CAVES AND SPELEOLOGY IN BULGARIA

Petar Beron   Trifon Daaliev   Alexey Jalov

Caves and Speleology in Bulgaria

Pensoft Publishers
Bulgarian Federation of Speleology
National Museum of Natural History - Sofia
2006

CAVES AND SPELEOLOGY IN BULGARIA
© Petar Beron, Trifon Daaliev & Alexey Jalov

Co-published by
© Pensoft Publishers
© Bulgarian Federation of Speleology
© National Museum of Natural History - Sofia

First Published 2006
ISBN-10: 945-642-241-X (Pensoft Publishers)
ISBN-13: 978-945-642-241-5 (Pensoft Publishers)

All rights reserved. No part of this publication may be reproduced, stored in a retrieval system or transmitted in any form by any means, electronic, mechanical, photocopying, recording or otherwise, without the prior written permission of the copyright owners.

Pensoft Publishers
Geo Milev Str. 13a, Sofia 1111, Bulgaria
pensoft@mbox.infotel.bg
Fax: +359-2-8704282
www.pensoft.net

Printed in Bulgaria, October 2006

CONTENTS

FOREWORD 7
ACKNOWLEDGEMENTS 9
HISTORICAL REVIEW OF BULGARIAN SPELEOLOGY — ALEXEY JALOV 11
EDUCATION OF CAVERS — ALEXEY JALOV 15
CAVE RESCUE IN BULGARIA — TRIFON DAALIEV 17
INTERNATIONAL CONNECTIONS OF BULGARIAN SPELEOLOGISTS — ALEXEY JALOV 20
BULGARIAN CONTRIBUTIONS TO THE EXPLORATION OF CAVES AND
    KARST IN OTHER COUNTRIES — PETAR BERON 22
SCIENTIFIC EXPLORATION OF BULGARIAN CAVES 39
PLANT AND ANIMAL LIFE IN BULGARIAN CAVES — PETAR BERON 40
    CAVE FLORA AND MYCOTA 40
    CAVE FAUNA 40
ARCHAEOLOGICAL RESEARCH IN BULGARIAN CAVES — MAGDALENA STAMENOVA AND ALEXEY JALOV 61
PALEONTOLOGICAL RESEARCH IN BULGARIAN CAVES — PETAR BERON 66
GEOLOGICAL, GEOMORPHOLOGICAL AND GEOPHYSICAL STUDIES OF THE KARST AND
    CAVES IN BULGARIA — ALEXEY JALOV 79
CAVE MINERALS IN BULGARIA — YAVOR SHOPOV 82
HYDROGEOLOGICAL, HYDROLOGICAL AND HYDROCHEMICAL STUDIES IN
    BULGARIAN CAVES — ALEXEY BENDEREV 85
STUDIES ON THE CAVE MICROCLIMATE IN BULGARIA — ALEXEY STOEV 94
MEDICO-BIOLOGICAL AND PSYCHOLOGICAL STUDIES IN CAVING AND SPELEOLOGY — ALEXEY JALOV 97
CAVE DIVING IN BULGARIA — ALEXEY JALOV 99
PROTECTION OF KARST AND CAVES IN BULGARIA — ALEXEY JALOV 101
KARST AND CAVES IN BULGARIA — ALEXEY BENDEREV 104
SOME IMPORTANT KARSTIC REGIONS IN BULGARIA — ALEXEY BENDEREV 104
THE LONGEST AND THE DEEPEST CAVES IN BULGARIA 127
POTHOLES AND CAVE SYSTEMS IN BULGARIA 128
CAVES AND POTHOLES INCLUDED IN THIS BOOK 129
SOME REMARKABLE BULGARIAN CAVES — TRIFON DAALIEV AND ALEXEY JALOV (FAUNA: P. BERON) 133
    CAVES IN THE DANUBE PLAIN 133
    CAVES IN STARA PLANINA AND THE PREDBALKAN 145
    CAVES IN THE TRANSITIONAL REGION 396
    CAVES IN RILA-RHODOPEAN REGION 408
    CAVES IN PIRIN 408
    CAVES IN THE RHODOPES 430
    CAVES IN THE STRANDJA AND SAKAR 480
WHO IS WHO IN BULGARIAN SPELEOLOGY — PETAR BERON 491
SELECTED LITERATURE 499

FOREWORD

It was late that the many thousands of Bulgarian caves attracted the attention of explorers. The First Bulgarian Caving Society was founded as early as 1929, but the first cave animals, fossils and archaeological artefacts in Bulgarian caves had been found and published even earlier, by the end of 19 Century. However, the sport Speleology, contributing so much to the study of potholes and of difficult cave passages, was developed much later – only after the restoration of the organized caving after 1958. By that time no more than 200 caves had been known (very insufficiently) in Bulgaria. Now their number is over 4500, most of them being well documented. Cave animals are known from more than 700 Bulgarian caves. As a result of the many years of hard work Bulgarian cavers discovered and surveyed 65 caves longer than 1000 m and 59 potholes, deeper than 100 m. So far no giant caves (over 100 km) of potholes (deeper than 1000 m) have been found in Bulgaria, but the search for them is still going on. They are in the hands of an army of enthusiastic cavers and of several speleologists – scientists of European renown. Bulgarian caving is now a well organized system of caving clubs, including more than 800 members. Courses for training of cavers and rescuers take place every year, experience is being exchanged with foreign colleagues.

Everywhere Speleology is a blend of sport and science, but especially for Bulgarian caving the strong participation of scientists has been typical since the very beginning. The first Presidents of the Bulgarian cave movement were Professors Stefan Petkov and Rafail Popov, Ivan Buresch, Member of Bulgarian Academy of Sciences. After the Second World War the tradition was followed up by Prof. Lyubomir Dinev and Dr Petar Beron. Other specialists too played an important role in the organized caving - Eng. Pavel Petrov, Eng. Radush Radushev, Dr Hristo Delchev, Dr Stoitse Andreev, Dr Alexi Popov, Dr Vassil Gueorguiev, Dr Vladimir Popov, Dr Vassil Popov, Eng. Panayot Neykovski, Prof. Dr Stefan Shanov, Dr Alexey Benderev, Dr Alexey Stoev, Eng. Konstantin Spassov, Dr Yavor Shopov and many others. A special place in this review is reserved for the speleologists who after 1958 played a key role in the restoration of the organized caving in Bulgaria, and, first of all, Petar Tranteev and Petko Nedkov. It would be difficult to enumerate here all sporting speleologists and divers, who proved that they could match any of their foreign colleagues, working in quite different conditions.

Very fit and well trained, Bulgarian speleologists are on their way to caves in many countries. In 40 years they organized the exploration of hundreds of caves in more than 45 countries, including independent expeditions to Austria, France, Spain, Italy, Greece, Cuba, China, Vietnam, Indonesia and participated in such complex international expeditions as the British Speleological Expedition to Papua New Guinea in 1975. Many new animal species were described from caves in far away countries by Bulgarian specialists or by their foreign colleagues on the basis of the material, collected by them. Being still within the modest East European frames, Bulgaria is already a leading country when its international speleological activities are concerned. This has been appreciated by

The authors – Trifon Daaliev, Petar Beron and Alexey Jalov – photo Valery Peltekov

the world speleological community. Prof. L. Dinev and Dr P. Beron have been twice Board Members of U.I.S., Trifon Daaliev – First Deputy President of the Cave Rescue Commission, Dr Yavor Shopov – President of the Commission for Hydrology and Hydrochemistry of Karst at the U.I.S. The Bulgarian Federation of Speleology initiated the creation of a Balkan Speleological Union (President: Dr P. Beron, Secretary General: Alexey Jalov).

One of the foremost achievements of Bulgarian speleology was the creation of the Main Card Index of Bulgarian caves, directed first by Eng. R. Radushev and then by Zdravko Iliev. In 1958 Dr Georgi Ikonomov published the first map of Bulgarian caves, in 1962 – the Catalogue of Bulgarian cave animals was published (V. Gueorguiev and P. Beron) and Bulgaria was subdivided into karst regions (Vladimir Popov). Several books about Bulgarian caves have been published (almost entirely in Bulgarian). Thanks to the prominent schoolteacher from Chepelare Dimitar Raichev a Museum of Speleology and Karst was created in his small Rhodopean town.

In spite of these good results, our assessment is that their analysis is yet to be made. Lately some booklets with descriptions of the deepest potholes in several regions have been published (T. Daaliev, A. Jalov, Z. Iliev, A. Benderev, D. Gospodinov, M. Mircheva, R. Pandev and others). With them the large-scale idea is to compile a complete catalogue of data about all Bulgarian caves, an endeavour started by Nenko Radev as early as 1926–1928. There are almost no written sources about Bulgarian caves in foreign languages, despite the obvious interest in them.

That is the reason for us to publish both in Bulgarian and English the concise descriptions of 260 of the most important caves and potholes in Bulgaria, together with an outline of some general subjects. The data have been put together by generations of Bulgarian cavers and "cave scientists". Some of them even gave their life for the sake of the Speleology. To all of them we owe this book.

The Authors

ACKNOWLEDGEMENTS

The idea to make such a book started in the distant years preceding the foundation of the First Bulgarian Caving Society. Heralds of the intention to put together the knowledge on Bulgarian caves were the books of the Škorpil brothers, and especially "Krazhski yavleniya" (Karstic phenomena, 1900). The beginning of the Catalogue of Bulgarian Caves was marked by Nenko Radev, who in 1926 and 1928 described 16 caves, in two papers, together with outlines and maps.

After the rebirth of the organized caving in Bulgaria the idea of preparing a Monograph "Caves in Bulgaria" was discussed, but the information was not enough. The milestones of this idea were the books of P. Tranteev, most of all "Caves objects of tourism" (1965), "The caves in Bulgaria" (1978, together with K. Kosev), as well as "Travel underground" (1982) by Vladimir Popov. The Catalogue of cave fauna of Bulgaria (Guéorguiev and Beron, 1962) and its supplements in 1967, 1973 and 1994 contained short descriptions of more than 700 Bulgarian caves. In 1968 P. Beron proposed to P. Tranteev the project of a three-volume monumental work called "The caves in Bulgaria". This project was approved by the Direction of the Institute of Zoology (Bulgarian Academy of Sciences), by this time the working place of both authors. Between 1969 and 1973 systematic investigations (village after village) were carried out in the caves of the districts of Vidin, Montana, Vratsa and Sofia. Later, for various reasons, this project faded away.

Gradually, together with the increasing activity of Bulgarian cavers and with the accumulation of new knowledge about the caves, the need of a general review of Bulgarian caves and speleology became obvious. Moreover, Bulgaria became one of the countries with developed speleology and caving, but remained very backward in terms of publications on karst and caves, especially in foreign languages. This was painfully obvious during international congresses and meetings, the Bulgarian presentation of books remaining inadequately poor. Lastly, several booklets and separate descriptions of particular karstic areas have been published. Among the most active authors of these publications were Trifon Daaliev, Alexey Jalov and Zdravko Iliev, having at hand the information contained in the Main Card Index of Bulgarian Caves. The President of the Federation Dr P. Beron wrote several generalized papers on the cave fauna in Bulgaria. He proposed to start work on the present book in two versions – in Bulgarian and in English – and edited both publications.

The data in this book are the result of the work of many people, part of them being mentioned in the descriptions of caves and in the scientific chapters. To them we all owe the chance to have such a general publication, and to these humble and qualified explorers goes our sincere gratitude.

The Authors are grateful to their teachers in the field of speleology, and especially to Dr Ivan Buresch, Member of the Bulgarian Academy of Sciences, Petar Tranteev and Radush Radushev, who laid the foundations of the modern caving and Speleology in Bulgaria. We also thank all colleagues and specialists, some of which wrote important chapters of this book: Dr Alexey Benderev, Dr Alexey Stoev, Mrs Magdalena Stamenova, Mr Petko Nedkov, Dr Yavor Shopov.

Two hundred fifty-eight (258) maps of caves, prepared for publication by T. Daaliev also found a place in the book. They were inked by miss Angelina Petkova, an enthusiast caver. To her and to the authors of the photographs (enumerated further below) goes our most cordial gratitude.

We should not omit the voluntary computer work on the colour maps of the distribution of caves and potholes in the whole of Bulgaria and in the particular regions. These maps we owe to the active caver Ivan Alexiev, with the expert help of Trifon Daaliev and Dr Alexey Benderev.

Our thanks also go to the colleagues who helped improve this book by reading critically the entire MS or parts of it. Especially helpful were the active veteran cavers Ivan Lichkov and Zdravko Iliev.

Special thanks are due to the persons and the institutions who made possible the publishing of the book, and especially to the Ministry of Youth and Sport (Minister mr. Vassil Ivanov-Luciano), the Ministry of Environment and Waters (Minister Mrs. Dolores Arsenova) and of Pensoft publishers (Dr Lyubomir Penev).

And last, but not least, we would like to offer our sincere thanks to the local people (shepherds, foresters, inquisitive teachers and local explorers), who showed us many caves or helped with their exploration.

HISTORICAL REVIEW OF BULGARIAN SPELEOLOGY

Bulgarian caves have been known since far back in time. Descriptions of them are to be found in old manuscripts dated after 12th century and in the works of Bulgarian and foreign authors. The entrance premises of some of these caves have been converted into churches and monasteries, and have been used as cells for the monks or for providing shelter for men and animals.

In many books describing travels in the past centuries, Bulgarian caves are mentioned. These protospeleological notes have been analysed by several authors, and mainly by Jalov (1999). The first karstic cave in Bulgaria, mentioned in the old writings, is the cave of St. Ivan Rilski by the village of Skrino, Kyustendil district. We find it in the anonymous biography of the Saint dated from 12th century.

The scientific interest in the country's caves began soon after Bulgaria's liberation from ottoman domination in 1878. At the end of the 19th Century the first Bulgarian prehistorians, naturalists and geographers embarked on their investigation with publications about the results obtained. This work of exploration continued in later years as well.

The first Bulgarian Speleological Society was founded on March 18, 1929. This was the result of the acknowledged necessity to set up a public organization which, under the conditions prevalent at that time, would begin a systematic investigation of caves, protecting them from destruction and setting the beginnings of cave tourism. The founders of the Society were eminent Bulgarian scientists, people active in the realm of tourism, and cave-exploration fans – office employees and workers. The foundation of the First Bulgarian Caving Society marked the beginning of a new stage in the development of Speleology in Bulgaria. Though not very numerous in its membership, and despite its limited financial capacities, the Society engaged in huge-scale and useful activities. It made a reappraisal of all that had been done until that time in cave investigation and in obtaining more knowledge about the country's caves. Organized trips and studies were carried out in

Eng. Pavel Petrov – founder of the First Bulgarian Caving Society

Prof. Stefan Petkov – President of the Caving Society in 1929-1931 and in 1933-1935

certain caves and karst regions of Bulgaria. The results obtained were published in the scientific publication of the Society – "Bulletin of the Bulgarian Speleological Society" – Volume One of which appeared in 1936. Active propaganda was also carried out for the protection of the caves. There was a useful and active cooperation between the Bulgarian Caving Society and the Bulgarian Tourist Union. The first provincial branches of the Society were founded in Rakitovo and the towns of Dryanovo and Lovech. The Bulgarian Caving Society became more active after 1947. New members entered the Society, which included university students as well. In 1948 and 1949 they took part with great enthusiasm in what were known as the cave brigades organized with the generous support of the Bulgarian Academy of Sciences. Detailed investigations were carried out during this period of the karst regions of Lakatnik, Karloukovo, Rabisha (near Belogradchik), and Zlatna Panega in Lovech District. The charts prepared and the materials collected constituted an important scientific contribution.

After 1949 the Bulgarian Caving Society ceased its activities over a brief period of time. A good deal of work was done by the speleologists in the town of Rousse and by university students organized in their caving club "Akademik" in Sofia. Amateur work continued, as well as the research initiated in this field by the various institutes of the Bulgarian Academy of Sciences and by the Sofia University. However, there was a keenly felt need for a speleological organization in the country.

Such an organization was necessary to unite the efforts of amateur speleologists and of specialists and to promote the development of this branch of science in Bulgaria. Favourable conditions to that effect were created after 1957 with the restoration of the Bulgarian Tourist Union. At that time a number of scientists and amateur speleologists, as well as active supporters of the tourist movement, approached the Central Council of the Bulgarian Tourist Union with a proposal to set up a Committee for Cave Tourism. On July 14, 1958, the Central Council decided on setting up a Commission on Speleology and Cave Tourism. Forty-two clubs were set up all over the country.

With the generous support of the Central Council of the Bulgarian Tourist Union, the new speleological organization rapidly grew in strength and became very active. Its objectives became clearly formulated and presented, and a number of enactments were passed. There were also a number of additional initiatives, such as plenary sessions, conferences, gatherings, and international expeditions. Penetration, charting, and survey work in the known caves and the

Petar Tranteev – Vice-President of Bulgarian Caving Federation until 1979

Prof. Lyubomir Dinev – President of Bulgarian Caving Federation from 1958 to 1985 discovery of new cave sites became more active and on a higher scientific and technological level. The Commission on Speleology and Cave Tourism took on the task of the development and popularization of cave tourism in Bulgaria.

After the Fourth Congress of the Bulgarian Tourist Union in 1972, the Commission on Speleology and Cave Tourism was transformed into the Bulgarian Federation on Speleology, which continues operating today. This new and higher form of organization furnished a fresh impetus to the development of Speleology in Bulgaria. Its operation became more active and efficient. New initiatives were undertaken with a view to improving the training and qualifications of the speleologists in the country. The speleological clubs obtained new equipment and their activity increased. International ties grew stronger and broader. The successes attained were largely due to the correct understanding and to the moral and material support, which was and is still rendered by the Central Council of the Bulgarian Tourist Union.

Today, the Bulgarian Federation on Speleology is the only national organization in Bulgaria whose role is essentially devoted to speleology. The Federation represents the cavers and speleologists in Bulgaria. In this capacity it organizes National Congresses, and within the International Union of Speleology it nominates the Bulgarian delegates at the International Congresses. The Federation co-ordinates both the activities of Bulgarian cavers abroad, establishing contacts with the corresponding Societies, and the activity of foreign cavers in Bulgaria. With 600 members in 2006, the Federation is organized on a national level (an executive committee, a board of directors, a head office in Sofia), and on regional level it has 30 speleo clubs. Many expeditions were carried out. The caves in all Karst regions of the country have been duly studied, described, and mapped. Detailed scientific research and investigations were carried out in more than 1000 caves. The participants in these cave expeditions collected abundant research material and turned it over to the respective institutes and archaeological museums. The Cave Museum in the town of Chepelare is a valuable achievement, while the construction of the Caver's Holiday Home in Karloukovo is expected to promote still further the research work carried out in the field of Speleology.

A Main Card Index of Bulgarian Caves was set up in 1974 at the Bulgarian Federation on Speleology. This Card Index is a very comprehensive pool of information, the result of the systematic and continuous work of the Bulgarian speleologists – a work, which goes on unabated. Up to date the Index contains information on some 5500 caves explored in Bulgaria.

Eng. Radush Radushev – founder of the Main Card Index of Bulgarian Caves

The Bulgarian Federation on Speleology is engaged in active organizational work as well. Its annual plenary meetings and discussions have adopted decisions on important issues of organization, of providing better facilities to visitors in the caves, ensuring safety, improving instruction and sports training necessary for penetration in caves, providing still better equipment to that effect, and of cave preservation. Dozens of international, national, city and club expeditions have been carried out. An important role in the training of the speleologists is played by the annual courses for instructors in speleology and for life-guards in caves. The courses end up with the award of the badge called "Speleologist" and competitions for the best club of speleologists and for making the most accurate chart of a particular cave. For the purpose of honouring the services rendered to promote speleology in Bulgaria, the Bulgarian Federation on Speleology instituted in 1974 the so-called "Bat Badge" in Gold, Silver and Bronze. A number of commissions have been set up at the Bureau of the Federation for the purpose of improving its standards of operation, as well as a Cave Rescue Team with more than 100 members. Bulgarian speleologists participate in solving a number of practical problems connected with the country's caves. Featuring prominently in the work of the Bulgarian Federation on Speleology are its international activities. Close ties have been set up to that effect with a large number of speleologists and national speleological organizations in many countries. Speleologists from many European countries have participated in international expeditions organized in Bulgaria. The big expeditions organized in some of the deepest and longest caves in the world: Jean Bernard, Gouffre Berger, Pierre Saint-Martin in France, Ilaminako Atek and Torka Urielo – Spain, Snezhnaja, Pantyuhinskaya – Geogia, Antro del Coria, Michele Gortani, Spluga de la Preta – Italy, Epos Chasm and Provatina – Greece, Ozernaja and Optimisticheskaya – Ukraina etc. have been of great significance to the development of speleology and to the qualifications of the Bulgarian speleologists.

Particularly important are the expeditions for the exploration of new caves in Austria, South China, Cuba, Vietnam, Syria, Spain and especially in Albania, where more than 10 expeditions have been carried out since 1991.

Dr Ivan Buresch receives the badge Zlaten Prilep (Golden Bat), April 17th 1974. On the picture P. Beron, P. Tranteev, Iv. Buresch and V. Beshkov.

The care for the development and popularization of cave tourism in Bulgaria has been entrusted to the Bulgarian Federation on Speleology. With the setting up of the Federation an end was put to the uncontrolled entries into the country's caves and to the damage caused by such entries. Cave tourism became organized, with many improvements introduced for visitors, electricity included. The overall work connected with these improvement is under the guidance of the Federation. No cave can become a tourist site before it has been thoroughly studied. All structures and improvements made must be of such a type as not to affect in any degree the natural environment and the natural conditions existing in the caves.

Attempts at introducing certain basic improvements in some caves had been made even before the Second World War, but the all-round work on making the caves safe and accessible to visitors actually began in the sixties, i.e. after the setting up of the Bulgarian Federation on Speleology. Nowadays there are eight caves in the country, which have been provided with the requisite facilities as tourist sites; Magoura near the town of Belogradchik (since 1960), Ledenika in the Vratsa Mountain near the town of Vratsa (since 1961), Orlova Chouka to the south of the town of Rousse (1961), Bacho Kiro near the town of Dryanovo (1964), Sâeva Doupka at the Bresnitsa Village, Teteven District (1967), Snezhanka near the town of Peshtera in the Rhodope Mountains (1979 and 1980), Dyavolskoto Gurlo, Yagodinskata Peshtera and Uhlovitsa Cave in the Rhodope Mountains (1979 and 1980). All these caves are set to operate as tourist sites. They are closed and visitors can avail themselves of the services of specially trained guides.

The membership of the cave clubs is not very numerous. Numerical growth has not been our policy, but rather the high and steadily improving skills and qualifications of the cave-club members. The work done by the Bulgarian Federation on Speleology and by the individual cave clubs is aimed at creating the necessary conditions for the development of cave tourism and its popularization.

The Bulgarian Federation on Speleology, with its members – research workers and amateurs, is the sole organization in Bulgaria working for the advance of speleology and of cave tourism, and it is also the foremost unit carrying out the requisite research and protection of caves in the country.

EDUCATION OF CAVERS

In the period of the Caving Society in Bulgaria (1929–1949), no system of education or training existed in the country, or of perfecting the qualification of cavers. This necessity arose only after the foundation of the Commission of Caving in 1958. One of the first tasks was to train the first instructors in caving. The idea was that these people would initiate the creation of a network of caving clubs. The first attempt to organize a specialized course took place in 1959, but only in 1961, under the direction of P. Tranteev, a course for "Guides, explorers and surveyors in caves" was realized. Many people, interested in caving, participated to this course and in 1962 the caving clubs in Bulgaria were already 24.

The Federation approved a two – grade structure of education: courses for beginners (above 18years of age) for obtaining the badge "Peshternyak" (Caver), organized by the clubs, and courses for obtaining the rank of "Junior Instructor". Programs have been elaborated for both grades to unify the study. In 1969 two more pedagogical grades were introduced – "Instructor" and "Senior Instructor". At that time 13 cavers attended the 14-day courses for instructors and 6 other passed the exams for the grade "Senior Instructor". Another 25 cavers obtained the grade "Senior Instructor" because of their long experience and proven capacity to teach.

In 1974, for the needs of the newly created Rescue Team within the Federation courses for cave rescuers started. These courses exist even today and help greatly in training volunteers to help cavers and citizen in an emergency.

In the period 1976–1979 the Federation experimented with courses for junior beginners (12 – 16 years of age). These courses found a place in the practice of the clubs, but not for long.

According to the changes in the theory and praxis of caving and with the new techniques for descending in potholes, the structure and the programs of the courses underwent changes several times. In the 80s it was experimented with two specialized courses: for technical perfection and for surveying, but they too did not last a long time.

Now the Federation has three grades of training: Beginners, taught in the clubs and obtaining the rank of "Peshternyak" (Caver), courses for "Junior Instructors", available for people already having the rank of "Peshternyak", and courses for instructors for those already having the rank of "Junior Instructor".

Participants to the First National Caving Assembly „Karlukovo – 1964" – photo Mihail Kvartirnikov

CAVE RESCUE IN BULGARIA

Three periods could be outlined in the history of cave rescue in Bulgaria

First period (1964–1973)

During the National expedition-convention, which was held from April 1 to 7, 1964 in Zadanen dol – Kunino, several clubs (Akademik and Ivan Vazov – Sofia, a group from Yambol) demonstrated different techniques for rescue, but these techniques required the use of many people and damaging of ropes. During the same year in Karlukovo, many clubs demonstrated their experience in rescue, also the use of trolley and some mistakes were assessed.

On September 8, 1965 during the exploration of the pothole Zmeyova dupka near Vratsa Petko Lazarov from Veliko Tarnovo died. That was the first fatal accident with a caver in Bulgaria.

This accident caused a decision by the then Republican Caving Commission (1966) to create rescue teams in all caving clubs in the country with the necessary equipment. Rescue demonstrations were included in all caving conventions. In 1967 Alexey de Martynoff, President of the International Rescue Commission, visited the demonstration in Karlukovo. No specialized cave rescue existed in this period.

Second period (1973–1985)

In 1972 specialized equipment was acquired, making possible in 1973 the first specialized rescue course with 14 participants (hand-

Kunino – 1965. Demonstration of a rescue technique

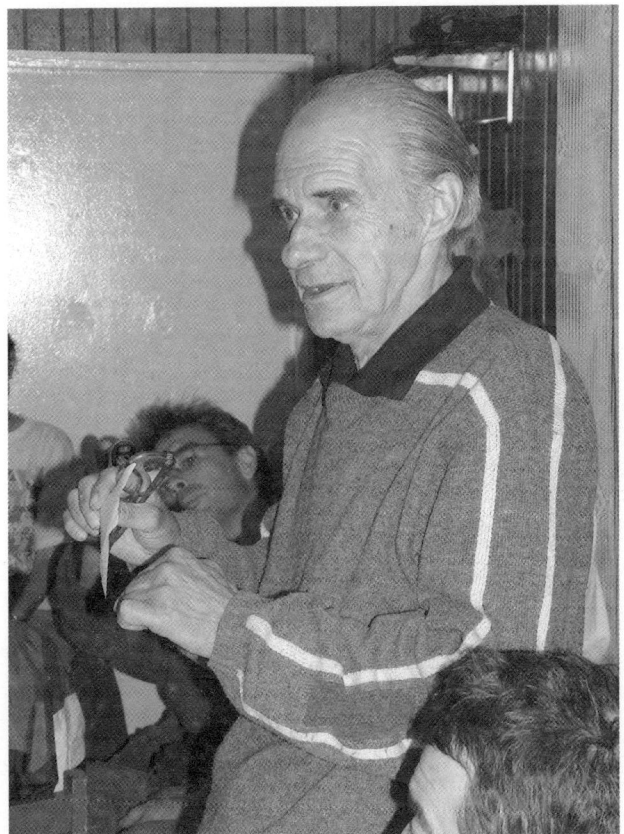

Petko Nedkov – Vice-President of Bulgarian Caving Federation until 1974 and Head of the Cave Rescue in Bulgaria from 1974 to 1989 – photo Valeri Peltekov

picked). In 1974 a second such course took place, again with 14 participants. The 28 trained rescuers formed the base of the Rescue Group. It included Trifon Daaliev, Nikolay Gladnishki, Vesselin Gyaurov, Kiril Yonchev, Trayan Georgiev, Kuncho Kunev, Petar Petrov and others. The start of the organized rescue in Bulgaria was on October 31, 1974, when a meeting initiated by Petko Nedkov took place in Sofia. A Board of the Group was elected: Petko Nedkov (Chairman) and Trifon Daaliev and Vassil Stoitsev (Deputies). Many other courses have been organized, rescue techniques were perfected, some of the rescuers penetrated some of the deepest potholes in the world (V. Nedkov, M. Dimitrov, V. Peltekov, O. Atanassov, S. Gazdov, V. Mustakov, K. Petkov, V. Oshanov). A constant effort was (and still is) made to follow the latest achievements in the world rescue practice.

In this period organized rescue existed, but without specialized stretchers. A Cassin stretcher was used, but it is more suitable for rock climbing rescue rather than in caves.

Third period – from 1985 till now

In this period the techniques and stretchers in use were entirely suitable for cave rescue. The stretchers have undergone changes three times. From 1985 to 1992 a French stretcher model (made in Bulgaria following the project of N. Gladnishki and T. Daaliev) was used. In the time from 1992 to 1999 another type of stretcher was in use, designed and made by Nikola Landjev. Now another stretcher, also made by N. Landjev, is being tested and will be introduced to Bulgarian cave rescue.

The modern methods applied by Bulgarian cave rescuers and their good training enhanced their authority among the members of the International Rescue Commission of the U.I.S.. This enabled Bulgaria to organize the first International Rescue School in 1988, later our country hosted a Conference on Cave Rescue (1991) with over 40 participants from France, Rumania, Italy, Belgium, Austria, Russia and other countries. The President of the International Commission André Slagmolen from Belgium also participated in these meetings. During the Congress of U.I.S. in Switzerland (1997) T. Daaliev was elected First Deputy-President of the International Commission for Cave Safety and Cave Rescue.

Meanwhile, Bulgarians helped organize cave rescue in several countries (Russia, Turkey, Macedonia, Georgia, Lithuania). Cave rescuers from these countries were trained in Bulgaria. Our rescuers exchanged experience with their French colleagues and participated in international conferenc-

Valeri Peltekov accompanies the stretcher in the vertical of the cave Drangaleshkata peshtera - 1997

30 years of cave rescue in Bulgaria. On the picture from right to left: Petko Nedkov, André Slagmolen, President of the Cave Rescue Commission of U.I.S., Vassil Stoitzev, Secretary of BFS, Trifon Daaliev, Vice-President of the Cave Rescue Commission of U.I.S.. Photo Valeri Peltekov

es on cave rescue. The Bulgarian Federation of Speleology has a contract for mutual assistance with the State Agency of Civil Defcnce.

During this 30-years period, over 170 cavers got training in specialized courses in cave rescue. Most of them are still active.

Participation in Rescue Conferences

In 1965 during the Congress of U.I.S. in Lyublyana (Slovenia), the International Commission for Cave Rescue, proposed by Prof. Anavy from Lebanon, was created.

In 1971 the Second International Cave Rescue Conference took place in Belgium. There were 50 participants from 17 countries, incl. Radush Radushev and Proyno Somov from Bulgaria. In 1973 Petko Nedkov was elected member of the Commission. Other sessions of the Commission followed: 1975 (Salzburg – Austria), 1979 (Zakopane – Poland, the Bulgarian participants were Hr. Deltshev, V. Gyaurov, N. Gladnishki and T. Daaliev), !983 (Aggtelek – Hungary, V. Peltekov, N. Landjev, K. Dimchev and T. Daaliev participated), 1987 (Triest – Italy), 1989 (Budapest – Hungary, participants P. Nedkov, V. Pashovski and T. Daaliev). Trifon Daaliev was elected Secretary of the International Commission for Cave Rescue. St. Tsonev and T. Daaliev participated in the meetings in 1988 and 1991 in Bulgaria, in 1992 in Belgium and Great Britain. During these conferences each country demonstrated the techniques and methods used for cave rescue. These conferences were very useful for exchange of experience and information.

Participation in Rescue Courses

In 1984 Nikolay Gladnishki and Trifon Daaliev attended a cave rescue course in France. The new techniques learnt there were successfully implemented in Bulgaria. In 1997 a similar course was attended by Nikola Landjev and Trifon Daaliev, in 1999 – by M. Dimitrov, N. Donchev, R. Vatev and K. Kolevski. The new methods, demonstrated during the courses, are now applied by most of the rescuers in Bulgaria.

Extraction with a balenceur – photo Trifon Daaliev

Transport of the strecher by troley – photo Vassil Balevski

INTERNATIONAL CONNECTIONS OF BULGARIAN SPELEOLOGISTS

The international connections of Bulgarian speleologists started at the very beginning of organized caving in Bulgaria and are most intensive in the domain of the biospeleological study of Bulgarian caves. In 1923 two prominent coleopterologists from Czechoslovakia (E. Knirsch and F. Rambousek) visited Bulgaria and, together with Iv. Buresch and N. Radev, undertook a thorough research on Bulgarian caves. In 1924 another Czech speleologist, K. Novak, visited several caves in Bulgaria and tried to reach the bottom of the pothole Bezdanniyat Pchelin near Yablanitsa (-105 m). Using a ladder, he descended "only" 85 m in this pothole. Dr Iv. Buresch established regular correspondence and active collaboration with prominent speleologists from France, Spain, Italy, Germany and other countries.

Later (1927), he participated in the Tenth International Zoological Congress and in the meeting of German, Austrian and Hungarian speleologists in Budapest, reporting on the cave fauna of Bulgaria.

We did not find data about international contacts during the existence of the Caving Society. The archives contain a letter from 1949, in which the President of the French national committee of Speleology R. Jeannel invites Bulgarians to participate in the international congress under preparation. Unfortunately, this participation was not realized.

The participants in the Conference on physicochemistry of karst in Stara Zagora. Two Presidents of UIS were present: Julia James from Australia and Derek Ford from Canada - photo Trifon Daaliev

The international activities of Bulgarian speleologists in the period 1958–2005 were in two main directions: collaboration with the sister-organizations, scientific institutions, speleological laboratories and specialists, as well as participation of Bulgaria in the activities of the International Speleological Union.

The first international connections after the founding of the Commission of Speleology in 1958 were realized in 1961, when two speleologists from Czechoslovakia (Peter Dropa and Moimir Kral) took part in the First International Expedition near Vratsa (Zmeyova dupka, Ponora and other caves). In 1962 contacts with the Polish National Commission of Speleology were established and three Polish cavers (Prof. K. Kowalski, K. Grotowski and V. Karch) were included in the Second International Expedition in the Rhodopes.

The exchange of speleological groups became a lasting practice also in the international activities of the Republican Commission of Caving and later of the Bulgarian Federation of Caving. Such exchange was most active with the sister-organizations from East Europe (GDR, Poland, Rumania, Czechoslovakia, Russia, Ukraine, Hungary, Yugoslavia), but also with the caving federations of Austria, Belgium, Great Britain, Turkey, West Germany, France and Switzerland. In the 80s contacts and also exchange of groups with Greece, Cuba, Vietnam and China took place and joint caving expeditions in these countries and in Bulgaria were organized. Ever since 1983, the Bulgarian Federation of Speleology has maintained very active connections with the French Federation of Speleology (contracts included in the International agreements between Bulgaria and France in the domain of Youth and Sport). Since 1991 there have been also intensive contacts with Albanian speleologists, allowing regular Bulgarian expeditions to this country. Later, in 1994 Bulgaria signed a special protocol for speleological collaboration with the official organizations of Rep. Macedonia and Albania, which gave rise to a trend of establishing contacts and interaction between Balkan countries. The peak of this trend was the creation in 2002 of the Balkan Speleological Union, including initialy the federations of Albania, Bulgaria, Bosnia and Herzegovina, Greece, Macedonia and Croatia, as well as clubs from Serbia and Montenegro.

The relations, established on a national level, are being multiplied by the caving clubs of these countries. These contacts facilitate the exchange of groups and the organization of joint expeditions in Bulgaria and abroad.

In 1965, during the International Congress of Speleology in Postoyna (Slovenia), the Republican Commission of Caving became a founding member of the International Speleological Union (ISU). Prof. Dr L. Dinev, Eng. R. Radushev, P. Tranteev and V. Gueorguiev were delegates. Bulgarian delegates participated also in the next speleological Congresses – Stuttgart, Germany (1969), Olomouc, Czechoslovakia (1973), Sheffield, Great Britain (1977), Bowling Green, USA (1981), Barcelona (1986), Budapest (1989), Beijing (1993), La Chaux de Font, Switzerland (1997) and Kalamos (2005). Bulgarian representatives and delegates contributed to the scientific sessions and participated actively in the work of the commissions of the Union and in the plenary sessions. As recognition of the achievements of Bulgarian speleology the long-lasting President of the Bulgarian Federation Prof. Dr L. Dinev was elected member of the Bureau of UIS (for two terms of office). Another Bulgarian – the actual President of BFS Dr P. Beron – was elected member of the Bureau of UIS. He too was a member for two terms (1989–1997).

Specialists in different fields, the members of BFS, have been members of different UIS commissions– Commission of Mapping (Eng. R. Radushev), of Bibliography (P. Beron), of Cave Rescue (P. Nedkov), of Underwater Studies (G. Mateev), etc. An honour for Bulgarian caving was the election of Y. Shopov as Chairman of the Commission for Physicochemistry and Hydrogeology of Karst. T. Daaliev became Vice-Chairman, and since 1997 he has been First Vice-Chairman of the Commission for Cave Rescue.

In September 1980 an important event took place in Bulgaria – the First European Regional Conference of Speleology. These conferences are held between the congresses and are being entrusted by UIS to be hosted by countries with developed speleology. Over 250 cavers from 17 countries took part in this Conference, and its excellent organization is still remembered by the participants. The Second European Conference was organized in 1992 in Belgium and Bulgarians participated actively in it.

The founders of the Balkan Speleological Union in 2002 – photo Trifon Daaliev

BULGARIAN CONTRIBUTIONS TO THE EXPLORATION OF CAVES AND KARST IN OTHER COUNTRIES

For a long time foreign caves have been the aim and dream of Bulgarian cavers and cave scientists. Before the Second World War there were so many unexplored caves in Bulgaria and so few cavers that it was out of the question to expeditions abroad. With the resurrection of the organized caving in Bulgaria (1958), the interest for studying caves of other countries increased, but only after 1963 Bulgarian speleologists started their long list of visits and expeditions to more than 45 countries. It is not possible to enumerate here all expeditions and visits, sometimes by clubs or individuals, but we want to contribute to the preserving of the history of Bulgarian caving abroad. Without underestimating anybody, we want to give prominence to the events, which have contributed substantially to our country's Speleology and to the enhancing of its image in the world.

EUROPE

Albania
1991

In the autumn the systematic Bulgarian research in Albania started. A team of three cavers

Karst plateau under Radohimës Peak (Alpet, N. Albania) – photo Trifon Daaliev

(A. Jalov, N. Gladnishki and N. Landjev) visited for the first time Albania, exploring and surveying 5 caves. Until 2005, altogether 14 Bulgarian expeditions with over 100 participants took place in Albania. Six of them were National Expeditions, led by: A. Jalov (5 expeditions) and V. Mustakov (1 expedition), 3 expeditions of the cavers of "Studenets" – Pleven Caving Club, led by S. Gazdov, and one expedition, organized by N. Landjev. In 2003 in the area of the village Boga a small club expedition of Pleven cavers took place, led by Orlin Kolov and succeeding to reach the bottom of the deepest pothole in Albania (610 m in the pothole BB-30). As a result of the long-term studies of Albanian caves 225 new caves have been discovered, studied and surveyed, the most important being BB-30 (-610 m) and Celikokave (-505 m), and rich information about the Geology, Hydrogeology, Morphology, the climate and the animals of Albanian karst has been gathered. All Bulgarian expeditions have been actively supported by our friend Prof. Gezim Uruci – President of the Albanian Speleological Association and also by other members of the Association.

Armenia
1988

P. Beron explored the fauna of the longest cave in Armenia – Archeri (The Bear's Cave).

Austria
1968

P. Tranteev and Hr. Delchev participated in an Austrian expedition to Plateneckeishöhle.

1974

Bulgarian – Austrian Expedition for exploring the pothole Bergerhöhle with 10 Bulgarian participants led by P. Tranteev.

1980–1983

Four Bulgarian expeditions were carried out for exploring new potholes in the Austrian Alps. Under the leadership of Petko Nedkov and Trifon

The entrance of the pot hole S-1 in Austria (-584 m) – photo Trifon Daaliev

Daaliev 93 new caves have been discovered, including S-1 (-584 m) and S-2 (-460 m Twenty three (23) cavers from 17 clubs (Nikolay Gladnishki, Vassil Nedkov, Valeri Peltekov, Milko Valchkov, Alexander Anev, Vassil Balevski, Nikola Landjev, Emil Todorov, Kamen Dimchev, Todor Todorov, Filip Filipov, Milko Paymakov and others) took part in these remarkable expeditions.

Belgium

Bulgarian speleologists visited several Belgian caves in 1967–1982, but mainly as tourists. Later P. Beron collected cave fauna in the longest Belgian cave Grotte de Han (10 693 m).

Czech Republic
1923

Dr Ivan Buresch visited several caves in Moravian Karst, including Macoha.

1961

A mission of the Republican Caving Commission (Prof. L. Dinev, P. Tranteev, Arch. P. Pâtskov, V. Gueorguiev and Hr. Mladenov) visited Czechoslovakia to study the experience in show caves. Among the caves visited were Dobshina, Belanska Cave, Demenovska Cave, Sbrashovska Cave and in Moravian Karst Punkva – Macoha, Balcarka, Sloupsko – Shushovska and Katerzhinska, also the "wild" cave Rudicko Propadanie.

1964

During the International Symposium in Brno P. Beron and P. Penchev visited some caves of Moravian karst.

1970

P. Beron explored the fauna of some caves of Bohemian Karst (Koneprusi and others) and visited caves in Moravian Karst. Part of the collected fauna has been already published.

In the period **1971–1989** groups of the Bulgarian Federation and the clubs in Vratsa, Russe, Pleven, Dryanovo, Peshtera, Blagoevgrad, Sofia, Botevgrad and other visited all show caves and the longest caves in Czech Republic.

France (continental)
1967

P. Beron received a French grant and spent 5 months in the Laboratoire Souterrain du C.N.R.S. in Moulis (Pyrenees), meanwhile visiting twice the caves of Corsica. In the Pyrenees Beron visited several caves, including the famed caves of prehistoric art in the estate of Count Beguen and collected cave fauna in other caves. Part of the data collected has been published and important experience has been obtained. The following year (1968) V. Gueorguiev also spent some time in the same Laboratory. During his stay P. Beron participated also in the training camp of French cavers at Vallon-Pont-d'Arc and visited some caves in Ardeche. Some scientific publications appeared from these stages.

1969

The first Bulgarian expedition to a cave deeper than 1000 m was organized by "Planinets" – Sofia Caving Club. The target was Gouffre Berger, by that time considered the deepest in the world (1122 m to the dry bottom). The daring action of the cavers of "Planinets" was carried out by 20 cavers, organized by N. Genov and Iv. Rashkov and led by Y. Strashimirov and P. Tranteev. A. Taparkova and P. Beron became the first Bulgarians to descend that deep, A. Taparkova marking a new world record for women. P. Beron became the first Bulgarian to cross the 1000-m point in a pothole. During this expedition for the first time Bulgarian cavers got Strumpf descendeurs and jumars, brought to the camp personally by the legend Fernand Petzl.

1973

Another expedition of "Planinets" – Sofia Caving Club, led by Y. Strashimirov, attacked the new deepest pothole – Pierre Saint Martin. Three out of the 23 cavers (A. Taparkova, V. Nedkov and K. Yonchev) reached the bottom. Ani marked a new world record, the expedition realized an interesting scientific program and the first Bulgarian speleofilm was shot in the giant pothole.

1974

Scientific expedition of "Planinets" – Sofia Caving Club explored the massif of the system Pierre Saint Martin and part of the system. Leader Y. Strashimirov.

1976

Scientific expedition of "Planinets" – Sofia Caving Club, NEK and the newspaper "Orbita" explored the massif of the system Pierre Saint Martin and part of the system. Leader Y. Strashimirov, participants: A. Tsvetkov, I. Parov, K. Sâbotinov, A. Bliznakov, P. Zlatarev, V. Gigov, A. Handjiyski. Geophysical observations were carried out by experimenting with the method "ULW" (Ultra Long Waves). In the area of Kokueta Cañon and under the Plateau Are d'Ani unknown cavities were discovered.

1982

A Bulgarian group (A. Jalov, V. Gyaurov, P. Saynov) participated in the course for instructors of the French School of Speleology. Several potholes (Trou qui souffle, Trou qui fume and Gournié) were penetrated.

1984

The new champion in depth – the pothole Jean Bernard, again in France, became the new challenge for Bulgarian cavers. The expedition of "Studenets" – Pleven with the Tourist Society "Kaylashka dolina", which consisted of 20 cavers led by Hr. Nedyalkov, headed for France on December 8th. The expedition lasted 20 days.

Eng. Radush Radushev and the famous French caver Norbert Casteret in 1973 – photo Nikolay Genov

Vassil Nedkov descends a waterfall in the pot hole Pierre-Saint-Martin in 1973 – photo Nikolay Genov

During this time the team overcame many difficulties and four cavers (Senko Gazdov, Krasimir Petkov and the brothers Valeri and Tsvetan Mirchev) reached 1358 m – a new deepest point for Bulgarians.

1987

This year will be remembered for the expedition of cavers from "Akademik" – Sofia Caving Club, led by St. Shanov, to the French part of the Pyrenees. Together with their colleagues from Nantes, Bulgarians realized traverses between two of the entrances (Mille-Penblanc and Coquille-Penblanc) of the system Felix Trombe – La Hen Morte, 81 km long and with denivelation of 1018 m. The first traverse, 12 km long, was overcome by four cavers, and the second, longer and more difficult, by V. Vassilev and N. Chobanov in ca. 22 hours.

1989

In July 10 cavers from "Akademik" – Sofia returned the visit of their French colleagues from the club "Saint Erblim" – Nantes. On July 13, V. Vassilev and B. Ivanov reached the siphon (-1122) of Gouffre Berger for ca. 17 hours. Yu. Atanasov, O. Stoyanov and P. Jivkov from the group led by S. Shanov penetrated the potholes Jean Nouveau (- 578 m, famous for its vertical entrance of 165 m), Soufleur (-640 m), Caldère (-667 m) and Romanesse (-205).

1989

In the same year V. Vassilev and B. Ivanov reached the bottom of Gouffre Berger for ca. 24 hours within the expedition of cavers from "Akademik" – Sofia and their colleagues from Nantes. Six other cavers, led by St. Shanov, descended into the potholes Jean Nouveau (-578 m), Souffleur (-640 m) and Caldère (-667 m).

1991

In July a group of seven cavers from Sofia (among them five from "Akademik", led by Tsvetan Lichkov) participated in the third joint expedition with colleagues from Nantes. The area of the expedition was NW of Montpelier, where they penetrated seven cave potholes.

The expedition, organized by "Galata" – Varna Caving Club, had the ambition, by overcoming the siphons at the end of Jean Bernard, to reach the final depth of 1494 m of the giant system. Six people reached the first siphon, one (the diver Hr. Raykov) managed to go beyond. Unfortunately, the narrow siphon stopped the second diver and the penetration was called off. The depth of 1250 m was reached by four other members of the expedition, including Diana Pencheva – this was her second descent below the 1000 m limit.

1993

From April 15th that year, Iv. Pandurski worked for 4 months in the underground laboratory in Moulis (Pyrenees). He explored the water fauna in some caves near the border with Spain, finding new Crustaceans. Ivan Pandurski visited and explored also:

a) caves in the region of Bordeau;

b) caves in the department Ariège, the region of Tarbes;

c) caves on the Vercors Plateau.

1993

The participants in the international expedition, organized by the caving club in Targovishte, P. Todorov, S. Milev, R. Tomov and T. Daaliev visited the cave Padirak, more than 10 km long and 235 m deep, as well as Gouffre du Sau de la Pucelle and Grotte Caniaque du Causse.

1993

A. Strezov, Y. Pavlov and K. Garbev participated in a contest in the Pyrenees, organized by the Olympic committees of France and Spain and won fifth place.

1994

In September several Bulgarian biospeleologists actively worked in the Laboratoire Souterrain de Moulis, studying the invertebrate fauna and the bats in many caves in the department of Ariège: Rumyana and Ivan Pandurski (two months), Hr. Deltshev, St. Andreev and V. Gueorguiev (one month).

1995

As a participant in a perfection course in France, G. Savov from "Helictite" – Sofia Caving Club took part in the biggest possible traverse between two of the entrances of Vernaud Cave System – the eighth longest in France. The distance of 8 km with a denivelation of 360 m was overcome for 18 hours. A. Jalov and G. Slavov from "Helictite" – Sofia Caving Club penetrated several potholes of the Vernaud System and the water cave Chauve Roche (7 km).

1999

From 1 to 20 August, a mixed group of cavers from Sofia (from the clubs "Akademik" and "Vitosha") carried out an expedition "Grenoble-99". Cavities visited: Trou qui Souffle, Gouffre Fumant, Gournie, Choranche, Pot 2, Trou de Spinette and others.

Corsica
1967

P. Beron visited the island twice and explored 21 caves, collecting many cave animals. Based on this collection, several specialists (H. Coiffait, J.-P. Mauriès, A. Vandel, J.-P. Henry, G. Magniez, M. Kolebinova) and P. Beron himself published articles with descriptions of many new species. In 1971 the Laboratory in Moulis published the general paper of P. Beron "Faune cavernicole de Corse".

Georgia
1971

A group of 6 cavers from "Akademik" – Sofia, led by Kliment Burin, participated in a joint expedition of "Speleolog" – Novosibirsk Caving Club in Arabika Massive, Western Caucasus. They descended in the potholes Ledyanaya, Karovaya (- 202 m) and Jubileynaya (-255 m).

1986

Two consecutive expeditions were carried out in the second deepest pothole in the world Snezhnaya (-1370 m) in Caucasus, Abhazia. The first was organized by four cavers led by Orlin Atanassov, all from "Aleko Konstantinov" – Sofia Caving Club. In 14 days they managed to penetrate from the lower entrance to the bottom of the pothole (denivelation -1335 m). So far this is the least numerous Bulgarian group to reach such a depth.

The second descent in 1986 was realized by an expedition of "Galata" – Varna Caving Club, led by R. Radulov and with the participation of cavers from Adler (Russia). The aim – to reach the maximum depth of 1370 m, using the upper entrance, was achieved by 6 cavers, including A. Taparkova – Pencheva and Diana Pencheva. Ani achieved a new personal record (for the third time). Diana became the second Bulgarian woman descending below 1000 m.

1988

First Bulgarian-Georgian expedition to Caucasus. Four Bulgarians (A. Mihov, S. Nenov, M. Tranteev and N. Valkov), led by T. Daaliev, had the task to penetrate, together with their Georgian colleagues, the end siphon of the pothole Vesennaya (-408 m). The diving team, including A. Mihov, penetrated the siphon, explored the next verticals and galleries and reached a second siphon, thus bringing the depth of the pothole to 555 m.

In the same time an expedition exploring the pothole Pionerskaya (-815 m) took place, organized by "Akademik" – Sofia Caving Club. The group (P. Saynov – leader and participants: Iv. Lichkov, V. Popov, I. Barova, V. Koleva, St. Hadjianastasov, B. Georgiev, Yu. Atanasov, P. Jivkov and P. Kolev) went there at the invitation of Crimean cavers. The pothole was a difficult one. The narrow passages and the rainy weather drained the strength of the cavers. Two teams managed to reach 500 m.

1989

24 August – 12 September: Joint Bulgarian-French-Georgian expedition in the region of the pothole Pantyuhinskaya – Caucasus; Organizer: BFS. Participants: T. Daaliev (Leader), B. Zahariev ("Planinets"- Sofia), V. Vassilev and A. Chobanov ("Akademik" – Sofia); K. Petkov ("Kaylashka Dolina" – Pleven); French participants: A. Guillon, B. Lips, J. Orsola and other. Georgian participants: V. Velisov and other. Results: equipment of Pantyuhinskaya up to 600 m for SRT.

The participants to the expedition Snezhnaya – 1986 in Caucasus – photo Trifon Daaliev

Germany
1926
Pavel Patev studied Rhizopoda (Protozoa) from the cave Salzlocher Höhle and published a paper on them.

In 1967–1982 Bulgarian cavers visited a number of caves in Germany.

Great Britain
1975 – During the preparations for an expedition to Papua New Guinea P. Beron explored the cave Easgill in Yorkshire.

1983 – The Yorkshire-Walles Expedition was organized by "Akademik" – Sofia Student's Caving Club. The leaders A. Grozdanov and I. Velichkova and the participants V. Popov, A. Strezov, F. Filipov were invited by the Speleo – Club of Sheffield University. The group penetrated four caves and one pothole in Derbyshire. Starting on 23 July, they took part in an expedition for the study of the longest system in Great Britain Easgill (52 km long, denivelation 173 m). During 10 joint penetrations, meteorological, biological and hydrogeological observations were conducted.

Greece
1968
P. Beron started the series of Bulgarian explorations in Greek caves. In January, together with his wife Kinka, he visited the cave Kutuki near Athens and the caves Kamilari, Melidoni, Kalamatu and Krionerida on Crete. Several scientific papers and new species were the result of this short visit. In the period 1968–2003 P. Beron visited Greece 11 times, collecting cave fauna.

1974
P. Beron and V. Beshkov explored the caves Dicteon Antron, Tzani, Sarkhos, Catholico Spilija, Arcoudas, Foli, Kamilari I et II and Trapezas on Crete, Zoodochos Cave on Santorin, the caves Eftamilos near Serres and Spilija Nycteridon near Petralona.

1979
The first Greek-Bulgarian caving expedition took place. Four Bulgarians led by Radoslav Rahnev explored, together with some Greek colleagues, caves in the mountain Erimanthos (Peloponnese). Nine new caves were mapped and another 3 visited and a good co-operation started.

1980
Vassil and Nadezhda Nedkov went to Greece on their personal expenses to penetrate, together with Greek colleagues, the ninth vertical pothole in the world – Mavro Skiadi (-347 m) on Crete. V. Nedkov made the descent in only 9 hours, thus becoming the first Bulgarian to have reached the bottom of one of the deepest shafts in the world.

1981
The third visit of P. Beron to Greek caves cantered on the cave of Zeus on Naxos and Aghios Joannis Cave on Iraklia Island.

1983
On their way back from Africa P. Beron and V. Beshkov visited the caves near Maronia, on the island Santorin, etc. and collected cave fauna.

1984
Descent into two of the deepest full verticals in the world (second and sixth), Epos and Provatina in Greece. In August, 6 cavers led by A. Jalov, together with Greek cavers, achieved this goal. The bottom of Epos (-451 m) was reached by 6 cavers, the bottom of Provatina (-392 m) – by three Bulgarian cavers, including L. Velichkova. Six cavers reached the bottom of Epos (-451 m) and three Bulgarian cavers, including L. Velichkova, reached the bottom of Provatina (-392 m)).

In the same year (April–May 1984) P. Beron explored the caves Ellinokamara and Stylokamara on Kassos I., two caves in the eastern part of Crete (together with K. Paragamian), Ideon Antron, the cave Katafyki on Kithnos I., the cave of Coufovouno in Thrace and collected cave fauna.

2000
In September, three Bulgarian Biospeleologists (Boyan Petrov, Stoyan Beshkov and Pavel Stoev) explored caves in Northern Greece (Western Thrace and Eastern Macedonia) and collected many cave animals.

2002
In September, a group of cavers from "Vitosha" – Sofia Caving Club, led by Ana Pencheva and with the participation of P. Beron and T. Daaliev visited the cave of Alistrati and the cave

Maara near Prosotsani. Some biological material was collected.

2003

In July, P. Beron, A. Jalov, M. Stamenova, J. Petrov, K. Kasabov, K. Stoichkov, together with Greek cavers, explored 6 caves in Othris Mountain, south of Volos. New caves and potholes were discovered (Titanospilya), some of the previously known caves were surveyed and the fauna of the caves Kokalya, Nero Spilya, Filaki, Metaxolaka and others was collected for the first time. New species were discovered in a completely unknown mountain (what concerns the cave animals). The expedition, organized by the Greek colleagues, explored in total 32 caves and marked the first joint exploration of cavers from the new Balkan Speleological Union.

Hungary
1964

P. Beron visited the huge cave Aggtelek and got acquainted with the underground laboratory there.

1982

Expedition of "Strinava" – Dryanovo Caving Club. Exploration of the cave Alba Regia (3 km) and of the pothole Czengö (-96 m).

1983

Expedition of "Protey" – Sliven Caving Club led by D. Panteleev. Visits to Szoymal cave (2 km) and the cave of Ferenz Hill (4 km), both of them in Budapest area, and to the potholes on the mountain Bükk – Istvan Kapo (3 km, 240 m) and the cave pothole Meteor.

1985

Expedition "Hungary–85", organized by "Protey" – Caving Club, Sliven, together with the Hungarian club VEAS. Nine participants, led by T. Toshev and M. Petrova. Visits to the cave Jozef higyi barlang on the hill Jozef in Budapest (2.4 km, denivel. 86 m), and to the following caves and potholes in the mountain Bükk: Fekete barlang (1 km – 152 m), Kiraly zsoumboy (-40 m), Istvan Lapoi Barlang (2 km – 240 m), Szepesi barlang (-165 m), Letrazi-vizes barlang (2.4 km – 86 m).

1987

Expedition "Hungary–85", organized by "Protey" – Caving Club, Sliven, together with the club VEAS. Bulgarian group: three cavers from the club "Prilep" with "Planinets" – Sofia Tourist Society and 8 cavers from Sliven, led by K. Aleksiev and M. Petrova. Visits to the cave in Matiasz Hill, Budapest (4.2 km). Joint exploration of the potholes in Aggtelek area: Szabo Lalas – 130 m Vechen Bükk – 245 m, Raytek (-72 m) and the cave Baradla – 9 km.

Italy (continental)
1972

Cavers of "Aleko Konstantinov " – Sofia tried to penetrate the then deepest Italian pothole Spluga della Preta (-878 m). Unfortunately, different problems hindered the action and after 144 hours of work the cavers reached "only" 420 m.

1979

The city caving section in Sofia organized an expedition to the new deepest Italian pothole Abisso Michele Gortani (-920 m). Nine cavers, led by A. Taparkova – Pencheva, realized the daring idea. The bottom was reached by 7 cavers, but "l'apetit vient en mangeant" and the cavers decided to penetrate Spluga della Pretta too. A. Pencheva and V. Peltekov reached the bottom of the second deepest Italian pothole.

1988

Led by Kr. Karlov, an expedition of "Veslets" – Vratza penetrated successfully the then deepest Italian cave system Abisso Figiera – Antro del Corchia (-1200 m). Eleven cavers, including Yu. Tsvetkova and A. Atanassova reached this depth. Thus, the number of Bulgarian women who had crossed the 1000-m limit increased to four.

1989

Wishing to test their skill in potholes deeper than Bulgarian potholes, nine cavers from "Iv. Vazov" – Sofia penetrated Abisso Emilio Comicci (-774 m) in Italy. On 21 September, 4 cavers reached the aim of the expedition.

1991 and 1992

Two consecutive expeditions of cavers from Pleven attacked the Italian pothole Emilio Comicci (-774 m) for further diving.

Sardinia
1980
P. Beron explored 4 caves. Among the material collected was the new Isopod *Oritoniscus beroni* Ferrara et Taiti, 1984.

Macedonia
1964
P. Beron studied the fauna of the cave Dona Duka at Rashche near Skopie.

1995
The first Bulgarian-Macedonian caving expedition took place with 5 participants from Bulgaria, including three divers, led by K. Petkov. The team was invited to explore several unknown siphons in the area of the canon Matka near Skopie. The total result was: 7 dives in 4 siphons, 280 m underwater galleries surveyed, 195 m of them in the karstic source "Vrelo". The end of the siphon was not reached, but nevertheless it became the longest siphon in the Republic of Macedonia.

Malta
1996
P. Beron collected cave animals in two small caves on Gozo I.

Poland
1968
Petko Nedkov and Konstantin Spassov became the first Bulgarians to descend into a really deep pothole (on the list of the 10 deepest in the world) – Sniezhna (-640 m) in Western Tatra mountains.

1969
Again a group of Bulgarian cavers attacked Sniezhna. In 32 hours five of the 10 cavers, led by P. Nedkov, reached the bottom of the pothole.

1973
Using the classical route, a joint team of BFS descended into Sniezhna, in 24 hours (V. Stoitsev, T. Daaliev, P. Evtimov, L. Mihaylov).

1974
The Troyan Caving Club also organized an expedition to Sniezhna – V. Markov, A. Borov reached 320 m.

1976
Expedition of "Edelweiss" – Sofia Caving Club, including V. Balevski from Troyan, managed to realize the most difficult itinerary in the system Nad Kotlinami and Sniezhna – the traverse between the two entrances. Three reached the maximum depth of 783 m.

1979
During the international rescue conference T. Daaliev and H. Gladnishki penetrated, together with Belgian and Polish cavers, in Bandjoha (- 558 m).

1980
Expedition "Tatri", organized by the caving club of "Prista" – Russe Tourist Society. Ten participants, inc. the leaders Iv. Ivanov and S. Guneshki. Descents into Czerna (4600 m; -150 m), Pod Vanta (-144 m), Velka Litworowa (-225 m) and Bandjoha (- 558 m), all in Tatra National Park.

1984
9–22 July. Joint expedition of cavers from Sofia with "Katovice" – Caving Club. Participants: P. Patev – leader, D. Angelov, Ya. Bozhinov, V. Naydenov and Hr. Delchev. Visits in Mariarova -110 m, Ptasha -230 m, Pod Vanta -144 m.

1987
Expedition "Sniezhna", organized by "Veslets" – Vratsa Caving Club. Leader Kr. Karlov. Seven cavers, including three women: A. Atanassova, K. Petkova and Yu. Tsvetkova, reached the bottom.

Milko Valchkov at the entrance Sniezhna in the Polish Tatra in 1976 – photo Vassil Balevski

1988

In June St. Hadjianastasov from "Akademik" Caving Club, together with Polish cavers, penetrated in Snejna until the first siphon at – 620 m and with the alpinist Svobodin Dimitrov – to the bottom of Velka Litvorova – 351 m.

1990

Expedition "Tatri", organized by the caving club of "Bacho Kiro" – Dryanovo and "Bobri" – a Polish Caving Club. Ten members, leaders Iv. Hristov and D. Tsanev. Descend into the pothole Velika Litworowa (-351 m) and Czerna (4600 m – 150 m).

Rumania
1963

Three Bulgarians (Vassil Guéorguiev, P. Beron and Georgi Velev) returned the Rumanian visit from the Institute of Speleology "E. Racoviþa" in Bucharest and Cluj. Together with V. Decu and L. Botosaneanu, they visited some caves (Ponikova, Topolniþa, the cave of Cloşani, Cioka cu Brebenei, Pozarul Policei, Geţarul de Scarişoara, Peştera Vîntului), exchanged experience and collected some cave animals. This was the first organized research of Bulgarians in caves abroad (without considering the visit of Dr Buresch and his group in Vetrena dupka in 1940).

1971

Expedition „Kluj '71", organized by the City Section of Speleology, Sofia. Leader Iv. Mishev, technical leader Eng. R. Radushev, A. Pencheva, V. Nedkov, T. Daaliev, N. Gladnishki and other. Visits of the potholes Parvi May (-220 m), Gheţarul de Scarişoara (600 m, depth 115 m), Peştera Vîntului. About two kilometres of new galleries were discovered and the length of P. Vîntului became 18 km.

1981

Expedition Königstein, organized by the caving club of "Prista" – Russe Tourist Society. Exploration of the area of the summit Grin, massif Königstein in the Carpatians. Leaders S. Guneshki and E. Zapryanov, 12 other participants. Visited the potholes Grin (-129 m), Podul Vilor II and Vladushka.

1981

Expedition Topolniþa, organized by the caving club of Prista Russe Tourist Society. Leader A. Evtimov. Explorations in the caves Topolniţa – 24 km and Epuran.

1982

Expedition Padizh, organized by the caving club of Bacho-Kiro Tourist society. Five participants visited the caves Barsa – 12 048 m, Shura Bosh – 400 m, Meles – 2070 m, Folkuviu – 950 m and the potholes Zapodie – 22 km, Zhemanta – 135 and Padizh – 52 m.

1982

Expedition Grin, organized by the caving club of Prista Russe Tourist Society and the club Avenul – Braşov. Six cavers, leaders K. Dimchev and S. Stoychev (Razgrad). The aim was to study new caves in the region of Königstein, summit Grin, Carpathians.

1990

A joint team of speleologists from Akademik – Sofia, Heliktit – Sofia and Cherni vrah Sofia Caving Clubs, together with the hydrogeological section of the Institute of Geology and Mining, participated to the International Scientific Conference for the Preservation of Karst.

1991–1992

Two consecutive expeditions of Edelweiss Sofia Caving Club, led by Iv. Alexiev. Exploration of the caves Cetatele Ponorului (8 km), Zaplodie (22 km), Barsa, Negra, etc.

Russia (entire)
1968

Participation of 5 members of the RCC in the All – Russia Caving Expedition in SW Caucasus (S. Kashev, S. Andreev, L. Popov, A. Grozdanov, Hr. Deltshev). Work in the pothole Rucheynaya and survey of the newly discovered Primusnaya (-200 m). Collecting of cave fauna (new species).

1969

Another group of 5 Bulgarian cavers: V. Gogov, V. Gyaurov, G. Milushev (Shumen), D. Uzunov (Varna), led by V. Stoitsev, participated to an international convention – expedition on Alek Ridge, SW Caucasus – work in the cave Vorontsovskaya and the potholes Rucheynaya (-150 m) and Oktyabrskaya (-420 m).

1971
August 9–23: – IV International Expedition in Caucasus (Alek Ridge). Bulgarian group included V. Stoitsev, At. Bliznakov (Planinets – Sofia), St. Stoykov, St. Milev (Targovishte), Z. Kachanov (Yambol). They penetrated the potholes Velichestvennaya (-260 m) and Geograficheskaya – 120 m.

1976
P. Tranteev, V. Stoitsev and T. Daaliev participated in an expedition of Russian cavers in Alek Range. T. Daaliev penetrated the potholes Rucheynaya (-510 m) and Nazarovskaya (-500 m).

1985
September 28 – October 7: – Expedition Caucasus, organized by BFC. Participants: A. Jalov (Sofia) and K. Petrov (Galata Varna Caving Club). Explorations: Ahunskaya Cave (603 m long, Ahunskiy Massif), Dolgaya Cave (1476 m long), part of the system Vorontsovskaya; the pothole Pechal'naya (-230 m, 900 m long), Dzahra Massif.

1986
From April 28 to May 8, 1986, T. Daaliev and I. Hristov from Galata Varna Club made an official visit to the Geographical Section of Sochi Department of the AS of the USSR to arrange a joint expedition in Snazhnaya. During the visit the participants penetrated the pothole Nazarovskaya (-500 m), together with cavers from Krasnodar and Sochi.

1988
Expedition of Aleko Sofia Caving Club in Altayskaya pothole (-240 m, 1250 m long).

1989
Expedition of Helictite Sofia Caving Club in the pothole Altayskaya (-240 m, 1250 m long). Leader: K. Danailov. In the same year an expedition of Veslets Vratsa took place.

1991
Expedition of Strandja Burgas Caving Club on Alek Ridge, SW Caucasus. Leader L. Zhelyazkov. Penetration in the pothole Osennaya (-500 m) – part of the cave system Nazarovskaya-Primusnaya – Osennaya.

Serbia and Montenegro
1940
Dr Iv. Buresch and his group of researchers from the Royal Museum of Natural History in Sofia explored the cave Vetrena dupka near Vlasi in the former Bulgarian Western Confines. Although formally in Serbia, we can hardly consider this an exploration of a "foreign" cave.

1997
In May, Bulgarian biospeleologists Ivan and Rumyana Pandurski, together with Milan Paunović, visited five caves in Eastern Serbia and collected cave animals.

2004
Expedition "Durmitor"- Montenegro. From 6 to 16 August a group led by Ya. Krachmarov (three cavers from Sofia Caving Club and two from Strinava – Gabrovo Caving Club of Bacho Kiro Tourist Society in Dryanovo) took part to the expedition in Durmitor, organized by the Student's Caving Club "ASAK" – Belgrade to the deepest pot hole of Serbia and Montenegro Yama na Vetrenom brdo (-775 m). The members of the Bulgarian team not only reached the bottom of the pot hole, but discovered and explored also several new vertical caves.

Slovakia
1961
A mission of the Republican Caving Commission (Prof. L. Dinev, P. Tranteev, P. Pâtskov, V. Gueorguiev and Hr. Mladenov) visited Czechoslovakia to obtain experience in show caves. Among the caves visited were Dobshina, Belanska Cave, Demenovska Cave and others.

1970
P. Beron explored the fauna of several caves. Part of the material collected has already been published.
The other visits in former Czechoslovakia are enumerated in the section of Czech Republic.

1983
Five Bulgarian cavers (A. Jalov, A. Apostolov, L. Adamov, P. Penchev and S. Guneshki) took part in the National Convention of Slovak Cavers. During their stay they penetrated all bigger show caves (Demenovkie caves, Lisitska Cave) and the pothole Stary uplaz (-86).

Slovenia
1909
Iv. Buresch visited the caves Postoyna and Skocianska yama.

1964
P. Beron visited the caves around Postoyna and collected some cave animals. He established a long standing connection with the speleologists from the Institute of Speleology in Postoyna.

2004
From February 19 to 23, at the invitation by the Slovenian Federation of Speleology, a joint team of Bulgarian cavers from Plovdiv, Russe and Sofia, together with three Slovenians, took part in the disequipment of the pothole Chehy 2. Lev Mironov reached the camp at 900 m, Theodor Kisimov, Georgi Staychev and Konstantin Stoilov crossed the 1000 m frontier and reached nearly 1100 m.

Postoyna and other famous Slovenian caves have been visited many times by Bulgarian cavers.

Spain (continental)
1986
The summer marked the successful return of the 22 participants in the expedition BU – 56 (-1338 m) in Spain – the fourth deepest in the world. These were the cavers of Studenets Pleven, led by P. Stoynov. Their aim was to increase the depth of the pothole by diving beyond the third known siphon. For this aim, never before attempted by Bulgarians, 14 people equipped the pothole and transported the diving gear to the "dry bottom" (-1325 m), to ensure the success of the divers I. Gunov and M. Dimitrov. Both crossed the three known siphons. I. Gunov dived on his own into the fourth, reached a dry gallery and returned. The exploration had to be stopped there due to the lack of equipment, but the pothole already ranked third in the world.

1987
For further study of the pothole BU – 56 was organized another expedition of Pleven cavers, led by P. Stoynov and S. Gazdov. The efforts were directed towards giving a chance to the divers I. Gunov, M. Dimitrov and V. Chapanov. They negotiated the known siphons and another two siphons, explored the 650-meter gallery after them and stopped before the seventh siphon. The survey from the beginning of the 4th siphon to the newly found parts showed that the depth was 1408 m. So, on September 11th, with the help of the Bulgarians the pothole, called Ilaminako Atek by the Basques, became the second deepest in the world.

1988
Third expedition of the cavers from Studenets Pleven Caving Club, leader S. Gazdov. Subject of exploration were 48 new potholes in the Larra Massif, in the area of the pothole BU-56, Pyrenees.

To descend the pothole Torka Urielo (-1017 m), an expedition was organized by 18 cavers from Planinets Sofia Caving Club, led by K. Sabotinov. Four members of the club and a representative of BFS (member of Etropole Caving Club) achieved this goal.

1993
Ivan Pandurski collected water animals in Casteret Ice Cave in the High Pyrenees.

Balearic Is.
1986
P. Beron joined the cavers from the speleo-club of Palma de Majorca and explored, together with them, 4 caves of Majorca.

Turkey (entire)
1971
P. Beron and V. Beshkov realized the first Bulgarian visit to Turkish caves (Barut Hane near Istanbul, Büük and Kücuk Maara near Boazkale)

The participants to the expedition BU - 56 in the Pyrenees in 1986

and collected biological material, part of which has been published.

1972
P. Beron, T. Michev and V. Beshkov visited the caves Barut Hane near Istanbul, Damlata near Alania and Insuyu near Burdur Lake and discovered new species of cave animals.

1993
T. Daaliev and P. Beron joined their colleagues from BUMAC – Istanbul to explore the deep pothole Düden Yayla and, after that, visited some caves around Beyshehir and collected cave animals. One month later another four Bulgarians participated in the expedition of BUMAK in Cukur Pinar (-1190 m), but had no chance to reach this depth.

2004
From 13th July to 15 August Bulgarian Federation of Speleology, together with speleologists from BUMAK – Istanbul, organized expedition for studying of the deepest pot hole in Turkey Mehmed Ali Yozsal-Evren Gyunay Düdeni. Bulgarian participants: 8 cavers from the caving clubs in Sofia, Plovdiv and Rousse, led by Tzvetan Ostromski. Turkish participants: 8–14 in the different stages of the study. Some results:

1. With the new parts, discovered and surveyed, the pot hole became the deepest in Asia, ranging 12th among the deepest pot holes in the World. Its depth "changed" from 1377 to – 1429 m.

2. Altogether six cavers (three from each side) reached the lowest pont of Asia (-1429 m). From Bulgaian side these are Teodor Kisimov and Svetlomir Stanchev from "Prista" – Rousse Caving Club and Konstantin Stoilov from "Akademik" – Sofia Caving Club. For the time being these are the "deepest" Bulgarians.

3. T. Kisimov, K. Stoilov, N. Kamenov, N. Tomov, Sv. Stanchev and Tzv. Ostromski reached the depth of – 1300 m for the first time – a chalenge for any caver.

4. The bones of Mehmed Ali Yosal who died at – 1270 m in 2001 during flood were brought to light, so the relatives were enabled to burry him and the authorities – to close the case.

Ukraine
1967
The first organized contact with the big caves of the world of Bulgarian cavers was during the expedition Podolie'67 in Western Ukraine, where five cavers of Akademik Sofia Caving Club, led by Kliment Burin, participated to the exploration of the cave Optimisticheskaya. This expedition seems to be the first Bulgarian club expedition outside Bulgaria. At that time the known length of the surveyed galleries had increased by 3184 m and with its 18 884 m the cave ranked second in the then URSS and 13th in the world. Now its length is already 205 km (second longest in the world).

1969
Again a team of 8 members of Akademik Sofia Caving Club, led by Radoslav Rahnev, participated in the study of Optimisticheskaya. Its length reached 43 500 m.

1971
Bulgarian cavers again attacked one of the longest caves. Five cavers, organized by A. Petkova and led by P. Tranteev, participated in the XXVI[-th] expedition of Ternopol cavers to study the cave Golubie ozera. With active Bulgarian participation (N. Gladnishki, Ch. Mitov, D. Mihaylov, T. Daaliev) 8 km of new galleries were surveyed, thus bringing the cave's length to 65 km. Now its length is more than 111 km. The caves Kristalnaya and Mlynki have been visited as well.

1973
In the period August 1-18, a group of cavers from Akademik Sofia led by Iv. Lichkov and participants P. Deltshev, A. Penchev, D. Stamenov, Iv. Shekerov and M. Koleva took part in the double jubilee of Cyclope Lvov Caving Club (10 years of the foundation of the club and the mapping of the first 100 km in the cave Optimis-

The participants to the expedition Golubie Ozera in Ukraine in 1971

ticheskaya). After the scientific conference in Lvov a surveying of the cave was carried out with many Russian clubs. The Bulgarians participated in the survey of 350 m of galleries and chambers, some of which bear Bulgarian names. The cave Kristalnaya was visited too.

1989

In July, A. Benderev and Ir. Ilieva from Akademik Sofia Caving Club, together with cavers from Lvov, explored the caves Optimisticheskaya and Zolushka.

2002

The cavers A. Jalov (Heliktit – Sofia), K. Petkov (Studenets – Pleven) and P. Todorov (Lucifer – Targovishte) participated in a Ukrainian expedition to Ozernaya Cave.

Several other Bulgarian expeditions to the caves and potholes in Crimea were carried out: 1978, 1980 – the cave Kizil Koba (Krasnaya) – 13 km on 6 levels, cavers from Akademik Sofia Caving Club, Protey Sliven CC and Mrak– Etropole CC; 1980, Akademik Sofia, the potholes Emine Bair Hisar, Bezdonnaya (-196 m), Kaskadnaya (-400 m); 1999, Edelweiss – Sofia, Hod Konem – (-213 m), Kaskadnaya, Bezdonnaya; 2001, Salamandar – Stara Zagora in the pothole Soldatskaya (-517 m).

AFRICA

Algeria
1989

With 1159 m the pothole Anu Iflis in Algeria is the deepest in Africa. Five cavers from Aleko Konstantinov Sofia Caving Club, led by S. Stoykov, attacked it, but at 500 m the pothole was too narrow for them. Four cavers reached the bottom of the second deepest pothole in Africa – Anu Boassili (-805 m). The first organized caving expedition of Bulgarians to African pot holes became a fact.

Kenya
1995

P. Beron and V. Beshkov visited some caves in Suswa Hills and near Mombassa.

Nigeria
1977

P. Beron explored the fauna of Pannini Cave near Mokwa.

Tanzania (continental part)
1983

P. Beron and V. Beshkov explored the fauna of Kulumuzi Caves near Tanga.

Zanzibar
1983

P. Beron and V. Beshkov visited the cave Pango Managole in the south of the island, collecting some animals.

Zimbabwe
1989

P. Beron explored the fauna of Sinoya Cave.

ASIA

China
1989

An expedition to China was organized by the Bulgarian Speleological Federation. Led by P. Beron, the expedition included T. Daaliev, V. Nedkov, N. Gladnishki and K. Spassov. Together with specialists – karstologists from the Geographical Institute of Kunmin University, the Sino-Bulgarian team studied some caves in connection with the building of a dam in the area of Menzi, Yunnan Province. Thirteen (13) caves and potholes with a total length of 8 km were surveyed and explored biospeleologically.

1990

Another important expedition started in February 1990. Five speleologists (K. Bonev, K. Garbev, P. Stefanov, I. Ivanov, led by A. Jalov) proceeded with the works of the First Sino-Bulgarian Expedition. Together with Chinese karstologists, geological, hydrogeological and geomorphological studies in the same area were realized. The Bulgarian team studied and surveyed nine caves and potholes (total length of 4 km).

1993

After the Speleological Congress in Beijing, P. Beron visited some caves near the capital, including the caves of the Prehistoric Man near Choukowdian.

Korea (North)
1982

Despite red tape, P. Beron entered one small cave and collected some animals.

India
1981

P. Beron visited a cave in Kashmir and explored its fauna.

1988

During the Sofia University Expedition to Gharval Himalaya Yavor Shopov discovered and explored several pseudo-karstic caves. The results have been published.

Indonesia
1994

P. Beron and V. Beshkov explored Indonesian caves: Ngalau Kamang on Sumatra, two caves on Nias, Gua Seplawan and Gua Kiskindo on Java, Gua Lawah on Bali, Gua Karangsari on Nusa Penida, Batu Chermin on Flores, four caves on Timor and four caves on Sumba. Many cave animals were collected.

1995

P. Beron, T. Daaliev and Teodora Ivanova explored some other caves on the islands Sumatra, Siberut, Java, Sulawesi. The speleological results obtained were: 13 caves explored, three new caves (850 m long) surveyed. Many new cave animals were collected.

Iran
1972

In December, P. Beron, T. Michev and V. Beshkov visited Shapur Cave in Zagros and explored the fauna. Some new species (beetles) have been described from this material.

Malaysia
Malaya
1994

P. Beron and V. Beshkov visited the Temple Cave in the complex Batu Caves near Kuala Lumpur, collecting cave animals.

Sarawak
1981

St. Andreev published the description of the new species *Cyathura chapmani* Andr. (Isopoda, Anthuridae), collected by the British Expedition in 1976.

Nepal
1981

P. Beron studied the fauna of Mahendra Gupha near Pokhara.

1984

P. Beron visited other small caves.

1987

During the expedition Ama Dablam Ivan Bonev studied the granite caves in the Himalaya and wrote an article about them (1989).

New Guinea
1975

P. Beron took part as a biologist in the British Speleological Expedition to Papua New Guinea – one of the most important speleological expeditions ever. This was the first participation of Bulgarians in the study of tropical caves. For five months about 200 new caves have been studied, including the over-20-km long Selminum Tem, then the longest cave in the Southern Hemisphere. After the end of the expedition, in the area of Telefomin, P. Beron visited on his own two caves in Chimbu (Simbu), two caves in New Britain and caves in New Ireland. A very important collection of cave animals was collected and is kept in the National Collection of Natural History in Sofia. Based on this collection, many new taxa have been described. P. Beron described this expedition in his book "Five months in New Guinea" (1986). Some materials on the caves of New Guinea and New Ireland (including new species) have been published also by the Bulgarian specialists V. Guéorguiev, K. Kumanski, St. Andreev and T. Marinov – one of the most important Bulgarian contributions to the study of the tropical cave fauna.

Syria
1972

P. Beron, Vl. Beshkov and T. Michev explored the underground channels dug in the limestone near Palmira and containing warm water.

1989

Expedition to Syria, organized by Sredets Sofia Caving Club, consisted of 12 cavers led by Kr. Garbev. Explored and surveyed were 9 caves, the longest of which – Joait (2600 m) becoming the longest Syrian cave. Karstological research was carried out as well.

Thailand
1984

P. Beron and Stoitse Andreev visited the cave Tham Chiang Dao in Northern Thailand and collected many cave animals, among which the new Isopod species *Exalloniscus beroni* Taiti et Ferrara, 1988. A cave near Phang Nga was also visited.

Turkmenia (Turkmenistan)
1990

A joint expedition of the clubs Studenets – Chepelare, Orpheus – Devin and Silivryak – Trigrad. Nine participants led by G. Raychev. Work on the Plateau Kugitag – penetrations in the cave Kap Kutan (38 km), Promezhdutochnaya (17.8 km), Gyulsherin and Hashima Oyug (8 km). At the same time St. Hadjianastasov and E. Tsaneva, together with cavers from Samarkand, Tashkent and Petersburg, penetrated the above-mentioned caves.

Uzbekistan
1987

Two members of the student's caving club (F. Filipov and St. Hadzhianastasov) penetrated in the then deepest pothole in Asia – Kievskaya (KILSI) in Uzbekistan (-990 m).

1989

June 18–July 10: – Expedition Samarkand'89 – Kirk Tau Plateau. Field work for 16 days. Organizer: Prilep – Gotse Delchev Caving Club. Participants: Dimitar Peychev (Leader), K. Stamatov, A. Hadzhiev, K. Milev, K. Zlatkov, K. Karavelov, I. Iliev (Blagoevgrad), I. Beshin, I. Perkov, V. Sarandev. Results: a new pothole – Nevrokopskaya (-150 m) was discovered; a prolongation to another pothole – Kulskaya (-83 m) was also discovered.

1990

Second Bulgarian expedition was organized by Helictite Caving Club with Ivan Vazov Tourist Society in the pothole Kievskaya. The team, led by V. Mustakov, included 9 cavers of Helictite, one caver of Edelweiss – Sofia and one from Kazanlâk. The caver from from Edelweiss and three cavers from Ivan Vazov – Sofia CC, including S. Krasteva, reached the bottom (-990 m). S. Krasteva remained only 11 meters short of becoming the fifth Bulgarian woman to cross the 1000-m limit.

Vietnam
1988

Expedition to Vietnam, organized to celebrate the 100 anniversary of the University of Sofia. Leader of the expedition was P. Beron and technical leader was A. Jalov. The big group included nine speleologists who explored 37 caves of total length of 5.7 km. At the same time the first Bulgarian biospeleological studies in Vietnam took place. Some new species were already published.

NORTH AMERICA

Mexico
1982

In January, P. Beron explored many caves in Tamaulipas, Chiapas and Tabasco and collected cave animals. Among the caves were Grutas de Cocona, Grutas de Quintero, La Trinitaria, Cuevas de San Cristobal, Cueva de la Capilla, Cueva de los Liones, Grutas de Loltun and other.

USA
1973

While specializing, Lyubomir Klisurov from the Laboratory for Underwater Research with the Institute for Marine Research in Varna visited several underwater caves in Northern Florida. He descended into the cave system Peacock-Olson (600 m), Little River, Devil's Eye.

1981

V. Beshkov visited Carlsbad Cavern.

1988

K. Burin visited the Mammoth Cave – the longest cave in the world.

1992

P. Beron participated in the Annual Convention of American Cavers in Indiana and visited four caves there. He also visited the Mammoth Cave.

1995

P. Beron participated in the Annual Convention of American Cavers in Colorado and explored some caves there.

1997-2000

St. Hadjianastasov visited several caves in different states: the Mammoth Cave, Carlsbad Cavern and others.

1999

Ts. Ostromski visited and explored the Mammoth Cave.

2001

B. Lilkov from Helictite – Sofia visited the Mammoth Cave and Diamond Cave in Kentucky.

SOUTH AND CENTRAL AMERICA

Bolivia
1979

P. Beron visited and explored the cave near Sorata in Cordillera Real.

Cuba
1981

The Student's Caving Club Akademik – Sofia, together with NEC (UNESCO) – Sofia, the Bulgarian Caving Federation and the Student's Tourist Council, organized a Bulgarian-Cuban expedition to Cuba. Its Cuban partner was the Club Martel de Cuba with President Prof. Antonio Nuñez Jimenez. Led by P. Beron, 9 Bulgarians studied the caves of Sierra de los Organos in Pinar del Rio Province. Explored and surveyed were 18 km of galleries, including 16 in Gran Caverna Fuentes. Scientific observations in the fields of Biospeleology, Hydrochemistry, Climatology and Mineralogy were carried out. This was the first Bulgarian speleological expedition outside Europe and its work was highly appreciated.

In March 1982 P. Beron explored the fauna of some "hot caves" near Santiago de Cuba.

Participants to the expedition Kuba – 81. From right to left: Tzvetan Lichkov, Alexander Filipov, Petar Beron, Ivan Lichkov, Ivan Tonchev, Panayot Neykovski, Konstantin Spassov, Petar Delchev, Ilcho Mitrashkov – photo Ivan Lichkov

1988

Important results were obtained by the National Speleological Expedition to Cuba, where 14 Bulgarians spent 2 months exploring (together with their Cuban colleagues) the karstic plateau Guaso in Oriente Province. The expedition, led by A. Jalov and T. Daaliev, discovered, explored and mapped 34 caves and potholes with a total length of 19 988 m, the longest being Cueva del Campanario (8382 m). At the same time, the complex karstological research aimed at finding additional water supplies for Guantanamo town.

Peru
1979

P. Beron visited and explored the cave near Tingo Maria in Peruvian Amazonia.

ANTARCTICA

Livingston I.
1997

During the Bulgarian Antarctic Expedition (January-February), Ivan Lichkov registered two ice caves. One was formed by a water current 250–300 m long, the height of the entrance – 2–2.3 m and of the exit – 1.2–1.5 m. The other cave was a dying one, 25–30 m long and with an entrance of 2.5-3 m.

In the 40 years since 1963, Bulgarian cavers have also visited or explored many other caves and potholes in different countries (more than

The 40-meter entrance of the cave El Campanario, explored by the members of the expedition Cuba – 88 – photo Trifon Daaliev

Ivan Lichkov at the entrance of an ice cave on Livingston Island (Antarctic)

45). We cannot compete with the leading countries, which have many more speleologists, longtime activities and resources incomparable to ours. However, we can compare ourselves to our colleagues from all neighbouring countries. None of them has such multilateral sporting or scientific activities abroad or in the caves of so many countries. Scientifically very advanced are the Rumanians, with their studies in Cuba and Venezuela and with the important scientific literature already published. Rumania and Slovenia are two of the countries where Speleology was born, they have Institutes of Speleology and old traditions. Nevertheless, no Balkan country can match the achievements of Bulgarian cavers abroad. This should lead us towards new horizons.

Cone karst in Sulawesi – Indonesia – 1995 – photo Trifon Daaliev

SCIENTIFIC EXPLORATION OF BULGARIAN CAVES

Bulgarian specialists have been very active in the scientific exploration of the caves and karst in Bulgaria. Almost all aspects of the cave science have found place in the country's scientific institutions, and it is important to remind that the founders of the Caving Society in 1929 and its Presidents (Prof. St. Petkov, Prof. R. Popov, Dr Iv. Buresch) were scientists. The tradition was followed with the resurrection of organized speleology in Bulgaria in 1958. The first leaders of the Bulgarian Federation of Speleology were Prof. L. Dinev and the geographer P. Tranteev, followed as President (since 1985) by Dr P. Beron. Many other scientists have contributed to various aspects of speleology: Alexander Filipov, Alexey Benderev, Alexey Jalov, Alexey Stoev, Alexi Popov, Boris Kolev, Boyan Petrov, Vladimir Beshkov, Vladimir Popov, Borislav Gueorguiev, Vasil Gueorguiev, Vasil Mikov, Vasil Popov, Georgi Antonov, Georgi Markov, Georgi Raychev, Dimitar Sabev, Dimitar Raychev, Geko Radev, Zlatozar Boev, Ivan Nikolov, Ivan Pandurski, Konstantin Spasov, Kamen Bonev, Mihail Kvartirnikov, Nenko Radev, Neno Atanasov, Neofit Cholakov, Nikolay Djambazov, Pavel Petrov, Pavel Stoev, Panayot Neykovski, Pencho Drenski, Petar Stefanov, Petar Petrov, Radush Radushev, Rumyana Pandurska, Stefan Draganov, Stefan Shanov, Stoitse Andreev, Stoyan Beshkov, Teodora Ivanova, Todor Stoychev, Christo Delchev, Yavor Shopov and others.

Important part of the knowledge on the caves is the description and the surveying of more than 5000 caves, done by many cavers over the years. We can mention (without being exhaustive): Alexandar Grozdanov, Alexandar Leonidov, Alexandar Popov, Aglika Gyaurova, Angel Penchev, Angel Velchov, Angelina Petkova, Alexandar Nikolov, Alexandar Siromahov, Alexandar Strezov, Alexandar Manov, Anna Taparkova, Antoniy Handjiyski, Atanas Bliznakov, Atanas Spasov, Bogdan Todorov, Bogdan Zachariev, Boris Kolev, Borislav Garev, Boyan Tranteev, Bozhan Marinov, Dimitar Dimitrov, Delyan Damyanovski, Filcho Filchev, Filip Filipov, Georgi Antonov, Georgi Chakalski, Georgi Markov, Ivaylo Ivanov, Ivan Alexiev, Ivan Zdravkov, Ivan Petrov, Ivan Lichkov, Ivan Rashkov, Iliya Nickoevski, Iren Ilieva, Ivan Tomov, Kamen Dimchev, Kiril Danailov, Kiril Stoyankov, Krasimir Petkov, Krasimir Karlov, Kuncho Kunev, Lyubomir Adamov, Lyuben Popov, Mariya Zlatkova, Mariyana Petrova, Maritin Tranteev, Milen Dimitrov, Metodi Metodiev, Mincho Gumarov, Nikolay Genov, Nikola Landjev, Nikolay Gladnishki, Nikola Gaydarov, Nikolay Milev, Pavlina Gyaurova, Petar Delchev, Petar Evtimov, Petar Tranteev, Petko Karabadjakov, Plamen Petkov, Pencho Todorov, Pencho Penchev, Radoslav Rahnev, Radush Radushev, Rayko Metodiev, Rosen Vatev, Sasho Alexiev, Senko Gazdov, Shterion Todorov, Stoyan Petkov, Stoycho Stoychev, Todor Todorov, Todor Troanski, Trayan Georgiev, Trifon Daaliev, Tsvetan Ostromski, Tsvetan Lichkov, Vasil Balevski, Velizar Botev, Veselin Gyaurov, Viliyan Vasilev, Valentin Valchev, Vasil Markov, Valentin Chapanov, Valeri Mirchev, Valeri Peltekov, Vasil Michaylov, Vasil Nedkov, Vasil Stoitzev, Veselin Mustakov, Venko Mustakov, Ventzislav Georgiev, Vladimir Gogov, Vladimir Pashovski, Yasen Vuchkov, Yordan Pavlov, Zahari Dechev, Zdravko Iliev, Zlatko Kachanov, and many many others. We ask the ones we have involuntarily omitted to be so kind to excuse us – the merits of everyone are acknowledged!

Traditionally, most of the scientific research on Bulgarian caves takes place in the National Museum of Natural History and the Institute of Zoology, BAS, but also in the Institute of Geography, Sofia University and other institutes.

PLANT AND ANIMAL LIFE IN BULGARIAN CAVES

CAVE FLORA AND MYCOTA

The first President of Bulgarian Caving Society Prof. Stefan Petkov published in 1943 his important paper on the cave flora in Bulgaria.

The Bacteria (Bacteriophyta) and the Blue Algae (Cyanophyta) are lower organisms on the verge of the vegetal kingdom. It has been found that Bacteria, belonging to the underground world (autochthonous) do exist. Their number varies between 10^5 and 10^7 in one gram of sediment or in one milliliter of water. They are chemiosynthetic, heterotrophic and oligotrophic.

There are more than 330 species of cave algae in the world, belonging to 6 classes: Chlorophyta, Euglenophyta, Chromophyta, Pyrrhophyta, Rhodophyta and Schizophyta. The blue algae (Cyanophyta) belong to Schizophyta, which contains more than one third of the genera and more than half (190) of the species of cave algae. It is followed by Chlorophyta, Chromophyta and the other rather insignificant ones. Troglobites are included in the groups Cyanophyta and Xanthophyta.

The research on the cave algae was carried out by S. Draganov and E.D. Dimitrova-Burin and the results were published in several papers (1968, 1977a, 1977b, 1980). At least 59 taxa of Cyanophyta, belonging to 3 orders, 7 families and 19 genera, have been established. Members of Chroococcales have been found almost exclusively in the entrance parts of the caves. What concerns the Hormogonales, 5 fam., 12 genera and more than 26 species have been found in the cave interior.

The lichens, the mosses, the ferns and the higher plants grow in the lighted or the semi obscure parts of the caves, as well as around artificial light ("Lampenflora", or "the green illness" of the caves). Blue and green algae (Cyanophyta and Chlorophyta) live on the wet cave walls around the lamps. In the caves of Germany only, more than 150 species of mosses from 60 genera and 20 families grow around the lamps. Among the ferns typical for the cave light sources is *Asplenium trichomanes*. Since the general publication of Stefan Petkov (1943), in Bulgaria no special studies on these important inhabitants of caves have been carried out.

Sometimes parasitic fungi (Ascomycetes, Laboulbeniales) are found on Diplopoda, Insecta or Acari. In Bulgaria these interesting parasites are almost unknown, but one species (*Rhachomyces middelhoeki* Banhegyi) has been described parasitizing the troglobite ground beetle *Pheggomisetes buresi* Knirsch.

CAVE FAUNA

The papers of Buresch (1924, 1926, 1927, 1936), Guéorguiev & Beron (1962), Beron & Guéorguiev (1967), Beron (1972, 1994), Delchev & Guéorguiev (1977), Beron (1986), Guéorguiev, Deltshev & Golemansky (1994) and Beron, Petrov & Stoev (in print) review all cave animals in Bulgaria, the literature on them and they outline the periods in the development of Bulgarian Biospeleology. From the first biospeleological publication (Frivaldzsky, 1879) to July 2006, more than 530

Acad. Dr Ivan Buresch (1885–1980), explorer of the cave fauna and Director of the Royal Museum of Natural History in Sofia. Honorary President of Bulgarian Federation of Speleology

articles and books on the cave animals in Bulgaria have been published. In them there is data about 780 animal species from 750 caves (the inhabitants of the psamal, of the soil, the hyporeic waters, and other non-cave underground habitats are not included). It is obvious that, despite the progress of Biospeleological studies in Bulgaria, less than one seventh of the almost 5000 caves, known in our country, have received (most of them only partially) some attention from Biospeleologists. Many years ago (Beron, 1986), we wrote "in Bulgaria there are more than 4000 caves whose fauna we know nothing about". This statement, unfortunately, remains valid even today. Nevertheless, the remarkable fact that in Bulgaria always existed many Biospeleologists, including young ones, is a guarantee that the research will continue.

Notes on the Groups of Cave Animals in Bulgaria

Protozoans (Protozoa). The free-living Protozoans in Bulgarian caves are not well known (Golemansky, 1983). Some epibionts from the genera *Vorticella* and *Tokophrya* (Ciliata), living on the cave crustaceans, have been recorded. Four endoparasitic **Gregarins** (Sporozoa) of the genera *Lepismatophila* (Stylocephalidae) and *Stenophora* (Stenophoridae) have been described. They live in *Plusiocampa bureschi* (Diplura) and in three species of Millipedes.

Flat worms (Plathelminthes). Some of the less known groups of animals in Bulgarian caves – **Planarians (Turbellaria)** – belong to this type. White planarians have been found in many caves, some of them certainly representing new species. The difficult collecting of these delicate animals in special liquids (they shrink in 70° alcohol) is the main reason for them to not have been studied in Bulgaria for so long.

Round worms (Nemathelminthes). Almost nothing is known about the multitude of tiny nematods, inhabiting cave substratum. Specialists need samples of cave clay in order to be able to extract the worms in a proper way.

Ring worms (Annelides). In the caves of Bulgaria the orders Oligochaeta (fam. Tubificidae, Haplotaxidae, Enchytraeidae, Lumbricidae) and Hirudinea (fam. Erpobdellidae) are known. Haplotaxidae is represented by the well known *"Pelodrilus"* (later *Haplotaxis*, now *De-laya*) *bureschi*. This large water worm lives not only in the cave Temnata dupka near Lakatnik, from where it was described, but also in other caves in Bulgaria, Rumania and exYugoslavia. Other such worms can be found after careful examination of the silt on the bottom of cave lakes. Their biology and zoogeography are particularly interesting.

Eight species from the **Lumbricids (Lumbricidae)** have been recorded so far in Bulgarian caves, but they are not of special interest, as these species live outside caves as well.

The **cave leeches (Hirudinea)** are, on the contrary, rather interesting. The large troglophile *Trocheta bykowskyi* lives in the caves near Komshtitsa and species of the genus *Dina* of unclear systematic position live in the caves near Gintsi and Druzhevo. Some of them live only in underground water (stygobites).

Molluscs (Mollusca). In Bulgarian caves the Mussels (**Bivalvia**) are represented only by one stygoxene species of the genus *Pisidium* (fam. Sphaeriidae). The snails (**Gastropoda**) however are numerous and include troglobites, stygobites and typical troglophiles. Due to the Polish specialist Prof. A. Riedel the family Zonitidae is well known in Bulgaria and on the Balkans. These are the transparent yellowish snails often found on the cave walls. One troglobite (?) belonging to this family has been described from the pothole Pticha dupka – the tiny white *"Lindbergia"* (now *Spinophallus*) *uminskii*. Many species of the closely related genus *Lindbergia* live in the caves of Greece, but so far *S. uminskii* is the sole species in Bulgaria. Other interesting taxa of cave snails (terrestrial and aquatic) should be expected in Bulgaria, taking into account the rich fauna of the neighbouring countries. The snails should be "drowned" in water (without air), then to be transferred to alcohol (the next day the diluted alcohol should be changed). If dropped simply in 70° alcohol, they shrink in their shells and their study becomes difficult, even in some cases impossible. So far 21 snail species have been recorded from Bulgarian caves, including 7 aquatic (fam. Hydrobiidae, the genera *Belgrandiella, Pontobelgrandiella, Saxurinator, Cavernisa* and *Insignia*). Many other species of these tiny snails, the shells of which we often find in the sand of the cave Temnata dupka near Lakatnik, are to be expected in Bulgaria. They are stygobites – live only in underground water. From our terrestrial snails, besides

Spinophallus uminskii, particularly interesting is the troglophile *Balcanodiscus frivaldskyanus*, living in the caves of the Eastern Rhodopes.

Crustaceans (Crustacea). Six orders (Ostracoda, Copepoda, Cladocera, Amphipoda, Isopoda and Decapoda) have been recorded so far in Bulgarian caves.

Ostracods are tiny bottom organisms looking like small bivalves. Only the stygobite *Pseudocandona eremita* is known from Bulgarian caves, but many other species are to be expected.

Copepods are also tiny, but planctonic organisms. During the last years, due to the Bulgarian researchers Ivan Pandurski and A. Apostolov many new species have been described from underground waters. The suborders Cyclopoida (33 sp.) and Harpacticoida (25 sp.) are represented in Bulgarian caves. The order Copepoda contains 25 stygobitic species found in Bulgarian caves (from the genera *Cyclops, Diacyclops, Speocyclops, Acanthocyclops, Elaphoidella, Stygoelaphoidella, Maraenobiotus, Nitocrellopsis*).

The **Cladocerans (Cladocera)** are not typical for the underground water (only accidental visitors are known in Bulgaria).

Almost every caver knows the **Amphipods (Amphipoda)**. Greyish, reddish or depigmentated, but always occulate, species of the genus *Gammarus* are found near cave entrances. Further in the dark, in the pools the white stygobites of the genus *Niphargus* live. Sometimes they are tiny (in the subterranean water other stygobite species of the family Gammaridae, still unknown in Bulgaria, live). Sometimes however, the *Niphargus* reach 2-3 cm (in the caves Leleshki dupki, Kamchiysko ezero nr. Bezhanovo). So far 7 species have been identified from Bulgarian caves, *N. bureschi* being the most widespread (W. Bulgaria). Besides it, also *Niphargus cepelarensis, N. pancici vlkanovi* and *N. pecarensis* have been described from Bulgaria, as well as species, captured in sources (*N. toplicensis, N. melticensis*). In Bulgaria both groups are studied by Dr S. Andreev.

The **Woodlice or Isopods (Isopoda)** are among the most interesting animals for a Biospeleologist. Three suborders are represented in our caves: Asellota, Flabellifera and Oniscidea. The **Asellots** include two species of the family Stenasellidae (*Protelsonia bureschi* and *P. lakatnicensis*), described from Temnata dupka near Kalotina and from the cave of the same name near Lakatnik. Other species are known from groundwaters.

The large Isopods of the suborder **Flabellifera** are among the most remarkable inhabitants of cave water in Bulgaria. There are two species of the family Cirolanidae – *Sphaeromides bureschi* from the caves near Tserovo, Iskrets and Beli Izvor and *S. polateni* from the source cave near Polaten, Distr. Lovech. These large water Isopods are relicts from ancient seas and it would be most interesting to find more of their localities. They are easily visible on the bottom of cave lakes and streams. The *Sphaeromides* should be collected in alcohol and it is necessary to remove the water from them with a dry tissue, or change the alcohol in a few hours, as the water dilutes the alcohol.

Isopoda Oniscidea. Terrestrial Isopods are among the most numerous and interesting troglobites in Bulgarian caves. They live on rotten wood, in clay and stalagmite surface. One of the most remarkable species (*Bureschia bulgarica*) is amphibious and occurs in big numbers around and in the lakes in the cave Temnata dupka near Lakatnik. As a whole, so far 45 species of terrestrial Isopods have been recorded from Bulgarian caves. At least 23 of them could be considered troglobites. Almost all of them are endemic. They live in a number of caves (so far Bulgarian caves are known to contain troglobitic Isopods), mostly in Stara Planina and the Rhodopes, but also in isolated localities in Russe and Blagoevgrad areas, in Strandja and Slavyanka. The troglobitic Isopoda belong to the family Trichoniscidae (*Hyloniscus, Trichoniscus, Balkanoniscus, Rhodopioniscus, Bureschia, Bulgaronethes, Bulgaroniscus, Tricyphoniscus, Beroniscus, Vandeloniscellus, Alpioniscus* (= *Ilyrionethes*) –at least 21 troglobites in total). Only one of the remaining 7 families of terrestrial Isopods, found in Bulgarian caves, contains troglobites – Styloniscidae with two species, the northernmost of the entire family.

The last order of Crustaceans inhabiting Bulgarian caves is **Decapoda** – the crayfish, well known in the sea and in the rivers, but rarely found in the caves of the Eastern Balkan Peninsula. One species of **Natantia** (*Austropotamobius torrentium*) sometimes enters the caves upstream the cave rivers. We found once a crab in the cave Samara in the Eastern Rhodopes. We still hope to find some day some of the white, blind, semi-transparent shrimps (Atyidae), living in the caves of France, Slovenia and the Transcaucasia. When looking in the cave rivers and lakes, our cavers should keep their eyes open for these 2-3 cm long shrimps – possible dwellers of Bulgarian caves.

The spider-like animals (**Arachnida**) are represented in Bulgarian caves by 7 orders: Palpigrada, Scorpiones, Pseudoscorpiones, Opiliones, Araneae and two orders of Acari. The **Palpigrads** are tiny, white, archaic animals. Their only localities in Bulgaria have so far been the caves Samuilitsa near Kunino, Mecha dupka near Salash and the pothole Randjolova Târsha near Prevala (Montana District). In Bulgaria they are unknown outside caves. **Scorpions** are rarely found in Balkan caves, mainly near the entrances. They are trogloxenes. Among the **Pseudoscorpions,** on the contrary, there are many troglobite species. So far 15 species have been recorded from Bulgarian caves, including 7 troglobites of the genera *Chthonius, Neobisium, Balcanoroncus* and *Roncus*. In the National Museum of Natural History – Sofia Pseudoscorpions found in many Bulgarian caves are preserved. The abundant material collected is still under study. Usually they are not numerous and some (Chthoniidae) are tiny and difficult to observe.

One of the most numerous groups of animals in Bulgarian caves, which are of biological importance are **spiders (Araneae).** They are well known in Bulgaria, due to Pencho Drensky (before the war) and now to Hr. Delchev and his team. So far about 80 species of spiders have been recorded from Bulgarian caves, but only six (*Centromerus bulgarianus, Trogloyhyphantes drenskii, Trogloyhyphantes* sp. n.*, Porrhomma microps, Protoleptoneta bulgarica* and *P. beroni*) are troglobites. The first is known from 4 caves (Sokolskata dupka near Lyutadjik, Belyar, Razhishkata dupka and Zidanka near Lakatnik). The second has been recorded only from Suhata dupka near Velingrad. The two species of *Protoleptoneta – P. bulgarica* and *P. beroni*, described respectively from Belimelskata peshtera (*P. beroni*) and from Djurdjina dupka near Erden and another 4 caves in the Western Predbalkan (*P. bulgarica*) are also among the many Bulgarian endemics. From the 17 families of spiders in Bulgarian caves the most numerous is Linyphiidae, containing ca. 30 species, including the two troglobites and many common troglophiles as the species of the genera *Centromerus, Troglohyphantes, Lepthyphantes, Porrhomma* and others. They are small spiders, often found among stones on cave floors and on rotten wood. The bigger troglophile spiders are most often found near the entrances. They belong to the genera *Tegenaria* (fam. Agelenidae), *Meta* (Araneae) and *Nesticus* (Nesticidae). Spiders play an important role in the food chain in cave communities.

The **Harvestmen (Opiliones)** look like spiders, but belong to a separate order. So far 19 species, out of 60 known in Bulgaria, have been recorded from Bulgarian caves. Four of them are troglobites, a fifth species from a cave in Slavyanka being under study. *Paralola buresi* from the caves near Lakatnik is the only representative of the family Phalangodidae and of the entire suborder Laniatores in Bulgarian caves. *Paranemastoma (Buresiolla) bureschi* is known from caves in Western Stara Planina. There are two members of the short-legged Opilion suborder Cyphophthalmi: *Tranteeva paradoxa* from the caves Rushovata near Gradeshnitsa and Toplya near Golyama Zhelyazna and *Siro beshkovi* from Haydushkata peshtera near Deventsi. The most common troglophile species is *Paranemastoma radewi* (Palpatores, Nemastomatidae) – a black long-legged opilion, living on rotten wood. The remaining species are occasional visitors to caves.

The large group of mites and ticks (**Acari**) is not well represented in caves. Many mites (**Acariformes and Parasitiformes**) live in the guano, but there are no troglobites among them. The only mites meeting to some extent the requirements for the category "troglobite" are some trombidiids from the Dinaric Karst and some species of Rhagidiidae (white mites, running fast on the cave clay). Flat, long-legged ticks are often seen on cave walls. They are the males of a specific bat parasite (not infesting man) – *Ixodes vespetilionis* (fam. Ixodidae). The bats harbour also other parasitic mites but they are not true cave dwellers. Bat parasites are host-specific and don't infest man.

The four classes forming the large group of **Myriapoda** are unevenly represented in caves. The small **Symphyla** and **Pauropoda** are almost unknown there, but the **Centipedes (Chilopoda),** and particularly the Millipedes (Diplopoda), are among the most important cave animals. Chilopoda are represented so far by 28 species from Bulgarian caves (2 from the order Geophilomorpha, 2 of Scolopendromorpha, 1 of Scutigeromorpha and 23 of the order Lithobiomorpha, all from the family Lithobiidae). Four *Lithobius* (*Lithobius tjasnatensis, L. stygius, L. lakatnicensis* and perhaps *L. bifidus*) are troglobites. Particularly widespread is *L. lakatnicensis,* known from 16 caves from Stara Planina and the Rhodopes. The large troglobite *Eupolybothrus andreevi* live in Vodnata peshtera near Tserovo.

The **Millipedes (Diplopoda)** are among the most interesting cave animals (together with the beetles, the woodlice and the pseudoscorpions). So far 54 species have been recorded from caves (more than half of the species of this Class living in Bulgaria), 17 of them being troglobites. From the small, white, rolling-in-ball representatives of **Glomerida** only *Trachysphaera orghidani lakatnicensis* from the caves near Lakatnik and from Tamnata dupka near Targovishte (Distr. Vidin) is considered troglobite, but *T. dobrogica* Tabacaru, described from Northern Dobrudja, could be expected in Southern Dobrudja too. Species of the order **Polydesmida** are often found in Bulgarian caves (so far 13 species and 4 subspecies). Troglobites are *Brachydesmus radewi* of Polydesmidae and two endemic subspecies of the family Bacillidesmidae – *Bacillidesmus bulgaricus bulgaricus* from the caves near Tsar Petrovo and Dolni Lom and *B. bulgaricus dentatus* from Drashanskata peshtera. Several interesting troglobites are known among the Antroleucosomatidae (**Chordeumida**, or Ascospermophora). Such are the inhabitants of the caves of Stara Planina (Ledenika and others) *Bulgarosoma bureschi* and of Rhodope caves (*Rhodoposoma rhodopinum, Troglodicus meridionale, T. tridentifer*), the species of *Anamastigona* (= *Prodicus*) – in the caves in Troyan area, Stoletovskata peshtera of Central Balkan, Lepenitsa in the Western Rhodopes, the caves near Paril and others. Each species contains the endemic genera *Stygiosoma* (*S. beroni* from Manailovata peshtera) and *Bulgardicus* (*B. tranteevi* from Bankovitsa near Karlukovo). The order **Callipodida** does not contain troglobites from our caves, but here belong four very large and typical troglophiles: *Balkanopetalum armatum* in the caves of Western Stara Planina and three species in the Rhodopes and in South Pirin. The *Typhloiulus*, particularly abundant in Temnata dupka near Lakatnik and in Vodnata pesht near Lipnitsa, belongs to the order **Iulida** (fam. Iulidae). Out of 6 species from the genus *Typhloiulus* in Bulgarian caves, 4 are considered troglobites. Similar to *Typhloiulus* is the genus *Serboiulus*, with one species *S. speleophilus* living in the caves of Northwest Bulgaria. Often the troglophile species of the genus *Apfelbeckiella*, and sometimes *Nopoiulus kochi* (= *N. venustus, N. pulchellus*), are abundant in bat guano. Other interesting troglobitic Diplopods are still under study by P. Stoev.

Many families and orders of insects avoid the cave environment. Among the insects inhabiting Bulgarian caves, we find troglobites only among the **beetles** (Coleoptera, fam. Carabidae, Cholevidae and Curculionidae), **Collembolans and Diplurans** (Diplura, Campodeidae). The remaining 12 orders (Thysanura, Ephemeroptera, Plecoptera, Orthoptera, Psocoptera, Homoptera, Heteroptera, Hymenoptera, Siphonaptera, Diptera, Trichoptera, and Lepidoptera) contain only troglophiles, trogloxenes and parasites. From more than 200 species of Collembola in the fauna of Bulgaria, 49 are known from caves, and 7 species are considered troglobites: *Onychiurus sensitivus, O. vornatscheri, Protaphorura beroni* (Onychiuridae), 3 representatives of Entomobrydae: *Pseudosinella bulgarica, P. duodecimocellata* and *P. kwartirnikovi* and one of Tomoceridae — *Tomocerus unidentatus*. We should proceed with collecting Collembola in our caves, as in the neighbouring Rumania their number is twice as big. More spectacular are the Diplurans (Diplura) — the frail white insects with long antennae and two cercae behind. Only 7 species of this order have been recorded (including 6 troglobites of the genus *Plusiocampa*), bur others are under study. The Diplurans must be collected with a wet brush and transported separately from the other animals in vials topped with alcohol, as they break easily. From the other family (Japygidae), so far there are no publications on troglobites in Bulgaria, but in the show cave Kutuki near Athens the most extreme troglobite of this group (new genus and species) has been found.

The cave beetles (**Coleoptera**) have been honoured as the first terrestrial troglobites to be described in Slovenia, Greece, Bulgaria and other countries, marking the rise of Biospeleology. In Bulgarian caves the family Carabidae is represented by 17 species of the genus *Duvalius*, 3 species and many subspecies of *Pheggomisetes* and one species *Rambousekiella* (*R. ledenikensis*), altogether 21 troglobite species, as well as by many other troglophiles and trogloxenes. Another remarkable family is Cholevidae (formerly Catopidae), represented so far by 26 troglobite species of the genera *Hexaurus, Beronia, Beroniella, Beskovia, Genestiellina, Netolitzkya, Radevia, Rhodopiola, Tranteviella, Bureschiana, Gueorguieviella, Vrat-*

zaniola and *Balkanobius* (Leptodirinae, formerly Bathysciinae), as well as by 8 "subtroglophile" species of the genera *Choleva*, *Catops*, *Nargus* and *Sciodrepoides*. Some of them (the *Beshkovia* in the caves near Tcherepish, the *Netolitzkya* in the caves near Arbanasi, the *Hexaurus* in Stoletovskata peshtera) are sometimes very numerous and could be collected for experiments without fear that this could destroy the population. Others are more rare and scarce in number. There are typical cases within the Carabidae. The *Pheggomisetes*, *Duvalius*, *Rambousekiella* and other cave beetles should not be collected excessively, as their number could decline very fast and their populations recover slowly. We can notice this in Dinevata Pesht near Gintzi. Most species are represented by few specimens, scattered and difficult to find without traps or bait. Cave beetles are important for zoogeography and they have biological peculiarities, completely unknown in Bulgaria. This is a vast field for further studies.

In the guano and litter in caves, other Coleopterans are also found, better known among which are the Staphylinidae. The bigger black Staphylinids usually belong to the genus *Quedius* and the tiny ones – to the genus *Atheta*. A total of 27 species of Staphylinids are recorded from Bulgarian caves, but this number does not contain troglobites. Other families of Coleoptera in Bulgarian caves are Dytiscidae, Hydrophilidae, Histeridae, Pselaphidae, Colydiidae, Endomychidae, Cryptophagidae, Curculionidae and Ptinidae. The Pselaphids and the weevils (Curculionidae) are of particular biospeleological interest. We could expect troglobites among the Pselaphids – such are known from Greece and ex-Yugoslavia. On the roots of plants in the caves, the careful observer may notice tiny weevils (*Troglorrhynchus*, Curculionidae). Other genera of this family, living underground are known from Greece.

The flying insects in caves are of limited interest. They are not troglobites and rarely the material collected contains new species. Sometime the large wasp – like **Hymenoptera** of the genus *Diphyus* (formerly "*Amblyteles*") (Ichneumonidae), hide among the stalactites. Often moth-like insects on the ceiling with wings like roof attract our attention. They are called **caddis flies (Trichoptera)**, 16 species of which are known in Bulgarian caves. Three genera of Limnephilidae: *Micropterna* (6 species), *Stenophylax* (3 sp.) and *Mesophylax* (1 sp.), accepted as regular trogloxenes, deserve particular attention. The remaining 6 sp. are occasional trogloxenes. Some **moths (Lepidoptera)** also regularly occur in caves. From the 17 species of the order Lepidoptera such are the two species of Geometridae *Triphosa dubitata* and *T. sabaudiata*, some Noctuids of the genera *Apopestes*, *Scoliopteryx* and others, representatives of Alucitidae (Orneodidae) and Tineidae. The last mentioned moths often live in cave guano and are inadequately known in Bulgaria, despite the good Catalogue of Cave Lepidoptera. They should be collected in dry tubes, the other Lepidoptera – in paper containers. The **flies (Diptera)**, among which 56 known and many other unrecorded species living in Bulgarian caves, can be collected in alcohol. More typical cave dwellers are the species of the families Muscidae, Helomysidae, Mycetophilidae, as well as the long-legged Limnobiids, often seen on cave walls.

The wingless parasitic flies of the family Nycteribiidae, easily seen on the fur of the bats, form a special group. Recently a species from another family of bat parasites (Streblidae) was added to the fauna of Bulgaria, due to the efforts of Bulgarian cave biologists (*Brachytarsina flavipennis* Mackwart).

The **Vertebrates** are not very numerous in Bulgarian caves. Sometime fish are captured deep inside some caves, but these are always species living in the surface water. No blind fish has ever been found in Europe. Frogs, toads or larvae of salamanders are rarely met in caves. At the entrances we can find nests of pigeons, swallows, the alpine chough (*Pyrrhocorax graculus*) nests in some deep potholes. This interesting bird is protected by law and its colonies need to be preserved, especially during the breeding season. Deep in caves, neter mice, dormice, foxes and other mammals can be found.

For the biospeleologist the most interesting cave mammals are the **bats (Chiroptera)**, 16 species of which have been recorded in Bulgarian caves (5 belonging to Rhinolophidae and 11 – to Vespertilionidae). Large colonies are formed by the Longwing Bat (*Miniopterus schreibersi*) and by both large species of the genus *Myotis* (*M. myotis* and *M. blythi*). Several protected winter and summer roosts have been observed, with colonies of *Myotis capaccinii* and horse shoe bats of European importance. Bulgaria enjoys the full list of European bats and some very numerous colonies, worth protection. Visits, especially of larger groups, in bat caves like Parnitsite near Bezhanovo, are very improper. The disturbing of breeding colonies, or in the winter roosts, could deprive our country of a national wealth. Some Central European countries are already facing this threat. Every caving club should take a special account of the bat caves in its region and protect them.

Troglobites and Stygobites in Bulgarian Caves
(some remarkable troglophiles are included)

Phyllum Annelides – Segmented Worms

Order Oligochaeta
Fam. Haplotaxidae
Delaya [*Haplotaxis*] *bureschi* (Michaelsen) – Cave Temnata Dupka (Lakatnik Railway Station). **Stygobite.**

Order Hirudinea – Leaches
Fam. Erpobdellidae
Dina absoloni Johansson – Cave Dinevata Pesht (Gintsi). **Stygobite.**
D. lineata arndti Augener - Suhata Jama (Druzhevo). **Stygobite.**

Phyllum Mollusca – Molluscs

Class Gastropoda – Snails

Order Basommatophora
Fam. Zonitidae
Spinophallus (before sub *"Lindbergia"*) *uminskii* (Riedel) – Ptichata Dupka (Cherni Ossam). Troglophile?
Balcanodiscus frivaldskyanus (Rossmaessler) – Caves Tilki Ini (Ostrovitsa), Hasarskata Peshtera (Gorna Snezhinka). Troglophile.
Fam. Hydrobiidae
Belgrandiella hessei A.J. Wagner – Cave Temnata Dupka (Lakatnik Railway Station). **Stygobite.**
B. bulgarica Angelov – The source cave Izvora (Polaten). **Stygobite.**
Pontobelgrandiella nitida (Angelov) – The source cave Izvora (Polaten). **Stygobite.**
Saxurinator bureschi (A.J. Wagner) – Cave Temnata Dupka (Lakalnik Railway Station). **Stygobite.**
S. copiosus (Angelov) – The source cave Izvora near Polaten. **Stygobite.**
Insignia macrostoma Angelov – The source cave Izvora near Polaten. **Stygobite.**
Cavernisa zaschevi (Angelov) – Cave Dushnika (Iskrets). **Stygobite.**
Iglica acicularis Angelov – Cave Dushnika (Iskrets). **Stygobite.**

Phyllum Arthropoda – Arthropods

Class Crustacea – Crustaceans

Order Ostracoda – Ostracods
Fam. Candonidae
Pseudocandona eremita (Vejdovsky) – Cave Temnata Dupka (Lakatnik Railway Station). **Stygobite.**

Order Copepoda – Copepods
Suborder Cyclopoida
Fam. Cyclopidae
Cyclops bohater ponorensis Naidenow et Pandourski – Cave Katsite (Zimevitsa). **Stygobite.**
Diacyclops chappuisi Naidenow et Pandourski – Caves Haydushkata Peshtera (Zemen) and Duhlata (Bosnek). **Stygobite.**
D. haemusi Naidenow et Pandourski – Cave Katsite (Zimevitsa). **Stygobite.**
D. strimonis Pandourski – Caves Duhlata and Zhivata Voda (Bosnek). **Stygobite.**
D. stygius (Chappuis) – Cave Dushnika (Iskrets). **Stygobite.**
D. clandestinus (Kiefer) – Caves Dushnika (Iskrets), Temnata Dupka (Lakatnik Railway Station) and Vodnata Pesht (Lipnitsa). **Stygobite.**
D. pelagonicus saetosus Pandourski – Cave Dushnika (Iskretz). **Stygobite**
Speocyclops rhodopensis Pandourski – Caves Sbirkovata Peshtera (Chepelare) and Lednicata (Gela). **Stygobite.**
S. d. demetiensis (Scourfield) – Caves Yalovitsa (Golyama Zhelyazna) and Brunoshushinskata Peshtera (Gortalovo). **Stygobite.**
S. infernus (Kiefer) – Caves Razhishkata Peshtera (Lakatnik), Ledenika, Belyar (Vratsa), 25 Years of Akademik (Gorno Ozirovo), Haydushkata Peshtera (Deventsi), Toplya and Yalovitsa (Golyama Zhelyazna). **Stygobite.**
S. lindbergi Damian – Caves Tsarkvishte, Marina Dupka and Travninata (Breze), Ezeroto (Gabare), Temnata Dupka (Berende Izvor), Bezimenna 22 (Karlukovo), Mechata Dupka (Zhelen), Haydushkata Peshtera (Deventsi) and Dushnika (Iskrets). **Stygobite.**

Acanthocyclops iskrecensis Pandourski – Caves Elata (Zimevitsa) and Dinevata Pesht (Ginci). **Stygobite.** Known also from karstic sources.

A. propinquus (Plesa) – Caves Peshterata na Bay Bore (Old Bore's Cave), Djeranica and Zhivata Voda (Bosnek). **Stygobite.**

A. radevi Pandourski – Caves Yamata and Vodnata Peshtera (Tserovo Railway Station). **Stygobite.**

A. balcanicus Naidenow et Pandourski – Cave Rushovata Peshtera (Gradeshnitsa). **Stygobite.**

Suborder Harpacticoida
Fam. Canthocamptidae

Elaphoidella cavernicola Apostolov – Cave Zidanka (Karlukovo). **Stygobite.**

E. karamani latifurcata Apostolov – Cave Orlova Chuka (Pepelina). **Stygobite.**

E. balcanica Apostolov – Cave 25 Years of Akademik (Gorno Ozirovo). **Stygobite.**

E. pandurskyi' Apostolov – Caves Akademik and Duhlata (Bosnek). **Stygobite.**

Stygoelaphoidella stygia Apostolov – Cave Vodnata Pesht (Lipnitsa) and Lyastovitsa (Glozhene). **Stygobite.**

S. bulgarica Apostolov – Cave Drashanskata Peshtera (Drashan). **Stygobite.**

S. elegans Apostolov – Cave Marina Dupka (Breze), Tsarkvishte and Dushnika (Iskrets). **Stygobite.**

Maraenobiotus bulbiseta Bassamakov et Apostolov – Cave Imamova Dupka (Yagodina). **Stygobite.**

M. parainsignipes Apostolov – Cave Dushnika (Iskrets). **Stygobite.**

Nitocrellopsis intermedia (Chappuis) – Cave Haydushkata Peshtera (Zemen). **Stygobite.**

Order Amphipoda – Amphipods
Fam. Gammaridae

Niphargus ablaskih georgievi St. et G. Karaman – Caves Popskata Peshtera and Urushka Maara (Krushuna), Stalbitsa (Karpachevo). **Stygobite.**

N. jovanovici St. Karaman – Caves Shokyovets (Cherkaski) and Sushica (Studeno Buche). **Stygobite.**

N. cepelarensis St. et G. Karaman – Cave Sbirkovata Peshtera (Progled). **Stygobite.**

N. pancici vikanovi St. et G. Karaman – Cave Zhivata Voda (Bosnek). **Stygobite**.

N. puteanus C.L. Koch – Cave Sbirkovata Peshtera (Progled). **Stygobite**.

N. bureschi Fage – Caves Falkovskata Peshtera (Falkovets), Golemata Mikrenska Peshtera (Mikre), Temnata Dupka (Lakatnik Railway Station), Temnata Dupka (Berende Izvor), Zadanenka (Karlukovo), Ptichata Dupka (Cherni Ossam), Lepenitsa (Velingrad), Dinevata Pesht (Gintsi), Drashanskata Peshtera (Drashan), Popovata Peshtera, Starata Prodanka and Moravata (Gabare), Golemi Pech (Varbovo), Skravenika (Karlukovo), Gornoto Ezero (Breste), Belyar (Vratsa), Kondjova Krusha (Chiren), Zvankova Dupka (Lilyache). **Stygobite.**

N. pecarensis St. et G. Karaman – Caves Golemi Pech (Varbovo) and Vodni Pech (Bashovishki Pech) near Oreshets Railway Station. **Stygobite.**

Order Isopoda – Woodlice
Suborder Asellota
Fam. Stenasellidae

Protelsonia bureschi (Racovitza) – Cave Temnata Dupka (Berende Izvor). **Stygobite.**

P. lakatnicensis (Buresch et Gueorguiev) – Cave Temnata Dupka (Lakatnik Railway Station). **Stygobite.**

Suborder Flabellifera
Fam. Cirolanidae

Sphaeromides bureschi Strouhal – Caves Vodnata Peshtera (Tserovo Railway Station), Belyar (Vratsa), Dushnika (Iskrets) and Toshova Dupka (= Kalna Matnitsa) (Stoyanovo). **Stygobite.**

S. polateni Angelov – The spring cave Izvora (Polaten). **Stygobite.**

Suborder Oniscidea
Fam Styloniscidae

Cordioniscus bulgaricus Andreev – Cave Boychovata peshtera (Logodash). **Troglobite.**

C. schmalfussi Andreev – Cave Shepran dupka (Belitsa). **Troglobite.**

Fam. Trichoniscidae

Bureschia bulgarica Verhoeff – Caves Vodnata Peshtera (Tserovo Railway Station), Temnata Dupka (Lakatnik Railway Station) and Belyar (Vratsa). **Troglobite.**

Balcanoniscus corniculatus Verhoeff – Caves Haydushkata Dupka (Karlukovo), Rushovata Peshtera and Djebin trap near Gradeshnitsa, Yamata (Tserovo Railway Station), Razrushenata Peshtera (Kunino Railway Station). **Troglobite.**

B. minimus Vandel – Cave Parnitsite (Bezhanovo), Danchova Dupka (Aglen). **Troglobite.**

Rhodopioniscus beroni (Vandel) – Caves Topchika, Ahmetyova Dupka and Yamata (Dobrostan), Shepran Dupka (Belitsa). **Troglobite.**

Trichoniscus bulgaricus Andreev – Yulen Ere 2 (Hristo Danovo). **Troglobite.**

T. rhodopiense Vandel – Caves Karagug (Peshterata) (Târnovtsi), Nadarskata Peshtera (Kremene), Boevskata Peshtera (Boevo), Samara (Ribino), Zlatnata Yama (Kremen), Zmiin Borun (Mostovo), Kraypatnata Peshtera (Smilyan), Ayna Ini (Ribino), Byaloto Kamene (Breze near Devin), Hasarskata Peshtera (Gorna Snezhinka), Bezimenna Peshtera (Svetulka), Rupata (Egrek). **Troglobite.**

T. valkanovi Andreev – Caves Sarpiyskata Peshtera, Kirechnitsata (Kosti). **Troglobite.**

T. anophthalmus Vandel – Caves Studenata Dupka and Ezeroto (Cherepish Railway Station), Mishin Kamik, Vreloto and Ayduchkata Dupka (Prevala), Tatarska Dupka (Replyana), Parasinskata Propast (Belimel), Toshova Dupka (Stoyanovo), Razrushenata peshtera (Kunino). **Troglobite.**

T. tranteevi Andreev (= *T. anophthalmus intermedius* Vandel) – Tâmnata Peshtera (Tabachka). **Troglobite.**

T. petrovi Andreev – Cave Shepran Dupka (Belitsa). **Troglobite.**

T. tenebrarum Verhoeff – Cave Andaka (Dryanovski Monastir), Golyama Podlistsa (Veliko Tarnovo), Devetashkata Peshtera (Devetaki), Gornik (Krushuna). **Troglobite.**

T. stoevi Andreev – Stoletovskata Peshtera (Shipka). **Troglobite.**

T. bononiensis Vandel – Caves Varkan (Druzhba), Suhi Pech (Oreshets), Rushkovitsa, Redaka 1, Prelaz and Mechata Dupka (Salash), Yame 3 (Targovishte), Propast (Oreshets), Krachimirskoto Vrelo (Stakevtsi), Zmiyskata Propast (Krachimir), Tsankinoto Vrelo (Granitovo), Haydushkata Propast and Neprivetlivata (=Gornata Propast) near Belogradchik, Tamni Pech and Golemi Pech (Varbovo), Pech (Gorna Luka), Yame 2 (Targovishte), Parnak (Oreshets), Yankulova Dupka (Prolaznitsa), Falkovskata peshtera (Falkovets). **Troglobite.**

T. garevi Andreev – Caves Sinyoto Ezero (Dragana), Alchashkata Peshtera (Bezhanovo), Popskata Peshtera (Krushuna), Tanyova Peshtera and Voditsata (Aglen), Skoka (Dragana), Gornik (Krushuna). **Troglobite.**

T. beroni Andreev – Cave Bratanovskata Peshtera (Kosti). **Troglobite.**

Beroniscus capreolus Vandel – Cave Parnitsite (Bezhanovo). **Troglobite.**

Bulgaroniscus gueorguievi Vandel – Caves Desni Suhi Pech (Dolni Lom), Zhivkova Dupka (Gorna Luka), Vreloto, Ayduchkata Dupka and Mitsina Dupka (Prevala), Neprivetlivata (Belogradchik), Yame 2 (Târgovishte), Tatarska Dupka (Replyana), Parnak (Oreshets). **Troglobite.**

Bulgaronethes haplophthalmoides Vandel – Novata Peshtera (Peshtera), Suhata Dupka. **Troglobite.**

Vandeloniscellus bulgaricus (Vandel)(= *Cyphoniscellus b.*) – Toshova Dupka (Stoyanovo). **Troglobite.**

Alpionicus (=*Ilyrionethes*) sp. – Starshelitsa (Goleshevo). **Troglobite.**

Tricyphoniscus bureschi Verhoeff – Caves Haydushka Dupka, Bezimenna 22 and Cherdjenitsa (Karlukovo), Vodnata Pesht, Kozarnika and Bankovets (Lipnitsa), Danchova Dupka (Aglen). **Troglobite.**

Hyloniscus flammula Vandel – Caves Magura, Magura 2 and Varnitsata (Rabisha). **Troglobite.**

Class Arachnida – Arachnids

Order Opiliones – Harvestmen
Suborder Cyphophthalmi
Fam. Sironidae
Siro beschkovi Mitev – Cave Haydushkata Peshtera (Deventsi). **Troglobite.**

Tranteeva paradoxa Kratochvil – Caves Rushovata Peshtera (Gradeshnitsa), Toplya (Golyama Zhelyazna). **Troglobite.**

Suborder Laniatores
Fam. Phalangodidae
Paralola buresi Kratochvil – Caves Kozarskata, Svinskata Dupka, Zidanka and Temnata Dupka (Lakatnik Railway Station). **Troglobite.**

Suborder Palpatores
Fam. Nemastomatidae
Paranemastoma (Buresiolla) bureschi (Roewer) – known from 31 Bulgarian caves: Dushnika and Otechestvo (Iskrets), Svardelo, Temnata Dupka, Zidanka and Yavoretskata Peshtera (Lakatnik Railway Station), Poroynata Dupka (Zasele), Sokolskata Dupka (Lyutadjik),

Shamak (Barlya), Propastta and Vodnata Peshtera (Tserovo Railway Station), Radolova Yama, Dinevata Pesht, Svetata Voda and Krivata Pesht (Gintsi), Mechata Dupka (Zhelen), Garvanets, Zmeyova Dupka, Reznyovete, Ledenika, Ledenishka Yama, Malkata Mecha Dupka and Golemata Mecha Dupka (Vratsa), Haydushka Dupka (Bistrets), Medenik (Eliseyna), (?) Saeva Dupka (Brestnitsa), Kitova Kukla (Druzhevo), Elata, Kolkina Dupka and Katsite (Zimevitsa), Radyova Propast (Milanovo). **Troglobite.**

Order Pseudoscorpiones – Pseudoscorpions
Fam. Chthoniidae
Chthonius troglodites Redikorzev – Kassapnitsite and Haydushka Dupka (Karlukovo Railway Station), Haydushkata Peshtera (Deventsi). **Troglobite.**
Fam. Neobisiidae
Neobisium (Heoblothrus) bulgaricum (Redikorzev)(syn. *Obisium subterraneum* Redikorzev) – Cave Yalovitsa (Golyama Zhelyazna). **Troglobite.**
N. (H.) beroni Beier - Cave Svinskata Dupka (Lakatnik Railway Station). **Troglobite.**
N. (Blothrus) kwartirnikovi Mahnert – Cave Duhlata (Bosnek). **Troglobite.**
Balkanoroncus bureschi (Redikorzev)(=*Obisium bureschi* Redikorzev = *Balkanoroncus praeceps* Ćurčić) – Cave Sâeva Dupka (Brestnitsa), Djebin Trap (Gradeshnitsa). **Troglobite.**
B. hadzii Harvey (= *Roncus bureschi* Hadzi) – Razhishka Peshtera (Lakatnik Railway Station). **Troglobite.**
Roncus mahnerti Ćurčić et Beron – Cave Vodna (Vodnata Peshtera) (Botunya). **Troglobite.**

Order Araneae – Spiders
Fam. Leptonetidae
Protoleptoneta beroni Deltchev – Belimelskata Peshtera (Beli Mel). **Troglobite.**
P. bulgarica Deltchev – Mecha Dupka (Lepitsa), Djurdjina Dupka (Erden). **Troglobite.**
Fam. Linyphiidae
Troglohyphantes drenskii Deltshev – Suhata Dupka, Lepenitsa (Velingrad). **Troglobite.**
Centromerus bulgarianus (Drensky) – Caves Zidanka and Razhishka Peshtera (Lakatnik), Sokolskata Dupka (Lyutadjik), Belyar (Vratsa). **Troglobite.**
Porrhomma microps (Roewer) – Kulina Dupka (Krivnya), Samar Daala (Voden), Prikazna (Kotel). **Troglobite.**

Myriapoda

Class Chilopoda – Centipedes
Fam. Lithobiidae
Lithobius tjasnatensis Matic (= *L. popovi* Matic) – Tyasnata Propast (Mramor). **Troglobite.**
L. stygius Latzel – Caves Lepenitsa and Suhata Peshtera (Velingrad), Yubileyna (Peshtera Town), Modarskata Peshtera (Modar), Lednitsata (Gela). **Troglobite.**
L. lakatnicensis Verhoeff – Caves Svinskata Dupka, Zidanka and Temnata Dupka (Lakatnik Railway Station), Novata Peshtera (Peshtera Town), Mechata Dupka (Zhelen), Grebenyo (Dolno Ozirovo), Yamata (Dobrostan), etc. **Troglobite.**
L. rushovensis Matic, 1967 (= *L. beschkovi* Matic et Golemansky) – Rushovata Peshtera (Gradeshnitsa) and others. **Troglobite.**
L. bifidus (Matic) – Cave Izvornata Peshtera (MIadezhko). **Troglobite** (?).
Eupolybothrus andreevi Matic – Vodnata Peshtera (Tserovo Railway Station). **Troglobite.**

Class Diplopoda – Millipedes

Order Glomerida
Fam. Doleriidae
Trachysphaera orghidani lakatnicensis Tabacaru – Caves Kozarskata Peshtera and Sedmovratitsa (Lakatnik Railway Station), Tamna Dupka (Targovishte). **Troglobite.**

Order Polydesmida
Fam. Polydesmidae
Brachydesmus radewi Verhoeff – Cave Promakinyalo (Dolna Beshovitsa), Chelovecha dupka (Vr 77). **Troglobite.**
Fam. Trichopolydesmidae (Bacillidesmidae)
Bacillidesmus bulgaricus bulgaricus Strasser – Mishin Kamik (Gorna Luka), Mladenovata Peshtera, Ponora (Chiren). **Troglobite.**
B. bulgaricus dentatus Strasser – Cave Drashanskata Peshtera (Drashan), Saeva dupka (Brestnitsa). **Troglobite.**
Fam. Anthroleucosomatidae
Bulgarosoma bureschi Verhoeff – Ledenika, Belyar, Malkata Mecha Dupka, Reznyovete (Vratsa), Kitova Kukla (Druzhevo). **Troglobite.**
Troglodicus meridionale (Tabacaru) – Imamova Dupka (Yagodina). **Troglobite.**
T. tridentifer Gulička – Chelechkata (Choveshkata) Peshtera (Orehovo). **Troglobite.**

Stygiosoma beroni Gulička – Manailovata Peshtera (Ribnovo). **Troglobite.**

Anamastigona lepenicae (Strasser) – Lepenitsa, Suhata peshtera (Velingrad). **Troglobite.**

A. delcevi (Strasser) – Art. galleries near Paril, Pothole Stoykova Dupka 1 (Goleshevo, Beron et Petrov leg., Hr. Deltshev def.). **Troglobite.**

A. alba (Strasser) – Pticha Dupka, Kumanitsa (Cherni Osam). **Troglobite.**

Bulgardicus tranteevi Strasser – Bankovitsa (Karlukovo). **Troglobite.**

Order Callipodida
Fam. Schizopetalidae

Balkanopetalum armatum Verhoeff – Studenata Dupka (Cherepish Railway Station), Vodnata Peshtera (Tserovo Railway Station), Kolibata (Beledie Han), Dushnika (Iskrets), Komina (Beledie han), Peshtereto (Lakatnik). Troglophile.

B. rhodopinum Verhoeff – Novata Peshtera, Yubileyna (Peshtera Town). Troglophile.

B. beskovi Strasser – Topchika, Druzhba, Yamata, Hralup, Pirkovskata peshtera (Dobrostan), Garvanyovitsa (Turen), Zmiin Borum (Mostovo). Troglophile.

B. petrovi Stoev et Enghoff – Samara (Ribino), Ogledalnata Peshtera, or Ayna Ini (Ribino), Zlatnata Yama (Kremen). Troglophile.

Balkanopetalum bulgaricum Stoev et Enghoff – Starshelitsa (Goleshevo). Troglophile.

Order Julida
Fam. Julidae

Typhloiulus bureschi Verhoeff – Mechata Dupka (Zhelen), Popovata Peshtera, Moravata (Gabare), Haydushka dupka, Svirchovitsa, Zadanenka, Zidanka, Bezimenna 22, Bankovitsa (Karlukovo), Sopotskata Peshtera (Sopot), Nikolova Yama (Dolno Ozirovo), Tsarkvishte, Padezh (Breze), Paraklisa (Bov), Toshova Dupka (Stoyanovo), Ponora (Chiren). Temnata dupka, Zidanka, Razhishkata peshtera, Svinskata dupka, Razhishka yama (Lakatnik), Vodnata peshtera, Mayanitsa (Tserovo), Promakinyalo (Dolna Beshovitsa), Yamata (Gintsi), Vodnata pesht, Bankovets (Lipnitsa), Sipo (Dolno Ozirovo), Toshova dupka (Glavatsi), Mandrata (Mikre), Shipochinata (Kunino). **Troglobite.**

T. georgievi Verhoeff – Futyovskata peshtera (Karpachevo), Popskata Peshtera (Krushuna), Toplya (Golyama Zhelyazna), Tsarskata peshtera (Belyakovets). **Troglobite.**

T. longipes Strasser – Belyar (Vratsa). **Troglobite.**

T. staregai Strasser – Prelaz (Salash). **Troglobite.**

Serboiulus spelaeophilus Gulička (=*S. popovi strasser*) – Shokyovets (Cherkaski), Mramornata Peshtera (Berkovitsa), Desni Suhi pech, Vodni Pech (Dolni Lom), Yame 2, Yame 3, Tâmna Dupka (Targovishte), Redaka 1, Redaka 2 (Salash), Tâmni Pech (Varbovo), Stanishina Dupka, Medjak Dupka (Replyana), Zmiyskata Propast (Krachimir), Falkovskata Peshtera (Falkovets), Zankovska Peshtera (Belotintsi), Randjolova Tarsha, Mishin Kamik, Vreloto, Pleshovskata peshtera (Prevala), Lisicha dupka, Parasinskata propast (Belimel), Pech (Gorna Luka), Propastta v Bucheto (Georgi Damyanovo), Lipova dupka (Mitrovtsi), Dupkata (Stubel), Propastta (Vladimirovo), Shtastie, Neprivetlivata (Belogradchik), Parnak (Oreshets). **Troglobite.**

Class Insecta – Insects

Order Collembola – Collembolans
Fam. Entomobryidae

Pseudosinella bulgarica Gama – Stoletovskata Peshtera (Shipka), Prikazna (Kotel). **Troglobite.**

P. duodecimocellata Handschin – Cherdjenitsa (Karlukovo), Ledenika (Vratsa), Vodopada (Krushuna), Imamova Dupka (Yagodina), Prikazna, Dryanovskata Peshtera (Kotel), Sbirkovata Peshtera (Chepelare), Prelaz (Salash), Murata (Ilindentsi), Padezh (Breze), Yavoretskata Peshtera, Temnata Dupka, Razhishka Peshtera (Lakatnik). – **Troglobite.**

P. kwartirnikovi Gama – Gurlyova Dupka, Tavancheto (Krushuna), Stoletovskata Peshtera (Shipka), Ovnarkata (Karlukovo), Kozarnika (Lipnitsa). **Troglobite.**

Fam. Onychiuridae

Onychiurus sensitivus Handschin – Ledenika (Vratsa). **Troglobite.**

O. vornatscheri Stach – Tsankinoto Vrelo (Granitovo). **Troglobite.**

Protaphorura beroni (Gruia) – Venetsa, Propast (Oreshets). **Troglobite.**

Order Diplura – Diplurans
Fam. Campodeidae

Plusiocampa bulgarica Silvestri – Lepenitsa (Velingrad), Yavoretskata Peshtera (Lakatnik), Hvoynenskata Peshtera (Hvoyna), Sbirkovata Peshtera (Progled), Haramiyskata Dupka (Trigrad), Dupkite (Chepelare), Imamova Dupka (Yagodina), Magura (Rabisha). **Troglobite.**

P. bureschi Silvestri (syn. *P. rauseri* Rusek) – Temnata Dupka, Razhishkata Peshtera (Lakatnik), Vodnata Peshtera (Tserovo Railway Station). **Troglobite.**

P. beroni Bareth et Condé – Magura (Rabisha), Varkan (Tsar Petrovo). **Troglobite.**

P. cf. *beroni* I Bareth et Condé – Novata peshtera (Peshtera Town). **Troglobite.**

P. cf. *beroni* II Bareth et Condé – Padezh (Breze). **Troglobite.**

P. gueorguievi Bareth et Condé – Toplya (Golyama Zhelyazna). **Troglobite.**

P. arbanisiensis Bareth et Condé – Lyaskovskata peshtera (Arbanasi). **Troglobite.**

P. vodniensis Bareth et Condé – Vodni Pech (Dolni Lom). **Troglobite.**

Order Orthoptera – Orthopterans (Grasshoppers and Crickets)
Fam. Raphidophoridae
Troglophilus neglectus Krauss – many caves in Western Bulgaria and in the Rhodopes. Troglophile.

Order Coleoptera - Beetles
Fam. Carabidae – Ground Beetles
Duvalius (Paraduvalius) balcanicus (J. Frivaldzsky) – A cave in Shipchensky Balkan. **Troglobite.**

D. (P.) beroni V. Guéorguiev – Toshova Dupka (Stoyanovo). **Troglobite.**

D. (P.) bulgaricus Knirsch – Zmeyovi Dupki (Hitrevtsi), Kumincheto (Genchevtsi). **Troglobite.**

D. (P.) bureschi Jeannel – Lepenitsa, Dupcheto (Velingrad). **Troglobite.**

D. (P.) garevi Casale et Genest - Sinyoto Ezero (Dragana). **Troglobite.**

D. joakimovi B. Guéorguiev – Stoykova dupka (not Golyama Stojkovitsa, as in the description !) (Goleshevo, Mt. Slavyanka). **Troglobite.**

D. (P.) kotelensis Genest – Prikazna (Kotel). **Troglobite.**

D. (P.) nedelkovi B. Guéorguiev - Western Rhodopes. **Troglobite.**

D. (P.) papasoffi Mandl – Temnata Dupka, Zidanka, Pyasachnata Dupka (Lakatnik Railway Station). **Troglobite.**

D. (P.) petrovi B. Guéorguiev – Zandana (Dolno Cherkovishte). **Troglobite.**

D. (P.) pirinensis B. Guéorguiev – Sharalijskata peshtera (Ilindentsi). **Troglobite.**

D. (P.) pretneri V. Guéorguiev – Mechata Dupka (Bov). **Troglobite.**

D. (P.) regisborisi Buresch – Yalovitsa (Golyama Zhelyazna). **Troglobite.**

D. (P.) zivkovi Knirsch – Ledenika, Malkata Mecha Dupka, Malkata Nevestina Propast, 25 Godini of Akademik (Vratsa). **Troglobite.**

D. (P.) legrandi Genest – Kalugerova Dupka (Arbanasi). **Troglobite.**

D. (P.) karelhurkai Farkac - Erkyupriya (Mostovo). **Troglobite.**

D. (Duvaliotes) beshkovi Coiffait - Mecha (Lisicha) Dupka (Stradalovo). **Troglobite.**

Pheggomisetes buresi buresi Knirsch – Ledenika, 25 Godini of Akademik (Vratsa). **Troglobite.**

Ph. buresi medenikensis Knirsch – Medenik (Eliseyna Railway Station). **Troglobite.**

Ph. globiceps globiceps Buresch - Dushnika (Iskrets). **Troglobite.**

Ph. globiceps breiti Mandl – Dinevata Pesht, Svetata Voda, Krivata Pesht (Gintsi). **Troglobite.**

Ph. globiceps georgievi Z. Karaman – Propast 30 (Karlukovo). **Troglobite.**

Ph. globiceps lakatnicensis Jeannel – Temnata Dupka, Zidanka (Lakatnik Railway Station), Radyova Propast (Milanovo), Kolkina Dupka (Zimevitsa), Golemata Mecha Dupka (Vratsa), Govedarskata Dupka (Chiren). **Troglobite.**

Ph. globiceps stoicevi V. Guéorguiev – Elata (Zimevitsa), Nevestina Propast (Vratsa). **Troglobite.**

Ph. globiceps karlukovensis Genest – Bezimenna 84 (Karlukovo). **Troglobite.**

Ph. globiceps ilandjievi V. Guéorguiev – Balabanova Dupka, Malata Balabanova Dupka, Granicharskata Propast (Komshtitsa). **Troglobite.**

Ph. globiceps mladenovi V. Guéorguiev – Malkata Mecha Dupka, 25 Godini of Akademik (Vratsa). **Troglobite.**

Ph. globiceps cerovensis V. Guéorguiev – Yamata, Peshterata, Propastta (Tserovo Railway Station). **Troglobite.**

Ph. radevi radevi Knirsch - Ledenika (Vratsa). **Troglobite.**

Ph. radevi tranteevi V. Guéorguiev – Suhata Yama (Druzhevo). **Troglobite.**

Rambousekiella ledenikensis Knirsch - Ledenika (Vratsa), Grebenyo (Gorno Ozirovo). **Troglobite.**

Fam. Cholevidae (Catopidae)
Beronia micevi V. Guéorguiev – Neprivetlivata (Gornata Propast), Haydushkata Propast

(Belogradchik), Dzhamiite (Bela), Venetsa, Propast (Oreshets). **Troglobite.**

B. andreevi Giachino et B. Guéorguiev – Ajduchka dupka (Prevala), Vodni Pech (Dolni Lom). **Troglobite.**

Beskovia bulgarica V. Guéorguiev – Studenata Dupka, Serapionovata Peshtera (Cherepish Railway Station). **Troglobite.**

B. tranteevi Giachino et Guéorguiev – Bezimenna 22 (Karlukovo). **Troglobite.**

B. beroni Giachino et Guéorguiev – Kozarskata peshtera (Lakatnik Railway Station). **Troglobite.**

Hexaurus merkli (J. Frivaldzsky) – the cave below Kurudja peak (?). **Troglobite.**

H. schipkaensis Zerche – Stoletovskata Peshtera (Shipka), Sokolskata peshtera (Gabrovo). **Troglobite.**

H. similis (J. Frivaldzsky) – A cave in Shipchensky Balkan. **Troglobite.**

H. paradisi Zerche – Han Maara (Ray Hut). **Troglobite.**

H. beroni Giachino et Guéorguiev – Duhaloto (Vidima). **Troglobite.**

Netolitzkya jeanneli jeanneli Buresch – Bacho Kiro, Andaka (Dryanovski Manastir). **Troglobite.**

N. jeanneli matroffi Jeannel – Lyaskovska Peshtera, Kalugerova Dupka (Arbanasi). **Troglobite.**

N. maneki maneki J. Müller - Zmeyovi Dupki (Hitrevtsi), Kumincheto (Genchevtsi), Bacho Kiro (Dryanovski Manastir). **Troglobite.**

N. maneki iltschewi Jeannel – Golyama Podlisca (Belyakovets), Troana, Bambalova dupka (Emen). **Troglobite.**

Radevia hanusi Knirsch - Ledenika, Bezimenna, Zmeyova Dupka I, Zmeyova Dupka III, Propast 13, Bulina Dupka, Golemata Mecha Dupka, Radyova Propast (Vratza). **Troglobite.**

Rhodopiola cavicola V. Guéorguiev – Sipeya (Bachkovski Manastir). **Troglobite.**

Vratzaniola pandurskii Dupré – 25 godini of Akademik, Barkite No 9 (Vratsa). **Troglobite.**

Tranteeviella bulgarica Pretner - Rushovata Peshtera (Gradeshnitsa), Varovit (Malka Zhelyazna). **Troglobite.**

Bureschiana drenskii V. Guéorguiev – Tilkiini (Ostrovitsa), Hasarskata Peshtera (Gorna Snezhinka), Maarata (Madrets). **Troglobite.**

B. thracica Giachino – Vodnata peshtera (Nedelino). **Troglobite.**

B. raitchevi Giachino - Fidjafkina dupka (Izbegli). **Troglobite.**

Balcanobius etropolensis V. Guéorguiev – Bezimennata Peshtera I, Zahlupena Dupka, Neykova Dupka (Etropole). **Troglobite.**

Genestiellina gueorguievi (Giachino) – Yalovitsa and Toplya (Golyama Zhelyazna). **Troglobite.**

Beroniella tetevensis Giachino et Guéorguiev – Dyado Draganovata Peshtera (Teteven). **Troglobite.**

Bathyscia raitchevi Casale, Giachino et Etonti – Imamova Dupka (Yagodina). **Troglobite.**

Gueorguieviella beshkovi Giachino et Guéorguiev – Razklonenata peshtera (Smiljan). **Troglobite.**

Fam. Curculionidae - Weevils

Troglorhynchus beroni Angelov - Inkaya (Tsvyatovo, Distr. Kardjali). **Troglobite.**

T. gueorguievi Angelov - Yalovitsa (Golyama Zhelyazna). **Troglobite.**

T. angelovi Gueorguiev - Zandana. **Troglobite.**

Zoogeographical Analysis of the Terrestrial Cave Fauna in Bulgaria

The intensive research on the cave and underground animals in Bulgaria since 1922 has accumulated rather complete information on the composition and distribution of most of the groups of underground living animals. Time has come to formulate hypotheses about the origin and the zoogeography of this fauna. Such hypotheses are due mainly to V. Guéorguiev, who analysed the terrestrial troglobites known from Bulgarian caves. His series of papers (1966 – 1977) were crowned by his monograph on the origin, formation and zoogeography of the terrestrial troglobites of the Balkan Peninsula (1977). This remarkable book was followed by his speleo-zoogeographical subdivision of Bulgaria (Guéorguiev, 1992, in Bulgarian). Other attempts to analyze the distribution of the terrestrial cave fauna in Bulgaria find place in the papers of Beron (1976, 1978) and in some articles on different groups of cave animals by Deltshev (1978, 1983), Riedel (1975).

All these papers do not deal with stygobites. Some of the stygobites have been found in caves, but the bulk is living in the stygal-hyporeic water or in other parts of the underground ecosystems. The distribution and the origin of the stygobites are subject to other regularities and have been analysed by several qualified Bulgarian stygobiologists (the late L. Tzvetkov, A. Petrova and the very active explorer of cave Copepoda Ivan Pandurski).

Many years have passed since the publication of the analyses of Guéorguiev in which new information had been accumulated. In his monograph Guéorguiev (1977) subdivided the Balkan Peninsula into four provinces (Dinaric, Egean, of Stara planina and Rhodopean). The Bulgarian territory falls within two of these provinces: the Province of Stara Planina (with two zones – Western and Eastern) and Rhodopean Province (also with two zones – Western and Eastern).

Beron (1976) subdivided the Province of Stara Planina into seven regions, some of them only provisional: Region of Eastern Serbia, Reg. of Ogosta, Reg. of Iskar, Reg. of Ossam, Reg. of Russe, Reg. of Kamchiya and Reg. of Dobrudja.

In his monograph Guéorguiev (1977) delineated also some special regions within his zones and subzones. In Bulgaria these regions (indicated on the map) are:

In the Western zone of Stara Planina:

Vrachanska Planina (the richest region in troglobites in the eastern part of the Balkan Peninsula) – up to 1977 there had been 29 terrestrial troglobites in this region, including 17 indicators: *"Cyphoniscellus"* [now *Vandeloniscellus*] *bulgaricus, Bulgarosoma bureschi, Typhloiulus longipes, Centromerus bulgarianus, Neobisium beroni, "Microcreagris bureschi"* [now *Balkanoroncus hadzii*], *Onychiurus sensitivus, Plusiocampa rauseri, Pheggomisetes radevi, Ph. r. ilcevi, Ph. r. tranteevi, Ph. globiceps mladenovi, Duvalius beroni, "D. deltschevi"* [now syn. of *Duvalius zivkovi*], *D. papasoffi, D. zivkovi, Radevia hanusi*. We can add *Vratzaniola pandurskii*.

Ponor Planina – 12 troglobites, incl. three indicators: *Eupolybothrus andreevi, Pheggomisetes globiceps globiceps, Ph. g. cerovensis*.

Golyama Planina – seven troglobites, incl. one indicator (*Duvalius pretneri*). Karlukovo Region – five troglobites, one indicator (*Tricyphoniscus bureschi*).

Vasilyovska Planina – seven troglobites, incl. five indicators: *Tranteeva paradoxa, Neobisium bulgaricum, N. subterraneum* [now syn. of *N. bulgaricum*], *Duvalius regisborisi, Tranteeviella bulgarica*.

Troyanska Planina – three troglobites, all considered indicators: *"Lindbergia"*(now *Spinophallus*) *uminskii, Trichoniscus bulgaricus, "Prodicus" albus* [now *Anamastigona alba*]. Now *"Lindbergia" uminskii* is no more considered troglobite.

In the Eastern Zone of Stara Planina:

Shipchenska Planina – three troglobites, all indicators: *Duvalius balcanicus, Haxaurus merkli, H. simile.* **New**: *Hexaurus paradisi, H. schipkaensis, H. beroni*.

Trevnenska Planina – three troglobites, all indicators: *Duvalius bulgaricus, Netolitzkya jeanneli, N. maneki*.

We can add now Kotlenska Planina as a new region, with the troglobite indicators *Duvalius kotelensis* and *Porrhomma microps*.

In the Western Rhodopes:

Batashki Rid (four troglobites, indicators *Troglohyphantes drenskii, Duvalius bureschi*). Ravnogor (two troglobites, indicator *Bulgaronethes haplophthalmoides*).

In the Eastern Rhodopes:

Chernatitsa – part of the Western Rhodopes (geographically) – two troglobites, indicator "*Bulgarosoma tridentifer*" (now *Troglodicus tridentifer*).

Dobrostan Massif – also geographically part of the Western Rhodopes – three troglobites, indicators "*Balkanoniscus*" [now *Rhodopioniscus*] *beroni* and *Rhodopiola cavicola*. **New**: *Nesticus beroni* (Araneae).

Trigrad Plateau – also geographically part of the Western Rhodopes – indicator "*Bulgarosoma*" [now *Troglodicus*] *meridionale*.

Zoogeographical Zones. Guéorguiev (1977) subdivided the eastern part of the Balkan Peninsula into the Zone of Stara Planina (in Bulgaria and in Serbia, East of Morava River) and the Zone of the Rhodopes (South Bulgaria and Northern Greece). There are very few genera and species-indicators in common between the two Zones and almost no genera and species in common between the Eastern and the Dinaric parts of the Peninsula. Such is *Lithobius lakatnicensis* (Chilopoda, Rhodopes and Stara Planina). Some genera, considered also to live both in Stara Planina and the Rhodopes, proved different and the Rhodopean species have been separated into new genera (*Rhodopioniscus, Rhodoposoma*).

Guéorguiev (1992) checked how his general zoogeographical subdivision of Bulgaria of 1982 fited the biospeleological subdivision. Besides the two zones, Bulgaria was subdivided into 7 regions and their troglobitic fauna was checked against the index of Tchekanovski – Sørensen. As the paper of Guéorguiev (1992) is in Bulgarian, it is worth summarizing its main conclusions. Moreover, this paper was submitted in 1989, so meanwhile several new troglobites and interesting troglophiles were added to the Bulgarian cave fauna.

The regions are characterized as follows:

1. **Danube Region.** The cave fauna consists of 26 troglophiles (trogloxenes not accounted for) and only one troglobite: *Trichoniscus tranteevi* Andreev (for *T. anophthalmus intermedius* Vandel, praeoccup.). According to the index of Sørensen – Tchekanovski, fifty percent similarity of the whole cave fauna (troglophiles and troglobites) with the Thracian Region only the troglobites – 54.2%. No similarity of the troglobites with any other region.

2. **Region of Stara Planina.** The richest cave fauna in Bulgaria (total 191 species, troglophiles 103, troglobites 88 species). Also the region richest in endemics. The troglobitic fauna bears insignificant similarity with the Rila-Rhodopean Region (5.7%), *Lithobius lakatnicensis* (Chilopoda), *Plusiocampa bulgarica* (Diplura) and *Pseudosinella duodecimocellata* (Collembola) being the only troglobites in common.

3. **Rila-Rhodopean Region.** In terms of cave fauna, this is the second richest region in Bulgaria. According to Guéorguiev (1992), here 46 troglophiles and 18 troglobites have been recorded. Since 1989 some other species have been added, but now (July 2006) the number of troglobites in the Rila-Rhodopean Region is still only 22: *Cordioniscus bulgaricus, C. schmalfussi, Trichoniscus rhodopiense, Rhodopioniscus beroni, Bulgaronethes haplophthalmoides, Alpioniscus* (= *Illyrionethes*) sp., *Nesticus beroni, Troglohyphantes drenskii, Lithobius lakatnicensis, Rhodoposoma rhodopinum, Troglodicus tridentifer, T. meridionale, Stygiosoma beroni, Anamastigona lepenicae, A. delcevi, Plusiocampa bulgarica, Pseudosinella duodecimocellata, Duvalius bureschi, D. nedelkovi, D. karelhurkai, Rhodopiola cavicola, Gueorguieviella beshkovi*. Among the troglobites, the most numerous are the Isopoda (6 sp.), the Diplopoda (6 sp.) and the Coleoptera (5 sp.). The endemism of this fauna also puts the Rila-Rhodopean Region in the second place with 21 endemic troglobites (Balkan, Bulgarian and local). As for troglobitic fauna, there is some similarity with the Thracian Region, the Danube Region and Region of Stara Planina. The only troglobite in common with the Thracian Region is *Trichoniscus rhodopiense* (Isopoda) and those in common with the Region of Stara Planina are *Lithobius lakatnicensis* (Chilopoda), *Pseudosinella duodecimocellata* (Collembola) and *Plusiocampa bulgarica* (Diplura).

4. **Region of Struma-Mesta.** Very few caves, mostly in Slavyanka Mt. Four cave animals are recorded: the troglobites "*Tranteevonethes gueorguievi*" (Isopoda, nomen nudum), *Duvalius joakimovi* and *Anamastigona delcevi* (Diplopoda) and the troglophile "*Lepthyphantes gueorguievi*" (Araneae, now *L. spelaeorum*).

Several other troglobites (Isopoda, Diplopoda, Araneae, Opiliones, etc.) are under study.

5. **Thracian Region.** Guéorguiev (1982) included also the Eastern Rhodopes in this region. The intensive research in the last years in this area (Beron, Petrov & Stoev, 2004) added new species to the list of the troglophiles and troglobites of the Eastern Rhodopes, an area insufficiently studied during the time of Guéorguiev. The actual list of the troglobites in the Eastern Rhodopes is as follows: *Trichoniscus rhodopiense* (Isopoda), *Lithobius tiasnatensis* (Chilopoda), *Duvalius petrovi, Bureschiana drenskii, Troglorhynchus angelovi, T. beroni* (Coleoptera). To this list we should add the troglobites found in the caves of the Greek part of the Eastern Rhodopes: *Alpioniscus thracicus* (Isopoda), *Maroniella beroni* (Coleoptera). Some interesting troglobites in the region are *Balkanopetalum petrovi, Apfelbeckiella trnowensis rhodopina, Rhodopiella beroni* (Diplopoda), *Balkanodiscus frivaldskyanus*, also *B. cerberus* (Mollusca) *in the* Greek part.

6. **Pontian Region** (in Guéorguiev: Region of the Black Sea coast). Very few caves, only one troglophile: the spider *Meta bourneti*, common with the regions or Strandja and Thrace.

7. **Strandja Region.** According to Guéorguiev (1992), the cave fauna consists of 21 troglophiles (6 of them endemics) and 4 troglobites: *Trichoniscus valkanovi, T. beroni* and "*T. tashevi*" (nomen nudum)(Isopoda) and *Lithobius* (*Monotarsobius*) *bifidus* (Chilopoda). All troglobites are so far Bulgarian endemics.

Endemics in the Cave Fauna of Bulgaria

Mollusca – the most important group of terrestrial Gastropoda living in Bulgarian caves belongs to the families Zonitidae. From ca. 33 sp. found in Bulgaria, one third (11 sp.) live in caves. Four of them can be considered endemic. Particularly interesting are *Balkanodiscus frivasdskyanus* (Rossmaessler)(relict) and *Schistophallus uminskii* (Riedel). Two slugs seem also to be endemic: *Litopelte bureschi* (H. Wagner) and *Milax kusceri* H. Wagner, but they are not troglobites.

Isopoda Oniscidea – From the 24 genera and 49 species (including 26 troglobites) of cave woodlice in Bulgaria, 7 genera (*Balkanoniscus, Rhodopioniscus, Bureschia, Bulgaronethes, Bulgaroniscus, Vandeloniscellus, Tricyphoniscus*, all belonging to Trichoniscidae) and 31 species are endemic for Bulgaria. Zoogeographically and from the point of view of cave evolution, the Isopoda terrestria are among the most important and interesting groups in Bulgarian cave fauna. With 32 species in caves (including 24 of all 26 troglobites), Trichoniscidae is by far the most important family among Bulgarian cave Isopoda. The only other troglobites (*Cordioniscus bulgaricus* Andreev and *C. schmallfussi* Andreev) belong to the family Styloniscidae. Both are Bulgarian endemics, resp. for Boychovata Peshtera on the border with the Republic of Macedonia and for the Rhodopes. From the 32 species of Trichoniscidae, 26 are endemic for Bulgaria (4 *Hyloniscus*, 14 *Trichoniscus, Alpioniscus* (= *Ilyrionethes*) sp., *Balkanoniscus corniculatus* Verhoeff, *B. minimus* Vandel, *Rhodopioniscus beroni* (Vandel), *Bureschia bulgarica* Verhoeff, *Bulgaronethes haplophthalmoides* Vandel, *Cyphoniscellus* (= *Bulgaroniscus*) *gueorguievi* Vandel, *Vandeloniscellus bulgaricus* (Vandel), *Tricyphoniscus bureschi* Verhoeff, *Beroniscus capreolus* Vandel, *Monocyphoniscus bulgaricus* Strouhal) and 3 are Balkan endemics (*Trichoniscus stankovici* Pljakić, *T. semigranulatus* Buturović, *T. rhodopiense* Vandel).

Some other families of Oniscidea also contain endemics (trogloxenes):
Ligidiidae – *Ligidium herzegowinense* Verhoeff – Balkan endemic
Trachelipidae – *Trachelipus bulgaricus bulgaricus* Verhoeff and *T. b. bureschi* Verhoeff – Bulgarian endemics
Philosciidae – *Chaetophiloscia hastata* Verhoeff
Porcellionidae – *Porcellium balkanicum* Verhoeff
Armadillidiidae – *Armadillidium elysii* Verhoeff

Pseudoscorpiones
Bulgarian endemics:
Chthoniidae – *Chthonius troglodites* Redikorzev – Western Predbalkan.
Neobisiidae
Neobisium (Heoblothrus) bulgaricum (Redikorzev)(syn. *Obisium subterraneum* Redikorzev) – Western Predbalkan.
N. (H.) beroni Beier – Western Stara Planina (Lakatnik). Endemic subgenus *Heoblothrus* Beier in Stara Planina.
N. (Blothrus) kwartirnikovi Mahnert – Vitosha (Bosnek).
N. (N.) intermedium Mahnert – Eastern Predbalkan (Prolazkata Peshtera)
Balkanoroncus bureschi (Redikorzev)(= *Obisium bureschi* Redikorzev = *Balkanoroncus praeceps* Ćurčić) – Central Predbalkan.
B. hadzii Harvey (= *Roncus bureschi* Hadzi) – Western Stara Planina (Lakatnik).
Roncus mahnerti Ćurčić et Beron – Western Stara Planina (Botunya).
Roncus parablothroides Hadzi is a Balkan endemic.

Opiliones – from the 22 species of harvestmen known to live in Bulgarian caves, 6 are Bulgarian endemics (including all 4 troglobites), and *Dicranolasma thracium* Starega and *Histricostoma drenskii* Kratochvil too. The species *Leiobunum rumelicum* Šilhavy, *Pyza bosnica* (Roewer) and *Rafalskia olympica* (Kulczynsky) are Balkan endemics. The troglophile *Paranemastoma radewi* (Roewer) is widespread in Bulgaria, but is also known from Bosnia and Northern Greece (Balkan endemic).

The four troglobites are confined to Stara Planina and the Predbalkan west of Troyan and are among the most remarkable cave animals in Bulgaria. the genera *Paralola* and *Tranteeva* are Bulgarian endemics. The only representative of Laniatores *Paralola buresi* Kratochvil is an inhabitant of the caves near Lakatnik Railway Station, west of Iskar River. The endemic Cyphophthalmid genus and the species *Tranteeva paradoxa* Kratochvil have been found so far in the caves Toplya and Rushovata Peshtera (Teteven District, Central Stara Planina). Among Bulgarian troglobitic Opilionids *Paranemastoma (Buresiola) bureschi* (Roewer) is the only species of the suborder Palpatores, which is predominant in Europe. It lives in many caves in Western Stara Planina and, most certainly, in the Western Confines (now in Serbia), as we found it at the very border. *Siro beschkovi* Mitov is a member of a Cyphophthalmid genus with many species on the Balkans and is known so far only from one cave of the Eastern Predbalkan.

Recently, another blind Nemastomatidae has been found in the Stoykova Dupka 1 in Slavyanka Mt. This is the first troglobitic Opilionid south of Stara Planina.

Araneae – from 80 species of cave spiders in Bulgaria (Beron et al, in print), 20 are considered here to be local, Bulgarian or Balkan endemics. According to

Deltshev (1996), *Antrohyphantes* is a genus related to *Fageiella* from the western part of the Balkan Peninsula and is therefore considered an ancient element (paleoendemic). Also according to the same author, "the group of cave endemic spiders has similar presence, 17 (39,43%) as the group of high altitude spiders".

Leptonetidae
Protoleptoneta beroni Deltchev – Bulgaria (Western Stara Planina)
P. bulgarica Deltchev – Bulgaria (Western Stara Planina)

Nesticidae
Nesticus beroni Deltshev – Western Rhodopes

Linyphiidae
Centromerus lakatnikensis (Drensky) – Balkan endemic (Western Stara Planina, Rep. of Macedonia)
C. milleri Deltshev – Balkan endemic (Eastern Rhodopes)
C. acutidentatus Delchev – South Pirin
?Lepthyphantes byzantinus Fage – Balkan endemic
L. istrianus Kulczynski (= *slivnensis* Drensky) – Balkan endemic
Pallidophantes trnovensis (Drensky) – Balkan endemic (Stara Planina, Rep. of Macedonia)
?L. jacksonoides Van Helsdingen – Western Rhodopes
Antrohyphantes sofianus (Drensky)(= *Lepthyphantes tranteevi* Miller) – Western Stara Planina
A. balcanica (Drensky) – Balkan endemic (Central Balkan and Republic of Macedonia)
A. rhodopensis (Drensky)(= *A. rodopicus* Dumitrescu) – Western Rhodopes
Troglohyphantes bureschianus Deltshev – Western Rhodopes
T. drenskii Deltshev – Western Rhodopes

Agelenidae
Histopona tranteevi Deltchev – Western Rhodopes
Coelotes drenskii Deltshev – Eastern Stara Planina
C. jurinitschi (Drensky) – Western Stara Planina

Amaurobiidae
Calobius balcanicus (Drensky) – Balkan endemic

Liocranidae
Mesioteles cyprius scopensis Drensky – Balkan endemic subspecies

Diplopoda – four genera (*Rhodoposoma, Troglodicus* and *Stygiosoma* from Western Rhodopes and *Bulgardicus* from the caves near Karlukovo in the Predbalkan) and all the 16 species of troglobitic Diplopoda in Bulgaria are (so far) Bulgarian endemics. The species are:

Order **Glomerida**
Doderiidae
Trachysphaera orghidani lakatnicensis Tabakaru – Western Stara Planina

Order **Polydesmida**
Polydesmidae
Brachydesmus radewi Verhoeff – Western Predbalkan
Trichopolydesmidae (Bacillidesmidae)
Bacillidesmus bulgaricus bulgaricus Strasser, *B. bulgaricus dentatus* Strasser – Western Stara Planina, W. Predbalkan

Order **Chordeumatida**
Anthroleucosomatidae
Bulgarosoma bureschi Verhoeff – Vrachanska Planina (Western Stara Planina)
Troglodicus meridionale (Tabacaru) – Western Rhodopes
Troglodicus tridentifer Gulička – Western Rhodopes
Anamastigona falcatus Gulička (troglobite ?) – Bulgarian endemic (Eastern Stara Planina)
A. lepenicae (Strasser) – caves near Velingrad (Western Rhodopes)
A. delcevi (Strasser) – Caves in Slavyanka Mt., (SW Bulgaria)
A. (Balkandicus) alba (Strasser) – Central Balkan (endemic subgenus *Balkandicus* Strasser)
Bulgardicus tranteevi Strasser – a cave near Karlukovo (Predbalkan)
Typhloiulus bureschi Verhoeff – many caves in Western Stara Planina
T. georgievi Strasser (troglobite ?) – Bulgarian endemic (Central Predbalkan)
T. (Inversotyphlus) longipes Strasser – Bulgarian endemic (endemic subgenus *Inversotyphlus* Strasser)
T. staregai Strasser – Western Stara Planina
Serboiulus speleophilus Gulička – 29 caves in Western Stara Planina

The following non-troglobitic Diplopoda also seem to be endemics:

Order **Glomerida**
Glomeridae
Glomeris balcanica Verhoeff, 1906 (= *G. bureschi* Verhoeff = *G. latemarginata* Attems) – Balkan endemic

Order **Polydesmida**
Polydesmidae
Polydesmus herzegowinensis Verhoeff – Balkan endemics

P. renschi Schubart, *P. zonkovi* Verhoeff, *P. bureschi* Verhoeff – Bulgarian endemics

P. tridens Attems – Balkan endemic

Brachydesmus herzegowinensis reflexus Strasser, *B. h. trifidus* Strasser, *B. h. confinis* Strasser *B. dadayi brusenicus* Gulička – Bulgarian endemic subspecies

B. cristofer Strasser – Bulgarian endemic

Mastigophorophillidae

Mastigona bosniensis (Verhoeff) – Balkan endemic

Haaseidae

Haasea (Histrosoma) vidinense (Strasser) – Bulgarian endemic species and subgenus

Paradoxosomatidae

Metonomastus pomak Golovatch et Stoev – Bulgarian endemic (Eastern Rhodopes)

Order **Chordeumatida**
Anthroleucosomatidae

Troglodicus rhodopinus Strasser – Bulgarian endemic genus and species (Rhodopes)

Order Julida
Julidae

Leptoiulus borisi Verhoeff – Bulgarian endemic

Typhloiulus kotelensis Jawlowski – Eastern Stara Planina

Megaphyllum rhodopinum (Verhoeff), *M. beroni* (Strasser) – Rhodopes

Balkanophoenix borisi Verhoeff – Stara Planina (Lakatnik, Western Stara planina)

Pachyiulus hungaricus gracilis Verhoeff – Rhodopes, end. subspecies

P. cattarensis (Latzel) – Balkan endemic

Apfelbeckiella trnovensis (Verhoeff) and the subspecies *A. t. rhodopina* Strasser and *A. t. deliormana* Strasser – Bulgarian endemics

A. byzantina Verhoeff – Balkan endemic

A. (Rhodopiella) beroni Strasser – Bulgarian endemic subgenus and species

Order **Callipodida**
Schizopetalidae

Balkanopetalum armatum Verhoeff – Western Stara Planina. Troglophile.

B. rhodopinum Verhoeff – Western Rhodopes. Troglophile.

B. beshkovi Strasser – Western Rhodopes. Troglophile.

B. petrovi Stoev et Enghof – Eastern Rhodopes. Troglophile.

B. bulgaricum Stoev et Enghoff – South Pirin. Troglophile.

Chilopoda – from the 44 species of this Class are found in Bulgarian caves, 14 are endemics (9 Bulgarian and 5 Balkan), as follows:

Order **Lithobiomorpha**
Lithobiidae

Lithobius tiasnatensis Matic (=*L. popovi* Matic) – caves in Central Predbalkan and Sakar

L. rushovensis Matic (=*L. beschkovi* Matic et Golemansky) – caves near Teteven (Central Predbalkan)

L. beroni Negrea – Balkan endemic (Bulgaria and Greece, non-troglobite)

L. lakatnicensis Verhoeff – Balkan endemic (Bulgaria and East Serbia)

L. bifidus (Matic) – five caves in Strandja

L. stygius Latzel – Balkan endemic

L. wardaranus (Verhoeff) – Balkan endemic

Harpolithobius folkmanovae Kaczmarek – Bulgarian endemic (SE Bulgaria, non-troglobite)

H. banaticus rhodopensis Kaczmarek – Rhodopes (endemic subspecies)

Eupolybothrus andreevi Matic – Cave Vodnata Peshtera near Tserovo (Western Stara Planina)

E. gloriastygis (Absolon) – Balkan endemic (Bulgaria and Bosnia). Known from two caves in Western Stara Planina.

Order **Geophilomorpha**
Himantariidae

Thracophilus beroni Matic et Darabantu – Western Rhodopes

Th. bulgaricus Verhoeff – Western Rhodopes and Sredna Gora (non-troglobite)

Order **Scolopendromorpha**
Cryptopidae

Cryptops croaticus Verhoeff – Balkan endemic

Diplura – so far three Bulgarian endemics:
Campodeidae

Plusiocampa bulgarica Silvestri – Western Rhodopes and Western Stara Planina.

P. bureschi Silvestri (syn. *P. rauseri* Rusek) – Western Stara Planina (Lakatnik and Tserovo).

Collembola – there are eight Bulgarian endemics in the caves of Bulgaria.

Neanuridae – *Bilobella digitata* Cassagnau (Eastern Stara Planina)

Onychiuridae – *Onychiurus bureschi* Handschin, *O. sensitivus* Handschin, *O. subgranulosus*

Gama, *Protaphorura beroni* (Gruia)(Western Stara Planina)

Entomobryidae – *Pseudosinella bulgarica* Gama (Central and Eastern Stara Planina), *P. kwartirnikovi* Gama (Central Balkan, or Stara Planina), *P. duodecimocellata* Handschin (caves in the Rhodopes and in Stara Planina)

Orthoptera – *Troglophilus neglectus* Krauss (Raphidophoridae) is Balkan endemic.

Coleoptera

Carabidae – There are 70 species of ground beetles known to live in Bulgarian caves. Two genera are Bulgarian endemics, both in Western Stara Planina: *Rambousekiella* Knirsch and (practically) *Pheggomisetes* Knirsch. *Pheggomisetes* was also found in one cave in East Serbia (the Western Confines of the former Bulgarian territory). The species and the subspecies are: *Pheggomisetes buresi buresi* Knirsch, *Pheggomisetes buresi medenikensis* Knirsch, *Ph. globiceps globiceps* Buresch, *Ph. g. breiti* Mandl, *Ph. g. georgievi* Z. Karaman, *Ph. g. lakatnicensis* Jeannel, *Ph. g. stoicevi* Guéorguiev, *Ph. g. karlukovensis* Genest, *Ph. g. ilandjievi* Guéorguiev, *Ph. g. mladenovi* Guéorguiev, *Ph. g. cerovensis* Guéorguiev, *Ph. radevi radevi* Knirsch, *Ph. r. iltschevi* Knirsch, *Ph. r. tranteevi* Guéorguiev. Another important genus is *Duvalius* Delarouzée, with 17 troglobites, endemic for Bulgaria, as follows:

D. (Paraduvalius) balcanicus (J. Frivaldzsky) – Central Balkan

D. (Paraduvalius) beroni Guéorguiev, *D. (P.) garevi* Casale et Genest, *D. (P.) papasoffi* Mandl, *D. (P.) pretneri* Guéorguiev, *D. (P.) zivkovi* Knirsch – Western Stara Planina and Western Predbalkan

D. (Paraduvalius) kotelensis Genest – Eastern Stara Planina

D. (Paraduvalius) bureschi Jeannel, *D.(P.) karelhurkai* Farkac – Western Rhodopes

D. (Paraduvalius) bulgaricus Knirsch, *D.(P.) regisborisi* Buresch, *D.(P.) legrandi* Genest – Central Predbalkan

D. (Paraduvalius) joakimovi B. Guéorguiev – Slavyanka

D. (Paraduvalius) nedelkovi B. Guéorguiev – Western Rhodopes

D. (Paraduvalius) pirinicus B. Guéorguiev – Pirin

D.(Paraduvalius) petrovi B. Guéorguiev – Eastern Rhodopes

D. (Duvaliotes) beshkovi Coiffait – mountains on the western frontier of Bulgaria

Three other species of *Duvalius (Paraduvalius)* have been described from the MSS and other localities outside the caves: *D. hanae* Hůrka from Central Balkan, *D. marani* (Knirsch) from Slavyanka and *D. rajtchevi* (Genest et Juberthie) from the Western Rhodopes.

Staphylinidae – Only two species (*Quedius troglophilus* and *Qu. gueorguievi*), described by H. Coiffait from Bulgaria, are so far considered endemic for this country. They are not typical cave animals.

Pselaphidae – only one (trogloxenic) species (*Bryaxis beroni* Z. Karaman) seems to be a Bulgarian endemic (Western Bulgaria).

Curculionidae – All three species of hypogeous weevils in Bulgaria are so far considered Bulgarian endemics – *Troglorhynchus gueorguievi* Angelov for Central Stara Planina, *T. beroni* Angelov and *T. angelovi* B. Gueorguiev for the Eastern Rhodopes.

Cholevidae (Leiodidae) – the highest endemism on generic level in the Bulgarian cave fauna is represented in this family. Far from being completely known in Bulgarian caves, there this family is represented by 34 species of 19 genera. From them,13 genera and 25 species are Bulgarian endemics:

All 13 genera – Bulgarian endemics – are very narrow endemics and they are as follows:

Beronia Guéorguiev – Northwestern part of Western Stara Planina (Vidin Distr.)

Beskovia Guéorguiev – the caves around Cherepish (Western Stara Planina)

Radevia Knirsch, *Vratzaniola* Dupré – Vrachanska Planina (Western Stara Planina)

Hexaurus Reitter – caves in the high Central Balkan (Stara Planina)

Balcanobius Guéorguiev – caves near Etropole

Beroniella Giachino et Guéorguiev – a cave in Teteven (Central Predbalkan)

Tranteeviella Pretner (= *Bulgariella* Z. Karaman) – caves in Teteven area (Central Predbalkan)

Genestiellina Giachino – caves in Teteven area (Central Predbalkan)

Netolitzkya J. Müller – caves in Tryavna area (Central Predbalkan)

Rhodopiola Guéorguiev – one cave near Bachkovo (Western Rhodopes)

Bureschiana Guéorguiev – caves in Kardjali area (Eastern Rhodopes)

Vratzaniola Dupré – Western Stara Planina
Gueorguieviella Giachino et Guéorguiev – Western Rhodopes

The species – Bulgarian endemics:

Beskovia bulgarica Guéorguiev, *B. andreevi* Giachino et Guéorguiev, *B. beroni* Giachino et Guéorguiev, *Beronia micevi* Guéorguiev, *B. tranteevi* Giachino et Guéorguiev, *Radevia hanusi* Knirsch, *Vratzaniola pandurskii* Dupré – Western Stara Planina

Hexaurus merkli (J. Frivaldzsky), *H. similis* (J. Frivaldzsky), *H. schipkaensis* Zerche, *H. paradisi* Zerche, *H. beroni* Giachino et Guéorguiev – the high Central Stara Planina

Beroniella tetevensis Giachino et Guéorguiev, *Genestiellina gueorguievi* (Giachino), *Tranteeviella bulgarica* Pretner, *Balcanobius etropolensis* Guéorguiev, *Netolitzkya maneki maneki* J. Müller, *Netolitzkya maneki iltschevi* J. Müller, *N. jeanneli jeanneli* Buresch, *N. jeanneli matroffi* Jeannel – Central Stara Planina and Central Predbalkan

Bureschiana drenskii Guéorguiev, *Gueorguieviella beshkovi* Giachino et Guéorguiev, *Rhodopiola cavicola* Guéorguiev – Rhodopes

All of them belong to the subfamily Leptodirinae (= Bathysciinae). All 24 troglobitic species of this subfamily in Bulgaria (plus 2 subspecies) are Bulgarian endemics.

Siphonaptera – Only subspecies (so far endemic for Bulgaria) of two species of fleas on bats have been described from Bulgarian cave: *Nycteridopsylla ancyluris johanae* Hůrka and *N. trigona balcanica* Hůrka.

Relics in the Cave Fauna of Bulgaria

The problem of the relictness and the antiquity of the troglobites still stays. For Jeannel (1944, 1960), followed by Vandel (1964) and Guéorguiev (1977), there was no doubt that the paleotroglobites are very ancient and have no relatives among recent animals living outside caves. On the contrary, neotroglobites still have relatives outside caves and are connected with them by intermediary forms. For Leleup (1965), the main lines of southeuropean troglobites have their origin in an orophilic prepleistocene fauna, very ancient and very rich, which lived in biotopes on land and had emerged in remote geological periods. For Vandel (1964), "Terrestrial troglobites are mostly descendants of a tropical fauna populating Europe and North America in the first half of the Tertiary". We have to keep in mind that the Paleogene (the first part of the Tertiary) covers a period between 67 and 25 million years.

In recent time Brignoli (1979) opposed this assertion, taken as axiomatic a long time ago. The early deceased prominent Italian specialist writes: "It is not true at all (or, at least, it is not sure) that the troglobites are ancient". And further: "The term (of) "relict" (or even (of) "living fossil"), so often applied to the troglobites, is for me completely meaningless". However, other prominent specialists do support the opinion of the ancient nature of troglobites. According to Beier (1969), "…the troglobite species show high degree of specialization and are without doubt to be considered (as) relicts from the Tertiary".

The present author also thinks that the assertions of Brignoli are exaggerated and that relicts do exist. Which troglobite is ancient and which is more recent is a matter of analysis.

Guéorguiev (1977) was a firm supporter of the theories of Jeannel and Vandel and his classification of the troglobites according to their origin will be resumed here, completed with some new data.

Descendants of Gondwanian Phyletic Lines

The woodlice of the genus *Cordioniscus* (Isopoda: Stylonisciidae) are to be considered among the most ancient Balkan relicts. The two species found in Bulgaria (*Cordioniscus bulgaricus* Andreev and *C. schmalfussi* Andr.) mark the northernmost localities of the family in Europe. The endemic genus and species of the short-legged Opilions *Tranteeva paradoxa* Kratochvil (Opiliones, Cyphophthalmi, Sironidae) from the caves of the Predbalkan are also to be considered Gondwanian relicts.

Descendants of Laurasian Phyletic Lines

Guéorguiev (1977) considered some spiders of the genus *Nesticus* and some Collembolans (*Acherontides spelaeus* Ionesco, which is now considered troglophile) Laurasian relicts. Bulgarian *Nesticus* species are also troglophyles. A typical Laurasian element, according to him, would be also the endemic genus and species of Isopoda Trichoniscidae *Bulgaronethes haplophthalmoides* Vandel, living in the caves near Peshtera (Western Rhodopes).

Descendants of Mesogeidean (Paleomediterranean) Phyletic Lines

The Isopods of the endemic genera *Balkanoniscus*, *Beroniscus* and *Bureschia* (all in Stara Planina and the Predbalkan) are considered descendants of a phyletic line populating in the Paleocene and early Eocene the land stretching from Cantabric Mountains to Caucasus and called Mesogeida. We should add here the genus *Rhodopioniscus* Tabacaru, described in 1993 for the Rhodopean *"Balkanoniscus" beroni* Vandel. Among the Diplopoda, such Mesogeidean relicts would be the Glomerids *Trachysphaera orghidani* Tabacaru and *T. dobrogica* Tabacaru (not yet recorded but certainly living in Bulgarian Dobroudja). Among the spiders (Araneae), such descendants are the *Troglohyphantes* species (in Bulgaria the troglobite *Troglohyphantes drenskii* Deltshev and the troglophile *T. bureschianus* Deltschev both in the Western Rhodopes, another *Troglohyphantes* from Slavyanka).

Mesogeidean origin is presumed also for the troglobitic Opilions *Paralola buresi* Kratochvil and *Paranemastoma (Buresiola) bureschi* Roewer. The endemic genus *Paralola* Kratochvil (Phalangodidae) and its only species *P. buresi* from the caves near Lakatnik in Western Stara Planina represent in Bulgaria the (mostly tropical) suborder Laniatores. Among the terrestrial troglobites, this strange creature gives impression of something really ancient and alien to the present European fauna. However, Martens (1972) wrote that the Laniatores "should not be considered any more as Tertiary relicts in the European fauna as they are widespread in the areas remaining outside the Pleistocene glaciation". Nevertheless, *Paralola* is beyond doubt a relict – its age is to be further considered.

As for *Buresiola*, it is no more considered a separate genus, but a subgenus of *Paranemastoma*. The only described species in Bulgaria is also an endemic of Western Stara Planina.

The pseudoscorpions of the subgenera *Blothrus* (genus *Neobisium*) and *Parablothrus* (genus *Roncus*) are also considered to be part of this category – Mesogeidean relicts. In Bulgaria, we know the species *Neobisium (Blothrus) kwartirnikovi* Mahnert from Bosnek (Vitosha).

According to Guéorguiev (1977), a third of all endemic troglobite genera of Cholevidae on the Balkan Peninsula are of Mesogeidean origin. In Bulgaria, the genera *Netolitzkya*, *Hexaurus*, *Radevia*, *Rhodopiola*, *Beronia*, *Bureschiana* also belong here.

Paleoegeidean (Protoegeidean) Relicts

Isopoda – the Haplophthalminae genera *Cyphoniscellus* and *Tricyphoniscus*, represented by the species *Cyphoniscellus gueorguievi* Vandel and *Tricyphoniscus bureschi* Verhoeff, belong here.

Diplura Campodeidae – the troglobitic species of the subgenus *Stygiocampa* (genus *Plusiocampa*), found in Bulgaria in Stara Planina and the Rhodopes, probably belong here.

Nordegeidean Relicts

Most terrestrial troglobites of the Balkan Peninsula belong to this category, due to the fact that major parts of former Yugoslavia, Bulgaria and Greece were situated in Northern Egeide for long periods during the Tertiary.

Many genera of different groups, which are listed by Guéorguiev (1977) in the categories of the descendants of Gondwanian phyletic lines (*Tranteeva*), Laurasian phyletic lines (*Bulgaronethes*) or Mesogeidean (paleomediterranean) phyletic lines (*Balkanoniscus*, *Bureschia*, *Buresiola*, *Paralola*) or are considered to be Paleoegeidean (Protoegeidean) relicts (*Tricyphoniscus*), are also considered by him to be Nordegeidean relicts. The assertion needs clarification. Furthermore, the following genera: Isopoda (*Hyloniscus*), Diplopoda (the Antroleucosomatids *Bulgarosoma*, *Stygiosoma* and the subgenus *Balkandicus*, and the Typhloiulids *Typhloiulus*, *Serboiulus* and *Apfelbeckiella*), Opiliones (*Siro*) are also considered Nordegeidean relicts. To them we can add the genus *Rhodoposoma*. The endemic subgenus *Heoblothrus* (genus *Neobisium*) from Stara Planina is also not far from this group. So are the species of the genus *Balkanoroncus* Ćurčić, not known to Guéorguiev by 1977. There are two species living in the caves of Stara Planina and the Predbalkan: *Balkanoroncus bureschi* (Hadzi) and *B. hadzii* Harvey. From the Carabidae, the species of the genus *Pheggomisetes* Knirsch belong here and from the Catopidae – the genera *Beskovia*, *Tranteeviella*, *Balcanobius*. We may add the newly described genera *Beroniella* and *Vratzaniola*. Nordegeidean relict is also the remarkable genus *Genestiellina* (Giachino, 1992).

Southegeidean Relicts

Gastropoda – the only Bulgarian "*Lindbergia*" (*L. uminskii* Riedel from Central Balkan) is now considered to represent a separate genus *Spinophalus*, but still of Southegeidean origin.

ARCHAEOLOGICAL RESEARCH IN BULGARIAN CAVES

The caves have been subject to Archaeology ever since it became a science in its own right. Man has used these natural phenomena during his cultural development since the earliest prehistoric times, in the Medieval ages and up to our time. Many caves in Bulgaria are sites of many layers, inhabited for millenia, and the depth of the cultural layer there is considerable. This fact is due to the crossroad situation of Bulgaria – since ancient times the Bulgarian land has been a contact region of different cultures and each of them has left traces, deciphered by archaeologists. In caves humans found shelter in troubled periods, built dwellings, created sanctuaries and monk cells.

The archaeological exploration of caves in Bulgaria has a history of more than one hundred years. That is why it is very difficult to enumerate and analyze all studies within such a short outline. Here only the most important of them will be listed.

The first known information about archaeological finds in a Bulgarian cave is due to the French traveller Dr K. Pouaye. In 1859, he mentioned that the cave near the village Topra Hisar (now Zemlen near Stara Zagora) was rich in archaeological remains.

The organized archaeological studies in Bulgarian caves started at the end of 19th century. The first explorers were not archaeologists. They were biologists, geologists, geographers, and part of them high school teachers. On May 20, 1890 Stephan Jurinich, a school-teacher in Gabrovo High School, started excavations in the cave Polichki near the Monastery of Dryanovo. In 1899–1900, the geologists G. Bonchev and Iliya Stoyanov carried out successive excavations in the cave Toplya near the village Golyama Jelyazna, Lovech District.

In the period up to World War II the explorations were due mainly to Rafail Popov (1876–1940), archaeologist and speleologist, an active member of the Bulgarian Caving Society and its President until his death. His contribution to the systematic study of Bulgarian caves laid the foundation of prehistoric archaeology in Bulgaria. He carried out his first excavations as a student in 1899, exploring Golyamata Peshtera and Malkata Peshtera near Veliko Tarnovo. Later, after specializing in European universities, R. Popov explored many other caves: Morovitsa (Glozhene), Mirizlivka (Oreshets), Bacho Kiro (Dryanovski Monastir), Duhlata (Veliko Tarnovo area) and others. Popov dated the material discovered back to the time of the Neolith (Malkata Peshtera), Halcolith (Tsarskata Peshtera, Golyamata Peshtera, Harlovata Peshtera, Popin Pchelin, Malka Listsa near Belyakovets, the caves near Madara). In 1924–1926, R. Popov partly excavated the cave Temnata Dupka near Karlukovo and discovered for the first time in Bulgaria and on the Balkan Peninsula human remains from the Late Paleolith.

Prof. Rafail Popov, Archeologist, Director of the National Museum in Sofia. President of the Caving Society in 1937-1938 and in 1940

In 1920, some other researchers joined the archaeological exploration of caves – N. Petkov (Mecha Dupka, Klisura), V. Atanassov and L. Filkov (Mirizlivka, Oreshets) and Vassil Mikov. Until 1929 Mikov had excavated 19 caves, and dated the artefacts found there as belonging to the Neolith (Devetashkata Peshtera, Tabashkata Peshtera, Lovech District); Halcolyth (Temnata Dupka, Popovi Pech, Lepenishki Pech and Musikalnata Peshtera, Vidin District; Averkovitsa, Svinskata Dupka, Haydushkata Dupka, Futyovskata Peshtera, Ladjenskata Peshtera, Hlevenskata Peshtera, Mikrenskata Peshtera and Draganchovitsa, Lovech Distr.); Iron Age (Ladjenskata peshtera, Lovech Distr.). In the 1930s and the 1940s, V. Mikov studied the caves Suhi Pech (1932) and Haydushkata Peshtera near Deventsi (1942).

In 1937, foreign specialists also participated in the excavations – D. Garod from the Institute for Prehistoric Studies in Chicago, together with R. Popov, excavated the cave Bacho Kiro near Dryanovo.

So, in the period until World War II the explorers did their best to lay the scientific fundament of the archeological study of Bulgarian caves. Due to the war and the death of R. Popov in 1940, this study was interrupted for several years.

Archeological studies in caves were restarted by the end of 1940-ties, with more systematic approach of some new researchers. Excavations in caves carried out G. Markov in the caves Emenskata Peshtera and Razhishkata Peshtera (1948–49); N. Djambazov excavated the cave Pesht near Staro Selo (1951–53); Ochilata (Aglen), Provartenka, Vodnitsa, Samuilitsa I and II near Kunino (1954), Tsakonichki Pech, Kalenskata Pesht and Morovitsa (1955), Tabashkata peshtera near Lovech (1952–1956) and the cave Snezhanka (1961), where traces from the Neolyth, Halcolyth, the Bronze and the Iron Ages have been discovered. N. Djambazov and V. Mikov carried out the studies of the cave of Devetaki (1950–1952), inhabited by man from the Neolyth to the Antiquity. In the cave Yagodinska Peshtera excavations were carried out by P. Detev (1956) and M. Vaklinova (1965–1966). At the end of the 1960s, A. Alexov from the Historical Museum in Plovdiv studied the cave Topchika near Dobrostan, where materials from the Bronze and the Iron Ages have been discovered.

It became clear that in Bulgarian caves cultural layers from the Middle Paleolyth (Moustier, 120 000–40 000 BP), Neolyth, Eneolyth, the Bronze and the Iron Ages, from the Antiquity and the Middle Ages had been recorded. Typical roughly worked tools from the Mousterian, as well as bones of animals – companions and food of the primitive people – have been found in the caves Bacho Kiro, Devetashkata Peshtera and Samuilitsa II. In some of these and some other caves (Temnata Dupka) remnants from the Late Paleolyth (Orignac) have been found. Well polished Neolythical tools, typical for the time 8000–4000 B.C., have been excavated in four caves: Devetashkata, Tabashkata, Toplya and Yagodinskata. In them vessels with natural paints (ochre etc.) have been found, as well as artefacts made of horn and bones of domestic animals. The Chalcolyth (Eneolyth or Stone – Copper Age, 4000–3000 B.C.) is characterized by artefacts of copper (needles, awls), ornaments and bone objects. Such artefacts have been found in Devetashkata Peshtera, Bacho Kiro, Andaka, Yagodinskata Peshtera, Morovitsa, Toplya, Emenskata Peshtera, Provârtenka.

Dwellings and fireplaces, as well as many artefacts from the Bronze Age (2800–1200 B.C.), have been found in Devetashkata, Tabashkata and Yagodinskata (Imamova Dupka) caves and also in Magura and Suhi Pech. Among the artefacts there are shining clay vessels, made with a pottery wheel, bronze axes, swords, sickles, etc.

The Iron Age (1200–2000 B.C.) is represented in Devetashkata cave, Magura, Tabashkata and Yagodinskata caves, in Provartenka, etc.

The cave Snezhanka had been used as shelter by the Besses (a Thracian tribe, inhabiting the Rhodopes during the period of Roman dominance).

With the creation of the Group for Paleolythical Studies at the Institute for Archaeology and the Museum of BAS (N. Sirakov, Sv. Sirakova, St. Ivanova, Iv. Gatsov and others) the systematic, long-term, regular excavations in caves started. In 1971–1973, together with archaeologists from the Yagelonian University in Poland, the group explored the cave Bacho Kiro near Dryanovo, where traces of a culture of the Middle and the Late Paleolyth were discovered (ca. 43 000 BP). Interdisciplinary methods of research found place in this research and helped achieve a better understanding of the life of the early inhabitants of our land.

At the same time other sites were explored: R. Katincharov studied the cave Magura, Vidin District (1971), with remains from the Halcolyth, the Bronze and Iron Ages. During the Strandja complex speleological expedition of Akademik – Sofia Caving Club (1975–1977), E. Teoklieva and P. Balabanov carried out excavations in Bratanova Peshtera and later also in the caves Leyarnitsata and Kaleto near the village Mladezhko, Burgas Distr. The artefacts found there were identified as belonging to the Bronze and Iron Ages. In the period 1976–1978, V. Gergov from the Historical Museum in Pleven led the archaeological excavations in the caves of Musselievo and the village Sadovets, Distr. Pleven. The material discovered was dated back to the Halcolyth and the Bronze Age. In 1978 V. Nikolov and V. Vassilev excavated the cave Golyamata Peshtera above the village Iliya near Kyustendil, with artefacts from the late Halcolyth and the Early Iron Age.

In 1981 the Group for Paleolythical Studies of the Archaeological Institute and Museum carried out initial excavations in the cave Toplya near the village Golyama Jelyazna, Lovech Distr. The material was dated as belonging to the second half of the Late Paleolyth. Further in the field of cave archaeology, there are other outstanding studies (Bulgarian – Polish – French) of Temnata Dupka near Karlukovo (1984–1994), under the leadership of N. Sirakov, Ya. Kozlowski and A. Laville. Traces of human settlements from the Middle and the Late Paleolyth (100 000–13 000 BP) were discovered. Mostly cavers of Iskar-Sofia and Studenets-Pleven Caving Clubs, as well as cavers from other clubs, participated in the research. In the same region, St. Ivanova did excavations in the cave Skandalnata near Kunino (1994), and found traces from the Late Paleolyth.

The archaeological studies in Rhodopean caves have been active as well: Hr. Valchanova from the Historical Museum – Smolyan and St. Ivanova carried out excavations in the caves Haramiyskata (1981–1983) and Partizanskata; M. Avramova carried out the research in Yagodinskata Peshtera (1983). In 1995 G. Nehrizov studies the cave Samara near Ribino. Considerable contributions to the cave archaeology of this region have also been made by Dimitar Raychev from Chepelare Caving Club, Mincho Gumarov from the Caving Club in Kardjali and Boris Kolev from the Caving Club in Haskovo.

After the removal of the petrol deposits in the cave Devetashkata Peshtera in the 90s, the archaeologists from the regional museums in Lovech and Pleven, directed by V. Gergov, carried out further studies there. This cave is a good example of the fact that some caves in Bulgaria have been used as human habitation for ten thousands of years uninterrupted – from the Middle Paleolyth to our days.

Since 1998 the Group for Paleolythical Studies, together with French scientists, have carried out archaeological explorations in the cave Kozarnika near Oreshets Railway Station under the direction of N. Sirakov and J.-L. Guadeli. So far in the cave layers traces of human presence, dating back to over 1.4 mill. years BP, have been found. The results indicate that Kozarnika appears to be one of the earliest populated caves in Europe.

Rock art in caves are also subject to prehistoric studies. I. Vardjiev and V. Mikov announced the existence of prehistoric rock paintings in the karstic canyon Chernelka near Pleven as early as the 1920s, but they were not well explored. Best known are the monochrome paintings in the cave Magurata, made with the use of various techniques (bat guano, ochre, graffiti, which are in several layers) and dating back to different times, the earliest being the Chalcolyth. The paintings have been explored many times by F. Birkner (1916), V. Mikov (1928, 1955), L. Filkov (1929), E. Anati (1971), but their most detailed study is due to T. Stoychev, V. Gerasimova and A. Stoev (1990-94). Another cave with rock paintings (disputed!) is the cave N° 2961 near the village Baylovo, Sofia Distr., studied in 1977–1993 by M. Zlatkova, T. Stoychev, V. Gerasimova, A. Stoev and others, The authenticity of these paintings has been strongly disputed by V. Beshkov.

Since the very beginning of archaeological studies in caves, explorers paid attention also to

Excavations in the cave Suhi Pech, where the oldest Man in Europe was found. Photo Nikolay Genov

caves with traces of later periods. Most sanctuaries discovered so far date back to the Iron Age and the Antiquity. Thracian sanctuaries are known from many caves, connected with springs or nearby water points. Such are the sanctuary of Asclepius and Hygia near the karstic source of Glava Panega, the Big Cave near Madara – a sanctuary of the nymphs and Heracles, Devetashkata Peshtera – a source, venerated by Thracians as sacred. These sanctuaries have been studied by many explorers.

B. Dyakovich studied the Thracian sanctuary near the cave Tsarna, Novosel (1905); V. Dobruski (1907), and later G. Katsarov (1910, 1925) studied the sanctuary in the cave near the karstic source Glava Panega, Lovech Distr.: N. Mushmov related on the sacred caves in Bulgaria (1924). G. Antonov wrote about the distribution of the cave-sanctuaries in Bulgaria and the historiography of their study (1977, 1979, 1981).

The use of caves as a cult space took place in later historical times. Such are the rock monasteries, crypts, monk cells and churches, covering a large time span – from 5th to 14th century and later. The monk dwellings were initially built in natural caves, later they were enlarged and adapted for the needs of the monk life. There are more than 1000 known Christian rock dwellings in Bulgaria. They are found almost over the entire territory of Bulgaria – the areas of Sofia, Vidin, Vratsa, Lovech, Pleven, Haskovo, Kârdjali, etc. Many of them are concentrated in Northeast Bulgaria (because of the type of karstification of the rocks there) – the Plateau of Shumen (Osmar, Troytsa), the Plateau of Madara and Provadiya, the regions of Targovishte, Razgrad, Silistra, Varna (best known is Aladja Manastir), Russe, South Dobrudja, in the area of Kamen Bryag village (Yaylata), cape Kaliakra and Tyulenovo. Such have been discovered in the area of Russenski Lom, in the gorge of Provadiyska Reka River from Provadiya to Kosovo Railway Station (studied by Ara Margos in 1979). Entire monastery complexes have been found on the cliffs of Dobrudja Black Sea Coast. They were surveyed by the brothers Škorpil as early as 18 and the results were published in Sbornik Narodni Umotvoreniya (a Collection of Popular Sayings).

Of particular importance are the rock monasteries in the area of the village Ivanovo, Russe,

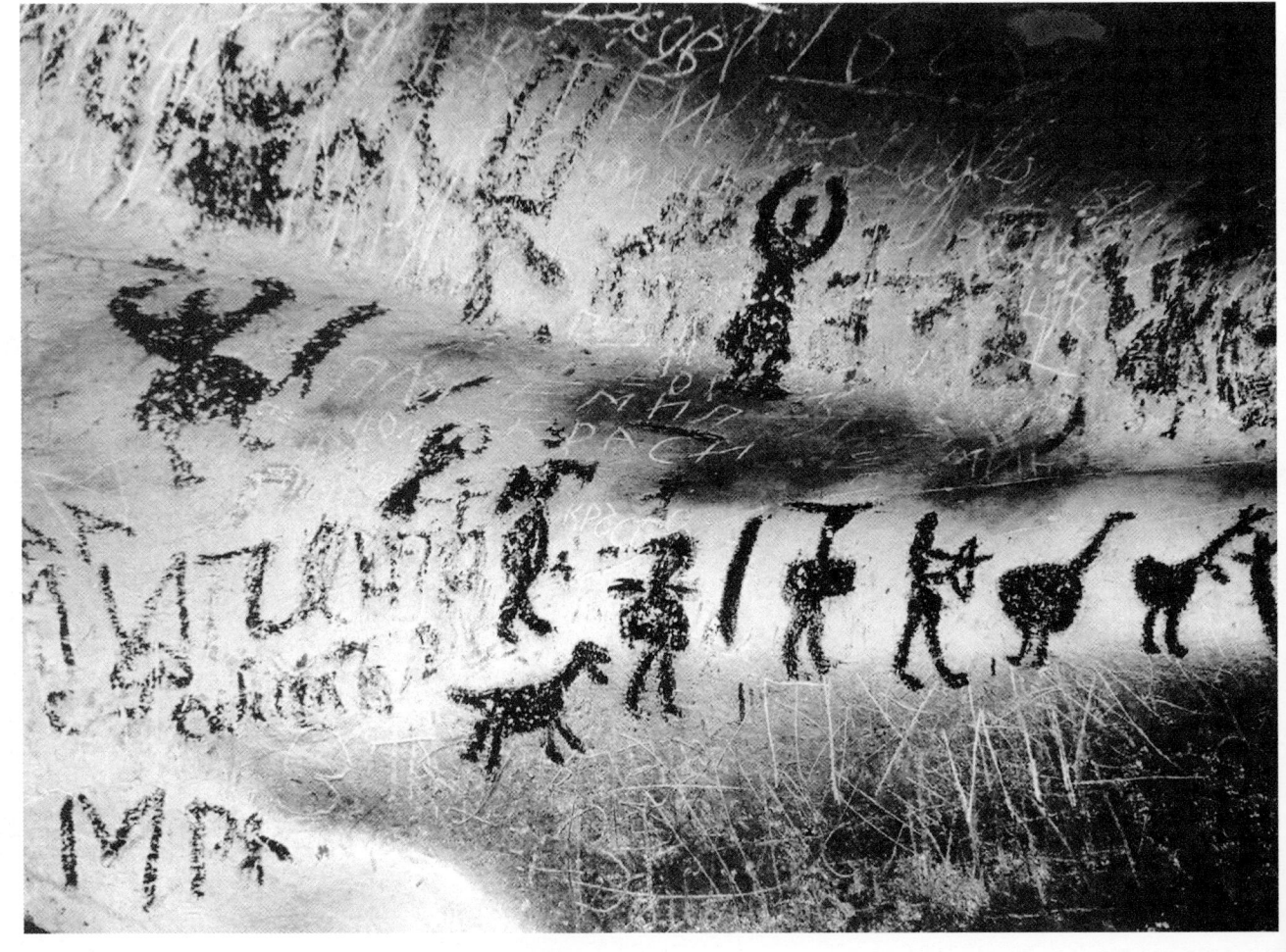

Paintings in the cave Magura – photo Mihail Kvartirnikov

District. They are remarkable with their polychrome wall paintings from 12–14th century. As objects of exceptional artistic value the monasteries of Ivanovo have been included in the World Cultural Heritage List of UNESCO.

The study of the rock dwellings started as early as the end of 19th century. In 1896 the book of K. and H. Škorpil "Primitive Men in Bulgaria" was published, where we find the first detailed descriptions of artificially elaborated rock monasteries in Dobrudja, Bulgarian North Black Sea Coast, the areas of Provadiya, Shumen and the valley of Russenski Lom.

Ivan Velkov studied the cave church Ipandi near the village Mihalich, Haskovo District (1933). Later, K. Miyatev (1936), P. Kamburov (1985), Maslev (1959, 1963), A. Margos (1970, 1976, 1983), Valov (1978), G. Atanassov (1984-1993) and others also explored the cave churches along Russenski Lom.

At the same time, many researchers paid attention to the Medieval graffiti drawings and signs. K. Škorpil (1904) documented the inscriptions from the rock monasteries near Krepcha, Targovishte Distr. K. Popkonstantinov (1977) dated them to the first half of 10th century (922) at the earliest. Of special interest are also "the royal inscriptions" from 13–14th century in the rock church of Royak near Varna (Margos, 1971).

A. Margos is among the most active researchers of the rock graffiti in the cave monasteries in the areas of Provadiya, Nikopol, Karlukovo, Beli Lom valley, Razgrad and others. Medieval graffiti from Northwest Bulgaria have been studied by M. Asparuhov (1984) in the caves near Debovo, Musselievo and Zhernov Pleven District, and the rock church near Nikopol. The graffiti in the caves near Provadiya have been studied also by I. Galabov (1966) and N. Panayotov (1977). P. Detev worked in the cave Topchika (1967). A generalization on the medieval drawings and graffiti, some of them from caves, was made by Ovcharov (1982).

The graffiti in the many caves and niches in the hills Sredniya Kamik and Govedarnika near the village Tsarevets, Vratsa Distr., are particularly interesting. Here, over 1000 anthropomorphic and zoomorphic pictures, solar signs and inscriptions, recording events from Medieval times have been discovered. This locality was recorded by the geographer M. Michev (1962) and was studied in turn by P. and B. Tranteev (1971-72), S. Bogdanov (1976), expeditions of Akademik – Sofia Caving Club and NEK – UNESCO (1979-80), V. Nikolov (1980), M. Asparuhov (1984), T. Stoychev and M. Zlatkova (1986–1987) and T. Stoychev (1988–2000).

It becomes clear from the brief survey that caves are important for the archaeological science in Bulgaria and their study is contributing a lot to the knowledge of the cultural heritage of Bulgaria. This makes it even more important and urgent to protect the caves for the future generations.

Rock art in the caves near Tzarevetz, Vratsa area – photo Nikolay Genov

PALEONTOLOGICAL RESEARCH IN BULGARIAN CAVES

Paleontological studies in Bulgarian caves started at the end of 19th century with the excavations carried out in the caves Polichki and Toplya by Yurinich (1891), Bonchev (1900) and Stoyanov (1904). The fundamental research in the Archaeology, as well as in the Paleontology of Bulgarian caves in the first half of 20th century is due to Prof. Rafail Popov (paleontological publications from 1904 to 1939). During these 35 years R. Popov recorded many species of large mammals in the caves Morovitsa, Temnata Dupka and other caves in Karlukovo area, Mirizlivka, Toplya, the caves near Dryanovo, on the Plateau of Belyakovets and Shumen, etc. Meanwhile, some other authors – Petkov (1926, 1958), Markov (1951, 1963), Bakalov & Nikolov (1964), undertook the study of the fossils of large mammals in Bulgarian caves. The essential collections of fossils of large mammals are housed in the National Museum of Natural History in Sofia. In this Museum Dr Ivan Nikolov worked until his death in 1982. Now the Museum has a Paleontological branch in Assenovgrad and boasts some of the most qualified specialists in large mammals from the Quaternary (Dr Nikolay Spassov) and birds (Dr Zlatozar Boev, DSc).

In his report before the Scientific Conference of BFS in 1976 Nikolov (1977) made a review of the finds of fossil mammal fauna in Bulgarian caves. According to him, 32 mammal species had been recorded from 36 caves. The list in his paper contains many obsolete, wrongly written Latin names (Ivan Nikolov was a mine engineer) and contains very few small mammals. The remains of birds, reptiles, fishes, molluscs, as well as of small mammals remained unidentified. The list of Nikolov is incomplete, as he obviously did not take into account the detailed list of Popov (1936). In another generalizing paper Nikolov (1983) recorded paleontological founds from 71 caves. The number of the mammals enumerated is different from the previous article of Nikolov – 30 species, 6 identified as "sp." and the group Primates.

Meanwhile, new research on fossils from Bulgarian caves took place on a modern basis. This is especially true for the small mammals. In their study substantial progress has been achieved thanks to the efforts and the high qualification of Dr Vassil Popov. Before his work some data on the fossil and subfossil small mammals have been published by Woloszyn (1982), Kowalski (1982); Kowalski & Nadachowski (1982), as result of the excavations in the cave Bacho Kiro by the expedition of the Jagiellonian University (Poland) and the Museum in Gabrovo in 1971-75. The book on this expedition (Kozlowski, Ed., 1982) contains also data on the fossil birds (Z. Boheński), Molluscs (E. Stworzewicz), Amphibia and Reptilia (M. Mlynarski), Insectivora (B. Rzebik-Kowalska), Lagomorpha (L. Sych), Carnivora (T. Wiszniowska), Rhinocerotidae (H. Kubiak), Equidae (A. Forsten) and Artiodactyla (H. Kubiak, A. Nadachowski). The most recent general survey of fossil mammals was done by V. Popov in vol. 27 (Mammals) of the series Fauna Bulgarica (Peshev, Peshev & Popov, 2004).

From 1983 to 2004 V. Popov recorded many species of rodents, insectivores and bats from caves. Part of the results obtained were analysed in his article on the proceedings of the Fourth National Conference of Speleology in Varna (1983). In this article 53 taxons of small mammals from 18 caves and niches are recorded. Some of them (Morovitsa, Temnata Dupka near Karlukovo, Mirizlivka near Oreshets) had been excavated first by Rafail Popov, but the small mammals were not studied. The cave Mechata Dupka near Bov (Zhelen) was explored also by Nedelcho Petkov. In the articles on the archeological results from the excavation of the caves near Lovech, Pesht near Staro Selo, Parnitsite, Morovitsa, Golyamata Mikrenska Peshtera and others the Archeologist N. Dzhambazov (1919–1982) also reported some remains of large mammals and other animals. Meanwhile, the well known school teacher and speleologist from Chepelare Dimitar Raychev and Y. Raicheva also carried out excavations in the caves in the Rhodopes and published some data on the fossils found there. Unfortunately, in many of the "bear caves" in Bulgaria, where large and impressive bones of cave bears are found, "excavations" by amateur cavers or chance visitors or by groups of "exploring" school children take place and damage the sites. One such indiscriminately excavated cave is Svinskata Dupka near Lakatnik. Often such "enthusiasts" are led even by their teachers who sincerely believe that by such excavations they are contributing to science. Actually, the result is just the opposite – the site is lost and the material collected, even though placed in a school exposition, is very soon destroyed. If not conserved properly in suitable liquids, the bones dry out, crack and are very soon disintegrated. Another damage is the disturbance of the layers, the throwing out of the "useless" earth, often containing even more important bones of small animals (some of them new for science). This earth should be sifted horizontally, layer by layer, and well documented. If fossils are seen in the cave clay, the correct thing to do is to inform the specialists and to help them, but always under their guidance.

Very substantial progress is also being achieved in the paleontology of bird remains in caves. The papers of Boheński (1982), Mlikovsky (1997) and, most of all, of Zlatozar Boev (from 1994) contain data on more than 160 bird species from the Pleistocene found in cave deposits (18 caves). According to the analysis of Boev (2001), the bird taxa from the Pleistocene have been found in Bulgarian caves as follows: Razhishkata peshtera (39 taxa), Toplya (3), Kozarskata peshtera (5), Tsareva Tsarkva (4), Mirizlivka (3), Mechata dupka (1), Filipovskata peshtera (30), Devetashkata peshtera (136), Karlukovo 4 Cave (2), Temnata dupka (38 – Late Pleistocene, 8 – Early Pleistocene), Cave 16 (52), Bacho Kiro (23), Kozarnika (43), Morovitsa (8), Cherdzhenitsa (3). Besides, many other mammals, birds and other vertebrates were found in clay filled karstic fissures, which are not caves (s.str.), or from destroyed caves (Muselievo, Varshets, Varbeshnitsa). These founds are not analysed here.

The results of the study of the fossil and subfossil animals, found in Bulgarian caves, allowed the specialists (most of all V. Popov, Z. Boev, N. Spassov) to reconstruct the paleoenvironment in many regions of Bulgaria and contributed to better understanding of the history of Bulgarian biota, especially durung the Pleistocene and the Holocene. Certainly, many other cave localities remain to be excavated and studied. Cavers can help the scientists in this task.

Quaternary mammals, known from Bulgarian caves (after Boev, 1994 to 2002, Djambazov, 1957 to 1981, Forsten, 1982, Garrod, 1939, Kowalski & Nadachowski, 1982, Kowalski, 1982, Kubiak, 1982, Kubiak & Nadachowski, 1982, Markov, 1951, 1963, Nikolov, 1977, 1983, Petkow, 1926, 1958, Popov R., 1904 to 1939, Popov V., 1983 – 2004, Raychev, 2002 and other publ., Raycheva, 1983, Stoyanov, 1904, Sych, 1982, Wiszniowska, 1982, Woloszyn, 1982).

Order Insectivora
Erinaceus concolor Martin – Mecha dupka (Bov), Strelite
Erinaceus sp. – Temnata dupka (Karlukovo)
Desmana sp. – Kozarnika, Futyovskata peshtera
Beremendia fissidens (Petenyi) – Kozarnika, Futyovskata peshtera
Sorex araneus L. – Mecha dupka (Bov), Morovitsa, Novata (Bosnek), Strelite, Borikovskata peshtera, Temnata dupka (Karlukovo), Kozarnika, Razhishkata peshtera, Bacho Kiro
S. cf. ***araneus*** L. – Morovitsa
S. ***subaraneus*** Heller – Morovitsa
S. cf. ***subaraneus*** Heller – Kozarnika
S. ***minutissimus*** Zimmermann – Cherdjenitsa, Morovitsa, Futyovskata peshtera
S. cf. ***minutissimus*** Zimmermann – Morovitsa
S. ***minutus*** L. – Mecha dupka (Bov), Morovitsa, Novata (Bosnek), Strelite, Borikovskata peshtera, Temnata dupka (Karlukovo), Kozarnika, Bacho Kiro
S. ex gr. ***runtonensis*** Hinton – Cherdjenitsa
Neomys fodiens Pennant – Mecha dupka (Bov), Novata (Bosnek), Strelite, Temnata dupka (Karlukovo), Bacho Kiro
N. ***anomalus*** Cabrera – Cherdjenitsa, Bacho Kiro
N. cf. ***anomalus*** Cabrera – Temnata dupka (Karlukovo)
Neomys sp. – Kozarnika
Crocidura leucodon (Hermann) – Cherdjenitsa, Razhishkata peshtera (Lakatnik), Mecha dupka (Stoilovo), Borikovskata peshtera, Temnata dupka (Karlukovo)
Crocidura suaveolens (Pallas) – Cherdjenitsa
Crocidura cf. ***zorzii*** Pasa – Morovitsa
Crocidura sp. – Temnata dupka, Morovitsa, Gininata peshtera, Bacho Kiro
Talpa europaea L. – Cherdjenitsa, Devetashkata peshtera, Morovitsa, Mecha dupka (Bov), Mecha dupka (Stoilovo), Novata (Bosnek),

Strelite, Borikovskata peshtera, Temnata dupka (Karlukovo), Mecha Kiro

Talpa cf. *europaea* L. – Mecha dupka (Bov), Kozarnika, Morovitsa

Order Chiroptera

Rhinolophus ferrumequinum (Schreber) – Bacho Kiro, TDK (Cave 15, Cave 16)

Rh. ex gr. *ferrumequinum* (Schreber) – Temnata dupka (Karlukovo), Futyovskata peshtera

Rh. euryale Blasius – Bacho Kiro

Myotis myotis (Borkhausen) – Bacho Kiro

M. blythi (Tomes) – Bacho Kiro

M. cf. *blythi* (Tomes) – Borikovskata peshtera

M. bechsteini (Kuhl) – Bacho Kiro, TDK (Temnata dupka – Karlukovo, includes Cave 15 and Cave 16)

M. nattereri (Kuhl) – Bacho Kiro, TDK (Cave 16)

M. mystacinus (Kuhl) – TDK (Cave 16), Razhishkata peshtera

M. brandti (Eversmann)/*daubentoni* (Kuhl) – TDK (Cave 16)

M. dasycneme (Boie) – Bacho Kiro

M. cf. *gundersheimensis* Heller – Kozarnika

Plecotus austriacus (Fischer) – TDK (Cave 15)

P. cf. *auritus* (L.) – TDK (Cave 15), Bacho Kiro

Barbastella leucomelas Cretzschmar – TDK (Cave 16), Bacho Kiro (sub "*B.* cf. *schadleri*" – see Woloszyn, 1982)

Nyctalus noctula (Schreber) – TDK (Cave 15, Cave 16)

N. lasiopterus (Schreber) – TDK (Cave 16)

N. leisleri (Kuhl) – Razhishkata peshtera, TDK (Cave 16)

Pipistrellus pipistrellus (Schreber) – TDK (Cave 15, Cave 16), Razhishkata peshtera, Bacho Kiro

Hypsugo savii (Bonaparte) – Razhishkata peshtera

Eptesicus serotinus (Schreber) – Bacho Kiro, TDK (Cave 16), Razhishkata peshtera

E. nilssoni (Keyserling et Blasius) – TDK (Cave 16), Razhishkata peshtera, caves near Karlukovo

Vespertilio murinus L. – Bacho Kiro, TDK (Cave 16), Razhishkata peshtera

Miniopterus schreibersi (Kuhl) – Bacho Kiro, TDK (Cave 16), Razhishkata peshtera

Order Lagomorpha

Lepus timidus L. – ? Emenskata peshtera, Karlukovskite peshteri and ? Vrazhi dupki near Lakatnik

Lepus capensis L. – Morovitsa

Lepus cf. *capensis* L. (= *L. capensis europaeus* Pallas) – Mecha dupka (Bov), Devetashkata peshtera, Mecha dupka (Stoilovo), Borikovskata peshtera, Temnata dupka (Karlukovo), Kozarnika, Bacho Kiro

Lepus sp. – Kozarnika, Futyovskata peshtera

Allilepus sp. – Temnata dupka (Karlukovo)

Hypolagus brachignathus Kormos – Temnata dupka (Karlukovo), Kozarnika, Futyovskayta peshtera

Ochotona pusilla Pallas – Temnata dupka (Karlukovo), Kozarnika, Mecha dupka (Bov), Bacho Kiro

Ochotona sp. cf. *pusilla* Pallas – Mecha dupka (Bov), Cherdjenitsa, Mecha dupka (Stoilovo), Kozarnika, Morovitsa

Ochotona polonica Sych – Temnata dupka (Karlukovo)

Order Rodentia

Sciurus vulgaris L. – Novata (Bosnek), Mecha dupka (Stoilovo)

Spermophilus citellus (L.) – Novata (Bosnek)

Spermophilus (= *Citellus*) sp., cf. *citellus* L.) – Golyamata and Malkata caves near Veliko Tarnovo, Malkata peshtera near Belyakovets, Cherdjenitsa, Mecha dupka (Stoilovo), Borikovskata peshtera, Temnata dupka (Karlukovo), Kozarnika, Morovitsa, Mecha dupka (Bov), Bacho Kiro

Spermophilus cf. *primigenius* (Kormos) – TDK (Cave 15)

Spermophilus cf. *nogaici* Topachevskyi – TDK (Cave 15), Futyovskata peshtera

Spermophilus sp. – Temnata dupka, Morovitsa, Mecha dupka (Bov), Haydushkata peshtera, Vodnitsata, Gornik, Gininata peshtera, TDK (Cave 15), Futyovskata peshtera

Marmota sp. – Kozarnika

Glis glis (L.) – Morovitsa, Mecha dupka (Bov), Haydushkata peshtera, Vodnitsata, Gornik, Gininata peshtera, Desni Suhi pech, Yagodinskata peshtera, Karnata, Borikovskata peshtera, Cherdjenitsa, Strelite, Mecha dupka (Stoilovo), Temnata dupka (Karlukovo), Kozarnika, Bacho Kiro

Dryomys nitedula (Pallas) – Morovitsa, Cherdjenitsa, Strelite, Temnata dupka (Karlukovo), Razhishkata peshtera, Bacho Kiro

Muscardinus dacicus – Temnata dupka (Karlukovo)

Muscardinus avellanarius L. – Mecha dupka (Bov), Strelite, Borikovskata peshtera, Morovitsa, Bacho Kiro

Myomimus sp. – Kozarnika

Myomimus sp. B

Castor fiber L. – Tamnata peshtera (Targovishte Village), Emenskata peshtera, Temnata dupka (Karlukovo), Bacho Kiro, Kozarnika

Hystrix sp. – Kozarnika, Temnata dupka (Cave 15), Futyovskata peshtera

Sicista subtilis (Pallas) – Temnata dupka (Karlukovo), Morovitsa, Mecha dupka (Bov), Cherdjenitsa, Mecha dupka (Stoilovo), Kozarnika, Bacho Kiro

Sicista sp. – Kozarnika

Allactaga jaculus – Temnata dupka

Allactaga major (Kerr) – Devetashkata peshtera, Novata (Bosnek), Temnata dupka (Karlukovo), Bacho Kiro

Nannospalax leucodon (Nordmann)(partly sub "*Spalax typhlus* Giebel") – Golyamata peshtera near Belyakovets, Mecha dupka (Bov), Cherdjenitsa, Temnata dupka (Karlukovo), Morovitsa, Haydushkata peshtera, Kozarskata peshtera, Vrazhi dupki, Desni Suhi pech, Vodnitsata, Gornik, Gininata peshtera, Borikovskata peshtera, Novata (Bosnek), Strelite, Mecha dupka (Stoilovo), Kozarnika, Bacho Kiro

Nannospalax sp. – TDK (Cave 15), Kozarnika

Nannospalax (Pliospalax) compositodontus (Topachevskyi) – TDK (Cave 15)

Nannospalax (Pliospalax) cf. *odessanus* (Topachevskyi) – Futyovskata peshtera

Prospalax cf. *priscus* (Nehring) – Kozarnika

Mus cf. *hortulanus* – Cherdjenitsa

Mus musculus L. – Bacho Kiro

Mus sp. – Temnata dupka (Karlukovo), Haydushkata peshtera, Gininata peshtera

Mus cf. *macedonicus* Petrov et Ružić – Mecha dupka (Stoilovo)

Mus cf. *spicilegus* Petenyi – Temnata dupka (Karlukovo)

Rattus rattus (L.) – Strelite

Sylvaemus mystacinus (Danford et Alston) – Kozarnika

Sylvaemus cf. *mystacinus* (Danford et Alston) – Mecha dupka (Bov), Futyovskata peshtera

Sylvaemus cf. *uralensis* (Pallas) – Temnata dupka (Karlukovo)

Sylvaemus ex gr. *sylvaticus – flavicollis* – Temnata dupka (Karlukovo), Morovitsa, Mecha dupka (Bov), Haydushkata peshtera, Vodnitsata, Gornik, Gininata peshtera, Futyovskata peshtera, Desni Suhi pech, Borikovskata peshtera, Yagodinskata peshtera, Karnata peshtera, Cherdjenitsa

Sylvaemus cf. *flavicollis* (Melchior) – TDK (Cave 15), Novata (Bosnek), Mecha dupka (Stoilovo), Borikovskata peshtera, Temnata dupka (Karlukovo), Kozarnika, Morovitsa, Mecha dupka (Bov), Bacho Kiro

Sylvaemus cf. *sylvaticus* (L.) – Kozarnika, Morovitsa, Mecha dupka (Bov), Bacho Kiro

Sylvaemus dominans Kretzoi – TDK (Cave 15)

Apodemus cf. *agrarius* Pallas – Mecha dupka (Bov)

Micromys minutus (Pallas) – Temnata dupka (Karlukovo), Morovitsa, Strelite

Cricetus cricetus (L.) – Morovitsa, Desni Suhi pech, TDK (Cave 15), Temnata dupka (Karlukovo), Kozarnika, Bacho Kiro

C. nanus Schaub – Kozarnika, Cherdjenitsa

C. runtonensis (Newton) – Temnata dupka (Karlukovo, incl. cave 15), Futyovskata peshtera, Kozarnika,

Allocricetus bursae Schaub – Morovitsa, Futyovskata peshtera, Cherdjenitsa, Kozarnika

Mesocricetus newtoni (Nehring) – Temnata dupka (Karlukovo), Morovitsa, Mecha dupka (Bov), Kozarskata peshtera, Vrazhi dupki, Haydushkata peshtera, Vodnitsata, Gornik, Gininata peshtera, Desni Suhi pech, Borikovskata peshtera, Cherdjenitsa, Novata Peshtera (Bosnek), Strelite, Mecha dupka (Stoilovo), Kozarnika, Bacho Kiro

Cricetulus migratorius (Pallas) – Mecha dupka (Bov), Cherdjenitsa, Borikovskata peshtera, Novata (Bosnek), Strelite, Mecha dupka (Stoilovo), Temnata dupka (Karlukovo), Kozarnika, Morovitsa, Bacho Kiro

Villania sp. – Mirizlivka, Temnata dupka (Karlukovo) (?)

Arvicola terrestris (L.) – Morovitsa, Mecha dupka (Bov), Kozarskata peshtera, Vrazhi dupki, Borikovskata peshtera, Cherdjenitsa, Novata (Bosnek), Strelite, Mecha dupka (Stoilovo), Temnata dupka (Karlukovo), Bacho Kiro

Arvicola cf. *chozaricus* Alexandrova – Morovitsa

Arvicola cf. *kalmankensis* Zazhigin – Morovitsa

Arvicola cantiana (Hinton) – Kozarnika

P. lenki (Heller) – Morovitsa

P. coronensis Mehely – Morovitsa

P. cf. *simplicior* Kretzoi – Cherdjenitsa

P. cf. *hungaricus* Kormos – Kozarnika

P. cf. *graecus* Mehely – Futyovskata peshtera
Ungaromys nanus Kormos – TDK (Cave 15)
Clethrionomys glareolus (Schreber) – Mecha dupka (Bov), Cherdjenitsa, Temnata dupka (Karlukovo), Kozarnika
Clethrionomys sp. – Morovitsa, Mecha dupka (Bov), Kozarskata peshtera, Haydushkata peshtera, Vodnitsata, Gornik, Gininata peshtera, Karnata peshtera
Mimomys pliocaenicus Forsith-Major – Temnata dupka (Karlukovo)
M. stenokorys Rabeder – TDK (Cave 15)
M. reidi Hinton – Temnata dupka (Karlukovo)
M. pusillus (Mehely) – TDK (Cave 15), Futyovskata peshtera, Cherdjenitsa
M. cf. *blanci* Van den Meulen – Kozarnika
M. pitymyoides Janossy et Van den Meulen – TDK (Cave 15), Futyovskata peshtera, Cherdjenitsa
M. tornensis Janossy et Van den Meulen – TDK (Cave 15)
Mimomys sp. – Temnata dupka (Karlukovo)
Villania sp.
Microtus arvalis Pallas – Bacho Kiro
Microtus arvalis/agrestis – Cherdjenitsa, Temnata dupka (Karlukovo), Kozarnika, Mecha dupka (Bov), Mecha dupka (Stoilovo)
Microtus arvalinus Hinton – Morovitsa, Kozarnika
Microtus cf. *agrestis* (L.) – Kozarnika
Microtus gregalis (Pallas) – Kozarnika
Microtus cf. *gregalis* (Pallas) – Morovitsa
Microtus arvalidens Kretzoi – Kozarnika, Morovitsa
Microtus oeconomus Pallas – Bacho Kiro
Microtus cf. *oeconomus* Pallas – Kozarnika
Microtus pliocaenicus (Kormos) – Futyovskata peshtera
Microtus burgondiae (Chaline) – Futyovskata peshtera, Cherdjenitsa
Microtus hintoni Kretzoi – Kozarnika
Microtus deucalion (Kretzoi) – TDK (Cave 15)
Microtus subterraneus (Selys – Longch.) – Mecha dupka (Bov), Cherdjenitsa, Razhishkata peshtera, Morovitsa, Novata (Bosnek), Strelite, Mecha dupka (Stoilovo), Borikovskata peshtera, Temnata dupka (Karlukovo), Kozarnika, Bacho Kiro
"*Pitymys*" *arvaloides* – Morovitsa, Temnata dupka (Karlukovo)
Dolomys dalmatinus Kormos – Kozarnika
Chionomys nivalis (Martins) – Mecha dupka (Bov), Cherdjenitsa, Razhishkata peshtera, Morovitsa, Temnata dupka (Karlukovo), Kozarnika, Mecha dupka (Stoilovo), Bacho Kiro
Borsodia hungarica (Kormos) – TDK (Cave 15)
B. petenyi (Mehely) – Temnata dupka (Karlukovo)
B. arankoides (Alexandrova) – TDK (Cave 15)
Kalymnomys major (Kuss et Storch) – TDK (Cave 15)
Lagurodon arankae (Kretzoi) – TDK (Cave 15), Futyovskata Peshtera, Cherdjenitsa
Lagurodon praepannonicus (Topachevskyi) – TDK (Cave 15), Futyovskata Peshtera, Cherdjenitsa
Prolagurus transylvanicus Tersea – Kozarnika, Futyovskata peshtera
Lagurus transiens Janossy – Kozarnika
Lagurus cf. *transiens* Janossy – Morovitsa
L. lagurus Pallas – Cherdjenitsa, Morovitsa, Temnata dupka (Karlukovo), Kozarnika, Bacho Kiro
Eolagurus luteus Eversmann – Temnata dupka (Karlukovo), incl. Cave 16
E. gromovi Topatchevsky – Morovitsa

Order Carnivora
Wolf (*Canis lupus* L.) – Bacho Kiro (Malkata peshtera), Svinskata Dupka, Mirizlivka, Malkata peshtera near Veliko Târnovo, Mladenova Propast, Toplya
Dog (*Canis familiaris* L.) – Bacho Kiro (Malkata peshtera), Razhishkata (Suhata) Peshtera, Golyamata Peshtera near Veliko Târnovo, Suhi Pech, Raztsepenata Peshtera, Popin Pchelin
[**Jackal** (*Canis aureus* L.) – Mladenova propast, Lednitsata (Dobrostan), Toplya. According to Spassov (1989), there are no remains of jackals in Bulgarian caves].
Fox (*Vulpes vulpes* L.) – Bacho Kiro (Malkata peshtera), Golyamata Peshtera near Veliko Târnovo, Mirizlivka, Golyamata Mikrenska peshtera
Polar Fox (*Alopex lagopus* L.) – Bacho Kiro
Red Dog (*Cuon alpinus* Pallas) – Bacho Kiro
Wild Cat (*Felis silvestris* Schreber (sub *F. catus* L.) – The caves near Krumovi Porti, Magura, Kozarnika (Suhi Pech), Toplya, Razhishkata peshtera, Yagodinskata Peshtera
Lynx (*Lynx lynx* L.) – Borikovskata peshtera
Leopard (*Panthera pardus* L.) – Triâgâlnata Peshtera, Bacho Kiro
Cave Lion (*Panthera spelaea* Goldfuss) – Temnata dupka (Karlukovo), Bacho Kiro, Svinskata Dupka (Lakatnik, Spassov, pers. com.)

Brown Bear (*Ursus arctos* L.) – from the caves Mirizlivka, Malkata Peshtera near Veliko Târnovo, Borikovskata Peshtera, Popin Pchelin, Malkata Peshtera near Belyakovets, Razhishkata (Suhata) Peshtera, Magura, Pesht near Staro Selo, Temnata dupka (Karlukovo), Bacho Kiro

Cave Bear (*Ursus spelaeus* Blumenbach) – Mecha Dupka (Bov), Bacho Kiro (Malkata Peshtera), Toplya, Temnata Dupka (Karlukovo), Devetashkata peshtera, Triâgâlnata peshtera, Borikovskata peshtera, Orlova Chuka, Boevskata peshtera, Potoka, Razhishkata (Suhata) peshtera, Svinskata Dupka, Golyamata peshtera near Veliko Târnovo, Malkata Peshtera near Veliko Târnovo, Mirizlivka, Kozarnika (Suhi Pech), Desni Suhi Pech, Mladenova Propast, Magura, Lucifer, Slivova dupka, Gorni Razh 1

Cave Hyaena (*Crocuta spelaea* Goldfuss) – Mirizlivka, Golyamata and Malkata caves near Veliko Tarnovo, Morovitsa, Magura, Bacho Kiro (= Malkata Peshtera) near Dryanovo, the caves near Karlukovo, Golyamata and Malkata caves near Belyakovets, Temnata dupka (Karlukovo), Borikovskata Peshtera

Badger (*Meles meles* L.) – Toplya, the caves near Krumovi Porti, Kozarnika (Suhi Pech), Razhishkata (Suhata) peshtera, Tsarskata Peshtera, Emenskata Peshtera, Magura, Popin Pchelin, Temnata dupka (Karlukovo)

Pine Marten (and/or *Martes foina* Erxleben) (*Martes martes* L.) – Toplya, Razhishkata (Suhata) Peshtera, Kozarnika (Suhi Pech), Bacho Kiro, Yagodinskata Peshtera

Polecat (*Mustela putorius* L.) – Golyamata and Malkata caves near Belyakovets, Temnata dupka (Karlukovo), Bacho Kiro

Stoat (*M. erminea* L.) – Bacho Kiro

Weasel (*M. nivalis* L.) – Devetashkata peshtera

Order Primates – Devetashka peshtera, Razhishkata (Suhata) peshtera, Svinskata Dupka, Snezhanka, Tsarskata peshtera, Emenskata peshtera, Magura, Mladenova Propast (ex-

Scull of cave bear – photo Atanas Russev

cept of genus *Homo*, no other primates have been identified from Bulgarian caves)

Order Proboscidea

Mammoth (*Mammuthus primigenius* Blum.) – Mladenova Propast, Golyamata Peshtera near Veliko Târnovo, caves in Karlukovo

Order Perissodactyla

Horse (*Equus caballus* L.) – Mecha Dupka (Bov), Temnata Dupka (Karlukovo), Magura, Emenskata peshtera, Mirizlivka, Golyamata peshtera near Veliko Târnovo, Mladenova Propast, Popin Pchelin, Parnitsite, Golyamata Mikrenska peshtera (the true nature of the horse remains from the older founds, published as "*E. caballus fossilis*" is not known)

Equus germanicus Nehring – Temnata dupka (Karlukovo), Bacho Kiro

Equus* cf. *stenonis – Temnata dupka (Karlukovo)

Wild Ass (*Equus hydruntinus* Regalia) – Temnata dupka (Karlukovo), Bacho Kiro

Rhinocerus (*Stephanorhinus*, sub "*Dicerorhinus*" cf. *hemitoechus* Falconer) – Mecha dupka (Stoilovo), Bacho Kiro

Rhinocerus (*Stephanorhinus kirchbergensis* Jäger = *Rhinoceros mercki* Kaup) – Mladenova Propast

Rhinocerus (*Coelodonta antiquitatis* Blumenbach = *Rh. tichorhinus* Cuvier) – Malkata Peshtera (Veliko Tarnovo), Mirizlivka

Order Artiodactyla

Wild Boar (*Sus scropha* L.) – Borikovskata Peshtera, Golyamata Peshtera near Veliko Târnovo, Kozarnika (Suhi Pech), Emenskata Peshtera, Magura, Mladenova Propast, Golyamata Mikrenska peshtera, Bacho Kiro, Razhishkata peshtera, Yagodinskata Peshtera

Red Deer (*Cervus elaphus* L.) – Yagodinskata Peshtera, Bacho Kiro ("Malkata Peshtera"), Toplya, Temnata Dupka (Karlukovo), Vodnitsata (Kirov Vârtop), Razhishkata (Suhata) Peshtera, Svinskata Dupka, Golyamata Peshtera near Veliko Târnovo, Emenskata Peshtera, Mirizlivka, Kozarnika (Suhi Pech), Mladenova Propast, Golyamata Mikrenska peshtera, Magura

Cervus* cf. *philisi Schaub – Temnata dupka (Karlukovo)

Fallow Deer (*Dama dama* L.) – Triâgâlnata peshtera, Borikovskata Peshtera

Giant Deer (*Megaloceros giganteus* Blumenbach) – Temnata dupka (Karlukovo), Bacho Kiro

Elk (*Alces alces* L.) – Temnata dupka (Karlukovo)

***Alces* sp.** – Bacho Kiro

Roe Deer (*Capreolus capreolus* L.) – Golyamata Listsa, Bacho Kiro (Malkata peshtera near Dryanovo), Magura, Golyamata and Malkata caves near Belyakovets, Prohodna, Borikovskata Peshtera, Orlova Chuka, Toplya and Andaka (Golyamata peshtera), Golyamata Mikrenska peshtera, Yagodinskata Peshtera

Reindeer (*Rangifer tarandus* L.) – Polichki

Chamois (*Rupicapra rupicapra* L.) – Temnata Dupka (Karlukovo), Malkata peshtera (Veliko Tarnovo), Bacho Kiro

Capricorne (*Capra ibex* L.)(partly published sub "*Capra aegagrus* Erxleben"; according to Spassov, 1982, all records of *Capra aegagrus* in Bulgaria represent actually *C. ibex*) – Temnata Dupka (Karlukovo), Razhishkata (Suhata) peshtera, Bacho Kiro, Emenskata peshtera, Mirizlivka, Kozarnika (Suhi Pech), Târmnata peshtera (Targovishte), Triâgâlnata peshtera (Borino)

***Capra* sp.** – Temnata Dupka, Borikovskata peshtera, Razhishkata (Suhata) peshtera, Emenskata peshtera, Kozarnika (Suhi Pech), Parnitsite

Sheep (*Ovis aries* Pallas) – Razhishkata (Suhata) peshtera, Golyamata peshtera near Veliko Târnovo, Tsarskata peshtera, Emenskata peshtera, Popin Pchelin

***Ovis* sp.** – Temnata Dupka (Karlukovo), Borikovskata peshtera, Razhishkata (Suhata) peshtera

***Bos* sp.** – Razhishkata (Suhata) peshtera

Cattle (*Bos taurus* L.) – Temnata Dupka (Karlukovo), Emenskata peshtera, Mirizlivka, Raztsepenata peshtera, Popin Pchelin, Golyamata Mikrenska peshtera, Bacho Kiro

Auroche (*Bos primigenius* Bojanus) – Golyamata and Malkata caves near Veliko Tarnovo, Magura, Borikovskata peshtera, the caves near Karlukovo, Golyamata and Malkata caves near Belyakovets, Harlovata peshtera, Tsarskata peshtera, Popin Pchelin, Temnata Dupka (Karlukovo), Mladenova Propast

Here is a review (according to districts) of the caves with paleontological excavations, with notes on these caves.

Burgas

Mecha dupka – cave near Stoilovo Village (Strandja Mt.). Founds: **mammals** – *Talpa* cf. *europaea*, *Mus* cf. *macedonicus*, *Ochotona pusilla*, *Spermophilus* cf. *citellus*, *Sciurus vulgaris*, *Glis glis*, *Nannospalax leucodon*, *Sicista subtilis*, *Rattus* sp., *Mus* sp., *Sylvaemus* cf. *flavicollis*, *Sylvaemus* cf. *sylvaticus*, *Mesocricetus newtoni*, *Cricetulus migratorius*, *Arvicola terrestris*, *Microtus subterraneus*, *Chionomys nivalis*, *Ursus spelaeus*, *Stephanorhinus* cf. *hemitoechus*

Gabrovo

Andaka (Golyamata Peshtera) – cave near Dryanovski Monastery. Length 4000 m. Founds: **mammals** – *Capreolus capreolus*

Bacho Kiro (Malkata Peshtera) – cave near Dryanovski Monastery. Length 3500 m. Excavations: D. Garod and R. Popov (1938), Polish expedition (1971-75). Founds: **birds** – *Anas platyrhynchos*, *Aquila chrysaetos*, *Circus aeruginosus*, *Lagopus mutus*, *Perdix perdix*, *Coturnix coturnix*, *Alectoris graeca*, *Gallus gallus*, *Rallus aquaticus*, *Porzana porzana*, *Gallinula chloropus*, *Bubo bubo*, *Ptyonoprogne rupestris*, *Delichon urbica*, *Lullula arborea*, *Alauda arvensis*, *Anthus campestris*, *Pyrrhocorax pyrrhocorax*, *P. graculus*, *Corvus monedula*, *C. corax*; **mammals** – *Sorex araneus*, *S. minutus*, *Neomys fodiens*, *N. anomalus*, *Crocidura* sp., *Talpa europaea*, *Rhinolophus ferrumequinum*, *Rh. euryale*, *Myotis myotis*, *M. blythi*, *M. bechsteini*, *M. nattereri*, *M. dasycneme*, *Barbastella leucomelas* [cf. *darjelingensis*], *Pipistrellus pipistrellus*, *Plecotus* cf. *auritus*, *Eptesicus serotinus*, *Vespertilio murinus*, *Miniopterus schreibersi*, *Lepus* cf. *capensis*, *Ochotona pusilla*, *Spermophilus citellus*, *Glis glis*, *Muscardinus avellanarius*, *Sicista subtilis*, *Allactaga major*, *Castor fiber*, *Cricetulus migratorius*, *Mesocricetus auratus*, *Nannospalax leucodon*, *Sylvaemus sylvaticus*, *S. flavicollis*, *Mus musculus*, *Cricetus cricetus*, *Clethrionomys glareolus*, *Arvicola terrestris*, *Microtus arvalis/agrestis*, *M. oeconomus*, *M. subterraneus*, *Chionomys nivalis*, *Lagurus lagurus*, *Ursus arctos*, *U. spelaeus*, *Alopex lagopus*, *Cuon alpinus*, *Canis lupus*, *C. familiaris*, *Vulpes vulpes*, *Ursus spelaeus*, *Crocuta spelaea*, *Mustela erminea*, *M. putorius*, *Martes martes*, *Panthera pardus*, *P. spelaeus*, *Stephanorhinus* cf. *hemitoechus*, *Equus germanicus*, *E. hydruntinus*, *Sus scrofa*, *Megaloceros giganteus*, *Cervus elaphus*, *Alces* sp., *Capreolus capreolus*, *Capra ibex*, *Rupicapra rupicapra*, *Bos taurus*

Polichki – Cave near Dryanovski Monastery. Length 46 m. Excavations: Yurinich (1890). Founds: **mammals** – *Ursus spelaeus*, *Rangifer tarandus*, *Bos* sp., "*Sus scrofa ferus*"

Lovech

Toplya – cave near Golyama Zhelyazna Village. Length 462 m. Excavations: M. Koychev, G. Bonchev and I. Stoyanov (1898, 1899, 1900). Founds: **mammals** – *Castor fiber*, *Canis lupus*, [*Canis aureus*], *Felis silvestris*, *Ursus spelaeus*, *U. arctoideus*, *Meles meles*, *Martes martes*, *Equus* sp., *Cervus elaphus*, *Capreolus capreolus*, *Bos primigenius*

Futyovskata Peshtera – cave near Karpachevo Village. Length 700 m. Founds: **mammals** – *Desmana* sp., *Beremendia fissidens*, *Sorex minutisimus*, *S.* cf. *subaraneus*, *Rh.* ex gr. *ferrumequinum*, *Hypolagus brachignathus*, *Spermophilus* cf. *nogaici*, *Hystrix* sp., *Nannospalax (Pliospalax)* cf. *odessanus*, *Sylvaemus* cf. *mystacinus*, *S.* ex gr. *sylvaticus/flavicollis*, *Cricetus runtonensis*, *Allocricetus bursae*, *Pliomys* cf. *graecus*, *Microtus pliocaenicus*, *M. burgondiae*, *Prolagurus transylvanicus*, *Lagurodon arankae*, *L. praepannonicus*, *Mimomys pusillus*, *M. pitymyoides*

Temnata Dupka – cave near Karlukovo. Length 215 m. Caves 15 and 16 are considered parts of the system of Temnata dupka. Excavations: R. Popov (1925, 1926), V. Popov (1980-1982), Bulgarian – Polish – French Expedition (since 1989). Founds: **birds** – *Aquila pomarina*, *A. chrysaetos*, *Accipiter nisus*, *Aegypius monachus*, *Circus cyaneus*, *Buteo* sp., *Falco tinnunculus*, *F. vespertinus*, *F.* cf. *subbuteo*, *Coturnix coturnix*, *Perdix perdix*, *P.* cf. *paleoperdix*, *Alectoris graeca*, *Phasianus colchicus*, *Tetrao tetrix*, *Lagopus lagopus/mutus*, cf. *Bonasa bonasia*, *Porzana parva/pusilla*, *Crex crex*, *Rallus aquaticus*, *Gallinula chloropus*, *Larus* cf. *ridibundus*, *Columba oenas*, *C. livia*, *Streptopelia turtur*, *Bubo bubo*, *Otus scops*, *Strix aluco*, *Asio otus/flammeus*,? *Aegolius*, *Apus apus*, *Dendrocopus* *minor, Alauda arvensis, Lullula arborea, Anthus trivialis, Delichon urbica, Ptyonoprogne rupestris/Riparia riparia, Lanius* cf. *collurio, Turdus viscivorus, T. merula, T. philomelos, T. torquatus, T. pilaris, Monticola* cf. *saxatilis, Oenanthe oenanthe, Erithacus rubecula, Bombycilla garulus, Cettia cetti, Lanius* cf. *collurio, Sitta* cf. *europaea, Carduelis cannabina, Fringilla coelebs, Loxia curvirostra, Luscinia* cf. *megarhynchos., Sturnus vulgaris, Pyrrhula pyrrhula, Emberiza* sp., *Passer montanus, Nuciphraga caryocatactes, Pica pica, Corvus monedula, C. frugilegus, C. corax, Pyrrhocorax graculus;* **mammals** – *Sorex minutus, S. minutissimus, S. araneus, Crocidura leucodon, Neomys fodiens, Talpa europaea, Rhinolophus* ex gr. *ferrumequinum, Ochotona* sp. cf. *pusilla, Ochotona polonica, Hypolagus brachignathus, Trischizolagus, Spermophilus* cf. *primigenius, S.* cf. *nogaici, Spermophilus citellus, Muscardinus avellanarius, Dryomys nitedula, Glis glis, Sicista subtilis, Allactaga major, Hystrix* sp., *Nannospalax (Pliospalax) compositodontus, Nannospalax leucodon, Sylvaemus mystacinus, Sylvaemus* ex gr. *sylvaticus/flavicollis, S. dominans, Cricetus cricetus, Allocricetus bursae, Cricetulus migratorius, Mesocricetus newtoni, Pliomys kretzoii, Clethrionomys glareolus, Lagurus lagurus, Microtus subterraneus, Microtus* gr. *arvalis/agrestis, Microtus* cf. *gregalis, Microtus (Allophaiomys) deucalion, Chionomys nivalis, Arvicola terrestris, Borsodia hungarica, B. arankoides, Lagurodon praepannonicus, L. arankae, Kalymnomys major, Mimomys pitymyoides, M. pusillus, M. tomensis M. stenokorys, M. reidi, Castor fiber, Ursus spelaeus, "Equus caballus fossilis", Equus* cf. *stenonis, Cervus elaphus, Cervus* cf. *philisi, Megaloceros giganteus, Alces alces, Rupicapra rupicapra, Capra ibex* [sub *"aegagrus"*], *Capra ibex, Ovis* sp., *Bos taurus, Bos primigenius*

Karlukovo 4 – cave near Karlukovo. Founds: **birds** – *Turdus* sp., *Pyrrhocorax graculus*; **mammals** – 28 spp.

Prohodna – cave near Karlukovo. Length 262 m. Founds: **mammals** – *Capreolus capreolus*

Devetashkata Peshtera – cave near Devetaki. Length 2442 m. Excavations: S. Ivanova (1996); Founds: **birds** – *Podiceps cristatus, P. griseigena, Anser* cf. *erythropus, Branta ruficollis, Anas platyrhynchos, A. crecca, A. penelope, A. acuta, A. querquedula, A. strepera, A. clypeata, Netta rufina, Aythya fuligula, A. nyroca, A. ferina, Bucephala clangula, Melanita nigra, Aquila pomarina, Circus aeruginosus, Buteo lagopus, B. buteo, Accipiter nisus, A. gentilis, Falco tinnunculus, F.* cf. *subbuteo, F. vespertinus, Tetrao tetrix, T. urogallus, Lagopus lagopus, L. mutus/lagopus, Tetrastes bonasia, Coturnix coturnix, Perdix paleoperdix, P. perdix, Phasianus colchicus, Alectoris graeca/chukar, Gallinula chloropus, Crex crex, Rallus aquaticus, Porzana pusilla, Tetrax tetrax, Otis/Tetrax, Vanellus vanellus, Recurvirostra avosetta, Himantopus himantopus, Philomachos pugnax, Pluvialis squatarola, Numenius phaeopus/tenuirostris, Actitis hypoleucos, Tringa totanus, T. ochropus, T. stagnatilis, T. nebularia, Calidris alba, Limosa limosa, Gallinago media, G. gallinago, Scolopax rusticola, Sterna hirundo, Sterna* sp., *Larus* cf. *canus, Larus* cf. *ridibundus, Chlidonias* sp., *Gelochelidon* cf. *nilotica,* cf. *Nyctea scandiaca, Athene noctua, Aegolius funereus, Asio otus, A. flammeus, ? Surnia ulula, Strix aluco, Strix* aff. *nebulosa, Glaucidium passerinum, Columba livia, Cuculus canorus, Apus apus, Dendrocopus major, D.* aff. *minor, Alauda arvensis, Lullula arborea, Eremophila alpestris, Ptyonoprogne rupestris/Riparia riparia, Hirundo rupestris, Anthus trivialis, Sylvia communis, Ficedula* cf. *albicollis, Turdus merula, T. viscivorus, T. iliacus, T. philomelos, Pyrrhula pyrrhula, Coccothraustes coccothraustes, Sturnus vulgaris, Garrulus glandarius, Corvus monedula, C. corone, C. frugilegus, C. corax, Nucifraga caryocatactes, Pica pica, Pyrrhocorax graculus, P. pyrrhocorax;* **mammals** – *Mustela nivalis, ? M. erminea, Lepus europaeus, Talpa europaea, Allactaga (?) major, Ursus spelaeus*, Primates

Cherdjenitsa – cave near Karlukovo. Length 45 m. Excavations: V. Popov (1985). Founds: **birds** – *Coturnix coturnix, Dendrocopus minor, Hirundo rustica, H. daurica, Ptyonoprogne rupestris, Riparia riparia, Lanius collurio, Cinclus cinclus, Turdus merula, T. philomelos, Monticola* cf. *saxatilis, Sylvia* cf. *atricapilla, Pyrrhula pyrrhula, Carduelis cannabina, Passer* cf. *domesticus, Anthus* sp., *Phylloscopus sibilatrix;* **mammals** – *Sorex minutissimus, S. runtonensis, Neomys anomalus, Crocidura leucodon, C. suaveolens, Talpa europaea, Beremendia fissidens, Ochotona* cf. *pusilla, Spermophilus* sp., *Nannospalax leucodon, Sicista subtilis, Sylvaemus sylvaticus/flavicollis, Allocricetus bursae,*

Cricetus nanus, Glis glis, Dryomys nitedula, Pliomys cf. *simplicior, Clethrionomys glareolus, Lagurodon praepannonicus, L. arankae, Mimomys pitymyoides, M. pusillus, Microtus pliocaenicus, M. subterraneus, M. burgondiae, M. arvalis/agrestis, Chionomys nivalis, Lagurus lagurus*

Parnitsite – Cave near Bezhanovo. Length 2500 m. Excavations: N. Djambazov (1960). Founds: **mammals** – *"Equus caballus", Bos* sp., *Capra* sp., *Cervus* sp.

Gornik – cave near Krushuna Village. Length: 1074 m. Founds: **mammals** – *Spermophilus* sp., *Glis glis, Nannospalax leucodon, Sylvaemus sylvaticus/flavicollis, Mesocricetus newtoni, Clethrionomys* sp.,

Golyamata Mikrenska peshtera – Cave near Mikre. Length 1921 m. Excavations: N. Djambazov (1960). Founds: **mammals** – *Vulpes vulpes, Bos taurus, Cervus elaphus, Capreolus capreolus, Sus scrofa, S. scrofa domestica, "Equus caballus".*

Morovitsa – cave near Glozhene. Length 3250 m. Excavations: M. Koychev (1909), R. Popov (1912), N. Djambazov (1955). Founds: **birds:** *Perdix perdix, Alectoris graeca, Coturnix coturnix, Alauda arvensis, Pyrrhocorax graculus;* **mammals** – *Sorex araneus, S. subaraneus, S.* cf. *minutissimus, S. minutus, Talpa europaea, Dryomys nitedula, Glis glis, Sicista subtilis, Nannospalax leucodon, Sylvaemus* ex gr. *sylvaticus/flavicollis, Clethrionomys* sp., *Microtus subterraneus, Chionomys nivalis,* cf. *Micromys minutus, Cricetus cricetus, Allocricetus bursae, Mesocricetus newtoni, Arvicola terrestris, Pliomys lenki, Lagurus lagurus, L.* cf. *transiens, Eolagurus luteus, Ursus spelaeus, Crocuta spelaea, Cervus elaphus, Capreolus capreolus, "Sus scrofa ferus", Bos taurus, Ovis aries*

Lovech Caves – caves in the park Bashbunar near Lovech (Tabashkata peshtera, Vassil Levski, Malkata peshtera and others). Excavation: N. Djambazov. Founds: **mammals** – *Lepus capensis, Arvicola* sp., *Ursus spelaeus, Crocuta* sp., *Canis lupus, C. familiaris, Vulpes vulpes, Martes* sp., *Bos primigenius, Bos taurus, Rhinocerus* sp., *"Equus caballus", "E. asinus", Cervus elaphus, Capreolus capreolus, Sus scrofa, S. scrofa domestica, Capra* sp., *Ovis aries,* Chiroptera; Aves, Pisces, Lamellibranchiata, Gastropoda

Vit Caves (Vachovata Dupka, Srednata Dupka and Ochilata) – caves near Aglen Vilage. Excavations: N. Djambazov (1959–1960).

Pazardjik

Snezhanka – Cave near Peshtera Town. Length 230 m. Founds: **mammals** – Primates

Pernik

Novata Peshtera – cave near Bosnek Village. Founds: **mammals** – *Talpa europaea, Sorex araneus, S. minutus, Neomys fodiens, Lepus* cf. *capensis, Spermophilus* cf. *citellus, Sciurus vulgaris, Muscardinus avellanarius, Glis glis, Nannospalax leucodon, Sylvaemus flavicollis, Mesocricetus newtoni, Cricetulus migratorius, Clethrionomys glareolus, Arvicola terrestris, Microtus subterraneus*

Filipovskata (Mislovishka) Peshtera – cave near Filipovtsi Village. Length: 83 m. Excavations and collecting: R. Pandurska, I. Pandurski (1995), Z. Boev (1993). Founds: **birds** – *Falco* sp. (gr. *tinnunculus), Lagopus* cf. *lagopus, Tetrao tetrix, Coturnix coturnix, Perdix* cf. *perdix, Phasianus colchicus, Athene noctua, Hirundo rustica, Sitta europaea, Prunella modularis, Erythacus rubecula, Carduelis* sp., *Pyrrhula pyrrula, Sturnus vulgaris, Corvus monedula, Pyrrhocorax graculus, P. pyrrhocorax, Pica pica;* **mammals** – *Glis glis, Lepus capensis, Vulpes vulpes,* Chiroptera

Tsareva Tsarkva Cave – cave near Zelenigrad Village. Excavations: Z. Boev (1994). Founds: **birds** – *Falco subbuteo, Coturnix coturnix, Corvus monedula, Pyrrhocorax graculus;* **mammals** – *Ursus spelaeus*

Zelenigradskata peshtera – cave near Zelenigrad Village. Excavations: Z. Boev (1993, 1994). Founds: **birds** – *Falco* cf. *tinnunculus, Perdix perdix, Gallus gallus domestica, Bubo bubo,* Fringillidae gen., *Pyrrhocorax graculus, Corvus monedula;* **mammals** – *Lepus capensis, Vulpes vulpes*

Vreloto – cave near Bosnek Village. Length: 5280 m. Hugh ammount of unidentified bones.

Pleven

Vodnitsata (Kirov Vârtop) – cave system near Bohot Village. Length 776 m. Founds: **mammals** – *Spermophilus* sp., *Glis glis*, *Nannospalax leucodon*, *Sylvaemus* ex gr. *sylvaticus-flavicollis*, *Mesocricetus newtoni*, *Clethrionomys* sp., *Cervus elaphus*

Gininata peshtera – cave near Sadovets Village. Length: 501 m. Founds: **mammals** – *Crocidura* sp., *Spermophilus* sp., *Glis glis*, *Nannospalax leucodon*, *Mus* sp., *Sylvaemus* ex gr. *sylvaticus-flavicollis*, *Mesocricetus newtoni*, *Clethrionomys* sp.

Russe

Orlova Chuka – caves near Pepelina Village. Length 13 437 m. Founds: **mammals** – *Ursus spelaeus*, *Capreolus capreolus*

Shumen

The caves near Krumovi Porti – Founds: **mammals** – *Bos taurus*, *Sus scrofa domestica*, *Felis silvestris*, *Meles meles*, *Mustella* sp.

Sliven

Lucifer – cave – pot hole near Kotel. Length 3200 m. Founds: **mammals** – *Ursus spelaeus*

Smolyan

Triâgâlnata Peshtera – Cave near Borino Village. Length 105 m. Founds: **mammals** – *Panthera pardus*, *Ursus spelaeus*, *Capra ibex*

Borikovskata Peshtera – Cave near Borikovo Village. Length 470 m. Excavations: D. Raychev (1981). Founds: **mammals** – *Talpa europaea*, *Sorex araneus*, *S. minutus*, *Crocidura leucodon*, *Lepus* cf. *capensis*, *Spermophilus* cf. *citellus*, *Muscardinus avellanarius*, *Glis glis*, *Nannospalax leucodon*, *Sylvaemus* cf. *flavicollis*, *Mesocricetus newtoni*, *Cricetulus migratorius*, *Arvicola terrestris*, *Microtus subterraneus*, *Lynx lynx*, *Ursus arctos*, *U. spelaeus*, *Crocuta spelaea*, *Sus scropha*, *Dama dama*, *Capreolus capreolus*, *Capra* sp., *Ovis* sp., *Bos primigenius*

Yagodinskata Peshtera (Imamova Dupka) – cave near Yagodina Village. Length 8501 m. Excavations: V. Mikov (1927), D. Raychev and others (since 1965). Founds: **birds**: *Tetrao tetrix*; **mammals** – *Glis glis*, *Sylvaemus* ex gr. *sylvaticus-flavicollis*, *Felis silvestris*, *Martes martes*, *Sus scrofa*, *Cervus elaphus*, *Capreolus capreolus*

Boevskata Peshtera – Cave near Boevo Village. Length 205 m. Founds: **mammals** – *Ursus spelaeus*

Karnata Peshtera – cave near Yagodina Village. Length 212 m. Founds: **mammals** – *Glis glis*, *Sylvaemus* ex gr. *sylvaticus-flavicollis*, *Clethrionomys* sp.

Tyovnata Dupka – cave near Zabardo Village. Length 179 m. Founds: **mammals** – *Ursus spelaeus*

Gorni Razh 1 – cave near Trigrad Village. Length 137 m. Excavations: D. Raychev and others. Founds: **mammals** – *Ursus spelaeus*

Gorni Razh 3 – cave near Trigrad Village. Length 78 m. Excavations: D. Raychev and others. Founds: mammals – *Cervus elaphus*

Slivova dupka – cave near Length 212 m. Excavations: D. Raychev. Founds: **mammals** – *Ursus spelaeus*

Potoka – Cave near Potoka Quart. Excavations: D. Raychev and others. Founds: **mammals** – *Ursus spelaeus*

Strelite – cave near Orphey Hut. Founds: **mammals** – *Erinaceus concolor*, *Talpa europaea*, *Sorex araneus*, *S. minutus*, *Neomys fodiens*, *Crocidura leucodon*, *Lepus* cf. *capensis*, *Muscardinus avellanarius*, *Dryomys nitedula*, *Glis glis*, *Nannospalax leucodon*, *Rattus rattus*, *Mus musculus*, *Micromys minutus*, *Sylvaemus* cf. *flavicollis*, *Mesocricetus newtoni*, *Cricetulus migratorius*, *Clethrionomys glareolus*, *Arvicola terrestris*, *Microtus subterraneus*

Sofia

Mecha Dupka – Cave near Zhelen Village (Bov Railway Station). Length 564 m. Excavations: N. Petkov (1948-1955) and others. Founds: **birds:** *Perdix perdix*; **mammals** – *Sorex araneus*, *S. minutus*, *Neomys fodiens*, *Talpa* cf. *europaea*, *Erinaceus concolor*, *Lepus capensis*, *Ochotona* sp. cf. *pusilla*, *Spermophilus* sp., *Glis glis*, *Muscardinus avellanarius*, *Sicista subtilis*, *Nannospalax leucodon*, *Sylvaemus* cf. *mystacinus*, *Sylvaemus* ex gr. *sylvaticus/ flavicollis*, *Apodemus* cf. *agrarius*, *Cricetulus migratorius*, *Mesocricetus newtoni*, *Arvicola terrestris*, *Clethrionomys glareolus*, *Chionomys nivalis*,

Microtus subterraneus, Ursus spelaeus, "Equus caballus fossilis"

Razhishkata (Suhata) Peshtera – Cave near Lakatnik Railway Station. Length 316 m. Excavations: G. Markov (1948, 1949), Founds: **birds** – *Anser* sp., *Anas* sp., *Tetrao tetrix, Bonasa bonasia, Perdix palaeoperdix, Perdix perdix, Coturnix coturnix, Crex crex, Tringa* cf. *stagnatilis, Athene noctua, Asio otus, Apus melba, Melanocorypha* sp., *Anthus* sp., *Parus major, Sylvia* sp., *Ptyonoprogne rupestris, Fringilla montifringilla, Loxia curvirostra, Coccothraustes coccothraustes, Carduelis chloris,* Carduelinae gen. cf. *Garrulus glandarius, Corvus monedula, C. corone, C. frugilegus, Pyrrhocorax graculus, P. pyrrhocorax., Petronia petronia;* **mammals** – *Sorex araneus, Crocidura suaveolens, Myotis mystacinus, Nyctalus leisleri, Pipistrellus pipistrellus, Hypsugo savii, Eptesicus serotinus, E. nilssoni, Vespertilio murinus, Miniopterus schreibersi, Meles meles, Martes martes,* Primates, *Ursus arctos, U. spelaeus, Canis familiaris, Capra ibex [*sub *"aegagrus"], Capra* sp., *Ovis aries, Ovis* sp. *Cervus elaphus, Bos* sp.

Svinskata Dupka – Cave near Lakatnik Railway Station. Length 240 m. Founds: **mammals** – *Canis lupus, Ursus spelaeus, Putorius putorius,* Primates, *Cervus elaphus.*

Veliko Târnovo

Golyamata Peshtera (Duhlata Peshtera, Podgolmelskata Peshtera) – cave about seven km north of Veliko Târnovo, near Preobrazhenski Monastery. With two entrances. Length ca. 120 m. Excavations by R. Popov in 1898, 1899, 1901 and 1905. Founds: **mammals** – *Spermophilus* sp. cf. *citellus, Mammuthus primigenius, Canis familiaris, Vulpes vulpes, Ursus spelaeus, Crocuta spelaea, "Equus caballus fossilis", Cervus elaphus, Sus scropha, Ovis aries, Bos primigenius.* Inhabited by Man.

Malkata Peshtera (Tonyuva Peshtera) – cave north of Golyamata Peshetra. Length 92 m. Excavated by R. Popov in 1899, 1900, 1905 and 1909. Founds: **mammals** – *Spermophilus* sp. cf. *citellus, Canis lupus, Ursus arctos, U. spelaeus, Crocuta spelaea, Meles meles, "Equus caballus fossilis", Coelodonta antiquitatis, Rupicapra rupicapra, Bos primigenius.* Inhabited by Man.

Raztsepenata Peshtera (Harlovata) – almost midway between Malkata Peshtera and Preobrazhenski Monastery. Length 15 m. Drilling by R. Popov in 1924. Founds: **mammals** – *Canis familiaris, Bos primigenius, B. taurus.* Inhabited by Man.

Tsarskata Peshtera – cave near Belyakovets. Length 80 m. Excavations: R. Popov (1906). Founds: **mammals** – *Meles meles, Canis familiaris,* Primates, *"Sus scrofa ferus", Ovis aries, Bos primigenius, B. taurus*

Popin Pchelin – cave near Belyakovets. Length 90 m. Drillings by R. Popov in 1906 and 1907. Founds: **mammals** – *Ursus arctos, Canis familiaris, Meles meles, Sus scrofa domestica, Bos taurus, Ovis aries, "Equus caballus"*

Emenskata Peshtera – cave near Emen Village. Length 3113 m. Excavations: G. Markov (1948-49). Founds: **mammals** – *Lepus timidus* (?) *Castor fiber, Meles meles,* Pimates, *"Equus caballus fossilis", Sus scropha, Cervus elaphus, Capra ibex [*sub *"aegagrus"], Capra* sp., *Ovis aries, Ovis* sp., *Bos taurus*

Vidin

Mirizlivka – cave near Oreshets Railway Station. Length 40 m. Excavations: V. Atanasov and L. Filkov (teachers from Vidin) in 1924 and 1929, R. Popov and V. Atanasov in 1931. Founds: **birds**: *Tetrao tetrix, Pica pica;* **mammals** – *Canis lupus, Vulpes vulpes, Ursus arctos, U. spelaeus, Crocuta spelaea, "Equus caballus fossilis", Coelodonta antiquitatis, Cervus elaphus, Capra ibex [*sub *"aegagrus"], Bos taurus*

Kozarnika (Suhi Pech) – cave near Oreshets Railway Station. Length 218 m. Excavations: N. Sirakov (since 1994). Founds: **birds** – *Anas crecca, Falco tinnunculus, F. vespertinus, Tetrao urogallus, T. tetrix, Lagopus lagopus, L. lagopus/ mutus, Tetrastes bonasia, Perdix paleoperdix, P. perdix, Alectoris graeca/chukar, Coturnix coturnix, Porzana* cf. *parva, Crex crex, Gallinula chloropus, Tringa stagnatilis, T. totanus, Apus apus, Nyctea scandiaca, ?Aegolius funereus, Athene noctua, Eremophila alpestris, Anthus trivialis, Hirundo daurica, Ptionoprogne rupestris, Riparia riparia, Lanius collurio,* cf. *Erithacus* sp., *Monticola saxatilis, Turdus merula, T. viscivorus, Carduelis carduelis, C. cannabina, Fringilla coelebs, Coccothraustes coccothraustes, Loxia curvirostra,* cf. *Pyrrhula pyr-*

rhula, Garrulus glandarius, Pyrrhocorax graculus, Corvus monedula, C. corone, C. corone/ frugilegus; **mammals** – *Desmana* sp., *Beremendia fissidens, Sorex* cf. *subaraneus, Crocidura leucodon, Myotis* cf. *gundersheimensis Lepus* sp., *Hypolagus brachignathus, Trischizolagus* sp., *Spermophilus* sp., *Marmota* sp., *Myomimus* sp., *Hystrix* sp., *Nannospalax* sp., *Prospalax* cf. *priscus, Cricetus nanus, C. runtonensis, Sylvaemus mystacinus, S.* cf. *flavicollis, Pliomys* cf. *hungaricus, Dolomys dalmatinus, Prolagurus transylvanicus, Mimomys reidi, M.* cf. *blanci, Canis familiaris, Felis silvestris, Ursus spelaeus, Meles meles, Martes martes, Sus scropha, Cervus elaphus, Capra ibex [sub "aegagrus"], Capra* sp., *Ovis aries, Ovis* sp., *Bos taurus*

Tâmnata peshtera – cave near Târgovishte Village. Length 276 m. Excavations: R. Popov (1931). Founds: **mammals** – *Ursus spelaeus, Canis lupus, Castor fiber, Capra ibex* [sub *"aegagrus"*]

Magura – cave near Rabisha Village. Length 2500 m. Excavations: G. Markov (1948, 1960-61). Founds: **mammals** – *Felis silvestris, Ursus spelaeus, Crocuta spelaea, Meles meles,* Primates, *"Equus caballus fossilis", Sus scropha, Cervus elaphus, Capreolus capreolus, Bos primigenius.* Inhabited by Man.

Desni Suhi Pech – cave near Dolni Lom Village. Length 591 m. Founds: **mammals** – *Sylvaemus* ex gr. *sylvaticus* – *flavicollis, Ursus spelaeus*

Vratsa

Mladenova Propast – cave near Chiren Village. Length 1723 m. Collecting: Nikolov in 1964. Founds: **mammals** – *Canis lupus,* [*C.aureus*], *Ursus spelaeus,* Primates, *Mammuthus primigenius, "Equus caballus fossilis", Stephanorhinus kirchbergensis, Sus scropha, Cervus elaphus, Bos taurus, Bos primigenius.*

Pesht – cave near Staro Selo Village. Excavation: N. Djambazov (1951-53). Founds: **mammals** – *Arvicola terrestris, Ursus arctos, Crocuta spelaea, "Equus caballus", "E. asinus"* (*hydruntinus ?*), *Bos primigenius, Cervus elaphus, Capra* or *Ovis,*; **birds, fish.**

Caves with unknown position

Slyapata dupka – *Bos taurus*

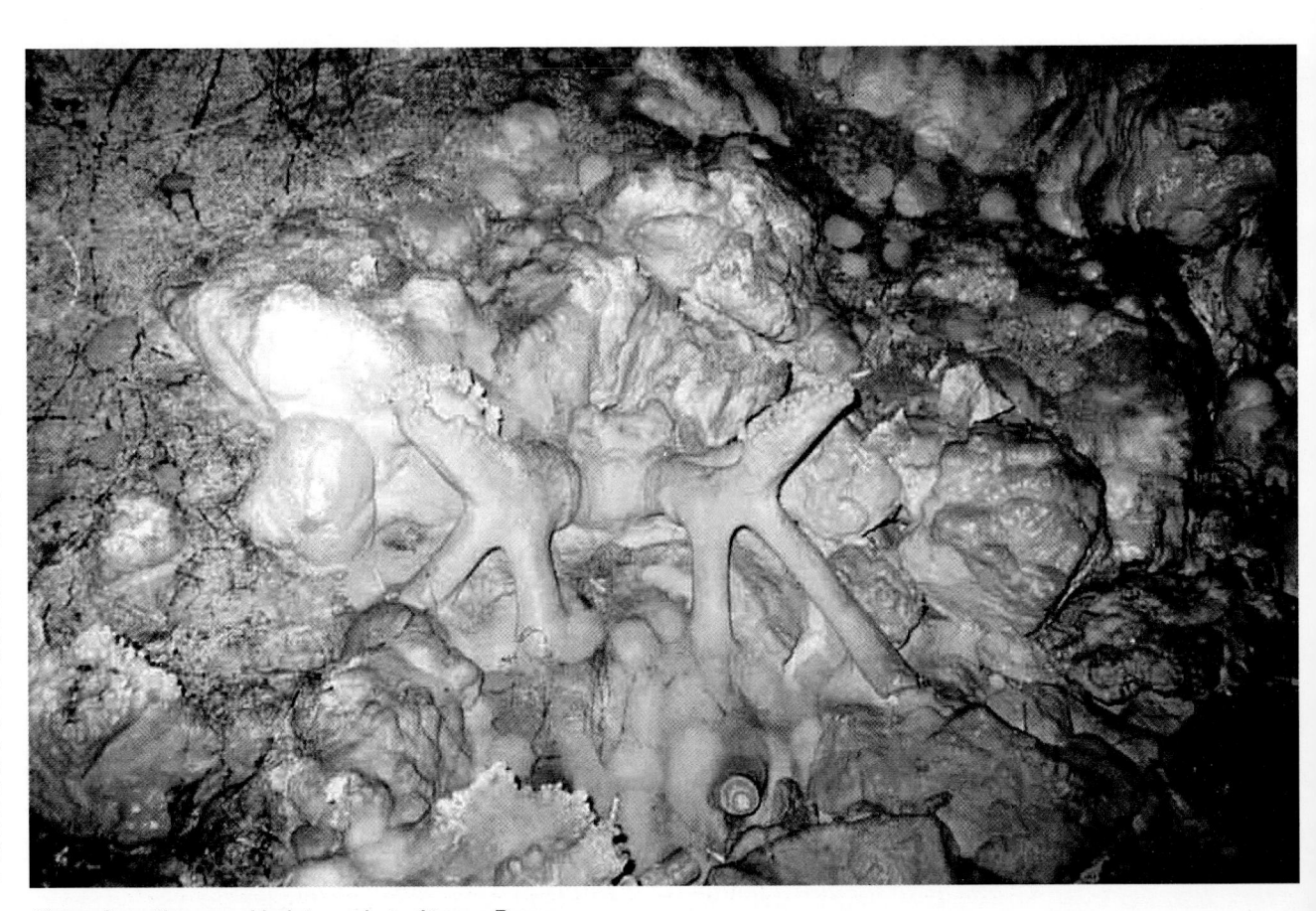

Horns from the cave Vreloto – photo Atanas Russev

GEOLOGICAL, GEOMORPHOLOGICAL AND GEOPHYSICAL STUDIES OF THE KARST AND CAVES IN BULGARIA

At the end of 19th and the beginning of 20th century karts and caves were not yet of special interest to scientists, but their work did require an accurate description, and geochronological dating of rocks, including those subject to karstification. During their field studies they described many features of the surface and underground karst – funnels, ponors, caves and karstic springs.

In this period the most productive researcher was Prof. G. Bonchev. In his publications there are data about caves in the areas of Dobrich, Veliko Târnovo, Kyustendil, Pleven, near Peshtera (Ushatovi Dupki), Kipilovo (the caves Kozya and Cheleshka), Golyama Zhelyazna (Toplya) and Oreshets Railway Station (Bashovishki Pech); about the rock bridges near Zabârdo and Belitsa; about sinks in Southern Dobroudja; about karst sources in the area of Shumen, Novi Pazar, Devnya, Kipilovo, Novi Pazar, Velingrad (Kleptuza), Dryanovo Village (Rhodopes), Aytos area, etc.

Besides his geological descriptions and notes, Prof. G. Zlatarski paid attention also to the caves near Cherepish, Iskrets (Dushnika), Lakatnik and Kunino, as well as to the funnels near Belogradchik, the karstic features in Ponor Planina. In 1900 the geologist Iliya Stoyanov carried out archaeological excavations in the cave Toplya near Golyama Zhelyazna, Lovech District. In 1904 the results of the excavations were published, the author paying attention especially to the geology of the karstic region and most of all to the morphometric description of the site (7 pages). Actually, this is the first detailed scientific description of a Bulgarian cave in the history of Bulgarian Speleology.

Geographers and Geomorphologists did more specialized studies on karst. In 1901, Prof. A. Ishirkov described in details the karstic features in the area of the source Glava Panega and the village Bâlgarski Izvor, and later (1905) described the sources of Devnya.

There is no doubt that the most important contribution to the study of karst in Bulgaria in this period was made by Prof. Zheko Radev. For four years (1911–1914) he explored in details Western Stara Planina, describing the many karst features in his monograph "Karstic forms in Western Stara Planina" (1915). This classic work contains a comprehensive morphographical analysis of the karstic regions in the mountain, as well as maps and descriptions of 12 caves. We can ascertain that the work of Zh. Radev is relevant even today and this publication deserves a foremost place in the karstological literature in Bulgaria. Its author could rightly be considered the founder of Bulgarian Geomorphology. The studies on karstic geomorphology at that time were completed by the paper of an unknown author ("X", 1903), which described in a professional way the surface karstic features in Dobrostan Massif, Persenk Section, and the village Gela (cave Lednitsata) in the Rhodopes.

One of the most dedicated explorers of caves, Nenko Radev published in his papers of

Prof. Zheko Radev – Geographer, explorer of karst and caves

1926 and 1928 maps and extensive morphological descriptions of 16 Bulgarian caves.

Eng. P. Petrov made the second serious attempt at a morphometric description of a cave in his publications on Devetashkata Peshtera in 1928 and 1929.

During the existence of the Caving Society (1929–1949), the studies on the geology and geomorphology of Bulgarian caves went on. Particularly intensive were the studies of P. Petrov on the development of karst in Devetashko Plateau and Dupnivrashko Plateau (now Druzhevsko), together with morphological characteristics of Hlevenska cave in Lovech area and Temnata Dupka near Lakatnik. The publication of V. Arnaudov on the karst of Gorna Djumaya (Blagoevgrad) also centered on geomorphology. The complete morphometric descriptions of the caves near Byala (Sliven District) are due to N. Atanassov.

In 1934, the citizen of Shumen V. Marinov published the book "Morphology of Shumen Highlands", in which the karstic terrain and its development were described in details, including the caves known at that time.

The scientific brigade „Todor Pavlov" considerably contributed to the geological and geomorphological exploration of karst and caves. Special studies were carried out in the areas of Zlatna Panega, Karlukovo, Lakatnik and Rabisha.

The geological preconditions for the development of the karst and caves in Bulgaria are discussed in details in the collection of papers "The Karstic Underground Waters in Bulgaria" (1959), under the editorship of D. Yaranov.

The early 60s of 20th century saw the start of some modern studies (mainly in the geomorphology of karst) and they were connected with the name of Assoc. Prof. Vladimir Popov. He explored in turn the mountain Vrachanska Planina and the morphology of the karst between the rivers Vit and Batulska (1962), the northern part of the Predbalkan between the rivers Yantra and Ossam (1965) and Iskâr and Vit (1969). In 1970, based on the morphostructural principle, V. Popov elaborated the subdivision of the caves in Bulgaria, which is in use even today. V. Popov worked actively on the genesis of the caves Magurata, Ledenika, Sâeva Dupka, Dyavolskoto Gârlo, Snezhanka, Bacho Kiro, as well as on some of the caves in Karlukovo karstic area.

Dimitar Sâbev made a considerable contribution to the geological exploration of karst, mostly in the Rhodopes. In the period 1964 - 1979, he published materials dealing with the geological basis for the development of karst and caves in Trigrad, Rozhen and Lâkavitsa karstic areas, Radyuva Planina, Mursalitsa and the Central part of the Rhodopes and others.

The newer geological and geomorphological studies are mostly of a regional type. A. Benderev, A. Radulov, G. Baltakov, D. Angelova, Z. Pironkova, I. Ilieva, I. Zdravkov, K. Kostov, M. Paskalev, P. Petrov, P. Stefanov have been and are still working in this direction. As a result of their work (and the work of others) the regions of Iskâr Gorge, Vrachanska Planina, Vitosha (Bosnek) and Ponor Planina have been researched and also the subregion of Ribnovo in the Rhodopes, the mountain Kotlenska Planina, Karlukovo Village, the upper part of Cherni Ossam River, the Plateaus of Devetaki and Shumen, the gorge of Russenski Lom, Golo Bardo, the morpho-structural features of Mezdra and other karstic areas. The development of paleokarst in Bulgaria was a subject of the papers of geologist Iv. Stanev.

At the same time, some caves have been described individually. Such are the papers of P. Stefanov "Morphometry and Morphography of the Cave Manailovska Dupka" (1982), "Taynite Ponori" and "Zandana" (2001); G. Raychev – "Morphology and Genesis of Yagodina Cave System (Imamova Dupka)" (1991); Y. Evlogiev – "The Geology of the Cave Orlova Chuka (1987) and "Study of the Genesis of the Crevices in the Cave Orlova Chuka" (1997) and others.

The use of geophysical methods for searching and discovering new underground cavities in Bulgaria started in 1970, when the Geophysical Methods of Exploration group at St. Ivan Rilski University of Geology and Mines carried out electrical experiments in the region of Chiren. The results, which proved the existence of an unknown connection between the caves Ponora and Mladenovata, were reported by S. Pishtalov, V. Ivanova, K. Spassov and others at the World Speleological Congress in Olomouc (former Czechoslovakia) in 1973 and were published in its Proceedings.

In 1968 in Bulgaria, the application of one of the known geophysical methods, that of ultra long waves (ULW), came into use. This method was tested for the first time in karstological research by Eng. A. Tsvetkov during the expedition Pierre-Saint Martin, France, carried out by Prilep Caving Club of Planinets, Sofia Tourist Society.

In 1979, the combined electroprophiling of V. Georgiev and Z. Andonov from the same club aimed to discover a prolongation of the cave Râzhishkata Peshtera near Lakatnik Railway Station. The results showed that 20 meters behind the huge blockage at the known end of the cave there was a large gallery. Despite the many attempts to clean up the blockage, the localized cavity has not yet been reached.

In the period 1981–1982, the cavers V. Georgiev from Prilep, Sofia Caving Club, and M. Zlatkova from the Student's Caving Club Akademik, Sofia, used several methods of electrical testing in order to find prolongation of the caves Dushnika near Iskrets and Givalaka near Zimevitsa, both in Ponor Planina (Sofia Distr.). A cavity has been found about 20 m from the final siphon of the first cave.

When studying the cave Dushnika in 1984, S. Shanov from Akademik Sofia Caving Club again applied the method of ULW. The study indicated the existence of unknown caverns in the proximity of the Cherna Voda Fault.

In the period 1984–1985, cavers from Akademik, Sofia Caving Club, led by Eng. M. Zlatkova, used four variants of electroprophiling in order to prove the existence of a walkable gallery beyond the karstic source Popov Izvor near Bosnek. The results were positive, but so far the speleologists have not managed to reach the underground cavity. Similar is the result of the exploration of S. Shanov and M. Zlatkova in the cave Orlova Chuka in 1985, when it was found that the underground cavities were filled with clay. In the same year, S. Shanov studied the karst near Karlukovo. By electroprophiling and the use of ULW and permanent current methods, the presence of large caverns between the caves Bankovitsa, Svirchovitsa and Prohodna was suggested. Later studies, carried out by P. Stefanov and students from the University of Geology and Mining, gave similar results (unfortunately unpublished). Encouraged by these data, for several years cavers from Helictite, Sofia Caving Club, had tried to reach the unknown caverns in Bankovitsa Cave. By using the method of symmetric profiling with a direct current, at the end of 1985, M. Zlatkova and cavers from Kupena, Peshtera Caving Club, looked for a prolongation in the show cave Snezhanka. Two extensions have been found so far, one of them proven directly.

In 1986, S. Shanov and M. Zlatkova and cavers from Vitosha, Sofia Caving Club, used various geophysical methods to search for underground caverns in the area of Ponor Planina. In both cases the cavers undertook cleaning in order to reach the caverns, but were confronted with considerable obstacles and dropped the idea. In 1987, M. Zlatkova, this time helped by cavers from Akademik, Veliko Tarnovo Caving Club, used the method of symmetric electroprophiling to search for caves between the villages Emen and Balvan near Veliko Tarnovo, but without result.

The last use of the methods of applied geophysics by Bulgarian speleologists was during the National Speleological Expedition "Guaso' 88" in Cuba. This time the aim was to measure precisely the depth of the karstic areas and to calculate how much of the caverns were filled with water, needed by the local people.

The cañon of the river Chudna between Kameno pole and Reselets Villages – photo Trifon Daaliev

CAVE MINERALS IN BULGARIA

Under the notion of "mineral species" we understand a natural chemical composite of a definite composition and crystal structure. One chemical composite can form several minerals of different crystal structure (solid phase). Till now ca. 3000 mineral species, subjects of speleomineralogy, are known. Out of them only 210 have been found in caves.

Speleomineralogical research started in Bulgaria in 1923. So far more than 53 mineral species are known from Bulgarian caves, 24 of them being new for Bulgaria. As a contribution of Bulgarian biomineralogy to the world science, we can mention the discovery in Bulgarian caves of 21 mineral species, previously not known from caves abroad. The development of speleomineralogy in Bulgaria is due mainly to scientists from St. Kliment Ohridski University of Sofia.

The first stage of speleomineralogic research in Bulgaria (up to 1974) is characterized mostly by scattered incidental studies by various authors. In 1928, Prof. G. Bonchev studied the morphology of calcite crystals from caves in Troyan, Teteven and Lovech areas. After more than 30 years, N. Dimitrov and V. Aleksiev studied the phosphorites from the cave Orlova Chuka near Russe, but their results remained unpublished. In some caves in the valley of Russenski Lom in 1960, Y. Yovchev found calcium nitrate.

More important was the contribution of Naydenova and Kostov, who did a modern study on these minerals in 1964. Prof. Iv. Kostov is the author of the famous evolutionary system of the shape of calcium crystals depending of the conditions ($t°$) and the stage of formation. In a cave (non-existing today), these authors found calcite, aragonite, gypsum, dolomite and celestine. They obtained some data, which suggested that in one sample of aragonite there were microinclusions of cerusite. It has been found so far in only one Bulgarian cave and it is connected with galenite. As a whole, this mineral is typical for the hydrothermal caves, probably because, being with low dissolution in the temperature of karstic caves, it cannot be transferred far from lead mineralization.

Cave pearls have been studied by N. Cholakov and P. Tranteev. V. Popov described the ice formations in Ledenika Cave near Vratsa.

The second stage (1974–1981) started with the acceptance of speleomineralogy as a course in the curriculum of the Department of Mineralogy and Crystallography at the Faculty of Geology and Geography of Sofia University. A considerable leap forward was noticed. Methods of the experimental mineralogy (M. Maleev, A. Filipov) and of the theoretical thermodynamic analysis of the formation of minerals were introduced in the speleomineralogical research in Bulgaria. The finding of brucite and saponite for the first time in caves by A. Filipov was a considerable success. The same author described from caves the minerals tarankite, hydromagnesite, apatite ardealite and brushite, unknown in Bulgaria till then.

Dimov & Dimitrov (1978) used the laser microspectral analysis to study the composition of the cave formations. Studies on the artificial acceleration of the growth and coloration of cave formations with their natural colorants started. At the same time,

Crystals in the cave Duhlata near Bosnek Village – photo Mihail Kvartirnikov

Bliznakov and Tokmakchieva from the Institute of Geology and Mining in Sofia and Iv. Bonev from the Institute of Geology, Bulgarian Academy of Sciences, studied the morphology of calcite formations.

With the foundation of a Cave Mineralogy Circle at the Department of Mineralogy and Crystalography of Sofia University by A. Filipov in 1981, the third stage was preconditioned. Then, A. Filipov studied minerals from Cuba and Y. Shopov was the first to start in Bulgaria research on the phosphorescence of minerals (from the cave Pepelyankata near Bosnek). This study laid the foundations of the physics of cave minerals in Bulgaria. Some studies started, which are still going on, on the synthesis of cave minerals with given admixtures for solving the direct problem of the luminescence (1982) of minerals and mineralogenesis, which were published later (1985).

The third stage started with the formation of the Speleology unit at the University of Sofia in 1982. Immediately after that, its President Y. Shopov formed a group called Physics and Chemistry of Karst and Cave Minerals, which grew in the years to follow into a large interdisciplinary group. The impact of this intensive research was reflected in the increased number of scientific publications and the extension of the scope of studies. Some theses on cave mineralogy were done.

Seven methods, expanding the capacity of the then-existing ones, and one method for field study of cave minerals were worked out, as well as six devices for applying these methods. The use of direct gamma spectroscopy allowed for the first in Bulgaria an EPR dating of cave formations (1985). For the first time, the Mösbauer's spectroscopy and the laser luminescent spectral analysis for studying cave minerals was used. During the studies, 19 modern physical methods were used. They made it possible to discover several rare minerals, which were difficult to identify. The finding of a hydrothermal cave by members of the Speleology Section and the applying of powerful physical methods made it possible to find minerals, which were new for science ($CaCO_3$-II), and over 20 minerals recorded for the first time in caves. Many of them were new for Bulgaria (Table). From the minerals shown in this table, only the almandine, the opal and the pyrite are of classical origin, e.g. cave minerals s. lato. All the others are hemogenic, autochthonous cave formations, or cave minerals sensu stricto. Among them, 19 are formed as a result of karstic processes in carbonate rocks, six are guanogenes (product of the interaction of the guano with mineralized solutions), four are hydrothermal, 23 are hypergenic (formed during processes of alteration of the chemical composition) of hydrothermal minerals.

The organization of the research was also made modern and complex. Of considerable importance was the publication of an Expeditionary yearbook of Sofia University under the care of the Speleology Section. A great deal of it concerns the speleomineralogy and the physics of karst (Table 1).

Table 1. List of the first findings of cave minerals in Bulgaria

No	Mineral	Year	Author	Formula
1	Ice			H_2O
2	Calcite	1923	Bonchev G.	$CaCO_3$
3	Nitrate	1960	Yovchev Yo.	KNO_3
4	Aragonite	1964	Naydenova, Kostov	$CaCO_3$
5	Gypsum	1964	Naydenova, Kostov	$CaSO_4 \cdot 2H_2O$
6	Celestine	1964	Naydenova, Kostov	$SrSO_4$
7	Dolomite	1964	Naydenova, Kostov	$CaMg_3(CO_3)_2$
8	Huntite**	1974	Maleev M.	$CaMg(CO_3)_4$
9	Tarankite**	1976	Filipov A.	$H_6K_3Al_5(PO_4)_8 \cdot 18H_2O$
10	Magnesite	1979	Filipov A.	$MgCO_3$
11	Hydromagnesite	1979	Filipov A.	$Mg_6(CO_3)_4(OH)_2 \cdot 4H_2O$
12	Brusite	1980	Filipov A.	$Mg(OH)_2$
13	Saponite	1980	Filipov A.	$Mg_3Si_4O_{10}(OH)_2$
14	Ardealite**	1980	Filipov A.	$Ca_2HPO_4SO_4 \cdot 4H_2O$
15	Brushite**	1980	Filipov A.	$CaHPO_4 \cdot 2H_2O$
16	Apatite	1980	Filipov A.	$Ca_5(PO_4)_3(F,Cl,OH)$
17	Todorkite**	1980	Filipov A.	$BaMn_3O_7 \cdot H_2O$

No	Mineral	Year	Author	Formula
18	Maghemite**	1983	Shopov Ya., V. Spassov	Fe_2O_3
19	Götite	1983	Shopov Ya., V. Spassov	á-FeOOH
20	Lepidocrocite	1983	Shopov Ya., V. Spassov	ã-FeOOH
21	Akaganeite*	1983	Shopov Ya., V. Spassov	â-FeOOH
22	Hematite	1983	Shopov Ya., V. Spassov	á-Fe_2O_3
23	Cerusite	1983	Shopov Ya. et al.	$PbCO_3$
24	Bylisyte**	1984	Shopov Ya. et al.	$K_2Mg(CO_3)_2.4H_2O$
25	Faterite**	1984	Shopov Ya. et al.	m-$CaCO_3$
26	Malahite	1984	Shopov Ya. et al.	$Cu_2CO_3(OH)_2$
27	Deviline*	1984	Shopov Ya. et al.	$Cu_4Ca(SO_4)_2(OH)_6.3H_2O$
28	Hydrozinkite	1984	Shopov Ya. et al.	$Zn_5(CO_3)(OH)_6$
29	Artinite**	1984	Shopov Ya. et al.	$Mg_2CO_3(OH)_2.3H_2O$
30	Djirdjeite**	1984	Shopov Ya. et al.	$Cu_5(CO_3)_3(OH)_4.6H_2O$
31	Azurite	1984	Shopov Ya. et al.	$Cu_3(CO_3)_2(OH)_2$
32	$CaCO_3$-II***	1984	Shopov Ya. et al.	
33	H.T. – FeOOH***	1984	Shopov Ya. et al.	
34	Galenite	1984	Shopov Ya. et al.	PbS
35	Djordjiosite**	1985	Shopov Ya. et al.	$Mg_5(CO_3)_4(OH)_2.5H_2O$
36	Dipingite**	1985	Shopov Ya. et al.	$Mg_5(CO_3)_4(OH)_2.5H_2O$
37	Hydrocerusite**	1985	Shopov Ya. et al.	$Pb(CO_3)_2(OH)_2$
38	Aurihalcite	1985	Shopov Ya. et al.	$CuZn_3(CO_3)_2(OH)_6$
39	Schrekingerite**	1985	Shopov Ya. et al.	
40	Sharpite**	1985	Shopov Ya. et al.	
41	Viartite**	1985	Shopov Ya. et al.	
42	Rutherfordin**	1985	Shopov Ya. et al.	
43	Rosasite	1985	Shopov Ya. et al.	
44	Celerite**	1985	Shopov Ya. et al.	$CaUO_2(CO_3)_2.5H_2O$
45	Pyroaurite**	1985	Shopov Ya. et al.	$Mg_6Fe_2(CO_3)(OH)_{16}.4H_2O$
46	Nakaurite**	1985	Shopov Ya. et al.	$Cu_8(SO_4)_4(CO_3)(OH)_6.48H_2O$
47	Poznyakite**	1985	Shopov Ya. et al.	$Cu_4(SO_4)(OH)_{16}.H_2O$
48	Tuyamunite**	1985	Shopov Ya. et al.	$Ca(UO_2)_2V_2O_8.8H_2O$
49	Anglesite	1985	Shopov Ya. et al.	$PbSO_4$
50	Bornite	1985	Shopov Ya. et al.	Cu_5FeS_4
51	Thermonatrite	1987	Shopov Ya. et al.	$Na_2CO_3.H_2O$
52	Natron	1987	Shopov Ya. et al.	$Na_2CO_3.10H_2O$
53	Gibsite	1987	Shopov Ya. et al.	$Al(OH)_3$
54	Aketamide	1987	Shopov Ya. et al.	CH_3CONH_2
55	Quartz	1987	Shopov Ya. et al.	á-SiO_2
56	Alophan	1988	Shopov Ya. et al.	$Al_2Si(OH)_{10}$
57	Nyuberite	1988	Shopov Ya. et al.	$MgHPO_4.3H_2O$
58	Purpurite	1988	Shopov Ya. et al.	$(Mn,Fe)PO_4$
59	Pyrolusite	1989	Shopov Ya. et al.	á-MnO_2
60	Epsomite	1989	Shopov Ya. et al.	$MgSO_4.7H_2O$
61	Hexahydrite	1989	Shopov Ya. et al.	$MgSO_4.6H_2O$
62	Depugeolasite	1989	Shopov Ya. et al.	$CaMg_4(SO_4)_2(OH)_6.3H_2O$

*** – minerals new for science
** – minerals new for Bulgaria
* – minerals, found by other authors in mines

HYDROGEOLOGICAL, HYDROLOGICAL AND HYDROCHEMICAL STUDIES IN BULGARIAN CAVES

The formation of karstic features and caves is impossible without water. That is why the knowledge on the penetration of waters underground, the character of their movement and the conditions of their resurgence (as karstic sources) is of considerable importance in the search of new caves. All caves are old or active waterways. Usually, in a karstic massive, it is very rarely possible to follow the way of an underground river from the sink to its resurgence. Such rare examples are the cave Parnitsite near Bezhanovo and some rock bridges (Chudnite Mostove in the Rhodopes, Bozhyia Most near Lilyache, etc.).

Beside the scientific problems, it is important for the cavers to know the conditions of the movement and the regime of waters in karstic massifs for better planning of the visits in caves and to avoid accidents.

Karstic waters and especially the waters of the caves are of practical importance too – most of the bigger karstic springs in Bulgaria are essential source for water supply in the settlements. They are studied by specialists – hydrogeologists.

The description and the research on karstic waters have started many years ago. Data concerning this research from the first records to the creation of the First Bulgarian Caving Society in 1929 are put together in the papers of Jalov (1999, 2000, 2002) on the history of Speleology in Bulgaria. According to them, the oldest record concerns the cave Mandrata, or Vodnata near the village Chavdartsi. The information is from the middle of 17th century (1640) and is found in the report of the first Bulgarian Archbishop Petar Bogdan to the Roman Congregation.

From 17th century are also the road notes of the most famous in this time Osman traveller Evliya Chelebi. In them is described his visit to the periodical source near Bosnek, known as Zhivata voda.

The following written sources are from 19th century and are due to Bulgarian and foreign travellers and explorers. More important among them are:

– in 1828 the French geographer and cartographer J.-G. Barbier de Bocage wrote about karstic sources and caves in the surroundings of Shumen, including the cave in Bunar, now known as Zandana (Biserna);

– in 1967 the French traveller Guillhomme Lejean wrote about Kayalashkata Peshtera near Pleven;

– in 1872 an anonymous author wrote about the cave near Mussina;

– in 1875 Iv. Slaveykov wrote about the caves and potholes near Chiren, Vratsa Distr., including the water caves known now as Tigancheto, Ponora and Bozhiya most near Lilyache;

– in 1877 the Englishman James Baker described for the first time the entrance of Devetashkata Peshtera;

– in 1973 the notes of the monk Pahomiy Stoyanov focused on the events at the monastery near Bachkovo and the water cave Andaka;

– in 1890 Captain A. Benderev wrote about the source Istoka and probably about the cave Mecha Dupka near Razlog;

– in 1896 Ivanov gave information about the karstic source Toplitsa (Toplika) south of Musomishta, Distr. Blagoevgrad;

– in 1887 Ts. Ginchev described the source Zlatna Panega.

In the second half of 19th century Bulgaria was explored by the well-known Czech scientist brothers K. and H. Škorpil. Their publications (1895, 1898, 1900, etc.) contain information about many caves, sinks and karstic sources, including interesting schemes showing the hydrological connections in the areas of Glava Panega and Vit, of Iskrets, on the Serbian-Bulgarian border, of Bezden near Sofia, Kotel, Mountain of Vratsa, Dobrostan, Narechen, Tri Voditsi and others. There are data about many of the major karstic sources in Bulgaria. They are divided into permanent, changing and drying out. One of the last papers of Škorpil brothers concerning Bulgaria appeared in 1910 and dealt with the sources of Devnya.

After the Liberation of Bulgaria in 1878 active work on the geological map of Bulgaria started. Some of the young geologists paid attention to the karst features and karstic waters, discovered during field studies. One of the first geologists was Prof. G. Zlatarski, who in 1884 described the karstic terrain around the source Zlatna Panega, and later wrote about other water caves like Dushnika (Iskrets), the cave near Lakatnik and others. In 1926 and 1928, the promising papers of N. Radev "Materials for Studying the Caves in Bulgaria" appeared, where several water caves found place, including Lednitsata near Gela, Sbirkovata near Chepelare, Temnata Dupka near Kalotina, Zhivata Voda near Bosnek, Andaka near Dryanovo, Marina Dupka and Zmeyova Dupka near Genchevtsi.

At the end of the 90s of 19th century the activities of the prominent geologist Prof. G. Bonchev started. He made drawings and did geological descriptions of the caves and karstic sources in the areas of Shumen, Novi Pazar, Devnya, Kipilovo, Velingrad (Kleptuza), Dryanovo (Rhodopes), Aytos, etc. He was also the first explorer of the cave Toplya near Golyama Zhelyazna (Bonchev, 1901). His hydrogeological research in the Ludogorie was important as he supplied it with water. Later (Bonchev, 1939), he indicated almost all karstic springs in Bulgaria (over 250) and described the general features of their formation, outflow and the water temperature.

A detailed exploration of the sources of Glava Panega and Bâlgarski Izvor was done by Prof. A. Ishirkov in 1901 (Ishirkov, 1902), also of the sources of Devnya (Ishirkov, 1902). To Yonchev (1902) we owe the first detailed description of one of the most interesting karstic sources in Bulgaria – the one near Iskrets.

One of the most profound studies of Bulgarian karst ever is the book of Zh. Radev "Karstic Forms in Western Stara Planina" (1915). In this geomorphological monograph many facts and conclusions about the water bodies in the region found place. All active or temporary sinks and sources were described and the probable connections between them indicated. The work includes descriptions of some water caves – Dushnika near Iskrets, Vodnata peshtera near Tserovo and others.

Eng. Pavel Petrov was the true founder of Bulgarian hydrogeology of karst. He studied geology and mining in Belgium, then in France he specialized in the area of mineral waters. Under his guidance many sources (including karstic) in Bulgaria have been piped. He was involved in the study of karstic waters also as a caver. He studied the water galleries and surveyed the caves Devetashkata Peshtera (Petrov, 1928, 1929) and Temnata Dupka near Lakatnik Railway Station

The book „Krazhski yavleniya" (Karst Phenomena) by K. and H. Škorpil

Alexander Strezov at the big karstic source Glava Panega – photo Nikolay Genov

(Petrov, 1936). Besides studying Devetashkata Peshtera, P. Petrov published some general information about the hydrogeology of the entire Plateau of Devetaki (1933).

After the Caving Society was found in 1929, the exploration of water caves in Bulgaria became more intensive and thorough, but almost results remained unpublished. Some karstic sources were piped and geological research on them took place. In 1948 the Caving Brigade worked also in the areas of Glava Panega and Temnata Dupka near Lakatnik.

With the setting of the Caving Commission and of many caving clubs (since 1958), the systematic study of Bulgarian caves started, which is now carried on under the care of the Bulgarian Federation of Speleology. Some major water caves were discovered and explored (Ponora near Chiren, Parnitsite near Bezhanovo, the caves near Krushuna and others). At the same time, many specialized studies were done for the needs of hydroenergy, mining and water supply. Most of them remained unpublished and are kept in the archives of the research organizations and in the State Archive. Many scientific articles on karstic water were published in this period. Especially important is the monograph of Hr. Antonov and D. Danchev "Underground Waters in Bulgaria" (1980).

As a whole, the research on waters in karstic terrains and caves has several aspects:

A. Research on the Karstic Basins (Massifs). The most important thing is the localization of the karstic massif, defining its boundaries, the depth of the karstifying rocks, the main karstic sources, the area of water catchment, etc. The knowledge about these particularities is important to assess the ways of the water in the karst and the chance to find caves and potholes. Some of the important studies on the karstic basins in Bulgaria (Dobrudja, Sadovets, the rivers Vit, Tazha and Panega, the area of Iskrets sources, in Kraishte, SW Pirin, Tri Voditsi, Madan, Nedelino, Strandja) were included in the monograph under the editorship of D. Yaranov (1959). In this book we find also the idea of the hydrogeological subdivision of Bulgaria. Some general regularities of the karstic morphology and hydrology were discussed by V. Popov, L. Zyapkov and P. Tranteev (1964) and by P. Penchev (1966). In the same year, N. Boyadjiev outlined 183 separate karstic regions in Bulgaria. This important scheme should serve the subsequent hydrogeological subdivision of Bulgarian karst. The essential hydrogeological regularities of the various karstic basins are treated in several publications:

– Antonov (1963) – about the karstic basins of Sofiyska Stara Planina;

– Popov, Penchev, Zyapkov (1965) – about the Predbalkan between the rivers Yantra and Ossam;

– Popov, Grozdanov, Vassileva (1971) – the area of Gabare;

– Penchev, Popov, Zyapkov (1971) – the northern parts of the Predbalkan;

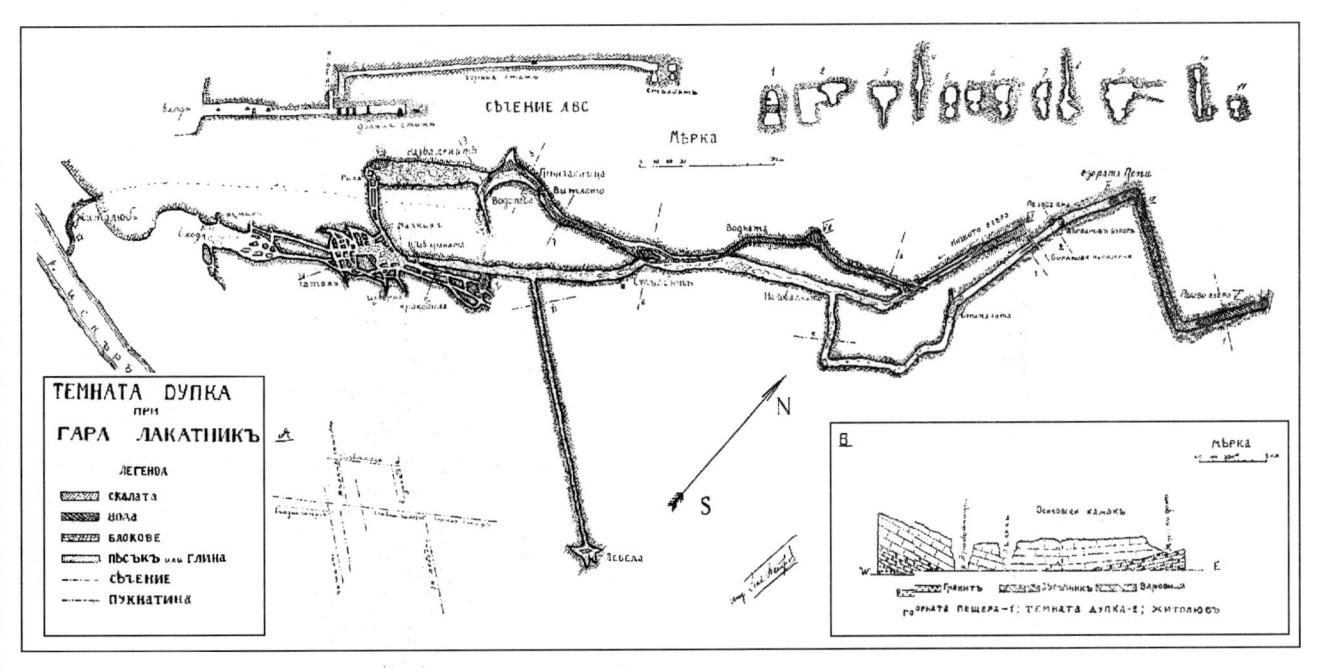

The first map of the cave Temnata dupka near Lakatnik Raylway Station by Eng. Pavel Petrov (1926)

– Spassov (1973) – about the Predbalkan of Vratsa;
– Benderev (1989) – about Ponor Planina;
– Gabeva et al. (1995), Benderev et al. (1997) – about Nastan – the Trigrad karstic basin;
– Angelova et al. (1995) – about the karstic basins of Iskar Gorge
– Benderev, Shanov (1997), Benderev, Angelova (2000) – about the karstic region of Bosnek;
– Spassov et al. (1998) – about the region of Vratsa;
– Benderev et al. (1999) – about the area of Chiren;

The mentioned publications do not include many other regional studies on the karstic water in Bulgaria. Most of the older data found place in the monograph "Underground Waters in Bulgaria" by Antonov and Danchev (1980). Many karstic regions were explored during the hydrogeological survey of Bulgaria.

B. Research of Karstic Springs. The situation of karstic springs, the change of the outflow, the temperature and the chemical composition are important in order to understand the degree of karstification and the chance of finding caves. As mentioned before, the first generalizing paper on the karstic sources in Bulgaria is the one of Bonchev (1939). In all hydrogeological papers describing karstic basins the karstic sources too are described. Since 1959 a National Network for Monitoring of Underground waters in Bulgaria has been created by the Institute of Hydrology and Meteorology of the Bulgarian Academy of Sciences. This network monitors all major karstic sources in the country. Part of the data obtained on the outflow, the temperature and the chemical composition have been published in the Hydrogeological Yearbooks (for the period 1959-1979), and the generalized reference books of Tsankov et al. (1993) for the period 1980-1991 and of Machkova, Dimitrov (1999) for the period 1980-1996. Data about the limits of the outflow of these sources are given in Table 2.

More detailed analysis of the changes in the regime of some karstic sources are contained in

Table 2. Data about the major monitored sources in Bulgaria

Name	Settlement	Rocks	Age	Minimum outflow, l/s	Maximum outflow, l/s	Important caves in the catchment area
Vreloto*	Krachimir	Limestone	J_3-K_1	11	1501	Krachimirsko Vrelo
Pech*	Dolni Lom	Limestone	J_3-K_1	5	1310	Suhi i Vodni Pechove
Kalna Matnitsa*	Stoyanovo	Limestone	K_1	22	949	Kalna Matnitsa
Bistrets	Bistrets	Limestone	J_3-K_1	0	5052	Barkite 14, Belyar and others
Berende	Berende	Limestone, dolomite	T_{2-3}	5	806	Temnata Dupka
Peshtta*	Iskrets	Limestone, dolomite	T_{2-3}	90	54900	Katsite, Dushnika
Peshterata*	Tserovo	Limestone, dolomite	T_{2-3}	1	348	Vodnata Peshtera
Zhitolyub*	Lakatnik	Limestone, dolomite	T_{2-3}	27	23470	Temnata Dupka
Glava Panega*	Zlatna Panega	Limestone	J_3	580	35700	Partizanskata, Morovitsa, Saeva Dupka
Toplya*	Golyama Zhelyazna	Limestone, dolomite	T_{2-3}	0.93	4780	Toplya
Maapata*	Krushuna	Limestone	K_1	3	5220	Vodopada, Boninskata, etc.
Peshterata*	Musina	Limestone	K_1	25	5650	Musinskata Peshtera
Kotlenski Izvor*	Kotel	Limestone	K_2^s	40	20061	Prikazna, Karvavata Lokva and others
Devnenski Izvori	Devnya	Limestone	J_3-K_1	2343	4270	
Marina Dupka*	Targovishte	Limestone	K_1	3	1529	Marina Dupka

Table 2. Continued

Name	Settlement	Rocks	Age	Minimum outflow, l/s	Maximum outflow, l/s	Important caves in the catchment area
Yazo	Razlog	Marble	Pz	440	2725	Vihren, Propast 9, Banderitsa and others
Kyoshka	Razlog	Marble	Pz	15	2770	Vihren, Propast 9, Banderitsa and others
Kleptuza	Velingrad	Marble	Pz	63	1209	Lepenitsa
Vrisa	Beden	Marble	Pz	132	3300	Drangaleshkata, Lednitsata and others
Nastanski Ribarnik	Nastan	Marble	Pz	40	2382	Drangaleshkata, Lednitsata and others
The 40 Sources	Muldava	Marble	Pz	57	1021	Ivanova Voda, Druzhba and others

Note: sources emerging from caves are marked with *

many publications of M.Machkova, D. Dimitrov et. al., T. Orehova et al., Bl. Raykova et al., T. Panayotov, A. Benderev et al.

C. Research on the Ways of Karstic Underground Waters. The direct research on cave rivers is an important part of the work of Bulgarian cavers. The penetration and the mapping of water caves allowed for the study of hundreds of caves with water currents, some of them of considerable length (Andaka, Vodnata Peshtera near Tserovo, the caves near Krushuna, Temnata Dupka near Lakatnik, Vreloto and many others). The waterfalls of Dyavolskoto Garlo and Bankovitsa have been studied. An underground stream can be followed along one of the deepest Bulgarian caves – Barkite 14. The basic hydrological scheme of the cave Duhlata, consisting of a main river and seven influxes, has been clarified. The exploration of the siphons is described elsewhere in this book.

In Bulgaria, many experiments with indicators for following the ways of the cave waters have been carried out (Table 3).

By means of these experiments the connections between the waters in the karstic basins and the sources draining them have been established, as well as between different caves. Several common water systems have been detected, like the ones of Boninskata Peshtera and Vodopada near Krushuna, Chuchulyan, PPD and Vreloto near Bosnek, Mladenovata Peshtera and Ponora near Chiren.

D. Hydrochemical Research on the Karstic Underground Waters. The study of the chemical composition of karstic waters has several aspects. Besides determination of macro and microcomposition of cave waters (made many times by different organizations), the studies concerning the hydrochemical regime of the karstic waters are also

The source Zhitolyub near Lakatnik Raylway Station – photo Trifon Daaliev

Table 3. Results of the indicating experiments for tracing of karstic underground waters

Date of release	Region	Place of release	Type and quantity of indicator	Place of arrival	Distance, m	Time of arrival, hours	Source
Not recorded	Chiren	Mladenova Propast Cave	Fluorescein	Cave Ponora			Data of BFS
20.08.1977	Chiren	End of Mladenova Propast Cave	Fluorescein	Waterfall in cave Ponora		480	Data of BFS
29.10.1968	Gabare	River in the cave Starata Prodanka	0.5 kg fluorescein	Gabare Source	500	75	Popov et al., 1968
21.08.1991		In the cave Aladjanska Peshtera	0.33 kg uranin	Baba Raditsa Source	850	36	Breskovski et al., 1991
20.08.1978	Devetaki	End siphon of Boninska Peshtera Cave	0.4 kg fluorescein	Cave Vodopada	400	22	Tranteev, 1978
				Source left of Vodopada Cave		35	
				At the entrance of Vodopada Cave		48	
23.08.1978	Devetaki	In Gornik Cave	0.25 kg fluorescein	Not established			Tranteev, 1978
	Devetaki	In Turska Cherkva Cave	0.25 kg fluorescein	Not established			Tranteev, 1978
19.02.1991	Devetaki	End siphon of Brashlyanska Peshtera	0.5 kg fluorescein	Sutna Bunar Source, Aleksandrovo Village	1250	28.5	Benderev, 1999
Not recorded	Emen	Estavela in the area of Visoka Gora, Dobromirka Village	6000 kg colored water	The lower source, Dobromirka Village	1340	130.00	Petrov, Botev, 1989
Not recorded	Emen	In the Lower source near Dobromirka Village (temporary)	Colorant	Nelaba Source		500	Petrov, Botev, 1989
29.04.1997	Shumen	Semi-siphon Starite Vodopadi in the cave Taynite Ponori	0.5 kg fluorescein	Water catchment of Beer Factory near Troytsa Village	2500	67.5	Stefanov et al., 1997
2.11.1955	Iskrets	Ponor Sink	6000 kg NaCl	Iskrets Source	9500	Ca. 24	Dinev, 1959
15.12.1955	Iskrets	Perachka Bara Sink	8000 kg NaCl	No result			Dinev, 1959
7.10.1955	Teteven	Sink in Yamite area	22 kg Engine oil	Source Glava Panega	6500	Ca. 250	Kovachev, 1959
16.10.1955	Teteven	Sink in	9000 kg	Source Glava	6500	286	Kovachev, 1959

Table 3. Continued

Date	Location	Injection Site	Tracer	Detection Site	Distance (m)	Time	Reference
		Yamite area	Na Cl	Panega			
25.08.1981	Iskrets	Studena Reka Sink	4 kg uranin	Iskrets Source	9500	200	Benderev, 1989
16.04.1982	Iskrets	Studena Reka Sink	600 kg PAV	Iskrets Source	9500	30	Benderev, 1989
20.04.1982	Iskrets	Studena Reka Sink	6 kg KBr	Iskrets Source	9500	50	Benderev, 1989
1.09.1982	Iskrets	Ponor Sink	5 kg uranin	Iskrets Source	9500	250	Benderev, 1989
May 1983	Iskrets	Sink of Zimevishka Reka below the cave Giva Laka	1 kg uranin	Source Skaklya	3000	160	Benderev, 1989
1966	Lakatnik	Zhekovo Ezero in the cave Temnata Dupka, Lakatnik	0.4 kg fluorescein	Source Zhitolyub	600	1h50 min	Tranteev, 1966
March-April 1967	Lakatnik	Proboynitsa River	0.25 kg fluorescein	Source Zhitolyub	2500	7	Bliznakov, 1967
26.05.1989	Teteven	the vertical of Maliya Sovat pothole	0.5 kg fluorescein	Malkata Voda Source	2300	9	Zlatkova, 1991
25.05.1967	Bosnek	The river near Ribkata, Duhlata Cave	Fluorescein	The fountains in Bosnek	850	2	Strezov, Saynov (in prep.)
11.03.1978	Bosnek	River in the cave Akademik	0.2 kg Fluorescein	The source near the fountains in Bosnek		Did not appear in the source	Strezov, Saynov (in prep.)
15.04.1978	Bosnek	River in the cave Akademik	Fluorescein	The source near the fountains in Bosnek		5	Strezov, Saynov (in prep.)
30.04.1978	Bosnek	River in the cave Chuchulyan	1.5 kg fluorescein	Vreloto		38.50	Strezov, Saynov (in prep.)
3.05.1978	Bosnek	River in the cave Akademik	Fluorescein	Fountains in Bosnek			Strezov, Saynov (in prep.)
1.05.1982	Bosnek	River near Ribkata, Duhlata Cave	Fluorescein	Fountains in Bosnek	850	2h10min	Strezov, Saynov (in prep.)
2.05.1982	Bosnek	River in the cave Akademik	Fluorescein	The source near the fountains in Bosnek		40	Strezov, Saynov (in prep.)
				The fountains in Bosnek		50	
11.11.1955	Velingrad	Sink of Chukura River	5000 kg NaCl	Source in the cave Lepenitsa	4600	9	Kerekov, 1959
				Source on the right bank of	7500	20.5	

Table 3. Continued

Date	Location	Site	Tracer	Source	Distance	Time	Reference
				Lepenitsa River Source on the left bank of Lepenitsa River	8500	22	
				Kleptuza	10500	18	
1955	Velingrad	Sink in Rakevo Dere	Salting and coloring	Toplika			Kerekov, 1959
				Sources in the area of Krivov Chark			
31.08.1954	Trigrad	Muglenska river, 3 km below Mugla	8950 kg NaCl	Source in front of Teshel Factory	11000	30	Yaranov, 1959
				Source S of Teshel Factory	11000		
				Near the former wooden bridge of Nastan	9000	45	
				Below the village Nastan	9000	680	
				Vrisa	7000	780	
				Malak Bedenski Izvor	7000		
3.08.1954	Trigrad	Cave Lednika	1000 kg NaCl	Did not appear			Yaranov, 1959
22.06.1999	Trigrad	Muglenska river, 3.5 km below Mugla	5 kg fluorescein	Nastan Source	8750	Not established	Machkova et al., 2000
				Nastanski Ribarnik	9250	50.00	
				Bedenski Ribarnik	7250	Not established up to 26.06.1999	
				Vrisa	7000	Not established up to 20.8.1999	

interesting, showing the connection between qualitative indicators and water quantity. Such are the studies of Iskrets source by Benderev (1989), as well as some studies of M. Machkova, D. Dimitrov, K. Spassov and P. Stefanov. Another field of hydrochemical research is the thermodynamic and kinetic studies of karst processes, based on the analysis of hydrochemical and hydrodynamic data. Such are the publications on the assessment of the coefficient of karstic water saturation concerning karstic waters with carbonates (the direction of karstic processes), made by B. Velikov, together with E. Stankova, A. Benderev, D. Danchev, etc. (about the source near Chepelare, the Iskrets source, the water along Russenski Lom and many others).

The first studies on the activity of karstic processes (the karstic denudation) in Bulgaria were done by Marcovitz et al. (1972) about the Balkan of Vratsa, Pirin and Zlatna Panega. The articles of Bl. Raykova et al.(1980, 1982) about Northeast Bulgaria, Benderev (1989) about Iskrets area, as well as the studies of Ya. Shopov,

K. Spassov, P. Stefanov, A. Benderev and others followed. We should pay attention to the major international project on karstic denudation (Bulgaria was represented by the Institute of Geography – group led by P. Stefanov). An important problem of the quality of karstic waters is the one of their pollution. Several major projects took place to that effect, including: a project concerning the aquifer from Sarmatian (Bulgarian-Spanish team, Bulgarian side represented by M. Machkova, D. Dimitrov, B. Velikov, P. Penchev and others); another one on the hydrochemistry and the pollution of Nastan-Trigrad basin – head B. Velikov, etc.; a project of Breskovski and others about the area around Petarnitsa and Gortalovo; about the Vratsa Balkan by V. Spassov, etc. A good deal of the results obtained have already been published.

The waterfall of the river Kumanitsa (The Big Cauldron) – „Steneto" Nature Reserve, Troyan Balkan – photo Vassil Balevski

STUDIES ON THE CAVE MICROCLIMATE IN BULGARIA

Cave microclimate is a function of the long-term regime of microclimatic elements (temperature, barometric pressure, relative humidity, speed of air current, etc.) in the atmosphere of caves and karstic systems.

The first known observations on cave microclimate in Bulgaria took place on July 19 and 16 September 1900. They were made by I. Stoyanov on the main axis of development of the cave Toplya near Golyama Zhelyazna Village. For the first measuring Stoyanov used the room thermometer of the village school, for the second – the 200° thermometer of the Institute of Geology and Mining at the High School in Sofia. A table of the measurements of temperature is attached to the article. At the same time, the author measured the temperature of the karstic source/spring. On January 29, 1901, the schoolteacher L. Radkovski similarly measured the air and water temperature (after the article of I. Stoyanov "The cave Toplya, v. G. Zhelyazna", 1904, p.114–115).

In 1921, the prominent scientist Rafail Popov published the measurements of the temperature in the main gallery of Tsarskata Peshtera near Belyakovets. Later (1923–1925), Nenko Radev measured the temperature of the following caves: Zhivata Voda near Bosnek, Zmeyova Dupka near Hitrevtsi, Kumincheto (now known as Marina Dupka). The data have been published by Radev (1926).

For a long time following this period no microclimatic data about caves appeared in the scientific literature. As late as 1956–1957, Dr G. Ikonomov published temperature measurements of the cave Orlova Chuka near Pepelina, Russe Distr. In 1949, the Cave Brigade studied the climate of several caves near Lakatnik, Brestnitsa, Karlukovo and Rabisha, but the report of the brigade does not contain results and they have never been published.

The expansion of caving in Bulgaria, after the formation of the Bulgarian Federation of Speleology in 1958, created favourable conditions for further research on the cave microclimate. In the 60s and the 70s, the caves Temnata Dupka near the village Berende Izvor (A. Grozdanov and V. Stoitsev), Razhishkata Peshtera near Lakatnik (P. Neykovski, P. Dechev and I. Mateev), the deep caves near Kotel and the cave Elata near Zimevitsa were climatically studied. Meteosurvey of caves was done also by cavers from Akademik, Sofia Caving Club, during their expeditions (published mostly by P. Neykovski, A. Grozdanov, V. Gruev and others in the four annual books of the club). All these microclimatic data are of historical value, only that they were not systematically collected and lack the necessary level of accuracy.

In the literature of this period, there are also some articles and data about the aerodynamic parameters of karstic caves (as a database). During the study of the ecology of the beetles of the genus *Pheggomisetes* in the cave Dinevata Peshtera near Gintsi, M. Kvartirnikov carried out one-year observations on the cave temperature and published them in 1970. Some other specialized microclimatic studies in the cave Magura by R. Mechkuev, B. Todorov, I. Popivanova and S. Toshkov (1983) proved the circadian and annual thermal regime of the cave. During the transformation of some caves (Ledenika, Saeva Dupka, Magura, Snejanka, Orlova Chuka and others) into show-caves, Vladimir Popov and his associates made some semi-stationary observations on the cave climate. Microclimatic observations in several Rhodopean caves were made in 1964–1985 by cavers from Chepelare and were published in Rodopski Peshternyak. Much later, Raychev (1991) analyzed the microclimatic observations in the cave Imamova Dupka near Yagodina.

During the period from the beginning of the 60s to the end of 20th century, many scattered data from separate non-system studies during expeditions in the country were collected in the Main Card Index of Bulgarian Caves.

In 1974 in Akademik, Plovdiv Caving Club, a specialized group for studying the cave microclimate was formed. A polygon for complex measuring of cave climate in Dobrostan Kartstic Area was set. Taking into account the fact that the microclimate of karstic caves is very different from the meteorology of the area, the following basic factors have been completely investigated:

– the morphological particularities of the cave or cave system;
– the altitude of the cave entrance and the karstic terrain, which includes it;
– the degree of fracture of the bed rock, building the karstic massif;
– the orientation of the entrance to the sun, the local air currents and the relief factors;
– the presence of ice and snow – firn bodies in the cave;
– the presence of intensive water dropping or of running water in the cave;
– the local intensity of the thermal field of the Earth in the area of the karstic massif.

The aerodynamic parameters, studied in the period 1974–1978, have been arranged in a meteorological data base, processed and presented as a diploma thesis "Thermodynamic Processes in the Caves of Dobrostan Karstic Region", presented by A. Stoev in 1979 at Plovdiv University. Later in a series of publications, again A. Stoev and a team presented an actual statistical and generalized information about several karstic regions (of Pirin, Vratsa, Kotel, Karlukovo, the karst in Rilo-Rhodopean area). The evolution of the meteoelements observed over the time, and their spatial structure, have been archived as polymeric tables. The decade and average monthly and yearly values have been calculated (mostly concerning measurements of the air, water and bedrock temperature, as well as the relative humidity of the cave atmosphere). For the first time one of the typical characteristics of the caves and the cave systems – the so-called Length of the Zone of Constant Temperature (LZCT) was applied. The thermodynamic processes, forming the ZCT, and its evolution depending on the outside climate and the telluric warmth of the Earth have been studied.

The expeditions to Karlukovo Karstic Region (1984–1987) for collecting complex data for the Encyclopaedia Lovech District studied 40 caves in stationary or semistationary regime. The data and the generalized article were published in the Encyclopaedia. For a detailed study of the special/spatial structure of the microclimatic fields in the caves and their interaction with the outer around-the earth atmospheric layer the following methods have been used: "compiling climatic maps" and "determining the vertical gradients of the value of meteorological elements". In 2002, these and many other studies on the karst regions in Bulgaria were put together in the report "Cave Microclimate in Bulgaria – Genesis, Evolution, Cycles and Territorial Distribution" by A. Stoev and P. Muglova. The paper used also data from many episodical, semistationary and stationary observations of the temperature of the air, rock and water, the relative humidity, barometric pressure and the speed of the air current. Through

Climatological observation by Alexey Stoev in the cave Drashanskata peshtera – photo Trifon Daaliev

diagrams the spatial distribution of the microclimatic fields in the caves and their connection to the outer, around-the earth, atmospheric layer was presented. Microclimatic maps of the various elements, depending on the season, the orientation and the height of the entrance and its situation were also made. These maps were compared with maps of the distribution of climatic parameters in the around- the earth atmospheric layer and the substratum in which the cave had developed. In 1981, at the [22] International School of Speleology, A. Stoev reported the idea and the first results of the research on paleoclimate by deciphering data about the solar activity in the "yearly circles" in stalactites, stalagmites and other speleological topics. The idea was undertaken and developed by Y. Shopov and his team, creating an interesting physical method (the so-called Lazer Luminescent Microzonal Analysis) for "reading" records in cave formations. It was shown in a series of articles that the cave formations (stalactites, stalagmites, stalactones, etc.) give information about the paleotemperature, paleosoils, seismic processes and quantity of precipitations in the past, dislocation of rocks, solar emissions, geomagnetic field, plant populations, chemical pollution, air composition, rising of the sea level and the level of subterranean waters, changes in the cosmic rays, obtaining of space-born isotopes and explosion of supernew stars. Secondary cave formations include a lengthy record of data with extremely high resolution, as they are formed, as a rule, under almost constant temperature and humidity and do not change over the time. Once formed, they conserve these data, which can be read by various methods – luminescence of cave formations, magnetometrics, laser luminescent microzonal analysis, contents of stable isotopes, etc. That is why the cave formations are among the best "archives" of the environment of the past.

In a series of publications at Bulgarian and international conferences, A. Stoev and his team presented thermodynamic models of the microclimate and the thermic field of the caves and the karstic massif in which they had developed, as well as studies on the resistance of the centuries-long value of the average monthly temperature of the cave atmosphere.

Vassil Stoitsev and Vassil Markov – photo Vassil Balevski

MEDICO-BIOLOGICAL AND PSYCHOLOGICAL STUDIES IN CAVING AND SPELEOLOGY

Caves, as a subject of exploration, have a complicated and varied relief – squeezes, verticals, underground shafts, lakes, rivers, waterfalls, siphons and others. Overcoming these obstacles is connected with serious physical efforts. Difficulties, danger and risk are combined with a considerable psycho-emotional tension, caused by unfavourable environment, characterized by high relative air humidity, darkness and relatively low temperature. The question of how caving activities affect the organism of cavers is important and interesting. Knowing the physiological and psychological changes, occurring in cavers during underground activities, we can manage their multilateral training.

The scientific research on the influence of caving on human organism started in 1962, when French speleologist M. Siffre descended in the pothole Scarson, remaining there for 62 days to study the big glacier in it, as well as the changes in an organism under complete social isolation. Later, in other countries 12 other similar experiments have been carried out.

The first studies on the physiological modifications in caving were done by B. Marinov and G. Shterev, while exploring the cave Gradeshnitsa (1964), and later by B. Marinov and P. Tranteev during the National Lepenitsa expedition (1970). The results have been published.

In 1971, some Bulgarians became interested in subterranean experiments. Initiated by Ivan Petrov from the District Caving Club in Plovdiv, together with the Medical Academy and the psychoneurological dispensary in Plovdiv, the first subterranean experiment in Bulgaria was organized. In the cave pothole Topchika, 60 m underground, the cavers G. Yolov, G. Trichkov, D. Zhishev and the organizer Ivan Petrov spent 30 days. Isolated from the outside world (except for the bilateral connection with the support team on the surface), the four effectuated a rich scientific program. The changes in different physiological and psychological parameters have been explored: skin, optical, auditory, palatal and other sensitivity, will activity, reaction speed, concentration and mobility of attention, psychical compatibility between the participants, the cardiovascular system, etc. The results indicated that in some of the participants the indexes studied showed a tendency towards worsening, but remained within the limits of the normal. Part of the results has been published.

In 1973, Dr M. Mihaylov did interesting psychological and medico-biological studies, unfortunately not published, on cavers during the expedition of Planinets, Sofia, to the pothole Pierre Saint Martin in France.

In March 1977, during the National Women's Expedition in the cave Orlova Chuka near Russe, A. Jalov accomplished some pedagogical and psychological research on the girl cavers. Part of the results (unpublished) showed decreasing of the level of physical fitness and of some psychological indexes after an everyday caving activity. In the same year (1977), the cavers A. Jalov, S. Tsonev and T. Stoychev from Aleko, Sofia Caving Club, proposed a new subterranean experiment. It was carried out in the cave Desni Suhi Pech (Vidin District) under the auspices of the Bulgarian Federation of Speleology and under the scientific guidance of the National Sport Academy. A. Jalov and S. Tsonev were participants in the experiment. They spent 62 days and nights in the cave in a total isolation (only transmitting information outside the cave). The program included a study of the changes in several physiological and psychological indexes. The results showed nonpathological changes in the observed spheres. The researchers observed some phases in the process of adaptation of the organism underground. Some of the results were published, others were used for the thesis of A. Jalov.

For a successful caving expedition, especially in difficult and deep potholes, the level of general capacity and psychical stability is decisive when extreme situations occur. The study of these two indexes is an important source of information about the status of the cavers during their training and about the final composition of the expedition. Along those lines were the medico-biological studies, carried out by Dr K. Mâlchinikolov from the District Dispenser of Sport Medicine in Pleven and by S. Gazdov from Studenets, Pleven Caving Club, while preparing for the descent into the world's deepest (at that time) pothole Jean Bernard in 1984. In the period 1986–1988, within the program „Man and His Brain" a series of studies took place, done by Dr G. Mateev, Dr T. Djarova and Dr D. Stefanova from the National Sport Academy, K. Krâstev from Pleven, K. Boyanov and A. Jalov during the preparation and the realization of the expeditions of Studenets, Pleven Caving Club, in Spain.

The results, most of them already published abroad as well, are of considerable practical value Hence they are of interest to the specialists. The generalized conclusion of the studies is that the functional aptitude of the cavers is important for enlarging of the organism's capacity to adapt under extreme conditions, which are encountered in every expedition. A. Jalov's studies on the psychology of the group proved the applicability of the sociometrical methods in speleology in terms of the composition of working groups.

Interesting studies have been carried out by B. Marinov and T. Daaliev on the participants in the expedition of Galata, Varna Caving Club, to the pothole Snezhnaya (Caucasus). They were directed to how to establish the changes in the accuracy of the muscular effort and strength endurance during the expedition. It was concluded that the activities of the cavers while penetrating this super deep pothole (-1370 m) influenced the indexes under study. These indexes can be used to assess the status of the cavers and to perfection the process of managing the preparation and the expedition itself.

In 1998, Emiliya Gateva from Ambaritsa, Troyan Caving Club, helped by the Sport Academy, carried out the next Bulgarian subterranean experiment. Unlike the previous explorers, E. Gateva remained alone in the pothole Ptichata Dupka in Troyan Balkan for 56 days. During her stay she had a two-way connection with the surface. A complex of physiological, and mainly psychological studies, were realized. Part of the results found place in the diploma thesis of Gateva in the National Sport Academy. The others are still unpublished.

Alexey Jalov and Stefan Tsonev after the end of the experiment in the cave Suhi Pech. Photo archiv BFS.

Ivan Petrov (the one being awarded), together with three other cavers from Plovdiv carry out the first experiment for long stay underground in Bulgaria.

CAVE DIVING IN BULGARIA

Until the beginning of cave diving the exploration of a cave had stopped where a siphon appeared. The first attempt to penetrate a submerged cave system in Bulgaria took place in October 1947 and was due to the naturalist Alexi Petrov. In a heavy diving suit, Petrov reached the bottom of the source lake Glava Panega (12 m deep), but it was not possible to enter the underground system. For many reasons real cave diving, requiring special equipment, started in Bulgaria much later than in other countries with developed Speleology. However, just in few years it achieved remarkable results.

First Organized Attempts to Penetrate Siphons

The real birth date of cave diving in Bulgaria was, however, November 7, 1959. The aim was to penetrate the siphon of the lake Mrachnoto Ezero in the cave Temnata Dupka at Lakatnik Railway Station. Seven cavers participated, two of which (Aleksandar Denkov and Tanyu Michev) tried, with primitive equipment, to discover parts of this big cave. After ca. 40 m, they had to return and, with considerable danger to their life, managed to join the group. Other attempts were made in the same year in the siphons in Temnata Dupka near Kalotina and Krivata Pesht near Gintsi, but without success, due to primitive equipment. In 1960, other attempts in the siphons of the pothole Djebin Trap and in the caves Lepenitsa and Rushovata Peshtera failed for the same reason.

In 1963 the Group for Cave Diving at the Republican Caving Commission already existed. The group included A. Denkov, T. Michev, M. Kanev, B. Antonov and Hr. Delchev. With some better equipment they managed to penetrate the siphon of the source Zhabokrek near Chiren (18-20 November 1963) and to discover a new chamber and a new 120-m gallery. The complete connection of Zhabokrek with the big cave Ponora was realized only in 2001 by a French-Belgian team, which overcame 6 more siphons (not possible in 1963).

In 1970, cave diving in Bulgaria took its first victims. In the cave Dyavolskoto Garlo near Trigrad (Smolyan Distr.) the divers S. Lyutskanova and E. Yonchev (with no underground experience) from Varna died.

In 1972, the cave divers T. Michev, A. Gyurov and V. Kiselkov tried to penetrate the source near Lakatnik Railway Station and the source Glava Panega. In Glava Panega T. Michev reached a depth of 29 m and V. Kiselkov (in 1978) – even 32 m. This was the end of the first Bulgarian cave diving group.

The New Generation In Action

Meanwhile, in the period 1976–1977, on the cave diving scene P. Petrov, V. Nedkov and, for a short time, P. Hristov appeared. They tried to penetrate various siphons in Lakatnik's Temnata Dupka, but most of all the siphons of the cave Katsite near Zimevitsa. In 1977, V. Nedkov penetrated the siphons of the difficult cave Katsite to a length of 2560 m and a depth of 220 m.

In 1978, A. Jalov overcame on apnea the short (2.5 m) siphon of the cave Gornik near Krushuna and discovered 627 m of new water galleries. Other discoveries were made after the siphons of the caves Brashlyanskata (Aleksandrovo), Mussinskata Peshtera (Mussina) and other caves.

In 1981, another tragic accident resulted in the death of the caver G. Antonov in the source Popov Izvor near Studena. Four years of standstill in the cave diving followed.

In 1985 a new period in Bulgarian cave diving started with the formation of the diving group of Studenets, Pleven Caving Club.

The divers I. Gunov, M. Dimitrov, V. Chapanov and A. Mihov started the exploration of the big cave Boninska Peshtera near Krushuna. Until 1987 they had overcome four siphons of 6, 6, 30 and 36 m and surveyed 461 m of new water

Vassil Nedkov – caver, alpinist and diver

galleries, thus bringing the cave's length to 4015 m. Later, in 1989, the divers from Pleven discovered another 515 m behind three new siphons in the same cave, bringing its length to 4530 m. Meanwhile, the group explored the siphons in many other caves, making considerable discoveries. In 1985, they returned to Glava Panega, but only in 1992 K. Petkov entered 230 m of inundated galleries, bringing the depth to 52 m. In Toplya the divers surveyed 178 m of new galleries.

Joy and Sorrow

During this period hundreds of new galleries were discovered by other cavers (G. Yordanov, D. Todorov, V. Pashovski, K. Georgiev) in the caves No7, Zâdanenka (Karlukovo), Krivata Pesht (Gintsi), in the source of Kotel and others. However, in May 1989 another tragedy shocked all Bulgarian cavers. Trying to penetrate the end siphon of the cave Urushka Maara near Krushuna, two of the best Bulgarian cavers and cave divers, V. Nedkov from Sofia and V. Chapanov from Pleven, died of suffocation in a gallery, full of (?) CO_2.

Despite the heavy losses, cave diving in Bulgaria did not cease to exist. In the autumn of 1989, divers from Sliven (A. Aleksiev, I. Zdravkov, G. Ganchev, D. Boyanov, S.Mihov) prepared an attack on the source in Kotel. On October 23, 1990, the team penetrated four siphons and discovered a chamber, which was 20–25 m high and 7 m wide. The exploration is not yet finished.

The divers from Pleven still active

In 1990–1991, the divers from Studenets, Pleven, explored many new siphons. In the pothole Golyamata Voda near Karlukovo, after three siphons of 8, 10 and 20 m, they reached a 400-m gallery and a fourth siphon at 105 m.

Although not so intensive, the siphon activities of the few cave divers are still resulting in new discoveries. In 2001, attempts to penetrate the submerged system of Andaka (Dryanovo) were undertaken by K. Petkov.

It is fair to give a due tribute to the considerable success of Bulgarian cave divers abroad, crowned by the penetration of the siphons of the giant Spanish pothole Bu-56. Thanks to the efforts of Pleven divers, on September 11, 1987 the depth of the pothole reached 1408 m and for some time the system was the second in depth in the world. We should remember this date, as well as the courageous men who have realized this exploit!

The authors of this book believe that cave diving is among the most dangerous human activities. We are proud that we know the people who dare see what is "behind" and who contribute to a better knowledge of caves. The problems in Bulgarian cave siphons are by no means over.

PROTECTION OF KARST AND CAVES IN BULGARIA

Karst and karstic phenomena are a natural entity with specific features, fauna and flora. At the same time, these regions and especially the caves in them contain a multilateral information concerning the development of the material and the spiritual culture of mankind, as well as about the animals and plants of the past of our planet. Taking this into account, it is certain that any excessive human activity, including the activity of cavers, could lead to disrupting the fragile balance and to irreparable damages to Nature, Science and Culture.

It is evident that if speleology did not exist, the human presence in the caves would not influence the natural environment and the negative changes would have been due only to external factors. The practising and the development of speleology should be carried out trictly respecting the rules for protection of environment. This is important also for anyone, so it is necessary for them to know what, why and how to protect what's in the caves and karstic areas.

The secrets of caves have attracted and still attract generations of Bulgarian cavers and the curiosity of many chance visitors. For many of them the animal world of the caves remains almost unnoticed. The caves offer a unique living environment – specific microclimate (almost permanent temperature and humidity), slight or absent daylight. This has provoked the subdivision of the animal world in caves into three essential groups: troglobites – animals, entirely adapted to life underground; troglophiles (cave loving) – animals, living in and outside the caves, but reproducing in caves, and trogloxenes – chance visitors in caves, sometimes quite regularly entering them.

So far 704 animal species (Invertebrates) have been recorded from Bulgarian caves. Many others live in other parts of the underground ecosystem (microcaverns of different type). At least 97 species are considered troglobites or stygobites (the name given to the water animals permanently living in caves). Usually, they are tiny white or semitransparent creatures, which could be seen on clay deposits, cave walls, flowstone formations, in pools, lakes or on layers of bat guano. Most troglobites are endemic species living only in several caves on a limited area, sometimes known only from one cave. Such are many cave beetles, Myriapods and Isopods. That is why the destruction of one cave could bring to the disappearance of entire animal species forever. All animals (invertebrates and bats) living in caves are protected.

Bats are the most typical vertebrates in caves. Almost all European species live also in Bulgaria (so far 30 species recorded, including two in caves). Bats are especially protected in all European countries and there are many restrictions for visiting the "bat caves" in Bulgaria, even for cavers. The local caving clubs should take care of these caves. Such caves of European importance are Parnitsite near Bezhanovo, Ponora near Chiren, Sedlarkata near Rakita, Morovitsa near Glozhene, Nanin Kamak near Musselievo, Devetashkata Peshtera near Devetaki, the old mine near Golak, etc. It is especially important not to disturb the bats during the breeding season, in winter roosts, to make fire in and near the caves or to chase the bats from their roosts.

Some birds also inhabit caves. They can be subdivided into two main groups: nesting in caves and taking shelter in them. The most typical representatives of the first group are the white-breasted swift, the Rock swallow, the alpine chough and the rock pigeon. Other birds (owls and many others)

Stalaktites, stalagmites and other flowstone formations in the cave Snezhanka – photo Valery Peltekov

also find shelter in the entrance parts of the caves, some of them even nesting there.

Among the birds which do not live in caves, but nest on the rocks or in niches around the cave entrances, are some of the rarest and most endangered birds: the rock eagle, the Egyptian vulture, some falcons, the eagle owl, the black stork and others. They are strictly protected and cavers should participate actively in their protection.

The natural beauty and the uniqueness of the caves is due to the stalactites, stalagmites and all other mineral formations. Depending on the geological and hydrogeological environment and the climate, the flowstone formations grow with different intensity and acquire various shapes, size and coloration. In all cases, the formation of secondary karst artefacts is a long process, while their destruction is a matter of minutes.

The karstic underground water is an important factor in the life of cave organisms. At the same time, it is often a resource for satisfying human needs (drinking and industrial water, irrigation, etc.). The pollution of underground water makes it dangerous for people and animals. If dumping of dead animals or other organic, chemical or mechanical pollutants in caves and potholes is detected, the Regional Inspectorate of Environment and the media should be informed.

Cave sediments contain material vestiges from the Paleolithic and the Medieval Man, and bones of many animals from different periods. Usually, the sediments conserve well these remains and protect them from erosion and other external factors. When not disturbed, the materials in the sediments, are usually in a chronological order. Each disturbance of the natural sedimentation makes impossible the dating of the finds and the proper archaeological, paleonthological, sedimentological and other scientific research. To preserve these extremely important sites from destruction there is a law concerning the monuments of culture and museums. According to the legislation, any excavation or digging in caves without the consent of the Archaeological Institute (Sofia) is illegal and prosecuted.

The Environment Protection Act provides for a special protection of caves of particular beauty or of high historical, scientific or cultural value. They may be declared **Protected Natural Sites** (PNS). Such are: the Natural Reserves, the National Parks, the Natural Parks, the Remarkable Sites, the Protected Areas and the Protected Plants and Animals. Depending on where/which category the caves belong, they get a different regime of protection and use.

Caves on the territory of Reserves are best protected. On their territory any economic activity or

Petar Beron collecting biological material in the cave Vodnata peshtera by the village Breste – photo Valery Peltekov

deeds, which endanger the natural habitat or ecosystem are prohibited. In Bulgaria, about 80 caves are protected within the Biosphere reserves Steneto and Boatin in Central Stara Planina, the reserves Vrachanski Karst, Bayovi Dupki – Djindjeritsa in Northern Pirin, Kupena and Kastrakl in the Rhodopes. Visiting and bivouacking in the reserves could take place only by A special permit of the Ministry of Environment and Waters (MEW).

One hundred forteen (114) Bulgarian caves altogether have been declared **Natural** or **Historic Sites**. They are protected, including the adjacent area, indicated in the Ministerial Decree, issued in Dârzhaven Vestnik (State Gazette).

In the decree stipulates that in the protected site it is prohibited to establish quarries nearby, to brake the flowstone formations, to leave graffiti on the walls, to light fire, to pollute and to do anything detrimental to the natural status of the site. It is prohibited to enter "bat caves" during their breeding season.

Protected sites could be areas with peculiar features of surface and subterranean karst. Well known protected sites in Bulgaria are the canyon of Chernelka River near Gortalovo and the Karst Valley near Petârnitsa, Pleven Distr.; the canyon of Negovanka River and the locality Ponorite near Mussina, Veliko Târnovo Distr., Zlosten near Kotel; the gorges of Trigrad and Buynovo in the Rhodopes. Four karstic sources are also declared Protected sites: the source in the locality Yamata near Stara Zagora, Kyoshkata near Razlog, Medvenskite Izvori near Medven and Zlatna Panega near the village of the same name, Lovech Distr. In the Protected Sites limited activities are permitted not damaging the natural landscape.

The **National Parks** and the **Natural Parks** are extensive protected areas. Caves or important features of the surface karst often fall within their territory.

The use of the natural assets, including caves, in the parks should not lead to a disbalance of the environment. Within the parks ca. 500 caves and potholes altogether are situated (and thus protected).

This makes the total number of Bulgarian caves, protected under the Environment Protection Act, about 750. Fines and other penalties are provided for by this Act and by other laws and decrees.

Some other Bulgarian caves are protected also by other laws (the Cultural Monuments Law or the Law on the Waters).

Devetashkata Peshtera (Monument of Culture of national importance), the rock monasteries near Ivanovo (Russe Distr.), the natural and archaeological reserve Yaylata near Kamen Bryag, Dobrich Distr., and others are protected as cultural monuments.

By virtue of the Bulgarian Constitution and the laws of the country, every Bulgarian citizen or organization can propose to the government institutions that some caves or areas be declared Protected Sites.

It is also possible to temporarily restrict all activities in one particular cave or groups of caves, by announcing the interested owners or organizations.

Declaring one natural site protected does not change its ownership but places its use under the restrictions of the law. National Parks and Reserves are Government property (Art. 18 of the Constitution).

Under the Law on the Waters there is a prohibition to penetrate the source caves, used for drinking, or the sanitary zones of the karstic sources. Some well known caves of this category are Tserovskata Peshtera near Tserovo, Garvanitsa near Kossovo, Vodni Pech near Dolni Lom, the sources Kotlenski Izvori in Kotel, the cave near Galata, Lovech Distr., the caves in the water collecting area of Kumanitsa karstic source near Cherni Ossam, etc. Visits in the cave sources of drinking water can be done only with a special permit by the authorized bodies.

To not leave behind rubbish is essential concern of the cavers – photo Trifon Daaliev

KARST AND CAVES IN BULGARIA

Karstic features are developed on 22.7% of Bulgarian territory (Popov, 1970a), or 25 171 km². According to Boyadjiev (1964) the karst areas are 15 778 km². As Popov (1970a) stated, this figure does not include buried karst. Together with this category, most of the karstified area belongs to the Danube Plain (66% of the karst in Bulgaria), the least is found in the Transitional Geomorphological Region (6%). After many years of research, in the period 1968–1978 Vladimir Popov subdivided Bulgaria into four regions and 50 districts.

The **Danube Plain** was subdivided into eight districts. Out of its extensive karst 70% is buried karst (Sarmatian sediments). From the Serbian-Bulgarian border to Pleven area several important caves are situated: Varkan near Druzhba Village (812 m), Bashovishki Pech near Oreshets Railway Station (over four km long), Sedlarkata near Rakita Village (1100 m), Gininata Peshtera near Sadovets Village (501 m), Kirov Vârtop near Bohot Village (776 m), Aladjanskata Peshtera (1083 m) and Haydushkata Peshtera (459 m) near Gortalovo Village. The region of the valley of Rusenski Lom River is of particular interest. There, on the steep slopes of the river valleys, near the villages Basarbovo, Ivanovo, Tabachka, Pepelina and others many caves, including Orlova Chuka (13 437 m), the second in length in Bulgaria, are formed. Kulina Dupka near Krivnja is 326 m long.

In Ludogorie and Dobrudja, there are many (over 350), but relatively small caves. More important are Stoyanova Dupka near Ruyno (357 m) and Dogoulite near Topchii (305 m). From the abrasive caves between Kaliakra and Shabla the longest is Tyulenovata Peshtera (107 m).

The karst of Shumensko Plateau harbours several caves including Zandana (2716 m) and the pothole Taynite Ponori (115 m deep and 1916 m long).

The Karstic Region of Stara Planina (the Predbalkan and the chain of Stara Planina) is subdivided into 19 districts. The karst of this Region covers 4980 km², or 19,2 % of its total area (Popov, 1970b). This Region is the richest in caves in Bulgaria. There, we find 19 of the 59 Bulgarian potholes, which are deeper than 100 m, and 46 of the 65 Bulgarian caves, which are longer than 1000 m. In all parts of the Regions there are big caves and potholes.

In Belogradchik District, built of limestone from the Mesozoic Era, there are such caves as Magura (2500 m), Vodni Pech near Dolni Lom Village (1300 m), Mishin Kamik near Gorna Luka (695 m), as well as Pleshovska Dupka near Prevala Village, which is 102 m deep.

In Salash District a noticeable cave is Rushkovitsa near Stakevtsi Village (450 m).

A classic karst area and one among the richest in Bulgaria is Vratsa District. In its thick Jurassic and Cretaceous limestone more than 500 caves and potholes have been discovered. In the higher parts of Vrachanska Planina the following potholes have been explored : Barkite 14 (-356 m denivelation, 2600 m long), Belyar (- 282 m Denivelation, 2560 m long), Barkite 8 (-190 m), Pukoya near Pavolche Village (-178 m), Yavorets (-147 m) and Panchovi Gramadi (-104 m) near Zverino and Haydushkata near Bistrets (-108 m). The water caves near Chiren (Ponora, 3497 m long; Mladenovata Propast, 1723 m long) are among the cavers' favourites. Other caves over 500 m long are Toshova Dupka near Stoyanovo (1302 m), Mizhishnitsa (885 m), Sokolskata Dupka near Lyutadjik (815 m), Gârdyuva Dupka near Zgorigrad (510 m). The longest of the 130 little caves near Cherepish is Studenata Dupka (623 m).

The caves near Lakatnik – the cradle of the cavers from Sofia – also belong to Vratsa Region. Here we find Temnata Dupka (over 7000 m), Kozarskata Peshtera (709 m), Razhishkata Dupka (316 m) and Svinskata Dupka (300 m).

More than 140 caves and potholes have been recorded from Ponor Karstic Region. Among them are the caves near Ginci, Komshtitsa, Gubesh, Iskrets, Zimevitsa and Tserovo. Some of them are well known to the cavers from Sofia: Dushnika (876 m), Katsite (-205 m, length 2560 m),

Golemata Temnota near Drenovo (-106 m, 2000 m length), Golyamata (4800 m) and Malkata (-125 m) Balabanovi Dupki, Radolova Yama (-88 m), Krivata Pesht (1500 m), Tizoin (-320 m, length 3599 m) and Saguaroto (-135 m, length 2217 m). Some of the longest caves in West Bulgaria are the caves near Tserovo Railway Station (Vodnata Peshtera, 3264 m, and Mayanitsa, 1426 m).

On the border with Serbia, the cave Temnata Dupka near Kalotina (Berende) marks the end of the present Bulgarian territory. Beyond the border, imposed by the Neuilly Treaty, the former Bulgarian caves near Odorovci and Vetrena Dupka near Vlasi remain.

Another classic karstic area is Kameno Pole – Karlukovo. In this majestic karst the National Caver's House Petar Tranteev has been built. In the valley of Iskar more than 600 caves have been discovered, including Bayov Komin (2196 m) and Drashanskata Peshtera (578 m) near Drashan; Starata Prodânka (628 m) and Popovata near Gabare; Golyamata Voda (-104 m, 612 m long), Zadânenka (1150 m), Bankovitsa (689 m); Stublenska Yama (-72 m, 562 m long); EC-20 (-94 m) and Tipchenitsa (-78 m), all near Karlukovo. In the majestic karstic landscape of Karlukovo, the cave Prohodna is also prominent, the lower entrance of which is 42 m high.

In the limestone, which is up to 450 m thick, in the valleys of the rivers Vit, Panega, Batulska and Yablanishka, the caves of the Panega Region are formed – Morovitsa near Glozhene (-105 m, 3200 m long), Vodnata Pesht near Lipnitsa (1100 m), the potholes Partizanskata (-107 m), Bezdânniyat Pchelin (-105 m), Yasenski Oblik (-108 m), Nanovitsa (-102 m) and others. One of the biggest karstic sources in Bulgaria – Glava Panega (max outflow 20 m^3/s) was explored by divers up to -52 m. However, it is justified to assume that amazing discoveries are yet to be expected in this huge underground system.

Large water caves are formed in Dragana – Bezhanovo Region of the Predbalkan: Parnitsite (2500 m), Gergitsovata (408 m), Sedlarkata (1100 m), Skoka (1000 m).

The caves of Teteven and Troyan regions are among the most interesting in Bulgaria. They belong to Vassilyov, Cherni Vit and Gorni Ossam areas. Some are particularly important, such as Raychova Dupka (Denivelation – 377 m, the deepest in Bulgaria, 3333 m long), Malkata Yama (-232 m), Borova Dupka (-156 m), Pticha Dupka (-108 m and 652 m long), Kumanitsa (-104 m and 1656 m long), Golyamata Gârlovina (-100) and Vurlata (1110 m), all in the area of the village Cherni Ossam. The water cave Trona (Duhaloto) near Apriltsi is 1040 m long.

In Lovech area many caves are known, including Golyamata (1921 m) and Malkata Peshtera (1295 m) near Mikre and Sopotskata Peshtera near Sopot (1225 m).

The subterranean karst in Devetaki Plateau forms a special cave area. It includes the famous Devetashka Peshtera, 2442 m and contains the largest cave chamber in Bulgaria, also the caves near Krushuna: Popskata, or Boninskata Peshtera (4530 m), Vodopada (1995 m), Urushka Maara (1600 m), Gornik (1074 m); near Gorsko Slivovo (Chernata Pesht, 741 m), Kârpachevo (Futyovskata Peshtera, 700 m), Aleksandrovo (Brâshlyanskata Peshtera, 608 m), Chavdartsi (Mandrata, 530 m). In this area also the potholes Kânchova Vârpina (-100 m, 463 m long) and Blagova Yama near Etropole (- 153 m) are situated.

Some of the remarkable Bulgarian caves are developed in the Apt-Urgon limestone of the plateau of Arbanasi and Belyakovets. Cavers are proud of the new discoveries near Emen: Russe (-100 m, 3306 m long), Troana (2750 m), Bambalovata (2923 m), as well as of the extension of the "old" Emenska Peshtera up to 3113 m. Other important caves are Genchova Dupka (265 m) near Malak Chiflik and the water cave Musinskata Peshtera (532 m) near Mussina.

Large caves have been found in Strazha-Debelets District. Such are Machanov Trap near Zdravkovets (1907 m), Izvora near Yantra (1400 m). The longest cave in Debeli Dyal Subdistrict is Marina Dupka near Genchevtsi (3075 m). A favourite meeting point of cavers are the big caves near Dryanovski monastery – Bacho Kiro (3500 m) and Andaka (4000 m). In the small cave Polichki in the same area out some of the first archaeological explorations in Bulgarian caves, were carried out, as early as the end of the 19th century.

The relatively small caves, high in Shipka Balkan (Stoletovskata Peshtera and others), contain interesting cave fauna. In the mountain of Elana – Tvarditsa only some 30 caves and potholes have been recorded, but they include the recently discovered pothole Magliviya Snyag (-146 m, length 3076 m) near Tvarditsa and the cave Dolnata Maaza near Byala (280 m), known since long time ago.

In the Cretaceous limestone of Kotel Mountain there are many caves and potholes, but the cave fauna is rather poor. The pothole Golyamata Yama near Kipilovo (-350 m, third deepest in Bulgaria) is worth mentioning. There are important caves in the area of Zelenich (Prikazna, 4782 m, with the richest fauna in the area; Lucifer, -130 m, length 3200 m; Kârvavata Lokva, -140 m, Bilernika, -80 m, Golyamata Humba, -94 m). Even during the time of Rakovski (the beginning of 19th century) the caves in the area of Zlosten were known – the potholes Lednika (-242 m, length 1367 m), Mâglivata (-220 m), Uzhasat na Imanyarite (Treasurehunters' Horror) (-160 m), Akademik (-158 m) and the cave sink Subatta (518 m). There are caves also near the village Medven (including the ice cave Lednitsata). The source cave in Kotel has an outflow of up to 2500 l/s.

In the mountain Preslavska Planina the most important cave is Prolazkata (Derventskata) Peshtera (569 m).

The Transitional Karstic Region comprises a long stripe stretching from the border with Serbia to the border with Turkey, between the regions of Stara Planina and the Rhodopes. In the mountains around Trân, Zemen and Treklyano the caves are small, but in the Trias dolomites and limestone on the western slope of Vitosha, near the village Bosnek, some of the most important caves in Bulgaria have been formed: Duhlata (18 000 m, the longest cave in Bulgaria), Vreloto (5300 m) and PPD (1020 m). In this belt from Sofia to the mountain Sakar, there are no caves of importance, but the cave source Chirpan Bunar near Belozem (220 m, 30–40 l/s) is to be noticed.

In the districts of Sakar and Dervent heigths, several dozens of caves in Trias marble-limestone are known. They include Bozkite (324 m) and Dranchi Dupka (257 m) near Mramor, Kirechnitsata (224 m) and Dranchi Dupka (-25 m) near Melnitsa.

In the Strandja District, 77 relatively small caves and potholes have been recorded, among which Bratanovata (384 m), Kaleto (302 m), Bâzât 1 (208 m), Bâzât 2 (208 m), Kerechnitsata (224 m) and Stoyanovata Peshtera near Kosti (150 m). The deepest among the few potholes in Strandja are Golyamata Vapa near Stoilovo (-125 m, length 450 m) and Tangarachkata Dupka near Bogdanovo (-74 m).

Rila-Rhodopean Region. There are 13 cave districts in the mountains Pirin and the Rhodopes, developed mainly in Proterozoan marble and (less) in limestone.

In Pirin there are mostly potholes, especially in Vihren Subdistrict. Right under the summit of Vihren the pothole Vihrenska Propast descends up to 170 m; in the circus Bayuvi Dupki is the pothole Chelyustnitsa (-103 m); in Banski Suhodol – "20 godini Akademik" (-118 m), Propast - 9 (-225 m) and Propast - 14 (-103 m). Among the caves in Sinanitsa Subregion the more prominent are Aleko (-130 m, length 600 m) Sharaliyskata (470 m), Rimskata (293 m) and Ruykovata (120 m). Near Razlog is the source cave Spropadnaloto (605 m).

The caves in South Pirin, Slavyanka and Stargach are few, but interesting. The deepest pothole is Garvanitsa (-60 m) near Gotsev Vrah, some interesting cave animals have been found in Rupite near Paril and Stârshelitsa near Goleshevo. In the marble of Dâbrash part of the Rhodopes more prominent is the cave Manailovata (Manoilovskata) Peshtera near Ribnovo (-115 m, length 2155 m).

In the Velingrad District, between Velingrad and Rakitovo, the cave Lepenitsa is situated. It is one of the first show caves, which however has been vandalized many times. Near the town of Peshtera the caves Vodnata Peshtera (1114 m),

Koncheto – part of the karstic ridge of Pirin – photo Valery Peltekov

Novata Peshtera (846 m), Yubileyna (814 m) and the show cave Snezhanka (276 m) are situated.

In the largest karst district in the West Rhodopes – Dobrostan District – we know over 200 caves and potholes. Important among them are Garvanitsa near Kosovo (897 m), Topchika (727 m), Hralupa (311 m) and Gargina Dupka near Mostovo (534 m). Particularly interesting are the potholes Lisek (-160 m) and Kutelska Yama (-88 m) near Dryanovo, Druzhba (-130 m) and Ivanova Voda (-113 m) near Dobrostan. The stone bridges Erkyupriya (Chudnite Mostove) near Zabârdo and Mostovo are to be noted.

In the marble of Trigrad District, which is thick up to 1600 m, there are majestic gorges with many caves. Among them are Yagodinskata Peshtera (Imamova Dupka -8501 m, the third longest in Bulgaria), Sanchova Dupka (888 m) near Yagodina, Izvora (2480 m) and Eminova Dupka (635 m) near Borino, the potholes Drangaleshkata (-255 m, length 1142 m) and Kambankite (-158 m) near Mugla, Ledenitsata near Gela (-108 m, length 1419 m). In Chepelare District, the longest cave is Samurskata Dupka (456 m).

Borikovskata Peshtera (470 m) and Goloboitsa (362 m) are caves in the marble South and Southeast of Smolyan. The deepest pothole in the area is Kladeto near Polkovnik Serafimovo (-147 m).

In Ardino District, 44 caves and potholes are known, including Vodnata Peshtera near Nedelino (203 m), Gyaurhambar near Stomantsi (112 m) and Karagug near Tyutyunche (105 m).

Among the small caves is Maazata near Mâdretsi (114 m). In the easternmost part of the Rhodopes, 35 caves have been explored, incl. Karangil near Shiroko Pole (490 m), Samara (327 m), Ogledalnata Peshtera (157 m) near Ribino and Belopolyanskata Peshtera near Belopolyane.

Caves in non-carbonate rocks. The volcanic caves in Bulgaria have been studied mainly by Boris Kolev. He has recorded only on the territory of the Eastern Rhodopes 77 caves (Kolev, 1987). They are short but spacious (Golyamata Peshtera near Byal Kladenets, 51 m, Ehtyashtata, 30 m and Prilepnata, 18 m).

Among the caves in sandstone, conglomerate and gneiss the ones worth mentioning are Lepenishki Pech near Prolaznitsa, Belogradchik area (35 m), Uske near Chetirtsi (88 m) and others.

The rocks of Cherepish – photo Trifon Daaliev

SOME IMPORTANT KARSTIC REGIONS IN BULGARIA

The Danube Plain
Region of Russenski Lom River and Its Tributaries

The karstic region of Russenski Lom River and its tributaries is one of the most interesting in the Danube Plain. The karst and the caves have been described in details by Krastev (1979), Radulov (2002) and others and are attached to the meandering valleys of the rivers Russenski Lom, Beli Lom from Senovo Town to the confluence with Cherni Lom; Cherni Lom from Tabachka Village downstream and Malki Lom near Svalenik Village. In these sectors, canyon-like valleys are formed, incised in some places by more than 100 m in the hilly relief of the Danube Plain (Altitude ca. 150–200 m). The karst and caves are formed in the oolythic, biogenous and detritus limestone of the so-called Russenska Suitee of Lower Cretaceous (Aptian) which is 50 to 300 m thick. To the south, they are gradually replaced by clay limestone and mergels and the karstification becomes weaker. Underneath, similar rocks from the Lower Cretaceous are found. There are overlaying sandy and clayish deposits from the Neogene and the Quaternary, represented mainly by loess material. The limestone layers are almost horizontal, slightly inclined to the north.

The cave formation is due mostly to river waters and less to precipitation, which in this region is ca. 580–600 mm/a. In the peneplenized parts between the valleys, permanent surface outflow is almost absent. Below the river level, the limestone is saturated with water, the direction of the movement of underground waters being towards the Danube River. There are also several karstic sources of local importance at the base of the limestone cliffs in the valleys.

The existing physicogeographical and geological situation is the reason for the type of karst – platforms, valleys and covered karst in the interriver massifs. The surface karstic forms are represented by different types of karren in the outcrops along the upper edges of the valleys. Several dolines and valogs covered by Quaternary deposits, have been registered. In the walls of the canyon-like valleys, there are niches, overhangs and other phenomena, due to river waters. The caves in the region are mostly horizontal, their entrances being situated on the slopes of the valleys, at a different relative distance above the river beds. Part of them are difficult to reach. About 50 caves have been registered, the most remarkable being the show cave Orlova Chuka, the second longest in Bulgaria. The rock churches and monasteries near Ivanovo Village are of considerable interest.

When visiting the caves in the area, it is recommended to keep in mind that part of the region is situated within the Russenski Lom Natural Park.

Stara Planina
Chiren Subregion

The Chiren Subregion is part of the so-called Vratsa Region thus named by V. Popov (1977). The reason for its separation is that this is an independent karstic basin, not connected and very different in type of karstification from the mountain Vrachanski Balkan. The subregion has been studied by Spassov (1973), Benderev et al. (1999) and others. It is situated in the lower Predbalkan, between the villages Lilyache, Mramoren and Banitsa. The relief is hilly, dominated by Milin Kamak. The altitude varies between 150 and over 400 m. The average precipitation is ca. 550–600 mm/a.

Geologically, the subregion covers the southern limb of the so-called Mramoren Syncline – a shallow tectonic structure directed E-W and long ca. 15 km with direction of the axis 100–110°. The northern limb is strongly dislocated and almost entirely destroyed. On the surface, alternating Lower Cretaceous carbonate and terrigenous materials appear and the layers dip northwards.

The Urgonian-type limestone of the so-called Cherepish suite is subject to karstification. Underneath, the mergels of the Mramoren suite are situated, the cover consisting of alternating terrigenous-carbonate rocks (limestone, sandstone-limestone, limestone-sandstone, aleurolythes, aleurites and sandstone-mergels) of the Lyutibrod Suite. The limestone complex çaëÿäa with an inclination of ca. 10–15° to the north and is split by a layer of mergels, 60–80 m thick. The lower part of the limestone (Lilyache Zone) appears as a rainbow shaped stripe 100–250 m thick and 1.5 km wide, between the villages Mramoren and Lilyache. The upper part (Banitsa Zone) is outcropping in the slope of the height Kaleto, between the villages Banitsa and Chiren and is 60–80 m thick and with a surface of 2.5 km^2.

Both zones are karstified, the karstification occurring mostly along the following fissure systems: 100–130° and 160–190°. There, karstic underground water is forming. The Lilyache Zone collects the waters from the zone of the limestone outcrops, as well as from the surfaces taken by terrigenous rocks. In the central part of the band, the surface water divide is expressed, separating this zone into two parts: western (of Lilyache) and eastern (of Mramoren), variously karstified. The natural draining of the zone is done via several sources; finally the waters of all sources merging in two rivers – Barata, heading west, and Mramorchitsa, heading east. Major sources are Zhabokrek, from which the water of Ponora cave, the source of Peshketo cave, the source Ezeroto and the sources in Mramoren Village appear. Banitsa Zone is drained by the group of sources Zankinoto in the center of the village Chiren with a capacity of 2.5 to 110 l/s and by the fountain in the west end of the village with a capacity of 5–6 l/s. In the village Banitsa, a subthermal source is situated.

Surface karstification is clearly expressed, especially in the Lilyache Zone. The karren – mostly east of the road Vratsa-Chiren, have large distribution. Two major surfaces of distribution of valogs and dolines are localized – one between the caves Ponora and Mladenovata, the other one between the villages Lilyache and Chiren. Dolines are of different shape and depth. Some of them are opened and give birth to potholes. Some others are filled up (such are most funnels between Lilyache and Chiren). The valley of Barata River, in some places a typical karstic canyon with steep cliffs, waterfalls, cauldrons of erosion and niches, also refers to the surface karst forms. Over this valley, the well-known natural rock bridge Bozhiyat Most, 443 m long, is formed. Towards the cave Ponora a blind valley is formed. In this region, 24 caves and potholes have been recorded. The potholes are few and represent wells, on the bottom of which there are horizontal parts.

The typical ponor caves prevail, some of them being of considerable dimensions. An exception is the cave Peshketo, the entrance of which is an old source and then an underground river is reached, surfacing ca. 50 m away. Another exception is Bozhiyat Most (a natural bridge with an ascending gallery). Generally, the caves can be localized in two zones. The first zone includes the valleys of Ponora and Zhabokrek – Barata. It is determined by the contact of the limestone of Lilyache Zone with the underlying rocks. Here, most of the major water caves are situated, the direction of the galleries parallel to the mentioned contact, while the direction of water movement is from E to W. One of the important caves is Mladenovata Peshtera, with a stream in it. An indicator experience/test it was proven that its waters join the water of the cave Ponora, the longest in the region. The water of Ponora reappears in the source Zhabokrek. From there, part of them enter the cave Tigancheto and reach Bozhiya Most.

Some major caves in the basin of Mramoren

No	Name	Site	Length, m
1	Ponora	Chiren Village	3497
2	Mladenovata Propast	Chiren Village	1723
3	Peshketo	Lilyache Village	630

More numerous, however smaller caves, are situated in the valley Suhata Dolina, parallel to the valleys Ponorska and Barata in the north. The shape of this valley is much more peneplenized. Here, there are many funnels, in some of which the entrances of caves and potholes appear.

The caves in the region are relatively accessible and are situated near some asphalt roads (between the villages Chiren, Lilyache and some others). In the eastern part, through the village of Mramoren, the road Vratsa-Oryahovo passes.

Vratsa Subregion

Within this subregion, most of the mountain Vrachnska Planina is situated – a band ortiented NE-SW. It rises sharply above the plain of Vratsa and the hilly area of Mezdra – bordering it on the northeast. The slopes are abrupt, rocky, about 600–700 m high. The valleys of the rivers Iskar and Botunya border the region on SE and NW, and the rivers Zlatitsa and Cherna – on SW and S. The valley of Leva River subdivides it into two parts: Bazovska (southeastern) and Stresherska (northwestern), with the highest summits respectively Buk (1394 m) and Streshero (1212 m). The valley of Leva River has formed the picturesque gorge Vratsata between Zgorigrad and Vratsa.

In general, plateau-like relief is typical for the subregion, with peneplenized surface, mainly between 1000 and 1200 m a.s.l. and abrupt slopes towards the bordering rivers, Vratsa Plain and Mezdra Lowland. The peneplenized area features a typical karst relief with residual tumuli, separated by poljes, blind dolines and blind valleys. Other surface forms are developed as well – funnels, karren, karren fields.

For the forming of karstic features here, especially important is the thick Upper Jurassic and Lower Cretaceous limestone. Underneath, there are rocks of Middle and Lower Jurassic (aleurolyths and aleuryths), missing in some places. A Trias carbonate complex (karstified limestone and dolomite) follows. In their base, they transit into Lower Trias sandstone and aleurolyths. The geological cross-section is best seen on the surface of the southwestern slopes and near Zgorigrad Village.

The basic tectonic structure here is the Zgorigrad Anticline, almost overlapping the Vratsa Subregion. From this anticlyne the arch and the northern limb are preserved, which determines the situation of the limestone layers. In their peneplenized parts, they are almost horizontal, on the northeastern slopes they are steep and in some places vertical. The pressure south – southwest determines the directions of the basic fissure systems. Contemporary karstification is due mostly to rainfall water, the average rainfall amount being about 800–1000 mm. In the peneplenized parts there is almost no constant surface outflow, as the entire amount of rainfall is taken by the surface karst forms. After its penetration into the massif, the water goes towards its periphery, surfacing as karstic springs. The most important ones are situated at the base of the northeastern slope – the source near Toshova Dupka, Stoyanovo Village (outflow 78–390 l/s), near Matnishki monastery (total outflow 138–430 l/s), near St. Ivan Pusti monastery (ca. 10–15 l/s), near Bistrets Village (127–780 l/s), near Pavolche Village (3–497 l/s), as well as the two sources in Iskâr Valley near Cherepish Railway Station (respectively 33–1010 l/s and 11–139 l/s). On the southern side, the two sources above Zverino Village and the source Chigoril (up to 40 l/s) are situated. In the central parts of the massif, in the valley of Leva River, there are the sources above Zgorigrad Village (total of 20 – 250 l/s) and the source Ludoto Ezero (8–220 l/s). According to the geological conditions the sources are mostly contact ones, surfacing on the contact between the

The karst of Vratsa – photo Valery Peltekov, Trifon Daaliev

The rocks near Lakatnik – photo Valery Peltekov

karstified and the non-karstified rocks. Only the situation of the sources near Cherepish Railway Station and Ludoto Ezero is determined by the base of erosion.

As a whole, in the sub region in 2002, more than 250 caves and potholes were known More numerous, although smaller are the caves in the eastern part. However, in the western part most of the bigger cave potholes are situated. So far, in Vratsa Subregion there are seven cave potholes and proper potholes over 100 m deep (some of them also of considerable length).

Besides the caves, the water caves, situated immediately close to the big karstic springs, are also interesting. Underground rivers, connected with springs, run in them. Such are Toshova Dupka near Stoyanovo Village, 1302 m long, Cherniyat Izvor near Mâtnishki monastery, 546 m long, Gârdyova Dupka near one of Zgorigrad sources (510 m long). Near Ledenika Hut we also find the interesting sink-cave Mizhishnitsa, 855 m long, as well as the pothole Zmeyova Dupka (-52 m, full of beautiful flowstone formations). In this subregion, the show cave Ledenika is situated, as well as the beautiful pothole N 13 (Vârteshkata) below the summit Rogo and others.

Some major caves in Vrachanska Planina

No	Name	Site	Depth, m	Length, m
1	Toshova Dupka	Glavatsi Village	63	1302
2	Cherniya Izvor	Glavatsi Village	+12	753
3	Belyar	Gorno Ozirovo Village	282	2500
4	Barkite 14	Gorno Ozirovo Village	356	2 600
5	Barkite 8	Gorno Ozirovo Village	208	above 1000
6	Mizhishnitsa	Gorno Ozirovo Village	85	885
7	Ledenika	Vratsa	-14+16	226
8	Gârdyuva Dupka	Zgorigrad Village	+25	555
9	Pukoya	Pavolche Village	178	48
10	Yavorets	Zverino Village	147	50
11	Panchovi Gramadi	Zverino Village	104	93
12	Varteshkata No 13	Zverino Village	74	158
13	Haydushkata Propast	Bistrets Village	108	20

When visiting the caves and the potholes in the area, we have to take into account that the subregion falls almost entirely within Vrachanski Balkan Natural Park.

Eastern Berkovska Planina and Ponor Planina

This region overlaps with the karstic region of Ponor, thus named by V. Popov. It is situated in the western part of the Main Balkan Range, between Iskar River and the border between Bulgaria and Serbia. The northern and the southern limits follow geological patterns – the northern limit coincides with the lithological limit between the Trias carbonate rocks and the Lower Trias sandstone, while the southern boundary runs over the Iskrets – Gubesh Dislocation. The region is divided by Ginska River into two. The western part is situated in the southern slopes of Berkovska Planina, while the eastern coincides with the mountain Ponor. The reason to merge the two parts is the undisturbed distribution of Trias limestone and dolomite. Comparatively better studied is the eastern (Ponor) section (Dinev, 1959, Stoitsev & Neykovski, 1975, Benderev, 1989).

The relief is typically mountainous, the altitude varying between 400 m in the gorge of Iskar to 1473 m at the summit Sarbenitsa in the Ponor part and to 1499 at the summit Meshin Kamak and the summit Biloto west of Ginska river. The slopes of the mountain, descending towards Iskar, are steep, in some places vertical. The higher part (900–1300 m) is plateau-like, with rounded hills, on which several smaller plateau rise, often surrounded by karstic cliffs. The mountain is cut by the valley of Gintska river, separating Ponor Planina from the eastern parts of Berkovska Planina. The slopes of the valley are gently falling down to the upper part of the valley and near the village of Gintsi they are steep to abrupt. The valley of Komshtitska River is of a similar type, situated near the state border with Serbia. Both rivers start from the higher parts of the Main Balkan Range, flow south and, entering the carbonate rocks, lose considerable amount of water. Many other rivers disappear completely, entering Trias limestone and dolomite.

The climate of the area is mountainous, with high precipitations. According to Koleva and

Peneva (1990), at the station of Petrohan, situated north of the region at 1400 m, the annual average rainfall is 1164 mm/a, while at the station of Gintsi it is (1050 m) – 846 mm/a. The lowest rainfall of 658 mm/a (altitude ca. 600 m) is observed in the lower parts, where the sources of the Ponor part of the region, near Iskrets Village, are situated.

Geologically, the region is built mainly of Mesozoic rocks, forming the following complexes of different level of karstification:

a) Lower Trias non-karstic complex, formed by sandstone, conglomerates and aleurolythes. This complex serves as foundation of aquitard of the karstic groundwater; Trias karstified complex, built of limestone and dolomite with a total depth of up to ca. 600 m;

b) Lower-Middle Jurassic non-karstic complex, represented mostly by terrigenous suites, with participation of agilities, aleurolythes, sandstone and mergels. In the lower part of the complex, Ozirovska suites is situated, built of limestone and limestone-sandstone, also karstified. This suites has a broader distribution, mostly in the western (Gintsi) part.

c) Higher Jurassic karstified complex ca. 100–150 m thick. Its structure consists mostly of limestone suite, a smaller part of it being covered by alternating limestone and mergels from the Lower Cretaceous.

The spatial setting and the relations between these complexes are determined by the tectonic situation in this part of Stara Planina. The region falls within the southern limb of Berkovska Anticline. The kern of Izdremets-Gubesh Syncline had been subject of intense tectonic impact, which led to the formation of a major dislocation. Along this dislocation, the terrigenous Paleozoic and Lower Trias rocks, building the northern limb of the Svoge Anticline, have been thrust in the western part and in the eastern part they are overthrust on Mesozoic complexes. This dislocation forms the southern limit of the karstic region under study.

The carbonate complex is subject to active karstification through rain and river waters. All rivers, starting from the upper parts of Berkovska Planina (in the western part) and from Koznitsa, form their outflow in Lower Trias sandstone and older rocks, feed the karst waters, mainly in Trias limestone and dolomite, and relatively fast emerge in karstic sources. The connection between the rivers in the Ponor section and the lowest situated source – the Iskrets source – was proven by indicator experiments. The time for emerging is 1–2 days in the period of high water and 7–8 days during the drier season. The situation of the source is structurally defined. Nearby is the front part of Iskrets Nappe, built of terrigenous Ordovician materials. The sources are characterized by their very variable outflow – from 280 to more than 50 000 l/s – the highest outflow of a source, measured in Bulgaria. When higher outflow occurs, the cave Dushnika takes part, appearing as a drain.

Other major sources in Ponor section are situated in the valley of Iskar River. Such are the two sources springing from Vodnata Peshtera cave near Tserovo and the sources under Skaklya waterfall (Bov Railway Station).

Some major sources in the Ponor section are Varo and Buchiloto near Zimevitsa Village and the sources near the villages Mecha Polyana and Dobravitsa. In the Gintsi part, the sources near Gintsi and Komshtitsa are situated.

Within the outcrops of carbonate rocks, the surface karstic forms are widespread. Zheko Radev described them in details as early as 1915. There are many valogs of considerable dimensions, as well as many funnels. The karren are partly covered by soil. Blind valleys, complicated by valogs and dolines, are typical for the surfaces in the northern part. The rivers Gintska, Komshtitska, Zimevishka, and indeed Iskar, have formed steep valleys, typical for karstic areas and sometimes – even canyon like.

In this region many caves, some of which quite big, have been explored. The difference in the altitudes of the higher parts of the mountain, where the cave entrances are, and the karstic

The area of Gintsi – photo Trifon Daaliev

sources, which are situated much lower, is the reason for the considerable denivelation of some of them. In the Trias complex, these are the caves Katsite in the Ponor part and in the part of Gintsi – Golyamata Balabanova, Malkata Balabanova, Granicharskata and others.

Some major caves in Ponor Planina and Eastern Berkovska Planina

No	Name	Site	Depth, m	Length, m
1	Golyamata Balabanova Dupka	Komstitsa Village	80	4800
2	Malkata Balabanova Dupka	Komstitsa Village	125	420
3	Granicharskata Propast	Komstitsa Village	80	224
4	Tizoin	Gubesh Village	320	3599
5	Saguaroto	Gubesh Village	135	2217
6	Krivata Pesht	Gintsi Village	+40	1500
7	Dinevata Pesht	Gintsi Village	10	451

Glozhene Region
Sectors 207 (Panega) and 209 (Vassilyov)

According to V. Popov's subdivision of karst in Bulgaria of, the karstic region of Glozhene includes the neighbouring sectors of Panega and Vassilyov, separated by the river Vit. Taking into account that the basin of the big karstic source Glava Panega includes parts situated on both sides of Vit, we consider reviewing both sectors together correct.

The Glozhene Region is situated in Central Predbalkan, on both sides of the Vit River and its tributaries Cherni Vit and Beli Vit. The region is limited on the north by the band Zlatna Panega-Brestnitsa-Boaza-the valley of Kalnik River up to the town of Yablanitsa. On the west the region almost reaches Yablanitsa town, on the east – the village Shipkovo. The southern limit is difficult to localize, it follows the northern slopes of Tetevenska Planina. The region includes the entire the mountain Vassilyovska Planina, parts of the ridges Debeli Dyal and Lestvitsa, as well as the massif Lisets. The relief is medium mountainous, while in the area between Vit River and Yablanitsa it is low mountainous. The main draining system is formed by Vit River and its tributaries – Gradezhnitsa, Cherni Vit, Beli Vit, as well as by some smaller streams falling into Kalnik River – Lessidrenska, Toplya. The slopes of their valleys are mostly steep and forested. The altitude changes from ca. 180 m near the source Glava Panega up to 1490 m at Vassilyov Summit. The average yearly rainfall changes from 760 mm (Ugarchin Station) to over 940 mm (Teteven Station). The highest rainfall was recorded in May-June and the lowest in September – October and February – March. The precipitation is mostly rainfall. The share of the snowfall increases from north to south, the average period of snow cover being 40–100 days yearly. The average year air temperature is 10 °C.

The geological conditions in Glozhene Region are extremely complex. Rocks, belonging to two carbonate complexes, have been subject to karstification and separated from each other by non-karstifying rocks. The first (upper) complex includes drab grey to pink Upper Jurassic limestone. It is no more than 120–130 m thick and is covered by terrigenic/terrigenous (mostly flishoid) rocks from the Upper Jurassic-Lower Cretaceous. It lies on a non-karstifiable complex of water impermeable argilite and aneurolite of Middle Jurassic sandstone mergels, sandstone, aleurolite and Lower Jurassic clay limestone. The terrigenous complex separates the Upper Jurassic limestone from the lower karstified complex of Trias age. The latter is built of limestone and dolomite and in its upper part also of limestone containing conglomerate and breccias – conglomerates.

Tectonically speaking, the region falls within the Teteven Anticlinorium – part of the Predbalkan structural zone. Two major anticlines are included in it – the Glozhene and the Teteven Anticlines, with the Cherven Anticline in between. They have a linear (East-West) orientation. The axis of the northern anticline (of Glozhene) crosses Glozhene village and ends south of Golyama Zhelyazna. Its nucleus is built of Permian and Trias rocks, the limbs are formed by Jurassic rocks. The southern limb is cut off from the Koynets fault. The axis of the Teteven Anticline passes south of Teteven and north of Vassilyov Summit. The axis of Cherven Syncline is oriented west-east, formed in Lower and Middle Jurassic sediments.

The complex tectonic structure and the fragmented relief are the reason for the fragmentation of the outcrops of the karstified carbonate complexes. For this reason, the whole region can be subdivided into several smaller karstic basins, developing independently and usually isolated from one another. In the Glozhene Anticline, there are three major basins in the Upper Jurassic complex and one in the Trias rocks. The most important among the Jurassic complexes is the westernmost – the Panega complex. Here, the karstification is due to rainfall, as well as to the draining water of Vit River. The entire amount emerges through the source Glava Panega, one of the biggest in the country (outflow from 0.5 to 35.5 m^3/s). The connection between the river and the sources was proven as early as 1959 by I. Kovachev, who used nine tons of salt. The distance of 6.5 km was overcome in 12 days. The next basin to the east is the one south of Bâlgarski Izvor Village. Important sources in it are found in Bâlgarski Izvor (outflow 26–320 l/sec) and Galata (10–15 l/s). The easternmost karstic basin is a band west of Lesidren Village, with small sources. The karstic basin, formed by Trias carbonate rocks, is a band in the west part of the nucleus of Glozhene Anticline, crossing Vit River near Glozhene. Part of this basin is crossed also by Gradezhnitsa River. Some important sources in this area are the two near the school of Gradezhnitsa Village (total outflow ca. 40 l/s) and the source of the cave Rushovata Peshtera (min. outflow 15–20 l/s). In the valley of Vit River, from this basin two sources emerge – one with a changing outflow, sometimes more than 200 l/s, the other with a more permanent outflow of ca. 20/25 l/s.

In the Teteven Anticline, the Upper Jurassic rocks form narrow bands along the western and eastern ends of the anticline. They are drained by small sources – the bigger being the ones near Koman Hut (outflow ca. 30 l/s). The Trias carbonate rocks form one interrupted ring-shaped basin. The more important karstic sources there are the one near Lesidrenska River, on the contact between limestone and dolomite (outflow up to 50 l/s), and the sources Toplya, draining the lower part of the basin. The outflow of these sources varies in a wide range – from 17 to more than 1000 l/s. There are also some other small sources in the area (outflow of up to 10 l/s).

The substantial rainfall and the geological, hydrogeological and relief features, especially of Teteven Balkan, are the reason for intensive

The source Glava Panega – photo Vassil Balevski

karstification in the area. Morphologically, the karst is best expressed in the western parts of the region (especially in the basin of Glava Panega). There, the surface karst formations are represented by karrens, dolines and valogs in the higher parts and by steep karstic valleys with sinks in the river valleys, crossing the limestone and the dolomites. The most typical sinks are found in the valley of Vit River near Assen area, along the valley of Suha River, Galatska River, etc. So far, in this area, 190 caves and potholes have been recorded, some of them of a relatively important size. One of them is a show cave (Sâeva Dupka). Other caves are of historical (the cave of Benkovski) or archaeological interest (the caves Toplya and Morovitsa). In this karstic region, two of the most interesting, not fully explored, siphons in Bulgaria are also situated – the one of Glava Panega and the source of Toplya. Six of the caves have been declared protected: Morovitsa, Rushovata Peshtera, Toplya, Lyastovitsa, the Cave of Benkovski and Sâeva Dupka.

Some major caves in Glozhene Region

N	Name	Site	Depth, m	Length, m
1	Morovitsa	Glozhene Village	105	3200
2	Maliya Sovat	Bryazovo Village	140	125
3	Nanovitsa	Yablanitsa	102	400
4	Bezdanniya Pchelin	Yablanitsa	105	20
5	Yasenski Oblik	Yablanitsa	108	10
6	Lyastovitsa	Glozhene Village	-	320
7	Partizanskata	Glozhene Village	107	711
8	Planinets	Glozhene Village	71	20
9	Rushovata Peshtera	Gradezhnitsa Village	+15	908
10	Djebin Trap	Glogovo Village	88	42
11	Kozya	Gradeshnitsa Village	50	
12	Nikolova Yama	Gradezhnitsa Village	70	
13	Toplya	Golyama Jelyazna Village	- 5 +10?	462
14	Stotaka	Bryazovo Village	104	347
15	Saeva Dupka	Bresnitsa Village	22	205

The caves of Glozhene karstic basin are often visited by cavers, due to the well-developed road network, especially in the western part of the area.

Region 211 (Gorni Vit – Cherni Ossam)

According to the subdivision of karst in Bulgaria (Popov), this region is an almost uninterrupted band of Mesozoic limestone along the northern slope of Stara Planina, immediately close to its ridge. It starts from Malak Iskar River and goes almost to Botev Peak. In both its ends the band widens, and there, sectors with interesting caves are formed. They could be taken also as two independent subregions – of Troyan (eastern) and of Etropole (western).

Troyan Subregion

It is situated in the mountain Troyanska Planina, in the upper part of the river Cherni Ossam. To the west, Beklemeto is accepted as a conditional boundary, to the east – the foothills of the peaks Levski and Golyam Kupen. The southern boundary runs below the main ridge of Stara Planina, while on the north the boundary reaches Neshkovtsi area. The relief is defined by the position of the subregion on the northern slopes of Stara Planina. They are inclined to the north and are fragmented by Cherni Ossam River and its tributaries. The valleys are deep, with steep slopes, almost canyon-like. The altitude changes from ca. 750 m in the valley of Cherni Ossam, above Neshkovtsi area, to ca. 1600–1700 m below the main ridge of Stara Planina.

The rocks, in which most caves have been formed, are Trias limestone and dolomite. They are ca. 300–330 m thick. In some places there are terrigenous rocks. In the eastern part, also Upper Trias red limestone and dolomite can be seen, alternating with mergels and conglomerate. To the north and in some places in the central parts the Triasrocks are covered with sandstone, sandstone-limestone, alevrolithes and argilites from Lower and Middle Jurassic, about 50–60 m thick. There follows a series of Middle and Upper Jurassic limestone, ca. 20 m thick. Along their contact with the underlying terrigenous rocks, the deepest cave in Bulgaria – Raychova Dupka has been formed. More to the north, there are flish rocks from Upper Jurassic – Lower Cretaceous. To the south, the Trias carbonate rocks come into contact with sandstone from Lower Trias and Paleosoic non-karstifiable rocks. In some places, Upper Cretaceous limestone lies over the dolomite. In the southwestern parts of the territory under consideration, there are gneiss and granitogneiss formations, transported from the south, superimposed over Trias rocks.

The Troyan Subregion is situated on the northern limb of the Cherni Ossam Anticline. The core of the fold is formed by Trias carbonate rocks, while the northern limb contains Upper Trias and the Jurassic sediments.

The altitude and the relief are the reason for the relatively high average rainfall (ca. 1000–1400 mm). Part of the rainfall form the temporary and the permanent surface outflow of Cherni Ossam and its tributaries, the remaining part (in the area of the carbonate rocks) is draining underground. Underground is draining also part of the water of the rivers, starting outside the karstifiable rocks. Water is sinking in Chaushov Dol, Borov Dol, but most of all in the locality Kazanite, from where downstream Kumanitsa River is dry in summer. Latter this water is crossing the cave Kumanitsa and reappearing in the source Kumanitsa (outflow from 300 to 1300 l/s). There is surfacing the main quantity of underground karstic water. The karst is situated in the crossing point of the valley of Cherni Ossam and Haydushki Thrust, forming a barrier for the underground water.

The big amount of rainfall, the geological and relief peculiarities are the cause of the intensive karstification of Troyan Balkan. The surface karst formations are represented by karrens, dolines and valogs in the higher parts and by steep karstic valleys with sinks in the river beds in limestone and dolomite. So far many caves and pot holes have been recorded there, some of them of considerable size (on Bulgarian scale). To be noted is also the unique natural bridge Krali Markova Dupka.

As a whole, the region is difficult to visit and is rarely visited by tourists and cavers. This is due to the rough landscape, far away from settlements. A good starting point is Neshkovtsi area, ca. 9 km from Cherni Ossam Village. The nearest tourist huts are outside the karstified rocks and are not suiteable as a starting point for the major caves.

Some major cavities in Troyan Subregion

N	Name	Denivelation, m	Length, m
1	Raychova Dupka	- 377 +10	3333
2	Malkata Yama	- 232	700
3	Borova Dupka	- 156	40
4	Pticha Dupka	- 108	652
5	Kumanitsa	- 104	1656
6	Golyamata Gârlovina	- 100	44
7	Vârlata	- 41	1110
8	Golyamata Yama	- 43 +18	459
9	Opushenata	- 32	232
10	Chernokozhevi Dupki	-24	213
11	Paisievi Dupki	-22	150

The whole karst subregion is situated on the territory of Steneto Biosphere Reserve, within the boundaries of Central Balkan National Park. A

The natural phenomenon Krali Markova Dupka – Steneto Nature Reserve, Troyan Balkan – photo Vassil Balevski

Steneto – the cañon of the river Kumanitsa – photo Vassil Balevski

special permit is needed to stay and work in the park. Besides, part of the valley of Cherni Ossam River falls within the guarded sanitary zone of the source Kumanitsa, used for piped water for several major settlements.

Karlukovo Gorge

After leaving the picturesque site Ritlite near Lyutibrod, Iskar River crosses the Syncline of Mezdra and follows its way to the north. Near Kunino, steep rock complexes appear again, a mighty cliff starting from Kunino and going as far as the site Provartenika north of Karlukovo. The Gorge itself is ca. 6 km long and in some places up to 3-4 km wide. If we add the karstified four kilometers band on the left bank of Iskar, the total surface of the karstic region will be ca 76 km^2. The region contains some of the most important karst formations in Northwestern Bulgaria. The average altitude of the Gorge with its cliffs is 350 m, the hypsometric level of the river Iskar being no more than 190 m.

Geologically, the Gorge of Karlukovo belongs to the northernmost parts of the Predbalkan and is built exclusively of Upper Cretaceous (Campan – Maastricht) limestone. This limestone has been a subject to tectonic tension for a long time and, due to the fact that it has long been on the sea level, it is strongly karstified. The karstification has been practically uninterrupted since the end of the Late Cretaceous to our time, as the Gorge has not been submerged neither by sea, nor by freshwater basins. These circumstances have determined the variety and the size of the karst features in the area.

Around Kunino, before the incision of river Iskar into Upper Cretaceous limestone, the river banks are built of Lower Aptian (Bedul) sandstone, underlying the rock complexes from Upper Cretaceous. The same sandstone could be observed ca. 5–6 km after the railway tunnel at Karlukovo Railway Station, where it constitutes the cliffs on the right bank of the river.

As we already indicated, the Gorge falls within the northern parts of the Predbalkan. South of it the northern limb of Lukovit Syncline is situated. Only the Maastricht limestone has been karstified, everywhere Campan clay limestone being the base. As the erosion cross section of Iskar River is not deep enough around Karlukovo Railway Station, the Campan limestone remains in the depth. It surfaces near Kunino Railway Station, south of lime furnaces, immediately above the sandstone of the Lower Cretaceous. The preserved depth of the karstified Maastricht limestone – sandstone is about 120–130 m, increasing to the northeast, towards Lukovit Town, where it participates again in the structures of Lukovit Syncline. Due to the heterogeneity of the Maastricht limestone, its karstification over time and space has been running unevenly, which is reflected in the dislocation of underground karst formations.

The Maastricht sediments form the mantle of several structures (anticlines and synclines). Among the negative structures, the most extensive is Karlukovo Syncline, being the east prolongation of Kameno Pole syncline, west of Iskar River. Actually, both synclines form a common structure. Karlukovo Syncline is widely spread and its axis is difficult to follow. Near Karlukovo Village it points to 110°. Southwest of the village near the hill Golyama Mogila, the Maastricht limestone of the south limb of the syncline sinks NNE (22°) and overlays the Lower Aptian sandstone of the Belene Anticline.

Rock cliffs above Iskar River near Karlukovo Village – photo Trifon Daaliev

To the east, Karlukovo Syncline widens and passes into Lukovit Syncline. The karstic water of this syncline, because of the general eastward sinking towards Vit Tectonic Lowering, flows smoothly to the east. That is why, in the eastern periphery of the syncline, the biggest Maastricht karst springs – Ezeroto and Temnata Dupka have been formed on the left bank of Panega River near Lukovit and Byalata Voda near Rumyantsevo Village.

The spring Ezeroto is situated immediately under Lukovit Power Station. Its outflow varies between 10 and 25 l/s, the water temperature – between 13.8 and 14 °C. Via a water-conduit for the water goes to Lukovit. The spring Temnata Dupka comes out of a small cave on the left bank of Panega (the locality Chervenata Stena, south of Lukovit Power Station). The outflow is 15.5 l/s, the water temperature is 13.3 °C. During river high waters the spring is flooded. The spring Byalata Voda is situated on the left bank of Panega near Rumyantsevo (outflow 25.6 l/s, temperature 13.2 °C (23.06.1967). The spring is connected with Skravenishki Valog, west of Rumyantsevo. In the west end of the valog runs a small source coming from a ponor cave in the east end. The outflow of Skravenishki Izvor is 3.4 l/s, the water temperature is 12.8 °C (30.06.1967).

On the right bank of Iskar Valley (the local base of erosion for part of the karst waters) several smaller karstic sources are situated. They are: the source Izvora with a permanent outflow of 4 l/s and temperature 15 °C, which is flooded during the high waters of the river; the source Patitsite, flowing underwater in the Iskar River, between the valley of Postoyna and Zadanen Dol; the source Averkovitsa, situated immediately under the cave of the same name, dry in the summer and quite big in the spring. There are also two other sources under the cliff ca. 1.8 km from the old Karlukovo Railway Station.

The surface karst in Karlukovo Syncline is represented by karren, karren fields, valogs, dolines and periodical ponors. Karren are most typical for the plain of denudation between 320–360 m altitude. They are relatively well preserved between Lukovit and the villages Petrevene and Karlukovo. The most widespread among surface karts forms are the funnels and among the underground forms – the potholes. Only in Karlukovo area they are more than 150. On the plain of denudation their number is insignificant. East of the hill Gerilitsa, in the place called Vlaykovi Livadi, there are eight funnels and in the place Skravenik there are another six, oriented NW-SE.

Around Karlukovo Village, the funnels and the potholes are formed at a lower level of denudation of 200-260 m. It is gulf-shaped and covers the areas of Gerdjikov Vrâh, Valchanov Gradezh and Koritska Rodina. Most funnels are found in the area of Valchanov Gradezh, where on 0.5 km^2 16 funnels (32 funnels on km^2) are situated.

East of the Iskar River, in the area of Karlukovo Village, several caves have been declared Natural Monuments: the caves Prohodna (with 1.5 ha around it), Temnata Dupka (area of 1.5 ha), Bankovitsa (area of 3.0 ha) and Svirchovitsa (with 2.0 ha). West of the Gorge is the cave Haydushkata Dupka (also a Natural Monument with 3.0 ha around). The caves Samuilitsa I and II and Galabarnika near Kunino also have the status of protected sites.

Some major cavities in Karlukovo area

Name	Length, m	Depth, m
Zadanenka	1150	26
Golyamata Voda	612	105
Bankovitsa	689	51
Stubleshka Yama	565	72
Svirchovitsa	231	39
Propadnaloto	205	40.5
Ogradite (S 20)		97
Dâlbokata		72

According to the subdivision of karst and caves in Bulgaria (Popov, 1976), the karstic area of Karlukovo falls entirely within Region 206 (Kameno Pole – Karlukovo).

Devetaki Plateau

According to the system of regions outlined by V. Popov, the plateau of Devetaki is situated between the rivers Ossam and Rossitsa and overlaps the Devetaki Region. Its altitude is from 380 to 470 m, and the base is 100–200 m. The relief is typically plateau-like, with a vast distribution of surface karstic forms (Petrov, 1933; Popov et al., 1965).

The slopes of the plateau are of different inclination. The western parts (towards Ossam River) are steeper. They are with valleys, often dry and with steep slopes. Climatically, the region falls within an area, characterized by a moderate continental climate. According to the Institute of Geography (Bulgarian Academy of Sciences), the average annual precipitation depends on the orographic conditions and oscillates between 600 and 700 mm. The average annual temperature is ca. 10 °C in the plateau-like part, while at the base of the plateau it is 2–3 degrees higher. Geologically, the Devetaki Plateau almost coincides with the area of the Devetaki suitee, built of organogenic limestone. The depth of Devetaki Suitee is ca. 200 m and is of a monocline bedding (inclination of the layers of 5 to 20 degrees to the north). The southern limit of the suitee is a lythologic one with the Smochanska Suitee, the northern and the western limits are tectonic and are controlled by the Krushuna and Ossam Faults. Hydrogeologically, the strongly karstified limestone of Devetaki Suitee appears as a principal collector of underground waters. Smochanska Suitee is a lower aquitard. The feeding of the waters is realized via precipitation in the peneplenized parts of the plateau (ca. 33.5% of them). Because of the many karst forms, it goes rapidly. Draining occurs through sources, situated at the base of the plateau, the majority of them being in the valley of Ossam River, at a lesser altitude. The most important ones are the Krushuna sources, the source of the cave Devetashkata Peshtera, the sources in Krushuna, Beliya (the White) and Cherniya (the Black) sources northwest of Devetaki, Sutna Bunar near Alexandrovo Village and others.

Surface karst forms are widespread and are described in details by Popov, Penchev & Zyapkov (1965). Because of the soil-covered rocks, the karren and karren fields are not widely distributed. There are 670 dolines and valogs recorded, some of them with colmatated bottoms, and in them small karstic lakes and swamps have formed. In the periphery of the region there are dry steep valleys without permanent outflow. Interesting formations are the travertine terrace and the waterfall near Krushuna Village.

In the Devetaki Plateau some of the biggest water caves in Bulgaria have been discovered. Their entrances are situated mostly in the periphery of the plateau and are of considerable volume and length. They are penetrated in boats.

Some major cavities in the Devetaki Plateau

No	Name	Site	Length, m
1	Popskata Peshtera	Krushuna Village	4530
2	Devetashka Peshtera	Devetaki Village	2442
3	Vodopada	Krushuna Village	1995
4	Urushka Maara	Krushuna Village	1600
5	Gornik	Krushuna Village	1074
6	Chernata Pesht	Gorsko Slivovo Village	741
7	Futyovskata Peshtera	Karpachevo Village	700
8	Brâshlyanskata Peshtera	Alexandrovo Village	608
9	Mandrata	Chavdartsi Village	530
10	Kânchova Varpina	Tepava Village	463

From the caves enumerated above, Devetashkata Peshtera is of particular interest, not only because of considerable underground chambers and interesting lakes, but also because of the archaeological and other discoveries, which have taken place in it. Part of the cultural layer was destroyed – despite being proclaimed a Protected Natural Monument, for a long period the cave has been used as a depository for fuel,. Another interesting cave in the region is Stalbitsa, near Kârpatshevo Village, a large chamber with a collapsed roof, from where it is possible to enter the cave using a metal ladder (made by local people to facilitate the access).

In the higher peneplenized part of the plateau, some of the funnels give access to potholes, the deepest being Kânchova Varpina, ca. 100 m deep.

Brashlyanski Dol – one of the valleys of Devetaki Plateau – photo Trifon Daaliev

The Devetaki Plateau is easily accessible. Along the northern boundary of the plateau the railway and the tarred road run, connecting Lovech and Levski. There are other asphalt roads to the villages Tepava, Devetaki, Brestovo, Krushuna, Kârpachevo and Gorsko Slivovo.

Kotlenska Planina Mt

Kotlenska Planina is situated in Eastern Stara Planina, from Stara Reka river to the pass Varbishki Prohod. On the North, its limits are the right hand tributaries of the rivers Stara Reka and Ticha and on the South –Luda Kamchiya Valley. It overlaps the Kotel Karstic Region, the region's name given by V. Popov (1976). This is one of the favoured karstic areas in Bulgaria, with many deep potholes. Many expeditions and explorations took place in this area, some of which were published by Neykovski et al. (1972). Paskalev (2001) published the geological setting, which allows for karstification of the region.

Kotlenska Planina is characterized by a mountainous relief and has a relatively peneplenized upper part with summits of 900–1000 m. The northern slopes are relatively steeper and more rocky than the southern. North of Kotel, the ridge lowers to 550–600 m (Kotel Pass). West of it, the highest summit of Kotlenska Planina – Razboina (1128 m) is situated. East of the pass, the highest point is Karaburun (1053 m). The river valleys are usually at 400–500 m.

The climate is moderate continental to mountainous, with an average annual precipitation of 810 mm/a.

The Balkan of Kotel in the area of Kipilovo – Camp of rescuers 2000 – photo Trifon Daaliev

Geologically, Kotlenska Planina falls within a region of extremely complicated tectonic setting. Here, the boundary between the tectonic zones of the Predbalkan and the Eastern Balkan runs, called Dislocation Chudnite Skali (Yovchev Ed., 1971). The karstified rocks – Maastricht (Upper Cretaceous) limestone are 250–280 m thick (Paskalev, 2001) and they are situated between the two tectonic structures. These rocks participate in the building of east-west syncline structures, complicated by smaller dislocations. The limestone outcrops form two non-interconnected sectors: western – between Kipilovo and Kotel and eastern – Zlosten. In the central part of the sector, between Kipilovo and Kotel, some of the carbonate rocks are covered by Paleocene limestone aleurolytes. Besides the two mentioned sectors, karstified carbonate rocks have been established also south of the front part of Kotel Nappe – in the areas east of Kotel, near Medven, etc.

Hydrogeologically, both main sectors are characterized as independent karstic basins. The alimentation of the underground waters is done mainly by rainfall and in the western sector also by Suhoyka, which runs across the karstic basin. The draining is done through karstic springs, situated in the periphery of carbonate rocks outcrops. In the sector Kipilovo – Kotel, the most important one is the Kotel Spring, situated in the easternmost part of the basin. The waters run out of a cave, the fluctuation of its debit varying widely – from 40 l/s to over 20 000 l/s. Besides this spring, the sector is drained also through other smaller springs, situated along the northern periphery of the region and of much smaller debit. Relatively more important are the springs, situated in the northeastern part – above Kipilovi Village – Studeniya Kaynak (51–126 l/s), Saygonitsa (6–43 l/s) (Antonov, Danchev, 1980) and others. In the eastern sector – Zlosten, the springs are also situated along the periphery of the karstified rocks, but are of smaller outflow. Usually, they are grouped together. One of these groups is situated at the base of Yurushkite Skali, north of Kotel Town, with a total outflow of 3 to over 20 l/s. The sources Suchakta, situated in Kayadere, at the base of Zlosten, have a similar outflow. About 2 km northeast, in the locality Shitlik, there is another group of several sources with an outflow of 1 to 10 l/s.

Both sectors are considerably karstified. A diversity of surface karstic formations exists. Because considerable parts of Kotlenska Planina are forested, karren fields exist only in places without forests (in the northern part of Zlosten). In some of the negative formations (funnels, dolines), we find the entrances of pothole caves, typical for Kotel Karstic Region. Along the periphery of limestone outcrops, cliffs and steep dry valleys are formed.

In the sector of Kotel – Kipilovo, one of the three deepest potholes in Bulgaria – Golyamata Yama, near Kipilovo, is situated. Some other well-known potholes in the region are Karvavata Lokva, Lucifer, Golyamata Humba. The biggest horizontal cave in this sector is Prikazna, about 6 km long. Cave divers explore the karstic spring near Kotel Town.

In the Zlosten sector, also pothole caves have been discovered – Maglivata, Akademik, Uzhasat na Imanyarite, Lednika and others. They are less horizontal caves and are not very long. The cave of Rakovski is of historical interest. There are many legends about hidden treasures, causing considerable activity of gold-seekers.

Some major cavities in Kotel Region

No	Name	Site	Denivelation, m	Length, m
1	Golyamata Yama	Kipilovo Village	- 350	740
2	Lednika	Kotel – Zlosten	- 242	1367
3	Mâglivata	Kotel – Zlosten	- 220	1244
4	Uzhasat na Imanyarite	Kotel – Zlosten	- 160	440
5	Akademik	Kotel – Zlosten	- 158	580
6	Kârvavata Lokva	Kotel – Zlosten	- 140	646
7	Lutsifer	Kotel – Zlosten	- 130	3200
8	Golyamata Humba	Kotel – Zlosten	- 96	
9	Subatta	Kotel – Zlosten	- 55	518
10	Prikazna	Kotel – Zelenich	- 37	4782
11	Karstoviya Izvor	Kotel – Zelenich	- 32	410
12	Ralena	Kotel – Zlosten	- 15	450
13	Kipilovskata Peshtera	Kipilovo Village	- 6	182
14	Golyamata Medvenska	Medven Village	- 16	314

Transitional Area
Bosnek Region

The karstic region of Bosnek is situated in the upper part of Struma valley, above Studena Dam, in Golo Bardo mountain and in the southern slopes of Vitosha. It was named after the village Bosnek, in the area of which most of the major caves and sources were found. It is a mountainous region with altitudes going from ca. 800 m to higher than 1400 m (the summits Petrus in Golo Bârdo and Mecha Mogila in Vitosha). The region has been shaped by the incision of Struma and of the nearby valleys.

Geologically, Bosnek Karstic Region consists of Triassic and partly Jurassic rocks, thrusted from the south above the pluton of Vitosha (Zagorchev et al., 1994). The karst and karst waters were formed in Trias carbonate. As a fundament, we see red Lower Trias conglomerate, together with alternating sandstone and aleurolites. Above them, a thick complex of limestone and dolomite gradually follows. Within them, there are layers of argylites and aleurolites, widespread east of Bosnek, around Struma valley. They are about 50 m thick and serve as a local aquitard, dividing the carbonate complex in this sector into two. In the northern end, these terrigenous rocks are missing.

The basic structural unit in the region is the Pernik Fault Zone. It is oriented at 120–140° and is 7 km wide. It is represented by a series of parallel faults, with horsts and grabens inbetween. In one of the horsts, Lower Trias surface terrigenous rocks divide Bosnek Karstic Region into two hydrogeologically independent parts. The boundary runs south of Bosnek Village. Most caves are situated in the northern part.

Surface karstic forms are poorly represented. So far, in Bosnek Karstic Region, 26 caves have been found, almost all of them formed in the slopes of the river valley and constituting old or actual ways of the river water.

The geological circumstances define the presence of two aquifers, each of them with a main karstic aquiferous system. One of the systems was formed in carbonate rocks, lythologically situated under the local water resistent, the other one – above it. Each system starts with the intrusion of the river in the rocks, ends in a source and is named after the most important cave – Vreloto or Duhlata respectively.

Both systems were formed essentially by the water of Struma and are presently fed by it, but rainfall on the limestone is also of some importance. The influence of rainfall is lesser on the Duhlata system because of the steep slopes and the smaller denuded area. In both systems, parts of the underground water way flow into caves. For the Vreloto system, in the feeding zone, such are the small stream in the cave Chuchulyan – a pothole, the bottom of which is situated ca. 40 m under the Struma river bed, and the river in the cave PPD, and in the draining zone – the cave Vreloto. The indicator experiments proved the connections between the water in PPD and Chuchulyan and the source. The time of arrival of the indicator is ca. 2-3 days. Also the maximum of the outflow after raining is delayed by several days. In Duhlata system, the underground water ways flow into Duhlata cave. After the end siphon of the cave, there is more infusing water, partly from Akademik, a horizontal cave 350 m long, which stream is also fed by the Struma river. Another interesting cave (without an underground stream), belonging to Duhlata system, is Pepelyankata – a 330 m-cave with beautiful (but unfortunately very damaged) flowstone formations.

Besides the main sources around Bosnek, there are also other sources. One of them is in the village of Bosnek and its outflow is rather constant. About 2.5 km north of the village is the pulsating source Zhivata Voda with an outflow rarely higher than 1 dm^3/s. The source Popov Izvor drains the slightly more karstified southern part of Bosnek Region.

Rila-Rhodopean Area
Northern Pirin

Northern Pirin is the major part of the so-called Vihren-Sinanitsa Karst Region (Popov, 1976, Lichkov, Delchev, 2002). It is situated along the main ridge of the mountain and north of it, between the huts Vihren and Yavorov. The relief is typically mountainous, of alpine type, forming a central ridge with high summits, narrow ridges, deep valleys, above which summits over 2500 m high are situated. The first summit from east to west is Vihren (2914 m), followed by Kutelo I (2908 m) and Kutelo II (2807 m), Banski Suhodol (2284 m), Bayuvi Dupki (2820 m) Kamenititsa (2726 m), Razlozhki Suhodol (2688 m). For the relief formation, the glaciations in Riss and Wurm were also important. During these glaciations the typical trog valleys and circuses were shaped. Because of the absence of water, they have got the local name of Suhodol (Dry valley). From SE to NW they are called Kazanite, Kutelo, Banski Suhodol, Bayuvi Dupki, Kamenititsa and Razlozhki Suhodol. In the upper parts of all of them relatively levelled areas, called "rigels" are formed, alternating with steeper sections. The valleys, starting from these circuses, descend towards Razlog Kettle, having an altitude of ca. 850–900 m.

The climate in the higher parts of Pirin is typically mountainous, while in the foothills (Razlog) it is of submediterranean influence. Precipitation, snow cover and the temperature have considerable influence on the formation of caves. Because of the considerable difference in the altitude, their values change very fast, the precipitation changing from 657 mm/a in Bansko to ca.1200 and more mm/a in the highest parts of the mountain. The considerable snowfall in the higher parts of the mountain and the long period of negative temperature contribute to the accumulation of a thick snow cover and to its slow melting in the spring. The perpetual snow formed by it remain up to early fall and serve as a source of very aggressive water, which also contributes to the karstification.

Along the main ridge of Pirin the water divide between the rivers Struma and Mesta goes. The karstic part is situated mostly in the water-collecting basin of the Mesta River and in it there are almost no surface waters. Almost the entire amount of rainwater, falling in this region, is immediately lost, re-appearing afterwards in

Panorama – Pirin Mountain – photo Valery Peltekov

the base area of the mountain as sources. Some rivers like Banderitsa and Bela Reka, starting outside this region, lose part of their water as soon as they reach the marble area. In the summer and autumn, smaller streams and rivers form out of the perpetual snow and are soon lost in some crevice or karstic form.

The karstic sector of Northern Pirin is build of Proterozoan age marble, massif or in bands, with muscovite-biotite and biotite gneiss, shists and amphibolites. In the contact areas with granites recrystalization and skarn formation are observed, which is the result of contact-metasomatic processes. To the west, the boundary of the suitee can be followed between the summit Okaden and Yavorov Hut, it runs east of Butin Summit, then south of the main ridge of the mountain to the south to Vihren Peak, then turns NE, crossing the valleys of the rivers Banderitsa and Damyanitsa and goes further towards Bansko Town. The northern boundary is a tectonic one and runs at the base of the steeper part of the mountain slopes. In this part also relatively recent and karstified marble breccias are widespread. These boundaries determine the surface taken by the karst. The thickness of the suitee reaches 1000 m. The marble are bedding steeply to the northeast. Its situation is controlled by the tectonic frame. The main structure here is the Pirin Horst-Anticlynorium, the karstified rock having taken its northern part. It is separated from the Razlog Graben through Predel Zone of faulting. The marble is additionally broken up and fissured.

Considering the underground water, here a karstic basin of monoclyne type is formed.

The underground water is formed in the marble of Dobrostan Suitee, which is broken up in blocks and is considerably karstified. The gneiss of the Lukovishka Suitee serve as водоупор (water-retention). To the south and south-east, the marble is in contact with Upper Cretaceous and Paleogenous intrusive rocks with fissure water in them. Northwards, the marble is covered with proluvial deposits in the outskirts of Pirin Mt. Underneath in the area of the sources, draining the basin, is the tectonic contact of the marble with the relatively water-resistant Pliocene material, filling the Razlog graben. The feeding of karstic underground water is done through rainfall. The amount of precipitation here is rather high, due to the high mountainous type of relief. Besides rainwater, karstic underground water is fed also through the river outflow, formed in the area of non-carbonate rocks. This is done in the peripheral sectors of the basin. In the western part, these are mainly the waters of Bela Reka river, in the eastern part – the water of Banderitsa River. Unconfined karstic flow, formed in the basin, moves from south to north – northwest, towards Razlog Kettle, surfacing along several parallel faults or through the proluvium. The impregnable barrier of Pliocene material predestine barraging of the karstic ground waters and presence of a saturated (phreatic) zone in the basin. The biggest karstic sources, draining the basin, are: Yazo with an outflow of 440 to 2750 l/s, Kyoshka (15–2770 l/s) and Kalugeritsa (43–386 l/s). The maximum of the water quantities is at the beginning of the summer period (June – July), when intensive melting of the snow cover in the higher part of the mountain takes place, as well as in the rainy period of the year. The minimum of the water quantities is in February – April, most often in April, when considerable part of the catchment area of the source is frozen and the snow cover is not melted and the feeding of the underground waters has practically stopped.

The karstification of marble is considerable. It is defined by the high degree of tectonic and weathering fissurization. In the circuses, series of funnels are formed, usually distributed in lines along faults. Most of them are filled with rock blocks and angular gravel, while in some others there are fillings of firned snow and ice – the so called snow ploughs. Only few of them are opened and are followed by potholes. The tectonic crevices are widespread and karstified in different degree. The major negative surface karst forms are Golemiyat (the Big) and Malkiyat (the Small) Kazan under the summit Vihren. Because of the intense weathering processes typical karrens are almost missing. Along the edges of karstified fissures and potholes there are rillenkarrens. The high degree of karstification, especially in the peneplenized sectors above the rigels, contributes to the outflow of the rainwater in the depth, as well as the water of perpetual snows, melting in spring and summer. There are almost reduced to streams running on the surface for more than 100 m. Even the water of Banderitsa River disappears when the river enters the marble part.

For Northern Pirin, the potholes are typical, due to the considerable difference between the zone of sinking (usually at an altitude of 2000 to 2400 m and higher) and the sources (at ca. 900 m), horizontally, the potholes being rather close to each other – up to 5–6 km. This is also the reason for hoping to discover in Pirin the deepest potholes in Bulgaria – over 1000 m. Until the end of 2003, in Northern Pirin more than 120 potholes and caves have been discovered.

Some important potholes and caves in Northern Pirin

No	Name	Site	Depth, m	Length, m
1	System 9–11	Banski Suhodol	-225	311
2	Vihren	Vihren	-170	396
3	Bânderitsa	Bânderitsa	-125	243
4	K-18 Ledenata	Circus Kamenititsa	-126	
5	20 Godini Akademik	Circus Banski Suhodol	-118	80
6	K-19	Circus Kamenititsa	-136	
7	Chelyustnitsa	Circus Bayuvi Dupki	-104	
8	No14	Circus Kamenititsa	-103	
9	Spropadnaloto	Razlog	-5	605

For most potholes, the presence of moving blockage and squeezes is typical, making it difficult to penetrate deeper. In some potholes (Vihrenskata Propast, Propast No9 – Banski Suhodol) there is permanent ice. One typical water pothole, containing waterfalls and ending in a siphon, is Banderitsa.

The biggest horizontal cave in the area – Spropadnaloto – is situated in the outskirts of the mountain, above the sources. In this cave part of the water, appearing in the resurgences, was followed. There are macadam roads to Vihren Hut and Yavorov Hut, situated along the periphery of the karstic region. An official authorization is needed to camp in the karstic region, as it is within the boundaries of Pirin National Park and the circus Bayuvi Dupki and the adjacent areas are within a Biosphere Reserve.

Dobrostan Karstic Massif

The Dobrostan Karstic Massif is situated in the northernmost parts of the Rhodopes, bordering the Upper Thracian Lowland. It is a karstic region, outlined by Popov (1976). It is separated from the Prespa part of the Rhodopes by the locality Inkaya (Yanev & Popov, 1980). On the west it is limited by Chepelarska River and by the lower parts of Yugovska River, on the south – by the steep valley of Mostovska Sushitsa River, on the east – by Topolova River and the pass Topolovski, and to the north its ridges descend to the Upper Thracian Lowland. The southern part of the massif rises, plateau-like, with an average altitude of ca. 1300 m. Here, the highest summits in the region are found – Stariya Kladenets (1517 m) and Chervenata Stena (1504 m). In the south, west and north, steep slopes are formed, complicated by valleys, descending towards the rivers and the Thracian Lowland. The climate in the higher part is mountainous, with sharp daily fluctuations of temperature and an average amount of precipitation above 800 mm/a.

The Dobrostan Massif is built almost entirely by Precambrian metamorphic rocks from the so-called Rhodopean Supergroup. The massive and layered marbles of Dobrostan suite, which is of the biggest spreading, are subject to karstification. Their thickness is over 1600 m, in some places containing thin layers of various schists, gneiss and amphibolites. Under the marble, gneiss and schists of the Lukovishka suite lie. Tectonically, the region forms the eastern ending (periclyne) of the Northrhodopean Antycline, that is why the marble formations are inclined to 30-50° NE and SE. In NE, through a series of faults, coinciding with the base of the Rhodopean Mts., the Dobrostan massif is separated from the Upper Thracian Graben, filled with nonbound Pliocene and Quaternary materials. Several faults run through the massif as well, roughly directed to west-east. The karst waters in the marble are formed mainly by rainfall and temporary surface water, disappearing in the higher peneplenized and karstified parts of the massif. Occurring underground, they form underground streams, surfacing along the periphery of the massif in sources (Troshanov, 1992). The biggest source – Chetiridesette Izvora (The 40 Sources) – is situated in the lowest part of the region, at the base of the Rhodopean slopes, and is determined by the contact of marble with the terrigenous sediments, filling the Upper Thracian Graben. Its outflow varies from 105 to 1439 l/s. Some other major sources are Varite (30–50 l/s) near Mostovo Village, Kluviyata (average 30 l/s) above Bachkovo Monastery, Kaptazha (6–456 l/s), Maykite (about 5 l/s) near Oreshets Village and others.

The surface karstic forms have developed mainly in the higher plateau-like part. They are represented by many dolines and valogs. There are also several blind valleys. Because of the soil layer, karren and karren fields are absent or rarely met. The river Sushitsa, delimitating the Dobrostan massif from the south, has formed a picturesque canyon-like valley. Near the village Mostovo, the river forms a natural rock bridge, from which the name of the village comes.

The Dobrostan massif is rich in caves and potholes – ca. 90 of them (Pandev, 1993, 1994). Most of them, including the deepest, are situated in its higher part.

Major caves and potholes in Dobrostan Massif and the adjacent territories in the south

No	Cave	Site	Length, m	Deniv., m
1	Topchika	Dobrostan	727	61
2	Ivanova Voda	Dobrostan	695	113
3	Gargina Dupka (Garvanitsa)	Mostovo	524	+38
4	Vodnata Peshtera	Mostovo	450	53
5	Hralupa	Dobrostan	311	10
6	Zmiin Borun	Mostovo	242	+22 -86
7	Ahmetyova Dupka	Dobrostan	44	14
9	Druzhba	Dobrostan		130

From the caves indicated the pothole Druzhba, with the biggest cave vertical in Bulgaria, is of particular interest. The cave Ahmetyova Dupka is a show cave, while in the cave Topchika interesting archaeological items and new species of animals have been found.

Most of the bigger caves are situated near Martsiganitsa Hut, which is connected to Dobrostan Village through an asphalt road and further to Assenovgrad. Another hut in the area is Bezovo, situated in the northern part of the massif. Assenovgrad is the exit/staring point for it. Some caves can be reached starting from the village Mostovo.

Within this karstic region Chervenata Stena reserve is situated.

Trigrad Region

The karstic region of Trigrad is one of the most extensive in the Rhodopes. It is situated in the upper part of the basin of Vâcha river and its tributaries. In the inter-river basins, the higher parts have plateau – like relief. The river valleys are canyon-like, with steep rocky slopes.

In this part of the Rhodopes, mostly metamorphic and magma rocks are represented. In terms of karstification, most important are the marble rocks, which are widespread and thick up to 1600 m. They lie on gneiss and schists of Paleosoic age. In the western part on the marble, there is riolith from the effusion of the Oligocene. The whole rock complex contains granite bodies of different size, transient to the west into the pluton of Barutin-Buynovo.

Tectonically, the region falls within the South Rhodopean Syncline. On the north, it is limited by the fault of Shiroka Laka, the southern block of which has subsided by about 5000 m. The syncline is a big graben, complicated by many faults and other smaller structures, mainly of northwest – southeast direction (the direction of the Shiroka Lâka fault). Some graben-like low places exist, filled by lakes from the Palcogene and by aluvium – proluvium deposits.

Trigrad Gorge and the entrance of the cave Dyavolskoto Garlo, Rhodopes – photo Trifon Daaliev

Trigrad Region is characterized by rainfall, which changes, according to altitude, from 692 to 1394 mm (average 953 mm – a value higher than the average for Bulgaria). In the marble terrains, one part of the rainfall drains superficially and takes part in the formation of the river outflow, while the bigger part infiltrates various karstic forms and replenishes the reserve of ground water. Besides, the rivers, which start outside the marble outcrops (Buynovska, Trigradska, Chairska, Muglenska – they all form Krichim River), enter karstified rocks and start losing water in sinks along the riverbeds. The most typical are the sinks of Muglenska River, where the entire outflow disappears. Entering underground, the water starts moving towards the sources (situated most often in tectonic faults), ponding of underground waters. The biggest karstic sources in the region are situated in the valleys of the rivers Krichim and Shirokolashka, near the villages Nastan and Beden. These are the lowest points of marble outcrop on the surface. Its place is determined by the contact between marble and the Proterosoan non-karstified rocks. These rocks serve as a threshold, behind which a large saturated zone is formed. Here, the largest karstic sources (near the villages Nastan and Teshel, the source Vrissa near Beden and other sources near Shirokolashka River) are situated. There are also sources along Krichim River and its tributaries. They are situated at river level. The more important among them are the sources at Mugla Village, along Chairska River, the sources Kaynatsite, situated along Buynovska River.

The varied geological structure has a considerable impact on the relief morphology in the region. Among the karstic formations worth noting are the karren and karren fields, dolines, valogs and the karstic valleys (dry, blind and gorges). Around the villages Yagodina, Trigrad, Zhrebevo and Chamla, typical uvalas are also formed. Here are the most picturesque karstic gorges in Bulgaria – these of Trigrad and Buynovo. The considerable vertical fragmentation of the relief and the presence of deep erosion base are the causes for some of the deepest potholes in the Rhodopes to have developed in Trigrad region. A total of over 85 caves and potholes have been recorded in the region, part of them quite sizeable.

Some important caves in Trigrad region

No	Name	Locality (village, area)	Length, m	Deniv., m
1	Yagodinskata Peshtera	Yagodina Village	8501	60
2	Lednitsata	Gela Village	1419	108
3	Izvora	Orpheus Hut	2480	
4	Drangaleshkata	Mugla Village	1142	255
5	Kambankite	Chamla Village		158
6	Dyavolskoto Gârlo	Trigrad Village	480	89
7	Partizanskata	Orphey Hut		80
8	Sanchova Dupka	Yagodina Village	888	12
9	Haramiyskata	Trigrad Village	495	48

Two of them are show caves and, together with the gorges in which they are situated, are particularly interesting for tourists. The pothole Dyavolskoto Gârlo is very impressive with its entrance 40-meter waterfall (the entire Trigrad River enters it, re-emerging some distance lower) and with its voluminous chambers. Nearby, the cave Haramiyska is situated (actually two caves, connected by 36 m vertical). Some of the caves of Trigrad Region contain important archaeological and paleontological sites.

The important settlements in Trigrad basin, except for Yagodina Village and Chamla Quarter, are situated in the river valleys.

The picturesque and almost virgin nature of Trigrad basin makes it attractive for tourism, together with the several mountain huts and hotels in Trigrad.

The natural phenomenon Ritlite of Lyutibrod – photo Trifon Daaliev

THE LONGEST AND THE DEEPEST CAVES IN BULGARIA CAVES LONGER THAN 1000 M (BY JULY 2006)

1. Duhlata, Bosnek Village — 18 000
2. Orlova chuka, Pepelina Village — 13 437
3. Imamova dupka (Yagodinskata), Yagodina Village — 8501
4. Temnata dupka, Lakatnik Railway St. — 7000
5. Vreloto, Bosnek Village — 5300
6. Golyamata Balabanova Dupka, Komshtitsa Village — 4800
7. Prikazna, Kotel Town — 4782
8. Boninskata (Popovata), Krushuna Village — 4530
9. Andaka, Dryanovo Town — 4000
10. Tizoin, Gubesh Village — 3599
11. Bacho Kiro, Dryanovo Town — 3500
12. Ponora, Chiren Village — 3497
13. Raychova Dupka, Cherni Osam, Village — 3333
14. Russe, Emen Village — 3306
15. Vodnata Reshtera, Tserovo Village — 3264
16. Lucifer, Kotel Town — 3200
17. Morovitsa, Glozhene Village — 3200
18. Emenskata Peshtera, Emen Village — 3113
19. Magliviyat snyag, Tvarditsa Town — 3076
20. Marina Dupka, Genchevtsi Village — 3075
21. Bambalova dupka, Emen Village — 2923
22. Troana, Emen Village — 2750
23. Zandana (Biserna), Shumen Town — 2716
24. Barkite 14, Vratsa Town — 2600
25. Belyar, Vratsa Town — 2560
26. Katsite, Zimevica Village — 2560
27. Magura, Rabisha, Village — 2500
28. Parnicite, Bezhanovo Village — 2500
29. Izvorat, Borino Village — 2480
30. Devetashkata cave, Devetaki Village — 2442
31. Saguaro, Bilin Dol, Gintsi Village — 2217
32. Bayov Komin, Drashan Village — 2196
33. Manuilovata, Ribnovo Village — 2155
34. Golyamata Temnota, Drenovo Village — 2000
35. Vodopada, Krushuna Village — 1995
36. Golyamata Mikrenska (Mandrata), Mikre Village — 1921
37. Tajnite ponori, Shumen Town — 1916
38. Machanov Trap, Zdravkovets Village — 1907
39. Mladenovata Propast, Chiren Village — 1723
40. Labirinta (A.Blazhev), Pepelina Village — 1697
41. Kumanitsa, Cherni Osam Village — 1656
42. Urushka Maara, Krushuna Village — 1600
43. Lepenitsa, Rakitovo Village — 1525
44. Krivata Pesht, Gintsi Village — 1500
45. Mayanitsa, Tserovo Village — 1426
46. Lednitsata, Gela Village — 1419
47. Izvorat (Padaloto), Yantra Village — 1400
48. Lednika, Kotel Town — 1367
49. Toshova Dupka, Glavatsi Village — 1302
50. Vodni pech, Dolni Lom Village — 1300
51. Bashovishki pech, Oreshets Railway St. — 1298
52. Malkata Mikrenska Peshtera, Mikre Village — 1295
53. Sopotskata, Sopot Village — 1255
54. Mâglivata, Kotel Town — 1244
55. Zadanenka, Karlukovo Village — 1150
56. Drangaleshka Dupka, Mugla Village — 1142
57. Vodnata Peshtera, Pesthera Town — 1114
58. Vârlata, Cherni Osam Village — 1110
59. Sedlarkata, Rakita Village — 1100
60. Vodnata Pesht, Lipnitsa Village — 1100
61. Aladjanskata, Gortalovo Village — 1083
62. Gornik, Krushuna Village — 1074
63. Duhaloto (Trona), Apriltsi Town — 1040
64. PPD, Bosnek Village — 1002
65. Skoka, Dragana Village — 1000

POTHOLES AND CAVE SYSTEMS IN BULGARIA DEEPER THAN 100 M (BY JULY 2006)

1. Raychova dupka, Cherni Osam Village — 387
2. Barkite 14, Gorno Ozirovo, Vratsa — 356
3. Yamata na Kipilovo (Golyamata yama), Kipilovo Village — 350
4. Tizoin, Gubesh Village — 320
5. Beliar, Gorno Ozirovo Village — 282
6. Drangaleshkata Dupka, Mugla Village — 255
7. Lednika, Kotel Town — 242
8. Malkata Yama, Cherni Osam Village — 232
9. Propast (Pothole) N9, Bansko Town — 225
10. Mâglivata, Kotel Town — 220
11. Katsite, Zimevitsa Village — 205
12. Barkite 8, Gorno Ozirovo Village — 190
13. Pukoya, Pavolche Village — 178
14. Vihren, Bansko Town — 170
15. Lisek, Dryanovo Village — 160
16. Uzhasat na imanyarite, Kotel Town — 160
17. Akademik, Kotel Town — 158
18. Kambankite, Mugla Village — 158
19. Borova dupka, Cherni Osam Village — 156
20. Blagova Yama, Etropole Town — 153
21. Yavorets, Zverino Village — 147
22. Kladeto, Rudozem Town — 147
23. Mâgliviyat snyag, Tvarditsa Town — 146
24. Malyat Sovat, Btyazovo Village — 140
25. Kârvavata Lokva, Kotel Town — 140
26. Propast K - 19, Razlog Town — 136
27. Saguaro, Bilin Dol, Gubesh Village — 135
28. Aleko, Ilindentsi Village — 132
29. Druzhba, Dobrostan Village — 130
30. Lucifer, Kotel Town — 130
31. Propast K - 18 (Ledenata), Razlog — 126
32. Malkata Balabanova Dupka, Komstitsa Village — 125
33. Golyamata Vâpa, Stoilovo Village — 125
34. Banderitsa, Bansko Town — 125
35. 20godini Akademik, Bansko Town — 118
36. Manoilovata dupka, Ribnovo Village — 115
37. Tyinite Ponori, Shumen Town — 115
38. PPD, Bosnek Village — 115
39. Ivanova voda, Dobrostan Village — 113
40. Pticha Dupka, Cherni Osam Village — 108
41. Haydushkata Propast, Bistrets Village — 108
42. Lednitsata, Gela Village — 108
43. Yasenski Oblik, Yablanitsa Town — 108
44. Partizanskata, Glozhene Village — 107
45. Golyamata Temnota, Drenovo Village — 106
46. Golyamata Voda, Karlukovo Village — 106
47. Morovitsa, Glozhene Village — 105
48. Bezdanniyat Pchelin, Yablanitsa Town — 105
49. Kumanitsa, Cherni Osam Village — 104
50. Panchovi Gramadi, Glozhene Village — 104
51. Stotaka, Brezovo Village — 104
52. Kamenititsa N14, Razlog Town — 103
53. Chelyustnitsa, Bayovi Dupki, Bansko Town — 103
54. Nanovitsa, Glozhene Village — 102
55. Pleshovskata pestera, Prevala Village — 102
56. Golyamata Garlovina, Cheni Osam Village — 100
57. Ruse, Emen Village — 100
58. Kânchova Varpina, Tepava Village — 100
59. Draganchovitsa, Glozhene Village — 100

CAVES AND POTHOLES INCLUDED IN THIS BOOK

20 Years of Akademik
Ahmetyova Dupka (Prokletata, Dobrostanski Biser)
Akademik
Aladjanskata Peshtera
Alchashkata (Alchanskata) Peshtera
Aleko
Akademik
Andâka (Golyamata, Vodnata Peshtera)
Bacho Kiro (Malkata Peshtera)
Balyovski Obruk
Bambalova Dupka
Bânderitsa
Bankovitsa
Barkite – 14
Barkite – 8 (25 Years Akademik)
Bashovishki Pech
Bayov Komin
Belyar
Bezdanniya Pchelin
Blagova Yama
Boninskata (Popskata) Peshtera
Borikovskata Peshtera (Chervena Dupka)
Borova Dupka
Boychovata Peshtera (Boychova Dupka)
Bozhkova (Kulina) Dupka
Bozkite (Babini Bozki)
Brâshlyanskata Vodna Peshtera
Bratanovata Peshtera
Brunoshushinskata Peshtera
Cheleveshnitsa
Cheleveshnitsa (Cheleveshkata, Choveshkata, Chilyashkata)
Cheloveshnitsa
Chelyustnitsa
Chernata Pesht
Cherniyat Izvor
Chudnite Mostove (Erkyupriya)
Dâlbokata (Propast N 27)
Desni Suhi Pech
Devetashkata Peshtera (Maarata)
Dinevata Pesht
Djebin Trap
Dolnata Maaza
Drânchi Dupka
Drangaleshka (Dangalashka) Dupka

Drânkalna Dupka
Drashanskata Pechtera
Druzhba (Sveti Hris)
Duhlata
Dushnika (Iskretskata Peshtera, Peshtta, Vodnata)
Dyavolskoto Garlo (Harloga)
Elata
Emenskata Peshtera
Eminova Dupka
Futyovskata Peshtera
Gabrika
Gâgla
Gârdyova Dupka
Gargina Dupka (Garvanitsa)
Garvanitsa (Kosovskata Peshtera)
Georgievata (Gyurgyovskata) Peshtera
Gergitsovata Peshtera
Gininata Peshtera
Golemata Voda
Golemiyat Burun (Oynyolar Marasa)
Golyamata Garvanitsa
Golyamata Humba
Goloboitsa 1 and 2
Golyamata Balabanova Dupka
Golyamata Gârlovina
Golyamata Maara
Golyamata Mikrenska Peshtera (Mandrata)
Golyamata Temnota (Yamkata)
Golyamata Vâpa
Golyamata Yama
Golyamata Yama (Yamata – 3)
Gorna and Dolna Peshtera
Gornik
Gradeshnishkata (Rushovata) Peshtera
Granicharskata Propast
Grimnena Dupka
Han Maara
Haramiyska Dupka
Haydushkata Peshtera
Haydushkata Peshtera
Haydushkata Peshtera
Haydushkata Propast
Haydushkata Propast
Hralupa (Hralup)
Imaneto

Ivanova Voda
Izvora (Izvora na Kastrakli)
Izvorat (Vodnata Peshtera, Vodopada, Padaloto)
Kalitsa
Kambankite (Propast M-4)
Kânchova Varpina
Karstoviya Izvor (The Karstic Source)
Kârvavata Lokva (Kladentsite)
Katsite
Kipilovskata Peshtera (Choveshkata, Chelovesht-
　　　nitsata)
Kirechnitsata (Kerechnitsata, Golyamata Mahara)
Kirov Vârtop (Vodnitsa, Vodnata Peshtera)
Kladeto
Kokalana (Stoletovskata Peshtera)
Kondjova Krusha
Krivata Pesht
Kumanitsa
Labirinta (Aleksander Blazhev Cave Complex)
Ledenika
Lednika
Lednitsata
Lepenitsa (Vodnata, Mokrata, Izvora)
Levi Shupli Kamak
Lipata System
Lisek (Pantyolova)
Lucifer
Lyastovitsa (Lastovitsa)
Kalenska Pesht
Kaleto
Kalnata Propast
Kamenititsa No 14
Karangil
Kozarskata Peshtera
Krachimirskoto Vrelo
Machanov Trap
Mâgliviyat Snyag
Magura
Maliya Sovat
Malkata (Malata) Balabanova Dupka (Yamata na
　　　Iskrechinata)
Malkata Mikrenska Peshtera
Malkata Yama
Mandrata (Lâdjenskata, Vodnata Peshtera)
Manuilovata Dupka (Manailovata Peshtera)
Marina Dupka (Parova Dupka, Kumincheto, Pro-
　　　pastite)
Martin 11
Mayanitsa
Mazata (Haydushkiya Zaslon)
Mâglivata
Mechata Dupka

Mishin Kamâk (Kamik)
Mizhishnitsa
Mladenovata Peshtera
Morovitsa
Mussinskata Peshtera
Nahodka – 13
Nanin Kamâk
Nanovitsa
Neprivetlivata (Gornata Propast)
Nevestina Propast
Novata Peshtera
Ogradite (S-20)
Opushenata ("Smoked")
Orlova Chuka (Orlovo Gnezdo)
Panchovi Gramadi
Parasinskata Propast
Parnitsite
Partizanskata Peshtera
Pavla (Vodnata Peshtera)
Peshketo
Peshterata s Dvata Vhoda
Pleshovska (Pleshova) Dupka
Ponora
Popovata Peshtera
PPD ("Introduction to Speleology")
Prikazna
Prilepnata Peshtera (Cave Complex Bozhiyat
　　　Most – Prilepnata Peshtera)
Prohodna
Prolazkata (Derventskata) Peshtera
Propada (Paramunska Yama)
Propast
Propast (Pothole) No 9 (Devyatkata, System 9-11)
Propast K-18 (Ledenata)
Propast No 35 (Cyclope)
Propast K -19 (Kamenititsa – 19, Bulgaria – France
　　　– Belgium)
Propastna peshtera (Cave – Pot Hole) N 29
Propastta na Benyo Ilyov
Pticha Dupka
Pukoya
Radolova Yama (Toshkova Dupka)
Ralena (Vodnata)
Raychova Dupka
Razhishkata Peshtera
Rushkovitsa (Prelaz, Partizanskata)
Russe
Sâeva Dupka
Saguaroto (Golyamata Peshtera v Bilin Dol)
Samara (Samarskata Peshtera)
Samuilitsa (Vassilitsa)
Sanchova Dupka

Sedlarkata (Mandrata)
Serapionovata Peshtera
Sharaliyskata Peshtera
Shepran (Shepra) Dupka
Shipkata
Sifona (Novata)
Sinyoto Ezero
Sinyoto Kolelo
Sipo I
Skoka
Snezhanka
Sokolskata Dupka (Peshtera)
Sopotskata Peshtera (Tâmnata Dupka)
Spropadnaloto (Propadnaloto)
Stâlbitsa
Starata Prodânka
Stotaka
Stoyanovata Peshtera
Stublenska Yama I
Studenata Dupka (Cherepishkata Peshtera)
Subattâ
Suhi Pech
Svinskata Dupka
Svirchovitsa
Tâmna Dupka
Taynite Ponori
Temnata Dupka
Temnata Dupka
Temnata Dupka (Izvorskata Peshtera)
Tigancheto
Tipchenitsa
Tizoin
Topchika
Toplya
Toshova Dupka (Kalna Mâtnitsa, Izvorna)
Troana
Trona (Duhaloto)
Tsakovska Pesht (Tsakonichki Pech)
Tsarkvishte (Tsarkveto)
Uhlovitsa (Ultsata)
Urushka Maara (Proynovata)
Uzhasa na Imaniyarite
Varkan (Vrkan, Peeshtata, Musikalnata Peshtera)
Vârlata
Vârteshkata
Venetsa
Vihrenska Propast
Vodnata Pesht
Vodnata Peshtera
Vodnata Peshtera (Borovskata Vodna Peshtera)
Vodnata Peshtera (Tserovskata Peshtera)
Vodnata Propast (Propust)
Vodni Pech
Vodnite Dupki
Vodnitsata
Vodopada (Maarata)
Vreloto
Yagodinska Peshtera (Imamova Dupka)
Yame – 3 (Srednitsa)
Yamkata (Yamkite)
Yasenski Oblik 2
Yavorets
Yubileyna
Yulen Ere
Zadânenka
Zandana (Biserna)
Zhabata
Zhivata Voda (Ayazmoto)
Zidanka
Zidanka
Zmeyova Dupka
Zmiyn Burun

CAVE MAPS:

Preparation of the maps for printing – Trifon Daaliev, Alexey Jalov, Zdravko Iliev
Inking of the maps reduced to A4 – Angelina Petkova
Scanning and other work on the maps – Trifon Daaliev
Geographical maps showing the situation of the caves – Ivan Alexiev, Trifon Daaliev and Alexey Benderev
Map showing the distribution of the karstified rocks and the situation of caves – Ivan Alexiev, Trifon Daaliev and Alexey Benderev

PHOTOGRAPHS:

Copyright remains to the authors

SOME REMARKABLE BULGARIAN CAVES

CAVES IN THE DANUBE PLAIN

Name: **ALADJANSKATA PESHTERA (139)***
Gortalovo Village, Pleven Distr.
Length 1083 m. Denivelation 41 m (-36,+5)

A cave 2.5 km from Gortalovo, in the protected area called Gushtera. Formed in limestone from the Upper Cretaceous (Maastricht). Cascade – descending, two-storey, branched, water cave. Entrance on the bottom of funnel, ponor-like. The main gallery descends up to a well, eight-meter deep, giving access to the lower storey, where there is an underground river. On the upper level, in one of the branches of the main gallery, a labyrinth with a total length of 105 m and surface 400 m^2 is formed. Thin stalactites in the entrance parts of the main gallery. The underground stream is fed by rain and condensation water, formed in the watershed-uvala 3x2 km. Discovered in 1979 and explored by cavers from Studenets – Pleven Caving Club in the period 1981–1986 by K. Petkov, S. Gazdov, B. Garev, M. Dimitrov, V. Chapanov and others. Protected together with the whole area.

Fauna: including not described troglobite Diplopod.

* The numbers in brackets pointed the location of the caves on the Maps of the location of the caves.

Aladjanskata peshtera

Gortalovo Village

Name: **BOZHKOVA (KULINA) DUPKA (172)**
Krivnya Village, Razgrad Distr.
Length 326 m

A cave 1200 m E of Krivnya, in the middle of the cliff between Krivnya and Senovo, in the western part of the region Ludogorie. Formed in limestone from Lower Cretaceous (Hotrivien-Aptien).

Bozhkova Dupka is the longest cave in the Ludogorie. It is horizontal, dry and branched, divided into two parts. In the first part, there are stones on the floor of the gallery. After 36 m, the gallery narrows and through a squeeze it is possible to enter the second part of the cave. There are several chambers, 7-8 m high, with small side galleries. All branches end in very narrow impenetrable places. The cave is poor in flowstone. In some places there are thick layers of bat guano (up to 120 cm).

The first graffiti in the cave (Deliorman) were done in 1924 by "tourists". First studied on May 7[th,] 1961 by S. Terziev, N. Stanchev and R. Dimitrov from the Caving Club at Buyna Gora – Razgrad Tourist Society. The cave was surveyed in details in 1976 by S. Ivanov, M. Krastev and D. Nenov from the same club. During the caving expedition "Ludogorie'77-78", P. Tranteev, Zdr. Iliev and Iv. Rashkov corrected the map. There is a legend according to which the cave had been inhabited by Bozhko Voyvoda, hence its name. In the entrance parts, fragments of ceramic vessels have been discovered, thus the assumption that the cave might have been inhabited.

Fauna: only 4 species of spiders have been identified so far.

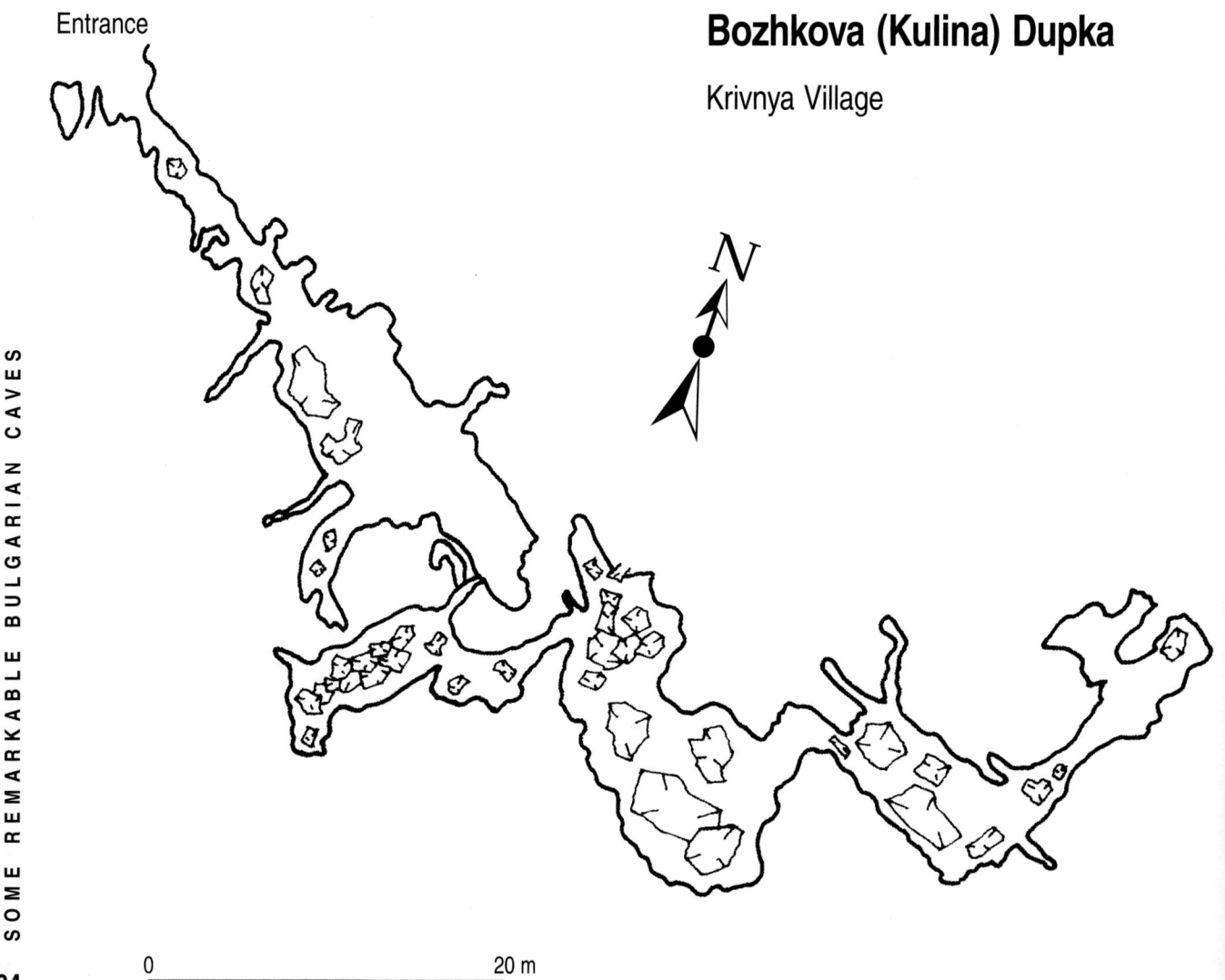

Bozhkova (Kulina) Dupka

Krivnya Village

Name: **KIROV VÂRTOP (VODNITSA, VODNATA PESHTERA) (143)**
Bohot Village, Pleven Distr.
Length 776 m. Denivelation -42 m

Branched cave pothole in the locality Buchi Geran. Water, one-storey cave, formed in Sarmatian limestone. Entrance in a funnel with diameter of 20 m. The galleries of Kirov Vârtop are 416 m long, but through a siphon of 1.30 m the cave is connected to Vodnitsata. During heavy rain the entrance also becomes a siphon, the river sinking into a similar river at the bottom of the cave. Where the cave is divided into two branches, there is an accumulation of guano and a bat colony on the ceiling. Bones belonging to a cave bear have been found in the cave (N. Spassov det.). The entrance of the cave is about 300 m away from the pumping station.

Vodnitsata is a two-storey cave with a lower water level. The upper two galleries are dry. The cave ends with a semidry water siphon. When the pumps at the station are not in use, the water rises and the siphon is closed. No cave flowstone formations.

The map and description of Kirov Vârtop were made by Studenets, Pleven Caving Club (K. Petkov, Ts. Hristov and B. Garev), on 15th October 1982.

Fauna: only guano-inhabiting invertebrates are known.

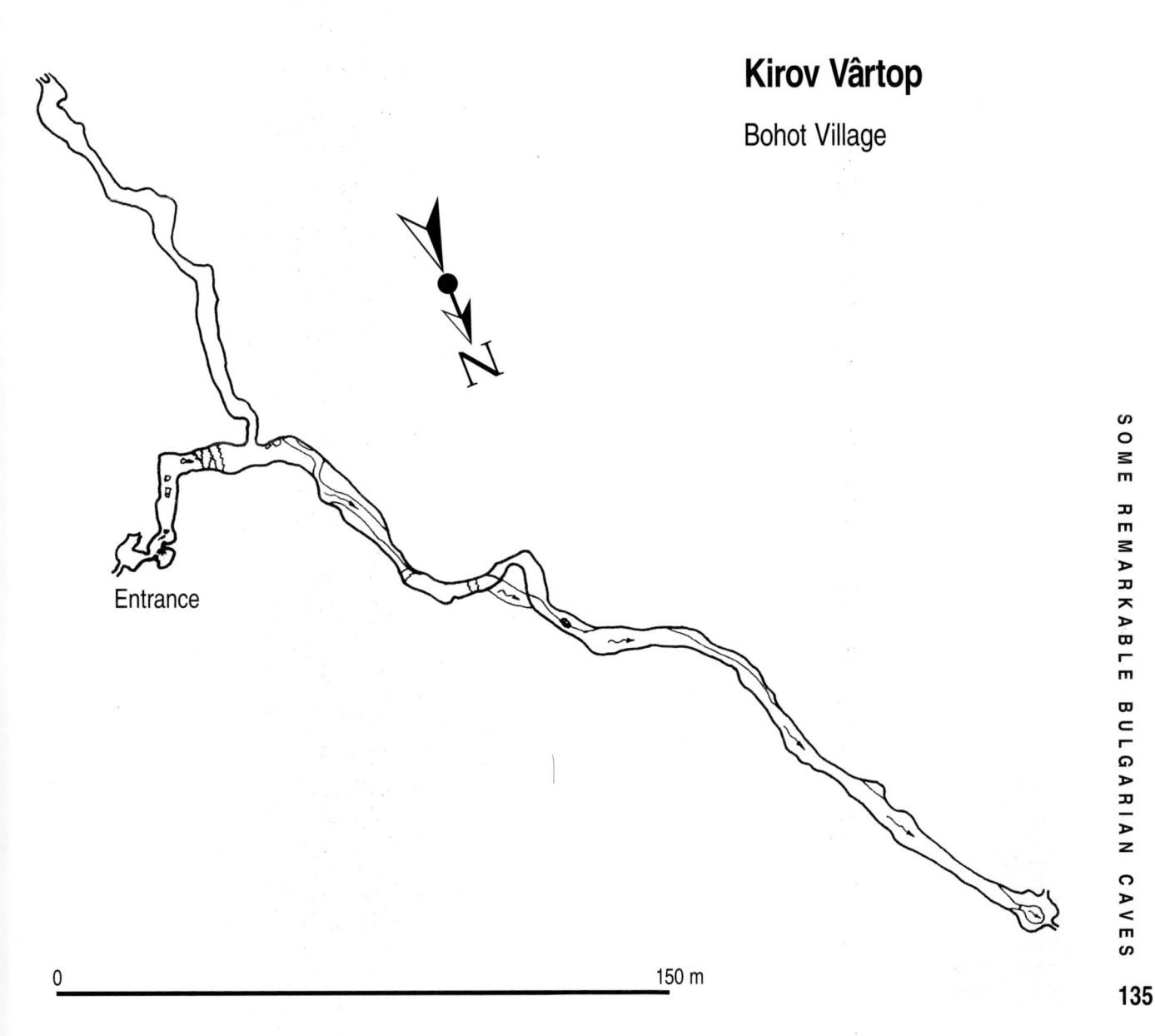

Kirov Vârtop

Bohot Village

Name: **LABIRINTA (Aleksander Blazhev Cave Complex) (170)**
Pepelina Village, Russe Distr.
Altitude 80 m. Altitude above the LBE: 45 m. Length 1697 m

A cave in the eastern part of a cliff facing Pepelina Village. There are several hidden entrances, situated about 50 m from a typical square rock niche, which are about 20 m higher than the niche and are situated 18 m from the edge of the cliff.

The cave is formed in Cretaceous (Apt) limestone and has developed in NE direction. It is horizontal, labyrinthine and narrow, with clay floor, and in some places with sinter plates and rock boulders. Crawling on wet clay, we reach an oblong chamber – this is the end of the "old" parts. The way to the new parts starts from the end of the chamber, through a meandering dug-out tunnel, and leads to many smaller chambers, connected by squeezes. The end is a labyrinth where one could be lost.

Discovered by cavers from Russe, the exploration of this difficult cave took many years. The first description was made by I. Kehayov from Haskovo – up to 328 m, the second and third – by K. Dimchev (Russe) – up to 1063 m, the fourth – by K. Stoyanov, Hr. Ivanov and T. Kisimov from Russe – up to 1697 m. A guide from Russe Caving Club is needed. The new parts were discovered at the beginning of 2000 by Hristo Ivanov, Nikolay Kovachev and Svetoslav Denev from Prista Caving Club. Cavers from Russe carried on digging. The survey was done by Kamen Stoyanov, Hristo Ivanov and Teodor Kisimov in September 2000.

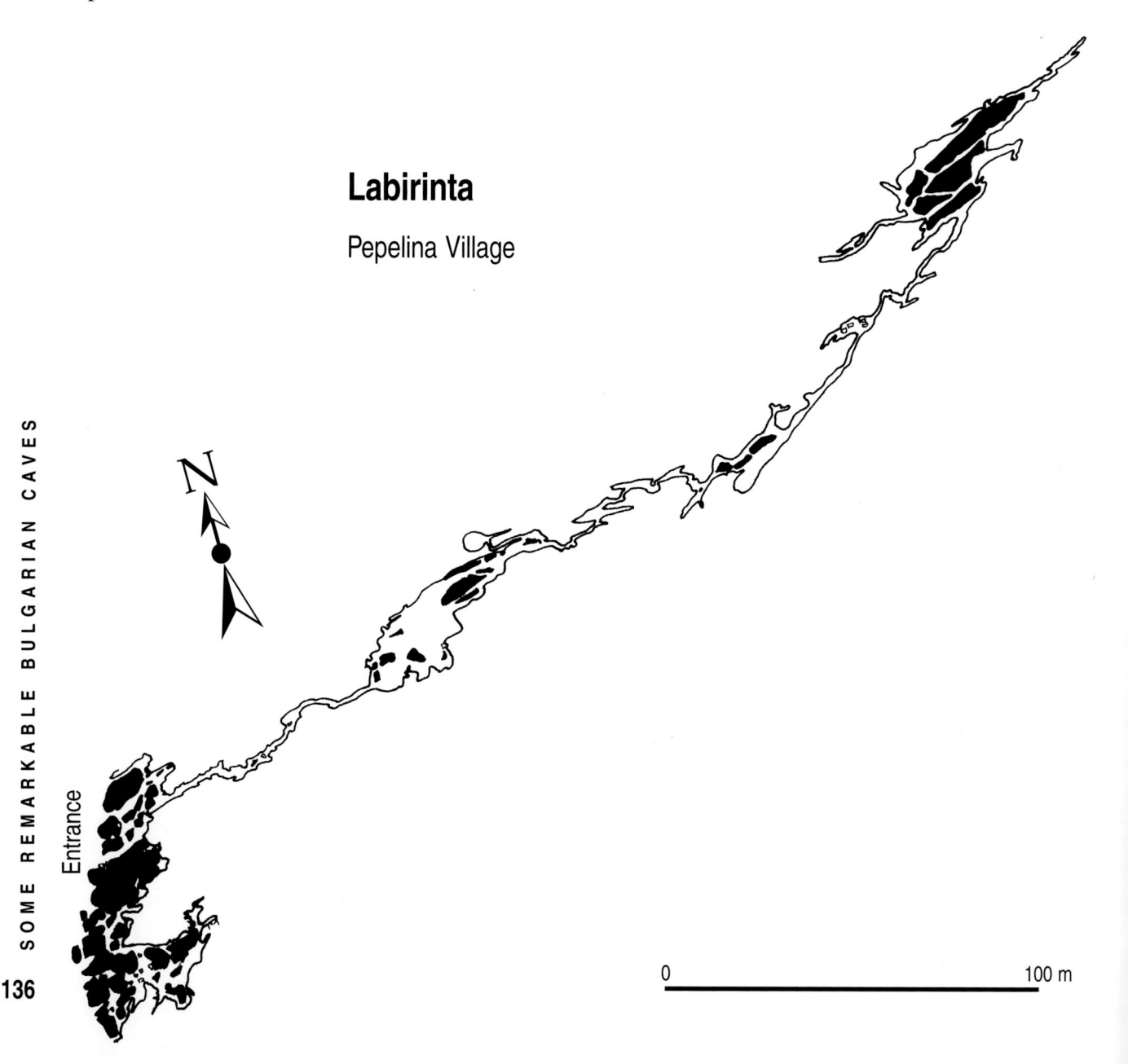

Labirinta

Pepelina Village

Name: **NAHODKA – 13 (173)**
Shumen
Altitude 470 m. Length 470 m. Denivelation -20 m

A cave in the southwest part of Shumensko Plateau, about 500 m SW of the Old Town. Entrance (1.5 x 1.5 m) in sinkhole. A 20-m gallery follows, transformed in a 20-m long diaclase. Two meters of vertical leads in a 25-m gallery with stalactites. Another 5-m vertical and a short gallery lead to the biggest chamber of the cave (16 x 20 m). On the floor blocks and clay, the chamber is poor in limestone formations. A 50-m gallery starts from the chamber. Explored in 1972 by cavers from Shumen Caving Club. Horizontal map by At. Spassov, St. Markov, M. Dinkova and E. Gerov. After 12 m of digging, another 150 m were added. In 1986, A. Spassov, V. Mirchev, V. Chapanov, M. Petrova and V. Kolev finally surveyed the cave.

Fauna: solitary bats.

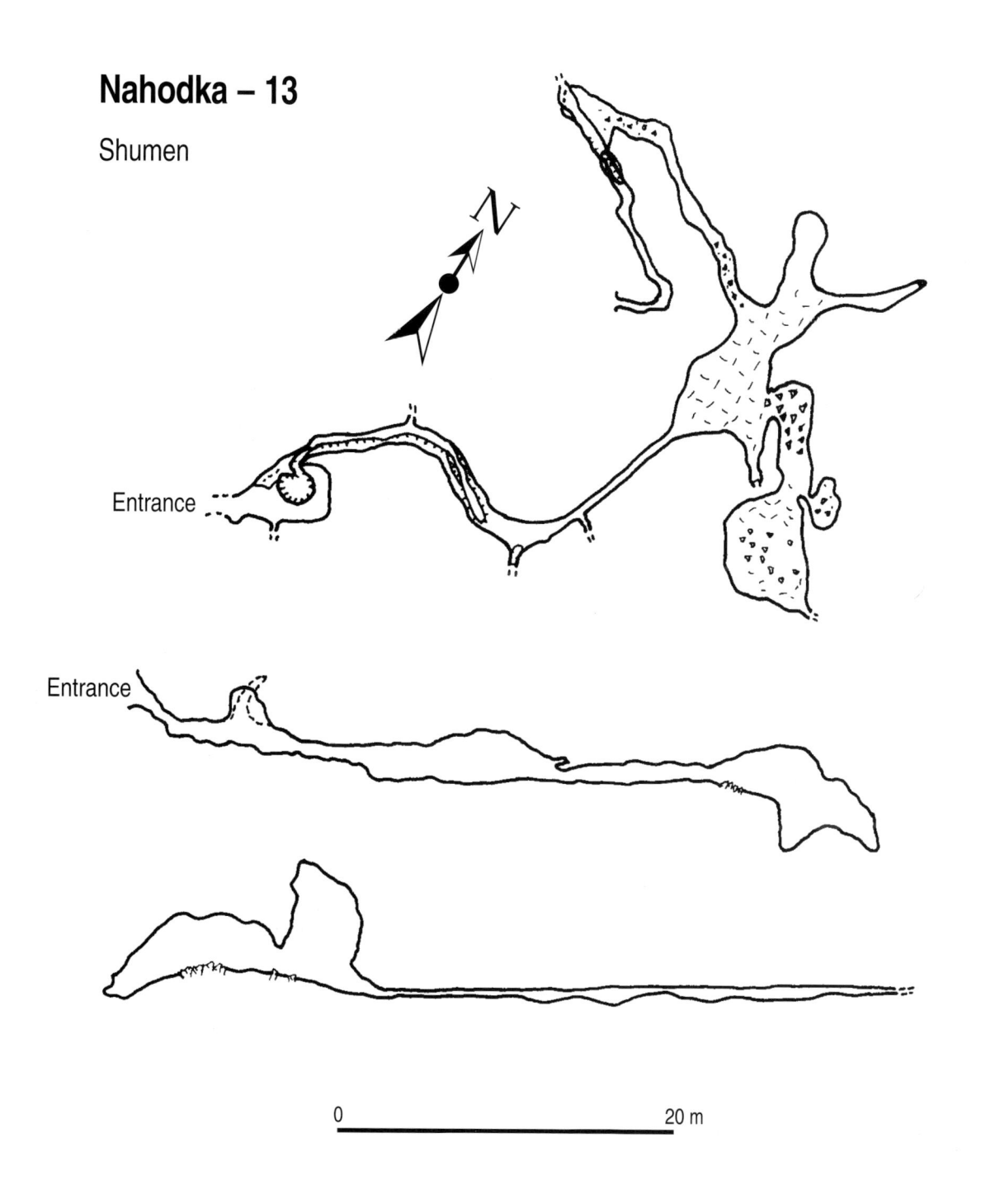

Name: **NANIN KAMÂK (144)**
Muselievo Village, Pleven Distr.
Length 178 m. Denivelation 8.5 m
Coordinates: N 43° 37' 42.9" E 24° 51' 24.8"

A cave in the locality Nanin Kamâk, about two kilometers SW of Muselievo. Formed in limestone from the Upper Cretaceous (Sarmatian). The cave is horizontal, branched, one-storey and dry. Two galleries start from the small entrance chamber. One is ascending and ends in an opening on the ceiling of the cave. The floor of a part of the main gallery is covered with bat guano, some colonies including rare species, thus being of considerable importance.

Further, the cave is divided into two more galleries, where bats live too.

The cave has been studied mainly by biospeleologists (P. Beron, T. Ivanova, V. Beshkov). In this cave, some systematic observations on bats have been carried out with the participation of German specialists.

Fauna: 13 animal species have been observed (guano-inhabiting animals), incl. *Rhinolophus mehelyi* and another five species of bats.

Protected Natural Monument with adjacent area of 14.1 ha (Decree 238/07.06.1996, gazetted 55/1996).

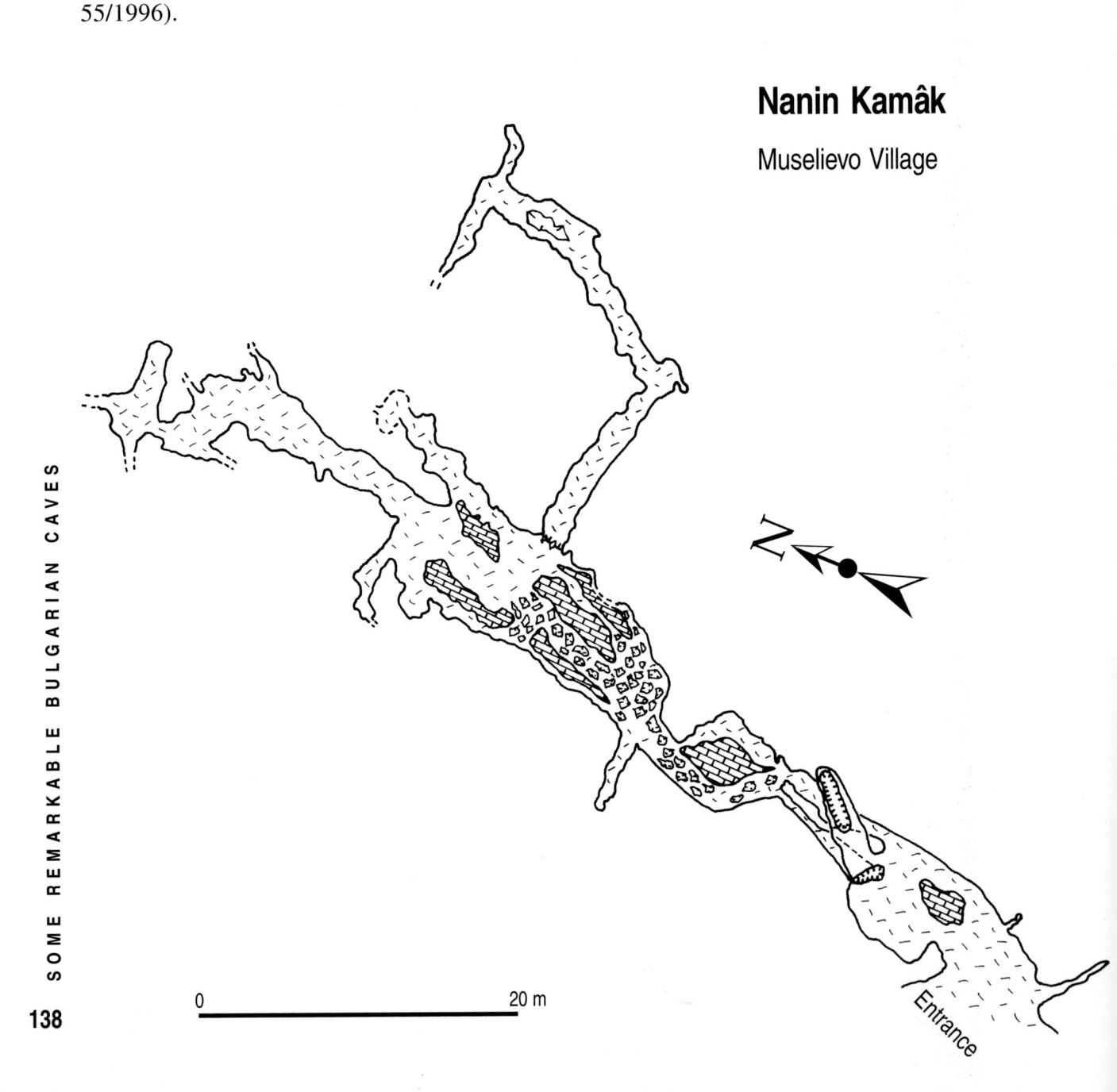

Nanin Kamâk
Muselievo Village

Name: **ORLOVA CHUKA (ORLOVO GNEZDO) (171)**
Pepelina Village, Russe Distr.
Length 13 437 m (second longest in Bulgaria). Denivelation 12 m
Coordinates: N 43° 35' 39,1" E 25° 57' 59,7"

A cave system 2.5 km NE of Pepelina, Dve Mogili municipality. Entrance on a river terrace, on the left bank of Cherni Lom, at an altitude of 80 m above the river level. Formed in sandy Aptian limestone. A labyrinth of dry, large and narrow, mainly horizontal galleries, connecting ca. 10 chambers of different size. Poor in speleothems. Widespread sediments thick up to 5.6 m. No running water. Average air t° = 10.8°C, relative humidity – 88%. Discovered in April 1941 by St. Spassov from Pepelina. First explored by Lokomotiv Tourist Society, Russe, under the guidance of T. M. Avramov. Iron gate installed and officially opened for tourists on June 7th, 1942. Later explored by the speleological section of the District Museum in Russe and by the Speleoclub of Prista Tourist Society in Russe. Until August 27, 1961, 530 m were known and on that day another 760 m of galleries were discovered. After the National Expeditions in 1970 and 1972, the length became already 7410 m, respectively 8796 m and in 1977 it became 11 483 m. New discoveries (still going on) increased the length to13 437 m. From 1956 to 1961, it was arranged as a show cave. In 1970 the show (lighted) itinerary was one km. Since 1994 the cave has been managed by the Speleoclub of Prista Tourist Society.

Declared a natural monument and issued in the State Gazette (together with 7.6 ha adjacent area)(Decree No 2810, published in 1963). Within Russenski Lom Natural Park.

Archaeological finds from the Middle Paleolithicum – East Balkan Musterien.

Paleontology: fossils of *Ursus spelaeus*.

Fauna: 15 species, including five species of bats.

Orlova Chuka

Pepelina Village

Entrance

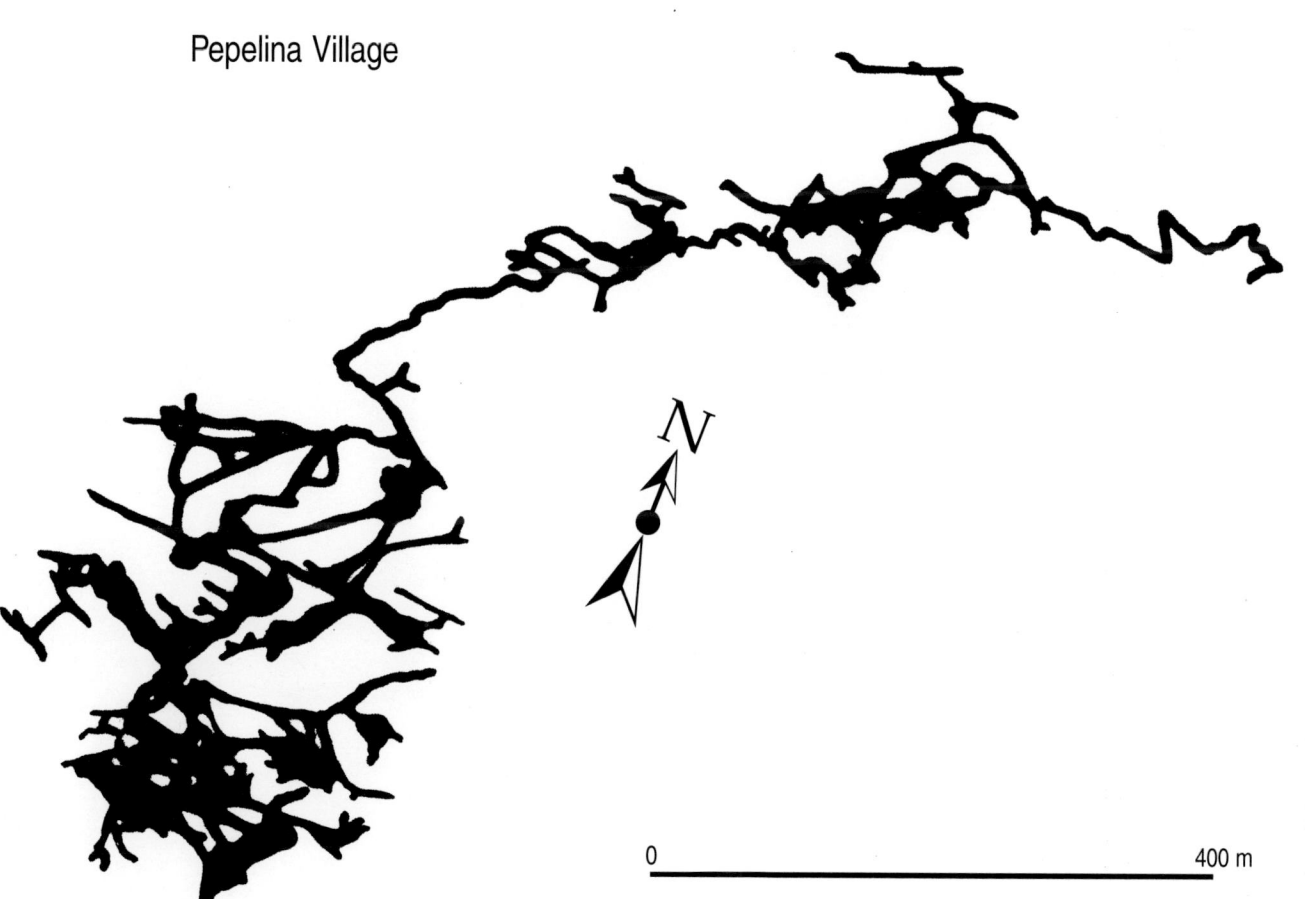

Stalagtons - Orlova Chuka – photo Vassil Balevski

Column – photo Trifon Daaliev

Cave decoration – photo Trifon Daaliev

Name: **TAYNITE PONORI (174)**
Shumen Town
Altitude 445 m. Total length 1916 m. Denivelation -115 m

A cave pothole on the Shumen Plateau, in the locality Bostanlaka. Discovered in 1983 through a ponor desobstruction. The pothole is a part of the system of the karstic source Bashbunar in Troishki Boaz, and is the deepest in NE Bulgaria. Consists of two main water galleries with permanently flowing streams. Four levels have been detected, at 20, 40–55, 60–70 and 90–100 m. The link between the storeys consists of verticals and internal pits. The water ejection is through two groups of karstic sources: Gorni, or Upper (in the locality Bashbunar, through sewers it goes to Shumensko Pivo Brewery), and Dolni, or Lower (by the northern part of the Troitsa), where Troishka Reka starts.

The cave is studied by the cavers of Shumen Caving Club.

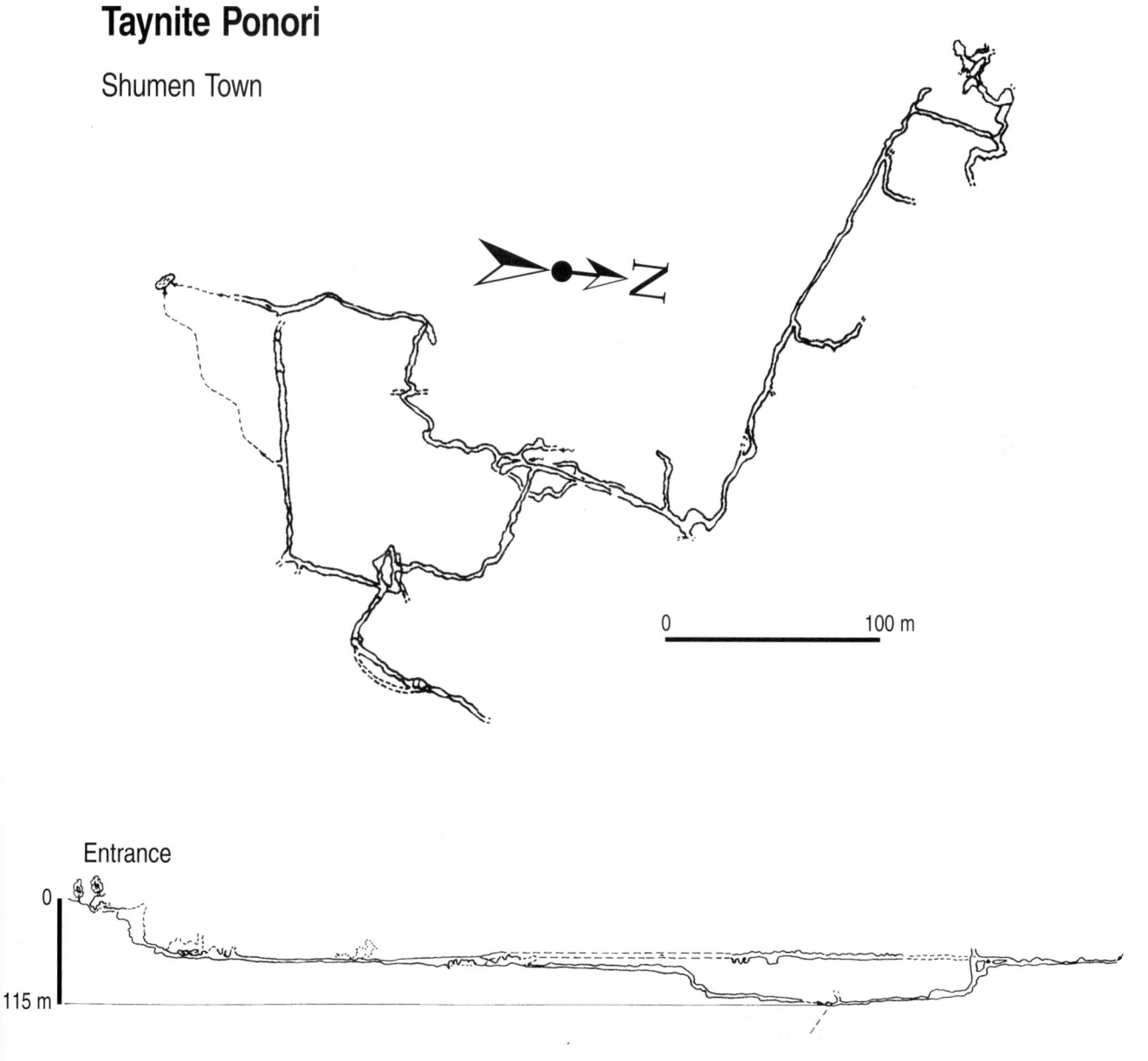

Taynite Ponori

Shumen Town

Name: **VARKAN (VRKAN, PEESHTATA, MUSIKALNATA PESHTERA) (1)**
Tsar Petrovo Village, Vidin Distr.
Length more than 800 m

The entrance of this big cave is small and narrow. During flood rain, in the semi blind valley, in which the entrance is situated, a huge amount of water is collected. As the entrance is small, the water is sucked inside and enters, creating a whirlpool with peculiar sounds. Hence, the Vlah name of the cave Vrkan (whirlpool), or The Singing Cave. The third name is a modern one.

The cave is horizontal, one-gallery and descending. An underground stream runs through it. The cave ends in a siphon. There are few secondary formations. The underground stream comes out ca. 300-400 m lower, as a karstic spring (description after P. Tranteev, 1978).

The cave has been known to local people for a long time. It was explored and surveyed in 1969 by P. Tranteev, P. Beron and Zdravko Iliev.

Fauna: so far 16 species have been identified (collected by P. Beron in 1969). Among them are the troglobites *Trichoniscus bononiensis* and *Hyloniscus flammula* (Isopoda) and *Plusiocampa beroni* (Diplura).

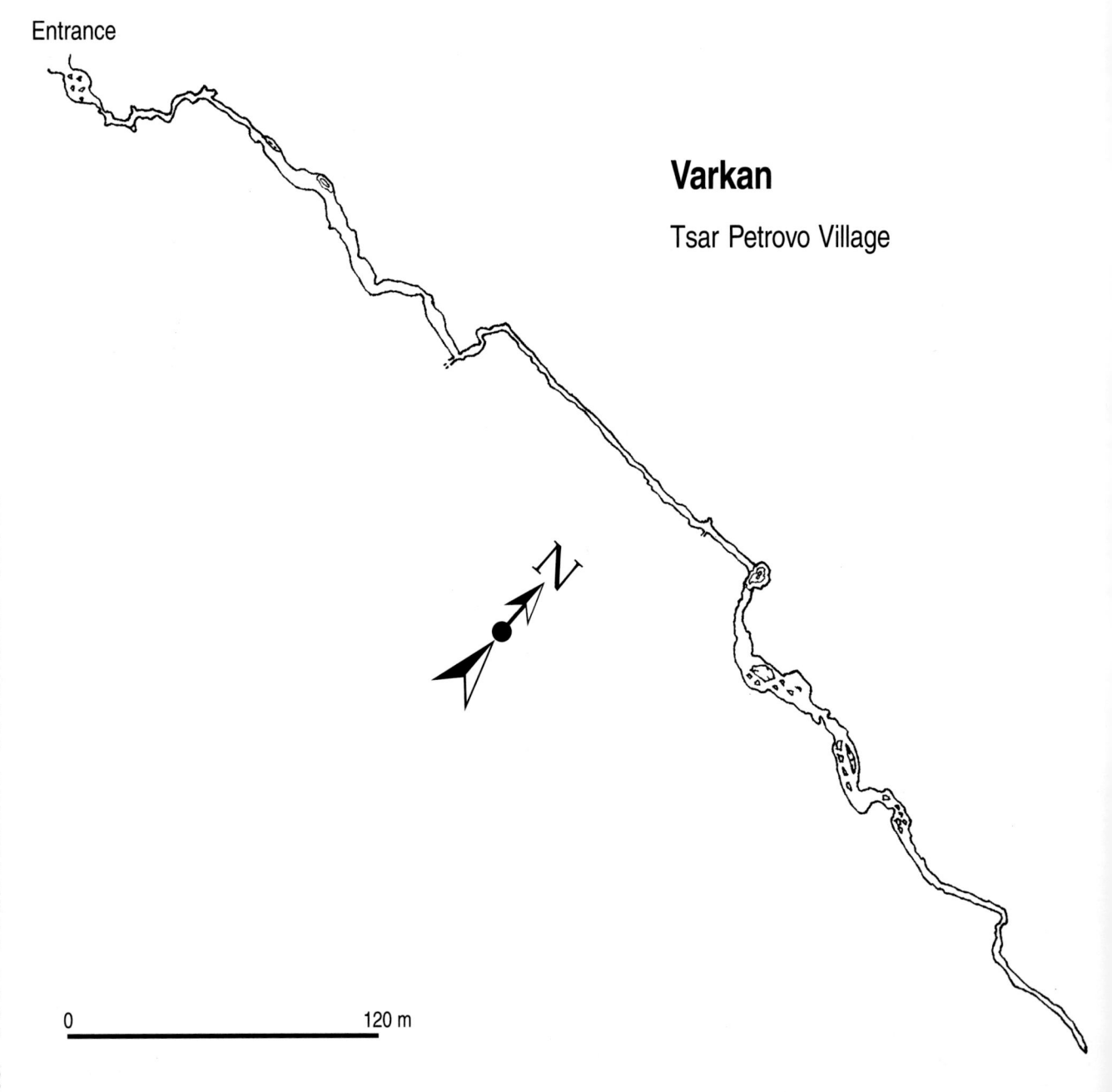

Varkan

Tsar Petrovo Village

Name: **ZANDANA (BISERNA) (175)**
Shumen Town
Length 2716 m. Denivelation 18 m. Surface 10 420 m²
Volume ~50 000 m³

A cave, one km west of Shumen, in the park Kyoshkovete. Situated on the NE slope of Shumensko Plateau. Formed in sugar-like yellowish limestone. The cave is ascending, two-storey, branched and with a permanently flowing river. Length of water gallery – 882 m. Dimensions of the cross-section of the cave: 5.5 m wide, 4 m high. Maximum dimensions: width 46 m, height 12 m. Rich in flowstone. Clay deposits, thick up to three meters. Boulder chokes. In the gallery an underground stream flows with an average outflow of 8 l/sec, min. outflow 4 l/sec, max. ~600 l/s.

This water cave has been a source for drinking water for Shumen since 1897. The first information about the cave is dated as early as 1828 (by the French geographer J.-G. Barbier de Bocage). The first modern exploration was done in 1968 by cavers from Shumen and Sofia. At that time 1440 m were surveyed. In 1971, K. Dimchev and Sn. Stancheva discovered an upstream continuation and the known length of the cave became 1800 m. For the transformation of Zandana into a show cave, A. Spassov made in 1979–1980 a detailed instrumental survey up to the present day length of the cave. In the period 1976–1985, the cave was converted into a show cave, but was never inaugurated for visiting. The research in 2003 resulted in the discovery of new galleries, bringing the length of the cave to 2716 m. Protected, within Shumensko Plateau Natural Park.

Fauna: five species of invertebrates are recorded and at least 11 species of bats.

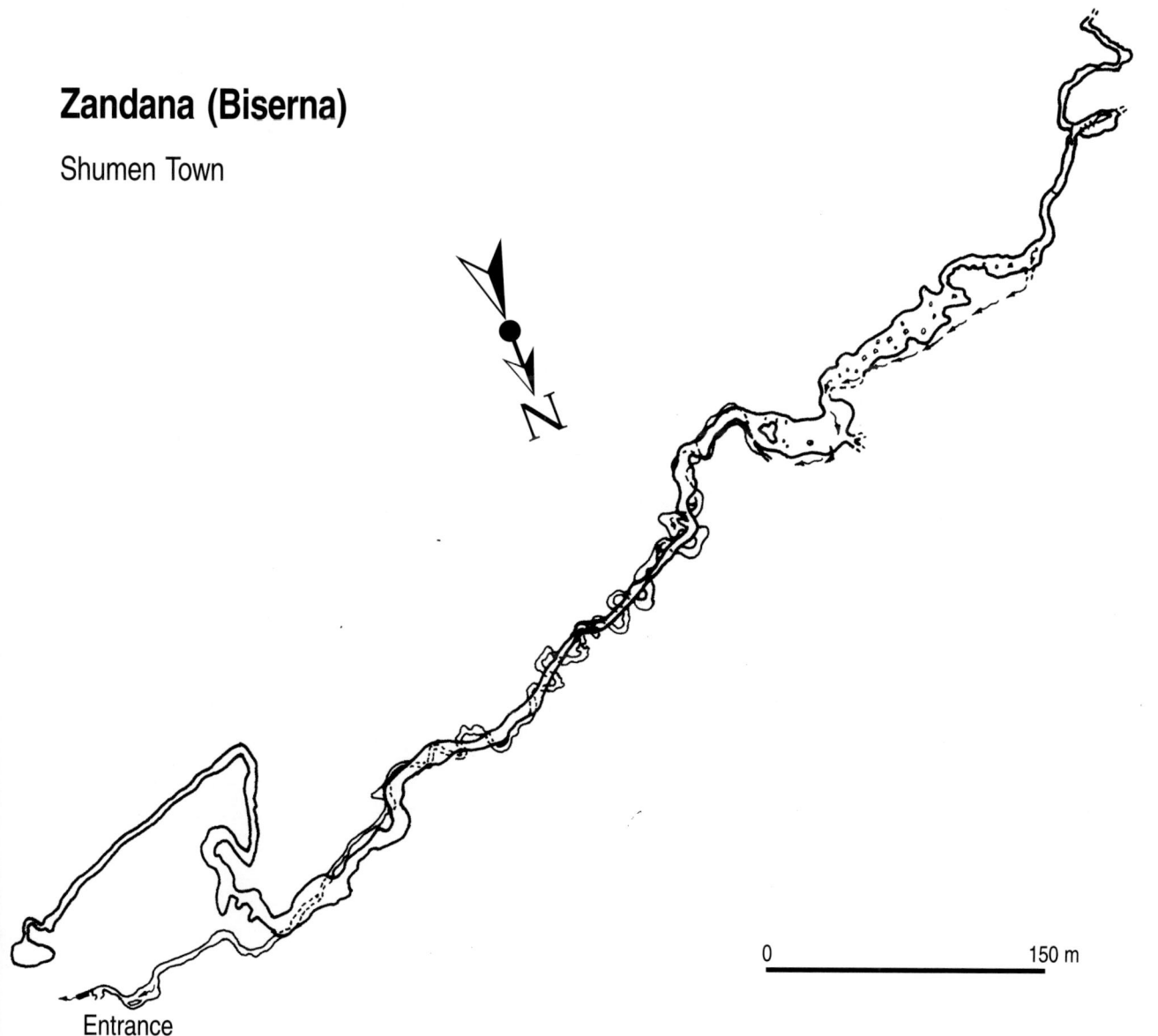

Zandana (Biserna)

Shumen Town

The Big Room – photo Trifon Daaliev

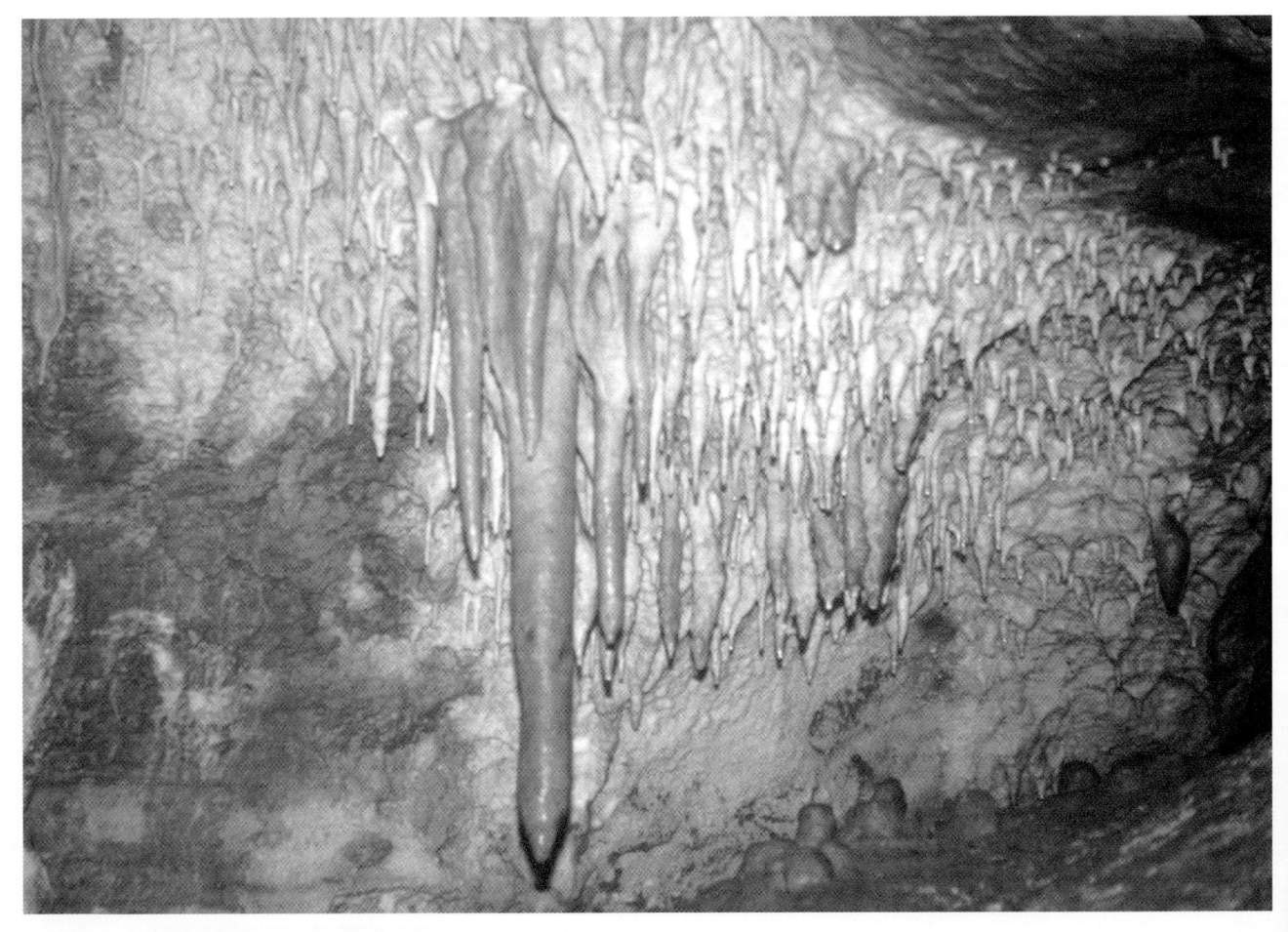

Zandana – Stalaktites – photo Trifon Daaliev

CAVES IN THE REGION OF STARA PLANINA AND THE PREDBALKAN

Name: **AKADEMIK (179)**
Kotel Town, Sliven Distr.
Altitude 705 m. Length 580 m. Denivelation -158 m
Coordinates: N 42° 56' 08,55" E 26° 30' 46,08"

A cave pothole in Senonian limestone in the locality Zlosten, 3.5 km northeast of Kotel Town. Length on the main axis – 300 m. Ten shafts of different dimensions, the deepest being 13 meters. In the middle, a small stream appears. Explored and surveyed in 1964 by cavers from Akademik, Sofia Caving Club, and later by cavers from Protey, Sliven Caving Club.

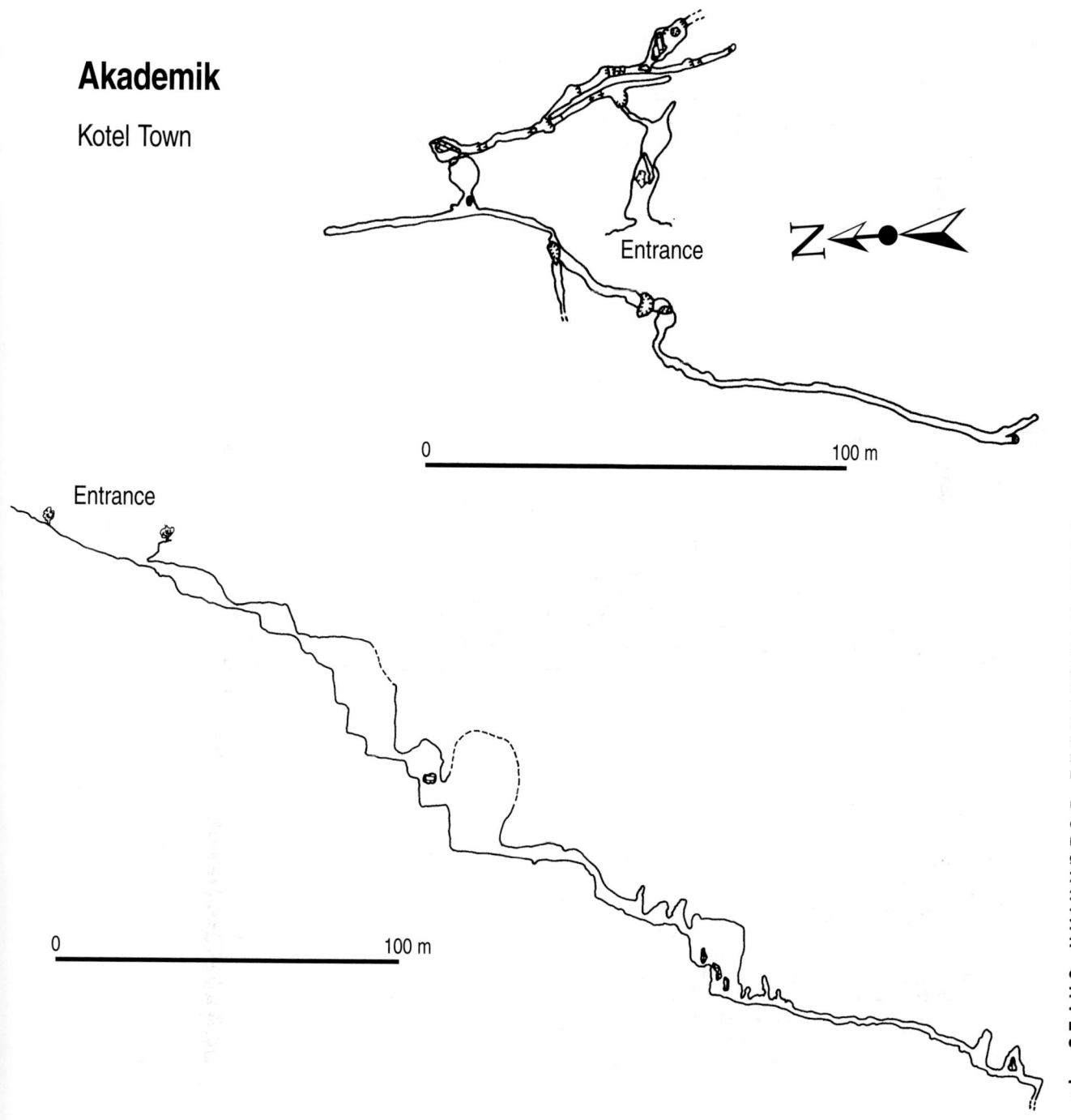

Akademik

Kotel Town

Name: **ALCHASHKATA (ALCHANSKATA) PESHTERA (102)**
Bezhanovo Village, Lovech Distr.
Length 227 m. Denivelation 2.4 m. Volume 296.6 m³

A cave 2.5 km from Bezhanovo, in the locality Gorna Turiya (Alchashkoto). The entrance is situated on the right bank (geographically) of Kamenitsa River (dimensions 6x3m). The cave is horizontal, dry diaclase, periodically humid. Formed in limestone from the Cretaceous (Maastricht).

The cave was surveyed in 1982 during a District Caving Expedition, which added another 40 m to its length. Air temperature at the entrance 12.4°C, at the bottom 12°C. Humidity is 56%.

Fauna: collected by B. Garev, but not fully identified.

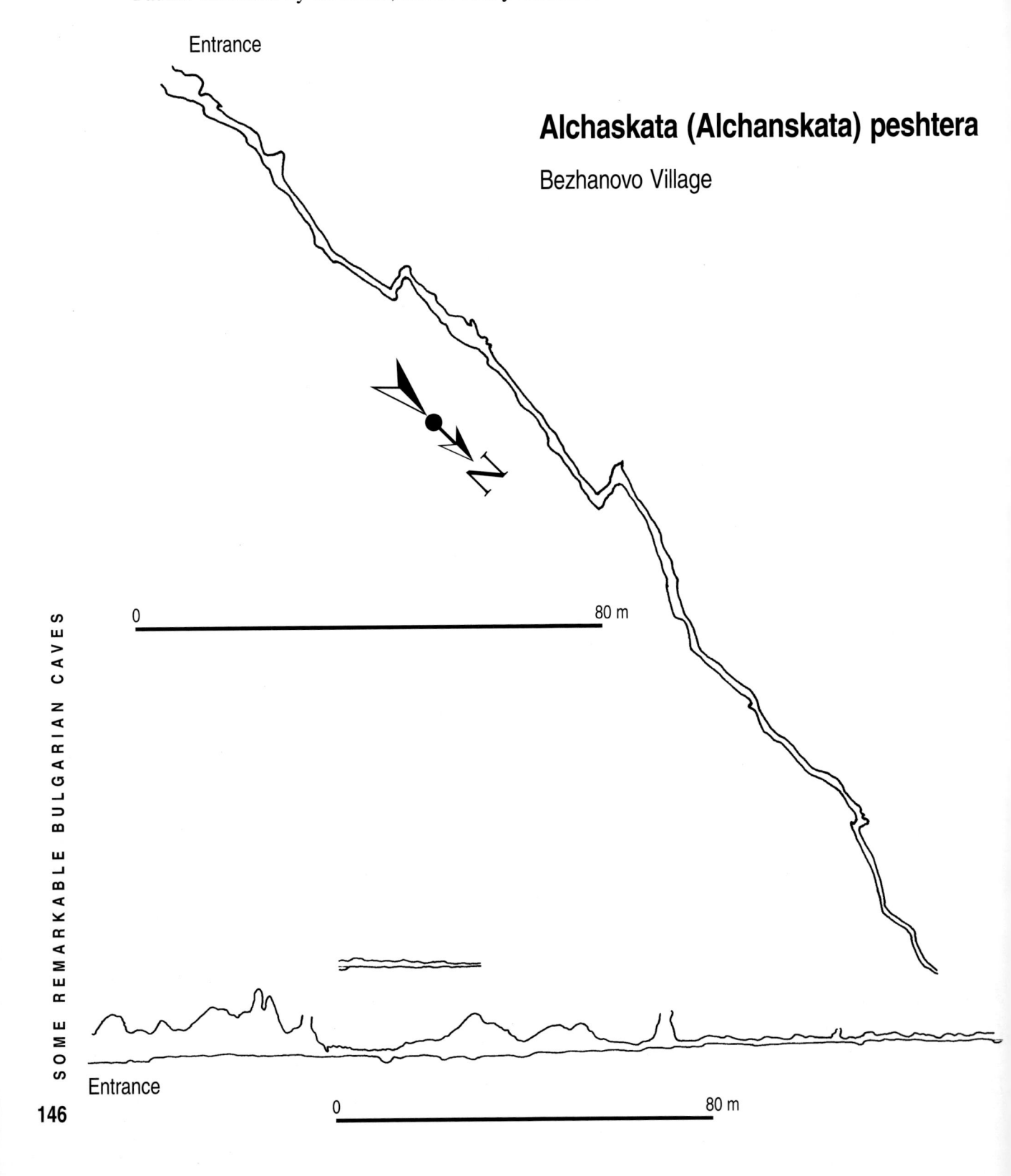

Alchaskata (Alchanskata) peshtera
Bezhanovo Village

Name: **ANDÂKA (GOLYAMATA, VODNATA PESHTERA) (166)**
Dryanovski Monastery
Altitude 280 m. Length ca. 4000 m
Coordinates: N° 42 56' 52,5" E 25° 25' 54,1"

A cave, 150 m west of the St. Archangel Mikhail monastery, in the eastern periphery of the karstic plateau Strazhata. Formed in Lower Cretaceous limestone.

A horizontal cave with large chambers through which an underground stream runs.

The cave is situated at the end of the ? canyon of Andaka River. The name of the cave and the river come from the large entrance (Andak = Hendek = pothole), which was closed for the construction of a water cleaning station in 1946. The entrance of the cave faces east (ca. 25 m high and ca. 40 m wide). In the entrance hall there is a water cleaning station, which provided drinking water to Dryanovo up to 1979. The cave has developed in Cretaceous limestone (Apt – Urgon – Barem complex) of the syncline karstic plateau Strazhata, the main direction being NE-SW. In the cave there is an underground river (min outflow ca. 15 l/s).

The connection of this cave with the caves near Dolni Varpishta Village (1 km S-SW, 475 m altitude) and with some ponors near Donino Village (7 km SW, 550–600 m altitude) has been proven, as well as with the karstic valley near Kostenkovtsi Village (2 km NW, 650 m altitude).

These indisputable, but yet unpenetrated connections, prove the existence of a huge cave complex. We could divide the cave in three parts, according to their discovery.

The Older part – up to the Second siphon, surveyed in 1974–1976, is on two levels, with labyrinthine galleries, an indisputable part of Bacho Kiro – Andaka System. Proven, but yet unpenetrated, is the connection with Vodnata (Kazana) in Dolni Varpishta Village. Three streams with different water temperature come from the left side and join the main river.

At the beginning the new parts represent low galleries with small branches, ending in three large chambers. The galleries have developed on the right and above the main river. Coming from the right (direction Kostenkovtsi), there are two streams running across the galleries, one of them being a sink of Andaka River. Further, the cave goes on with high galleries, giving access to the main river in a wide and low chamber with siphons on both ends.

The newest parts, also on the right side of the river, are still under study and survey.

The cave is poor in flowstone formations. In some chambers, there is a high content of CO_2 up to 4 % (description of Vanyo Stanev).

First mentioned by St. Yurinich in 1891. Excavations at the entrance by R. Popov in 1920 and in 1937. First registered research by N. Radev on October 28, 1923. In May 1924, Iv. Buresch and D. Ilchev collected cave fauna. N. Radev published (1926) a survey of the first 100 m. In 1966, Planinets, Sofia, surveyed another 1800 m of the cave. Under the care of Strinava, Dryanovo Caving Club, the known length of the cave increased to 3200 m in 1977 and to 4000 m in 1984.

Fauna: 16 species, including the troglobite *Trichoniscus tenebrarum* (Isopoda).

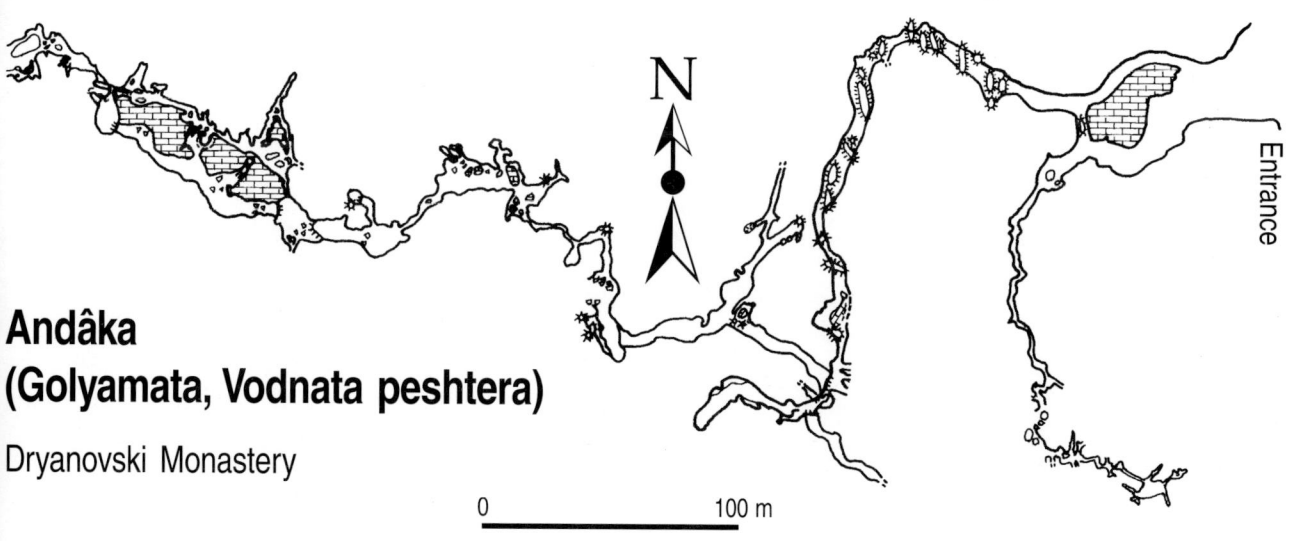

Andâka
(Golyamata, Vodnata peshtera)
Dryanovski Monastery

0   100 m

The entrance of the cave Andaka – photo Trifon Daaliev

The stone waterfall – photo Trifon Daaliev

V. Peltekov on the cement water-line – photo Trifon Daaliev

Name: **BACHO KIRO (MALKATA PESHTERA) (167)**
Dryanovski Monastery
Length 3500 m. Denivelation 65 m
Coordinates: N 42° 56' 50,5" E 25° 25' 54,7"

A cave, 150 m from Dryanovski Monastery and 50 m from the cave Andâka. Formed in Lower Cretaceous limestone. A complex labyrinth of galleries, developed in four main levels and connecting ca. 12 chambers, the biggest and the highest of which is Srutishteto (25x18 m), reaching its maximum height in Dâzhdovnata Zala. Minimum dimensions along the profile of the main galleries – 0.30 x 0.60 m, maximum – 12 x 30 m. Main direction NE-SW. Diversity of flowstone and other cave formations. On the floor of the lower level galleries of the cave there are thick clay formations, on the upper storeys – mainly river gravel from the earliest stages of the cave formation – ca. 1.8 million years ago. Archaeological excavations in the cave revealed artefacts from the following cultures: Middle Paleolythicum – Moustier – 47 000 years ago; Late Paleolythicum – Orygnac (over 43 000 years ago); Neolythicum and Eneolythicum. Remains of *Homo sapiens* dating back to more than 43 000 years ago have been discovered and also fossil fauna (bones of 21 species of small mammals from the Upper Pleistocene and from 11 species of large mammals). The cave was first recorded by H. Škorpil (1884). Explored on May 19, 1890 by St. Yurinich – up to 143 m. First archaeological explorations in 1920 and 1924 by R. Popov. Map of entrance parts (60 m) published by N. Radev (1926), who visited the cave in 1923 and 1925. In 1937 and 1938 other explorations were done by R. Popov and by the American D. Garod. Survey of the cave in 1966 by cavers from Planinets, Sofia Caving Club (up to 2400 m). Research has been carried out also by cavers from Strinava, Dryanovo Caving Club, bringing the cave to its present length. In 1971–1976, Polish-Bulgarian archaeological excavations took place. Geomorphologically studied by V. Popov and Iv. Vaptsarov (1972). Partly electrified in 1937 (the first Bulgarian show cave). In 1964 the tourist circuit was increased to 450 m.

Declared Natural Monument with the adjacent 0.5 ha in 1962. Declared Monument of Culture in 1971.

Fauna: 28 species, including the troglobite *Netolitzkya jeanneli jeanneli* (Coleoptera).

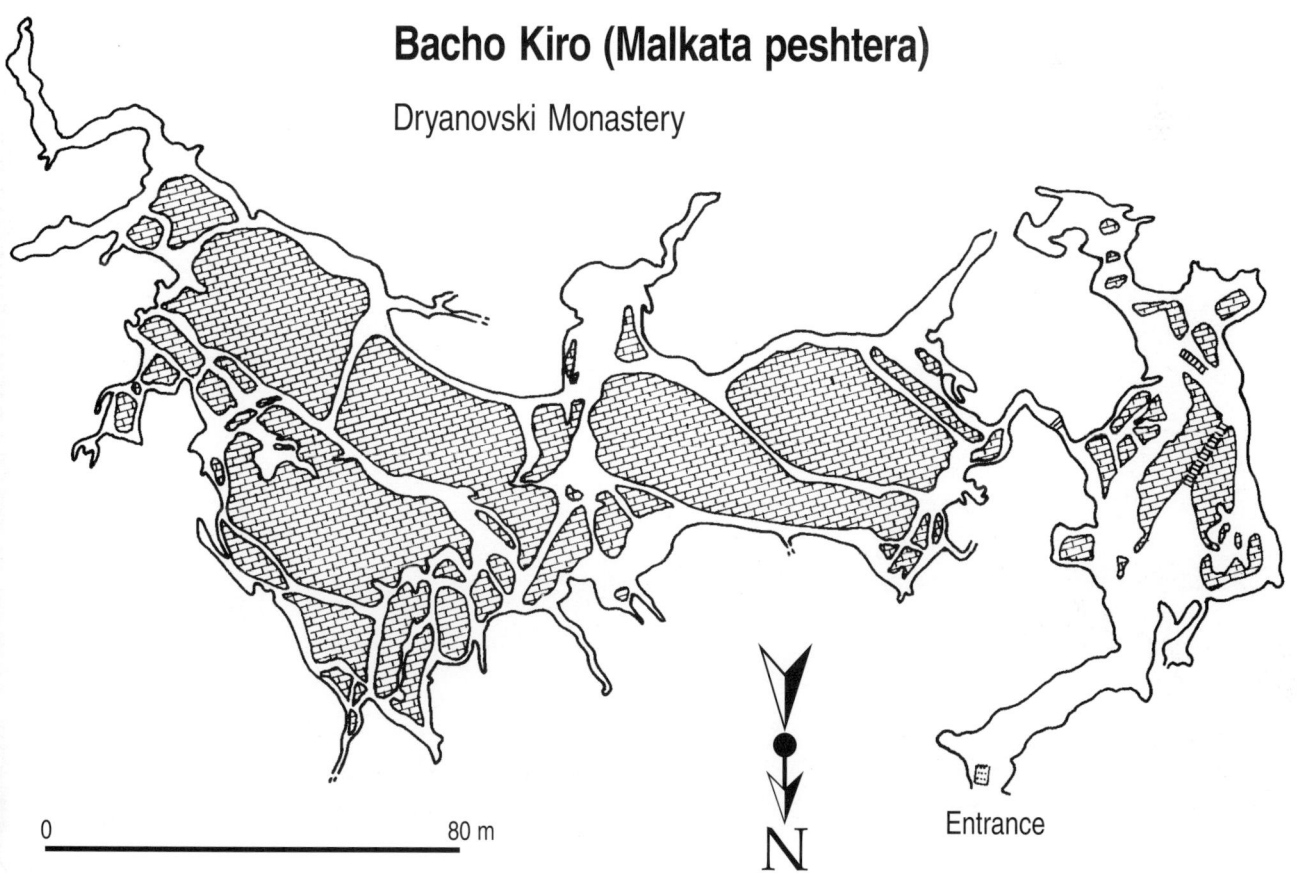

Bacho Kiro (Malkata peshtera)
Dryanovski Monastery

The opening ceremony after reconstruction of the cave Bacho Kiro – photo Trifon Daaliev

In the last room of the tourist part – photo Trifon Daaliev

Column – photo Trifon Daaliev

Name: **BALYOVSKI OBRUK (158)**
Krushevo Village, Gabrovo Distr.
Length 517 m. Denivelation -48 m

A cave 1.5 km from the camping site Burya in the locality Razboynik, 300 m northeast of the fountain. The cave starts with a funnel. The first 30-40 m are wide and low. Then they change their character, becoming narrow and high. By the middle of the cave there is a 10–15 m vertical. From the bottom of the vertical a water gallery starts, in which there is more mud than the water.

V. Botev, T. Boteva, B. Drandarov and R. Rachev, cavers from Viten, Sevlievo Caving Club explored the cave in 1975. In 1982 they made the map and the description.

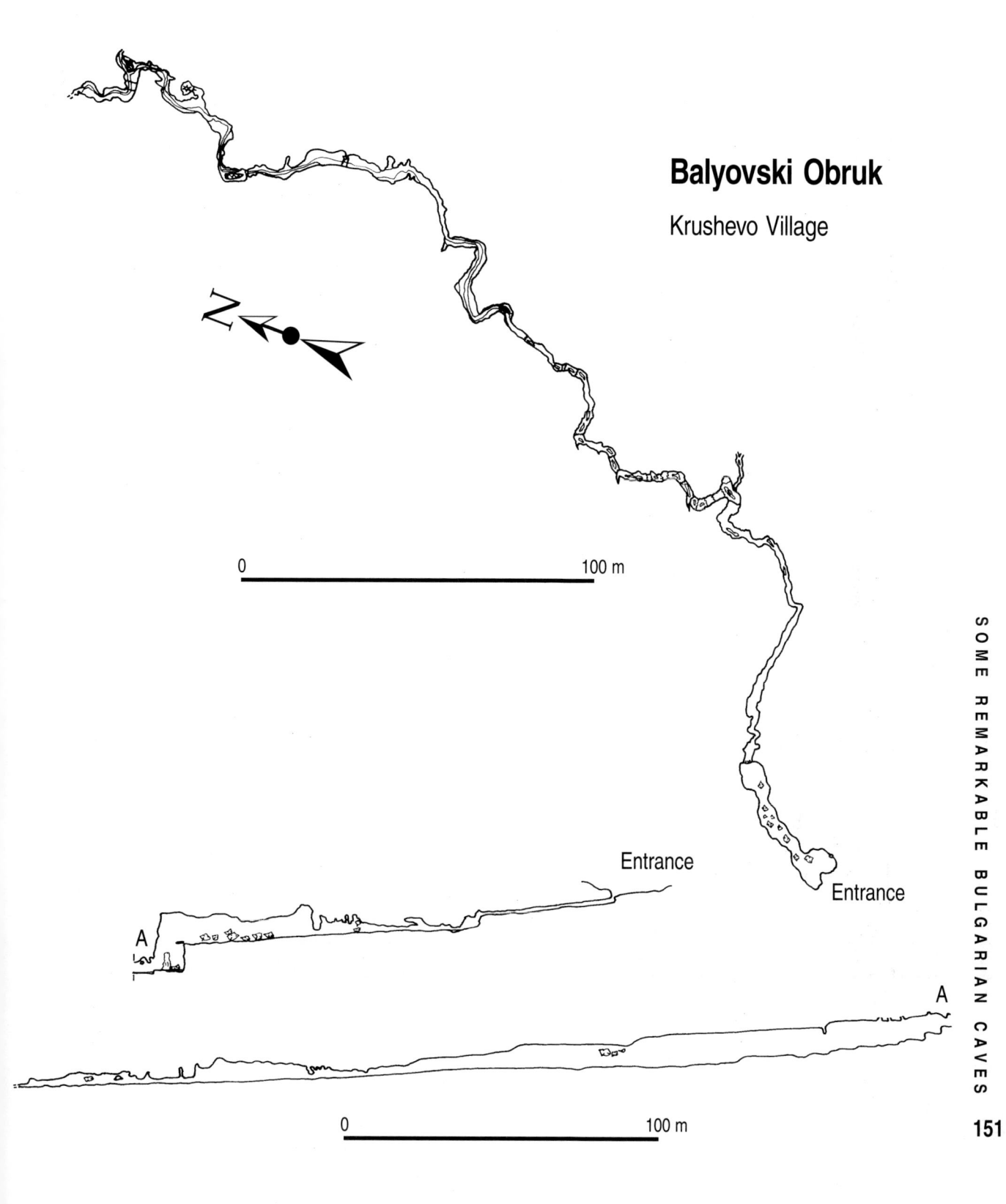

Balyovski Obruk

Krushevo Village

Name: **BAMBALOVA DUPKA (160)**
Emen Village, Veliko Tarnovo Distr.
Length – 2923 m. Denivelation -20 m
Coordinates: N 43° 07' 26,2" E 25° 19' 30,4"

    A water cave, 2,7 km SW of Emen. Formed in Aptian-Urgonian limestone. A descending, two-storey, branched water cave. The entrance serves as a periodical sink of surface water. High and narrow galleries. Two rivers flow in the cave, they appear from underground sources and sink in siphons. The rivers are fed by rainwater in the cave basin (surface 5 km^2). Found by the geographer P. Petrov in 1980. Entrance parts studied by V. Nedkov and T. Daaliev. Detailed survey and exploration by Prista, Russe Caving Club. Up to 1984, 1211 m had been explored, in 1980 – another 1323 m, in 1985 – another 389 m.
    Fauna: the troglobite *Netolitzkya maneki iltschewi* (Coleoptera).

Bambalova Dupka

Emen Village

The entrance of the cave Bambalova Dupka – photo Alexey Zhalov

Name: **BANKOVITSA (87)**
Karlukovo Village, Lovech Distr.
Altitude 240 m. Length 430 m. Denivelation -50 m. Surface 4150 m². Volume 27 667 m³
Coordinates: N 43° 10' 24" E 24° 04' 44"

A pothole system, formed in Upper Cretaceous (Senonian) limestone, 1.5 km North of Karlukovo Village. The entrance is a pothole, 17.5 m deep (incl. 15 km vertical). Dimensions of the entrance 22 x 12 m. A horizontal gallery follows with a recurrently opening mud siphon. When open, it leads into a small chamber, a much bigger chamber follows, with a lake with dimensions 10 x 14 m and depth of 1.5 m. Everywhere there are huge masses of clay. The prolongation after the siphon was discovered in May 1958 by the participants in a caving course, conducted by P. Tranteev. N. Korchev and P. Beron were the first to enter the big chamber. In the lower part, underground rivers flow. Surveyed many times, the actual map was made in 1988 by V. Mustakov, I. Yordanov, K. Danailov, D. Lefterov, cavers from Helictite, Sofia Cavining Club.

Fauna: among the many animals recorded are the troglobites *Bulgardicus tranteevi* (endemic genus and species, described from this cave) and *Typhloiulus bureschi* (Diplopoda) and the troglophiles *Trichoniscus bureschi* (Isopoda), *Porrhomma convexum* (Araneae), *Paranemastoma radewi* (Opiliones) and *Tomocerus minor* (Collembola).

Bankovitsa

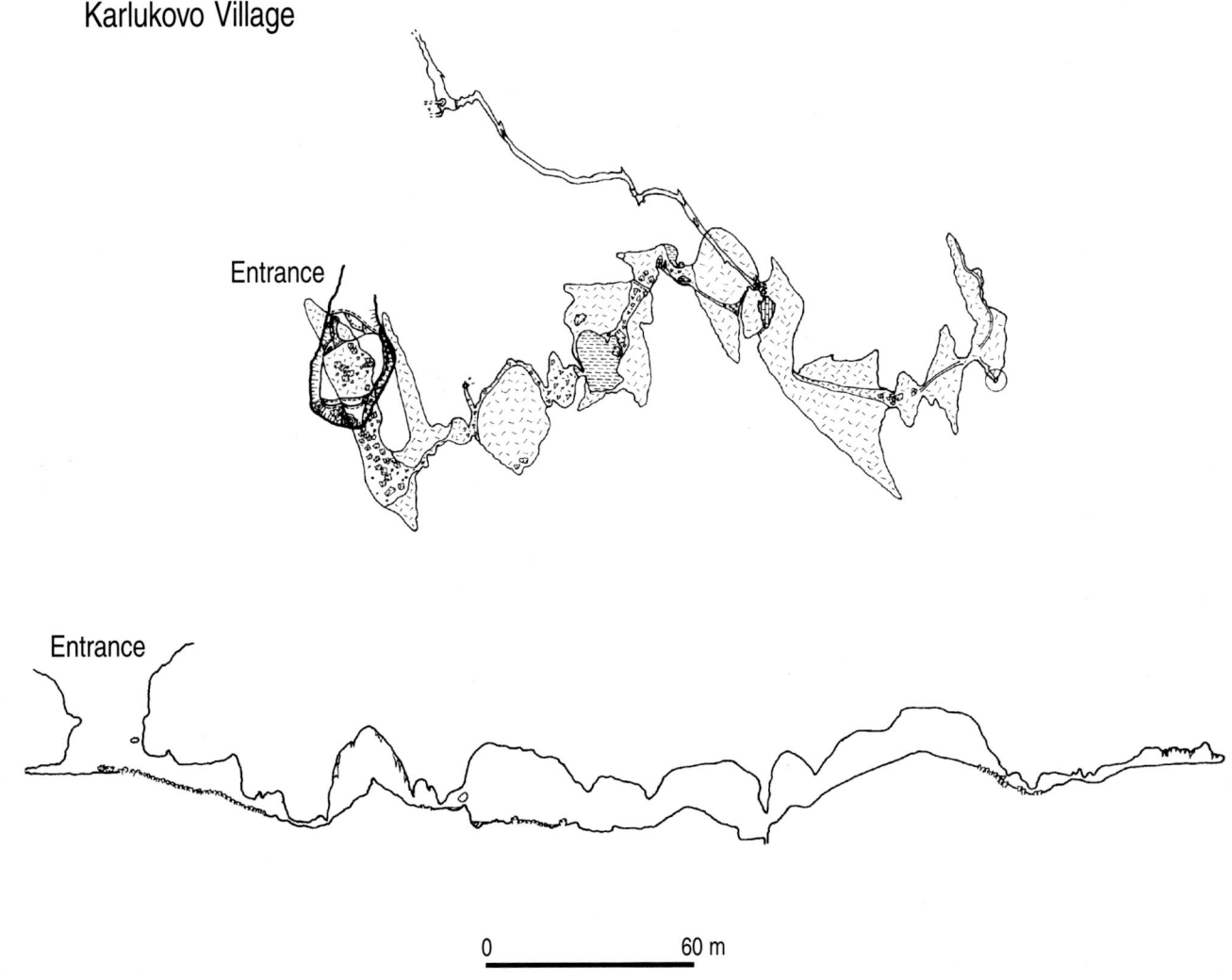

Karlukovo Village

Name: **BAYOV KOMIN (82)**
Drashan Village, Vratsa Distr.
Length 2196 m

    A water cave, 2.5 km W of Drashan, in the outskirts of the hill Kirkovoto. Formed in Upper Cretaceous (Maastricht) limestone. A horizontal, branched cave, developing E/NE, with a side-elongated profile of galleries of average dimensions of the cross section: width 4.6 m; height 1.4 m and min/max resp. 1.4–8 m and 0.4–2.8 m. Sinter formations. The entrance is at the end of an adoline and periodically takes in its waters. The main water input comes from a left branch, 70 m from the entrance. The water of the underground stream is drained through the cave Drashanska Peshtera. Discovered in 1968 (information from local people) by the cavers from Veslets, Vratsa Speleoclub. By 1977 it had been completely explored and surveyed by K. Karlov, S. Dyankov, G. Chakalski and K. Genadiev.

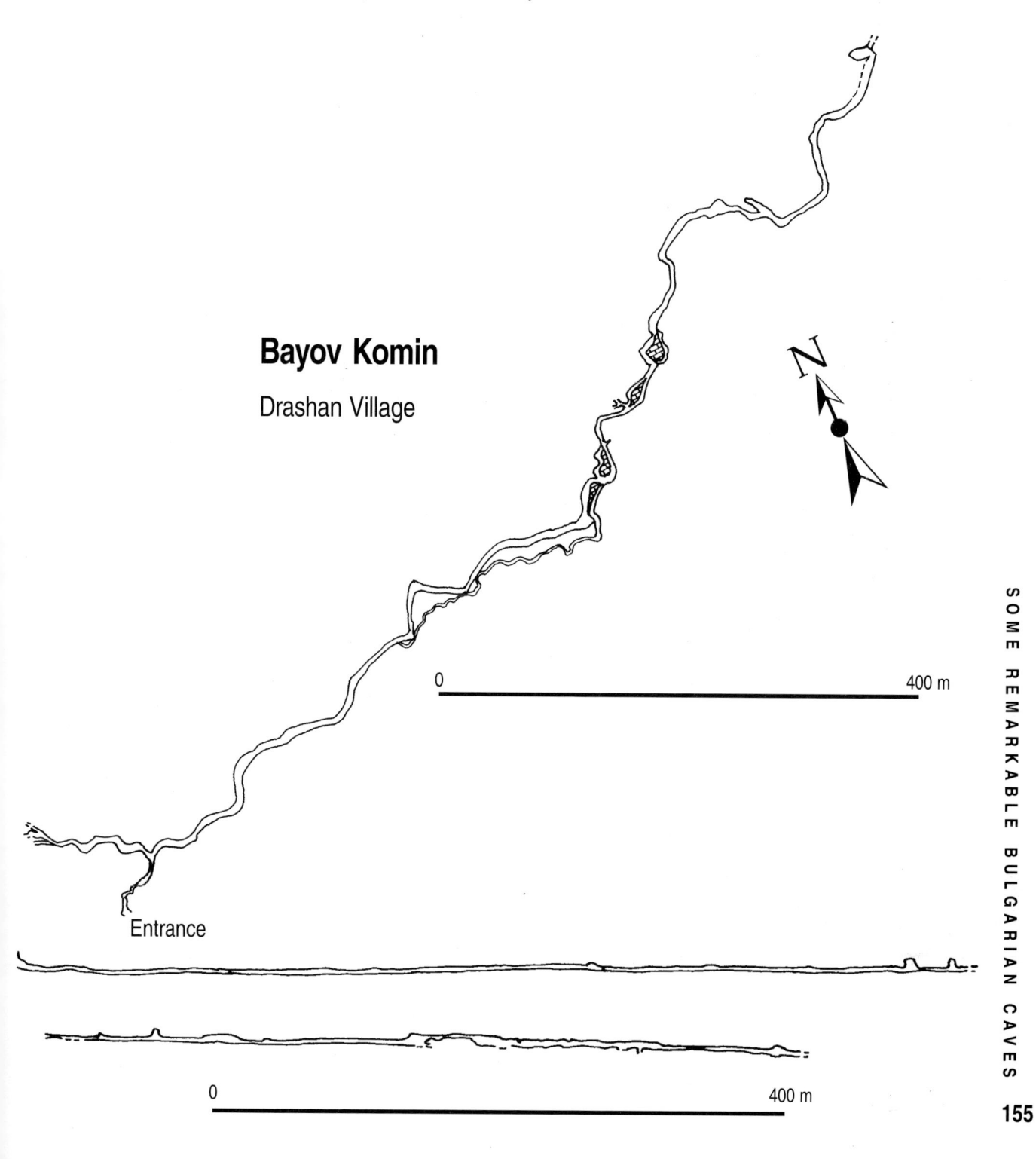

Bayov Komin

Drashan Village

Name: **BARKITE – 14 (37)**
Gorno Ozirovo Village, Montana Distr.
Altitude 834 m. Length 2600 m. Denivelation -356 m
Coordinates: N 43° 13' 15" E 23° 27' 29,7"

The entrance of the deepest pothole in the park Vrachanski Balkan is situated near a place called Studenite Korita, ca. four kilometers NW of Ledenika Hut. The pothole is built of Upper Jurassic – Lower Cretacous limestone.

From the entrance (0,5 x 0,5 m) a descending gallery starts, going on for 76 m. Its length is 200 m. A very narrow opening follows and a vertical of nine meters. Further on, the gallery goes into a squeeze with water in it. After 450 m, a vertical of seven meters is reached (this point is at 145 m denivelation). A lake (2 x 3 x 0,5 m) and an upper storey follow. After 350 m, we find a narrow semisiphon and a 130-m horizontal part. After several verticals of six, seven and ten meters, we pass a 13-m meander, followed by an eleven-meter vertical, twelve meters after which we reach the biggest vertical in the cave. It is 18 m, divided into two parts of six and twelve meters. The last vertical of eight meters leads to the end of the siphon. There is an underground stream.

The participants in the International Ralley Ledenika 68 T. Tranteev, Z. Iliev, T. Daaliev, P. Beron and others (up to 70 m) were the first to penetrate the cave. After more than 15 years, cavers from Akademik – Sofia Caving Club, led by M. Zlatkova, enlarged the squeeze and discovered the new parts. M. Zlatkova and M. Stoyanov reached the siphon at – 140 m in May 1982. One year later the siphon was passed by Tz. Ostromski and M. Zlatkova. The bottom was first reached in early November 1983 by M. Zlatkova, M. Stoyanov, O. Stoyanov, Ya. Bozhinov and V. Shekerdjieva. The depth reached was 350 m, and the length 2560 m. In 1988 the cavers of Studenetz – Pleven Caving Club organized diver's expedition and underground bivouak for exploring the end siphon. The divers V. Chapanov and M. Dimitrov dived in the siphon and the result, after all these efforts, was that the system became longer by 24 m (to 2584 m), and the depth increased by 6 m.

Fauna: known stygobites are *Diacyclops languidoides*, *D. bisetosus* (Copepoda)

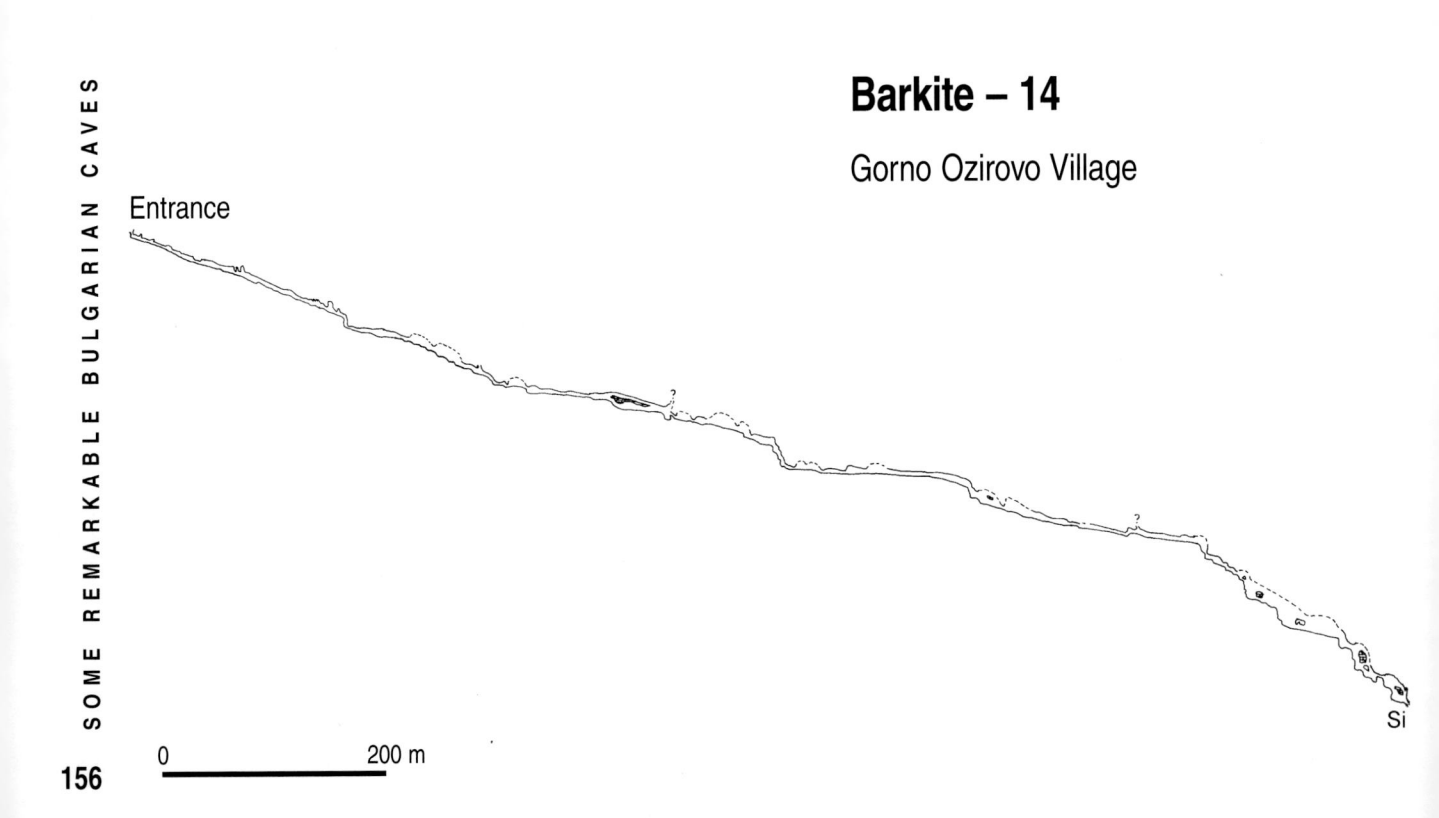

Barkite – 14

Gorno Ozirovo Village

Name: **BARKITE – 8 (25 YEARS AKADEMIK) (33)**
Gorno Ozirovo Village, Montana Distr.
Altitude 830 m. Explored length 733 m. Denivelation -190 m
Coordinates: N 43°13' 18.7" E 23°27' 24.6"

A water cave in Upper Jurassic – Lower Cretacous limestone, on the territory of Vrachanski Balkan National Park. Discovered in 1982 by cavers from Akademik, Sofia Caving Club.

The first part is relatively narrow and descending, then it becomes wider and more abrupt. There is only one vertical of six meters near the end of the cave. When in flood, it is better to equip the vertical with a twelvemeter rope or ladder, in order to avoid the water. There is a small stream.

In the middle of the cave there is a dry, ascending, side gallery.

Map – 1985 SPK Akademik, Sofia. Tz. Ostromski, M. Zlatkova, D. Shekerdjiev, T. Medarova.

Fauna: includes the troglobites *Pheggomisetes globiceps mladenovi, Ph. buresi* and *Duvalius zivkovi* (Coleoptera, Carabidae).

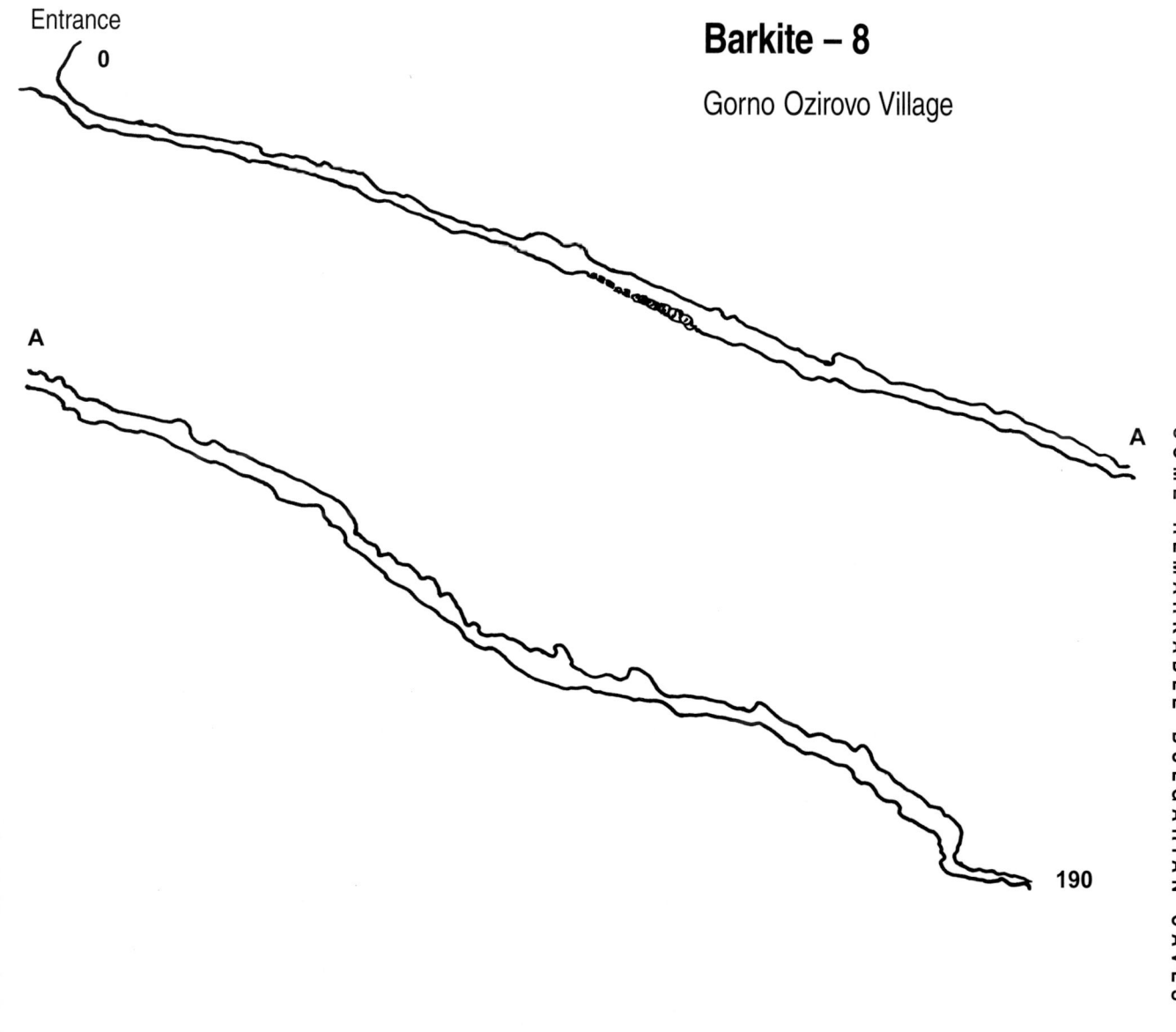

Barkite – 8

Gorno Ozirovo Village

The beauty of the gallery – photo Trifon Daaliev

The decoration of the last part – photo Trifon Daaliev

Stalagton – Barki 8 – photo Vassil Balevski

Name: **BASHOVISHKI PECH (5)**
Gorno Ozirovo Village, Vidin Distr.
Altitude 300 m. Total surveyed length 1298 m. Explored length ca. 3300 m
Denivelation ca. 10 m

A water cave about 1500 m W-NW of Oreshets Railway Station near the bent of Vidin – Sofia Railway. Its entrance is situated on the eastern slope of Pâkina Glama, with a surface of 10 m^2 and dimensions: 7 m wide and 5 m high, 3/4 of it taken by huge boulders due to blasting.

Immediately after the entrance, there is a big chamber, periodically filled by water. A cave is formed in it. About 200 m from the entrance, there is a siphon, open for only 3-4 months every year. Another 200 m further, the gallery narrows heavily and is practically obstructed by sinters. From the inside, a stream with an outflow of 0.250 l/s runs. Immediately before the sinters, the gallery takes a branch on the right. Is walled by a one-meter concrete dam, under which the water passes through the deposits. It is possible to pass through a small opening in the wall. A gallery, about 1900 m long, follows. The first 80 m are inundated, the water being 1–1.5 m deep. Another 700 m of dry gallery follow, after which there are sinter lakes, which are dry in some periods. A gallery of ca. 300 m follows, ending in a siphon, circumvented by passing over to the second storey of the cave. The cave goes on for another 400 m and ends in a chamber. After the chamber, the gallery narrows heavily and it becomes impossible to continue. It is possible to look for a continuation.

About 800 m from the dam, down on the left, there is a branch. About 10–15 m from its entrance, we reach the bottom of a vertical of 10 m, from which an ascending gallery goes for some 400 m. It is devoid of flowstone formations, very narrow and with cutting edges on the wall. The sector goes further on with horizontal galleries and a height of ca. 30–40 cm.

The left gallery is a water table, about 400 m long. A low, ascending gallery, about 30–40 m long, follows, from which we reach a huge chamber, its length about 70–80 m and covered with huge blocks. Further on, we can either go down by the water, or up, over the blocks. The gallery goes on for another 500 m, lowers and ends in a siphon, not yet surmounted. The stream in the left gallery has an outflow of ca. 7–8 l/s. Its total length is ca. 1000 m. The entire cave is poor in flowstone formations, except for sinter lakes. In the right gallery, after the dam, for some 500 m there are clay deposits, which are about one meter thick.

The cave is a recurrent source with a maximum outflow, until December 3, 1985, of 3000 l/s. During heavy rains and snow melting water emerges from there. Known since the 19th century, its water being used in water mills and for irrigation.

In 1961, the cave entrance was blasted during Bashovitsa quarry works. Opened again on June 24, 1979 by cavers from Bononia, Vidin Caving Club (initiative of Svetoslav Gavrilov and Bogdan Todorov). In the same year, the first 200 m were explored. In the following years, cavers explored and described some 3300 m of galleries and surveyed ca. 1300 m of them.

Fauna: in 1959, the stygobite *Niphargus pecarensis* (Amphipoda) was described from this cave. The fauna includes also the troglobite Isopod *Trichoniscus bononiensis*.

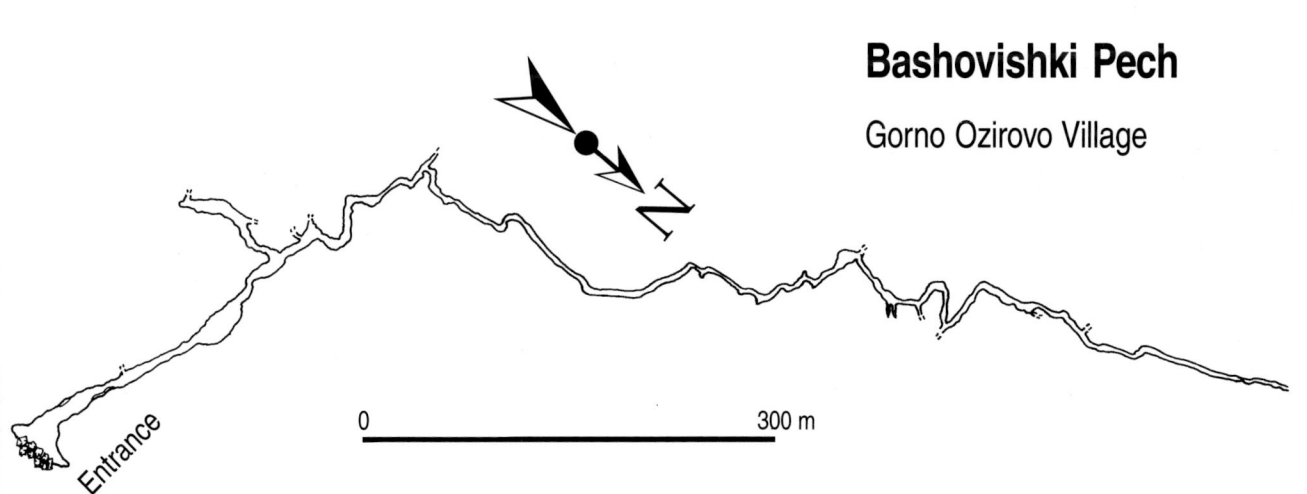

Bashovishki Pech

Gorno Ozirovo Village

Name: **BELYAR (34)**
Gorno Ozirovo Village, Montana Distr.
Altitude 860 m. Total length 2560 m. Denivelation -282 m
Coordinates: N 43° 13' 26.58" E 23° 27' 16.27"

A cave pothole, NW of Ledenika Hut. Formed in Upper Jurassic – Lower Cretaceous limestone.

The 18-m entrance vertical leads into a chamber with a mud floor, ending with huge boulders. A small stream runs through the cave. The end used to be a siphon, which was drained by cavers from Sofia, and the lowest point was brought to 282 m.

In the spring of 1968, local people showed the cave to V. Velkov and G. Valchinov (from the staff of Ledenika Hut). In October the same year, P. Tranteev and R. Rahnev surveyed 236 m of the cave (by that time, the second deepest cave in Bulgaria). In 1984, G. Markov, M. Zlatkova, Ts. Ostromski and other cavers from Akademik, Sofia Caving Club, drained the end siphon and discovered new parts. The cave is within the Natural Park Vrachanski Balkan.

Fauna: known troglobites and stygobites are *Speocyclops infernus* (Copepoda), *Niphargus bureschi* (Amphipoda), *Bureschia bulgarica, Sphaeromides bureschi* (Isopoda), *Centromerus bulgarianus* (Araneae), *Bulgarosoma bureschi* and *Typhloiulus staregai* (Diplopoda)

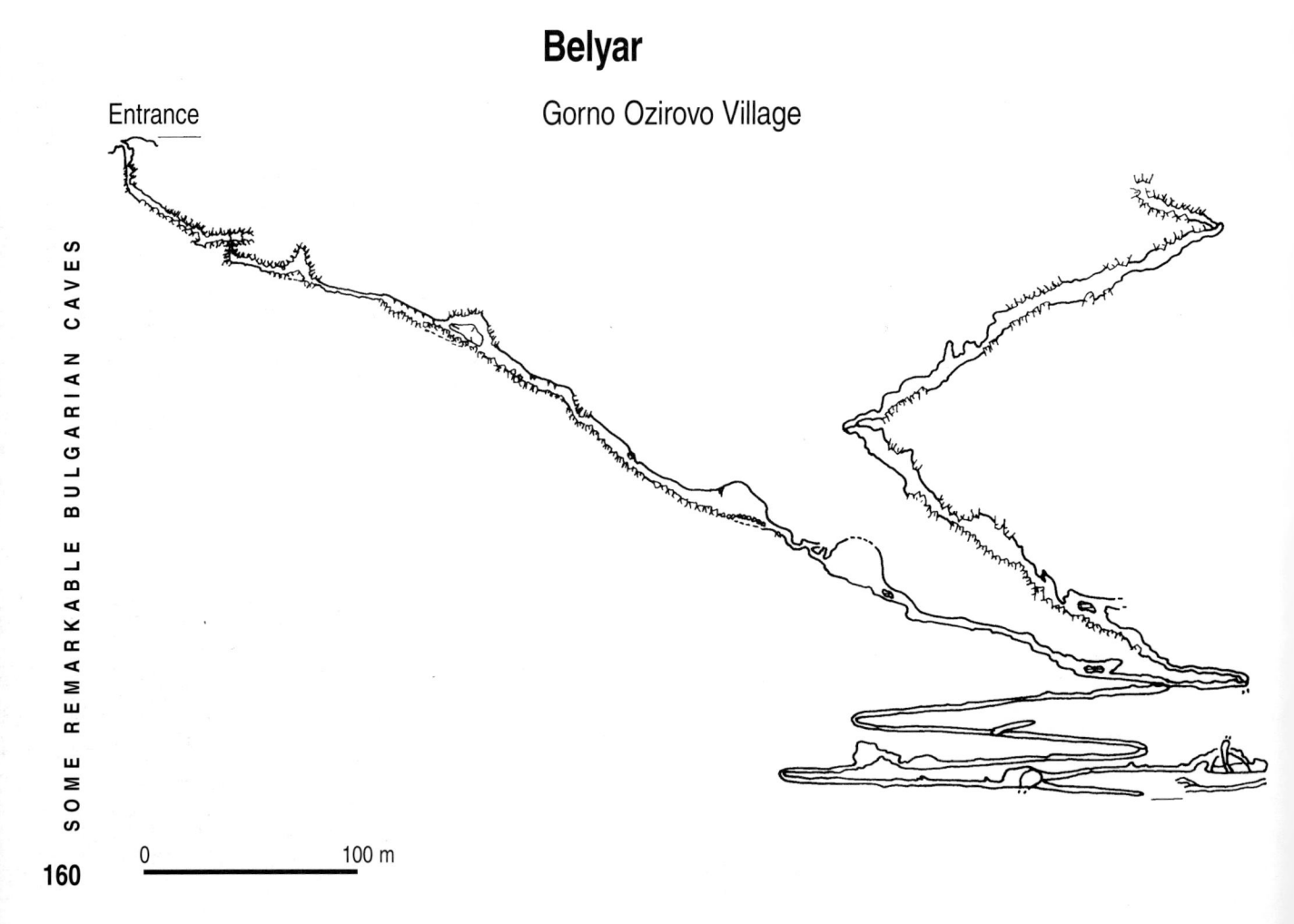

Belyar
Gorno Ozirovo Village

A. Manov waiting for the stretcher – photo Trifon Daaliev

The stone waterfall in the old bottom of the cave Belyar – photo Trifon Daaliev

Name: **BEZDANNIYA PCHELIN (109)**
Yablanitsa Town, Lovech Distr.
Denivelation -105 m
Coordinates: N 43° 04' 46.932" E 24° 16' 56.292"

A pothole, six kilometers NE of Yablanitsa, near a limestone quarry. Formed in Upper Jurassic limestone (Titonian). A huge opening (50 x 30 m). A vertical part of 90 m, followed by a scree of 15 m. It ends with a filled ponor. No cave formations.

The first attempt to penetrate the pothole was made by the Czech caver Novak in 1924. He used a ladder and surveyed the pothole up to 88 m. On April 5, 1931, N. Radev and P. English descended the pothole and assessed its depth at 146 m. In 1948, another group, led by A. Petrov, descended again the pothole and considered it being 127 m deep. Until 1957, Bezdânniya Pchelin was thought to be the deepest pothole in Bulgaria. In 1967, the Caving Commission of the City Council (Sofia) of the Bulgarian Tourists Union organized an expedition to descend and correct the map of the pothole. Members of the Sofia Caving Clubs of Edelweiss, Planinets and Vitosha – V. Nedkov, A. Bliznakov, T. Daaliev, N. Genov, L. Popov, R. Radushev and others took part in the expedition.

Bezdanniya Pchelin

Yablanitsa Town

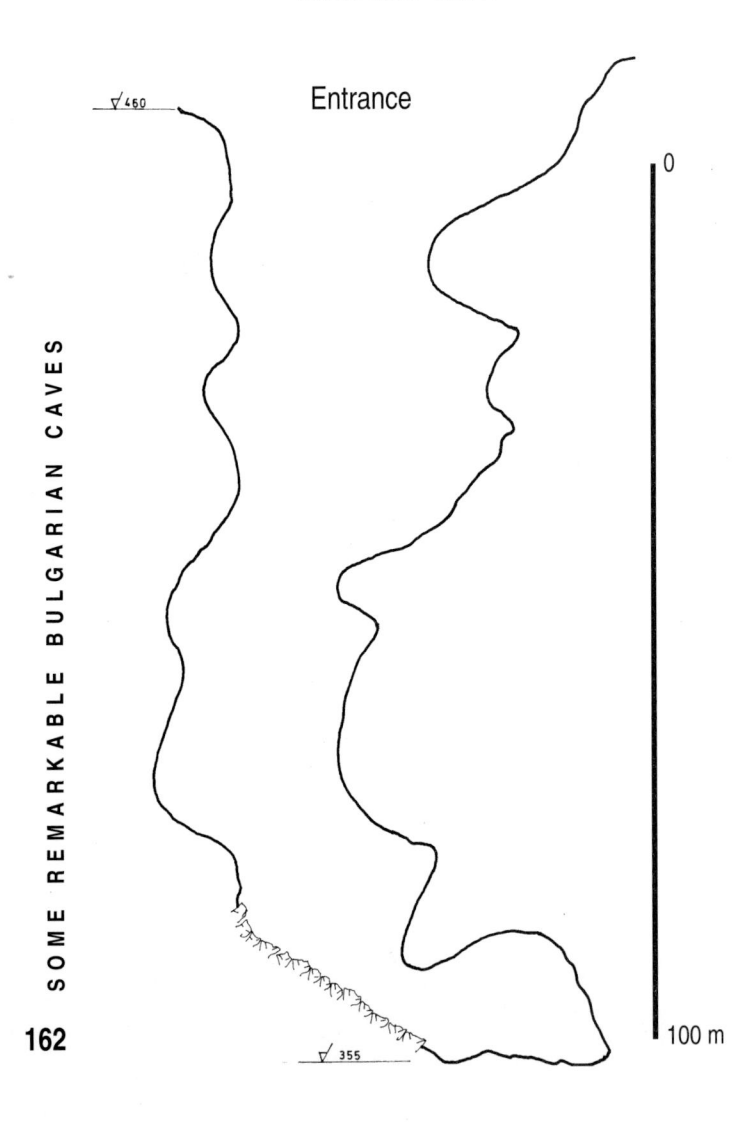

The bottom of the shaft – photo BFS archive

Name: **BLAGOVA YAMA (73)**
Etropole, Sofia Distr.
Altitude 750 m. Length 284 m. Denivelation -153 m

A pothole in Trias limestone and dolomite not far from Etropole, in the locality Blagovoto. A cave pothole with a 60-meter internal vertical (diameter 10 m) and several smaller verticals. They are separated by muddy passages and are sometimes very wet, even with running water. The pothole was discovered in 1959 by Dyado (Grandfather) Blago, an old man who inhabited the house at that time, while trying to dig a well. First penetrated by D. Ilandjiev in 1966 (reached a depth of 55 m). In 1969, L. Popov and cavers from Al. Konstantinov, Sofia, reached 112 m. In 1979, another 63 m of a narrow horizontal meander was overcome by cavers from Mrak, Etropole Caving Club. In 1987, after hard work the same club reached 153 m.

Fauna: only the troglophile spider *Porrhomma convexum* was identified.

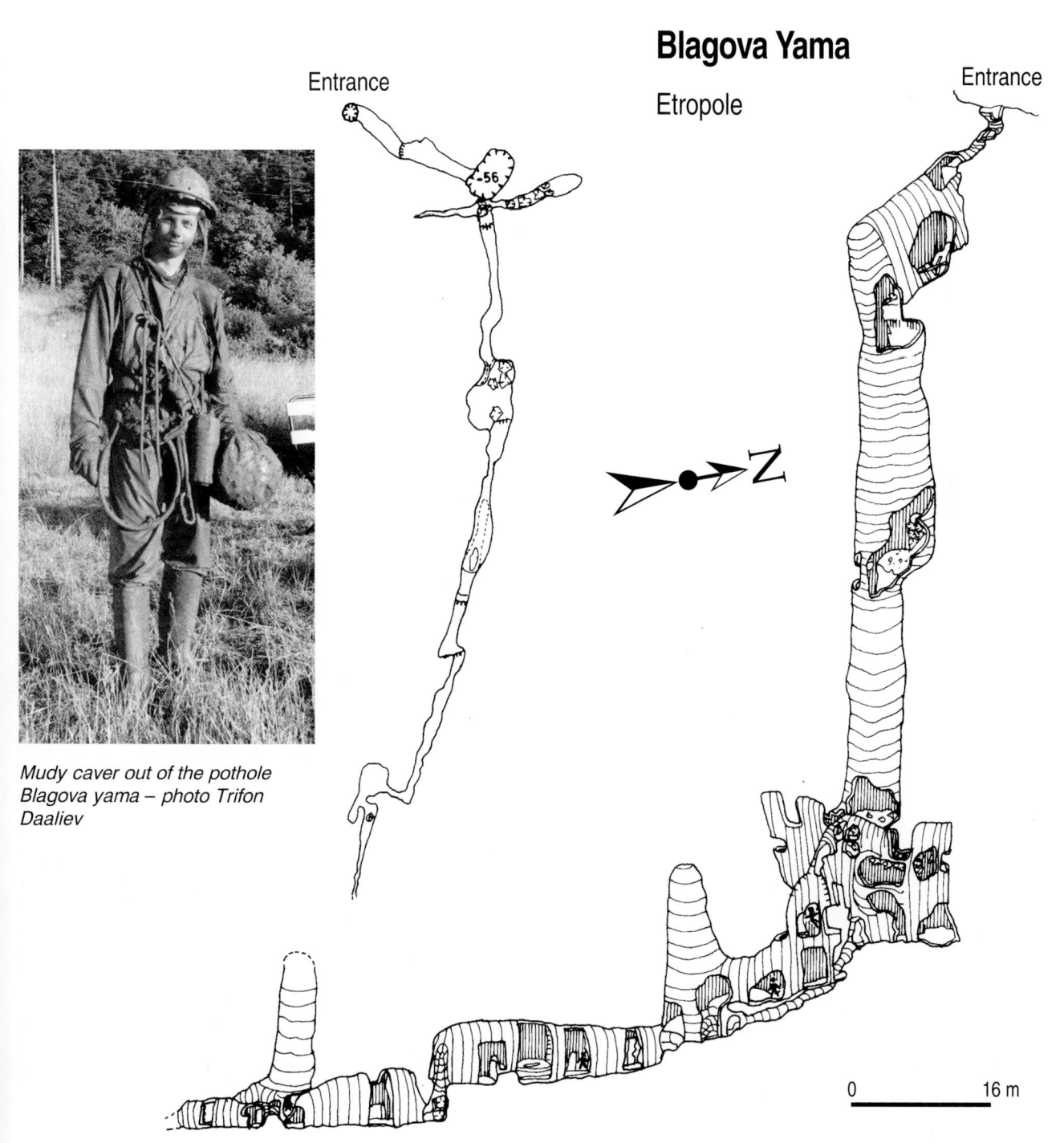

Mudy caver out of the pothole Blagova yama – photo Trifon Daaliev

Name: **BONINSKATA (POPSKATA) PESHTERA (154)**
Krushuna Village, Lovech Distr.
Altitude 279 m. Length 4530 m. Denivelation -22 m
Coordinates: N 43° 14' 25.4" E 25° 02' 53.8"

A water cave, 2,5 km SE of Krushuna, on the southern slope of Slivov Dol, the north periphery of Devetaki Plateau. Formed in Aptian limestone (Lower Cretaceous). Entrance – 7,5 x 3,8 m, followed by a dry horizontal gallery, 280 m long. An underground stream in the main gallery, about 800 m E and 80 m W of it. Both ends finish with siphons. After the siphon, there is an upstream gallery, which is 515 m long. It goes downstream with one half-siphon and three siphons, which are 5, 20 and 4 m long respectively and are followed by a 1100-m gallery and a new siphon. Rich in flowstone. The river goes further through the source cave Vodopada (average outflow 10 dm^3/sec). Until 1972, only the river part was known. In the period 1974-75, the cavers of Studenets, Pleven Caving Club, explored and surveyed 2753 m. In 1978, during the International expedition Druzhba'88, another 133 m, downstream from the semi-siphon, were mapped. In 1985-87, divers from the same club penetrated the three siphons downstream and brought the length of the cave to 4015 m. After 1990, the same team overcame the upstream siphon, adding another 515 m and the cave reached its actual length.

Fauna: so far 14 species known, including the stygobite Amphipod *Niphargus ablaskiri georgievi* and the troglobites *Trichoniscus garevi* (Isopoda) and *Typhloiulus georgievi* (Diplopoda).

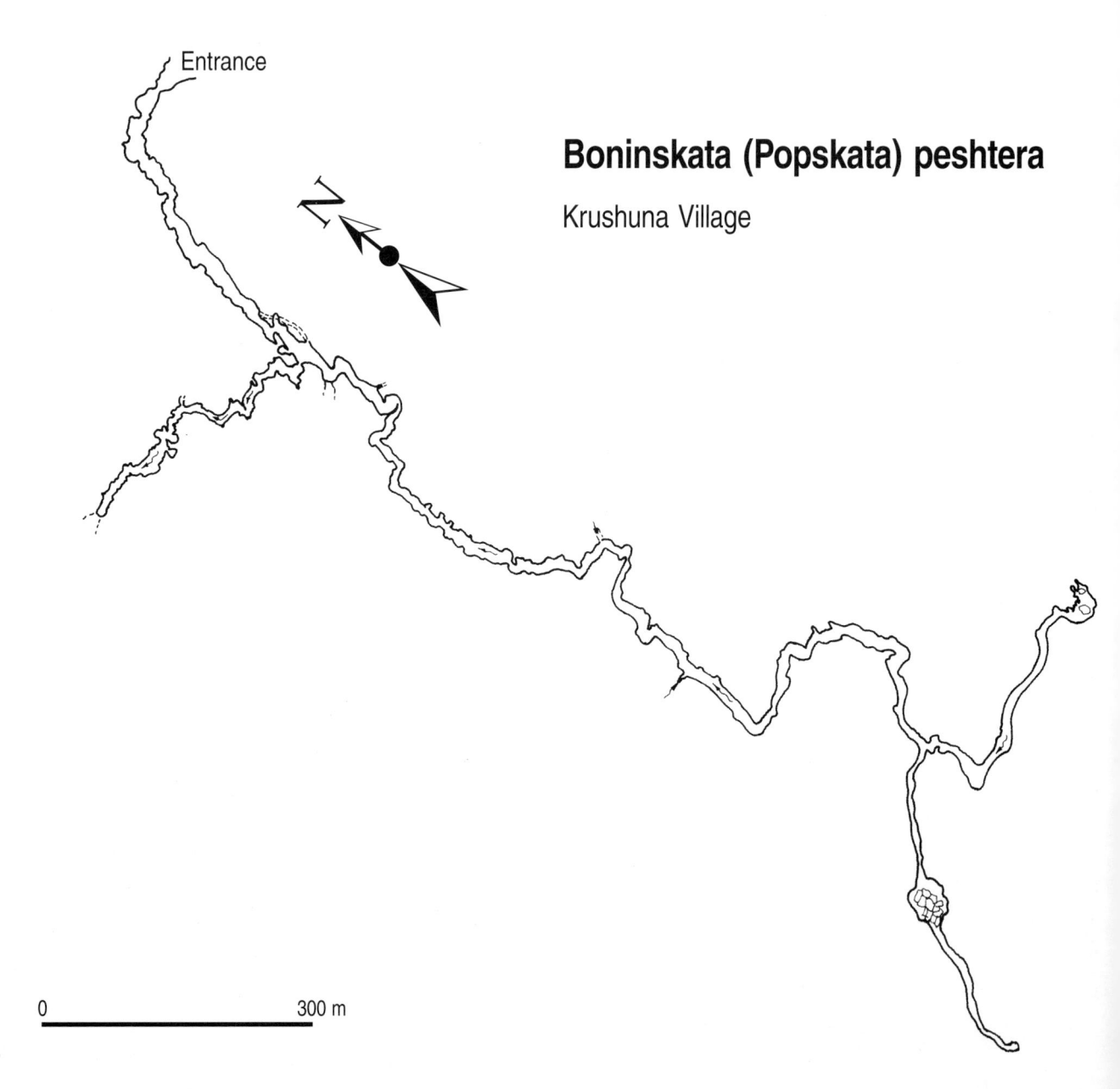

Boninskata (Popskata) peshtera

Krushuna Village

The cave entrance – photo Vassil Balevski

The entrance of the cave Boninska Peshtera from inside – photo Vassil Balevski

Name: **BOROVA DUPKA (125)**
Cherni Ossam Village, Lovech Distr.
Altitude 1257 m. Length 240 m. Denivelation -156 m

    A pothole in Borov Dol, Troyan Balkan, in the Steneto Biosphere Reserve. Formed in Trias limestone and dolomite. The pothole starts with three shafts of 13, 22 and 15 m. Two shafts of 30 and 60 m start from the bottom of the last one. From there, an inclined part follows up to 140 m, where an underground stream appears.
    Surveyed in 1973 by P. Nedkova, A. Taparkova, N. Nedkova, V. Balevski, G. Hitrov and other cavers from Sofia and Troyan. An authorization from the headquarters of Central Balkan National Park is needed to enter the pothole.

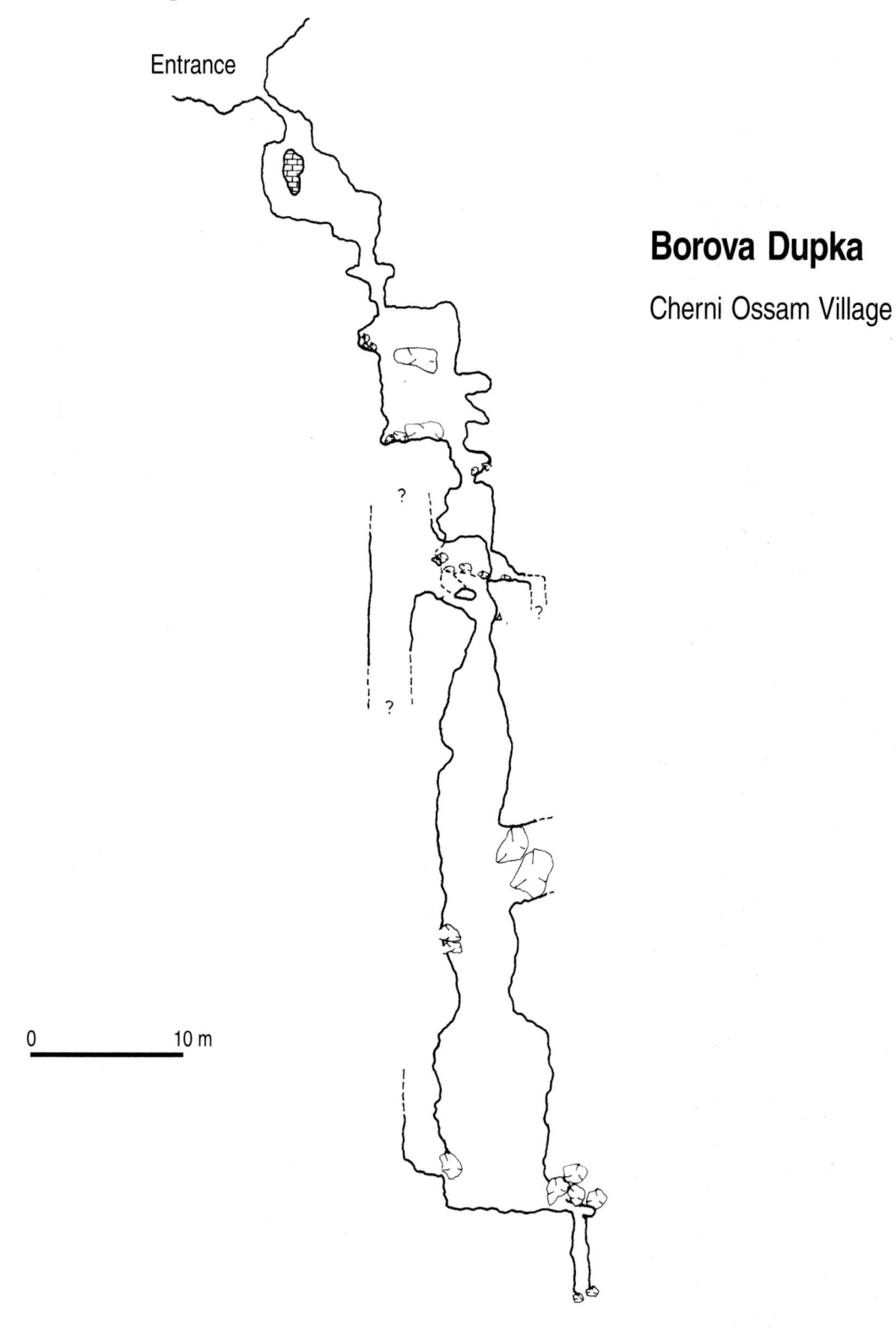

Borova Dupka

Cherni Ossam Village

The participants in the first exploration (1969) – photo Nikolay Genov

V. Balevski and C.Hitrov (1969) – photo Nikolay Genov

P. Beron is looking for cave fauna (1969) – photo Vassil Balevski

The first exploration was made by using of ladder (1969) – photo Vassil Balevski

Name: **BRÂSHLYANSKATA VODNA PESHTERA (145)**
Aleksandrovo Village, Lovech Distr.
Altitude 175 m. Length 608 m + ca.200 m of siphons. Denivelation -10 m

A water cave situated on the left bank of Ossam River and on the southern slope of the valley Brâshlyanski Dol.

The entrance part of the main gallery is dry, with a stony floor. About 30 m from the entrance, the water part starts and, through three consecutive lakes, goes on till an upstream siphon is reached. On the left side (entry direction), there is a short and narrow branch from which sometimes water comes out. When in flood, the current is reversed and part of the water runs outside through the entrance.

Downstream, the cave ends with a wide siphon at the end of its left (southern) branch. The water runs along a descending gallery, 120 m long, with a bottom of interconnected lakes. At the branch of the gallery, there are ruins from an old Roman (?) installation, which possibly served to divert the water and make it run through the entrance (skewered for the needs of the nearby castle ?). Along the stream, there are some small dams, the highest being 3.9 m (at the end of the gallery).

To establish the direction and localize the outlet of the underground water of Brâshlyan Cave, on February 19, 1991, (12.30 h), 0.5 kg of fluorescein were released in its end siphon (downstream).

With a distance of 1250 m between the place of release and the place of its reappearing, the maximum flow speed of the water underground is 43 m/h. The main water quantity flows at a speed of 31 m/h.

Thus, the hydraulic connection between Brashlyanskata Peshtera and the source Sutna Bunar was proved. The speed of water in a straight line should be relatively low (ca. 30–40 m/h).

The hydrometric measurements indicate that the water of Brâshlyanskata Peshtera forms less than half of the outflow of the source Sutna Bunar.

The cave has been known since Antiquity. The first exploration took place in 1966–1968, carried out by the cavers D. Trifonov, V. Markov during district caving expeditions. Surveyed in 1972 by V. Markov.

During the international expedition Krushuna in 1978, V. Nedkov, V. Balevski and T. Daaliev had to correct the cave map. Then V. Nedkov overcame several semi-siphons upstream and two siphons of length ca. 200 m. T. Daaliev and V. Nedkov climbed the entrance chimney and found another gallery, 30 m long. Another check was done by divers (M. Dimitrov and A. Mihov and some cavers from Gotse Deltshev).

Hydrogeological studies were done in 1991 by A. Benderev, A. Jalov and N. Landjev.

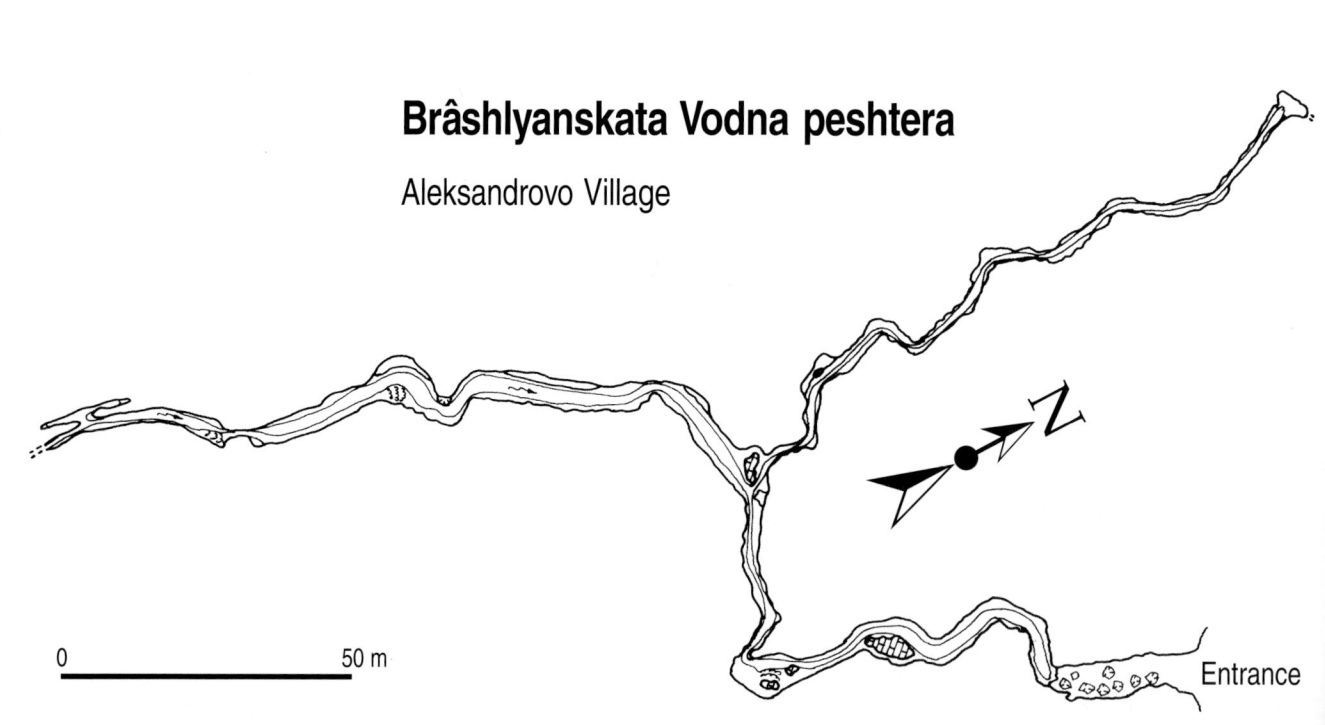

Brâshlyanskata Vodna peshtera

Aleksandrovo Village

The cave entrance – photo Vassil Balevski

Roman period wall – photo Vassil Balevski

Name: **BRUNOSHUSHINSKATA PESHTERA (140)**
Gortalovo Village, Pleven Distr.
Length 555 m

A cave, three km from Gortalovo, upstream the river Pârchovitsa, on its left bank. The entrance is 8–10 m above the local base of erosion, at the base of a cliff. A trapezoid entrance, three meters high and 12–15.5 m wide. A horizontal, wet, one-storey, branched cave. From the entrance hall, two galleries start, the main (on the right) being 55 m long and 0.5–3 m high.

Fauna: stygobite is *Speocyclops d. demetiensis* (Copepoda).

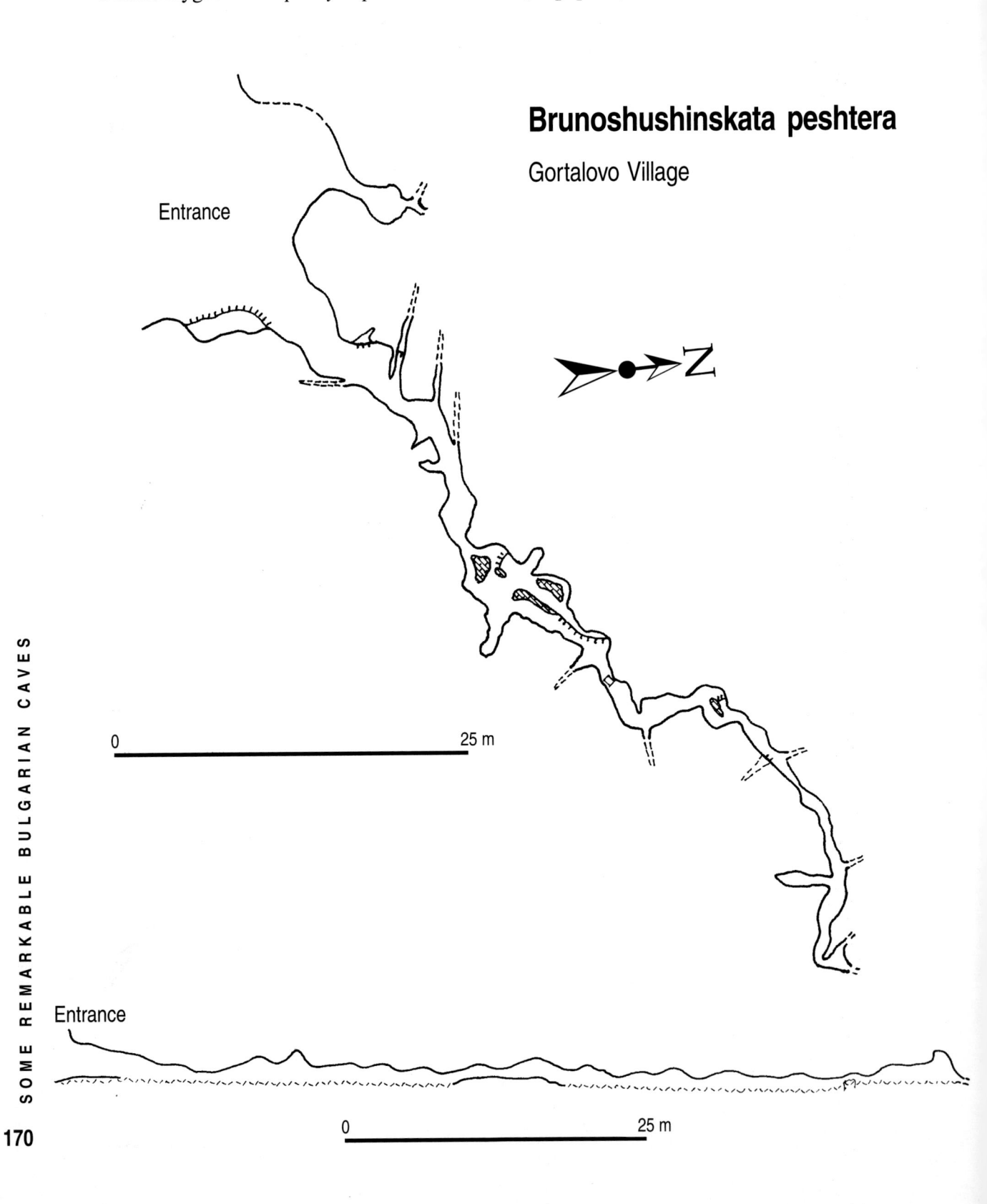

Name: **CHERNATA PESHT (62)**
Gorsko Slivovo Village, Lovech Distr.
Length 741 m. Denivelation 3.2 m

A cave, 1550 m SW of Gorsko Slivovo in the NE part of Devetaki Plateau, Middle Predbalkan. The entrance is 6.5 m wide and 2.5 m high. A stream runs out of it (outflow 0.200 l/s in summer and autumn and 5–6 l/s in April–May). About 40 m from the entrance, there is a lake siphon, which explorers call Tranteev's Siphon. To penetrate it, the water should be taken out. A gallery follows with a triangular profile and dimensions: width 1–1.5 m and average height 1 m. About 260 m from the entrance, a second siphon is reached, penetrated as the first one. Further on, the gallery gets higher (up to 4 m). Its floor is covered with clay. Gradually, a third siphon is reached (8 m long). To overcome it, it is necessary to take out the water. The cave ends in a blockage of small stones. In some places sinter lakes and some flowstone formations are seen.

In 1967, P. Tranteev explored and surveyed the cave up to the first siphon. Later, cavers from Akademik, Sofia Caving Club, took over the exploring of the cave (in the summer of 1982). Definitely explored in November 1982 by a joint expedition of several clubs from Sofia.

Description after A. Grozdanov, G. Markov, I. Lichkov. 2000. Expeditions of Akademik, Sofia Caving Club, in the area of Devetaki Plateau (MS).

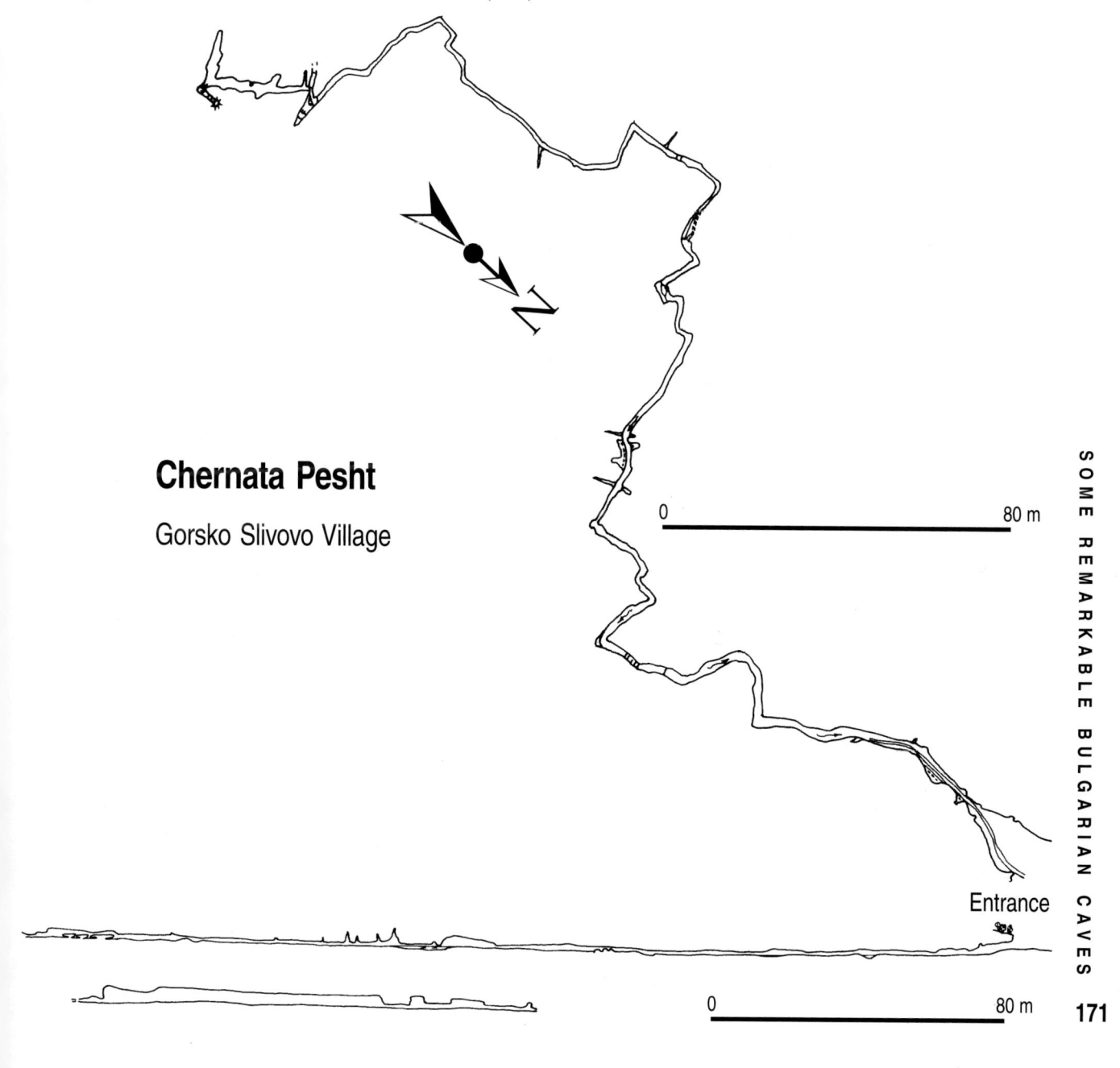

Chernata Pesht

Gorsko Slivovo Village

Name: **CHERNIYAT IZVOR (31)**
Glavatsi Village, Vratsa Distr.
Length 550 m. Denivelation 12 m

St. Nikola monastery is in the northeastern part of Vrachanska Planina, in the foothills of the summit Golyama Lokva. About 300 m east of the monastery is the karstic source Beliya Izvor (the White Spring), skewered for the needs of the nearest villages (11–180 l/s). Above the source, in the cliffs, there is a recurring karstic source called Cherniya Izvor (the Black Source), a high narrow fault in the cliff with a small lake. When in flood, up to 50–60 l/s run out of it.

The two sources are at ca. 500 m from each other and the denivelation is 60 m. They are formed in Aptian limestone. Some cavers managed to penetrate inside the Black Source through a hole in the middle of the cliff. The cave contains several stagnant lakes, some of which interconnected by siphons. After ca. 300 m of a low, semi-inundated gallery from the last lake, one enters a cave of a different type, already with running water. This is the water of Beliya Izvor. After another 150 m upstream, a siphon is reached. After 50 m downstream, the stream enters a narrow fissure. As a rule, the upper storeys have nice flowstone formations and the lower ones are washed out, with typical evorsion formations. If the upstream siphon could be overcome, more galleries are to be expected.

Explorers of the cave: Ts. Ostromski, M. Zlatkova, Zdr. Iliev, St. Petkov, T. Ivanov, Yu. Petrov and others from Sofia and Iv. Tomov and A. Velchev from Mezdra. Description after the article of Maria Zlatkova.

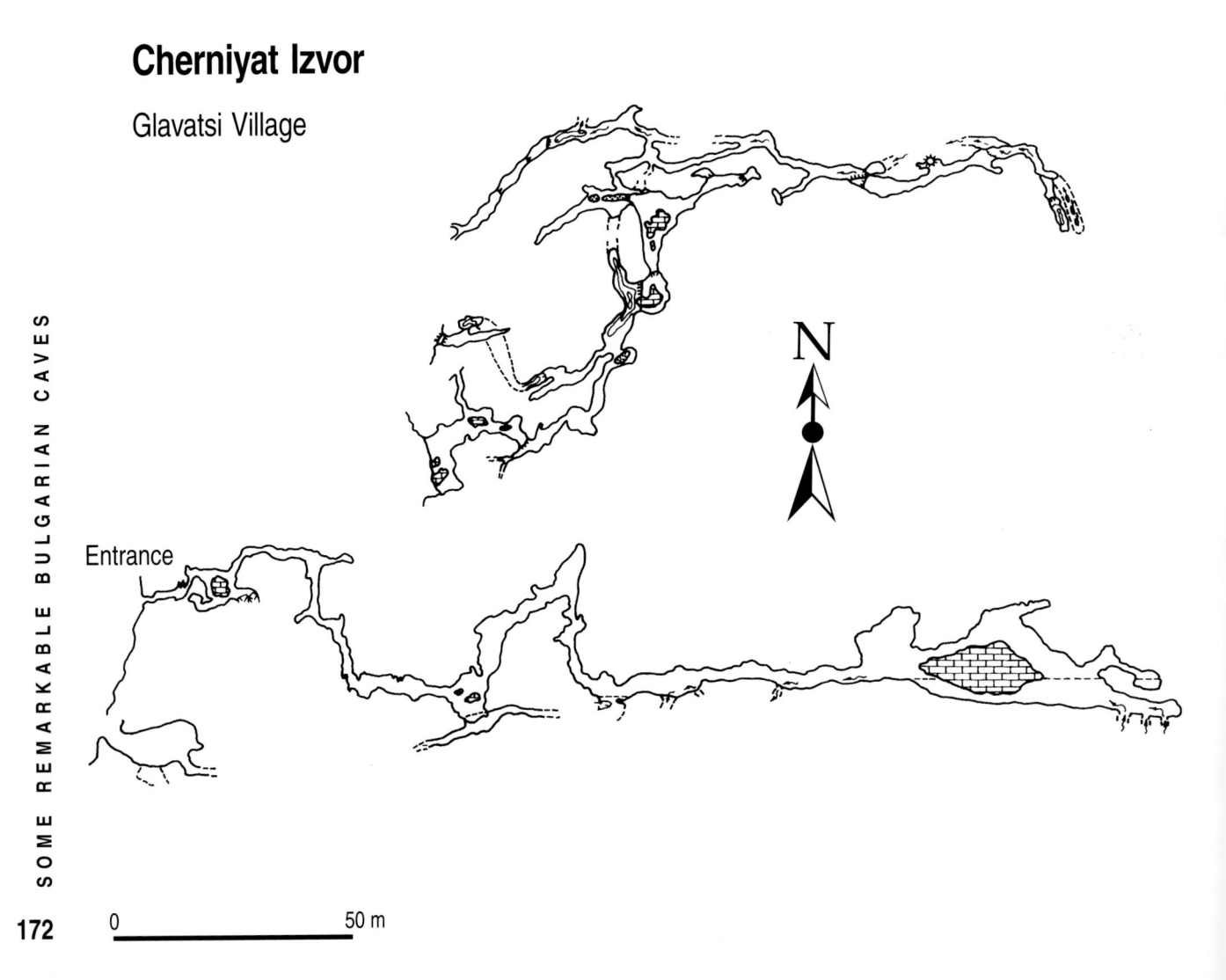

Cherniyat Izvor

Glavatsi Village

Name: **DÂLBOKATA (Propast PIII N 27) (89)**
Karlukovo Village, Lovech Distr.
Altitude 252 m. Altitude above the LBE: 126 m. Length 113 m. Denivelation -72 m
Coordinates: N 43° 10'37" E 24° 05'47"

A pothole, 1250 m ENE of the karstic sump Lilov Vir. The entrance is in the center of a funnel with trees (diameter 4.5 m). A shaft, deep 18 or 22 m (depending on the attachment point) follows, leading to a level bottom. In its NE part, a gallery opens, leading (after steep descent over boulders) into an entrance chamber, 16 m long and ca. 5 m wide. It is followed by a gallery, 12 m long and with a 2x2 m cross section. Under a sinter vertical of 2 m, sometimes a small lake is formed. Then, the cave starts developing in NE-SW direction, with a horizontal gallery, which is 28 m long and 1 m wide. The main gallery leads to a chimney – 8 m high. Narrow places give access to a 22-m vertical with a stony bottom. The wider of the two branches on the left leads to a second pitch, 19 m deep and 2 m wide. The cave ends at its bottom.

Explored many times. Surveyed mainly in July 1976 by V. Gechev, A. Siromahov, Al. Alexiev, V. Vassilev and Iv. Tsvetkov.

Fauna: six species have been identified.

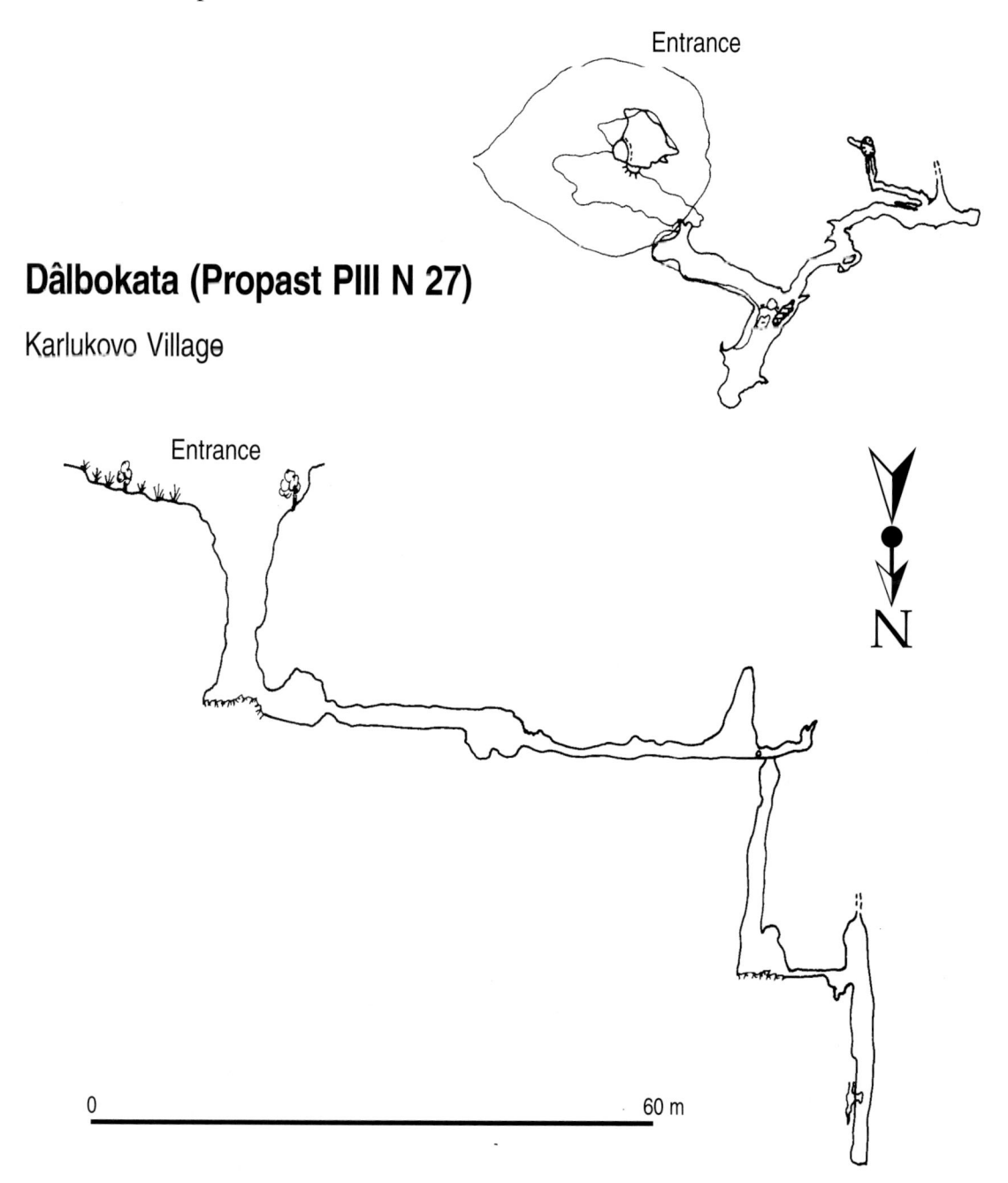

Name: **DESNI SUHI PECH (14)**
Dolni Lom Village, Vidin Distr.
Altitude 500 m. Length 591 m (the entire system Suhite Pechove is 799 m long) Denivelation -27 m
Coordinates: N 43° 29' 40.9" E 22° 46' 58"

A cave complex (together with the neighbouring Levi Suhi Pech) in Middle Trias limestone of Berkovitsa antycline. An entrance chamber with excavations done by gold-seekers, followed by a gallery with several small branches, many sinter lakes and water pools.

Air T° 11.2 °C, humidity 96 %. In 1977, A. Jalov and S. Tsonev carried out an experiment remaining underground for 62 days (Jalov, Tsonev, 1980).

Fauna: studied by P. Beron since 1960.

Identified troglobites and stygobites in the whole system of Levi and Desni Suhi Pech: *Niphargus pecarensis* (Amphipoda), *Bulgaroniscus gueorguievi* (Isopoda), *Bacillidesmus bulgaricus, Serboiulus spelaeophilus* (Diplopoda)

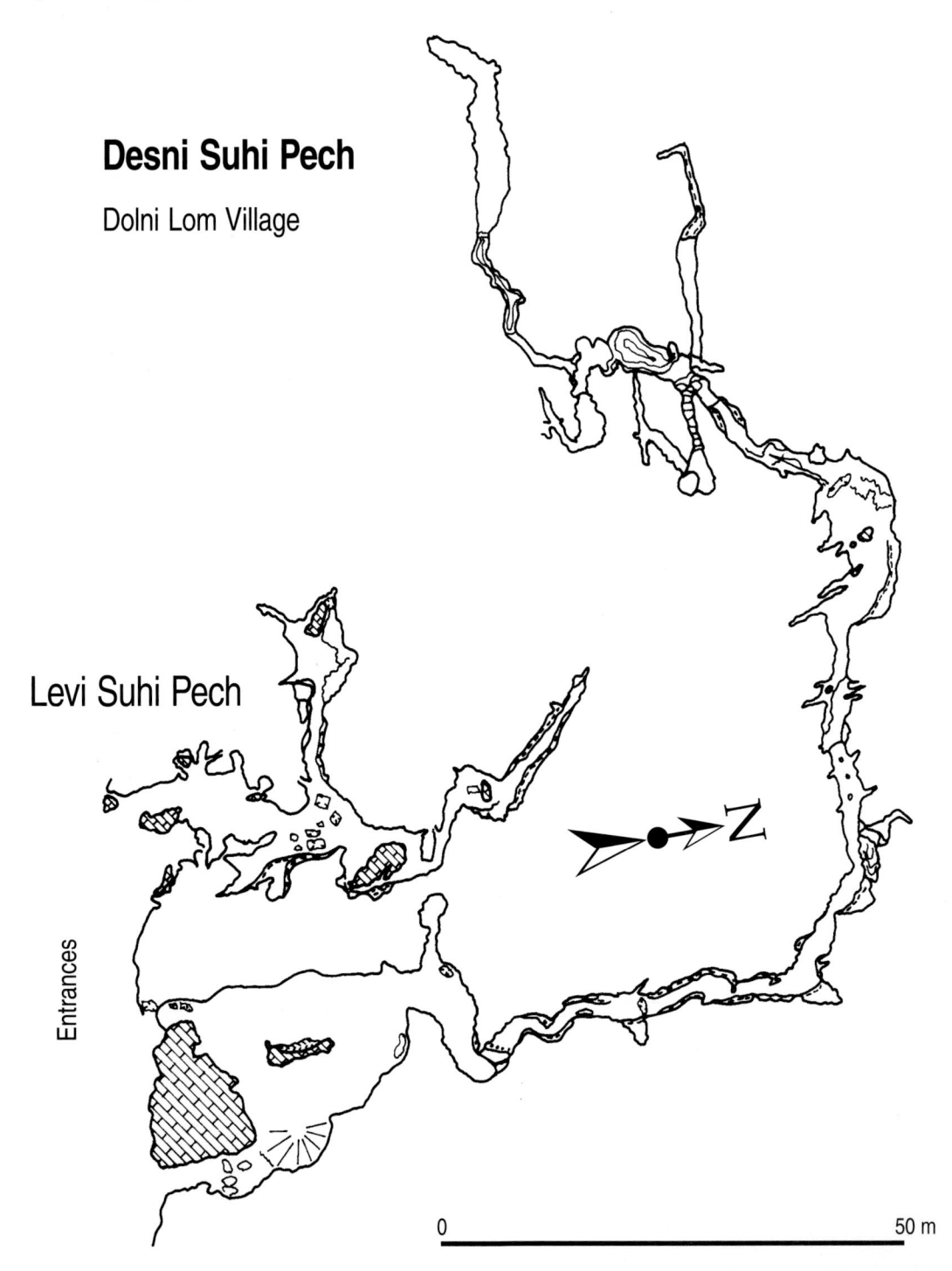

Desni Suhi Pech
Dolni Lom Village

Levi Suhi Pech

Entrances

0 _____ 50 m

Name: **DEVETASHKATA PESHTERA (MAARATA) (146)**
Devetaki Village, Lovech Distr.
Length 2442 m
Coordinates: N 43° 14' 05.2" E 24° 53' 23.2"

A cave, which is 1,5 km NW of Devetaki on the right bank of Ossam and on the northern slope of Devetaki Plateau. Formed in Aptian-Urgonian limestone. Entrance with a semi-elliptic shape of 30 x 35 m. The gallery widens into a colossal chamber (25 000 m^2) with height of the ceiling up to 58 m. The chamber is lit by seven holes, called "okna", the biggest (cave) one (Kilika) being 73 x 48 m. The volume of the chamber is 800 000 m^3, making it the biggest cave chamber in Bulgaria. A stream runs in the cave, forming deep pools and huge sinters.

First studied in 1921 by G. Katzarov, who discovered artefacts. In 1925-1928, several times explored by Pavel Petrov. He sailed the cave river on wooden boats and made hydrogeological observations, surveying the cave. In 1950-1952, the archaeologists V. Mikov and N. Djambazov excavated the entrance hall and discovered cultural layers from many periods – Middle and Late Paleolith and Neolith, the Iron and Bronze Age and from Roman times.

The cave is considered to be the richest and most representative site of different prehistoric cultures in Southeast Europe. In 1967, the cave was declared a Monument of Culture of National Importance (Decree 238/07.06.1996 with the adjacent 14.1 ha).

Fauna: 21 known species, including the Isopod *Trichoniscus tenebrarum*, the Amphipod *Niphargus bureschi*, and the Centipede *Lithobius tiasnatensis (= popovi)*(Chilopoda).

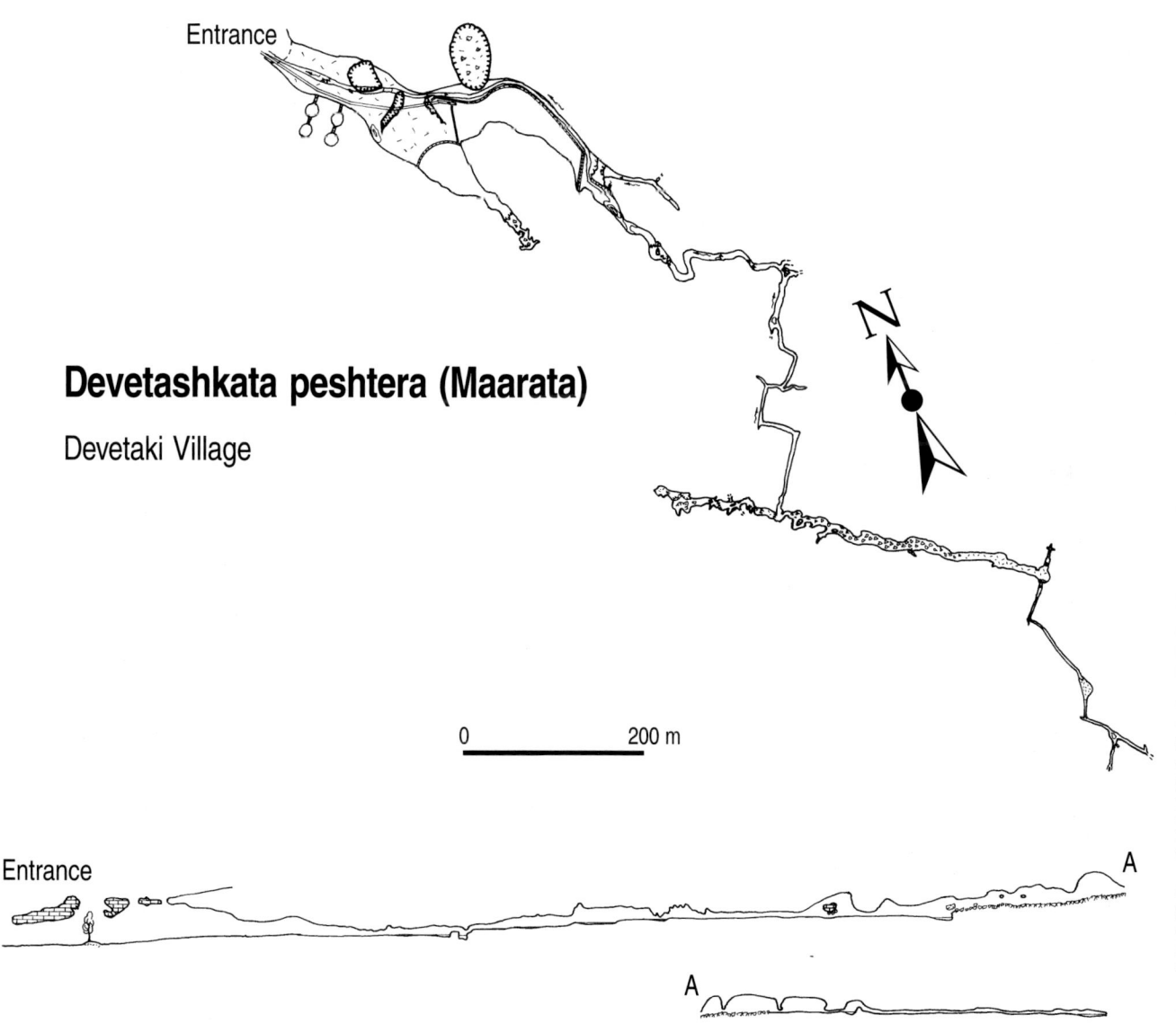

The biggest cave room in Bulgaria with its stone windows – photo Vassil Balevski

D. Daalieva in front of the giant cave entrance – photo Trifon Daaliev

D. Daalieva in the first caves lake – photo Trifon Daaliev

The waterfall – photo Vassil Peltekov

Name: **DINEVATA PESHT (64)**
Gintsi Village, Sofia Distr.
Length 451 m. Denivelation -10 m
Coordinates: N 43° 04' 00.8" E 23° 06' 23.1"

A cave on the right slope of Ginska Rcka (Nishava River), at the base of the cliff Zaskoko. Formed in Middle/Upper Jurassic limestone. A horizontal, branched, dry cave. The cross section has a triangular profile and average dimensions of 5 m (width) and 7.5 m (height). At the end of the cave there is a chamber 20 m long, 10 m wide and 15 m high. Poor in flowstone. Thick mud deposits. Periodically, the dripping water forms lakes in the chamber and in the only branch on the left.

The first known study was carried out on June 30, 1939 by a group of members of the Bulgarian Caving Society, led by Dr Iv. Buresch (P. Drenski, K. Tuleshkov, N. Atanasov, Iv. Julius and Bruno Pittioni were present also (survey and biospeleological collection)). Another survey by P. Beron in the 60s. In the period 1967–1968, fauna studies were carried out by M. Kvartirnikov (research on *Pheggomisetes*) and later by V. Beshkov and other bat researchers. Also systematic all-year climatic observations have been carried out. Studied and surveyed also in 1978 by Z. Iliev and K. Lyubenov from Edelweiss, Sofia Caving Club.

Fauna: 37 species known, including the leach *Dina absoloni*, the Amphipod *Niphargus bureschi*, the harvestmen *Paranemastoma* (*Buresiolla*) *bureschi,* the beetle *Pheggomisetes globiceps breiti.* The Crustacean *Acanthocyclops iskrecensis* (Copepoda) is also a stygobite. The cave is inhabited also by at least 16 species of bats, but not by large colonies.

A natural monument, as part of the protected karstic complex Zaskogo, together with the adjacent 88 ha (Decree 1141/15. 12. 1981).

Dinevata Pesht

Gintsi Village

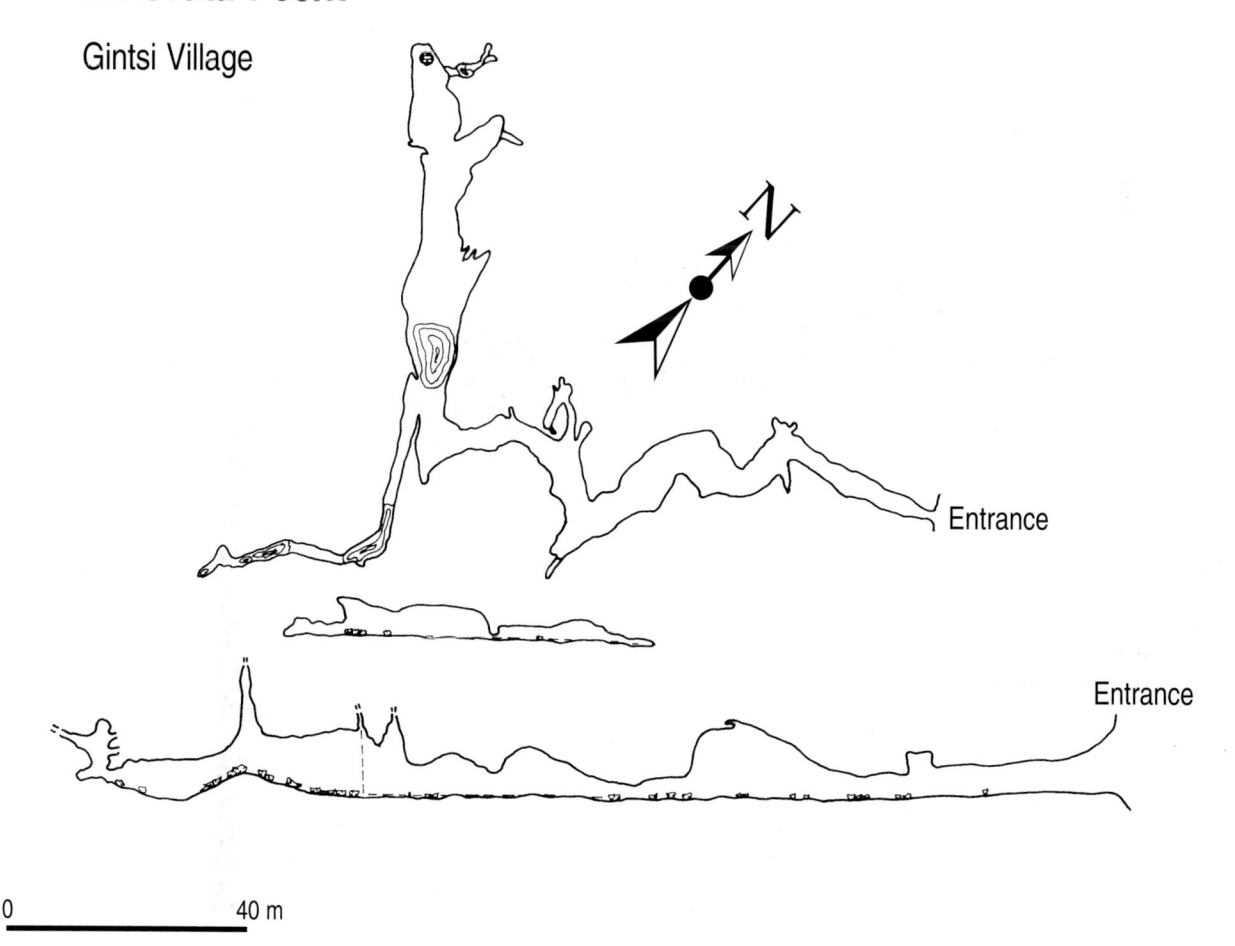

Name: **DJEBIN TRAP (116)**
Gradeshnitsa Village, Lovech Distr.
Denivelation -88 m
Coordinates: N 42° 58' 55" E 24° 16' 53.4"

The entrance of the pothole Djebin Trap (50 x 80 cm) is situated ca. 400 m north of Glogova Mahala and ca. 120 m above the river, under the house of Usin Avin. His sons and grandsons dammed the stream and used the water for household needs.

The cave was known to local people for a long time. The first attempts to penetrate the cave were made by P. Tranteev, V. Beshkov and M. Kwartirnikov in 1956. They overcame 45 m. From February 26 to 28, 1960, an expedition of the Republican Caving Commission, led by P. Tranteev, explored the pothole. P. Beron and T. Michev were the first to reach the bottom. They described the pothole in a special article. In 1968, cavers from Sofia (A. Pencheva, A. Kotev, A. Handjiyski, N. Genov, P. Gyaurova and T. Daaliev) also reached the bottom and made a new map and description of the pothole. In 1982, N. Gladnishki, N. Genov, V. Nedkov and T. Daaliev equipped the pothole for SRT.

Djebin Trap is one vertical, intersected on different levels by small and bigger platforms.

The small entrance leads into a relatively narrow and strongly twisted vertical (ca. 20 m). Through a crevice in the wall of descent, there is running water with variable outflow (in the spring up to 10–20 and more l/s).

The pothole goes on with a vertical of 25 m. When the pothole is flooded, the person descending gets entirely wet. The opposite wall is covered with crystals (dendrites) from the splashes of the waterfall.

About 45 m from the entrance, the water runs over a stone bridge – this part of the cave is very beautiful. On the left-hand side, the water disappears into a narrow channel, open for some 20 m.

From the platform verticals of 22 and 8 m follow, the latter leading to the large bottom. In the narrow gallery where the water disappears, it is possible to attempt and find a way further on.

Fauna: known troglobites are *Balkanoniscus corniculatus* (Isopoda) and *Balkanoroncus bureschi* (Pseudoscorpiones).

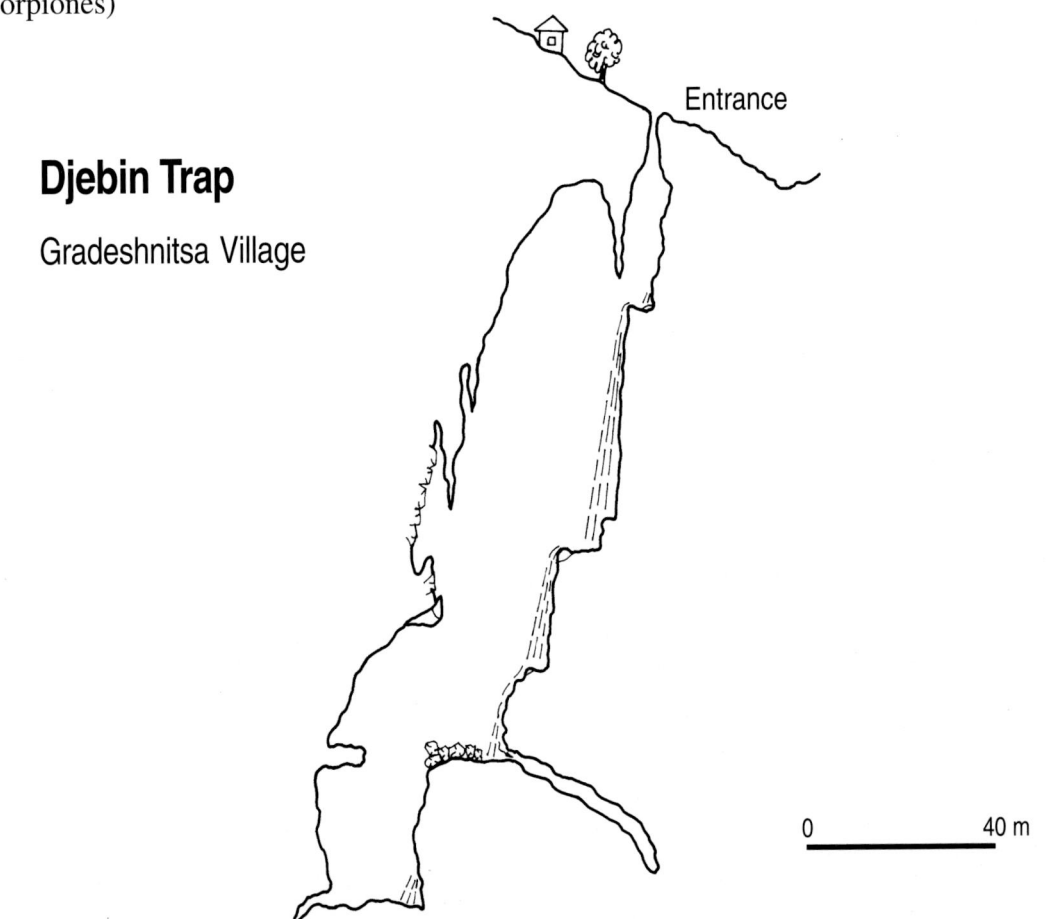

Djebin Trap
Gradeshnitsa Village

A. Apostolov in front of the entrance of Djebin Trap – photo Trifon Daaliev

Rocks in the area of Gradeshnishka Reka River – photo Trifon Daaliev

Name: **DOLNATA MAAZA (192)**
Byala Village, Kachulka Coal Mine
Altitude ca. 800 m. Length 280 m. Denivelation -52 m

Cave in Trias rocks on the southwest slope of the summit Kachulka (1082 m), Central Stara Planina, in the system of the coal mines Kachulka. The entrance is facing southwest and is 1.50 m high and 3.50 m wide. The gallery descends and at the seventh meter it bends to the north/northwards, narrowing up to 0.65 m and getting at a height of 0,70 m.

Further on, we reach a small, but high chamber with walls, covered with flowstone formations. After overcoming large stone blocks, we reach an upper branch. At the beginning, it is 3.75 m wide and 7–10 m high. We descend into the first big chamber, which is up to 10 m high and 17 m wide. The lower gallery widens until forming an elliptic chamber (10 x 4 x 2 m). Over stone blocks we reach the mouth of a 38-m deep vertical.

About 38 m away from the entrance, the gallery gets narrow up to 0.70 m. By the 42^{-th} meter it widens up to 2 m. Fifty seven (57) m from the entrance, the gallery is split in two. The right side branch leads into three consecutive interconnected chambers. The first chamber is bigger and it has high halls with small stalactites. The hall narrows, until we arrive to the second chamber.

The third chamber is the most beautiful in the cave, with two small lakes at the end. The left-hand branch has branches also. Both end in chambers. One has dimensions 5 x 2.5 m, and the other, called Elephant Ears, is 4.5 m long, 6.5 m wide and 1.50 m high. At the end, the chamber narrows into a confined gallery with a 7 m vertical at the end. At its end, we find the most spacious chamber in the cave, called The Candles (20 m long, 14.50 m wide and 10–15 m high). The chamber bears its name after the many stalagmites. From there, two galleries start which later merge. The right-hand gallery leads to the chamber, called The Clubs (Bozduganite). From this chamber, we descend to the lakes. The first lake (6 m long, up to 3.50 m wide and 1.50–2 m deep) is adorned with beautiful stalactites. Around the second lake (5 m long, up to 2 m wide and up to 1–1.50 m deep) is one of the most beautiful places in the cave, with a variety of stalactites. After the lake, the gallery bends, lowering and getting narrow, and finally ends (description by A. Jalov after Neno Atanassov).

The cave was first explored in 1926 by tourists from Yambol, led by L. Brânekov and St. Zafirov. In the period June 1–5, 1935, Neno Atanassov, accompanied by the discoverers of the cave – Iv. Târpanov and B. Monchev from Yambol, studied in details the cave and collected biospeleological material. Later, cavers from Kabile, Yambol Caving Club, and Protey, Sliven Caving Club, made other maps.

Fauna: 22 animal species known, but no troglobites.

Dolnata Maaza

Byala Village

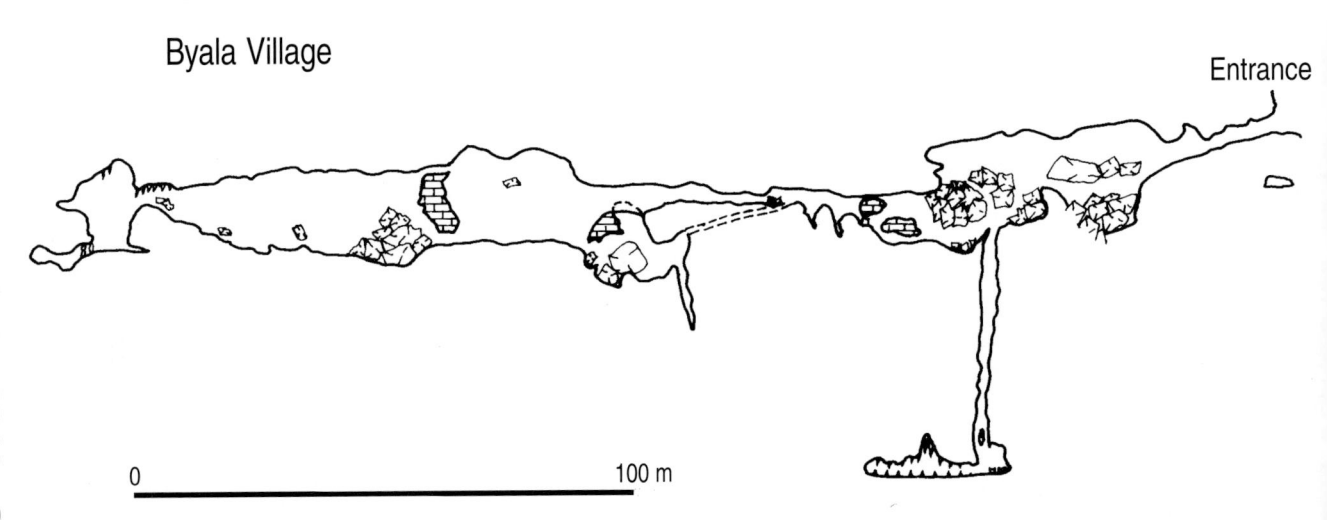

Name: **DRASHANSKATA PESHTERA (83)**
Drashan Village, Vratsa Distr.
Altitude 390 m. Length 578 m
Coordinates: N 43° 15' 18" E 23° 54' 27,7"

A cave in Maastricht limestone, at the base of a cliff south of Drashan. The main gallery is not branched. Along its whole length runs an underground stream with small, but apparently steady outflow. In some places it is not possible to follow the stream because of the low ceiling. Several bigger chambers with secondary formations (stalactites of various types). Sinter lakes at the end of the cave.

Fauna: stygobites are *Stygoelaphoidella bulgarica* (Copepoda), *Niphargus bureschi* (Amphipoda); troglobite is endemic subspecies of Diplopoda *Bacillidesmus bulgaricus dentatus*.

Drashanskata peshtera

Drashan Village

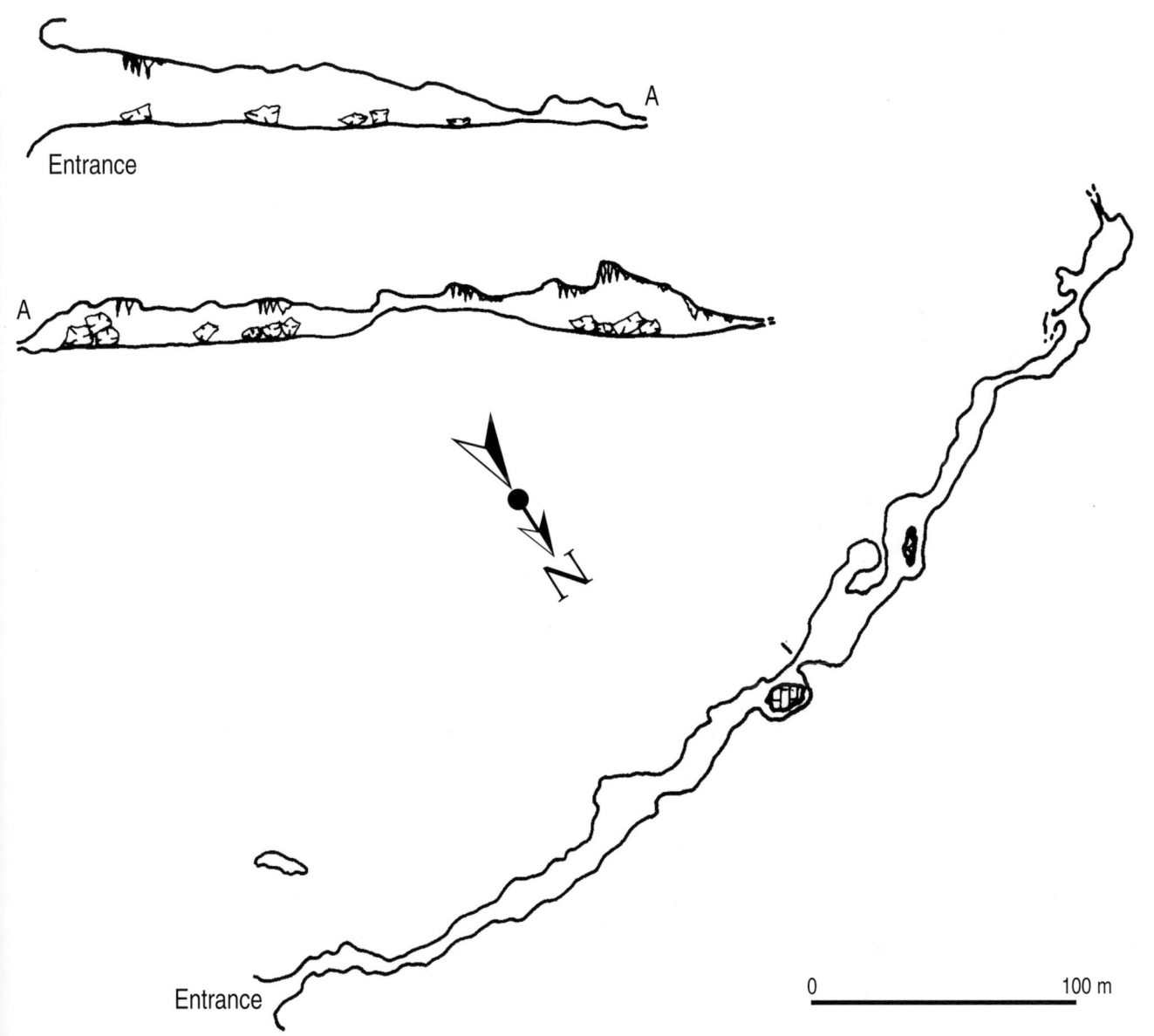

The entrance of Drashanskata Cave – photo Trifon Daaliev

Photo team in the cave Vodnata peshtera near Breste – photo Valery Peltekov

Name: **DRÂNKALNA DUPKA (35)**
Dolno Ozirovo Village, Montana Distr.
Length 78 m. Denivelation -83 m
Coordinates: N 43° 15' 01.2" E 23° 21' 34.4"

    A cave pothole in the locality Kreshtta, ca. 60 m above the river. The entrance is under the wall of Kaleto, which is difficult to find. A strongly twisted 40-m vertical follows. A descending gallery on the bottom leads to the next vertical of 5–6 m. Here a ladder is needed. The path follows a descending gallery to its end. Another climb and a descent between boulders (5–6 m) lead to a small chamber with stalactites and stalagmites. A descent of a diaclase of 12–15 m leads to a big lake siphon, which takes the whole bottom of the pit.

    The pothole was explored in January 1971 by members of Planinets, Sofia Caving Club. Principal surveyor – Vassil Nedkov. On June 14, 1970, P. Beron and V. Beshkov also descended the pothole and collected cave fauna. Actual map - Ina Barova.

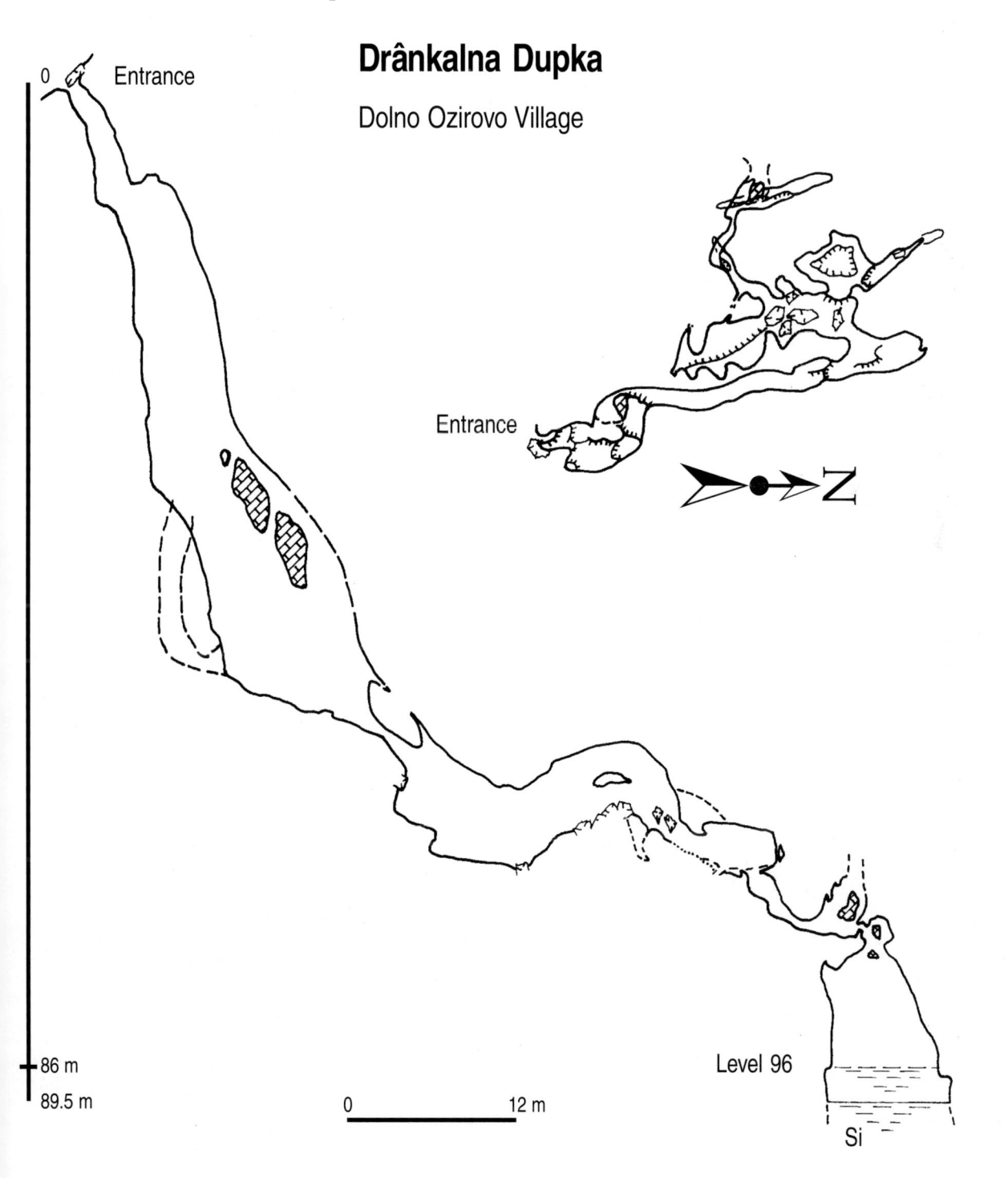

Drânkalna Dupka

Dolno Ozirovo Village

Name: **DUSHNIKA (ISKRETSKATA PESHTERA, PESHTTA, VODNATA) (61)**
Iskrets Village, Sofia Distr.
Length 876 m. Denivelation -12 m
Coordinates: N 43° 59' 53.1" E 23° 14' 09.9"

A water cave on the left bank of Brezenska river, ca. two kilometers NW of Iskrets. The road to Breze was built over the entrance chamber of the cave.

A horizontal, labyrinthine cave with four entrances. Water T° 11 °C. Developed in Trias limestone. In spring most of the cave is inundated, except for the entrance part. The hydrological system of the cave gives rise to the biggest dynamic karstic spring in Bulgaria (ca. 30 000 l/s).

The distance between Ponor Planina and the spring is ca. 7.5 km as the crow flies, the denivelation being over 650 m. The huge water basin indicates that if one day the end siphon is penetrated, the cavers may find themselves in a complex and very long cave system.

Known to local people since time unrecorded. First written information given by Zlatarski in 1904. Zheko Radev surveyed 225 m in 1915. During the period 1923-25 the cave was visited many times by Dr Buresch and his team for the study of the cave fauna. The club Akademik, Sofia, added in 1967–1969 some more information and brought the known length to 567 m. The latest map of the cave (1988, Z. Iliev, A. Jalov and others) indicated a total length of 876 m.

Fauna: Stygobites are *Cavernisa zaschevi, Iglica acicularis* (Gastropoda), *Diacyclops pelagonicus saetosus, D. stygius, D. clandestinus, Speocyclops lindbergi, Maraenobiotus parainsignipes, Stygoelaphoidella elegans* (Copepoda), *Sphaeromides bureschi* (Isopoda). Troglobites: *Paranemastoma (Buresiolla) bureschi* (Opiliones), *Pheggomisetes globiceps* (Coleoptera). Also known are many other troglophiles (incl. *Balkanopetalum armatum*) and trogloxenes.

Dushnika (Iskretskata peshtera, Peshtta, Vodnata)

Iskrets Village

The first map of the cave from Prof. Zheko Radev 1915

Entrance

0   200 m

Entrance

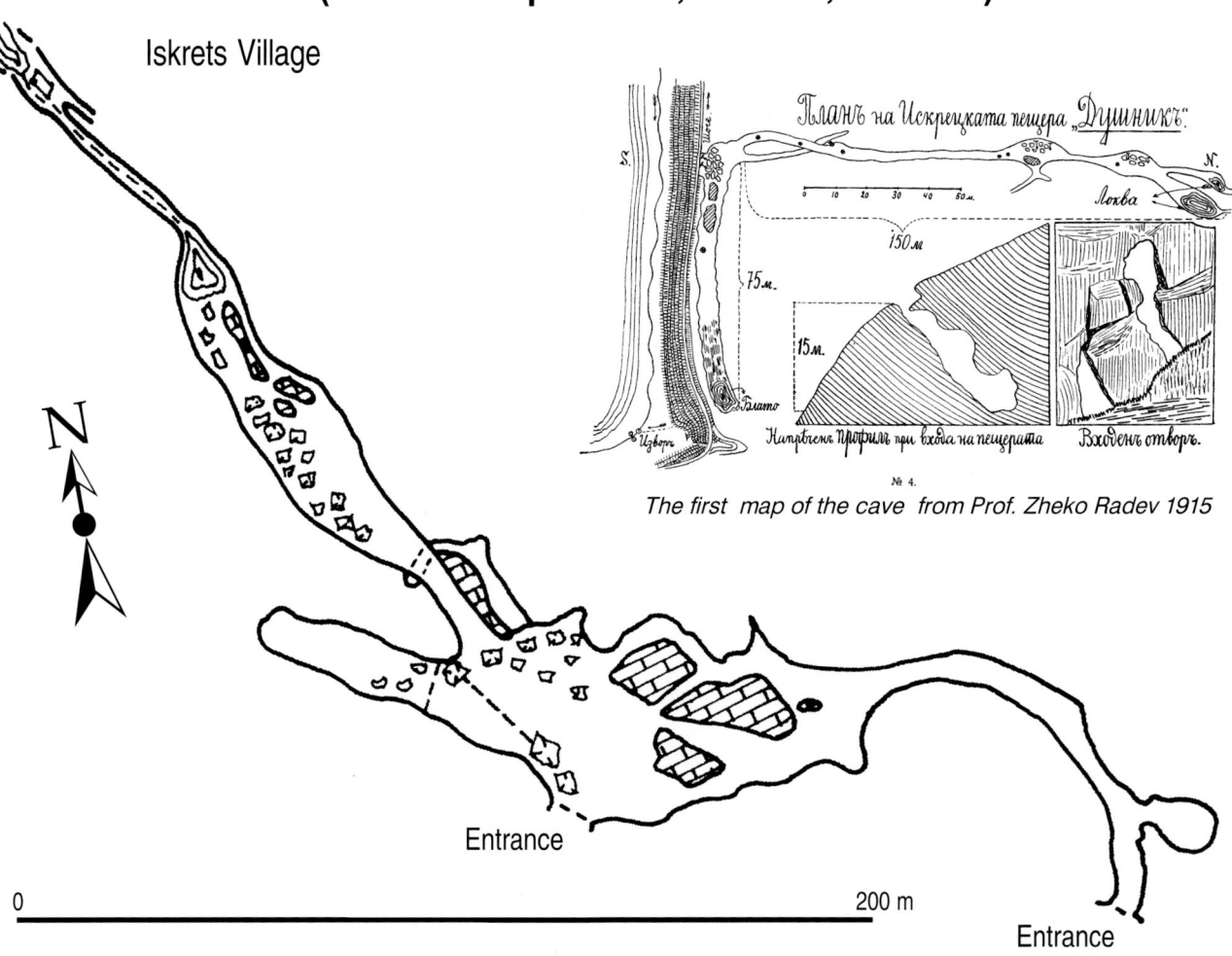

The upper entrance of the Dushnika Cave – photo Trifon Daaliev

The sump – photo Trifon Daaliev

The lake of the cave Dushnika (Dr. Ivan Buresch) – photo Dr K. Tuleshkov (16.10.1940)

Name: **ELATA (56)**
Zimevitsa Village, Sofia Distr.
Length 176 m. Denivelation -64 m
Coordinates: N 43° 02' 06.3" E 23° 17' 33"

    A cave in Jurassic limestone, two kilometers South of Zimevitsa. Entrance pitch of 18 m. Descending gallery, 4–5 m high. After 123 m the end siphon is reached. Many and beautiful flowstone formations, stalagmites, stalactites and others. No running water, sinter lakes.
    Discovered and explored by P. Beron, V. Beshkov and T. Michev in 1960. Surveyed by cavers of Akademik, Sofia Caving Club.
    Fauna: includes the following troglobites and stygobites: *Acanthocyclops iskrecensis* (Copepoda), *Paranemastoma (Buresiolla) bureschi* (Opiliones), *Pheggomisetes globiceps stoicevi* (Coleoptera, Carabidae)

Elata

Zimevitsa Village

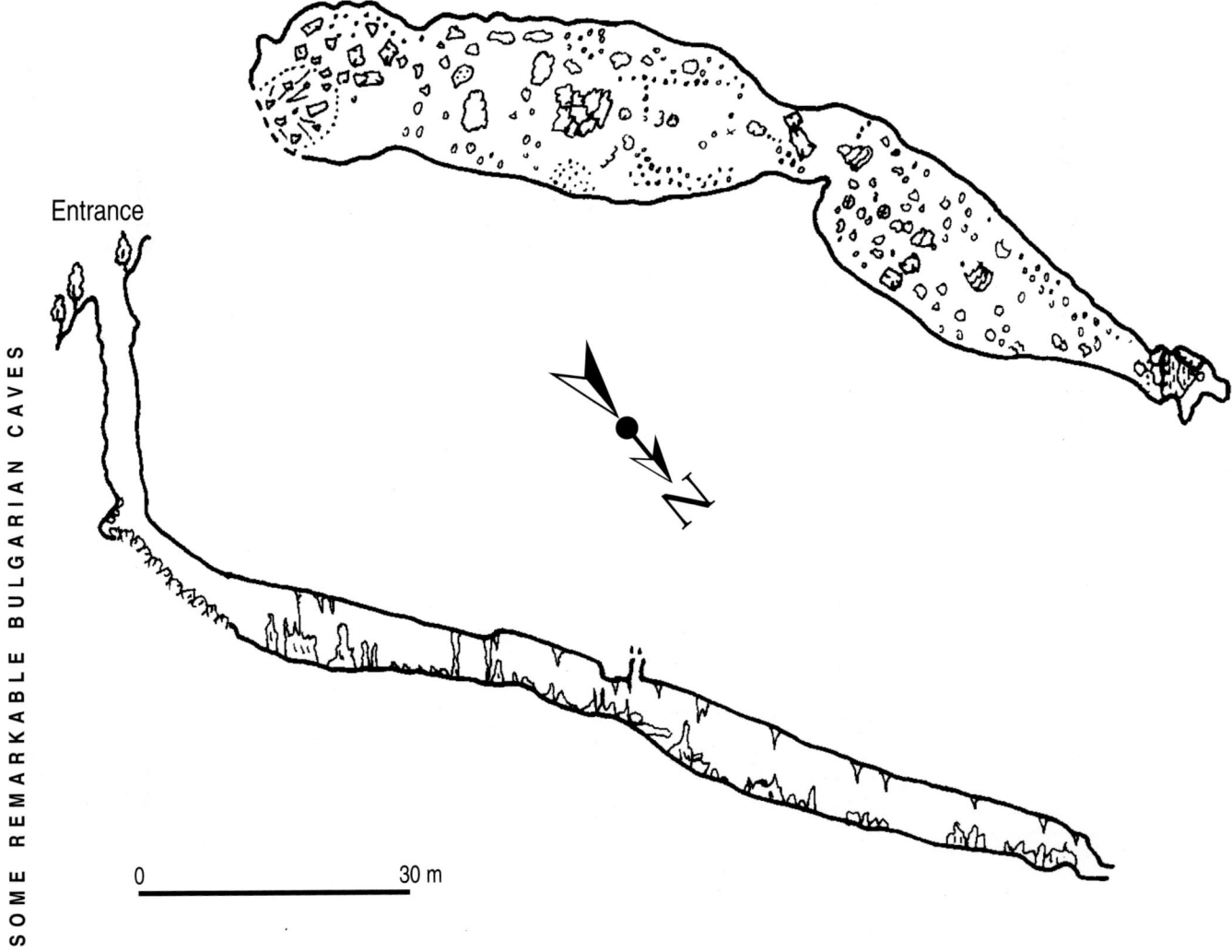

The walls and ceiling are well decorated – photo Nikolay Genov

Stalactites, stalagmites and columns – photo Nikolay Genov

Name: **EMENSKATA PESHTERA (161)**
Emen Village, Veliko Târnovo Distr.
Length 3113 m. Denivelation -40 m
Coordinates: N 43° 08' 07.4" E 25° 21' 58.6"

A cave at the beginning of Emen Canyon, on the left slope of Negovanka River, 17 m above the river bed, Middle Predbalkan. Formed in Aptian limestone. A horizontal, two-storey, branched, periodically inundated cave. Before 1959 there were thick clay deposits in the entrance parts, which were later removed. First studied and mapped (approx. 230 m) in 1883 by Karel Škorpil. Explored also in 1924 by the archaeologist V. Mikov who discovered artefacts from the Neolith. Paleontological excavations in 1948–1949 by G. Markov. In 1959 archaeological excavations under the guidance of N. Angelov and Ya. Nikolova from Veliko Târnovo Museum of Archaeology. The Neolithic material was confirmed, but also artefacts from the Bronze Age, the Early and Late Iron Age and from the Middle Ages were discovered. In 1959, it was turned into a place for cultivation of champignons, then into a cheese factory. Later it was used for military purposes and finally it became a restaurant. By 1960 only 236 m of the cave were known. In 1976–1977, cavers from Strinava, Dryanovo, studied and surveyed the cave up to 1200 m. then in 1979, cavers from Planinets Sofia Caving Club reached the length of 1713 m. In 1982–1983, the intense research of cavers from Prista Russe Caving Club resulted in another 1.5 km of new galleries, thus the cave reached its present length.

A Natural Monument (together with Emen Canyon)(Decree 880/25.11.1980).

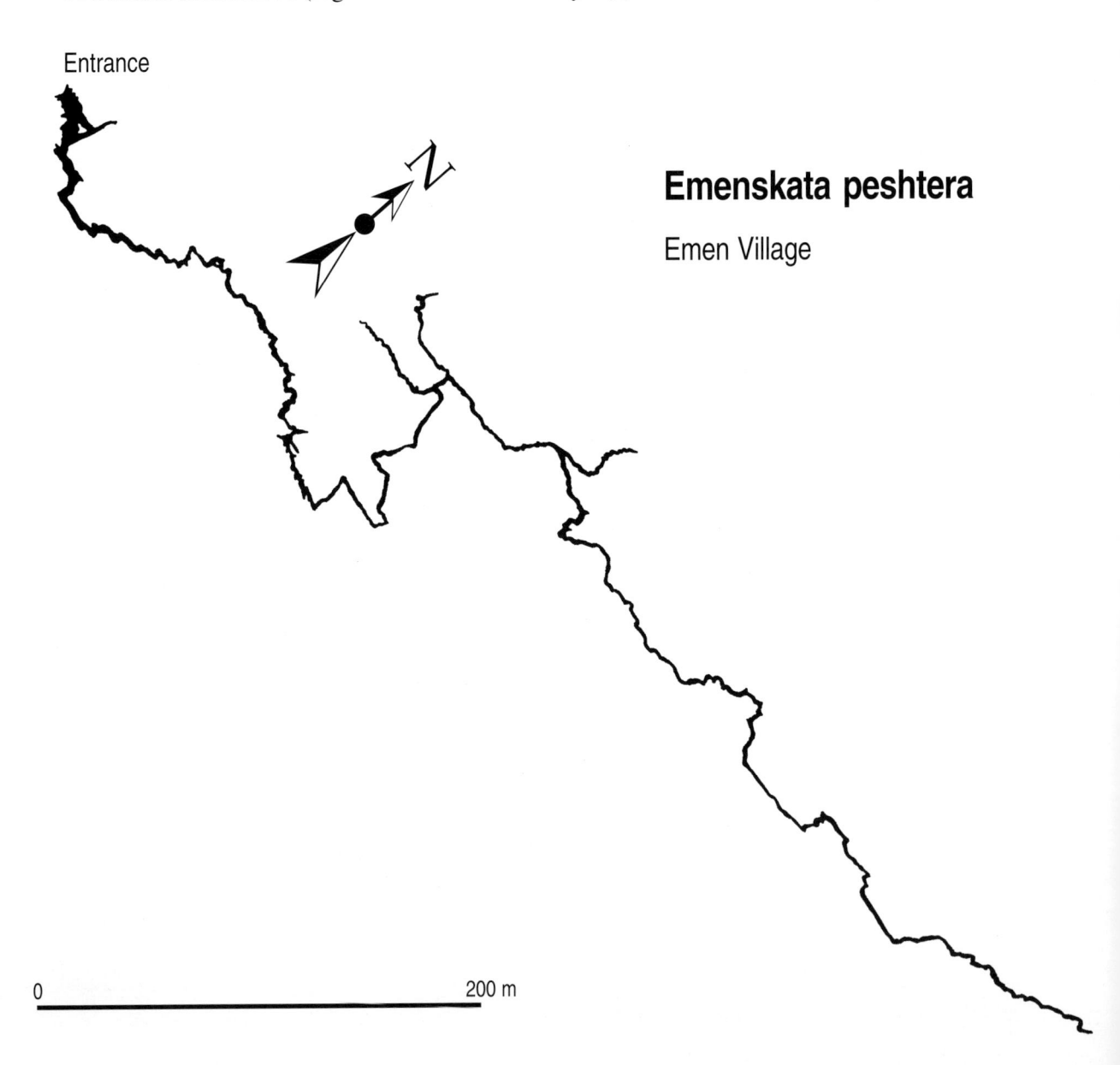

Emenskata peshtera

Emen Village

Entrance of Emenskata cave – photo Nikolay Genov

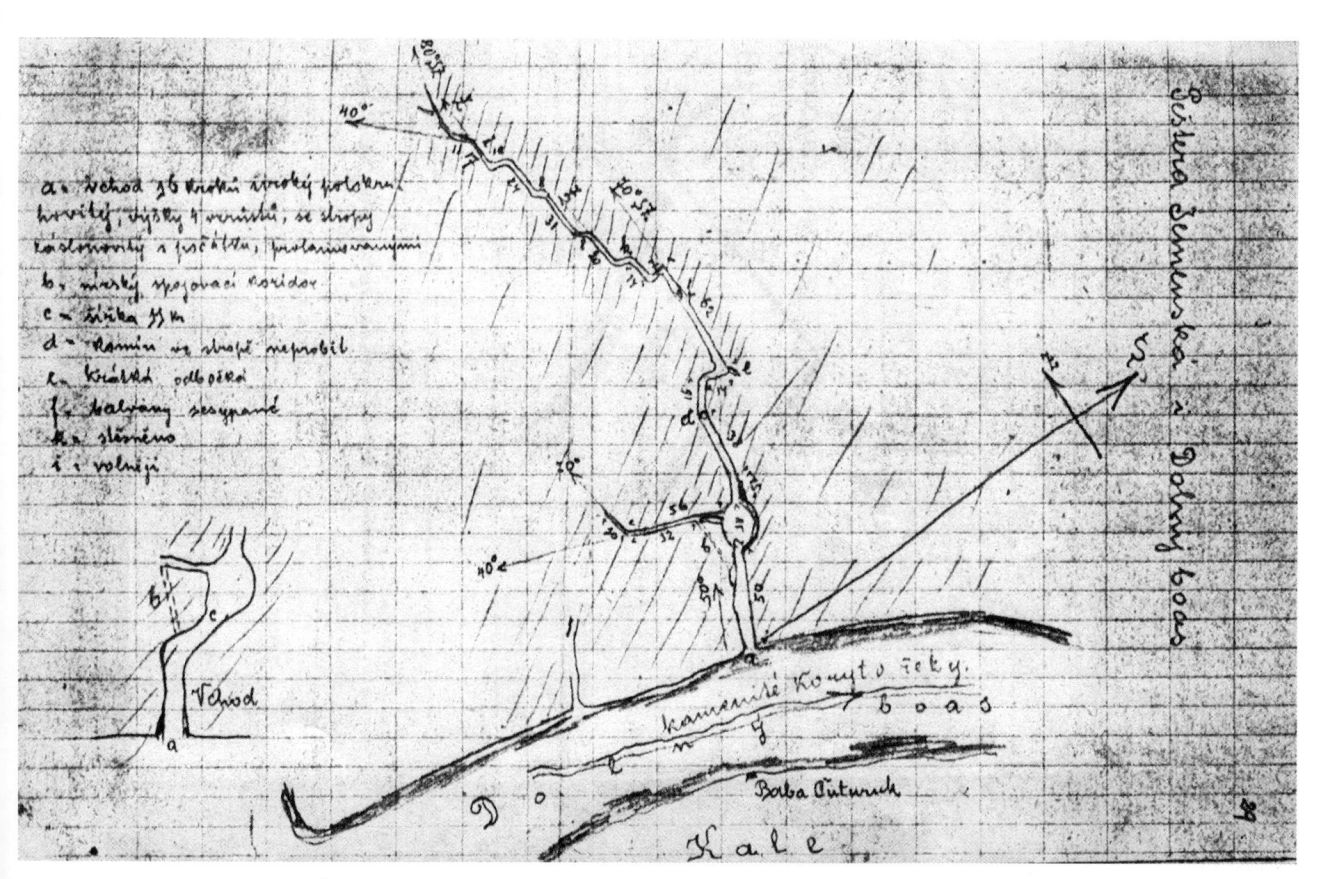

The first map is made by K. Škorpil in 1883

Name: **FUTYOVSKATA PESHTERA (149)**
Karpachevo Village, Lovech Distr.
Length 700 m. Denivelation -10 m

About 1.5 km southwest of Karpachevo the entrance of this large cave is situated, facing south. The cave is easily walk able, with an even floor covered with clay, with pools in some places. After ca. 400 m it branches out and becomes difficult to walk. In places, big evorsion cauldrons are seen. Local people call them "gyozyove".

The cave was well known to local people. It was surveyed in 1968 by V. Markov, Ts. Hitrov and T. Daaliev.

Fauna: 13 recorded species, including the troglobites *Lithobius tiasnatensis* (Chilopoda) and *Typhloiulus georgievi* (Diplopoda). In autumn, many bats are seen, including *Miniopterus schreibersi*, several species of *Rhinolophus* and others.

Futyovskata peshtera

Karpachevo Village

Name: **GABRIKA (136)**
Mikre Village, Lovech Distr.
Length 783 m. Denivelation 25 m

A horizontal cave with two entrances near Mikre. After the entrance, we reach an underground stream. The cave develops up and downstream. Downstream, the gallery goes for some 150 m and ends in a narrowing. Upstream, the gallery is winding and is about 600 m long. There are only two branches – one of 20 m at the beginning and one 40-m long in the middle of the gallery. The dimensions are: 1–2 m wide and 1–3 m high. In some places the underground stream has formed small gourds. The cave ends in a 14-m chimney. Poor in flowstone formations.

Explored and surveyed by S. Nedkov, V. Chapanov, M. Dimitrov, S. Gazdov and K. Petkov from Studenets Pleven Caving Club in 1983.

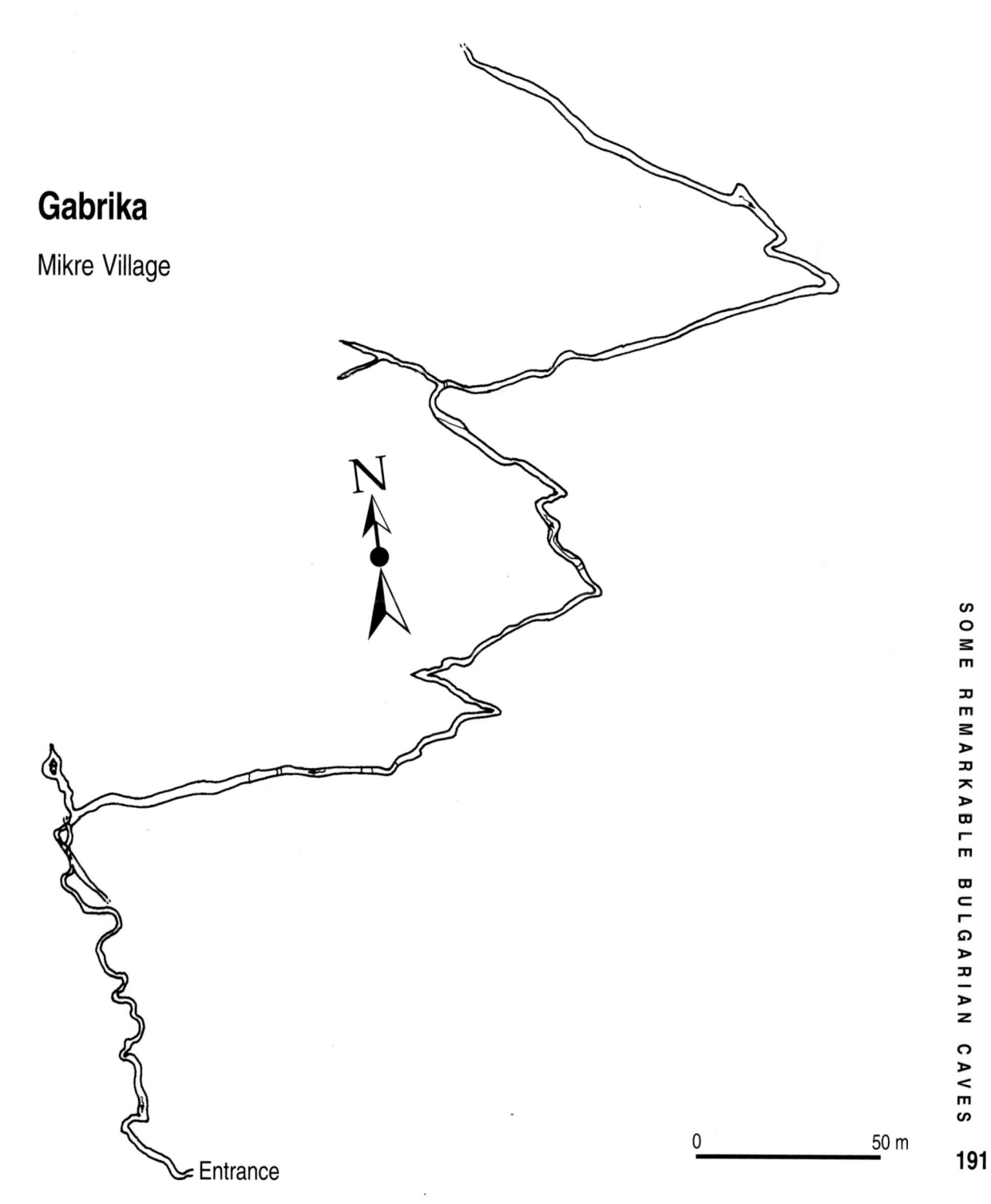

Gabrika

Mikre Village

Name: **GÂGLA (77)**
Gabare Village, Vratsa Distr.
Altitude 280 m. Denivelation -64 m

A pothole in Maastricht limestone ca. 4 km East of Gabare. The road from Gabare to Breste passes about 300 m from the cave entrance (with dimensions 1.5 x 5 m).

After the opening, the habitus of the pothole remains the same, with the same dimensions (1.5–5 m). The walls are clean, sculptured in vertical ribs. After overcoming the first half of the vertical, the pitch becomes larger and its elliptic shape has dimensions of 2 x 8–10 m until the very end. The bottom is covered with stones and is of elliptic shape (3–10 m). A vertical crevice goes for another 5 m downstairs and ends in a small chamber covered with a thick layer of clay. Here, when rainy, the water forms a siphon.

The depth of 64 m (including 50 m of full vertical) is the biggest depth in this area (description according to Hr. Delchev).

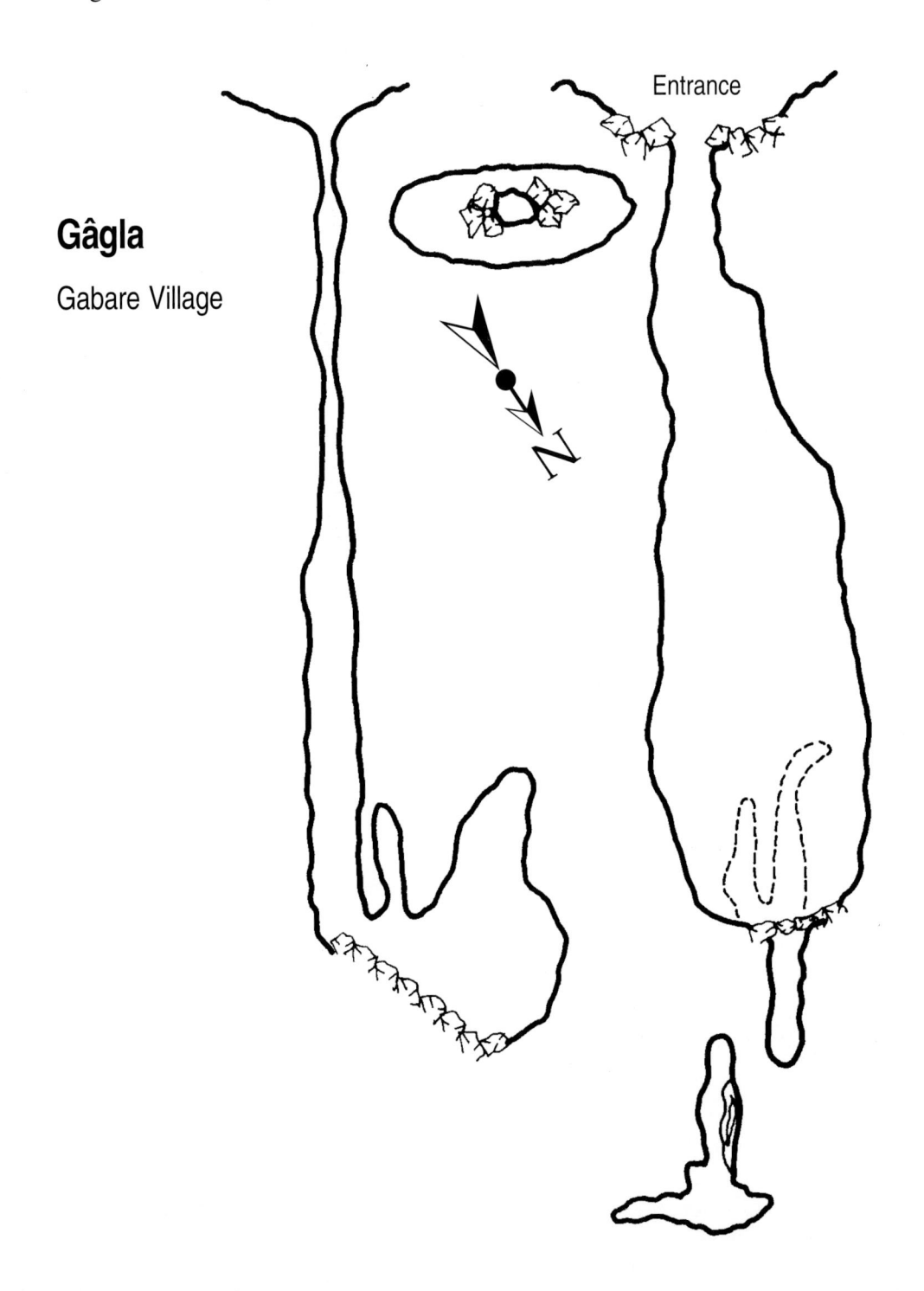

Gâgla

Gabare Village

Name: **GÂRDYOVA DUPKA (43)**
Zgorigrad Village, Vratsa Distr.
Length 510 m

A cave above Zgorigrad, on the western slope of the hill. In the upper side of the village, by the river there is a very big iron post. If we go upstream, we shall reach the first valley on the right. Following it, after 400 m, a piped source is reached. About 100 m above it, on the right, we reach the cave entrance.

Through a low and dry part a small chamber is reached, left of which a narrow labyrinth starts. On the right, the first stream is reached. Upstairs, the path follows terraces above several lakes, one of them being 1 m deep. Climbing a diaclase, we reach another horizontal part with a temporary pool. After 20 m, a branch is reached – the second stream, disappearing into a ponor. Upstream, by a boat, we reach a wide and deep lake siphon. Through another chamber, the second stream is reached again. At the chamber end, there is a very deep siphon. Downstream, a siphon part is reached. At the western end of the chamber, through narrow galleries, a deep lake is reached.

The cave name, according to the local tradition, has to do with the popular hero Gârdyu Voyvoda.

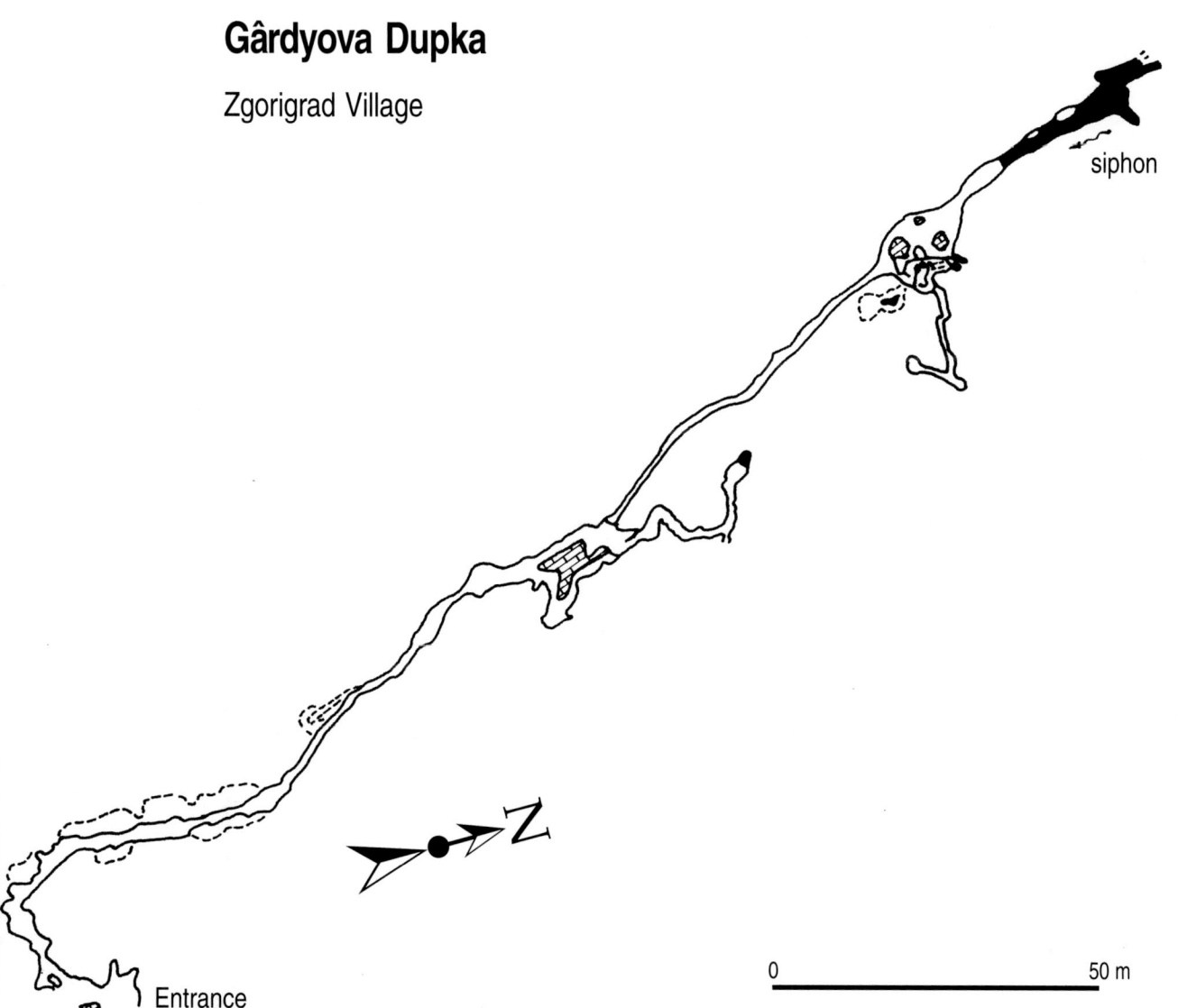

Gârdyova Dupka

Zgorigrad Village

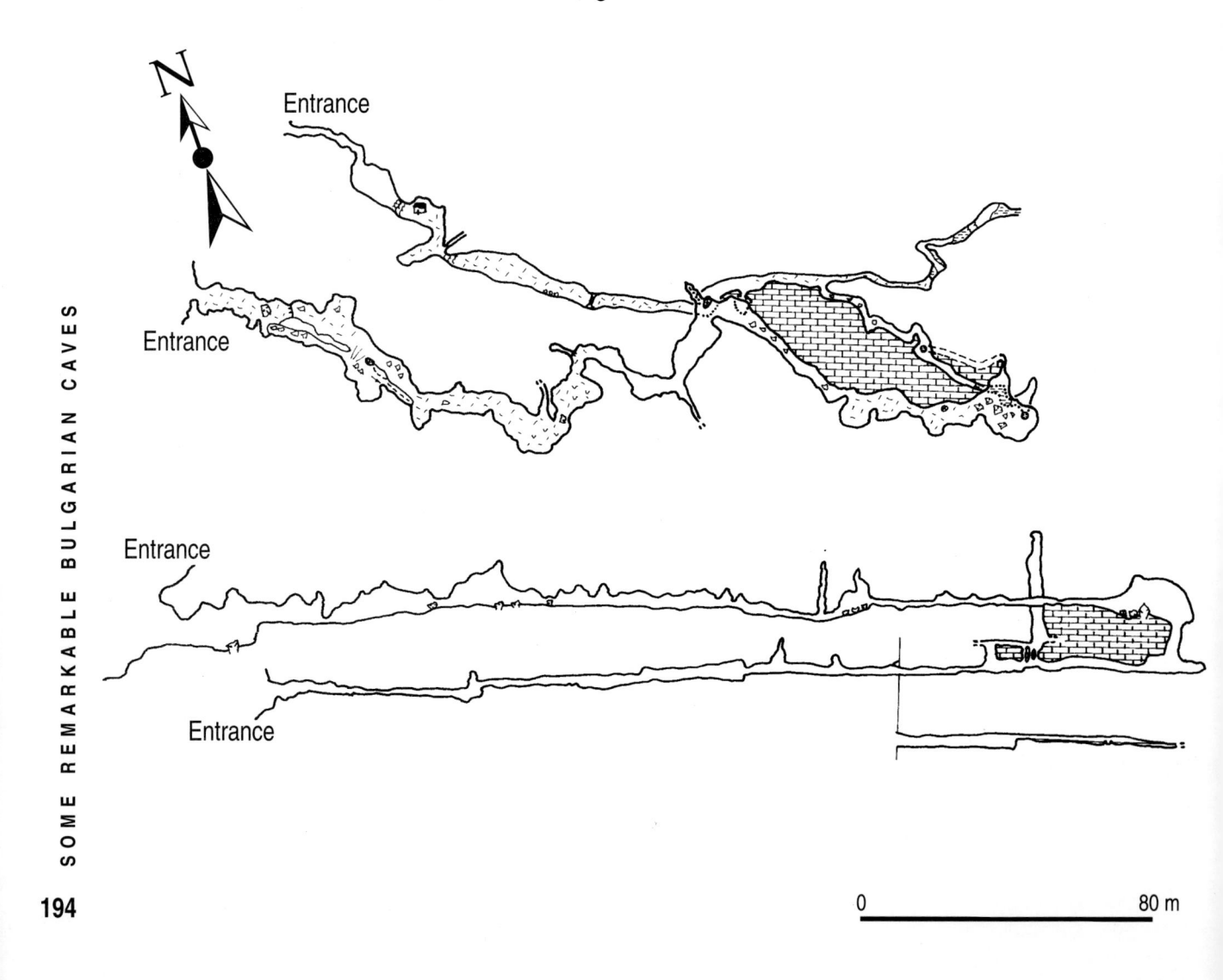

Name: **GERGITSOVATA PESHTERA (104)**
Bezhanovo Village, Pleven Distr.
Length 628 m. Length on the main axe 190 m

A cave, 3 km S of Bezhanovo, in the locality Smesite, 10 m from the river Kamenitsa, 20 m north of a karstic source. The entrance is situated at the base of a cliff, dimensions – 4 x 4 m. The cave is rich in stalactites, stalagmites, curtains, sinter lakes. One lake, which is 227 m from the entrance and has dimensions 3 x 4 m, ends with a siphon. Many bats.

Gergitsovata peshtera

Bezhanovo Village

Name: **GININATA PESHTERA (100)**
Sadovets Village, Pleven Distr.
Length 501 m. Denivelation 20.4 m
Coordinates: N 43° 17' 19.6" E 24° 22' 34.6"

A cave, three km SW of Sadovets, on the left bank of Vit river, with five entrances, four of them situated in a cliff, 50–60 m above the local base of erosion.

The cave is an ascending diaclase, branched, four-storey and dry.

The entrance to the first storey is up to seven meter high and 6 meter wide. The main axis of the first floor is 58 m long and 5 to 0.5 m high.

The second storey has three entrances and three parallel galleries. The right gallery is 34 m long and the left one – 114 m. After 65 m, the main gallery gives access to a chamber, about 35 m long and up to 12 m wide. After about a hundred meters, there is a narrowing, leading into a small labyrinth. After 180 m, the gallery comes to an end.

The third storey is 60 m long. It contains a chamber, up to eight meters wide, 30 m long and 3–10 m high. The cave is poor in flowstone formations. About 80 m deep into the second storey, more than 1000 bats were counted.

Fauna: troglophiles and trogloxenes.

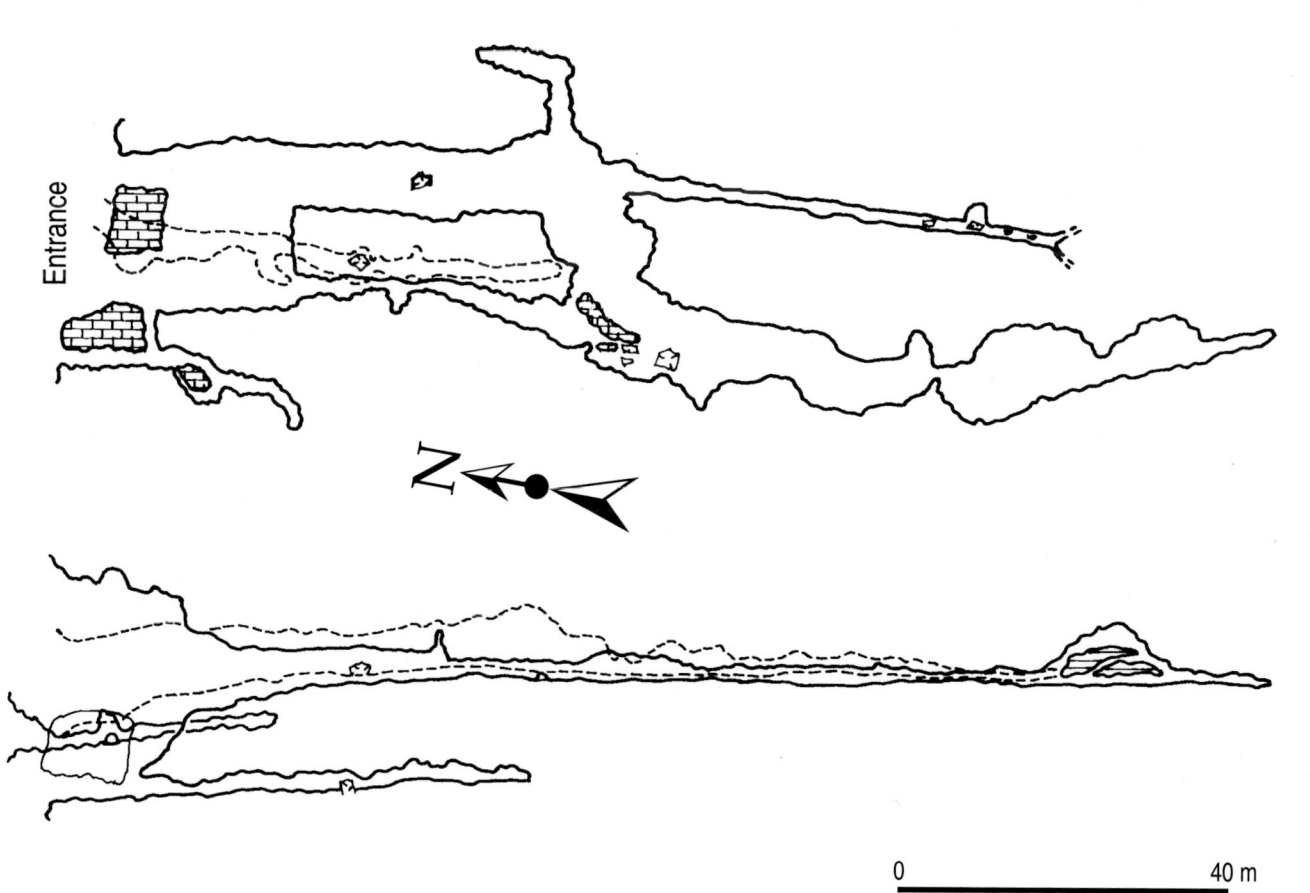

Gininata peshtera

Sadovets Village

Name: **GOLYAMATA BALABANOVA DUPKA (69)**
Komshtitsa Village, Sofia Distr.
Length 4800 m. Denivelation -80 m
Coordinates: N 43° 08' 06.3" E 23° 02' 31.5"

    A cave in Trias limestone, seven km SE of Komshtitsa, on the NE slope of Bilo ridge, Berkovska Planina. A labyrinthine, four-storey cave, relatively poor in sinter forms. Up to 160 m from the entrance parts, in colder months, a glacier is formed and remains until April. During high waters the cave represents a ponor. The cave was studied first in 1960 by D. Ilandjiev up to 260 m. In the same year, V. Beshkov, T. Michev and P. Beron too studied the cave. Later, explored and surveyed by cavers from Edelweiss Sofia Caving Club. In 1983, during a speleological expedition of Sredets Sofia Caving Club, J. Pavlov discovered an extension of the cave. Until 1985, the cavity had been completely explored and surveyed to its present length.
    Fauna: eleven known species, including the beetle troglobite *Pheggomisetes globiceps ilandjievi* (Carabidae). Another interesting inhabitant is the leech *Trochaeta bykowskii* (Hirudinea).

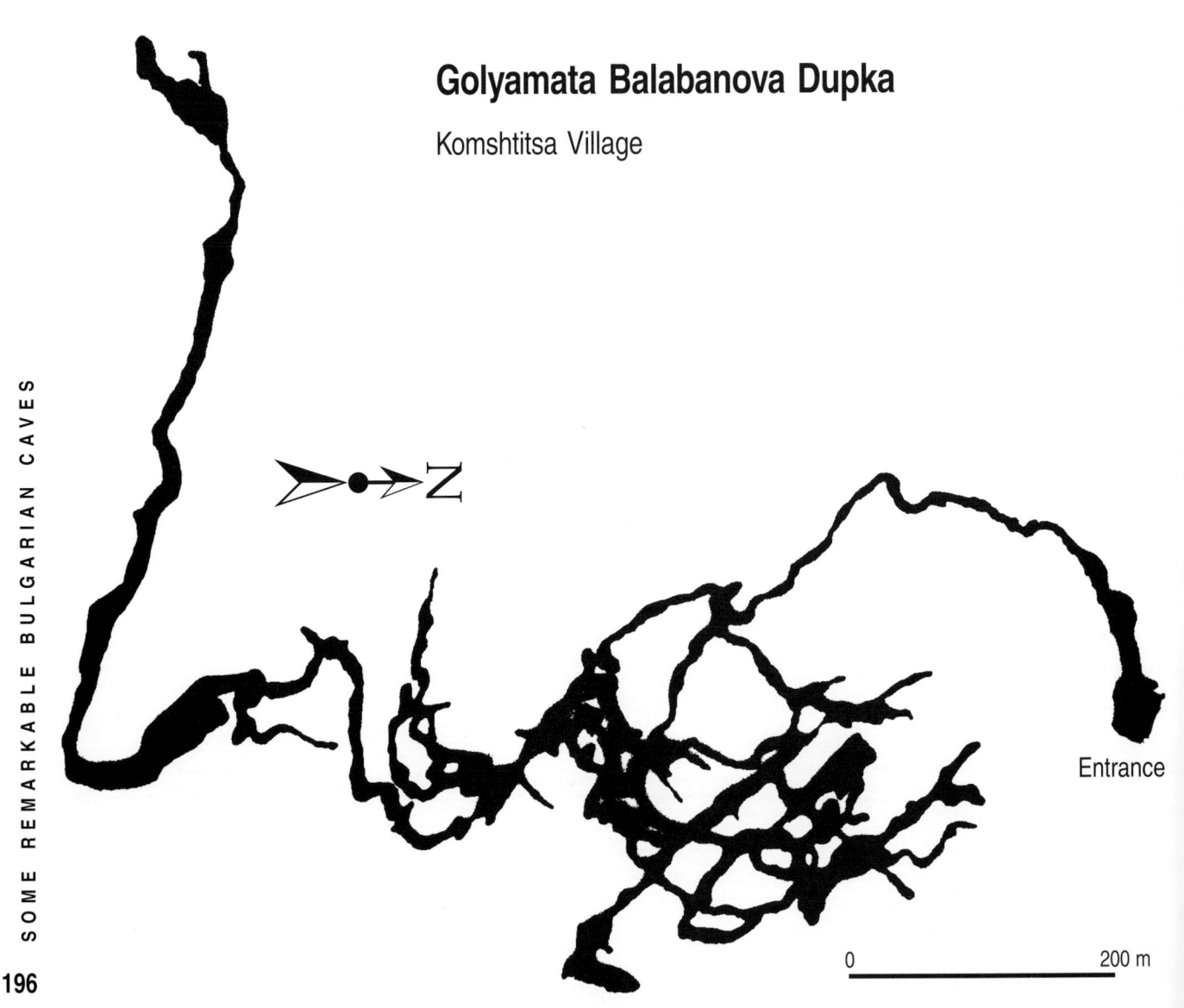

Golyamata Balabanova Dupka

Komshtitsa Village

The entrance of Golyamata Balabanova Dupka

Cavers after penetration – photo Trifon Daaliev

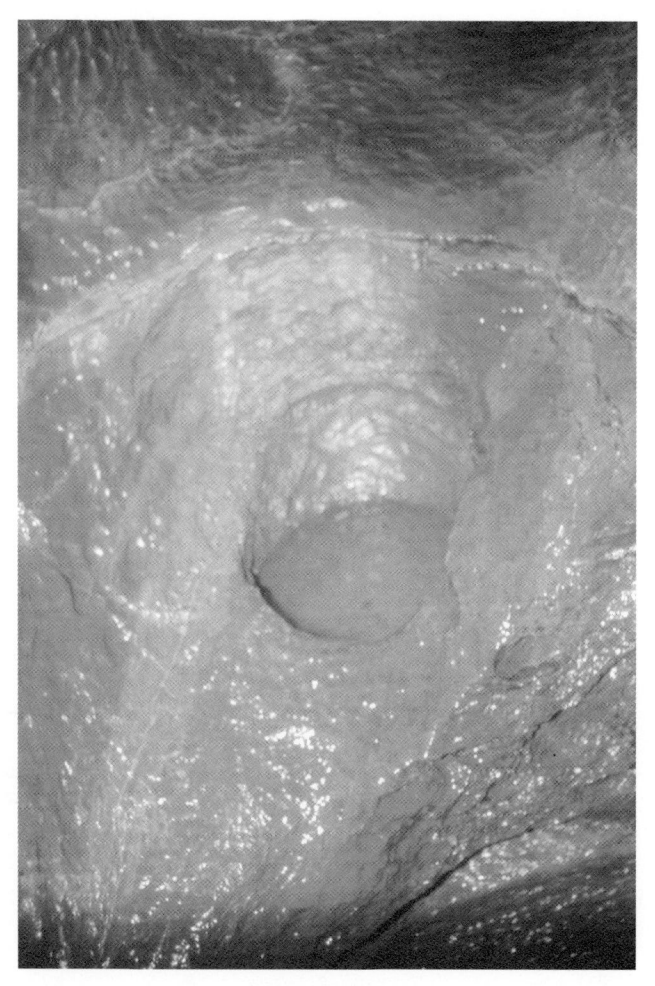

Evorsion pit – photo Trifon Daaliev

Name: **GOLEMIYAT BURUN (Oynyolar Marasa) (177)**
Varbitsa Town, Shumen Distr.
Altitude 950 m. Length 359 m. Denivelation -76 m

The cave pothole Golemiyat Burun is situated ca. 5 km NW of Chernookovo Village, in the locality Oytyolar (the Playground).

On the top of the ridge, there is a concrete triangular point and about 120 m E-SE of it the opening is situated. A five-meter vertical, a narrowing and a chamber start from the entrance (4 x 5 m). Three descending galleries are joint on the bottom of the cave (total length of 359 m). The left one is dangerous, because there is a scree moving towards the bottom. Excluding several stone waterfalls, the cave is poor in flowstone formations.

The cave pothole Golemiyat Burun has been explored many times. It was discovered and surveyed in 1975 by Shumen cavers from Madarski Konnik Tourist Society (main surveyor Atanas Spassov). Explored and surveyed in 1982 by Prista – Russe Caving Club (surveyor Kamen Dimchev). In 1986, during a District Caving Expedition, the cave was re-surveyed by P. Stefanov from Planinets, Sofia Caving Club. Description after A. Spassov, K. Dimchev, P. Stefanov.

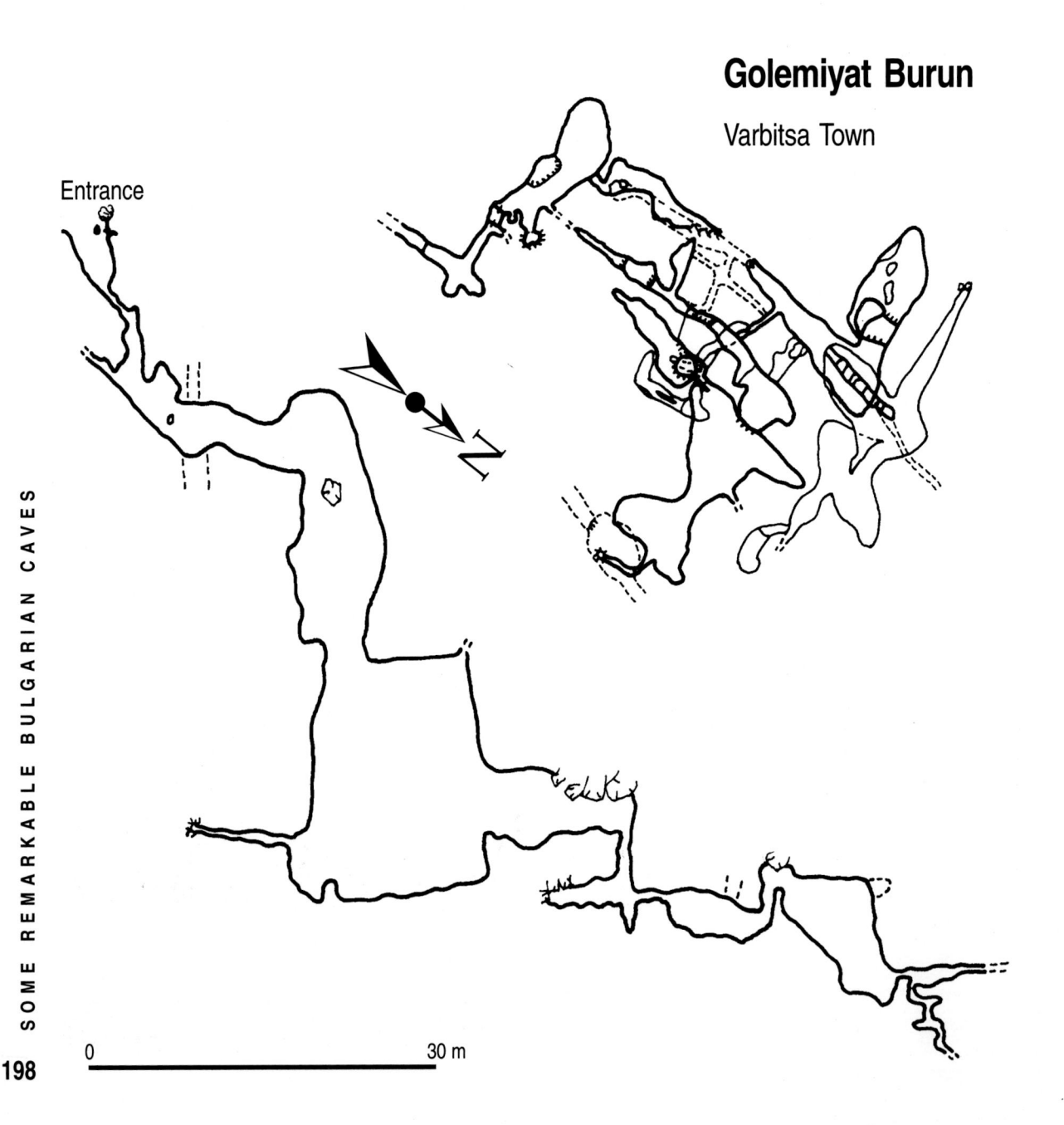

Golemiyat Burun
Varbitsa Town

Name: **GOLYAMATA GÂRLOVINA (127)**
Cherni Ossâm Village, Lovech Distr.
Altitude 1242 m. Length 44 m. Denivelation -100 m

A pothole in the eastern part of Troyan Balkan, on the territory of Steneto Biosphere Reserve. The entrance is situated in the upper end of the middle third of the valley Borov Dol. Formed in Trias limestone and dolomite.

The entrances of the potholes Golyamata Gârlovina and Borova Dupka are situated very close to each other. The altitude of the entrance of Golyamata Gârlovina is about the same as that of Pticha Dupka.

Golyamata Gârlovina is the fifth deepest pothole in this area. It is almost without a horizontal part. It starts with a narrowing (0.4 x 0.4 m). After the entrance, the gallery gets wider and steeper. A -1,5 m vertical follows, then the heavily descending gallery enters a small chamber. Further, a blockage causes bifurcation of the way. The climb over the blockage is wider and more comfortable. a steep descent over big blocks follows, ending in a small vertical. The small platform after the vertical is covered with a sinter plate and many flowstone formations. Here, a well (83 m deep) starts, with a 10-m diameter along its entire depth. The walls are covered with a thick calcite plate, the niches are full of flowstone formations (white and brown, with clay layers). This creates problems while equipping, as it is difficult to find a suiteable place for a spit. The bottom is slightly inclined, with a blockage and in some places covered with sinter plates. There are some shallow sinter lakes. The vertical goes further as a chimney (the ceiling is not seen). It would be possible to open an upper entrance by digging in a small funnel near the present one.

This 100-meter deep pothole was discovered on July 25, 1999, by Tsvetan Ostromski, Konstantin Stamov and Neda Daskalova. The same day the group equipped its 83-meter vertical, at that time the biggest in this area, and made the first attempt to reach the bottom. The survey, the description and the attachment to the nearest pothole in the area (Borova Dupka) was done by another small expedition in the autumn (October 9–10, 1999). Beside the discoverers, the group included Milen Bratanov and Georgi Georgiev. Description after Ts. Ostromski.

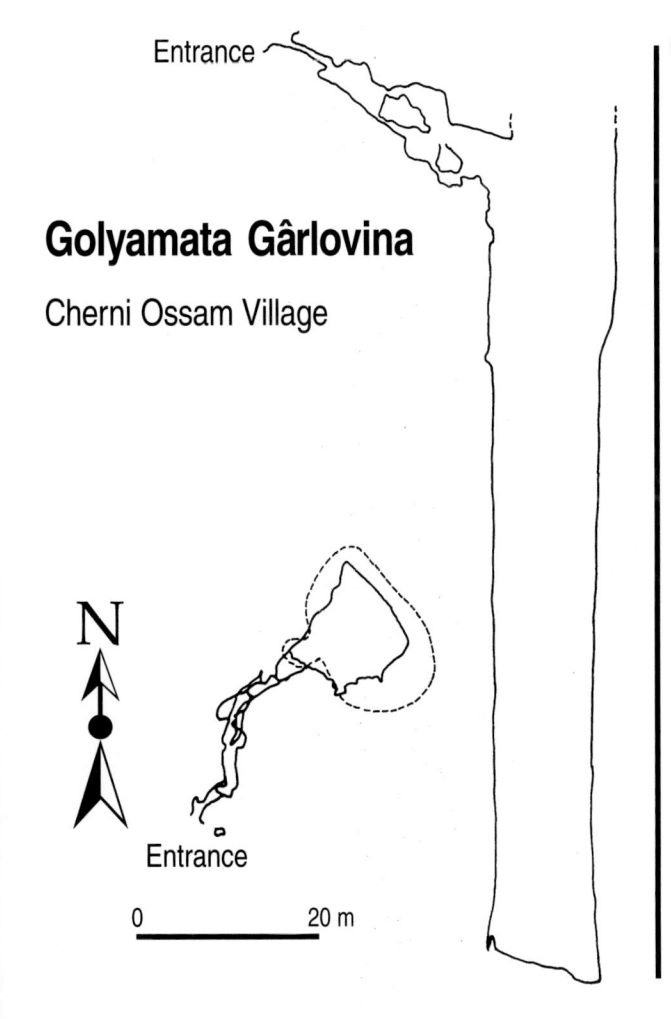

Golyamata Gârlovina

Cherni Ossam Village

The entrance – photo Tsvetan Ostromski

Name: **GOLYAMATA GARVANITSA (150)**
Gorsko Slivovo Village, Lovech Distr.
Length 273 m. Denivelation -50 m
Coordinates: N 43° 12' 43.2" E 25° 02' 56.7"

A pothole, 1.2 km NW of Gorsko Slivovo, in the NE periphery of Devetaki Plateau.

The entrance opens at the bottom of an elliptical funnel (34 x 17 m). A shaft of the same shape and dimensions follows and a depth of 34 m. Bottom covered with stones. In the SE part, there is a short gallery, ending in a lake siphon. An inclined gallery goes NW (17 m wide and with a height of up to 15 m). At the 35th meter, the gallery is transformed into a chamber (35 x 45 x 6 m). On the NE the chamber gradually lowers into a gallery, which after 30 m becomes wider. A pyriform chamber follows which has the following dimensions: length 60 m, width up to 30 m and height up to 19 m. The floor of the galleries is covered with thick clay deposits, angular stones and separate stone blocks. The cave is poor in flowstone formations. During rain and snow melting, from the lake siphon comes a considerable amount of water, flowing along the galleries and draining in the end chamber.

The cave was explored and surveyed first by G. Yanchev and S. Penchev in 1966 during a National Cave Expedition. In the 80s, at the initiative of P. Petrov and P. Stefanov of the Institute of Geography (BAS), cavers from Viten Caving Club, R. Mazalat Sevlievo Tourist Society, mounted a metal staircase and platforms, allowing access to the bottom of the entrance shaft. At the same time, the entrance was fenced and on the bottom a station for complex karstological research was installed.

Golyamata Garvanitsa

Gorsko Slivovo Village

Name: **GOLYAMATA HUMBA (180)**
Kotel Town, Sliven Distr.
Altitude 980 m. Length 20 m. Denivelation -94 m

A pothole, about 6.5 km NW of Kotel, on the southern slope of the hill Suhi Dyal, Kotel Balkan, Eastern Stara Planina. The entrance is quadrangular and is situated ca. 20 m below the ridge of the hill. From there, we descend a vertical with an oval cross section and 5-m diameter. About 30 m below the entrance, there is a small platform, then the vertical goes on, slightly elliptic and enlarged, for another 40 m. A strongly inclined bottom follows, gradually becoming a lowering gallery (3 m wide), going east. At the end of the gallery, there is another vertical (13 m), with a stony bottom. The air temperature on the bottom, measured on August 9, 1970, was 9.4° C.

The pothole had been known to local people for a long time, but was explored for the first time in 1962 by cavers from Plovdiv (the bottom reached by Stoycho Stoychev). Surveyed during an expedition of Akademik Sofia Caving Club in August 1970 (description of D. Dimitrov).

Golyamata Humba

Kotel Town

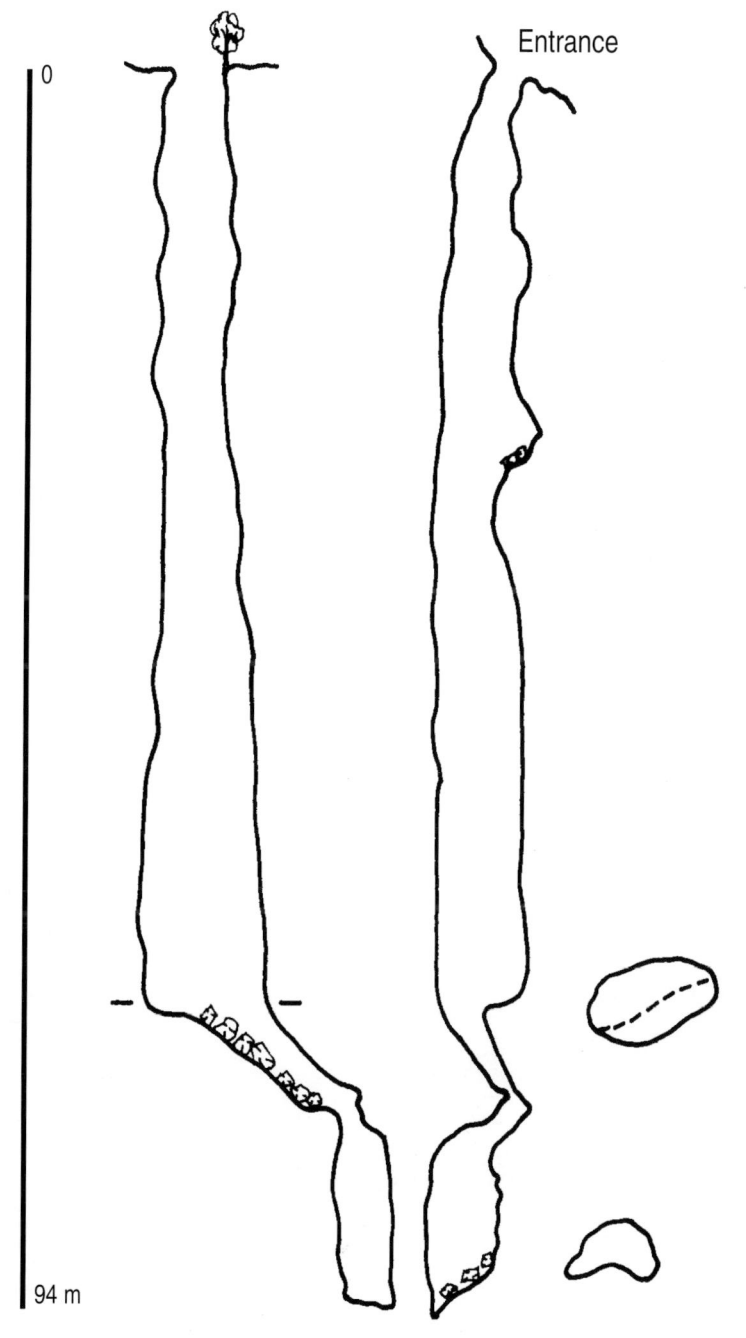

Name: **GOLYAMATA MAARA (178)**
Medven Village, Sliven Distr.
Length 314 m
Coordinates: N 42° 34' 46.38" E 26° 35' 01.39"

A cave near Medven, about 1300 m NE of Lednitsata, near the summit of Sreden Kayriak. Entrance 0.6/1 m. A spacious entrance chamber with sinters, in northern direction there is a main gallery with a floor of terra rossa. After a hole called Prilepna Dupka with a small bat colony, the main gallery goes on among boulders and sinter lakes. The gallery ends with a clay downing and a chamber. Explored in 1967 by Akademik Sofia and in 1981 by Protey Sliven Caving Clubs.

A Protected Natural Monument with adjacent area of 17.8 ha (Decree 1537/ 02.09.1968, issued in State Gazette 33/25.04.1969).

Golyamata Maara

Medven Village

Entrance

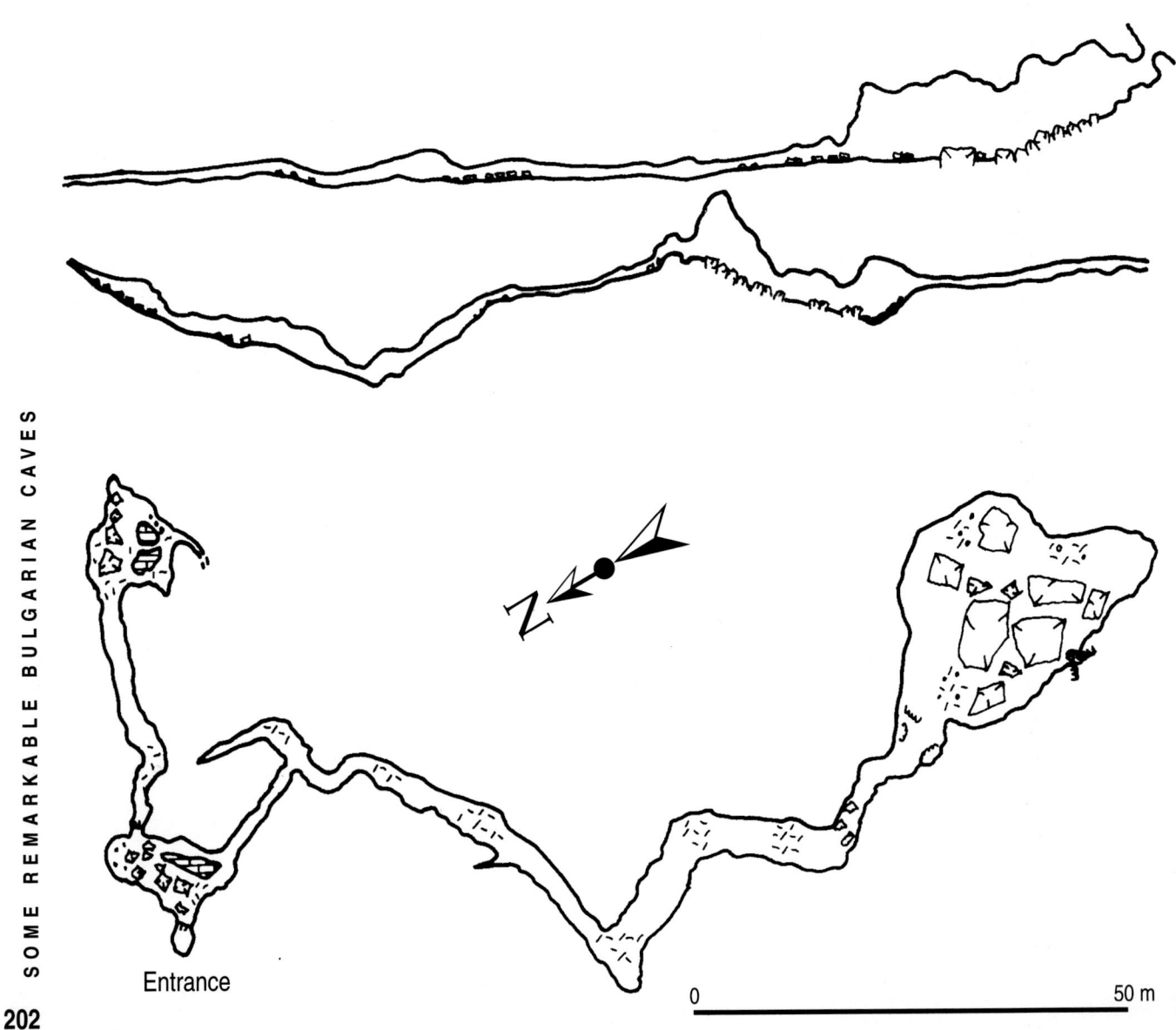

0   50 m

Name: **GOLYAMATA MIKRENSKA PESHTERA (MANDRATA) (137)**
Mikre Village, Lovech Distr.
Length 1921 m. Denivelation -9 m

A water cave north of Mikre, on the left bank of river Kamenitsa (Kamenka). Formed in Maastricht limestone (Lower Cretacetous). An asymmetric, labyrinth cave. Width of galleries 0.5–22 m, height 0.3–7 m. Well developed sinters.

A stream with a variable outflow comes out of the cave. Discovered by D. Iltchev (shown by local people), and on September 4, 1924, studied up to 200 m. In 1926, the archaeologist V. Mikov found artefacts from the Eneolith. Excavated also by N. Djambazov in 1960. Later used for cultivation of champignons, then for cheese making, hence the name Mandrata. In 1983, explored by S. Gazdov, Tsv. Hristov, K. Petkov from Studenets Pleven Caving Club and surveyed to its present length.

Fauna: eight species, including the stygobite *Niphargus bureschi* (Amphipoda).

Golyamata Mikrenska peshtera (Mandrata)

Mikre Village

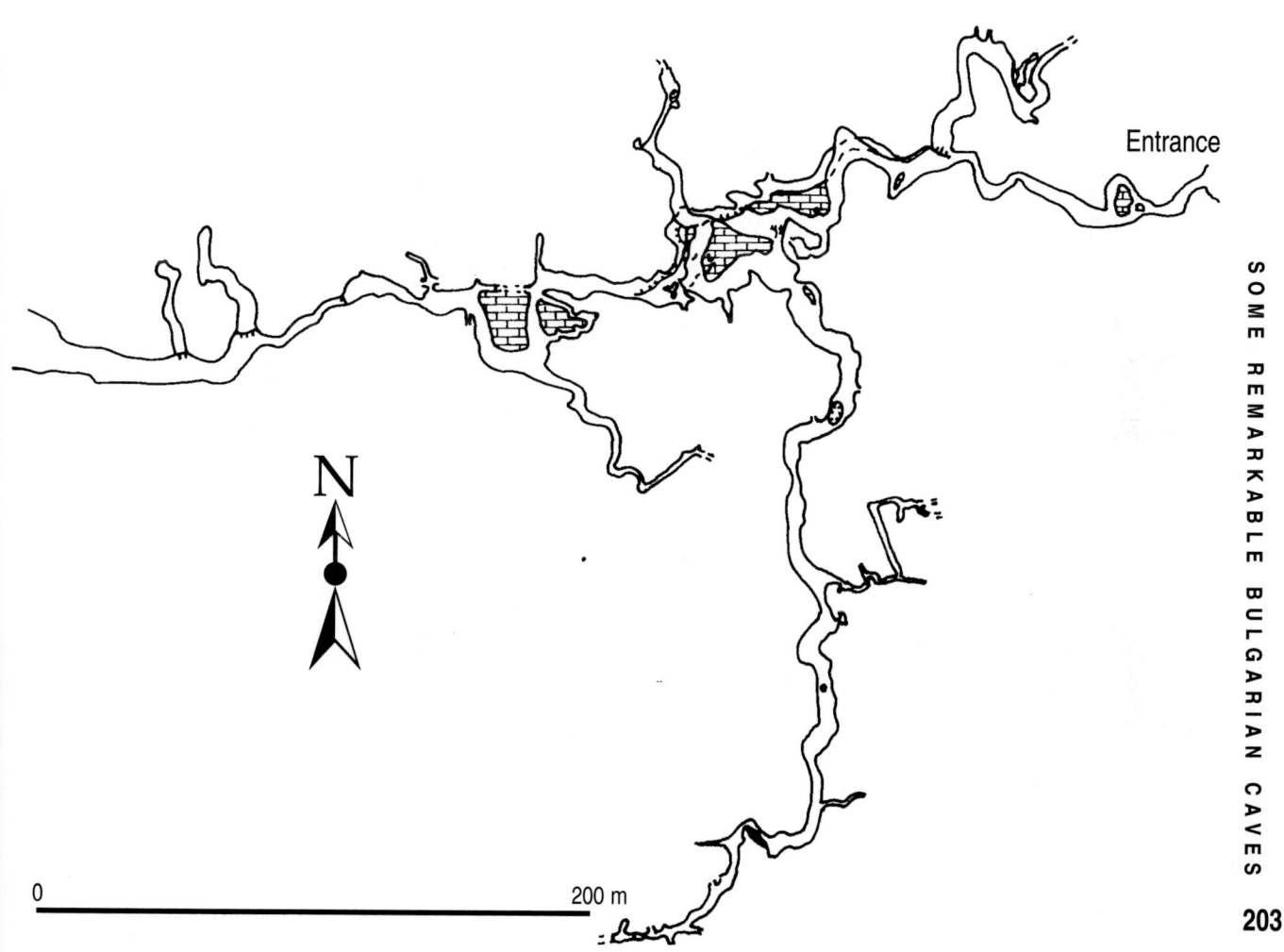

Name: **GOLYAMATA TEMNOTA (YAMKATA) (63)**
Drenovo Village, Sofia Distr.
Total length 2000 m. Denivelation -106 m

A cave in Trias limestone and dolomites, 35 km from Sofia, near the quarter Stephanovtsi of Drenovo, in the locality Razdoltsi. Entrance in a triangle, fortified by wooden beams. A two-storey, branched cave. A narrow squeeze, 18 m from the entrance, eight meter long. Upper storey – 350 m long and very beautiful, leads again to the river. After ca. 100 m, a siphon marks the end of the cave. An underground stream, used for drinking water in Drenovo.

After two weeks of hard digging and fortifying of the entrance, on January 16, 1994, the first researchers M. Zlatkova, St. Petkov, D. Angelov and others entered the cave.

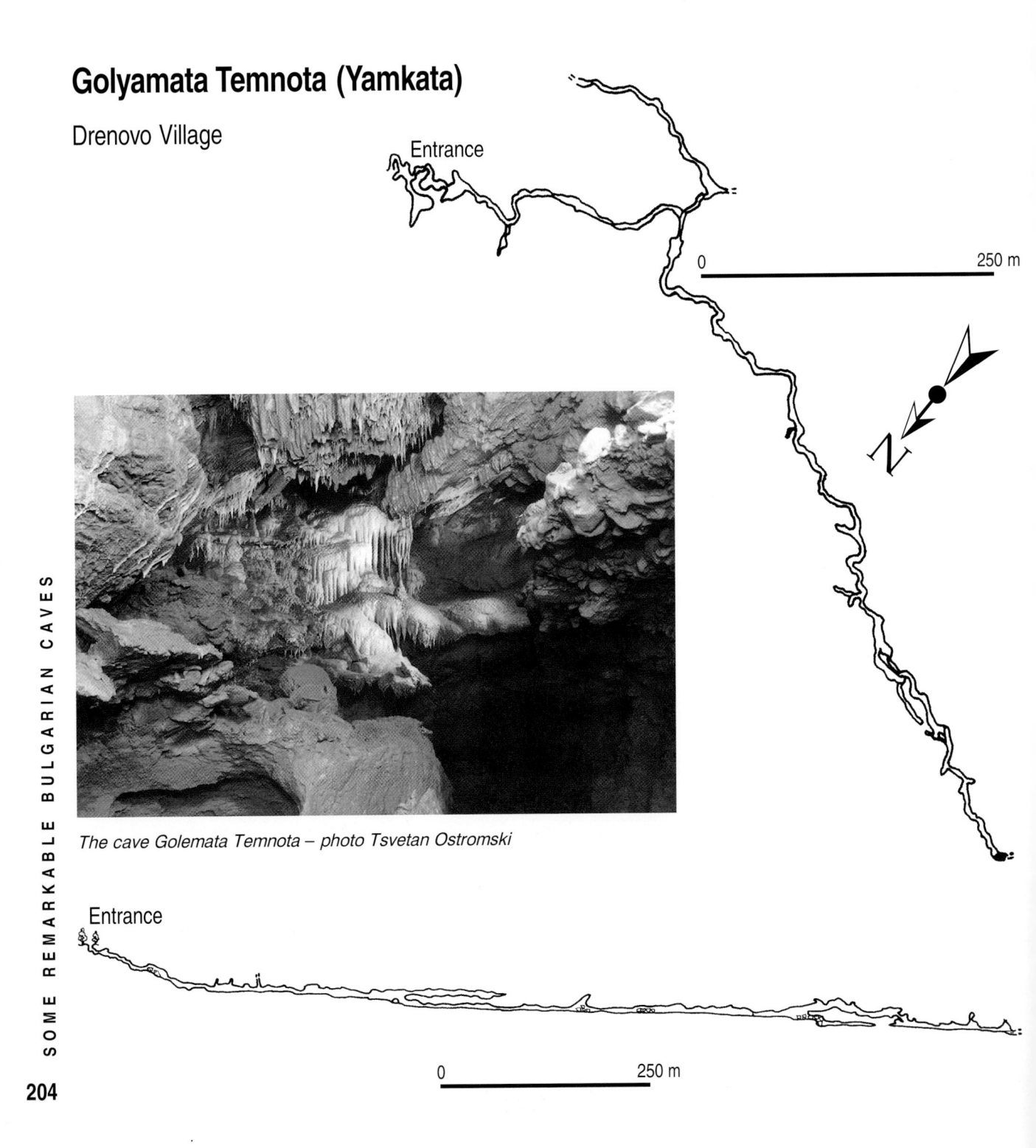

Golyamata Temnota (Yamkata)

Drenovo Village

The cave Golemata Temnota – photo Tsvetan Ostromski

Name **GOLEMATA VODA (88)**
Karlukovo Village, Lovech Distr.
Altitude 255 m. Above the LBE: 129 m. Length 612 m. Denivelation -105 m
Coordinates: N 43° 09' 36" E 24° 04' 37"

A cave near Karlukovo, 25 m on the right from the road to Rumyantsevo and 210 m SE of the village mill. A narrow entrance leads to a two-meter descent, then three interconnected verticals of 14.7 and 35 m, with a total depth of 56 m, are descended. A chamber, 15 m long and ca. 5 m wide, follows. A stream (3 l/s) runs in the chamber. Upstream, a hall, 10 m high, is reached by crawling and on the SE side a gryphon is formed from which water runs out. On NE, from the bottom of the vertical, the stream forms an 8-m siphon, penetrated by I. Ivanov from Studenets Pleven Caving Club on April 4, 1991. After a 30-m long water gallery and a second siphon of 10 m there is a third siphon of 20 m (overcome by divers from Pleven). The cave is descending, a large gallery, 65 m long, leads into an elliptical widening, 30 m long and 18 m wide. Three galleries follow. On the right, there is another stream (5 l/s) coming from an ascending, large and high gallery (not fully explored), on the left there is a 170-m dry, ascending gallery, wide ca. 4 m. After ca. 100 m, the middle gallery ends in a siphon.

In 1990, M. Zlatkova and A. Drazhev discovered the cave by digging. The survey of the dry parts was done by the discoverers and by L. Zhelyazkov from Burgas. The siphons and the parts after them were mapped by M. Dimitrov, Kr. Gigenski and Iv. Ivanov from Studenets Pleven Caving Club.

Golemata Voda

Karlukovo Village

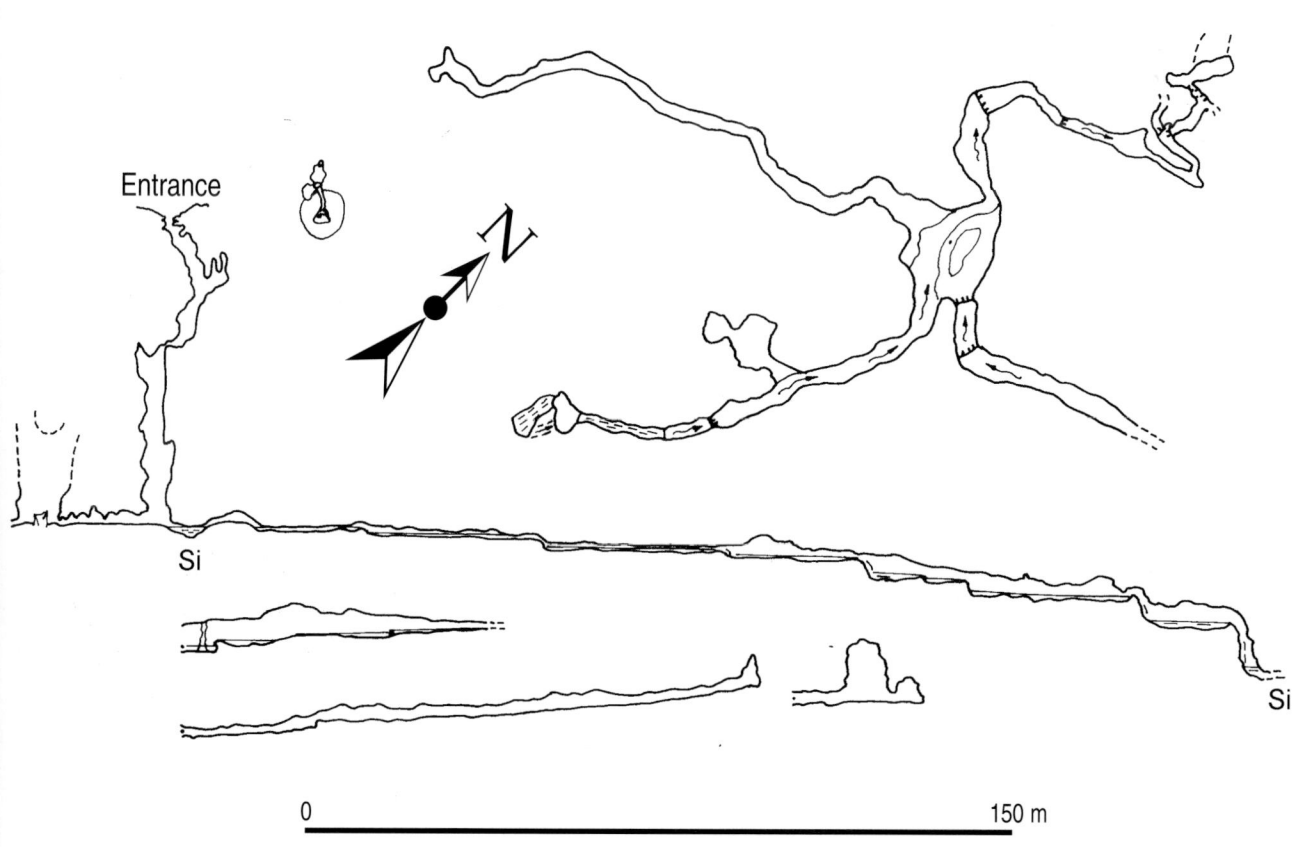

Name: **GOLYAMATA YAMA (YAMATA – 3) (190)**
Kipilovo Village, Sliven Distr.
Altitude 920 m. Length 740 m. Denivelation -350 m

The cave pothole Yamata – 3 is developed in Senonian limestone in the locality Gyoldjukleri. The cave is a diaclase, branched and periodically watered.

One 220-meter long diaclase cascade, intersected by wells, 10, 70, 50, 21 m deep, leads to Temerut chamber. The dimensions of the chamber are 50 x 100 m. This is the biggest chamber in the cave. Huge boulders are scattered around, among them runs a stream. A descending clay gallery follows, then it widens and forms an oblong chamber, ending in stone blocks. The floor is covered with dolomite flower. Here, the cave is about to crumble. Yamata -3 is very rich in secondary formations. The chamber with the lake is particularly beautiful. There, everything is snow white. The most typical formations start after the 35-meter vertical. In the chamber, after the 11-m vertical, there are cave pearls. The bottom is impenetrable, narrowing at 350 m. There is a stream in the cave, presumably emerging at the karstic spring in Kipilovo, T° 10 °C.

In 1964, cavers from Akademik Sofia Caving Club explored the entrance parts of the pothole. In 1978, cavers from Protey Sliven Caving Club found an extension and after many expeditions managed to explore and survey the third deepest cave in Bulgaria. In the following years, ascending verticals were found, and wells emerging from Temerut Chamber and reaching the surface.

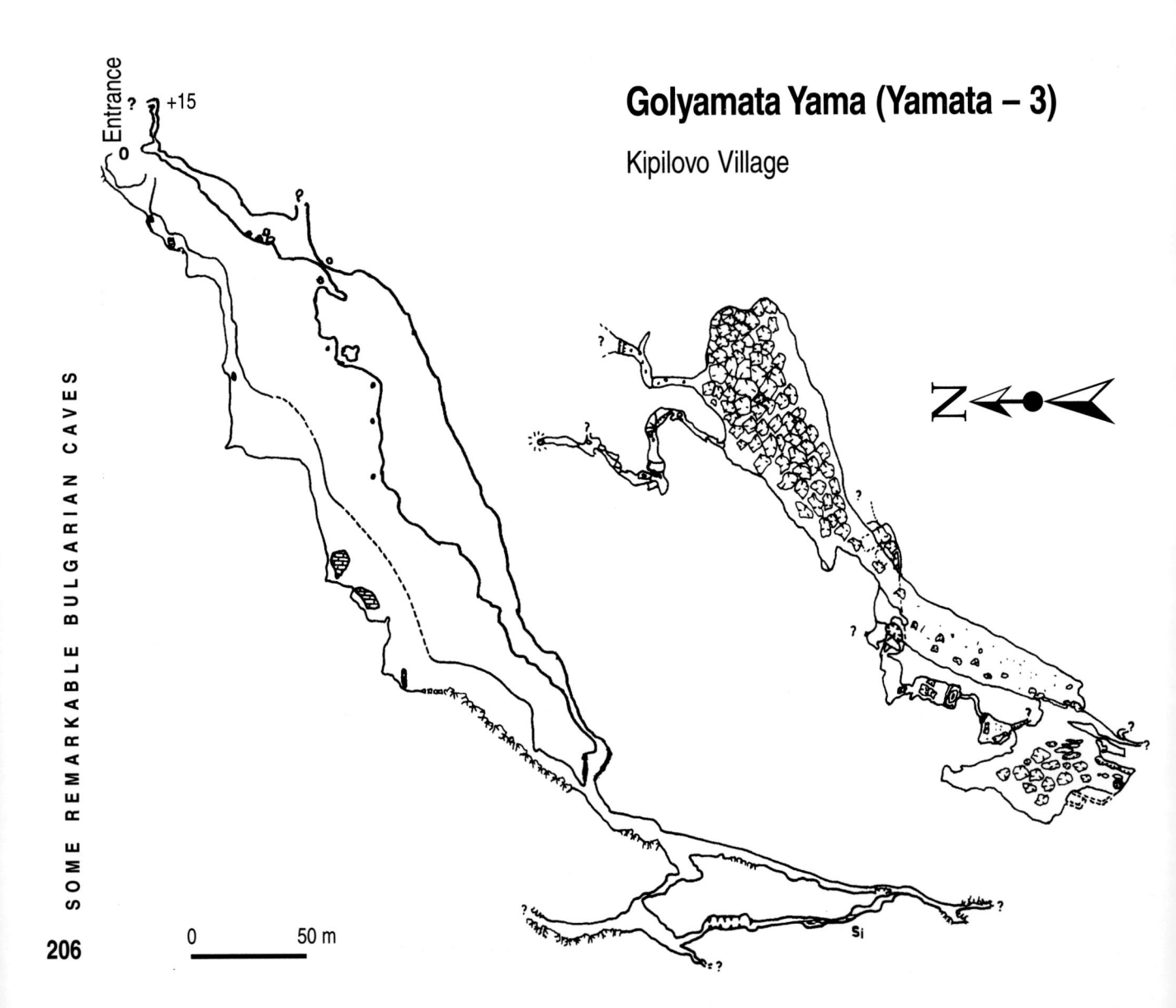

Golyamata Yama (Yamata – 3)

Kipilovo Village

Name: **GOLYAMATA YAMA (128)**
Cherni Ossam Village, Lovech Distr.
Altitude 1343 m. Length 459 m. Denivelation -43 + 18 m
Coordinates: N 42° 45' 51.6" E 24° 40' 53,5"

A water cave in Chaushov Dol, in the park Steneto, with an occasional stream. The cave is labyrinthine, with many storeys. There are three streams, running in it, and many sinks. When in flood, the inside of the cave is entirely under water.

At 392 m, on the main axis, the cave ends in a muddy siphon. At the beginning of the central gallery there are big boulders. After a vertical of 2.5 m, on small gravel, we go on until reaching a siphon, which was penetrated in 1975. At +4.5 m, an upper storey is formed. Through a system of narrow passages, we reach the Big Diaclase (description according to Rumyana Sirakova and Panayot Neykovski).

The cave had been known to local people, but the first exploration took place in 1961, done by an expedition of cavers from Akademik Sofia, led by P. Neykovski. They surveyed ca. 300 m. In 1972–1974, the exploration was carried out by V. Balevski and V. Markov from Ambaritsa, Troyan, and N. Gladnishki and T. Daaliev from Edelweiss, Sofia. New galleries and storeys were discovered. The latest map of the cave was made by K. Dimchev and S. Popov from Russe in 1986. The cave was not completely explored. A gallery of some 200–300 m, discovered by V. Balevski, was obstructed again by silt, carried by the water.

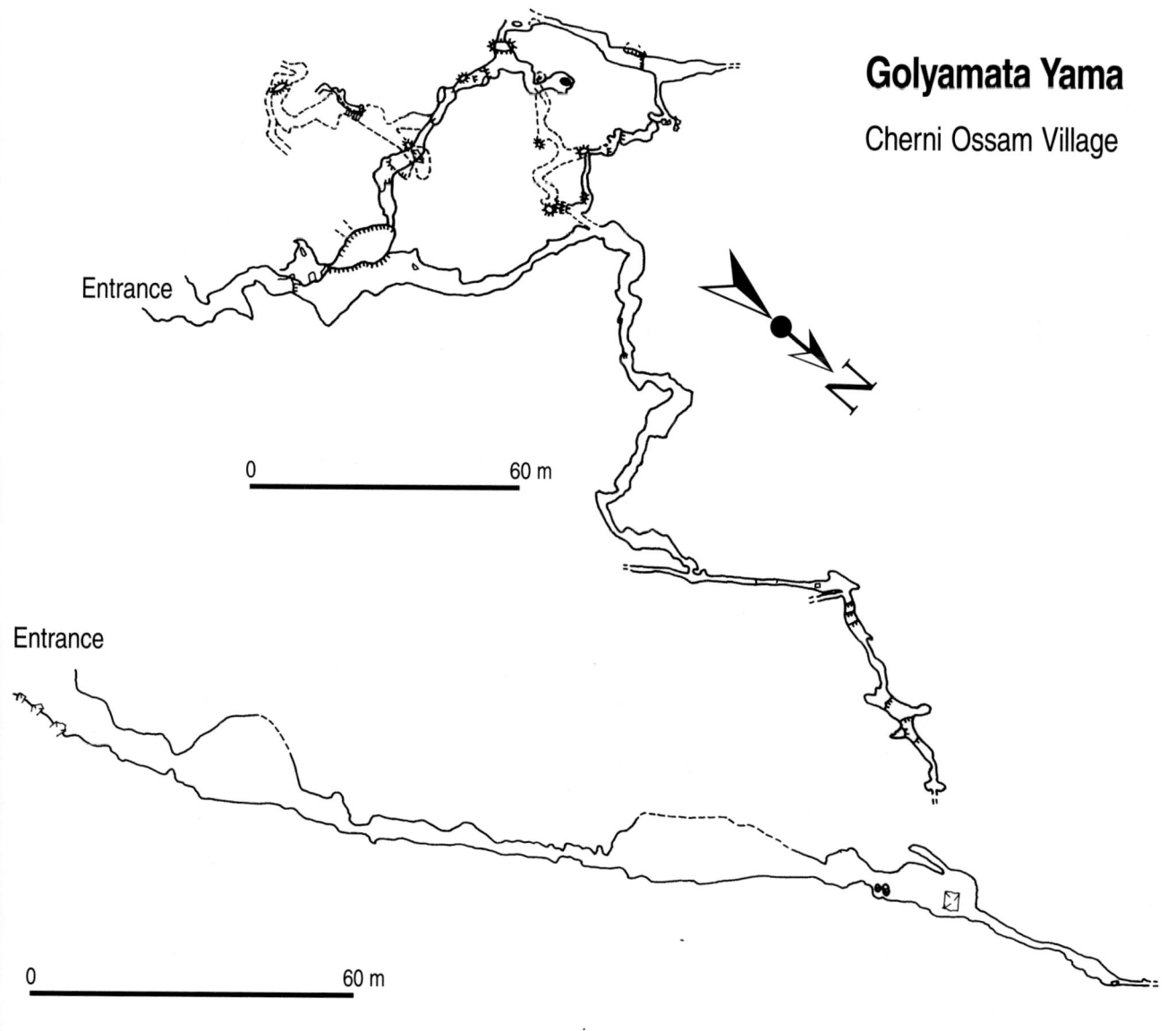

Golyamata Yama

Cherni Ossam Village

Entrance – photo Vassil Balevski

The valley Golyam Chaushov Dol – 200 m downstream is the entrance of the cave Golyamata yama – photo Trifon Daaliev

Name: **GORNA AND DOLNA PESHTERA (107)**
Zlatna Panega Village, Lovech Distr.
Altitude 180 m. Length 660 m. Denivelation -41 m

The entrance (3 x 3 m) is situated in the southeastern confines of Zlatna Panega, in the ridge Panezhki Rid. The two entrances are seen in a small cliff, immediately near and above the siphon of Glava Panega, from where the river runs out. A small stone staircase leads to the entrance of the first level. An iron bar separates the footpath from the siphon lake.

The entrance gallery, 30 m long, leads to a big chamber, open towards the siphon — this is the other entrance to the cave. The dimensions of the chamber are 100 x 30 x 15 m. It ends in a blockage, from which several narrow galleries start. On the lower level there are no flowstone formations.

The second floor is situated highly, about 100 m from the restaurant. A wide dirt road leads to it. The entrance of the Upper Cave is right above the big entrance of the lower cave. There is an iron door, as the cave is used to store blasting material for the quarry. The entrance part of the cave is dry, warm and without formations. It is possible to climb from one level to another. The third floor ends in a blockage.

The cave has been well known to local people since ancient time. It was described first by the brothers K. and H. Škorpil in 1895. Many prominent explorers of Bulgarian caves (P. Petrov, P. Tranteev, Iv. Buresch and others) have discussed this interesting lake and its caves. In 1948, a team called "Todor Pavlov" Scientific Brigade worked in the cave.

In 1973, cavers from Iv. Vazov Sofia Caving Club, led by P. Petrov, explored the cave, looking for a connection with the siphon.

In December 1975, cavers from Studenets Pleven and Planinets Sofia Caving Clubs made two competing maps of the cave.

Gorna and Dolna peshtera

Zlatna Panega Village

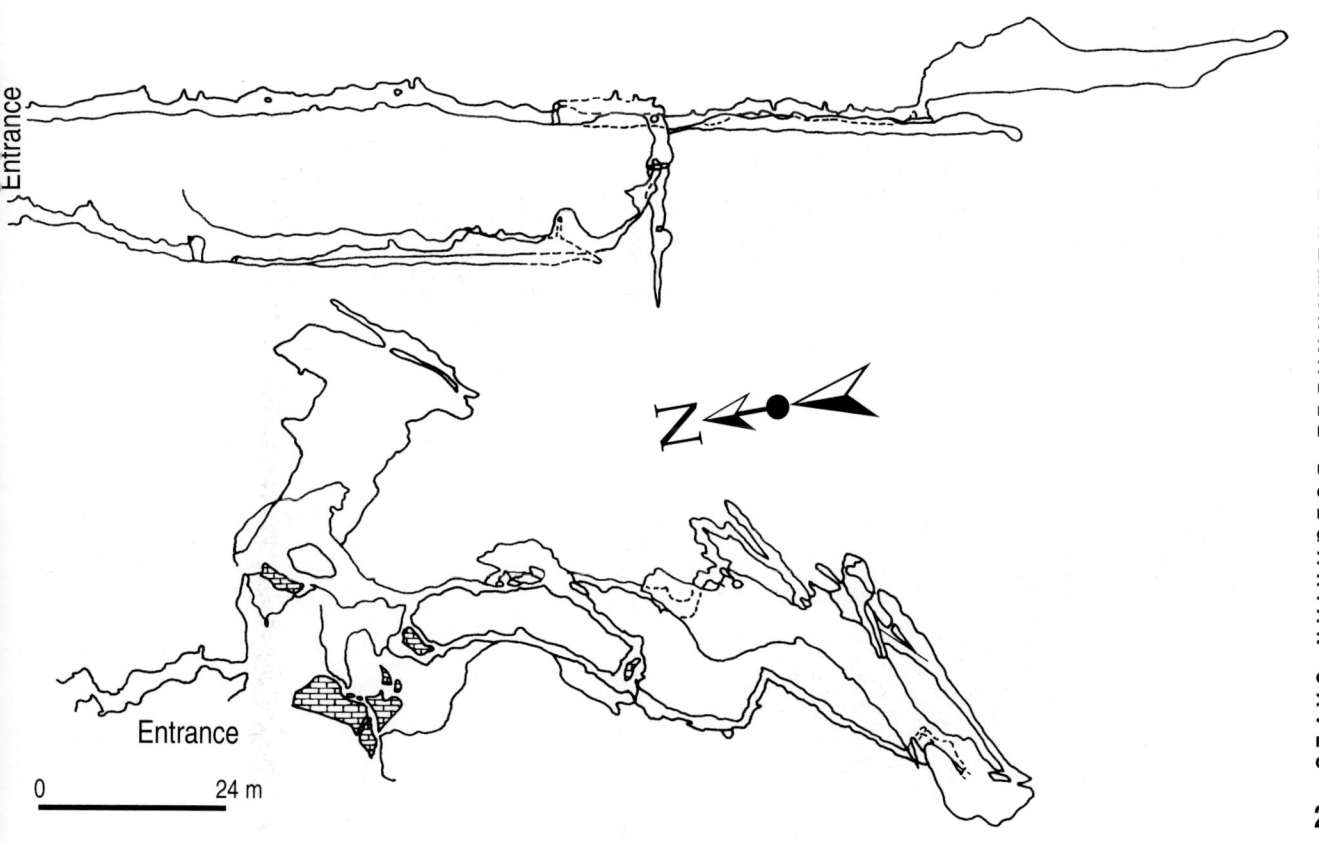

One of the entrances of the Lower Cave is looking to the source Glava Panega – photo Vassil Balevski

Name: **GORNIK (155)**
Krushuna Village, Lovech Distr.
Length 1074 m. Denivelation -4 m

A water cave, 2.5 km SE of Krushuna in the locality Lipichkite, in the northern periphery of Devetashko Plateau, Middle Predbalkan. Formed in limestone from the Lower Cretaceous (Aptian). A one-gallery cave. Entrance – 6 x 2 m. A descending dry gallery follows, which 370 m further opens in the wall of an active water gallery, three meters above the water level. About 70 m upstream a 1.5-m siphon is formed. Another 560 m of meandering gallery end in a siphon. Before the siphon, the gallery has a triangular cross section and average dimensions: 1.8 m wide, 1.4 m high. After that the shape becomes different and the dimensions respectively 2.30 x 2.95 m. Many flowstone formations, especially behind the siphon. On August 25, 1978, the outflow of the river was three l/sec.

Explored and surveyed in 1965 by Tsvetan Lichkov. During the International expedition Druzhba' 78, Alexey Jalov penetrated the siphon at 447 m. The entire cave was surveyed in February 1982 by P. Hristov, S. Tsonev, D. Nanev, I. Ivanov, cavers from Aleko Sofia Cavinig Club.

Fauna: only one species of caddis fly is known.

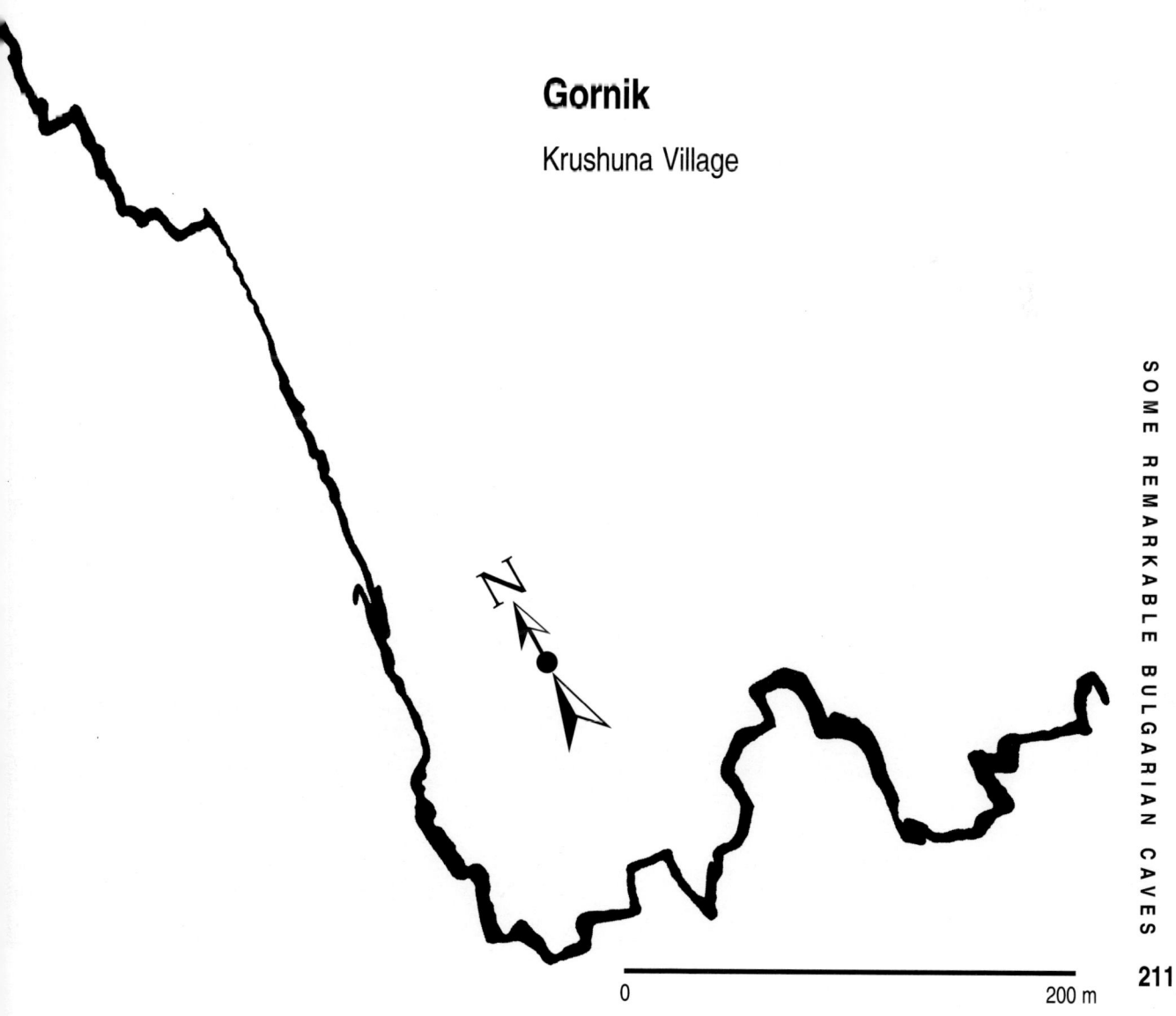

Entrance

Gornik

Krushuna Village

0   200 m

Name: **GRADESHNISHKATA (RUSHOVATA) PESHTERA (115)**
Gradeshnitsa Village, Lovech Distr.
Length 908 m
Coordinates: N 43° 15' 46" E 24° 15' 46"

The entrance of this cave is situated several meters above the level of the right bank of the river about 100 m from Gradeshnitsa, in the direction Glogova mahala. The cave is formed in Titonian (Upper Jurassic) limestone. A horizontal water cave with several branches in the end parts. There is rich adornment in the second half. Through most of the cave an underground river runs with an outflow of 30–40 l/s. The water comes from the end siphon, collecting the drained water of the surface river.

Known to the local people since time unrecorded. Announced in 1948 by P. Tranteev and surveyed by him later.

Known troglobite fauna: *Balkanoniscus corniculatus* (Isopoda), *Tranteeva paradoxa* (Opiliones, Cyphophthalmi), *Tranteeviella bulgarica* (Coleoptera).

Gradeshnishkata (Rushovata) peshtera

Gradeshnitsa Village

Rushovata peshtera – photo Trifon Daaliev

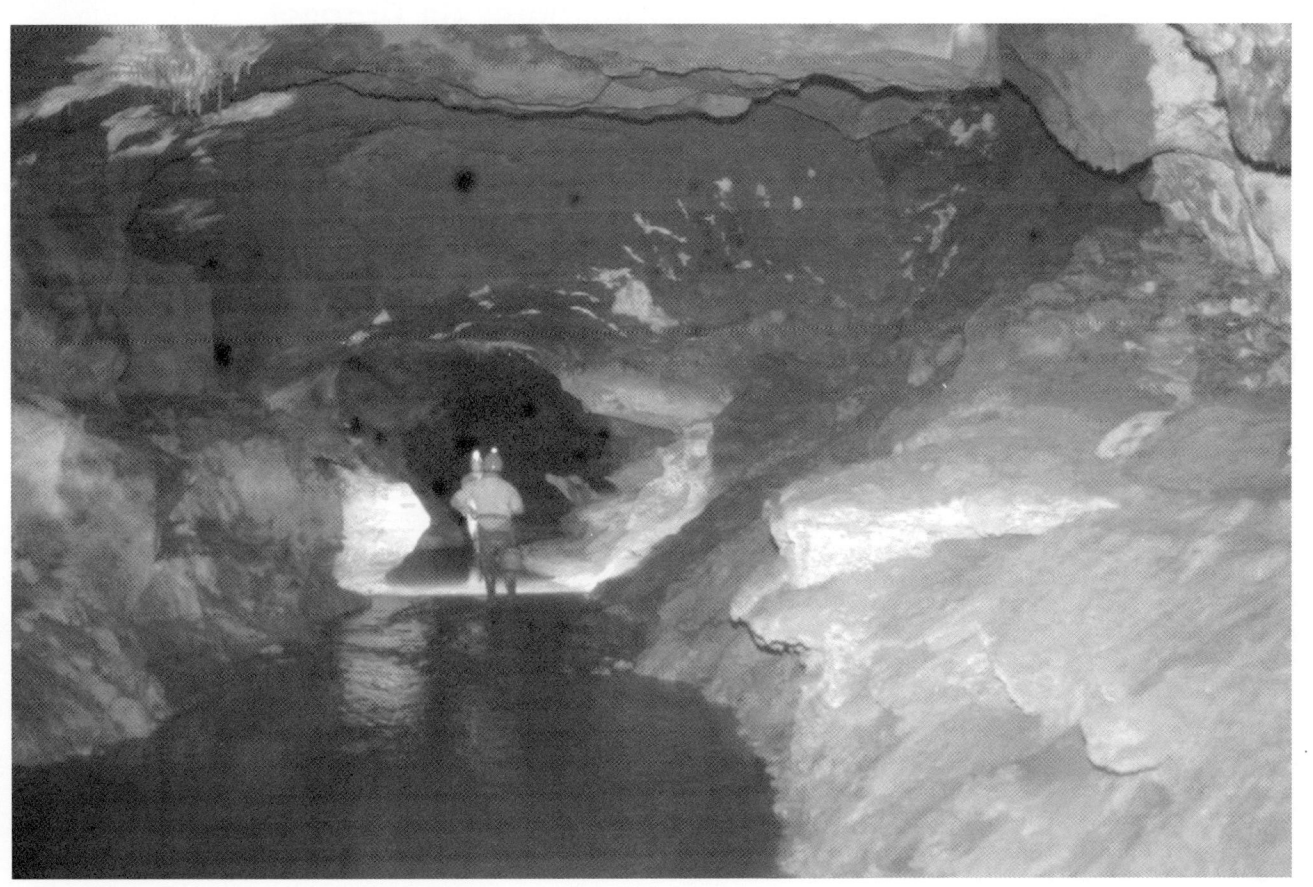

The underground river – photo Trifon Daaliev

Name: **GRANICHARSKATA PROPAST (71)**
Komshtitsa Village, Sofia Distr.
Length 224 m. Denivelation -80 m

A cave pothole ca. 6 km north of Komshtitsa, on the left bank of Komshtitsa River, ca. 200 m from the border post, immediately by the road in the locality Lissina.

The small opening of this cave is situated 3–4 m above the river level. An inclined part of 5–6 m and a 4-meter vertical follow, then an almost horizontal gallery, in which the underground water from the river Komshtitsa (Vissochitsa) comes. After ca. 30 m, we reach a 20-meter vertical, washed by water. Next is a 50–60 m long inclined gallery, on the walls of which there are flowstone formations. In some sinters there are sugar-like pears. At the end, there is a 6-7 m vertical. Through the cave an underground stream runs with an outflow of ca. 7 l/s (2.7.1974), forming a 20-meter waterfall. Up to the 15th meter, there are ice formations and frost weathering. The cave was formed along diaclase, by means of water coming through the side sinks of Komshtitsa River. Its exploration has not yet been completed.

The name Granicharskata (Border Guard) was given by the participants in the exploration in 1974, after border guards had shown them the cave. Some local people call it Hladilnika (The Fridge).

N. Gladnishki and T. Daaliev were the first to enter the cave on June 28, 1974. They reached ca. 80 m. On July 2, 1974, N. Gladnishki, A. Anev and T. Daaliev surveyed the cave, P. Beron collected cave animals and P. Tranteev and E. Svilenova made scientific observations.

In this pothole, on November 7, 1975, Yancho Manuilov from the Caving Club of Ivan Vazov Sofia Caving Society died. At the end of that year a road construction started and the opening was closed until 1994. Then, some cavers, led by Z. Iliev, dug out and reopened the cave entrance.

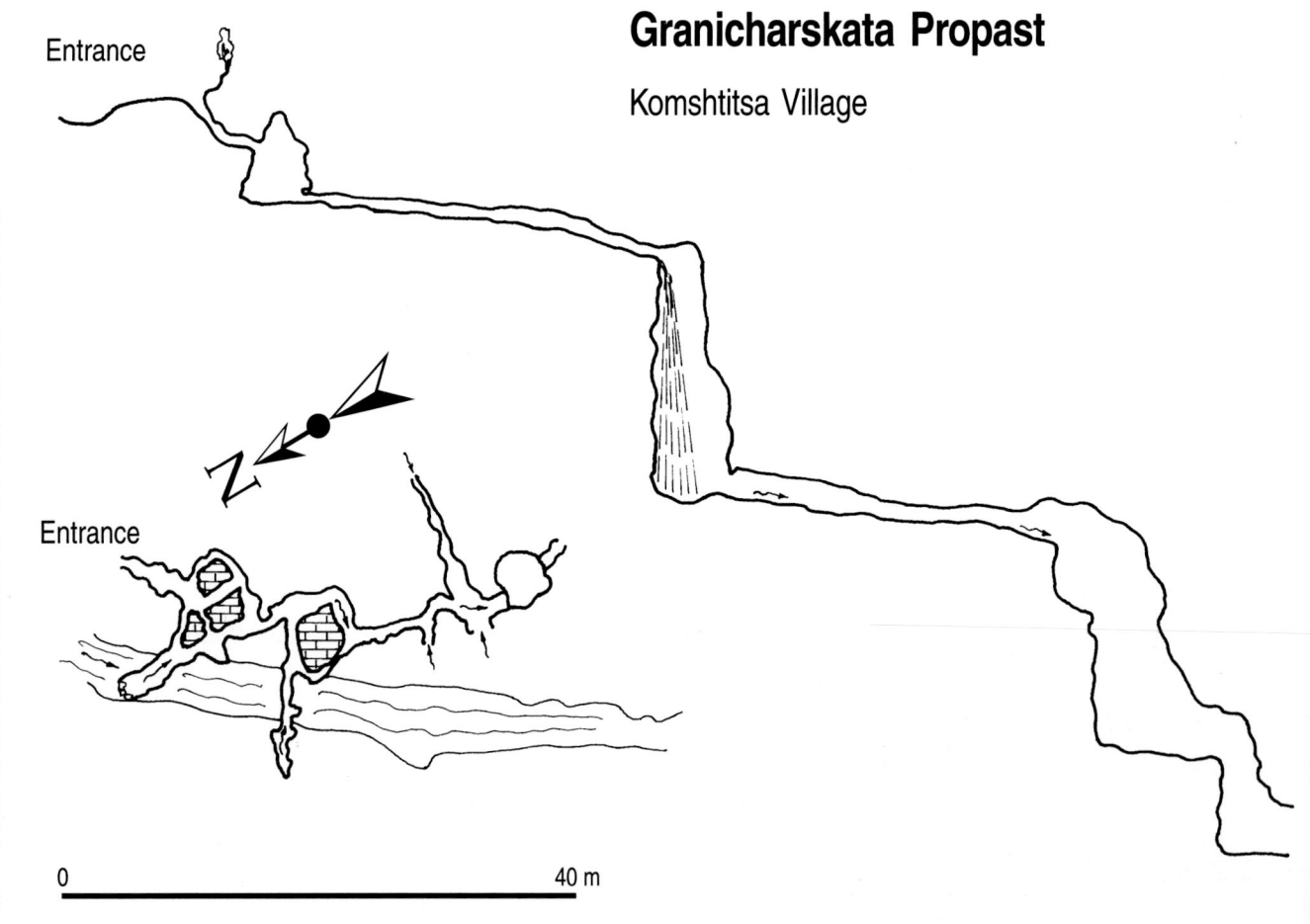

Granicharskata Propast
Komshtitsa Village

Name: **GRIMNENA DUPKA (32)**
Cherkaski Village, Sofia Distr.
Length 351 m. Denivelation -28.6 m
Coordinates: N 43° 14' 29.9" E 23° 15' 57.5"

A dry cave with two entrances. The upper entrance is 6 m wide and 2.8 m high. The lower entrance (situated 12 m under the upper one) is 5 m wide and 1.8 m high. The cave is descending, branched (in some parts a labyrinth), two-storey. During excessive rain, flood water enters the lower storey. There is one eight-meter shaft. Studied many times, surveyed by V. Georgiev and S. Chobanov by Pâstrinets Montana Caving Club in 1983.

Fauna: three species of spiders and seven species of bats are known.

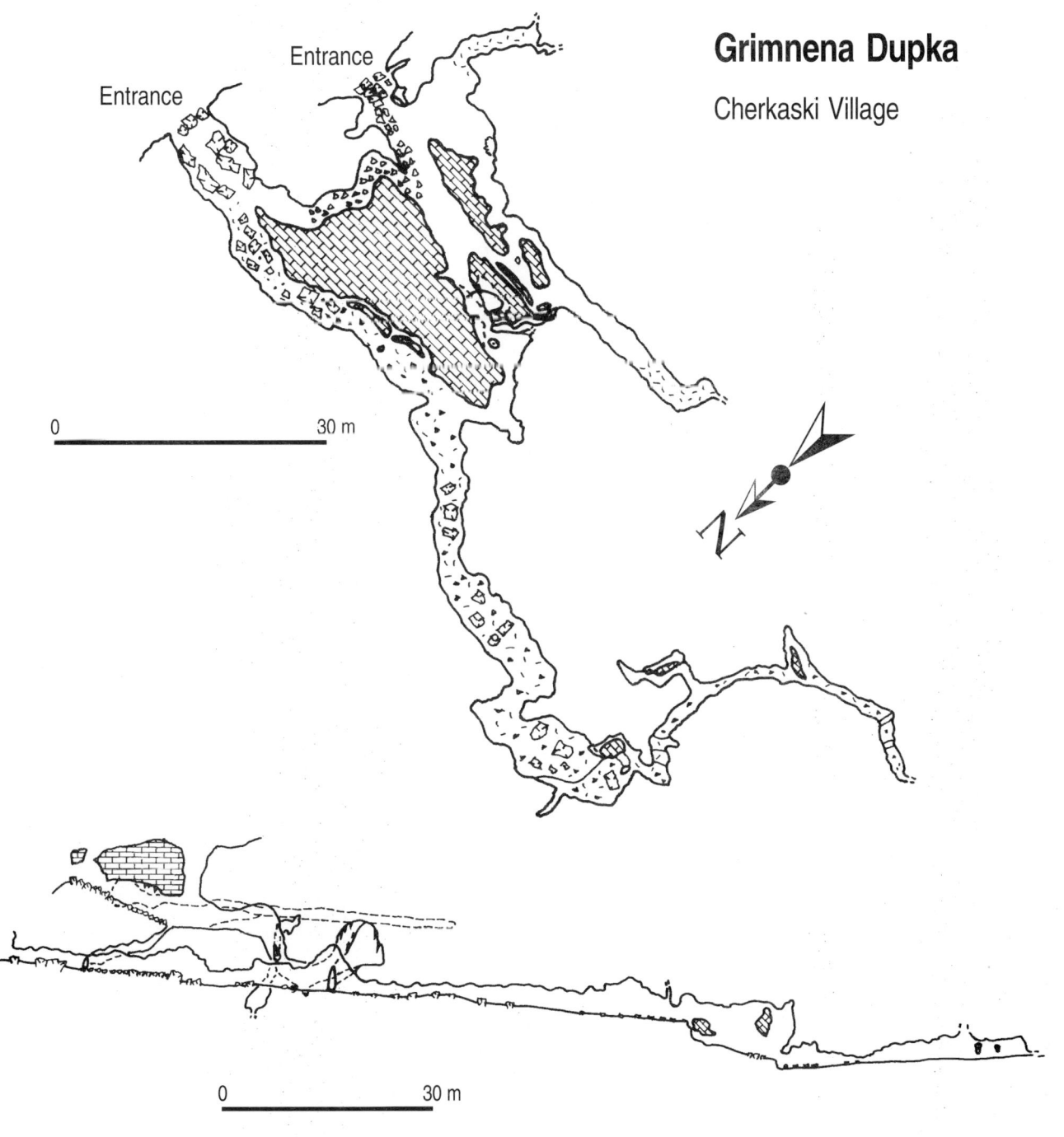

Name: **HAN MAARA (122)**
Ray Hut, Plovdiv Distr.
Length 182 m. Denivelation +3, -1 m

A cave 400 m west of Ray Hut, on the left of the footpath leading to Levski Hut and Bashmandra, under the waterfall Rayskoto Prâskalo. The entrance is facing south and gives beginning to two main galleries. The right side gallery leads north. It is dry, ca. 80 m long. The other after 15 m reaches a river fed by the water of Prâskaloto. After ca. 30 m upstream, we reach a low and narrow gryphon.

The caves were explored and surveyed in 1975 by a team of cavers from Akademik Varna Caving Club (D. Zdravkov, M. Marinov and A. Stoyanov).

The fauna includes the endemic troglobitic beetle *Hexaurus paradisi* (Cholevidae).

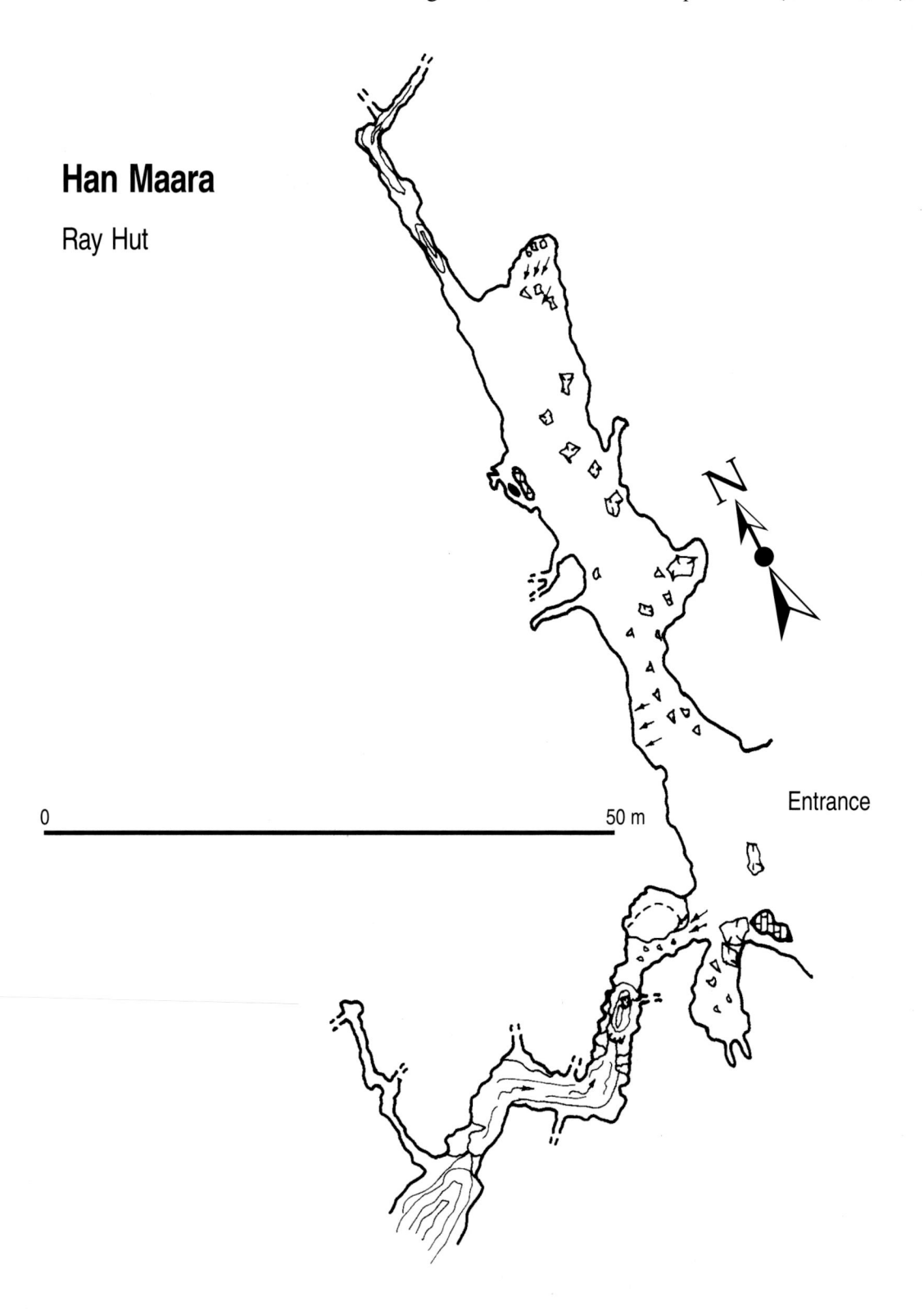

Han Maara

Ray Hut

Name: **HAYDUSHKATA PESHTERA (141)**
Gortalovo Village, Pleven Distr.
Length 433 m. Denivelation 2.2 m
Coordinates: N 43° 18' 02.8" E 24° 07' 23.1"

A cave in Maastricht limestone, 3 km from Gortalovo, upstream the river Pârchovitsa, on its left bank. The cave has 3 entrances, situated about 10 m above the local base of erosion and about 5 m from the base of the cliff. The uppermost entrance has dimensions 0.8 x 2.5 m. The water entrance is on the left of the main entrance, it is round and has a diameter of 1 m. The third one is small. The cave is branched, horizontal, one-storey, permanently with water. Up to 170–180 m, two parallel galleries cross each other repeatedly. The height of the water gallery in places goes down to 0.3 m below the water surface, the sinter lakes are deep up to 0.7 m. There are two more important side galleries – one 40-m long, 60 m from the entrance, and one on the right, 90 m from the entrance, with a length of 50 m and connected with Brunoshushinskata Peshtera. The cave ends with a small labyrinth and an impenetrable siphon.

The cave was surveyed in 1981 during the expedition „Chernelka–81" by E. Zapryanov, G. Vergilov, S. Guneshki, S. Gazdov and B. Garev.

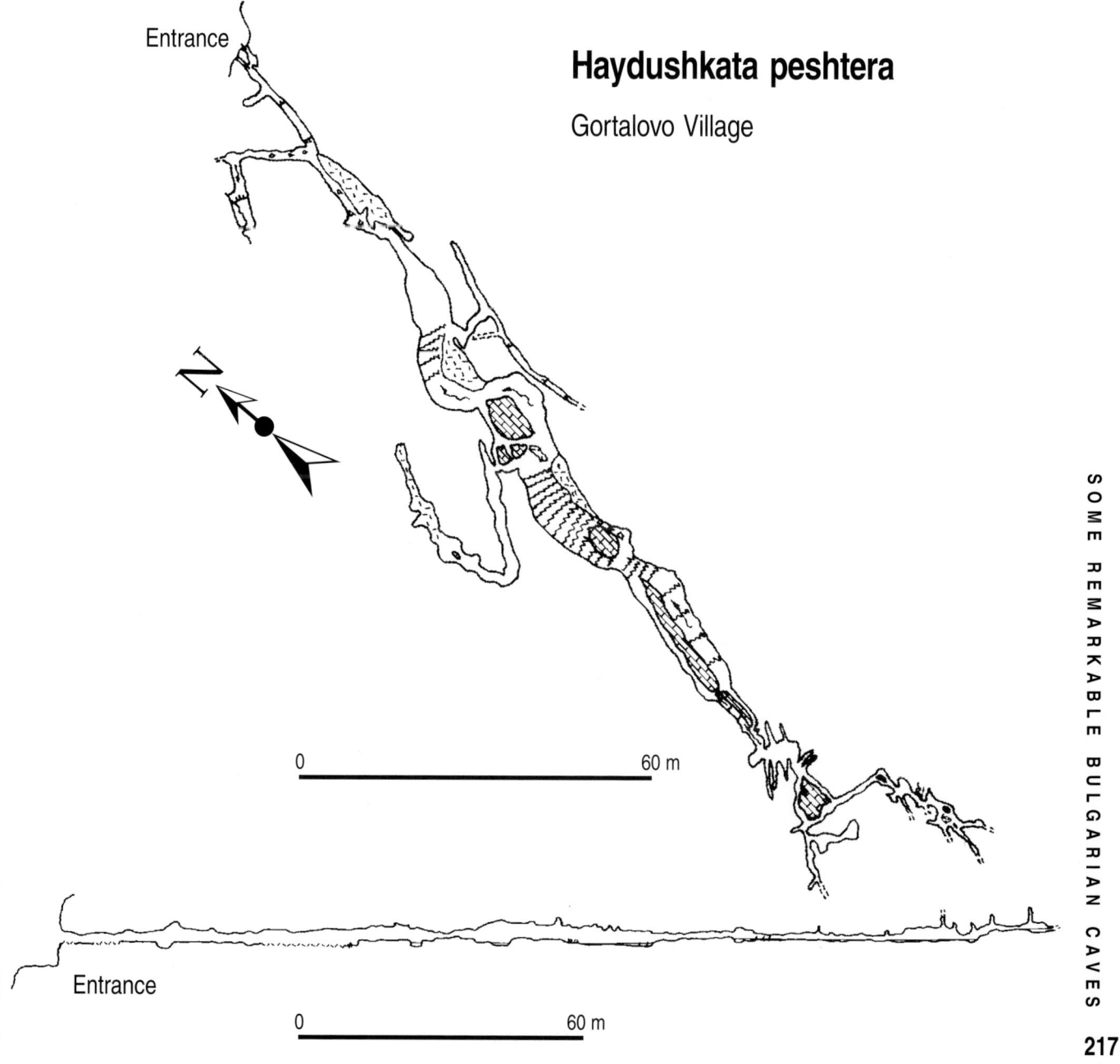

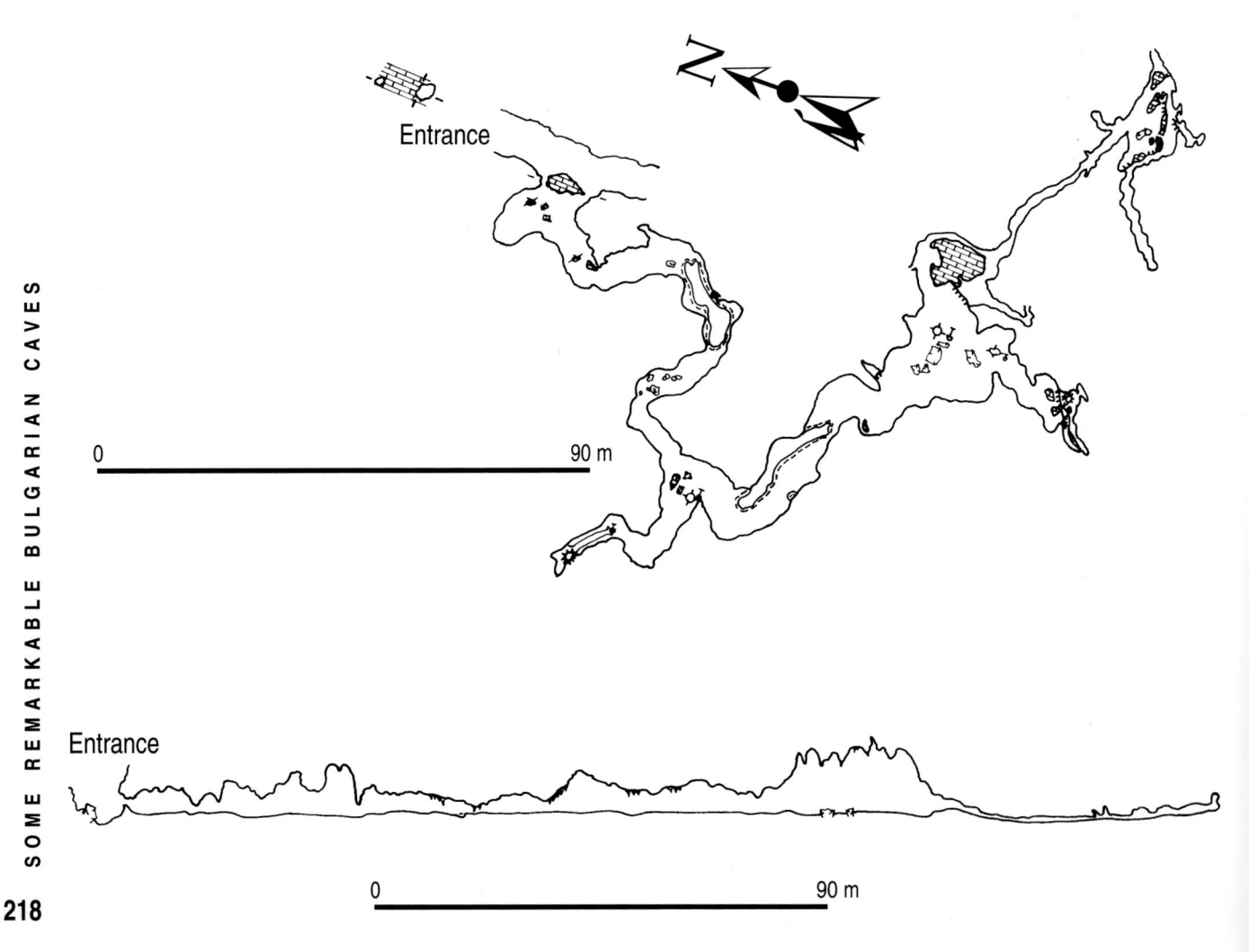

Name: **HAYDUSHKATA PESHTERA (99)**
Deventsi Village, Pleven Distr.
Altitude 156 m. Length 459 m. Denivelation 20 m
Coordinates: N 43° 18' 02.8" E 24° 07' 23.1"

A cave in the locality Skoka, SW of Deventsi. Two entrances, facing NE, in one of the steep valleys on the northern slope of the hill Markova Mogila. Denivelation from the local base of erosion – 56 m, length on the main axis – 218 m.

Formed in limestone from the Cretaceous (Maastricht). Horizontal, one-storey, with branched galleries. The entrance hall is high up to 7–8 m. The first stalactites appear at the 56th meter. The gallery leads into a large chamber, a wide gallery, 31 m long, follows on the right. Going left, we reach the biggest chamber in the cave with a ceiling high up to 12–14 m, and a floor covered with bat guano. So far, the length is 158 m. Several other galleries follow, some intersected by columns, giving impression of a labyrinth. In places – dripping water, many sinters and anemoliths.

Fauna: so far 33 invertebrates known, including the troglobites *Siro beschkovi* (Opiliones Cyphophthalmi, endemic to this cave) and *Chthonius troglodites* (Pseudoscorpiones) and the stygobites *Speocyclops infernus* and *S. lindbergi* (Copepoda). Bats are often found (*Rhinolophus hipposideros, Rh. ferrumequinum, Rh. euryale, Myotis myotis, Miniopterus schreibersi*).

Haydushkata peshtera

Deventsi Village

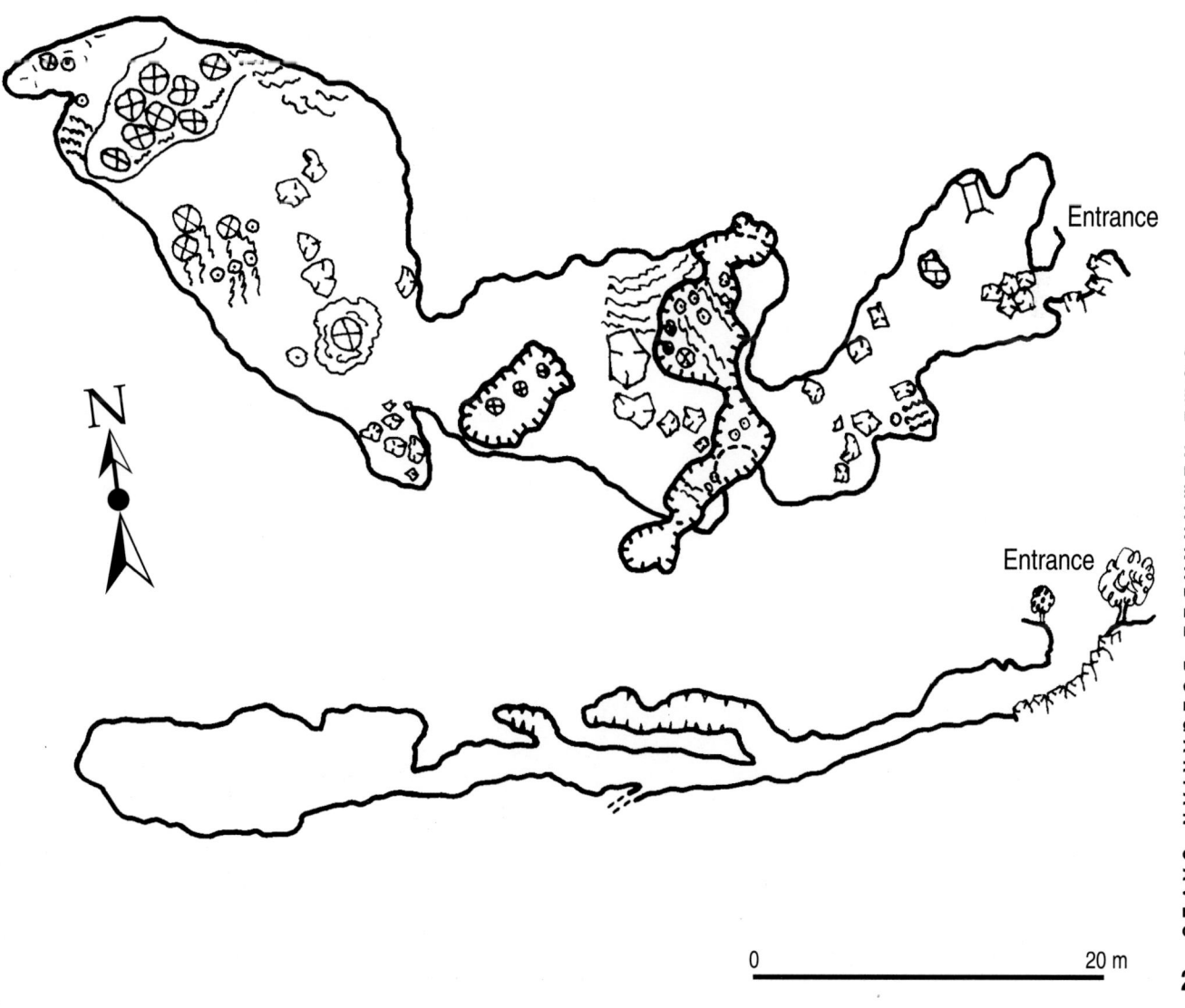

Name: **HAYDUSHKATA PESHTERA (98)**
Karlukovo Railway Station, Lovech Distr.
Length 69 m. Denivelation -10 m

A horizontal, two-storey cave. First studied by V. Mikov (excavations Eneolith ceramics), also by members of the Speleological Society in 1930–1936. Surveyed in July, 1948 and again in 1978.

Declared a National Monument (Decree 2810/ 1963), together with three ha of the surrounding area.

Fauna: 28 species, including the troglobites *Balkanoniscus corniculatus* and *Tricyphoniscus bureschi* (Isopoda), *Typhloiulus bureschi* (Diplopoda) and *Chthonius troglodites* (Pseudoscorpiones).

Haydushkata peshtera

Karlukovo Railway Station

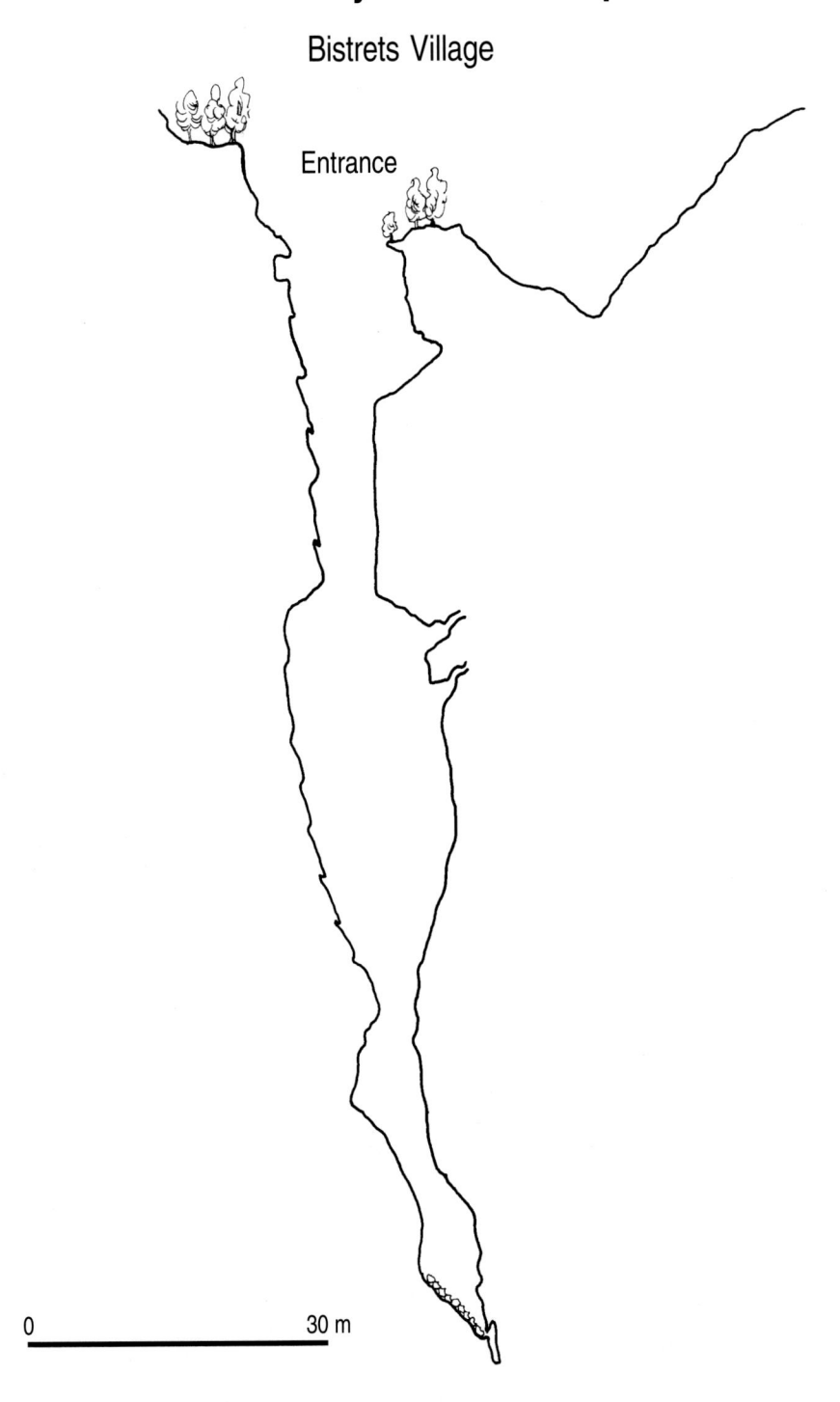

Name: **HAYDUSHKATA PROPAST (29)**
Bistrets Village, Vratsa Distr.
Altitude 720 m. Denivelation -108 m

A pothole (a pitch) in the upper third of Haydushki Dol – the third valley between the village Bistrets and the monastery Sveti Ivan Kassinets (Sveti Ivan Pusti), east of Kârni Vrâh. Developed in limestone from the Upper Jurassic – Lower Cretaceous.

The pitch has 95 m of full vertical. The opening is under a small cliff, dimensions: 6 x 10 m. Fifteen (15) m from the edge of the pitch, there is an inclined platform, from there to the bottom the descending caver is in a bell. Wet, sometimes with dripping water. First penetrated by the cavers A. Siderov from Plovdiv and I. Aleksandrov from Vratsa in 1962.

Haydushkata Propast

Bistrets Village

Name: **HAYDUSHKATA PROPAST (8)**
Belogradchik Town
Altitude 750 m. Length 87.5 m. Denivelation (-32, 4.5 m).

A cave near Belogradchik, in the area called Venetsa. Stalactites, clay, one chamber. Discovered (first entered) on February 7, 1960, by a group led by Alexander Leonidov and Nikola Korchev. The person who entered it first (P. Beron) found an old human skeleton in the entrance gallery and 4 other skeletons down the pitch. There were no vestiges of clothes or other objects on them.

The actual map was made in 1979 by A. Leonidov, V. Nenov and N. Gaydorov, cavers of Bel Prilep Belogradchik Caving Club.

Fauna: includes the troglobites *Trichoniscus bononiensis* (Isopoda), *Beronia micevi* (Coleoptera).

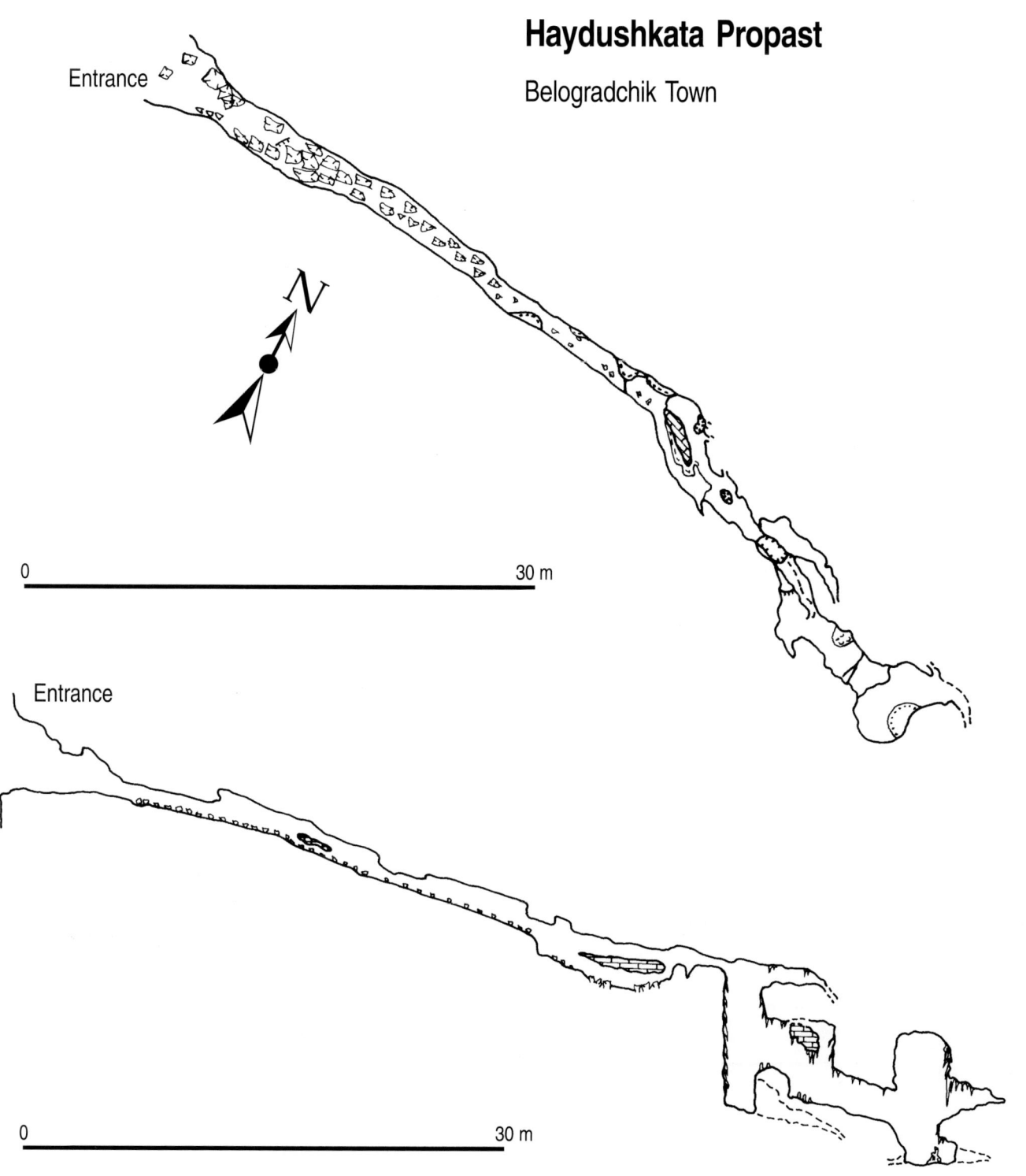

Haydushkata Propast

Belogradchik Town

Name : **IMANETO (92)**
Karlukovo Village, Lovech District
Length 275 m. Denivelation -24 m
Coordinates: N 43° 11' 9.9" E 24°03' 25.5"

    A two-storey, dry and branched cave in the area of Karlukovo, Central Predbalkan. On the left (geographically) bank of Iskar River, ca 2 km from the old railway station. The cave has two entrances, 17 m from one another. They lead into strongly descending to vertical galleries, which at some point merge. Further, the cave goes on as a 2-m wide and even gallery, up to 4 m high. Its initial direction is NW, after a turn it goes south, then west, and ends in a soil siphon. Above the NW sector, the cave has an upper level, rich in flowstone formations, connecting the underlying gallery by several openings. The bottom of the lower level is entirely covered with clay deposits. In some places there are small sinter lakes and other flowstone formations. The cave is formed in Cretaceous (Maastricht) limestone. Explored several times. The actual map was made in 1996 by a team of Helictite Sofia Caving Club, with main surveyor V. Mustakov.

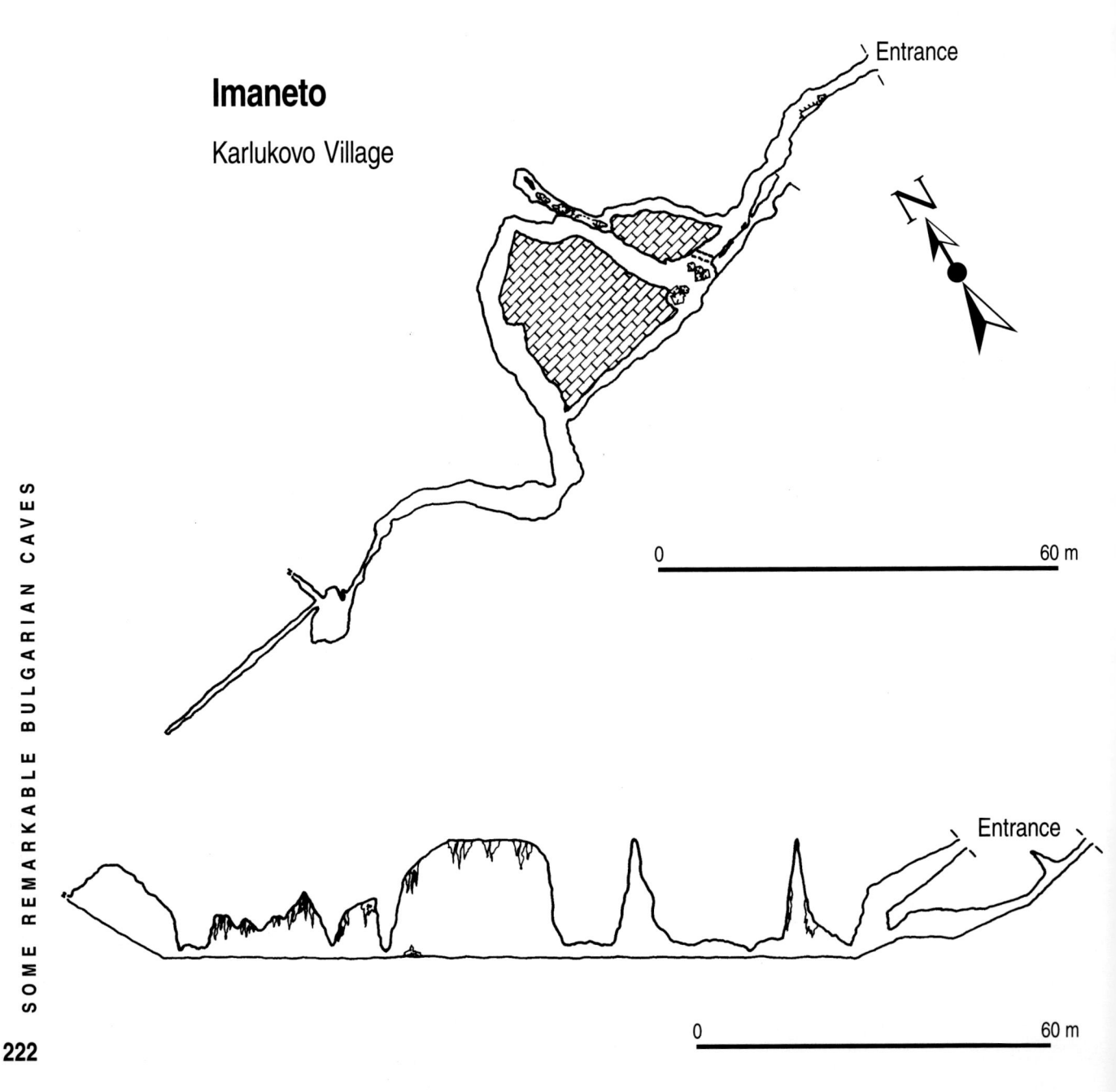

Name: **IZVORA (VODNATA PESHTERA, VODOPADA, PADALOTO) (164)**
Yantra Village, Gabrovo Distr.
Length 1400 m

A source cave, 2.5 km SE of Yantra Village. Formed in Aptian-Urgonian limestone. A horizontal, two-storey, branched cave with permanent water. Entrance 6 x 3.5 m. It is followed by a 80-m gallery, which reaches a 12-m width and is divided by rock pillars. After a bent to the north, the gallery leads into the biggest chamber of the cave (36 x 12 x 6 m). A meandering gallery with a cross section of 2.5 x 6.5 m follows. Initially, the last 120 m of the main gallery retain the shape of the cross section, but then they become lower and wider (6 x 1.2 m), up to the end siphon.

An active water cave without flowstone formations. Thick clay deposits in the entrance parts. Average outflow of the source – 50 l/s.

First exploration in December 4–8, 1962 by St. Andreev, Al. Grozdanov, N. Prodanov, N. Korchev and T. Nenov, members of Akademik Sofia Caving Club. The first three of them surveyed 810 m of the cave. Further research by members of Strinava Dryanovo Caving Club.

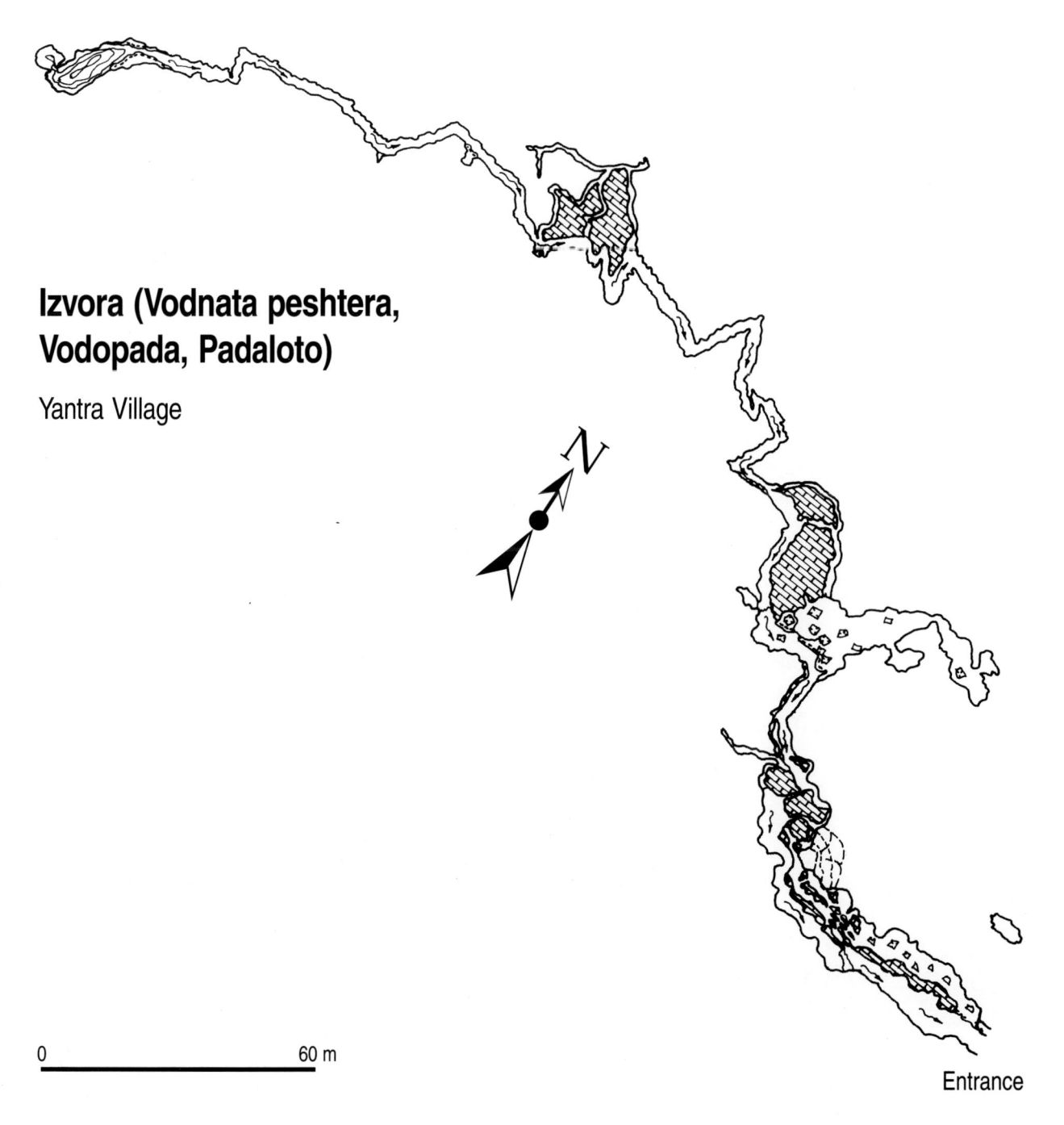

Izvora (Vodnata peshtera, Vodopada, Padaloto)

Yantra Village

0 — 60 m

Entrance

Izvora - the cavers before entering – photo Vassil Balevski

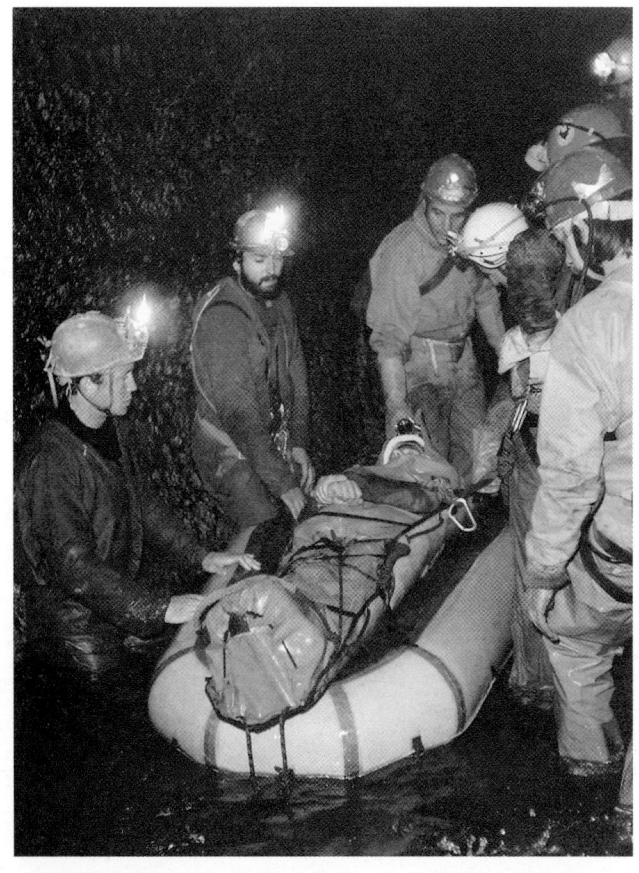

Transport along the water gallery of the cave – photo Vassil Balevski

The entrance of the cave Izvora (Yantra) from inside – photo Vassil Balevski

▲ Borneo – The mountain Kinabalu with the highest peak Low (4101 m)

▼ Conic karst in the park Bantimurung (Indonesia, Sulawesi, 1995)

▲ Lake in the cave Campanario (Guaso Plateau, Cuba, 1988)

▼ Limestone formations in the cave Campanario (Guaso Plateau, Cuba, 1988)

▲ Laundry in the atrium of a cave in Indonesia (Jawa, 1995)

▲ The last and most beautiful chamber in the cave Campanario (Guaso Plateau, Cuba)

▼ Shi Lin (The Stone Forest) – China

▲ The Swallow Cave (China)

▼ Milen Dimitrov and the Albanian Alps (Albania)

▲ The inauguration of the show cave Bai Long Tong (The Cave of the White Dragon) – China

▲ In the Albanian Alps Bulgarian cavers made 14 expeditions, explored 225 caves and potholes, including the deepest ones (-610 m in BB-30, -505 m in Celikokave)

▼ Preparation for exploring the potholes on the plateau above Boga Village (Albanian Alps)

▲ F. Habe, T. Daaliev, V. Stoicev, P. Tranteev (the second, fourth, fifth and sixth from left to right) and Vladimir Ilyuhin (the second from right to left) in the cave Novoafonskaya (Abhasia, 1976)

▲ French and Bulgarian cavers overcome by boat the Lake of Rain in the cave Padiraq

◀ Ice column

▲ The Plateau Tenengebirge in the Austrian Alps

▼ Transparent stalactites

▲ Hedgehog-like formations in a cave

▲ Bulgarian and Greek cavers in the cave Ton Limnon (Greece, 1979)

▼ Transparent stalactites and stalagtones, cobweblike helictites

▲ V. Nedkov in the cave Ton Limnon (Greece)

▶ Carrying out of a "casualty" from the pothole Drangaleshkata (Rhodopes, Bulgaria)

▲ Horizontal transporting of a stretcher in the cave Svirchovitsa (Bulgaria)

◀ Cave rescue action in the cave Yantra (Bulgaria)

▲ National Caving Convention
"Karlukovo – 1984"

▼ National Caving Convention
"Karlukovo – 1984" –
demonstration of rescue technique

▲ National Caving Convention
"Karlukovo – 2004" – 75 years of
organized speleology in Bulgaria

▲ Cavers of several generations commemorating 80 years from the birth of Petar Tranteev (Hera) – Museum Men and Earth, January 2004

▼ 75 years of organized speleology in Bulgaria – April 2004

▲ Sinter floor of the cave Barki 8 (Bulgaria)

▶ Gallery in the cave Saguaro (Bulgaria)

▼ Stalagtons in the cave Barki 8

▲ Stone waterfall (The First lake) in the cave Devetashkata peshtera (Bulgaria)

▼ The biggest cave chamber on the Balkans
(Devetashkata peshtera)

▼ Cave waterfall

▲ Play of Colours – the Big Hall of Pticha Dupka (Bulgaria)

▲▲ Courtains in the Big Hall of Pticha Dupka

▶ The Big Hall of Pticha Dupka

▼ Big stalagmites in the last chamber of the cave Pticha Dupka

▲ Winter view in front of the cave entrance of the cave Vodnite dupki (Apriltsi)

▼ Gallery in the cave Rushovata peshtera

▲ Cave Ledenika – the Concert Hall

▼ Ice formations in the Small Hall (Ledenika)

Name: **KALENSKA PESHT (75)**
Kalen Village, Vratsa Distr.
Altitude 440 m. Denivelation 15 m. Total length 826 m. Length on the main axis 528 m

A cave, 2600 m E – NE of Kalen Village, at the base of a cliff. Kalenska Pesht is a two-entrance, one-gallery, slightly ascendant, horizontal water cave. Its entrance has the shape of an irregular semiellipse (20 x 5 m) and is facing West. The second entrance is a small opening, which after 3 m joins the main gallery. The initial part of the cave is a large chamber (45 x 22 x 10 m). In the left lowest part of the Big Chamber, there is a stream and the floor is covered with mud, sheep dung and in places with bat guano. There are some stalactites on the ceiling. In the winter, in the Big Chamber (up to 35 m) ice formations of considerable dimensions are accumulated.

Three galleries start from the chamber, merging again after 20 m. The main gallery is mostly of NE and NW orientation. After the chamber, the gallery is at the beginning 5 m wide and 2 m high, but gradually it changes and becomes 10–15 m high, the width decreasing up to 1.6 m, up to the first siphon. Two hundred and eighty (280) m from the entrance there is a second level. The first siphon is the end of the known part of the cave, with a total length of 520 m and 380 m on the main axis. After the siphon, the gallery, which is 16 m high, sharply descends up to 1.2 m after the 25th meter and after another 20 m it reaches the Waterfall Chamber, full of roaming and water dust. The waterfall, 3.7 m high, leads into an upper level.

After the Waterfall, the water gallery bifurcates. Above the Waterfall there is a stream, coming from the left gallery. The ceiling of this gallery is sharply lowering and stops before the lake siphon, which is 3 m long. Further on, the gallery is low, half inundated, 30 m long. It leads to a two-meter waterfall and after 50 m it ends in a siphon.

On the right of the waterfall, the main gallery is 3–4 m wide and 8–9 m high. One hundred (100) m on the left, there is a short, water branch, 11 m long, ending in a squeeze. After 50 m a short, one-meter, siphon appears, which leads to a siphon, from where the main stream starts.

The explored and surveyed galleries have a total length of 826 m, out of which 520 m belong to the galleries, up to the first siphon, and 306 m are after it.

After the Waterfall, the stream runs over a slight slope and comes out of the cave. Its outflow varies in the span of 10–50 l/s. The water is collected at the end of the cave by three underground streams. The depth of the water in the cave is not more than 1.3 m. The floor of the galleries is covered with sand, gravel, stones, clay, calcite cover and bat guano. Flint nuclei are seen everywhere in the cave. The water cave of Kalen is rich in secondary flowstone formations. After the first siphon, the formations are intact. Their colors vary from milk-white to dark-brown, red and lead-black. The highest stalagtone is 6.2 m high, while the biggest stalactite is 3 m high.

In 1955, In the entrance parts of the cave, the archaeologist N. Djambazov discovered remains of the Iron Age. The cave was explored in September 1959 by P. Tranteev, P. Beron, M. Kvartirnikov and Vl. Beshkov. Explored again and surveyed in 1970–1976 by Petar Hristov from Kalen. On June 2, 1971, after the 6.5 m siphon (the First Siphon), a continuation was discovered (description by T. Daaliev, after P. Hristov).

The water of Kalenskata Pesht was piped and entering the cave is prohibited.

Several bat species live in the cave, white Isopoda and Amphipoda have been collected by P. Beron (11.06.2006).

Kalenska Pesht

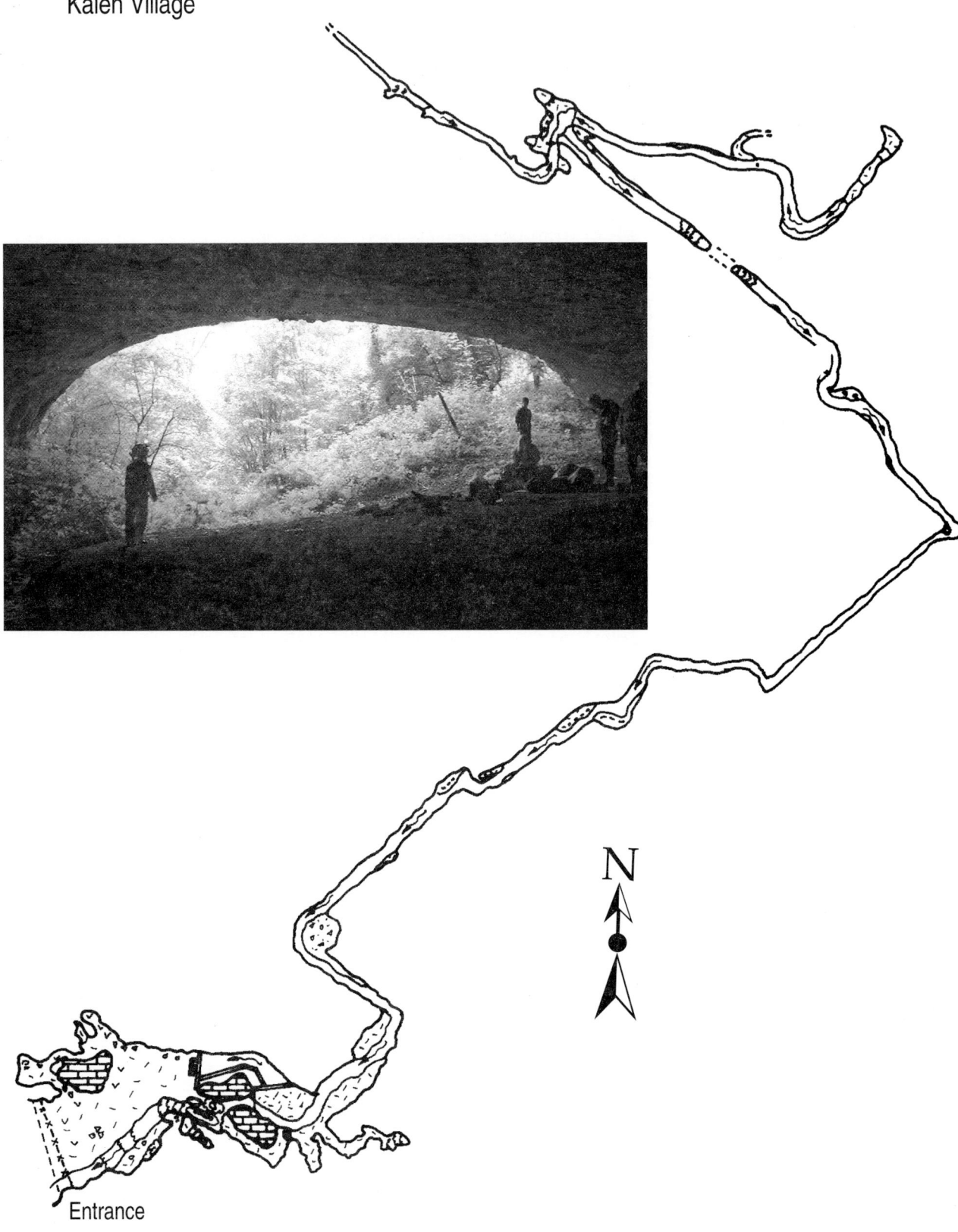

Kalen Village

Entrance

0   100 m

Name: **KALITSA (156)**
Krushuna Village, Lovech Distr.
Total length 430 m. Denivelation 5 m.

A cave about 3 km on the road between Chavdartsi and Krushuna. On the left side, at about 500 m is the karstic source Kalitsa, from where the cave starts.

The entrance is facing north and is 1.40 m wide and 1 m high. This is a horizontal water cave with an underground stream with an outflow of 3 l/s and water $T° 12.3°C$ (25.08.1982). The cave is formed in fissured limestone from the Upper Jurassic (with admixture of clay and flint concretions).

The first 120 m represent a lake, which is 1–2 m deep. The gallery is 1–2 m wide, further on it reaches 3 m and more. There is a side stream by the 350th meter. From the 250th meter to the end, there are stalactites, curtains and other flowstone formations. Along the last 50 m (the most beautiful part), the gallery is over 4 m wide and is with many stalactites and stalagtones.

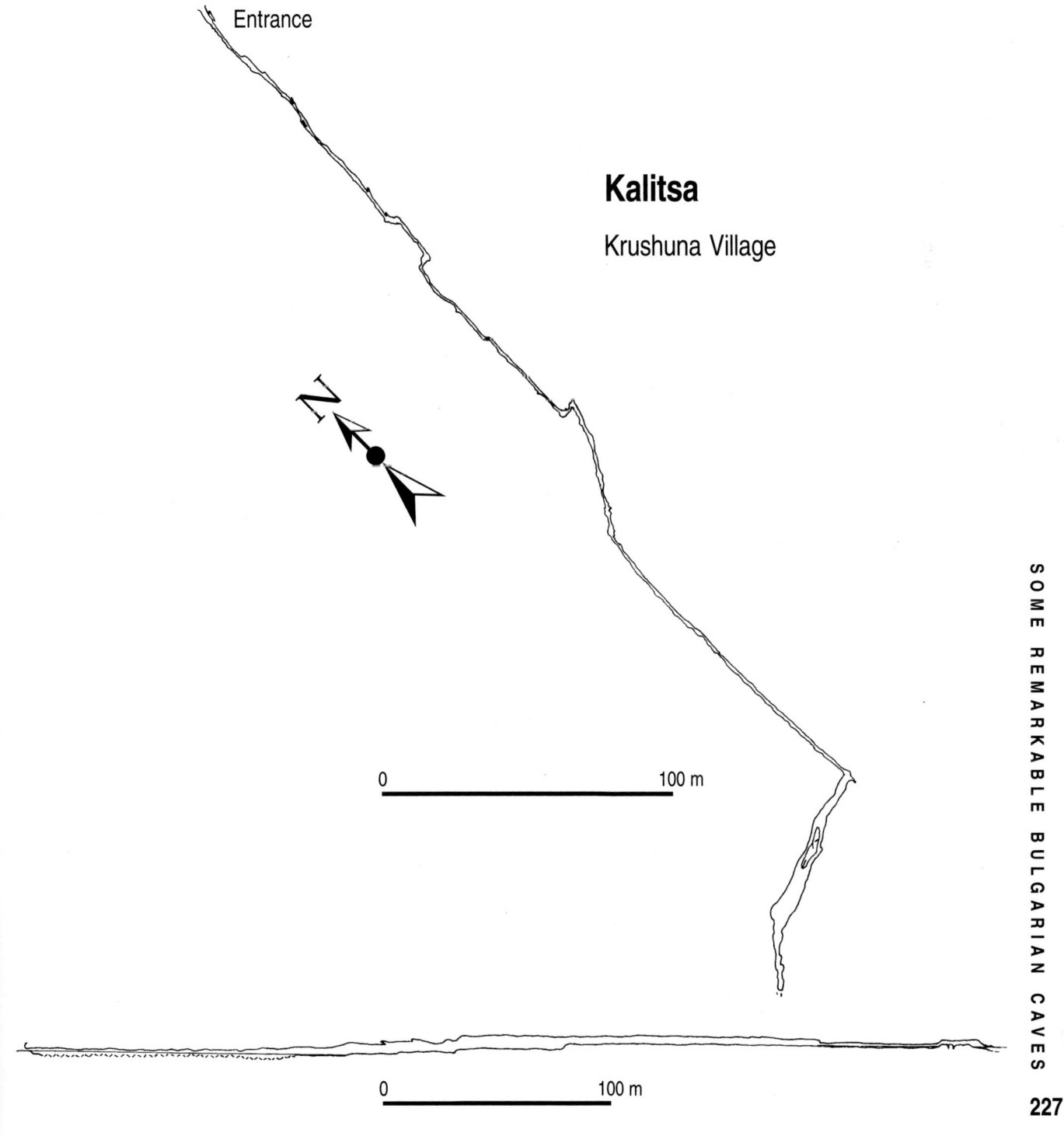

Name: **KALNATA PROPAST (42)**
Vratsa
Altitude 970 m. Length 32 m. Denivelation -86 m

A pothole in the locality Ibishov Valog, 45 minutes walk from Ledenika Hut. It is a cascade with interconnected verticals, with a general denivelation of 85 m. At the 52 meter there is a squeeze, which could be penetrated only by very slim cavers. The pothole is not rich in secondary formations.

Discovered and explored in 1968 by a team of cavers from Sofia (A. Petkova, A. Bliznakov, V. Nedkov, Sh. Bassat and T. Daaliev) and surveyed by the discoverers up to the squeeze at 52 m. The actual map was made by Tsv. Georgiev and Ya. Bozhinov from Aleko Sofia Caving Club in 1975.

Fauna (collected by P. Beron in 1968): three species known, including the troglobite *Pheggomisetes buresi* (Col., Carabidae).

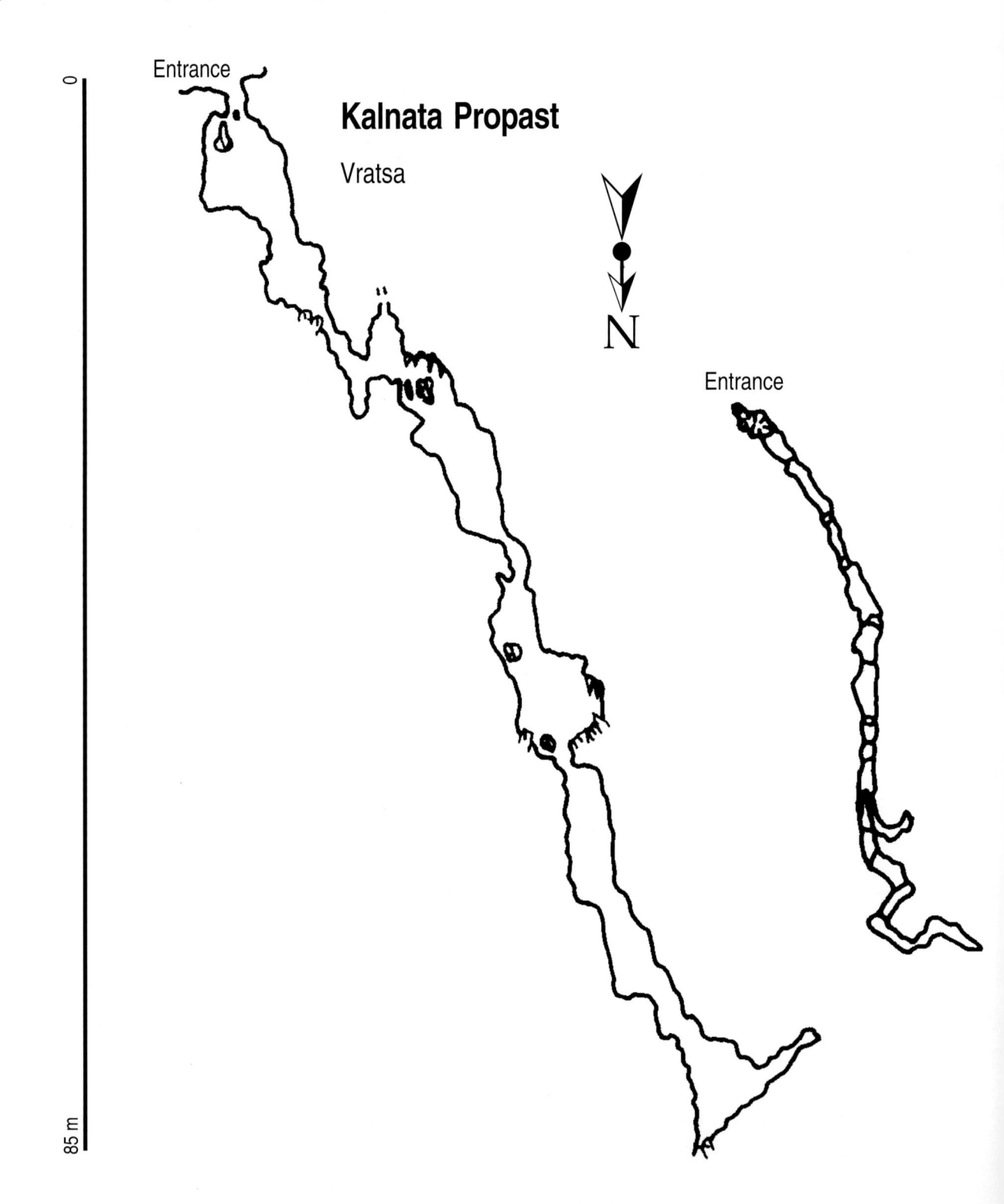

Name: **KARSTOVIYA IZVOR (THE KARSTIC SOURCE) (42)**
Kotel Town, Sliven District
Total length of the dry and inundated galleries 410 m. Denivelation 32 m

The karstic source near Kotel is in the western confines of the town, in the outskirts of Suhi Dyal Hill. The source consists of three lakes, which have a diameter of 1.5 – 2.5 m and are rather deep (8–9 m), of west-east situation. The outflow of these sources vary within the limits 50–2500 l/s. The water is piped for the needs of Kotel.

There had always been interest in these sources, but the first attempt to explore them took place in 1968. In very dry years, the water was pumped and the caver from Kotel Ivan Dragandjikov penetrated the system and found a cavern, developing to the north. Later on, divers from Varna again penetrated the cavern and reached a chamber above the water, through which a stream runs. Above it, in an old river bed tree roots have been noticed.

In 1984–1985, Sâbcho Dimitrov, Kuncho Solakov and other cavers from Kotel managed to open another (dry) entrance to the cavern, then they reached a siphon. In 1989, again divers from Varna penetrated through a second siphon to the end of this chamber and, after 61 m reached another branching off chamber, ca. 6–7 m wide and 30 m long. The left branch was very narrow and was not explored. The right branch ends in a siphon, about 5 m long. They crossed it and got into a third L shaped chamber, about 35–40 m long and ending in a siphon (fourth).

At the end of the same year, a group consisting of Alexey Alexiev, Ivan Zdravkov, Doncho Boyanov and Simeon Mihov started another exploration of the cave. The aim was to follow the underground stream, to measure the outflow at several points and to establish the possible drains of water in some parts of the cave. In 1990, after a long preparation, the fourth siphon was overcome. Special equipment was secured to allow work in difficult conditions.

The water temperature is 9.8 °C, the transparency in the siphons is ca. 25–30 cm. The new dry entrance is ca. 10 m higher than the level of the lakes. The length of the axis of the cavern is 78 m, the average width – 7-8 m. There are two floors – the upper one is dry, the lower is a river bed. In two places the two floors are interconnected. In the northern part there are several smaller galleries – parts of old river beds. The dry part helps circumvent the first siphon. The second siphon is 70 m long and 3 m deep. There are two air bells in it. A 60-m long gallery and a chamber, which is 6–7 m wide and 30 m long, follow. The measured outflow in the three places is ca. 70–75 l/s.

The second and the third chambers are poor in flowstone formations. The walls are covered by black deposit with clay underneath. Everywhere there are shields – remnants of layers with a higher content of silicates.

The third siphon is the smallest (10 m long and 3 m deep). The fourth is the biggest siphon in the cave (150 m long and 32 m deep). After it a diaclase follows (20–25 m high, the ceiling cannot be seen). It is ca. 7 m wide, there are 3–4 lakes on the floor.

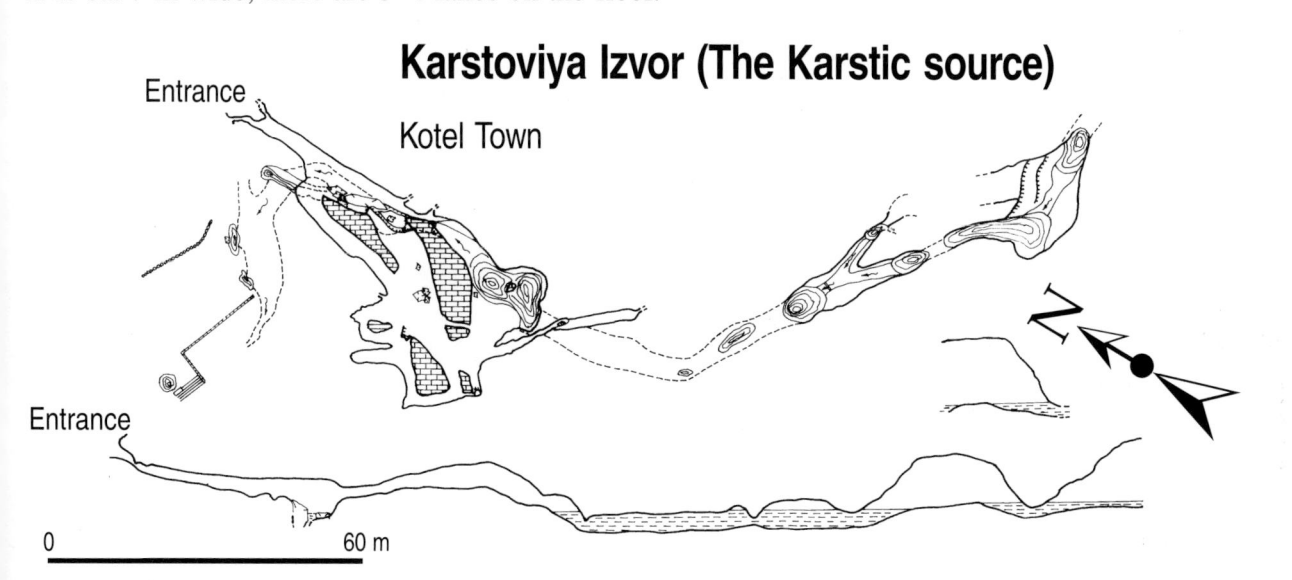

Karstoviya Izvor (The Karstic source)
Kotel Town

Name: **KATSITE (57)**
Zimevitsa Village, Sofia Distr.
Altitude 1190 m. Total length 2560 m. Denivelation -205 m

A cave in Trias limestone, descending, cascade, branched, broken by verticals, the highest being 26 m. Two siphons – at 26 and 160 m. Underground river with water $t° = 7°$ C. The dying of the waters in the sink Ponor has shown, that they spring from the Iskrets karst spring. The distance between Katsite and Iskrets source is 7.5 km, the denivelation – 650 m. The huge water basin suggests the existence of a complex cave system with length of over 100 km.

The cave had been known to local people since time unrecorded. First researched in 1960 by L. Popov and other cavers from Sofia. On December 11, 1960, P. Beron and V. Beshkov also penetrated the cave and collected cave fauna. In 1963, denivelation of 26 m was reached and a total length of galleries – 250 m. In August 1977, cavers from Edelweiss Sofia Caving Club penetrated the First siphon and reached another one at 160 m. In the autumn of 1977, the Second siphon was overcome and a boulder chock at 200 m was reached. In 1979, the Third siphon was reached at 205 m. The survey and most of the research were done by cavers from Edelweiss, with participation of cavers from Cherni Vrâh, Planinets, Akademik and other clubs from Sofia.

In August 1977 during expedition of Edelweiss – Sofia Caving has been negotiated the First Siphon (P. Petrov and V. Nedkov) and, after a series of verticals (the deepest being 26 m) a second siphon was reached at – 160 m (N. Landjev and M. Valchkov), Z. Iliev suveyed the big vertical. After one and half month the same cavers overcame the Second Siphon and reached a blocage at – 200 m. During the several following years many cavers participated to the mapping of over 2000 m of galleries, climbing of ascending chimneys up to 75 m. In 1979 A. Apostolov and M. Daaliev passed through the squeezes in the blocage. Follows a wide gallery, 250 m long, and a third siphon at -205 m. The survey and most of the reasearch was done by cavers from Edelweiss – Sofia Caving Club.

To penetrate the cave, a dry-type caving hydro-suite is needed, as one remains almost all the time in water with T° 7 °C.

Fauna: Known troglobite and stygobite fauna: *Cyclops bohater ponorensis, Diacyclops haemusi* (Copepoda), *Niphargus* sp. (Amphipoda), *Paranemastoma (Buresiolla) bureschi* (Opiliones)

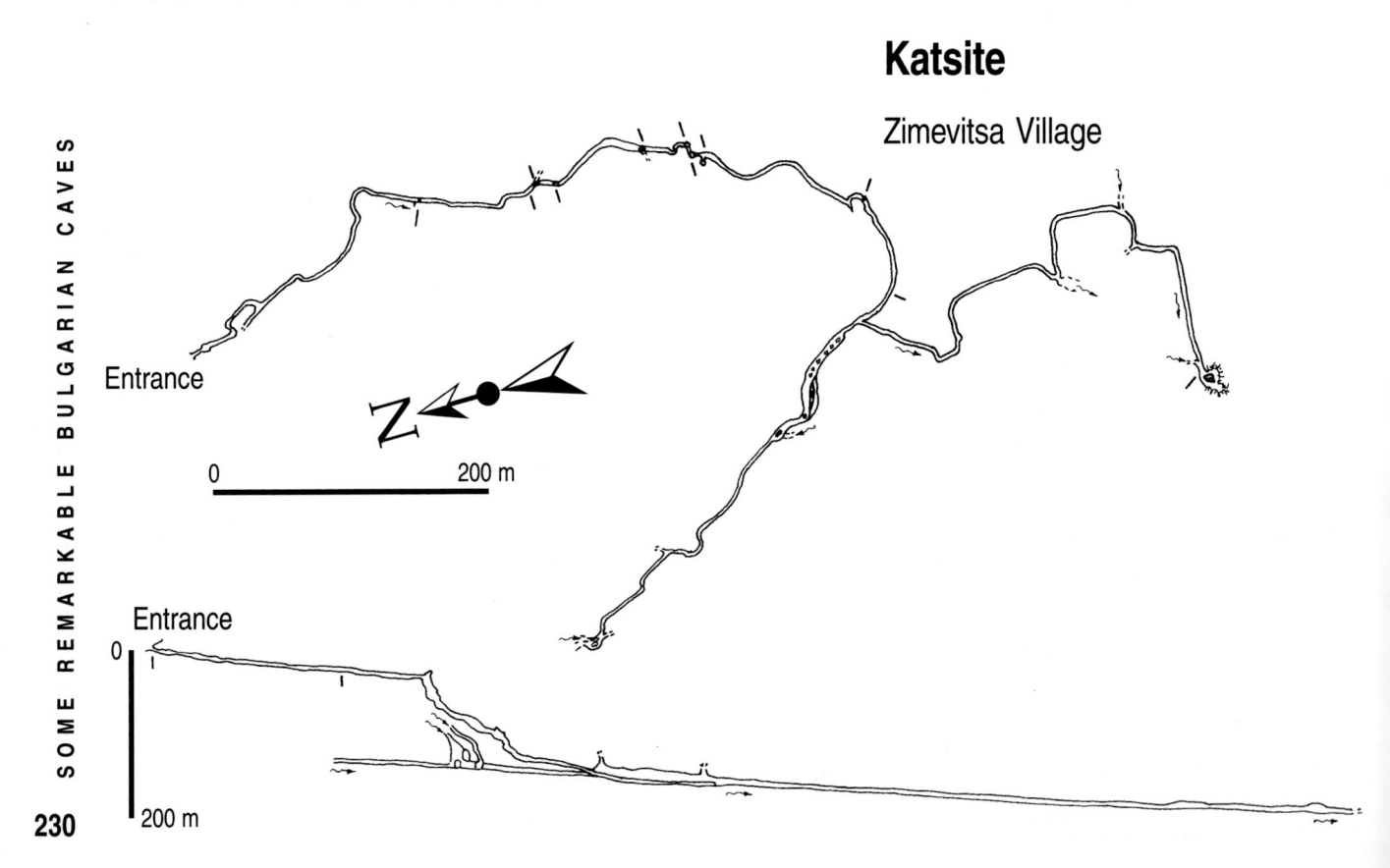

The entrance – photo Nikola Landzhev

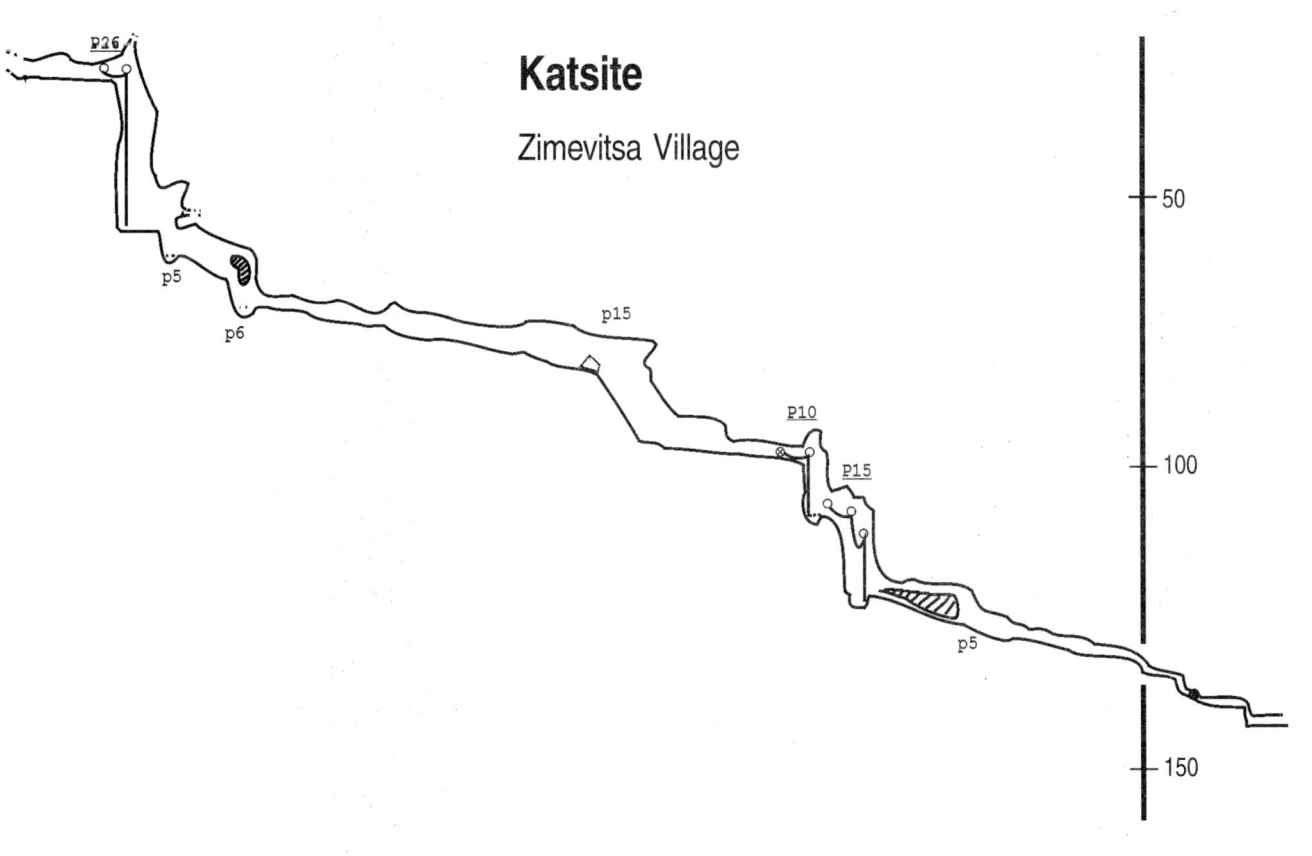

Katsite

Zimevitsa Village

Name: **KÂNCHOVA VARPINA (147)**
Tepava Village, Lovech Distr.
Length 463 m. Denivelation -100 m
Coordinates: N 43° 11' 15.2" E 24° 54' 13.9"

A pothole on Devetaki Plateau, East of Tepava, in the oak forest between Tepava and Devetaki, about 100 m from a karstic swamp. The opening of the pothole is on the bottom of a huge funnel with denivelation of 12 m.

Entrance – 1 x 1.5 m. At the beginning, the gallery is low and after the 17-th meter it reaches 2 m height. On the left, a 30-m long winding gallery follows. At the 22-nd meter, an upper storey begins, going up to the 140th meter. Up to this place, the verticals are small. After the 140th meter the verticals are from 4–5 to 10–15 m. Up to the 290th meter no gear is needed, except for short ropes. Bigger and smaller chambers follow. The bottom of the gallery is stony and dry. At the 290th m from the entrance, there is a bell-like vertical, which is 47 m deep.

The pothole is equipped for SRT. For the 47 m vertical, a rope of 60 m is needed.

The pothole was explored and surveyed by cavers from Aleko Konstantinov Sofia Caving Club (Lyuben Popov, Tsvetan Georgiev, Lyubka Dimitrova and Alexander Ivanchev). In 1978, L. Petrov, V. Nedkov, V. Balevski and T. Daaliev measured the pothole with two altimeters to 100 m. In the 80s the depth of 100 m was confirmed by using a hydrolevel.

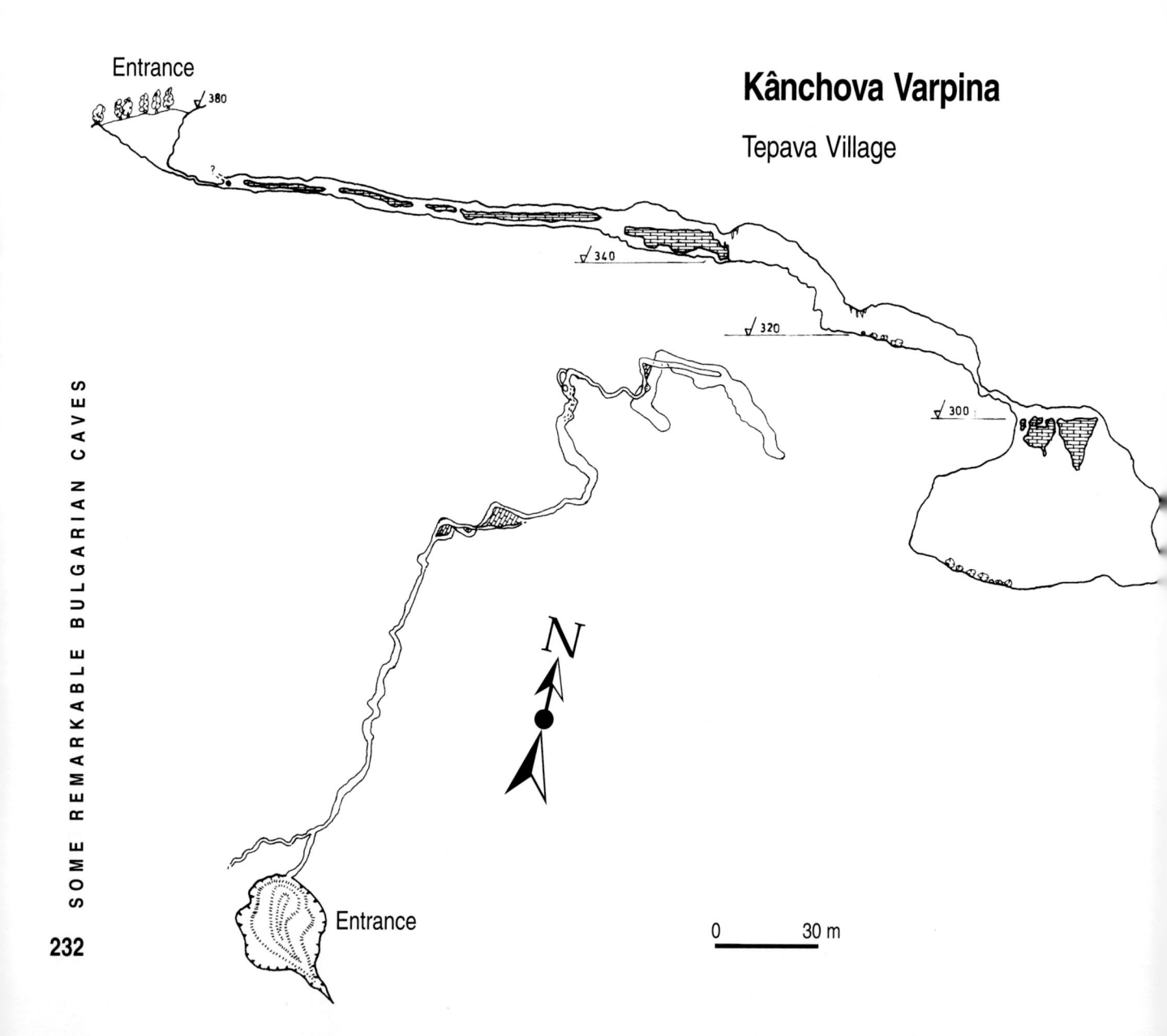

Name: **KÂRVAVATA LOKVA (KLADENTSITE) (182)**
Kotel Town, Sliven Distr.
Length 646 m. Denivelation -140 m
Coordinates: N 42° 52' 49" E 26° 21' 33.1"

A pothole near Zelenich shelter, on the way from Kotel to Kipilovo. Formed in Senonian limestone. Cascade cave with verticals of 15, 15, 38, 5 and 17 m. Siphon on the bottom. In the horizontal galleries there are beautiful speleothems. A stream with small outflow.
First exploration – during the expedition Kotel 1966 of Akademik Sofia Caving Club, up to 44 m. Second research – in 1969 by cavers from Yambol. They reached another shaft and called it Kladentsite, making a cave map up to 103 m. In 1971, again cavers from Yambol reached a depth of 140 m.

Kârvavata Lokva (Kladentsite)
Kotel Town

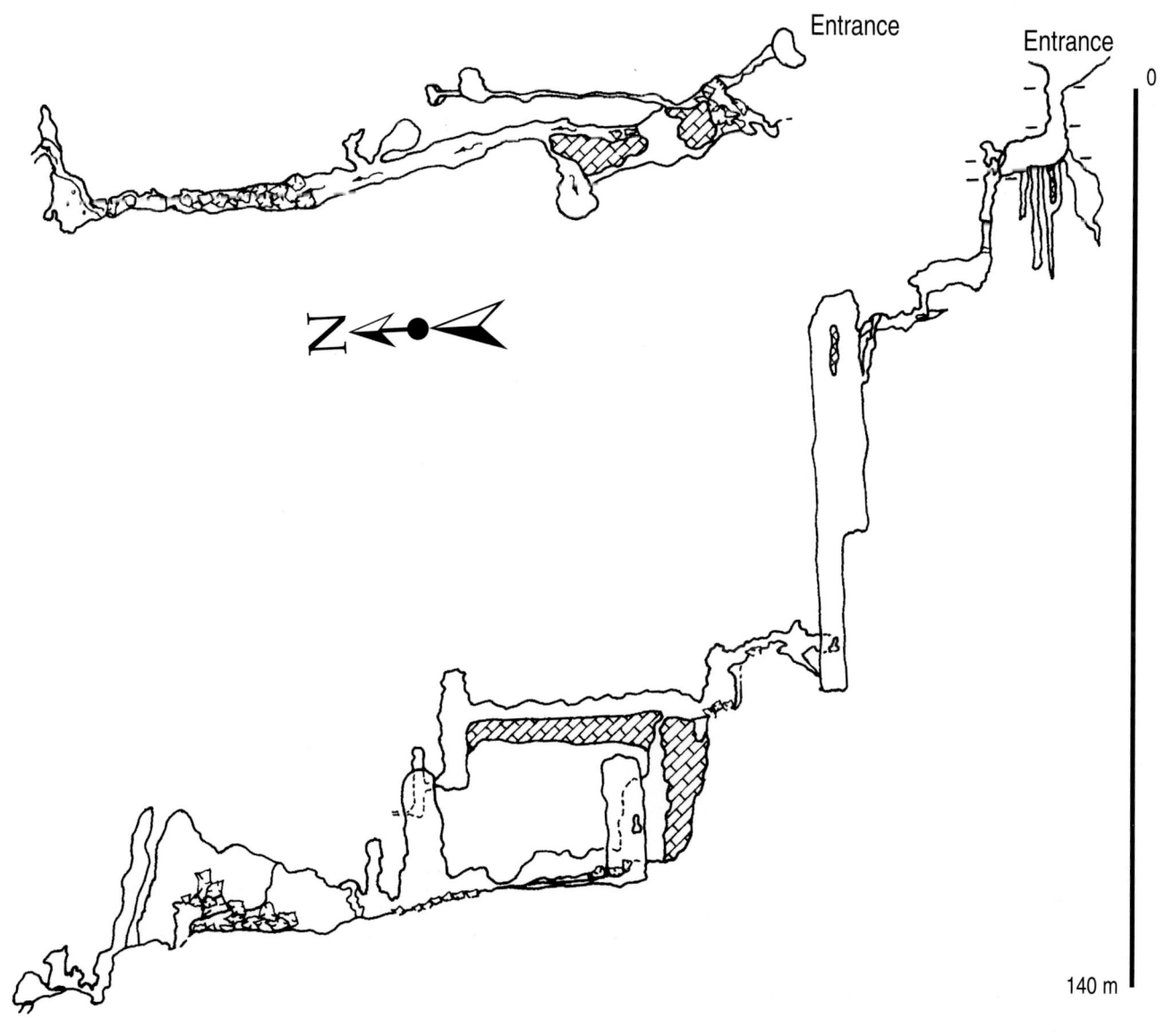

Name: **KIPILOVSKATA PESHTERA (CHOVESHKATA, CHELOVESHTNITSATA) (191)**
Kipilovo Village, Sliven Distr.
Length 182 m. Denivelation -6 m

A cave, 2.5 km from Kipilovo, about 100 m from the road and the river. A descending, branched, one-storey, wet cave. In the middle part there are many stalagmites, stalactites and several bats.
Other fauna: troglophile and trogloxene flies, spiders and other animals.

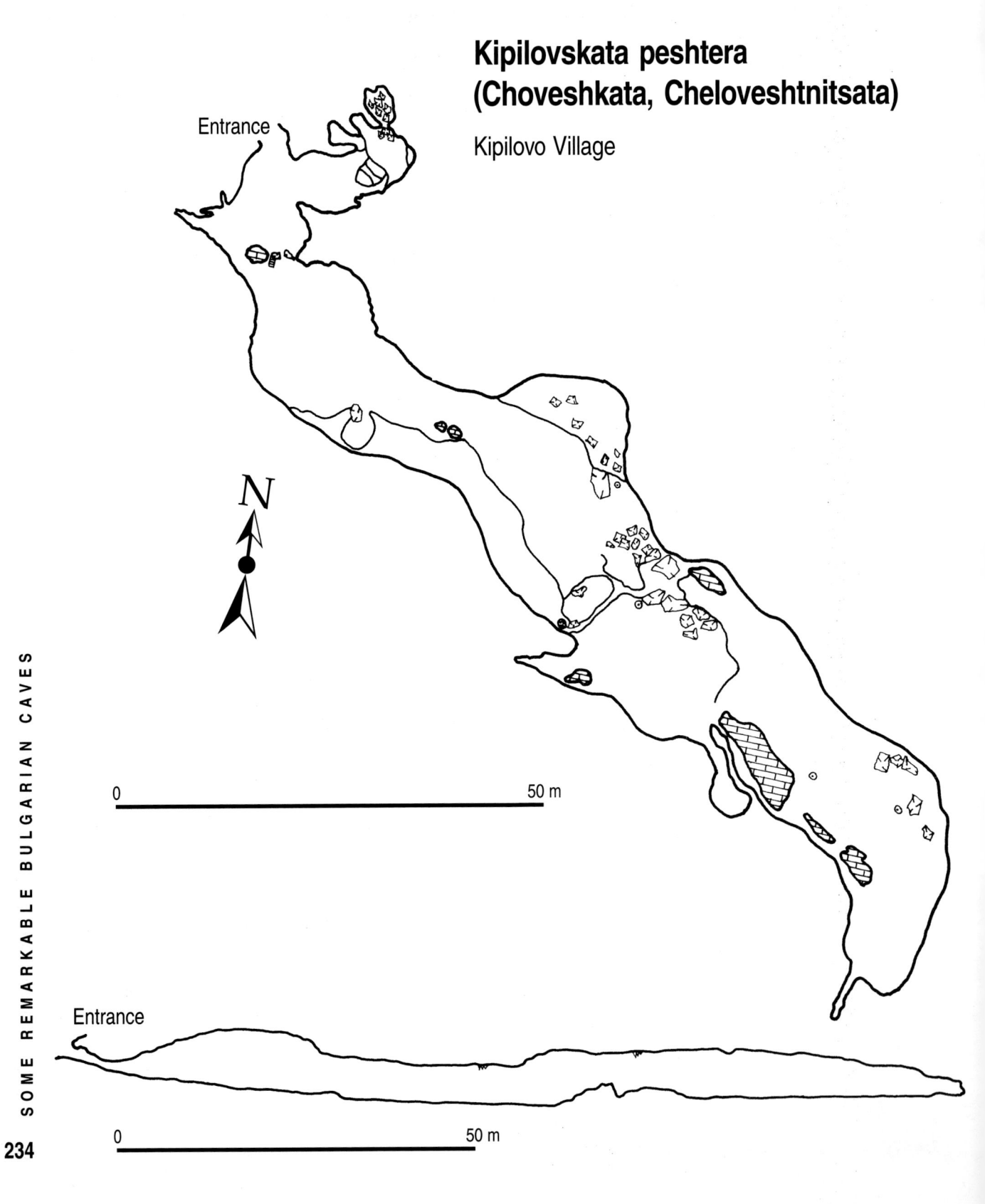

**Kipilovskata peshtera
(Choveshkata, Cheloveshtnitsata)**

Kipilovo Village

Name: **KOKALANA (STOLETOVSKATA PESHTERA) (169)**
Shipka Pass, Gabrovo Distr.
Length 114 m. Denivelation -12 m

A cave under the summit Stoletov, 150 m from the gas station on the road to Gabrovo. Entrance – 60 cm high, one meter wide, near a tall tree. The cave and the area are named Kokalana after the many bones of soldiers, killed there during the Russian-Turkish war in 1877. Developed along a fault (diaclase) inclined at 30°. Azimuth of the entrance – 324°. Floor in flowstone, dendrites from the 8th to the 30th meter. A narrow squeeze leads into a semicircular chamber, called Zala na Srutishteto, having a diameter of ca. 17 m. Boulders, sinter lake – 5 x 6 m.

Fauna: 11 invertebrates known, including the troglobites *Pseudosinella bulgarica* and *P. kwartirnikovi* (Collembola), *Hexaurus schipkaensis* (Coleoptera, Catopidae, endemic species, described from this cave) and *Trichoniscus stoevi* (Isopoda).

Kokalana (Stoletovskata peshtera)
Shipka Pass

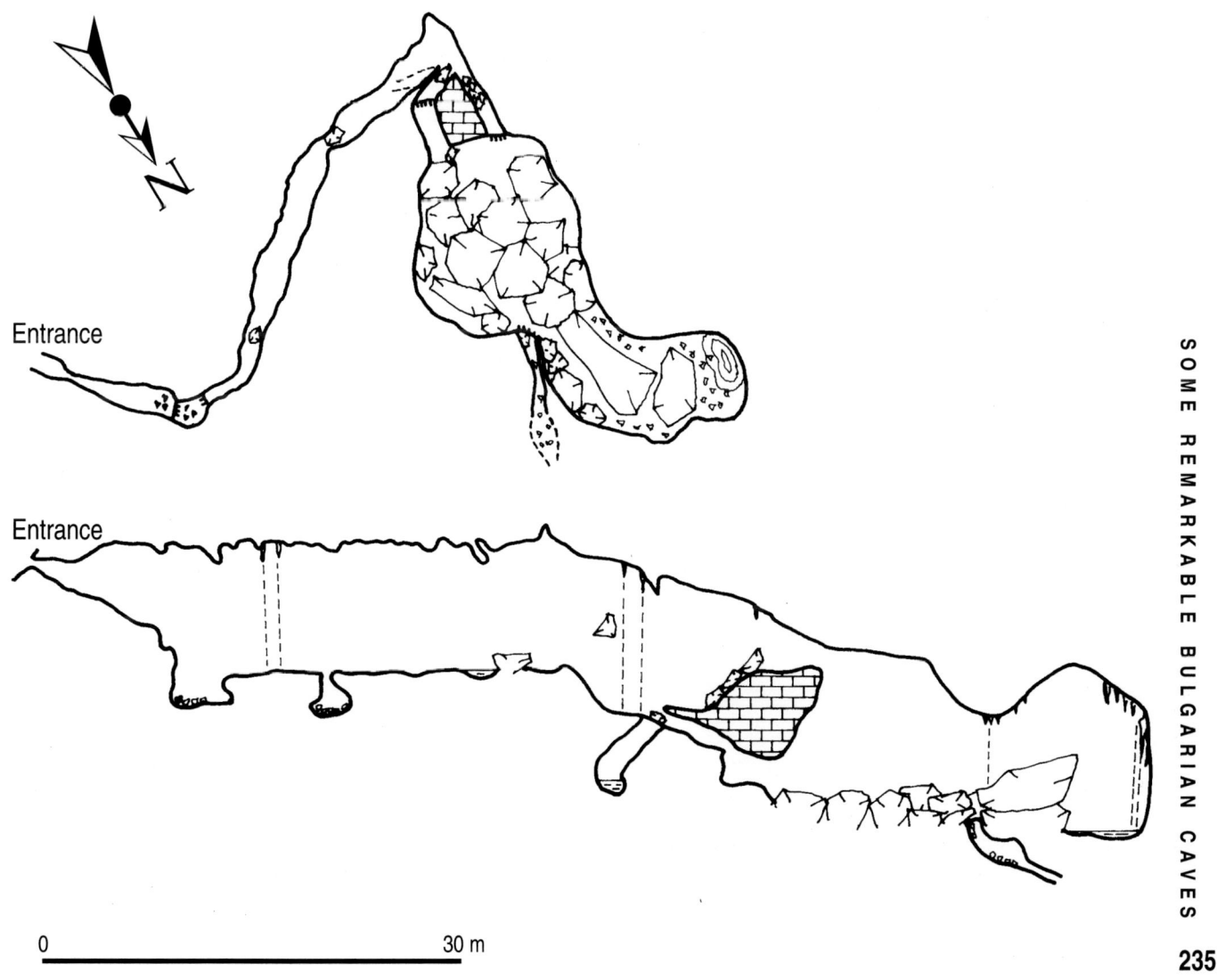

Name: **KONDJOVA KRUSHA (22)**
Lilyache Village, Vratsa Distr.
Length 552 m. Denivelation -38 m

A cave pothole, west of Lilyache, in the locality Mladjovoto, south of the cave Ponora. There is a karstic swamp about 1 km from Ponora. About 100 m from it there are two funnels. The pothole entrance is in the smaller one.

The start is an 18-m vertical, leading to a low and narrow gallery, along which runs a stream. The cave is developing in two directions. About 40 m downstream, the ceiling lowers to 50 cm in a semi siphon. After negotiating it (when in flood the semi siphon becomes a siphon!), 40 m of gallery follow, ending in a sinter wall of 2 m. From here, there is a wide and high gallery with sinter lakes up to 25 cm deep. From the 180th m starts the gallery, called The 20 Sinters, which is about 60 m long. In a nice white-and-rose chamber, we find the formation called The Jar (about 2 m sinter wall). The gallery is descending and ends in a small chamber with a mud cone on its left. From there, the cave aspect changes into a high canyon with a stream running on its bottom. Over huge boulders in a big chamber, we follow a winding gallery for another 30 m to reach the end chamber. In its first part there are boulders and clay, in the other – a siphon lake (10 x 6 m). Up to this place the length is 480 m.

Upstream from the entrance there is another gallery, low and narrow, 72 m long.

The cave entrance was known to the local people. The cave itself was explored and surveyed during the National Expedition in 1971 by L. Popov, Tsv. Georgiev, St. Tiholov and other cavers from Aleko Konstantinov Sofia Caving Club (description by L. Popov, Journal Tourist, 1971).

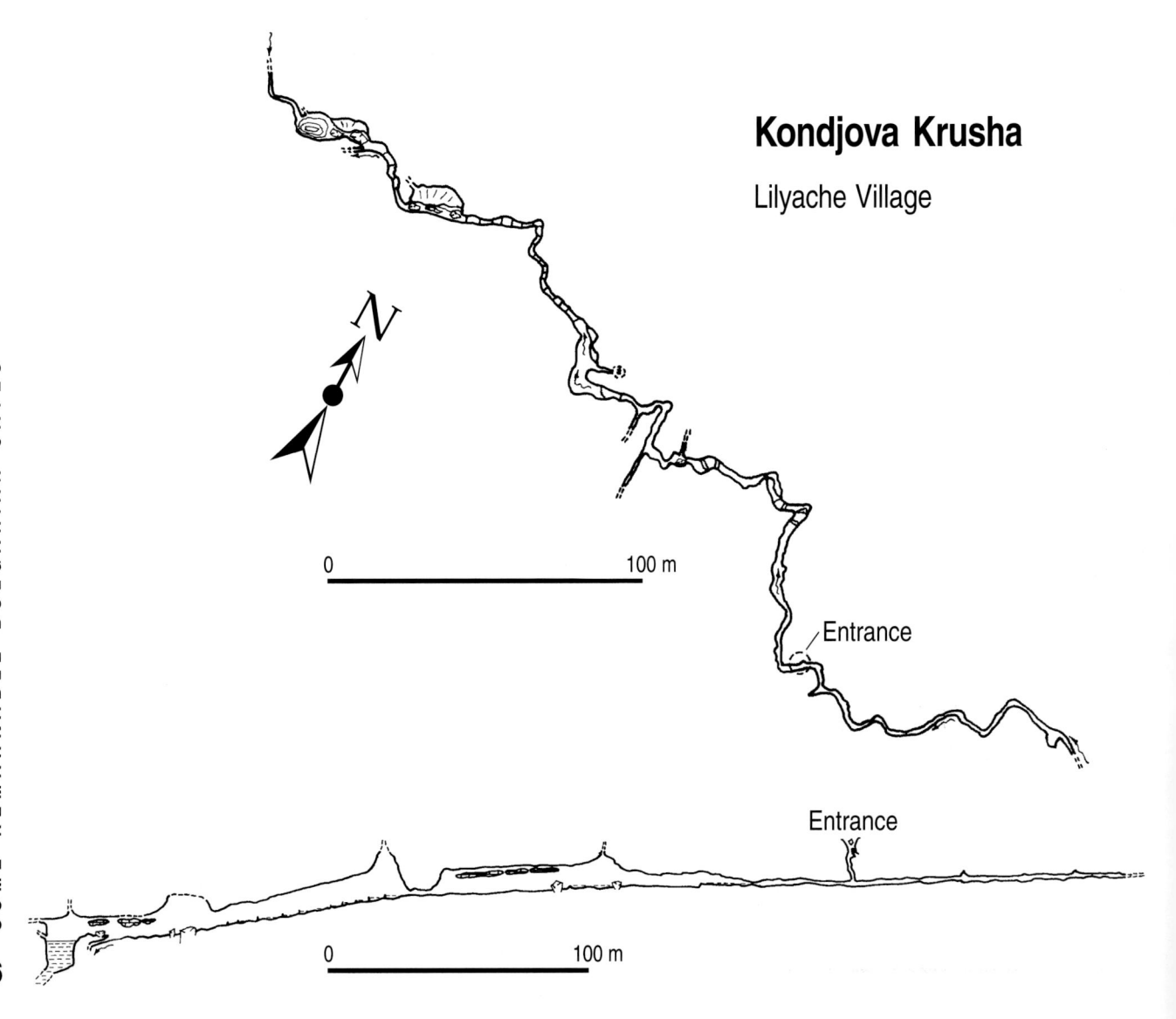

Kondjova Krusha

Lilyache Village

Name: **KOZARSKATA PESHTERA (54)**
Lakatnik Railway Station, Sofia Distr.
Altitude 395 m. Total length 709 m. Denivelation 12 m.
Coordinates: N 43° 05' 04.9" E 23° 22' 16.6"

A cave, 1.125 km SW of Lakatnik Railway Station, in the cliff on the right bank (geographically) of Proboynitsa River, Western Stara Planina.

The cave is formed in dolomitized limestone from the Jurassic. Developed along two main systems of fissures (NE-SW and NW-SE). There are 6 entrances. The cave is a system of galleries with a semi elliptic, upright general shape and average dimensions: width 2 m and height 3 m. Deposits – clay, smooth and angular gravel. The primary cave forms are represented by facets, levels of denudation, etc. Most types of secondary formations are also present. The cave is a site, where bones of a cave bear have been discovered (description: A. Jalov).

The first exploration and descriptions were realized by P. Tranteev at the beginning of the 60s of the 20th century. The cave was described and surveyed in details by cavers from Cherny Vrâh Sofia in 1978. Main surveyor: V. Vulchev.

Fauna: several cave animals have been found, among which the troglobites *Paralola buresi* (Opiliones) *Beskovia beroni* (Coleoptera) and *Trachysphaera orghidani* (Diplopoda).

Kozarskata peshtera
Lakatnik Railway Station

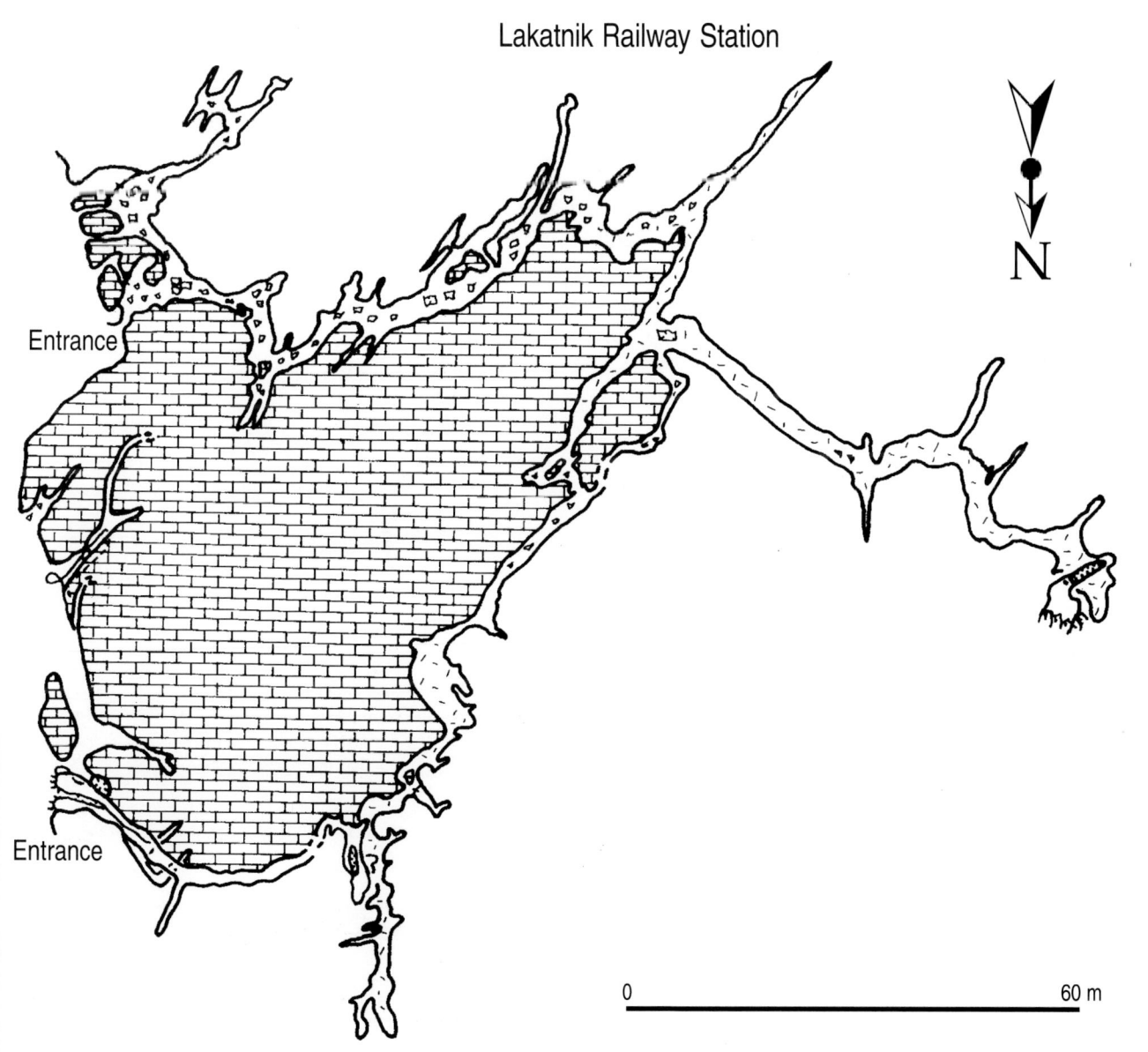

Name: **KRACHIMIRSKOTO VRELO (11)**
Krachimir Village, Vidin Distr.
Altitude 445 m. Length 333 m. Denivelation 22 m. Surface 925 m^2. Volume 4589 m^3

The entrances of this cave are situated in Zelyani Dol in the foothills of Kaleto Summit, 250 m from the branch to Krachimir and to Stakevtsi, behind the former guard post. There are 4 entrances, facing east. Through the lowest, an underground river runs out, piped for the needs of the guard post. The first part is a labyrinth. About 30 m along the water gallery, before the semi siphon (actually a siphon, after the damming), the galleries from the four entrances merge. About 6 m from the water entrance, a 45-m long lake starts with a maximum depth of 3 m. After a small vertical, another lake follows, which is 50 m long and has a maximum depth of 5 m at its end. A 25-m long dry gallery follows, with branches merging before the 14-m vertical. After the vertical, a 10-m long, very inclined gallery follows, ending with a 4-m vertical. The bottom of the vertical is the last lake, which is 35 m long and has a maximum depth of 5 m, from which an underground stream runs. There is a siphon at the end. This is a horizontal, permanently water cave, rich in flowstone formations. The deposits in the dry galleries are clay, sand and angular gravel.

In 1969, St. Tashkin, Z. Iliev, Chavdar Mitov, N. Gladnishki, Yu. Velinov and other cavers from Edelweiss Sofia Caving Club explored the cave, discovered new galleries and surveyed the initial parts. Accurate surveying was done in 1985 by A. Leonidov, N. Gaydarov, V. Nenov and S. Nenov.

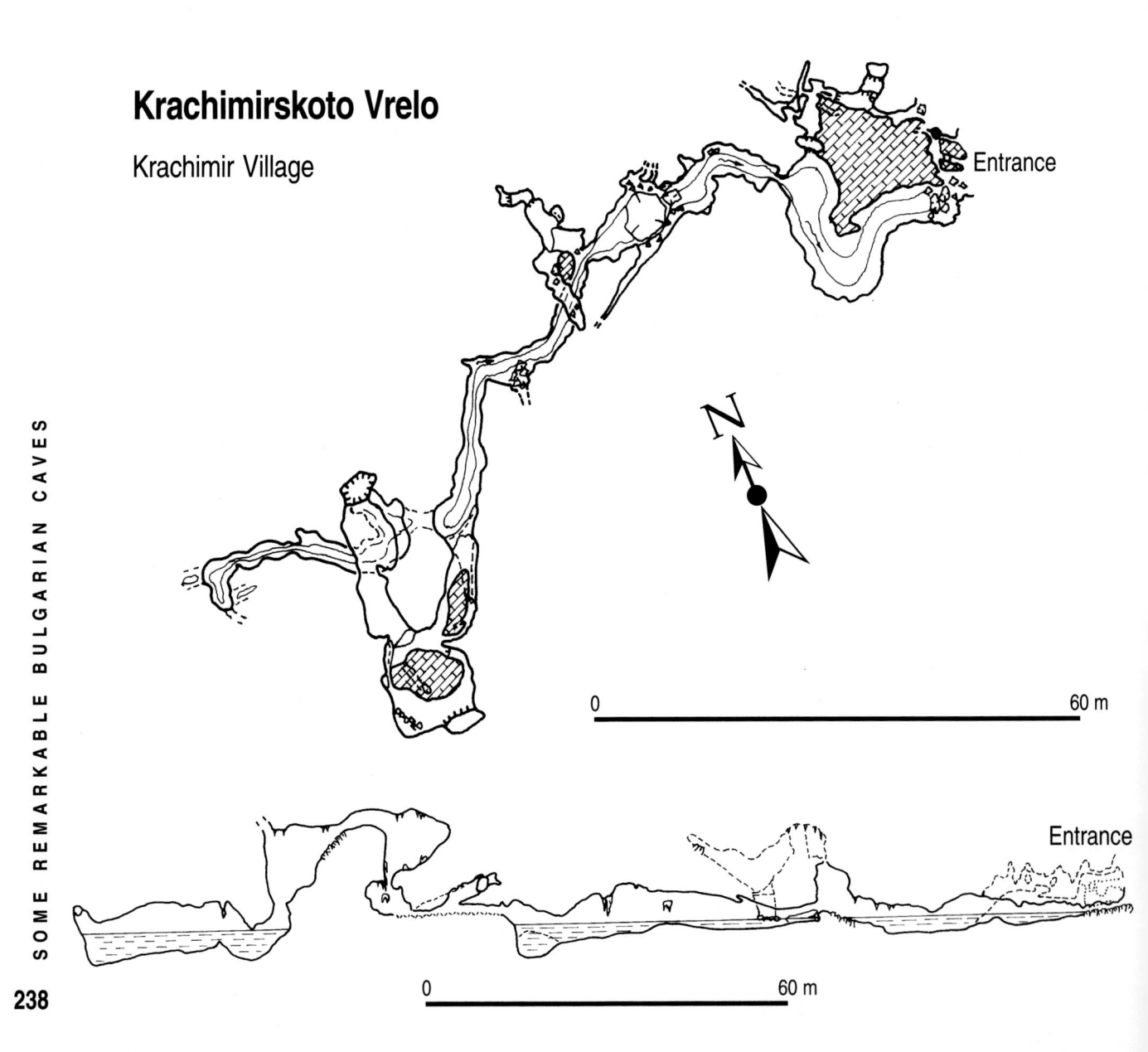

Name: **KRIVATA PESHT (65)**
Gintsi Village, Sofia Distr.
Length 1500 m
Coordinates: N 43° 04' 00.8" E 23° 06' 22.8"

A cave on the right slope of Gintska Reka (Nishava), in the locality Megerova Kukla. Formed in Middle-to-Upper Jurassic limestone. A horizontal, periodically watered cave. Deep layers of river sand after the 500th meter. There is a stream, periodically springing from them. A siphon follows, then galleries and chambers with dimensions 5 x 10 x 20 m. From an inaccessible opening, situated high up in a chamber, a stream, which forms a small waterfall, runs out.

First known exploration on June 30, 1939, by a scientific expedition led by Dr Iv. Buresch. The zoologists from the Royal Museum of Natural History in Sofia described the first 500 m and collected cave animals. Surveyed in details by cavers from Edelweiss Sofia Caving Club in 1978–1986. In January 1987, the siphon was overcome by D. Todorov and V. Pashovski and the end chamber reached.

The cave is part of the protected rock complex Zaskoko (88 ha, Decree 1141/ February 15, 1981).

Fauna: 15 species known, including the troglobites *Paranemastoma (Buresiolla) bureschi* (Opiliones) and *Pheggomisetes globiceps breiti* (Coleoptera, Carabidae).

Krivata Pesht

Gintsi Village

At the entrance of the cave Krivata pesht – 30.06.1939. On the photo: Dr Iv. Buresch, K. Tuleshkov, N. Atanassov, P. Drenski, Iv. Yulius, B. Pitioni

Entrance

0   120 m

Name: **KUMANITSA (129)**
Cherni Ossâm Village, Lovech Distr.
Length 1656 m. Denivelation -104 m
Coordinates: N 42° 44' 58.2" E 24° 41' 39.4"

The cave is formed in Trias limestone, in the eastern part of Troyan Balkan in the karstic canyon Steneto.

The entrance (2,5 x 2,5 m) is 1 m above the river Kumanitsa and 500 m from Kozi Brod. Two main galleries with flowing rivers of 300/400 l/s outflow. A downing cave with cataracts. Bats observed. At the beginning, there is a chimney, not fully explored. The river ends in a siphon. Many karst forms, flowstone, large chambers. In rainy weather, the exploration is dangerous and often impossible. The water is piped. The cave needs further exploration.

Penetrated and partly studied by P. Beron (August 29, 1960). Exploration and survey by Akademik Sofia Caving Club in 1962 – Hr. Delchev, St. Andreev, Tsv. Lichkov, P. Neykovski, Al. Grozdanov and others). Later on, new galleries have been discovered by V. Balevski, V. Nedkov and T. Daaliev (in 1977), A. Jalov and S. Tsonev (in 1979), E. Gateva and D. Marinova (in 1990) and others.

Fauna: the troglobite *Anamastigona alba* (Diplopoda)

Protected Natural Monument, together with 15 ha of adjacent land – Decree 2810/63 (56/19.07.1963). Situated within Steneto Biosphere Reserve and Central Balkan National Park.

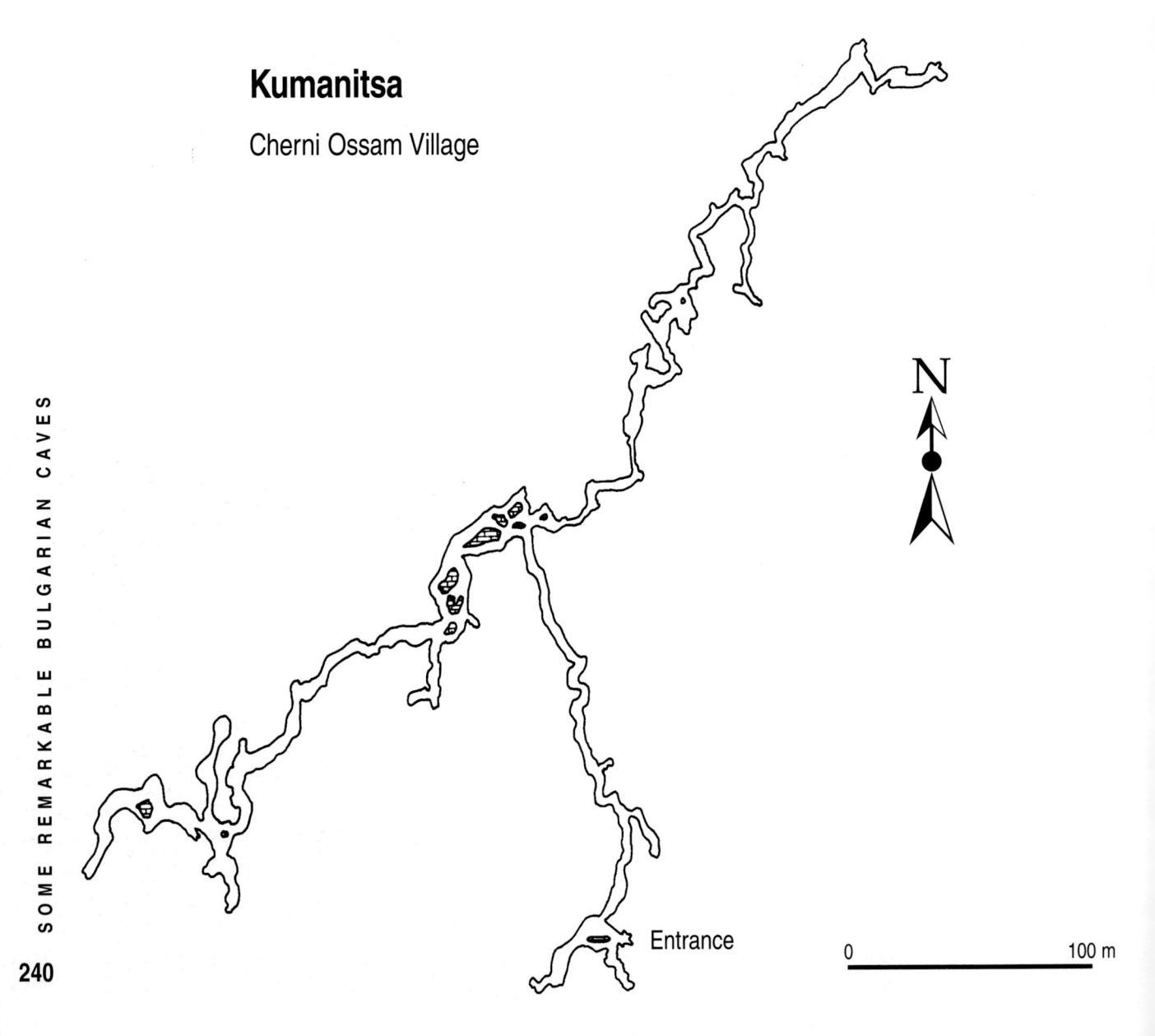

Kumanitsa

Cherni Ossam Village

The entrance of the cave – photo Tsvetan Ostromski

The waterfall before the end siphon of the cave Kumanitsa – photo Vassil Balevski

Name: **LEDNIKA (39)**
Kotel Town, Sliven Distr.
Altitude 736 m. Length 1367 m. Denivelation -242 m
Coordinates: N 42° 56' 00.9" E 26° 30' 46"

A cave in Zlosten region, 4.3 km NE of Kotel. Formed in Senonian limestone. A cascade pothole, with the biggest vertical of 18 m. One of the beautiful Bulgarian caves with stone waterfalls, sinter lakes, stalactites and stalagmites. There is an underground stream.

The pothole was descended first by P. Beron in 1961 (up to 130 m). The first exploration with a survey took place on July 16, 1964. It was carried out by S. Kazheva, Sv. Videnova, N. Popov, P. Gantchev, V. Stoitsev, Tsv. Lichkov, A. Grozdanov, B. Nedelchev.

Protected as a National Monument (Decree 2810/10.10.1962, with five of adjacent area).

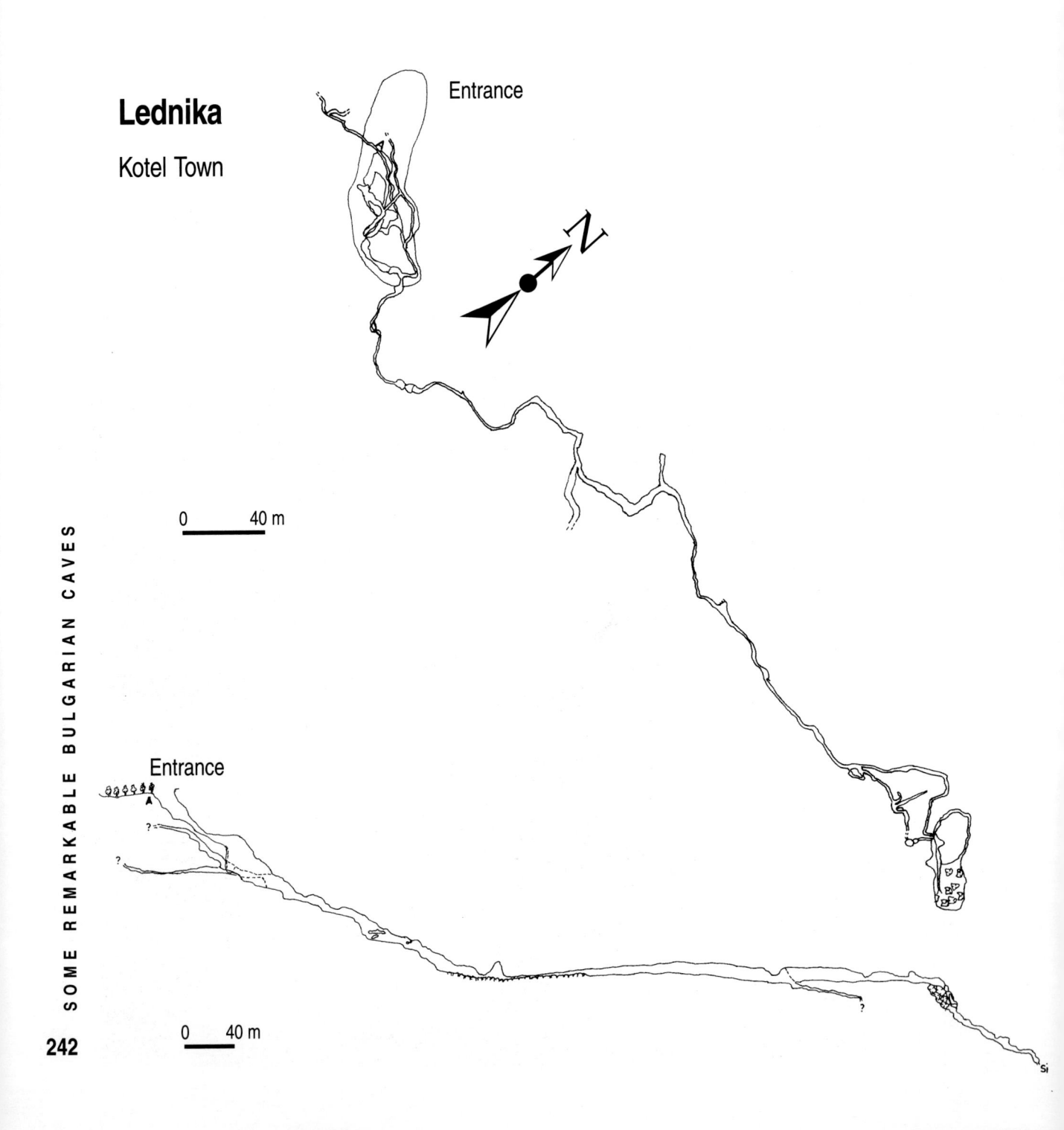

Name: **LEDENIKA (183)**
Vratsa Town
Altitude 830 m. Length 226 m. Denivelation -14 m, 16 m
Coordinates: N 43° 12' 35.6" E 23° 29' 34.8"

A cave in Upper Jurassic limestone Jurassic, 16 km from Vratsa, in the Vratsa Predbalkan. A dirt road goes to the cave, which is situated on the bottom of a huge uvala. The whole area is with such uvalas, flanked by a beech forest. There is a hut (Ledenika), 300 m from the cave, with 115 beds. In 1987, immediately in front of the cave, a small building was built for the visitors. The visitors' center of the park Vrachanski Balkan is situated in it.

The cave has two entrances. A stone staircase leads to the first chamber (Predverieto, 21 x 3 m). The lowest part of the hall, called Valchata Dupka (The Wolf's Den), is also the lowest part of the cave. A low gallery, 11 m long, called Plaznyata, connects Predverieto with Malkata Zala. Malkata Zala (The Small Chamber) has dimensions 21 x 17 m and a height of 5 m.

After a 6-m gallery, we reach the biggest and most beautiful chamber of the cave – Konzertnata (Golyamata) Zala. Its dimensions are 60 x 45 m and the height is 23 m. In it, there are huge and majestic flowstone formations.

In the highest part of the chamber, behind the flow, another hole is situated, called Hladilnika, with dimensions 21 x 10 m and a height of 4 m. It is also filled with beautiful formations and a small sinter lake.

From the Konzertna Zala, an ascendant gallery leads to "Malkata Propast", 13 m long and 9 m deep. At the highest point of the ascendant gallery, the temperature changes from 7–8° in Malkata Propast to 9,6° in the highest part of the cave, called Sedmoto Nebe. A narrow gallery connects Malkata Propast and Golyamata Propast (15 m long and 12 m deep). The gallery of the Curtains follows, which is 27 m long, 15 m wide and 17 m high. This gallery is also full of stalactites and other flowstone formations of different dimension and shape. After Prohoda na Greshnitsite (The Pass of Sinners), the White Chamber (18 x 15 m and 17 m high) and Sedmoto Nebe (the end part of the cave) follow. They are also richly adorned.

The cave had been known to the local population since time unrecorded. The shepherds used to keep milk in its cool entrance part. In the winter, from November to March, when the temperature in these parts drops below 0°C, in the Atrium and in the Small hall, transparent and milk-white masses of ice accumulate. They have been studied by the geomorphologist Vl. Popov since 1962. Data about the cave can be found in the publications of H. and K. Škorpil (1895) and Zh. Radev (1915).

Protected by virtue of Decree 2810 /1963. Show cave since September 17, 1961.

Fauna. In this cave, the first systematic research on the cave fauna of Bulgaria started (by Dr I. Buresch and his associates in 1922). So far, 53 species have been identified, including 10 troglobites or stygobites: *Speocyclops infernus* (Copepoda), *Bulgarosoma bureschi* (Diplopoda), *Tricyphoniscus bureschi* (Isopoda), *Paranemastoma (Buresiolla) bureschi* (Opiliones), *Onychiurus sensitivus, Pseudosinella duodecimocellata* (Collembola), *Pheggomisetes buresi, Ph. radevi, Rambousekiella ledenikensis* (Coleoptera, Carabidae), *Radevia hanusi* (Col. Cholevidae).

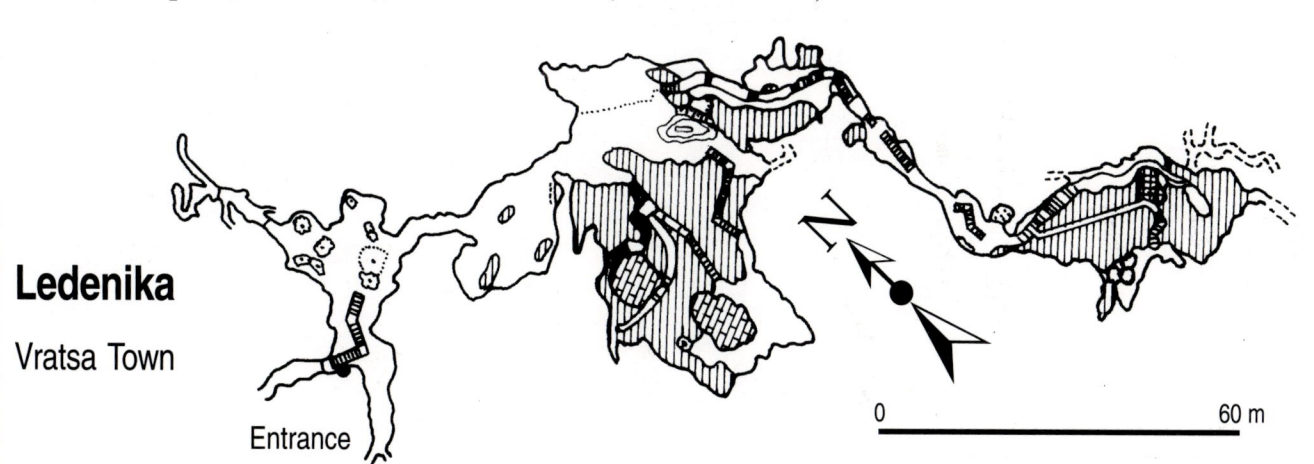

Ledenika

Vratsa Town

Entrance

0 — 60 m

The entrance of Ledenika – photo Valery Peltekov

The ice concretions in the entrance-hall – photo Trifon Daaliev

The concert Hall – photo Trifon Daaliev

The Visitor's Center – photo Trifon Daaliev

Name: **LEVI SHUPLI KAMAK (117)**
Prevala Village, Montana Distr.
Total length 342 m. Denivelation -20 m

Levi Shupli Kamak is a cave in the locality Shupli Kamak near Prevala. About 1.6 km on the road from Prevala to Gorna Luka, in the abrupt rocks, a triangular cave entrance is seen, ca. 3 m high. Here, a stream, coming from the entrance, joins the river. The stream leads into a natural rock tunnel through which the stream runs for ca. 15 m. In front of the tunnel, the stream comes out of two holes. The triangular entrance of Shupli Kamak is about 20 m high, facing east, and with dimensions 0.5 x 0.5 m.

The cave is inclined, branched at the beginning. The main gallery is horizontal, diaclase. Along the gallery, a stream runs toward the entrance (outflow of ca. 2 l/sec). The stream is permanent and comes out ca. 20 m under the entrance, joining Prevalska Ogosta. The floor of the water gallery is covered with sand and clay. In places, bigger rocks are seen. In some places in the cave, there are small white flowstone formations (stalactites on the ceiling and ribs on the walls). The cave ends with a big boulder jam, after which a continuation is possible (according to V. Georgiev).

The cave was known to local people and visited often by them (mainly by treasure hunters). Explored, surveyed and described in 1993 by cavers from Pastrinets Montana Caving Club.

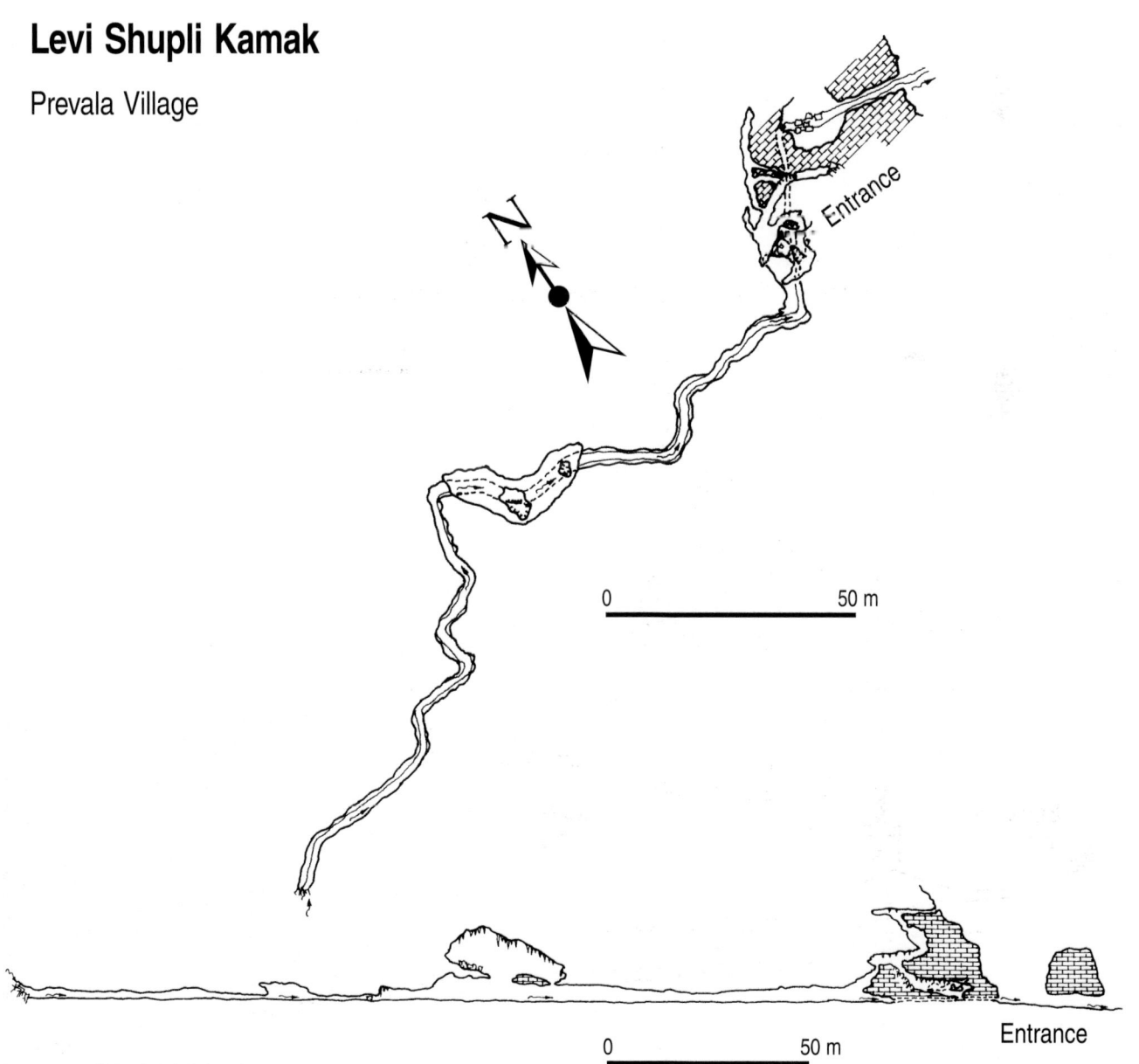

Levi Shupli Kamak
Prevala Village

Name: **LIPATA SYSTEM (79)**
Gabare Village, Vratsa Distr.
Length 657 m. Denivelation -21 m

Situated 2.83 km SE of Gabare, in the locality Lipata, Western Predbalkan. Formed in thick layered limestone from the Cretaceous (Maastricht), Gabarevo Syncline.

The system Lipata includes 6 funnels, four of which have no way further. The biggest funnel has three cave openings. One of them is the beginning of an not branching cave, 77 m long. The other two openings and the vertical in the westernmost opening are the three entrances into the system. They merge further by means of a labyrinth of low and narrow galleries covered with mud and of E-W direction mainly. The cave is poor in flowstone formations.

The first exploration was conducted by an expedition of Akademik Sofia Caving Club (leader Hr. Delchev), carried out from April 30 to May 2, 1968. In November that year, another expedition of the club (leader A. Popov) took place. This expedition connected the already surveyed cave potholes Lipa 1 and Lipa 2, discovered another entrance and proved all entrances as belonging to one system. The cave was surveyed by the same club (under the leadership of L. Lilov) also during the Republican Caving Expedition Gabare '86 by a team of cavers, which consisted of Yu. Atanasov, O. Ognyanov, V. Georgiev, Z. Zlatkov, A. Georgiev and S. Spasov. Description after A. Popov, A. Grozdanov and L. Vassileva.

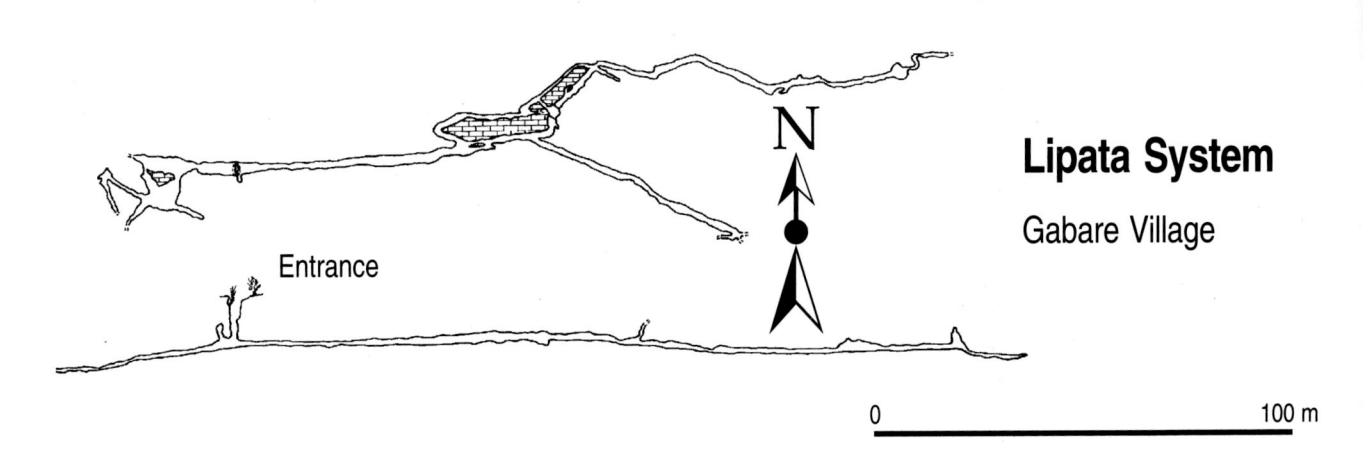

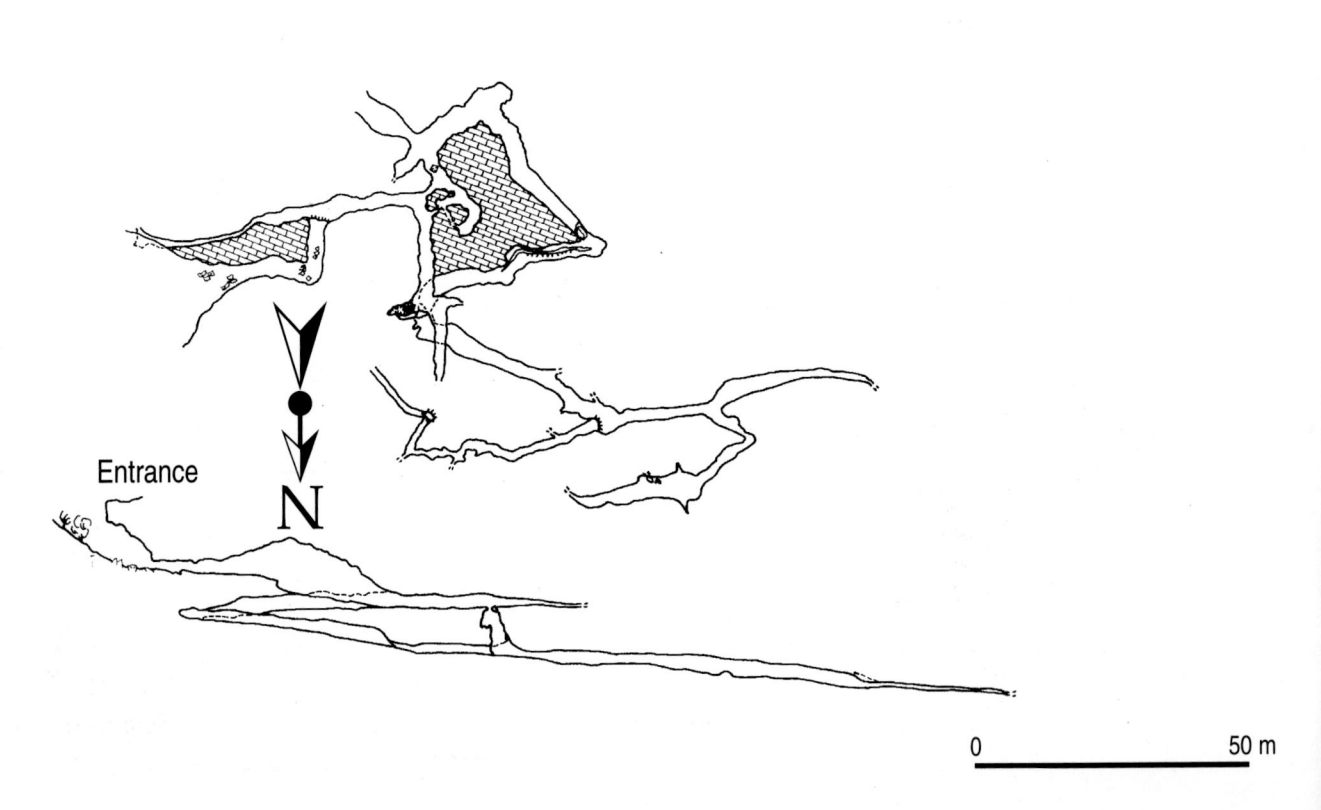

Name: **LUCIFER (189)**
Kotel Town, Sliven Distr.
Length 3200 m. Denivelation -130 m

A cave pothole, ten kilometers west of Kotel, in the locality Zelenich, on the left bank of Suhoyka River. Formed in Cretaceous (Maastricht) limestone. It starts with a 37-m shaft, leading into a high and narrow, descending and meandering gallery, broken by small verticals. This part ends in a water siphon. Several meters before it, a low and narrow dry gallery starts. On the upper floor, bone fossils from *Ursus spelaeus* (Cave bear) were found. During extreme rains and snow melting, the surface waters enter the cave, forming an underground river along the central gallery which sinks into the end siphon.

The cave was discovered by cavers from Lucifer Târgovishte Caving Club in 1992, following information from local people. In 1993, after digging, a branch at the end of the central gallery was found. Up to 1995 it had been completely surveyed and explored by the same club.

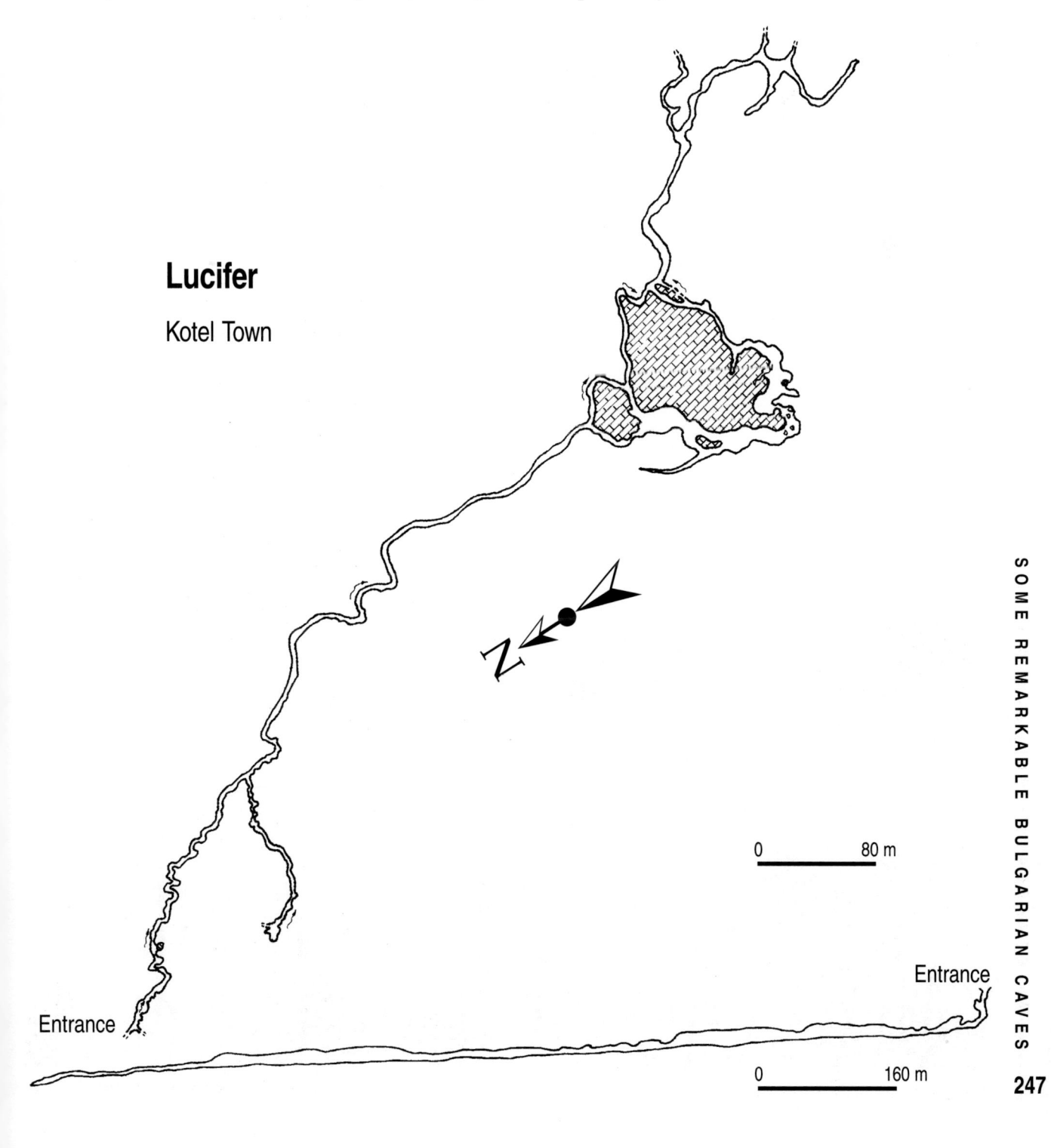

Lucifer

Kotel Town

The entrance of the cave Lucifer – photo Trifon Daaliev

Well deserved rest – photo Valery Peltekov

The bottom of the entrance vertical – Ascent – photo Valery Peltekov

Name: **LYASTOVITSA (LASTOVITSA) (112)**
Glozhene Village, Lovech Distr.
Length 320 m. Denivelation -2 m.
Coordinates: N 43° 01' 44,04" E 24° 10' 44.04"

A cave, 7 km from Glozhenc, in Lozeto quarter, not far from a dirt road. "Lyastovitsa" in Bulgarian means "swallow", but the legend has it that Lastovitsa was the name of a local feudal lord.

The cave entrance is a 3-meter diaclase. From there, the gallery follows two opposite directions. From the beginning, the first anemoliths are seen on the ceiling. After 20 m, the gallery becomes horizontal and passes by large stalagmites and boulders. By 80th meter, there are sinters, and through a squeeze we reach many pearls of caverns – china-like and sugar-like, some of them quite big (nut-size).

In some places, water drips from the ceiling. By the 130th meter, there are curtains on the ceiling. After a 7-meter climb, we reach small chambers with snow-white formations and sinter walls (30–40 cm high). There are heaps of bat guano. In 1967, a colony of some 500 bats was observed on the ceiling. At the 175th meter, we arrive to a narrow place, after passing by walls of different colour – from brown to snow white. The gallery is called Dreamland after this place. Nice poplar-like stalagmites are formed there, together with almost all possible flowstone formations. In some places, we can see pepper-like formations (rare in the caves). At the 175th meter, helictites replace the other formations. There, a siphon full of mud stops further penetration.

The cave had been known to local people since time unrecorded. It was explored and mapped in 1967 by N. Genov, V. Nedkov, P. Vassileva, A. Taparkova and other cavers from Planinets Sofia Caving Club.

Fauna: one stygobite known – *Stygoelaphoidella stygia* (Copepoda).

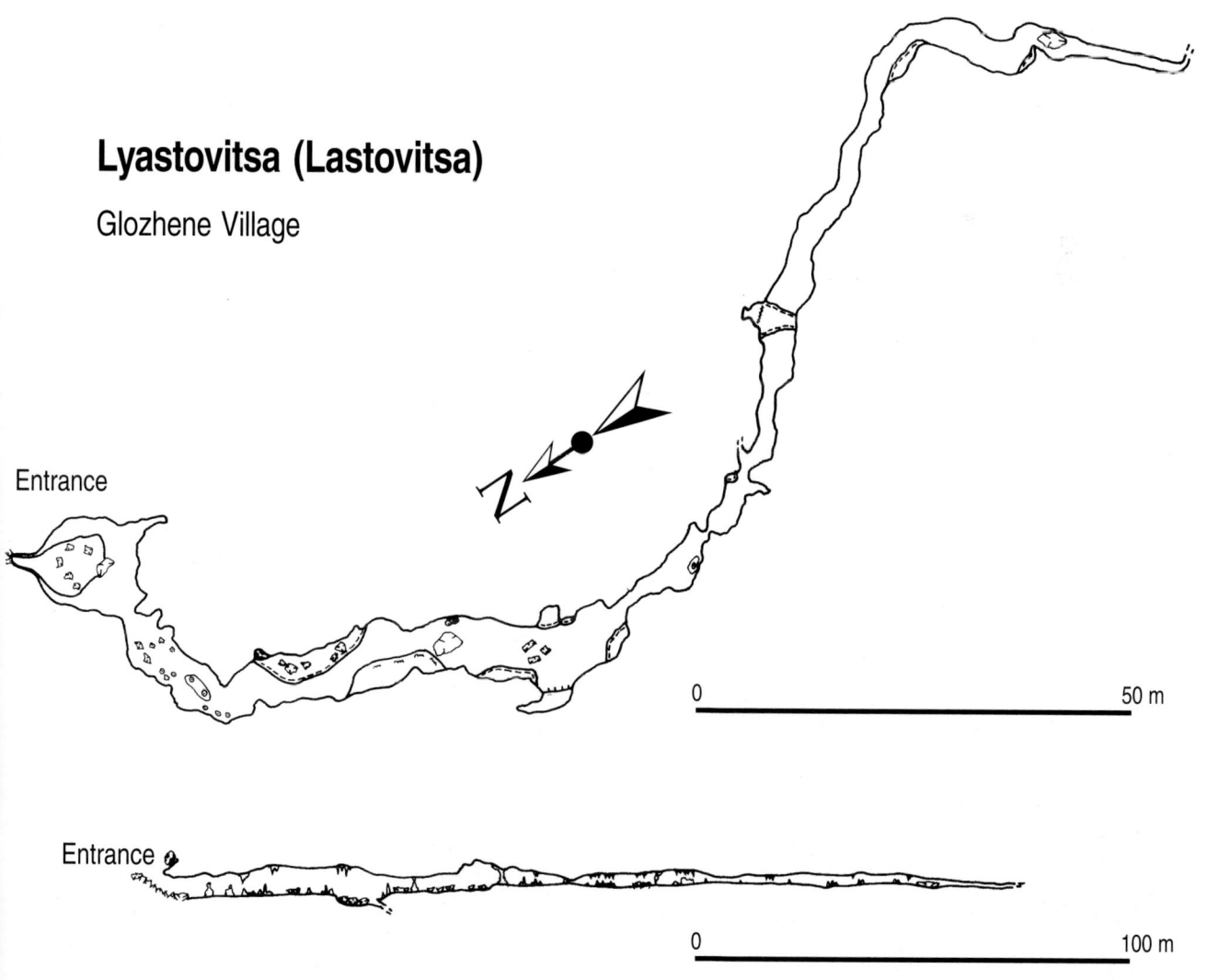

Lyastovitsa (Lastovitsa)

Glozhene Village

V. Peltekov in front of the entrance – photo Trifon Daaliev

The drop of water – creator of the fairy cave decoration – Lyastovichata peshtera Cave – photo Vassil Balevski

Columns – photo Trifon Daaliev

Name: **MACHANOV TRAP (165)**
Zdravkovets Village, Gabrovo Distr.
Length 1907 m. Denivelation -27 m (-21; 6 m). Surface 8000 m². Volume 15 000 m³

A cave, 0.5 km W of Zdravkovets, in the area called Bucheto, NW part of Strazhata Plateau, Middle Predbalkan. Formed in the contact zone between Lower Cretaceous sandstone and mergels and the under laying Aptain limestone. A branched, four-storey, water cave. Different shape of the cross sections: triangular, trapezoid, rectangular. Average dimensions of the galleries: width 3.65, height 2.37 m. Five bigger chambers, the most spacious one is 39x10x10 m. Many sinters and flowstone. Boulders in the lower levels. A permanent stream entering the cave (ponor). Four side streams sink together into a siphon in the lowest part. Water inflow measured in the period November 6 – December 17, 1983, between 0.18 and 0.9 l/s.

The cave was known to the local people. Studied and surveyed by cavers from Viten Speleoclub (Rossitsa – Mazalat) in the period 1981–1995.

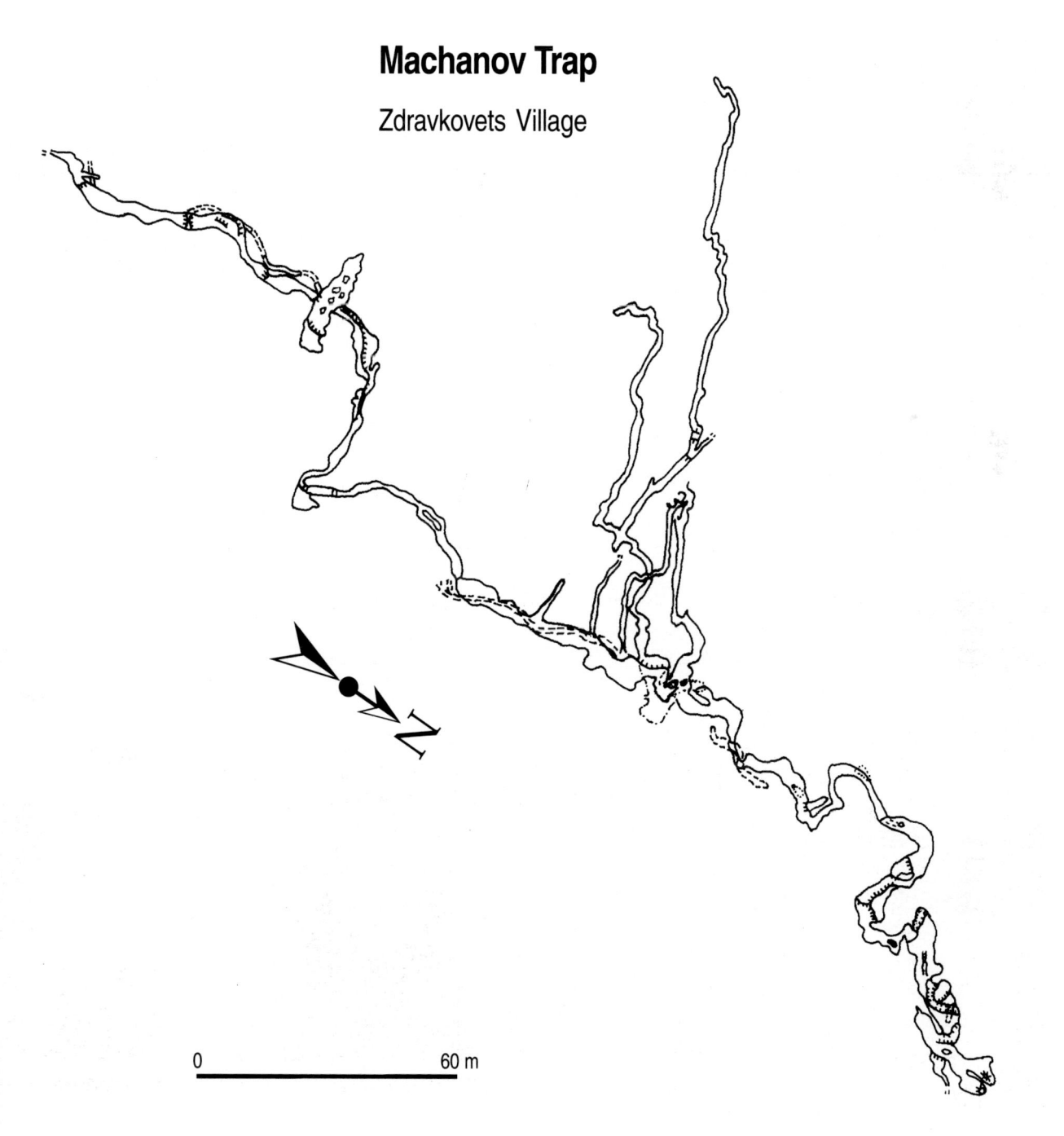

Machanov Trap

Zdravkovets Village

Name: **MAGLIVIYAT SNYAG (193)**
Tvarditsa Town
Length 3076 m. Denivelation -146 m
Coordinates: N 42° 46' 09.5" E 25° 54' 48"

A pothole in the locality Doksa, in the mountain Tvardishki Balkan. Entrance vertical, spacious, 5 x 10 m, -84.5 m deep, in two parts. Most of the year its walls are wet, T° 4 °C. On the walls there are stalactites, long up to 3-4 m, and several bats. Through the cave runs a stream, ending in a siphon. The way in goes through many narrow places, verticals and water. The entrance vertical is divided into two by a small platform, about 30 m from the entrance. The entrance has a stony bottom. From it the pothole splits into two parts, called Starite (The Old) and Novite (The New) parts. The old parts are reached through a small opening among the blocks. By descending among the blocks, the caver arrives to a gallery, on the bottom of which runs a stream. A narrow sector follows, usually with a waterfall.

Further on the gallery is diaclase shaped, gradually getting wider. After ca. 100 m there is a dry branch from the right, ca. 50 m long, ending in sand siphon. The left wall of the branch is covered by crystals and helictites. The main gallery goes further, following the stream. Another low gallery, ca. 40 m long, ends in 8 m vertical. The way through an opening 3 m high on the wall follows a 30 m gallery, ending in 4 m vertical. After a 10 m squeeze a large 80 m long gallery with nice formations is reached. Follows a 5 m vertical (rope is needed). The stream runs on the bottom of a gallery 20 m long. A branch is reached with another stream coming from the right. A siphon marks the lowest point of the cave, which is – 146 m. After climbing the stream is reached again. After 63 m there is a waterfall 2 m high. The gallery goes this way for another 128 m to a final narrowing.

The way to the "New parts" is a chamber lefts from the entrance vertical. After a 7 m vertical (usually equipped with rope) a spacious chamber is reached (35 m long, 20 m wide and 10 m high.Than the configuration is complex. Many chambers, passages and small verticals have to be negotiated before reaching a chamber 5 m wide, 8 m long and 6–7 m high. A 2 m squeeze leads in chamber 20-25 m high, than to 24 m vertical. After several narrow passages 26 m vertical is reached. Running water appears on the bottom. After several other passages and small verticals a small chamber is reached at 119 m. Than we reach a chamber, 35 m long, 20 m wide and 10 m high. On the left, a steep slope leads to the entrance pitch and it is suiteable for an intermediate camp. Behind a big boulder there is a vertical of 18 m with washed walls. Following the main direction, we enter a chamber 6–7 m Through the boulders on the left side, we reach a vertical of 14 m. Gradually, the gallery becomes 4 x 4 m, with boulders on the floor. On the left, a 3m vertical is reached, from where three branches start.

We then follow the main axis – a diaclase 64 m long, high in some places 20–25 m., wide up to 2 m, with wet floor covered with gravel.

On the right side of the chamber, after 20 m, a clearly shaped gallery is reached with a floor of small gravel. Some bats were observed there. For some 130 m the gallery does not changing shape. There are some nice flowstone formations. Some more complicated galleries lead to the bottom.

The pothole was discovered in 1995. The first exploration was carried out by Salamander Caving Club, Stara Zagora. Explored and surveyed entirely during the national caving expeditions in 2000 and 2001 and the expedition of Salamander Caving Club in 2002.

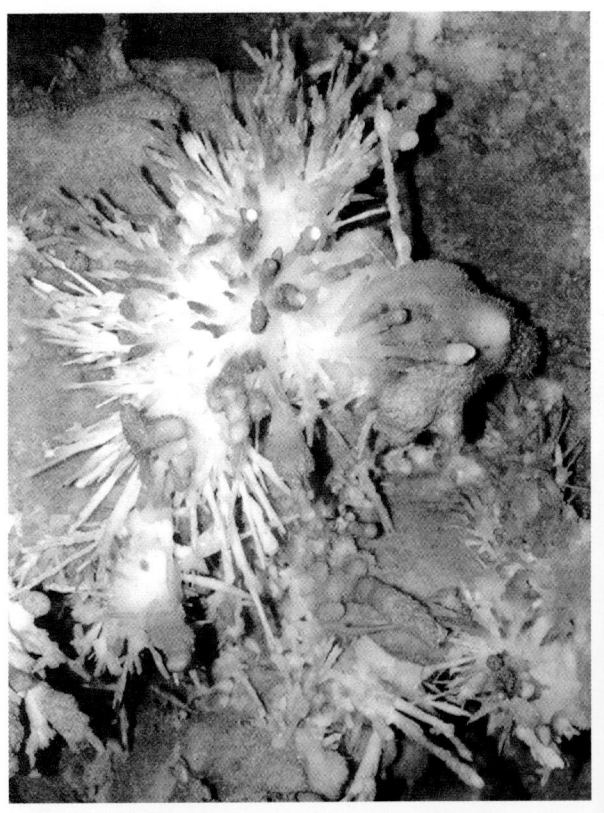

The crystals are one of the cave's beauties - photo Alexey Stoev

The entrance pit is 90 m deep – photo Vassil Balevski

Magliviyat Snyag

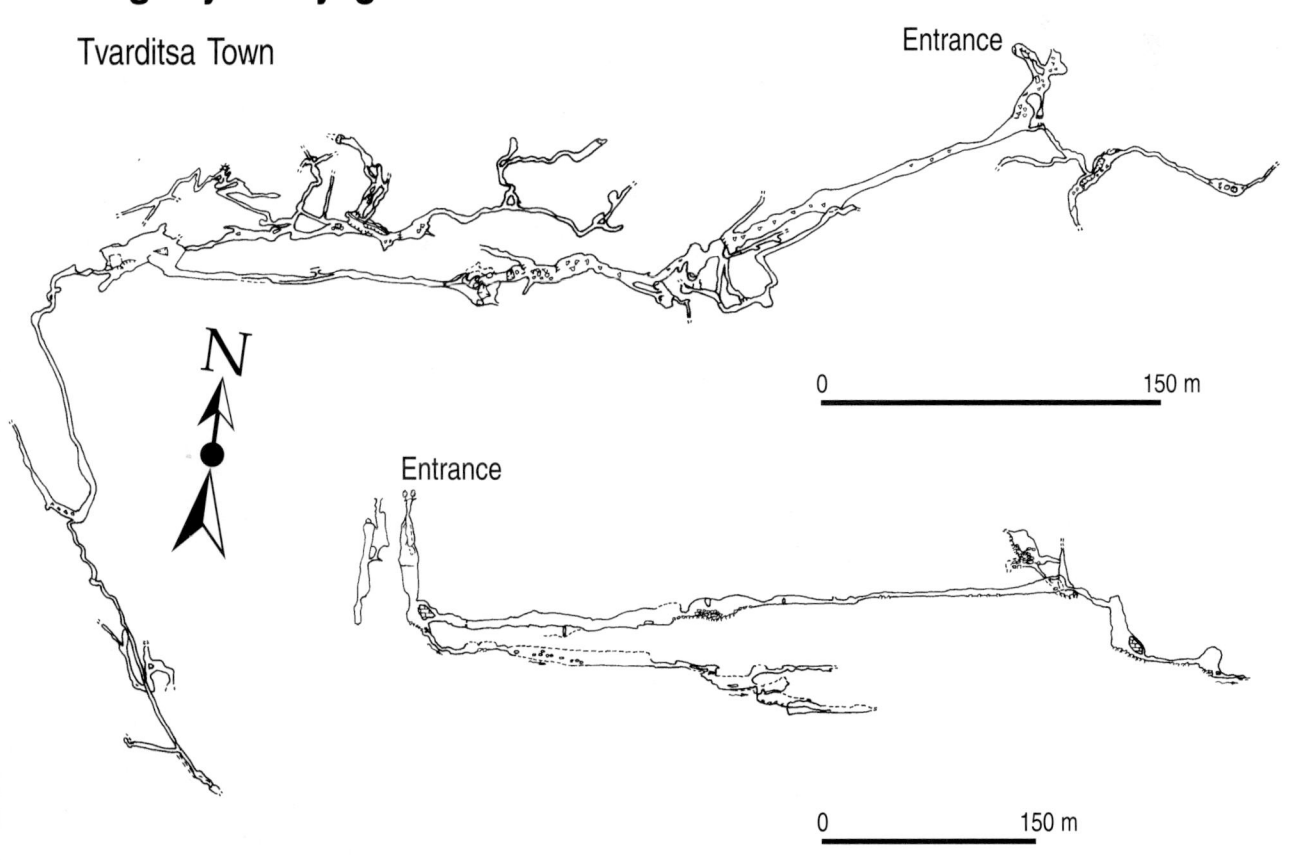

Tvarditsa Town

Name: **MAGURA (2)**
Rabisha Village, Vidin Distr.
Length 2500 m. Denivelation -56 m. Surface 30 000 m². Volume 220 000 m³
Coordinates: N 43° 43' 41.1" E 22° 34' 54.8"

A cave, 1.5 km NW of Rabisha, in the hill Rabishka Mogila (Magurata), Western Predbalkan, in Aptian (Urgonian) limestone.

A horizontal, branched cave with six chambers. The biggest one (Triumfalna Zala) has 5720 m², with dimensions: 128 m long, 58 m wide and 21 m high. There are eight other smaller chambers in the side galleries. The SW branch is of special interest because of the cave paintings on the walls (over 1000, made with ochre and bat guano). The cross section of the chambers is asymmetric with semivertical SW walls and NE inclined parts, defined by the position of the layers. Massive flowstone and river deposits.

The cave formation started in Tortonian (15 millions years ago). Fossils from seven species of Quaternary birds, fish and molluscs have been found in the deposits.

The cave was first discovered partly by Marinov (1887). F. Birkner published further archaeological data in 1916 and Dr B. Bonchev from Vidin published a small booklet. In 1928, R. Popov attracted the attention of Bulgarian and foreign scholars to the cave paintings. They were created in various periods, the earliest – in the Epipaleolith and Neolith, the prevailing part – in the late Eneolith – beginning of Early Bronze Age. This is the most important site with monochrome rock art in Southeast Europe.

The cave was studied in details by T. Pavlov Caving Brigade in 1948. Regular archaeological excavations were carried out in 1971–1975. The cave had been inhabited from the beginning of the Bronze Age to the Early Iron Age. In 1974, attempts to use the cave for asthmatic patients were made. Modern research on the rock paintings was carried out in 1988-93. Among the drawings in Svetilishte Chamber, records of a prehistoric sun calendar were identified and decoded.

Show cave since 1961 (the first show cave in Bulgaria). Later, a wine cellar was built in the first western branch. Declared a National Monument in 1960.

Fauna (studied by P. Beron in 1960) represented by 37 known species, including the troglobites *Hyloniscus flammula* (Isopoda), *Plusiocampa bulgarica* and *P. beroni* (Diplura).

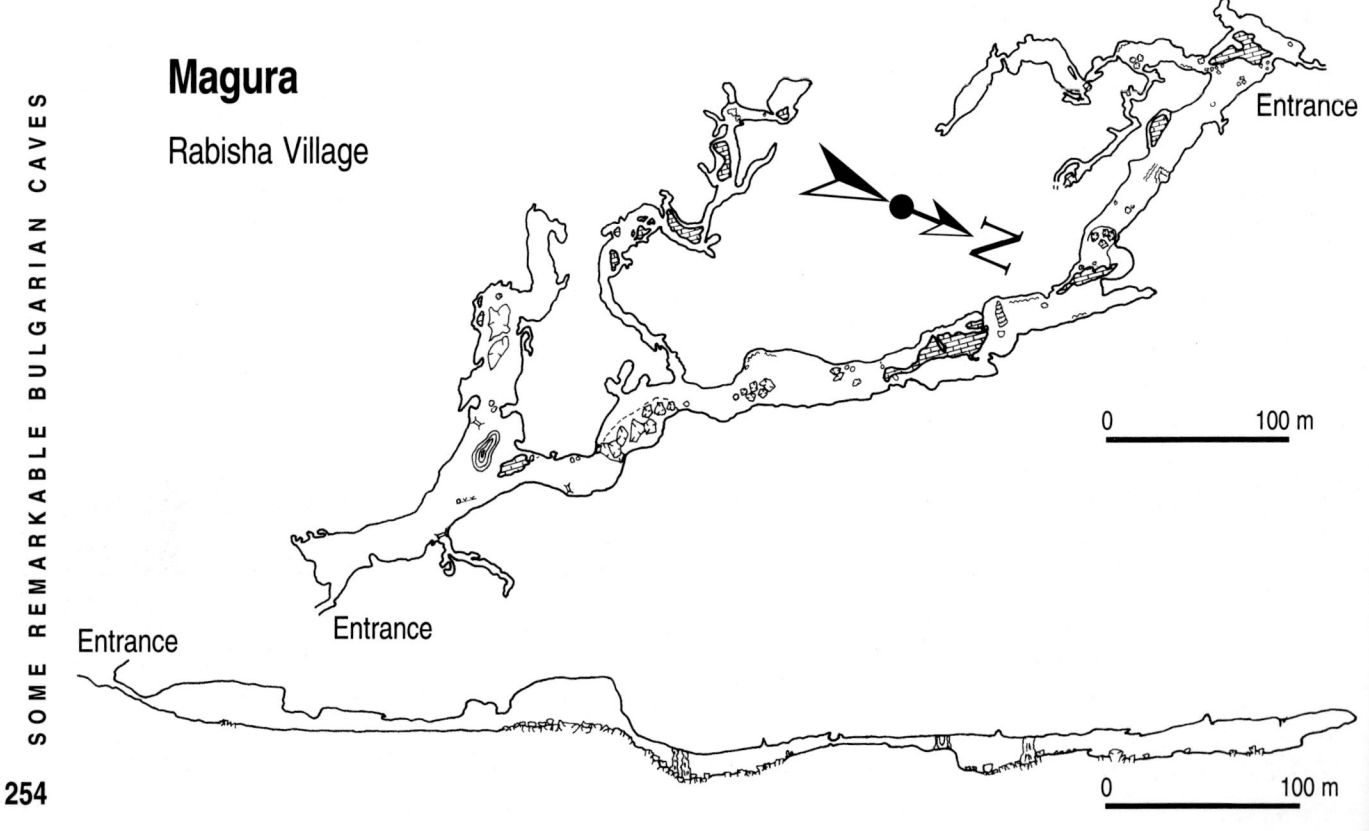

Monochrome paintings from Bronze age – photo Mihail Kvartirnikov

The entrance gallery – photo Valery Peltekov

A huge 10 meter high stalagmite – photo Trifon Daaliev

Stone waterfall – photo Trifon Daaliev

The Concert Hall (The Big Hall) – photo Valery Peltekov

Name: **MALIYA SOVAT (119)**
Bryazovo Village, Lovech Distr.
Length 125 m. Denivelation -140 m

The pothole Maliyat Sovat is ca. 7 km (2 hours of walk) from Bryazovo, on the road between Teteven and Ribaritsa, in a valley of the same name.

Formed in Upper Jurassic limestone.

The pothole has two entrances. One is small, situated on the bottom of the valley. When raining, the water is pouring in it. After a vertical of 8 m, we reach the floor or an inclined gallery. The second entrance is 20 m away from the first and is bigger. Through it, we enter a strongly inclined gallery with a floor, covered with blocks.

The beginning of the gallery is lit by the light, coming through the first entrance. After ca. 40–50 m, a 22-m vertical is reached. From a platform, with a diameter of 7–8 m, a vertical of 52 m starts, leading to the bottom of the pothole. The vertical is inclined and covered with secondary flowstone formations. The bottom of the pothole is slightly inclined and covered with clay and other material carried by the surface water.

After 20 m over a strongly inclined slope, we reach a narrowing, above which, at a height of 10 m, a 2-m wide opening 2 m wide, not yet checked. A narrow gallery with clay has been tried several times for prolongation, but without success. The water, passing through the pot hole, is surfacing in karstic well. When raining the surface water in the valley enter directly the pothole.

Only well trained cavers should visit this pothole.

The pothole is known to local people. In 1972 V. Nedkov, caver from Planinets Sofia Caving Club, was the first to reach the bottom, followed by N. Genov, Iv. Rashkov, Iv. Parov and others. N. Genov and Iv. Rashkov surveyed the pot hole and named is after the valley. In the years to follow many attempts to discover continuation failed. The pothole was equipped for the SRT technique by V. Nedkov, N. Genov, P. Nedkov, M. Valchkov and T. Daaliev in November 1980.

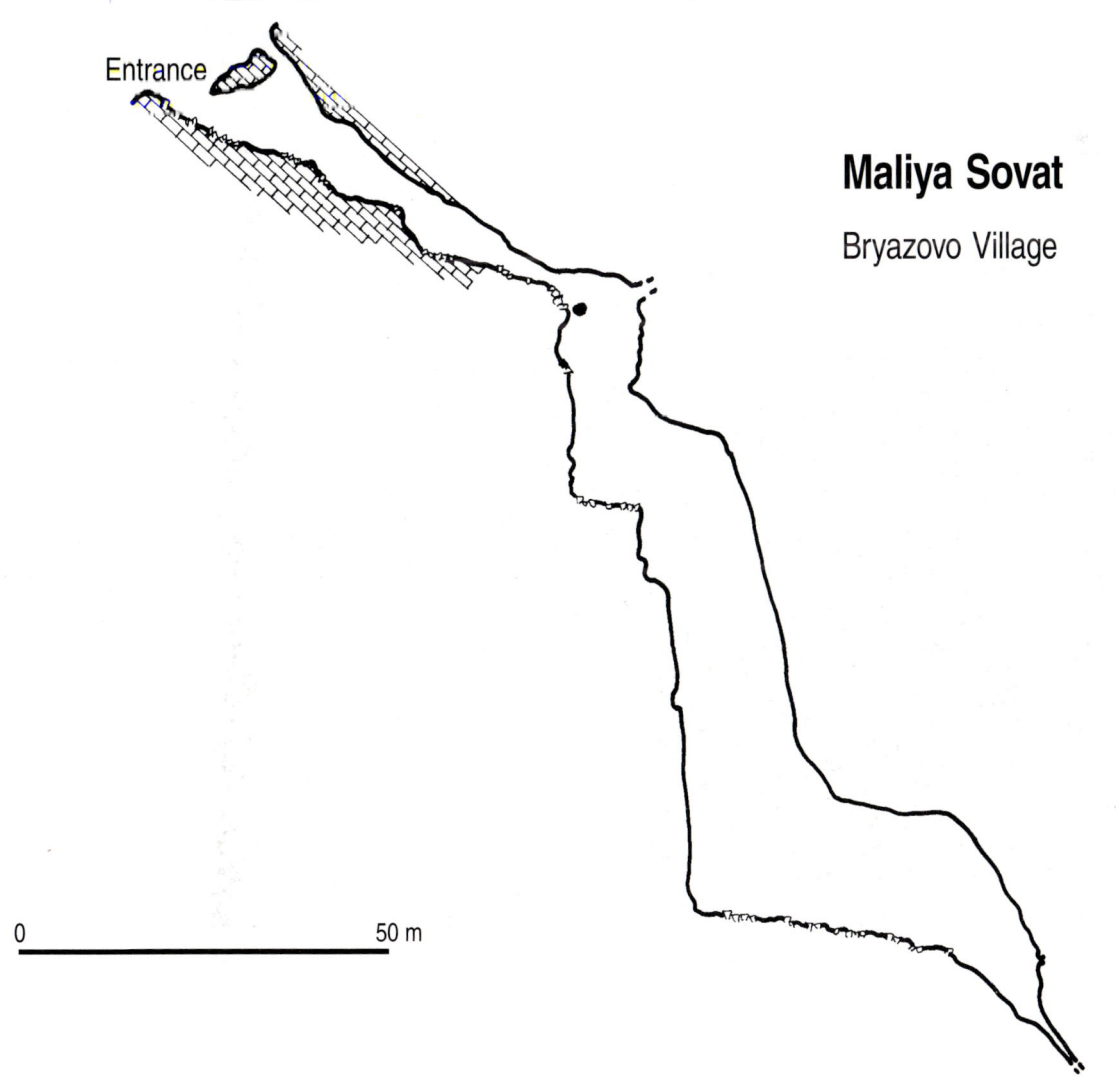

Maliya Sovat

Bryazovo Village

Maliya Sovat – ascending the second pitch – photo Nikolay Genov

Name: **MALKATA (MALATA) BALABANOVA DUPKA (YAMATA NA ISKRECHINATA) (70)**
Komshtitsa Village, Sofia Distr.
Length 400 m. Denivelation -125 m
Coordinates: N 43° 08' 00.5" E 23° 02' 32.2" (1-st entrance) * N 43° 07' 58.6" E 23° 02' 32.4" (2-nd entrance)

Malkata (or Malata) Balabanova Dupka had been known to local people for a long time. Its entrance part was first explored by Dimitar Ilandjiev in 1963. He announced the importance of the unexplored cave. In October 1971, T. Daaliev, N. Gladnishki, A. Anev, M. Vlachkov and N. Anachkov made a reconnaissance trip for discovering new caves and for penetration into Malata Balabanova. A local shepherd showed them a pothole (immediately) near the locality Iskrechinata, so the name "Yamata na Iskrechinata" appeared on the maps. They were the first to penetrate the pothole. They surveyed the cave up to 83 m denivelation. Then, they penetrated further down to the bottom (ca. – 130 m). They returned to this area in the spring of 1972, accompanied by V. Gyaurov and P. Nikolova. V. Gyaurov fined the small triangular entrance of Malata Balabanova Dupka and then they realized/established a horizontal connection with Yamata na Iskrechinata. At the beginning, there is a 10-meter vertical, leading to the big chamber at 83 m. The same cavers finished the survey at 125 m.

The entrance of Yamata na Iskrechinata is egg-shaped (2 m diameter), on a steep slope, about 30 m above the entrance of Malata Balabanova Dupka. The first vertical of 20 m leads into an inclined gallery, ending in a 10-meter vertical. From the Big Chamber, through flowstone formations, we reach a small lake, covered with transparent crust. Next, there is a small (4 m) sinter waterfall (with water). Then, from the bottom we follow an inclined gallery. The walls are black, there is running water on the floor. After ca. 50–60 m we arrive in the middle of a chamber, which is even bigger than the first one. After a descent over beautiful flowstone formations, the bottom of the chamber is reached at 83 m. The gallery after the chambers goes down in a conglomerate. After 70–80 m, on the right of the main gallery, water falls from 7–8 m. The gallery narrows and proceeds for another 30–40 m up to a 10-m vertical. There is running water also in a parallel gallery. The cave is very beautiful.

Description of Malkata Balabanova Dupka. Its small triangular entrance is situated in a valley about 300 m before reaching the huge entrance of Golyamata Balabanova Dupka. After several small verticals and a gallery of 10 m, a vertical, 10 meter high and 7–8 m wide, is reached. An inclined gallery follows, ending in a 10-m vertical, where the water falls into the big chamber at 83 m. Above the 10-meter vertical, there is a gallery, leading into diaclase galleries, which are 30–40 m long. At the end of these galleries is the connection with Yamata na Iskrechinata at this level.

At the end of September 1997, T. Daaliev, Atanas and Kiril Russev, together with French cavers (J. Orssola and O. Vidal), made a traverse above the last vertical of Malkata Balabanova Dupka and the pothole became deeper by several dozens of meters. In 1999, K. Stoyankov, N. Donchev, K. Dudlen and T. Daaliev realized the first acoustic connection between Malkata and Golyamata Balabanova Dupka. The connection was confirmed in 2000 by K. Stoyankov, V. Peltekov, M. Dobrichev, Y. Pavlov and others.

Malkata (Malata) Balabanova Dupka (Yamata na Iskrechinata)

Komshtitsa Village

In the secondary hall of the cave – photo Trifon Daaliev

Sinter lake – photo Trifon Daaliev

Name: **MALKATA MIKRENSKA PESHTERA (138)**
Mikre Village, Lovech Distr.
Altitude 377 m. Length 1295 m. Denivelation 5,4 m
Coordinates: N 43° 06' 46.65" E 24° 32' 33.74"

A cave 100 m E of Golyamata Mikrenska Peshtera, also on the left bank of Kamenitsa. Formed in Maastricht limestone (Lower Cretacetous). An asymmetric, labyrinth cave. Horizontal, one-storey, dimensions of the galleries: width 0.3–3 m, height 0.3–3,2 m. Rich in flowstone. A stream following from the entrance, 0.1–0.2 l/s in dry periods, 20–30 l/s after rain. Discovered by D. Ilchev with indications from local people (September 4, 1924). Studied to ca. 50 m. In 1980–1990, explored and surveyed by Studenets Pleven Caving Club up to its present length.

Fauna: four animal species have been identified.

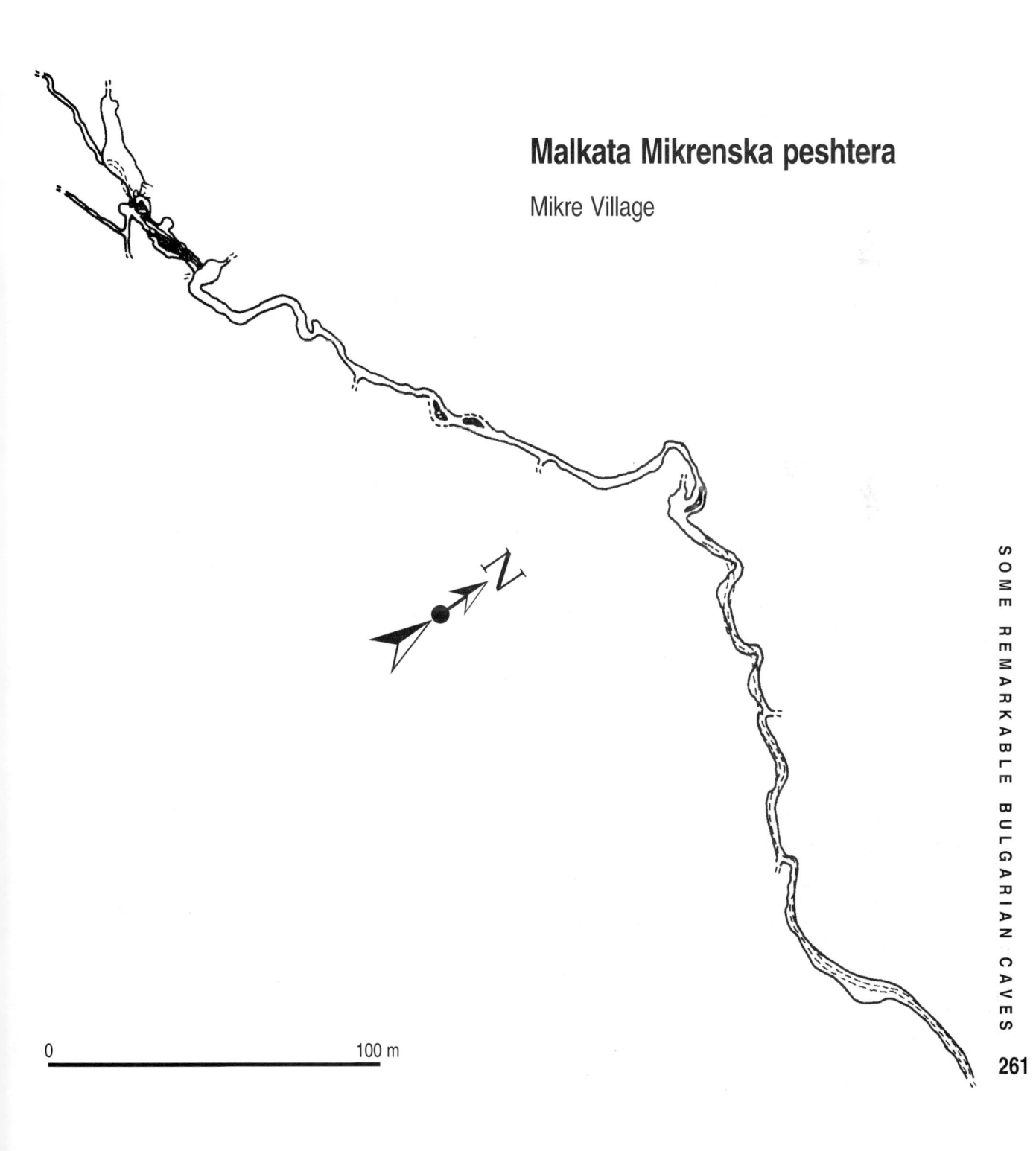

Malkata Mikrenska peshtera
Mikre Village

Name: **MALKATA YAMA (130)**
Cherni Ossam Village, Lovech Distr.
Altitude 1333 m. Length 700 m. Denivelation -232 m
Coordinates: N 42° 46' 03.2" E 24° 41' 02"

A cave pothole in the Steneto Biosphere Reserve. Entrance on the right bank of Malak Chaushov Dol. Formed in Trias limestone and dolomites.

A cascade pothole cave with two bigger verticals of 15 and 36 m. At 127 m, there is a huge chamber, over 50 m high, called The Big Hall. It is inclined and full with boulders, some of them having a volume of over 400 m^3.

An underground stream with an outflow of 1–40 l/s. T° 7 °C.

Discovered in 1969 by Troyan cavers V. Balevski and A. Borov. Surveyed and described by T. Daaliev, N. Gladnishki, V. Balevski and A. Pencheva during the International Caving Expedition Troyan'72.

Malkata Yama

Cherni Ossam Village

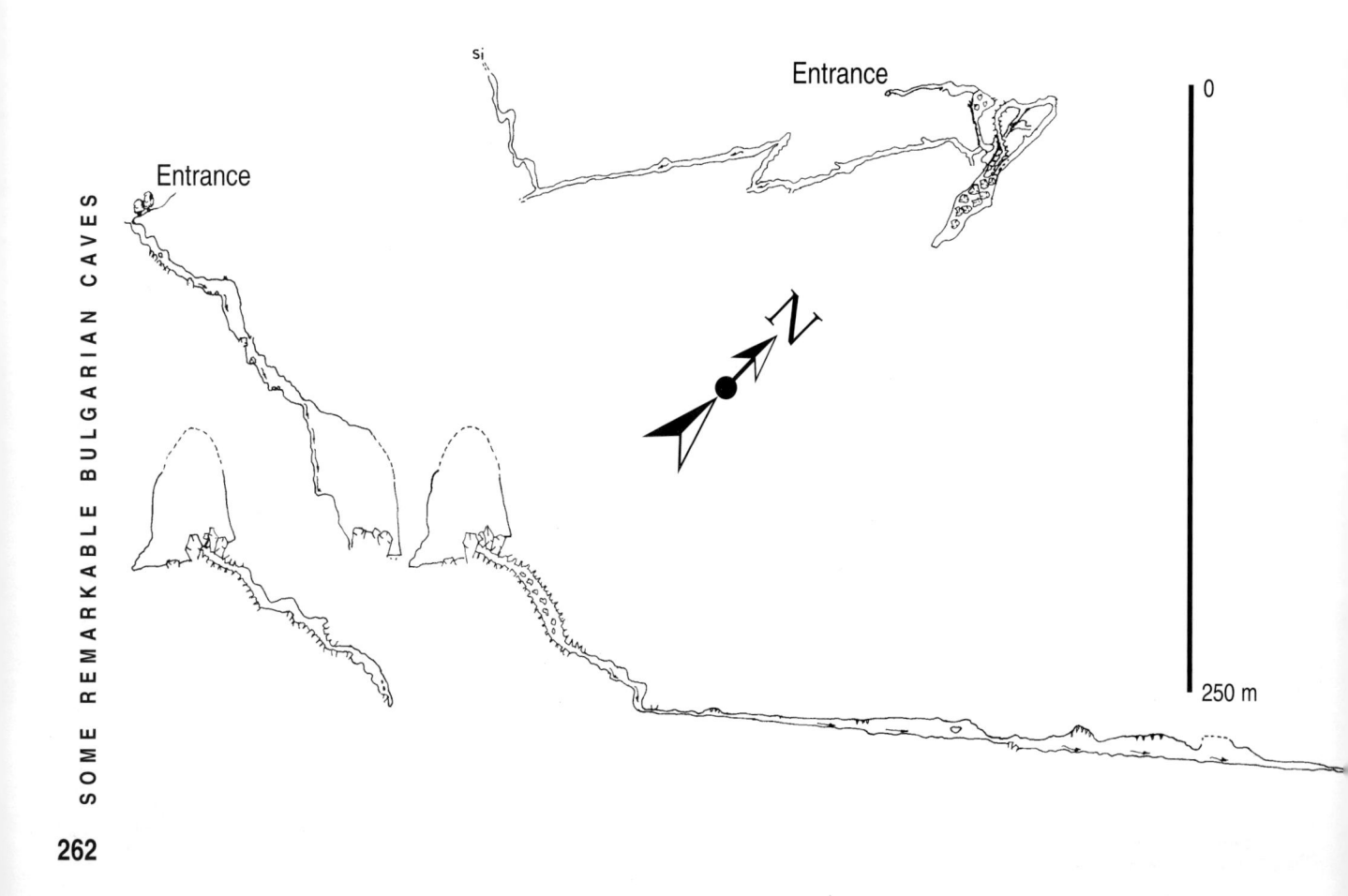

The ascending is not easy – photo Vassil Balevski

Name: **MANDRATA (LÂDJENSKATA, VODNATA PESHTERA) (152)**
Chavdartsi Village, Lovech Distr.
Length 530 m. Denivelation – 2 m
Coordinates: N 43° 14' 32,2" E 24° 58' 08,8"

A cave 2 km west of Chavdartsi (Lâdjene), in the NW periphery of Devetaki Plateau, Middle Predbalkan. Formed in limestone from the Lower Cretaceous (Aptian). A horizontal, branched, source cave, developed along a fault directed N/NE. Dimensions of the entrance: 12 m wide, 7.5 m high. Entrance chamber 81x 35x 15 m, after which two galleries follow: to the left – a river gallery, 165 m long, and to the right – a dry gallery, 103 m long. Denudation forms prevail. Deposits of clay and sand, no sinters. Measured outflow on October 15, 1931 – ca. 5 l/sec.

The first known exploration was done by Nenko Radev (September 18, 1926). Archaeological excavations were carried out by V. Mikov in 1926 and 1929 and artefacts dated from the Eneolith and the Iron Age were found. Detailed description and survey by P. Petrov in 1931 and later by national expeditions in 1965-78. At that time it was used as cheese factory, hence the name Mandrata.

Fauna: 13 species of Invertebrates.

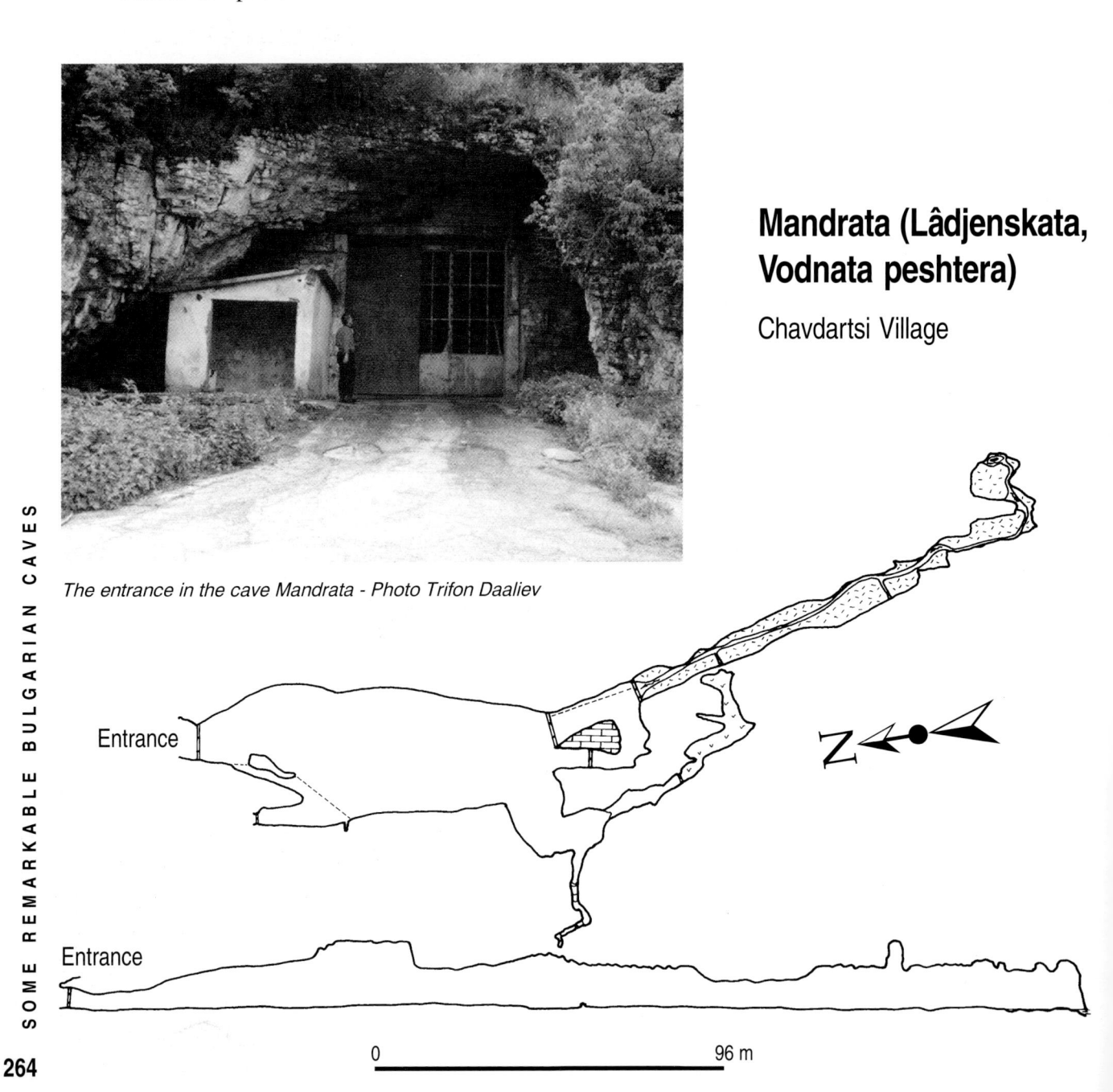

The entrance in the cave Mandrata - Photo Trifon Daaliev

Mandrata (Lâdjenskata, Vodnata peshtera)

Chavdartsi Village

Name: **MARINA DUPKA (PAROVA DUPKA, KUMINCHETO, PROPASTITE) (168)**
Genchevtsi Village, Gabrovo Distr.
Length 3075 m. Denivelation -45 m

A water cave, 4 km NW of Tryavna, in Genchevtsi Municipality, Trevnenska Planina. Formed in limestone from the Lower Cretaceous. A two-storey, branched cave with two entrances – the Northern horizontal, the Southern vertical (- 38 m). About 80 m from the horizontal entrance is the only bigger hall (22 x 16 m), from where the cave branches off. The SE branch ends in a siphon, in which the river sinks. After 30 m, the NW gallery takes the water of two streams on the right and forms a second river, which flows inside. After 140 m, it joins a third river, coming from the NW. The water flows further into a 300-m gallery, ending in a siphon. The NW branch forms the longest part of the cave, in which there are about 50 ascending verticals, 0.3–2.2 m. After 200 m upstream from the water joint, the gallery branches off. In both branches streams flow. The average width of the gallery is 2.5 m. Thick clay deposits and big boulders in the entrance part.

First studied by N. Radev and Iv. Buresch in September and October, 1923, and by Radev again in 1924–1925. In 1926, he published the first survey of the cave (150 m long). In 1970, Planinets Caving Club surveyed the cave up to 1498 m, the next year – up to 2932 m. Further studies have been carried out by Strinava Dryanovo Caving Club, bringing the cave to its present length.

Fauna: five species known, including the troglobite beetles *Netolitzkya maneki* and *Duvalius (Paraduvalius) bulgaricus*.

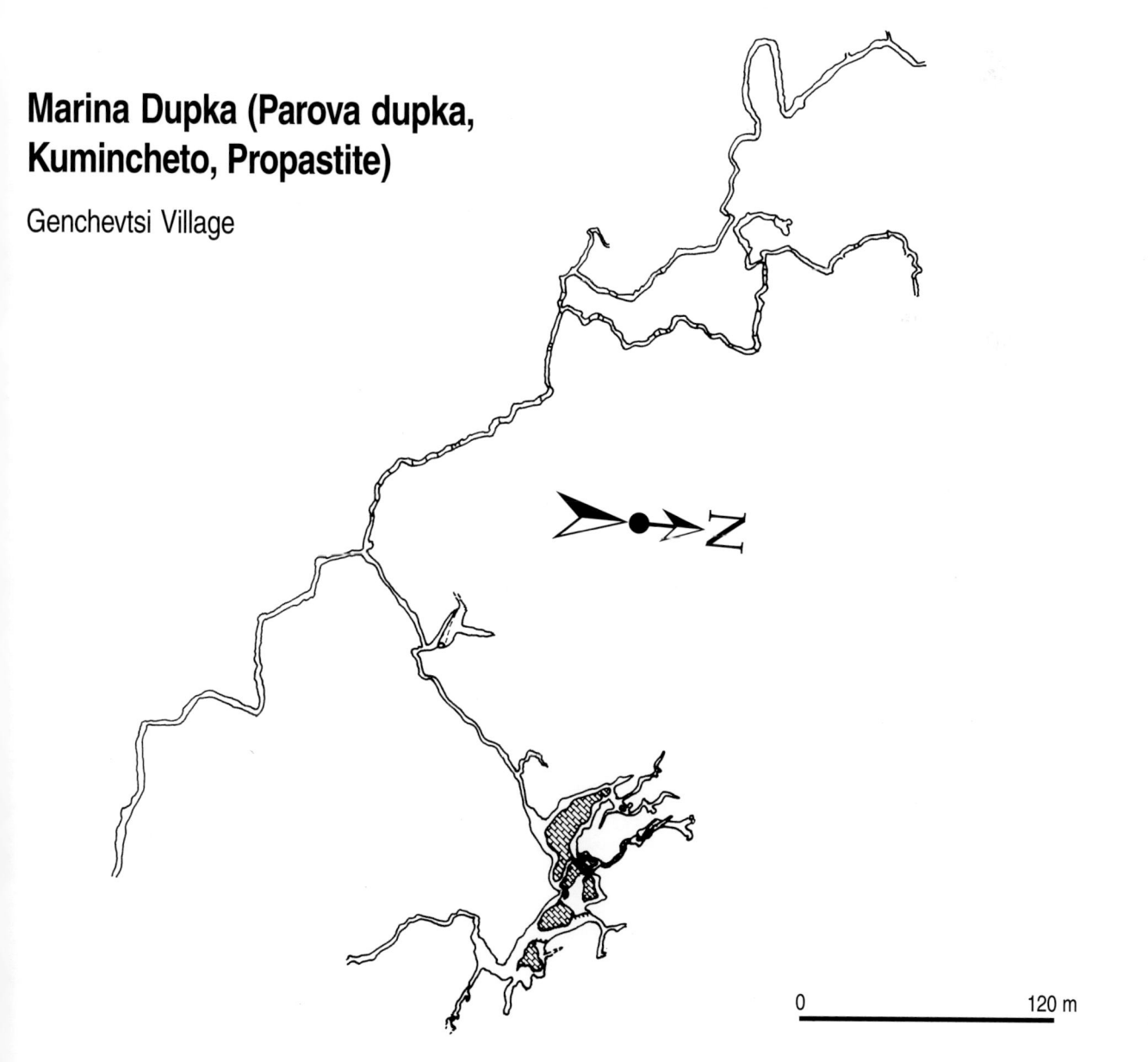

Marina Dupka (Parova dupka, Kumincheto, Propastite)

Genchevtsi Village

Name: **MARTIN 11 (10)**
Between the villages Prauzhda and Varbovo, Vidin Distr.
Denivelation -91 m

The cave pothole Martin-11 is situated on the southern rib of Belogradchik Anticline, between the villages Prauzhda and Varbovo, in one of the typical local hills "glami", called Klepats. It is an area between Prauzhda and Vârbovo, Belogradchik municipality. The name was given by members of Bel Prilep Belogradchik Caving Club after the expedition of 2000. Before that, the pothole had been unknown (during the first exploration, no traces of other visits were seen). About 300 m north of its entrance, in the monoclinal part of the Anticline, is the entrance of the better known Premenska Dupka.

The approach to the cave follows the road Belogradchik – Stakevtsi. The cave is about 10 km down this road, in a branch on the right to Prauzhda. In this place, on the right there is a dirt road, leading to the localities Proboynitza and Pragove. It is possible to drive on this road for another 1 km, then in 40 minutes the cave entrance is reached. If necessary, a guide from Bel Prilep Belogradchik Caving Club could be negotiated. The locality Pragove cannot be mistaken, as this is the lowest point between the two "glami" (limestone hills) Klepats and Golema Stena and here is the all-year source of Proboynitsa River.

After passing by the locality Pragove, about 100 m on the right, a cliff is formed. In the middle of the cliff, there is a footpath and about 2 m from the upper end of the footpath is the entrance of the pothole.

The pothole has 4 verticals and several inclined platforms. It is wet, but without running water. The entrance is facing north and is about 170 m above the local base of erosion. Immediately after the entrance there is a 24-meter deep vertical and we reach a chamber, covered with flowstone formations. After the chamber, by means of a vertical of 8.50 m, we get into the fist large chamber. After another vertical, a second big chamber is reached. Through a squeeze and a vertical we reach a third chamber and finally, after stone blocks, the bottom (siphon jammed by boulders and mud) is reached. The cave is rich in flowstone formations. This is the deepest pothole in the area. It is not equipped for SRT. The verticals are: I – 24 m, II – 8.50 m, III – 15 m and IV – 9 m.

No scientific observations have ever been carried out in this pothole, there is no information about its flora and fauna, as well.

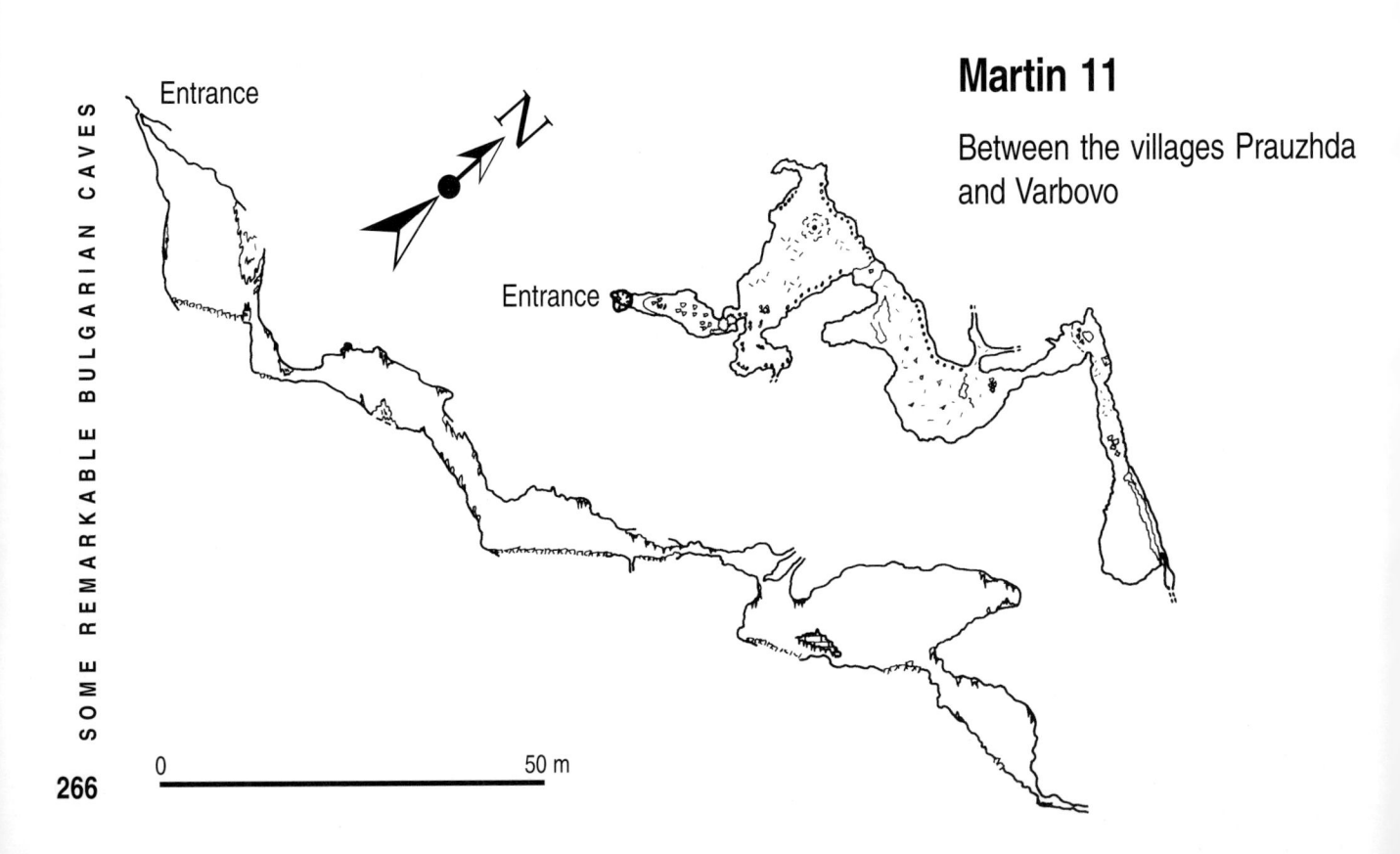

Martin 11

Between the villages Prauzhda and Varbovo

Name: **MAYANITSA (60)**
Tserovo Village, Sofia Distr.
Length 1426 m
Coordinates: N 43° 00' 02" E 23° 21' 35.6"

A water cave, SE of Zhelen Quarter on the right bank of the Iskar river, at the base of Tsarni Kamâk massif. Formed in limestone from the Middle Trias.

A horizontal, two-storey, branched cave. Cross section of the galleries: width 0.5–9 m, height 0.3–7.5 m. Rich in accumulative formations on the floor-sinters and others. An underground river with a permanent debit of ca. 2 l/s. and maximal ca. 30 l/s. runs through the cave. The waters bifurcate in the cave – both branches are piped). The cave water is used and the cave could be visited only with a permit. During the survey of the cave in 1991 the caver K. Valtchev discovered a way further. The cave was explored and surveyed by the middle of 1992 by M. Zlatkova, T. Medarova, V. Mladenova, St. Petkov, Ts. Ostromski and Z. Ivanov.

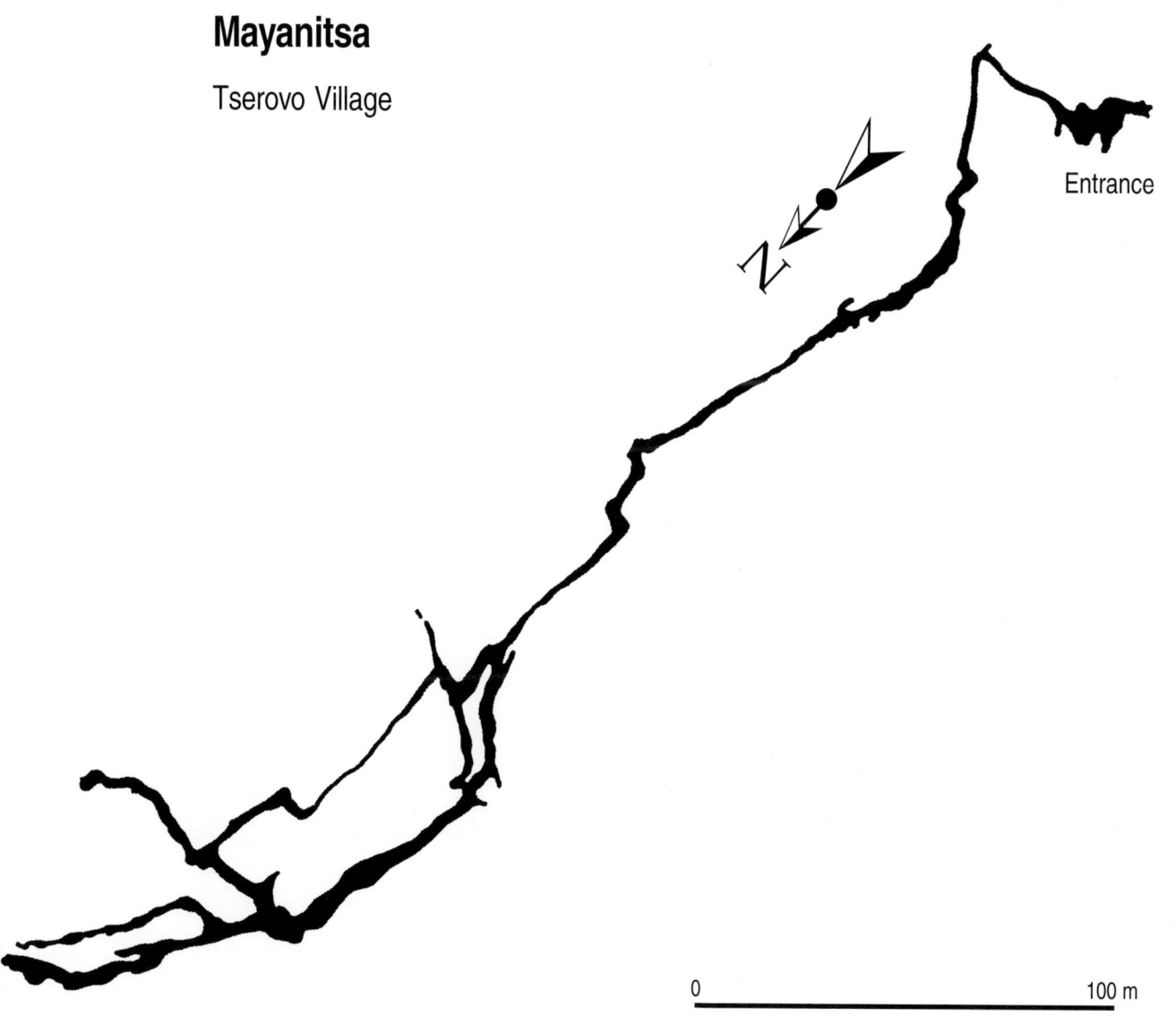

Mayanitsa

Tserovo Village

Name: **MAZATA (HAYDUSHKIYA ZASLON) (121)**
Hristo Danovo Village, Plovdiv Distr.
Length 367 m. Denivelation -11 m
Coordinates: N 42° 45' 59.7" E 24° 36' 01.4"

A cave in Troyan Balkan, 2 km west of Troyan Pass.

The entrance (3 x 8 m) is in a cliff. The cave is horizontal, with permanent water. The galleries are 1–3 m wide and 1–8 m high. In some places the cave is branched, at the 70th meter there is a second floor, about 40 m long. There are about 20 sinter lakes. The floor of the galleries is covered with clay and small gravel, in some places there are also bigger boulders. The end parts of the cave are the richest in flowstone formations (stalactites, stalagmites, drums, sinter lakes). The cave ends in a narrowing.

The name of the cave has to do with a local legend, according to which the cave served as a shelter to Velcho Vrazhaliyata and other local "hayduti" (kind of guerrilla).

The first study and survey of the map were done in 1971 by A. Ivanova and P. Vassileva from Planinets Sofia Caving Club. In 1976 M. Krastev and N. Dichev from Stratesh Lovech Caving Club made a horizontal and vertical survey of the cave and recorded the legends connected with it.

Fauna (explored by P. Beron and T. Ivanova): among the animals collected are *Lithobius rushovensis* (Chilopoda) and several species of bats (Chiroptera).

Protected Natural Monument, together with 0.4 ha of adjacent land (Decree 206 /March, 23,1981).

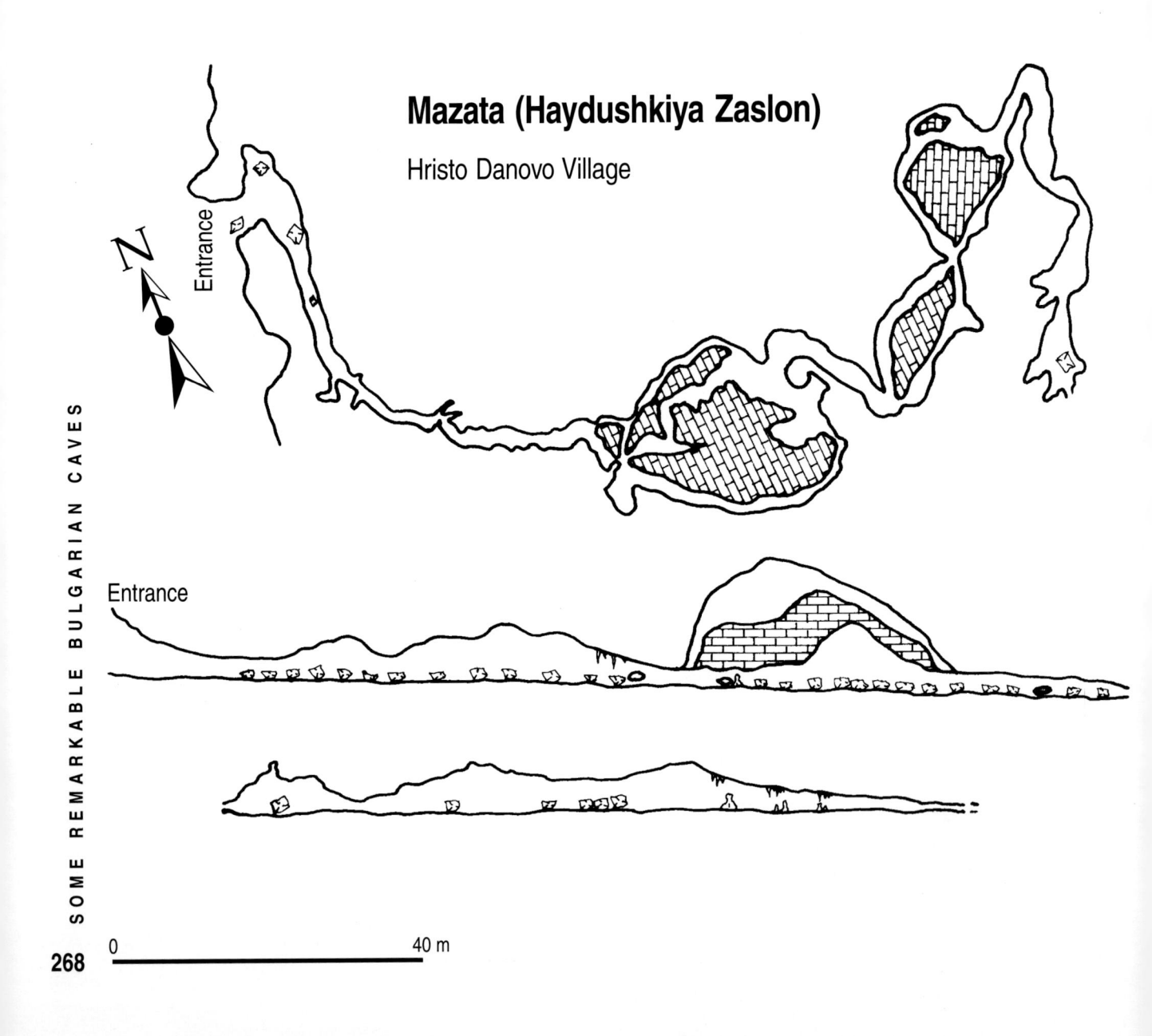

Name: **MÂGLIVATA (184)**
Kotel Town, Sliven Distr.
Total length 1244 m. Denivelation -220 m
Coordinates: N 42° 54' 54" E 26° 29' 38"

A cave pothole in Senonian limestone in the locality Zlosten, 3.3 km northeast from Kotcl. Length on the main axis 630 m. A cascade cave with eleven verticals of different depth, the biggest being 30 m. Poor in flowstone. Outflow of the underground stream 5–6 l/s. T° 10 °C. A siphon in the lowest part of the cave.

Discovered and surveyed up to 48 m by Akademik Sofia Caving Club in 1963/1964. In 1979 the new parts of the cave were found. Explored by Caving Clubs from Aytos and Sliven.

Protected by Decree 234/4.04.1980, together with an adjacent area of three ha.

Fauna: five species known, no troglobites.

Mâglivata

Kotel Town

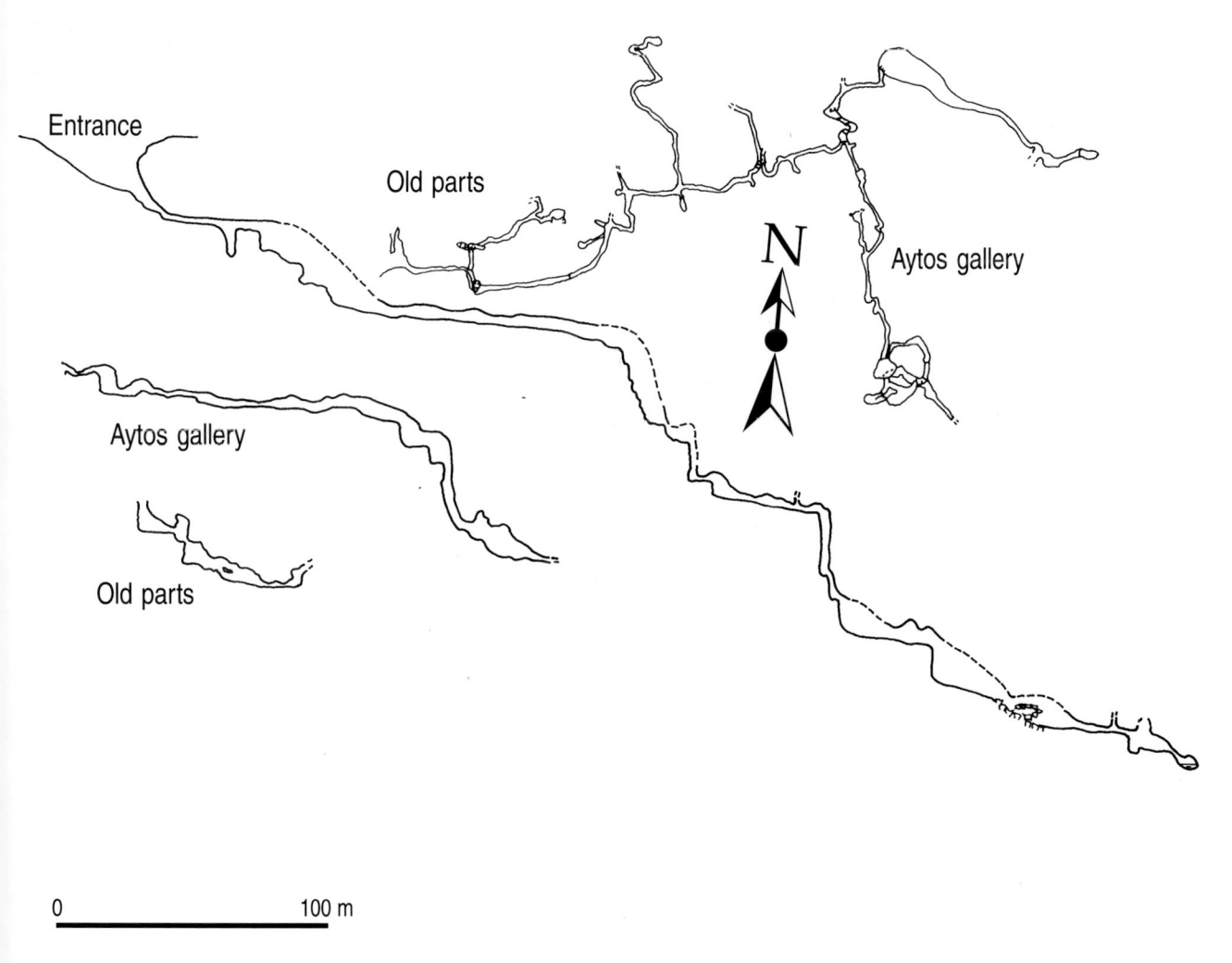

Name: **MECHATA DUPKA (58)**
Zhelen Village, Sofia Distr.
Altitude 910 m. Length 564 m. Denivelation -11 m
Coordinates.: N 43° 00' 18.4" E 23° 24' 40.9"

A cave 3.2 km NE of Zhelen, at the base of a cliff in the upper part of the massif Koev Kamak, Golyama Planina, within Treskavets Protected Area. The cave is formed in Trias limestone of the Svoge Anticline. Developed along a N – S fault.

There are 4 entrances facing N-NE. Usually, one enters via the largest entrance (irregular rectangle 3 x 2 m). It leads into a gallery, 16 m long and ca. 5 m wide, with two branches (first on the right, then on the left), then follows the first and the largest chamber in the cave (27 m long, 9 m max. width, 6 m max. height). In the north end of the chamber there is an elliptic sinter lake (4 x 5 m). Beyond the lake, a spacious but very low chamber with several branches starts. Many sinter basins and cascades and two small lakes are typical for it. Two galleries, ending in two of the cave entrances, start from the northern part of the chamber. Another narrow gallery, high up to 1 m, leads to a chamber, which is 20 m long, 5.5 m wide and 4 m high. The upper storey is reached through a wide crevice and a labyrinth of low and narrow galleries, ending in mud siphons.

In the middle of the left branch of the entrance gallery, there is a small triangular opening, leading into a narrow gallery. Then, we reach a small chamber and, through a crevice on the floor, after a 5-meter descent, we arrive in a beautiful small chamber. A small stream is seen, it's the water running through the cave Izvornata peshtera, 10 m lower.

Fossils from *Ursus spelaeus* have been found in the cave by N. Petkov.

The cave was known to local people. It was explored in the 50s, when biospeleological studies took place. The first map and description were published by N. Korchev and M. Chakârov in 1960 (Tourist, 7: 20–22). The actual map of the cave was done by T. Stoychev, St. Tsonev and A. Jalov from Aleko Sofia Caving Club in 1983 (description by A. Jalov, based on own observations and data from N. Korchev and M. Chakarov).

Fauna: so far, 23 animal species have been recorded from this cave, including the troglobites *Duvalius pretneri* (Coleoptera), *Lithobius lakatnicensis* (Chilopoda), *Typhloiulus bureschi* (Diplopoda), *Paranemastoma (Buresiolla) bureschi* (Opiliones) and the stygobite *Speocyclops lindbergi* (Copepoda).

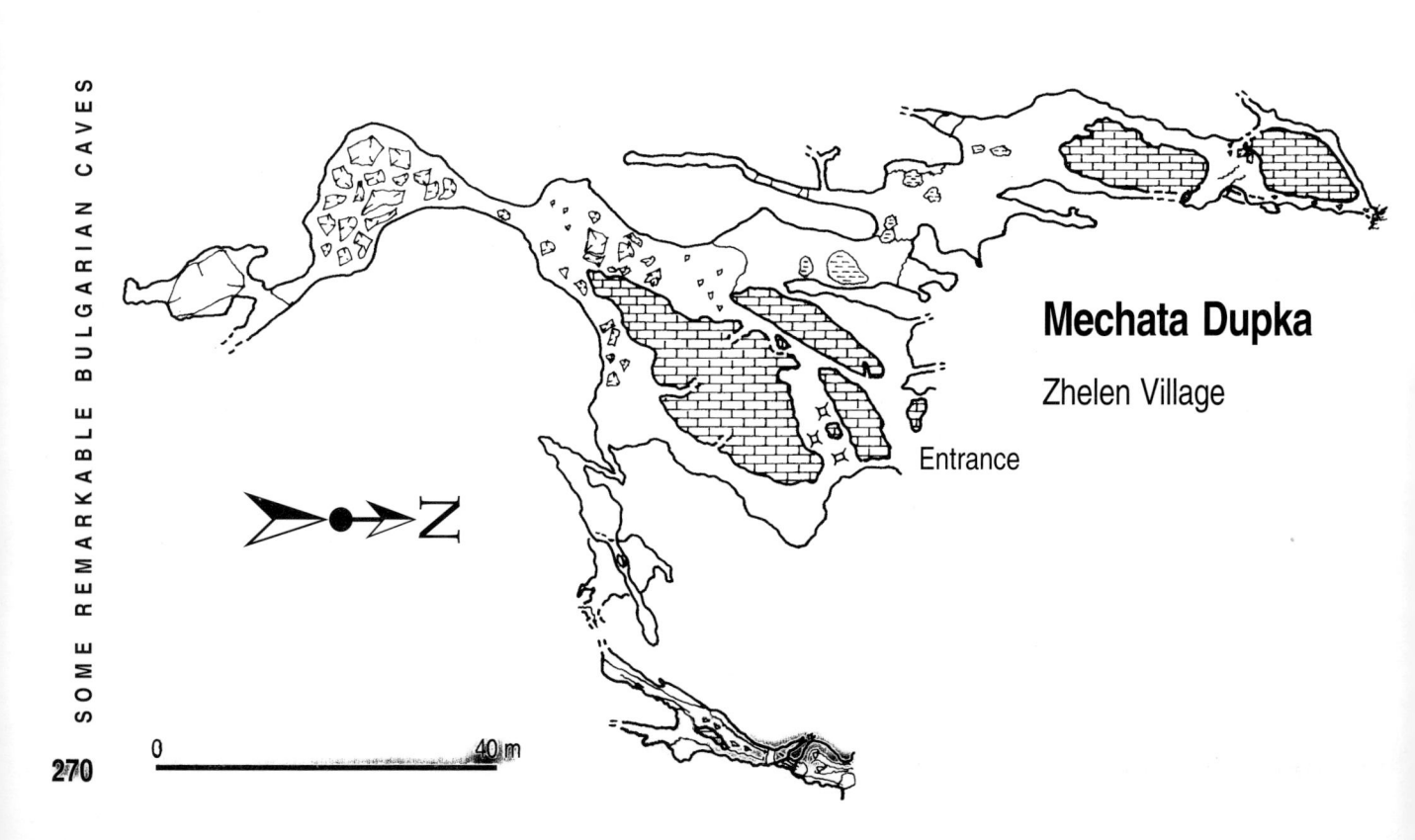

Name: **MISHIN KAMÂK (KAMIK) (20)**
Gorna Luka Village, Montana Distr.
Length 695 m. Denivelation 19 m (+ 4.2; -14.8 m)
Coordinates: N 43° 27' 46.4" E 22° 53' 11"

A cave on the right bank of the river Prevalska Ogosta, Western Predbalkan. A horizontal, two-storey, branched cave. There is an 80-m long entrance gallery, 8-9 m wide and 4–7 m high. In the middle of the gallery the entrance to the lower parts opens. The southern branch is long 155 m and wide from 3.5 to 85 m; its average height is 2.5 m, and it is divided by a massive stone.
Natural Monument (Decree 2634/21.09.1962), with 0.5 ha adjacent area.
Fauna: 25 species known, including the troglobites *Trichoniscus anophthalmus* (Isopoda) and *Serboiulus spelaeophilus* (Diplopoda).

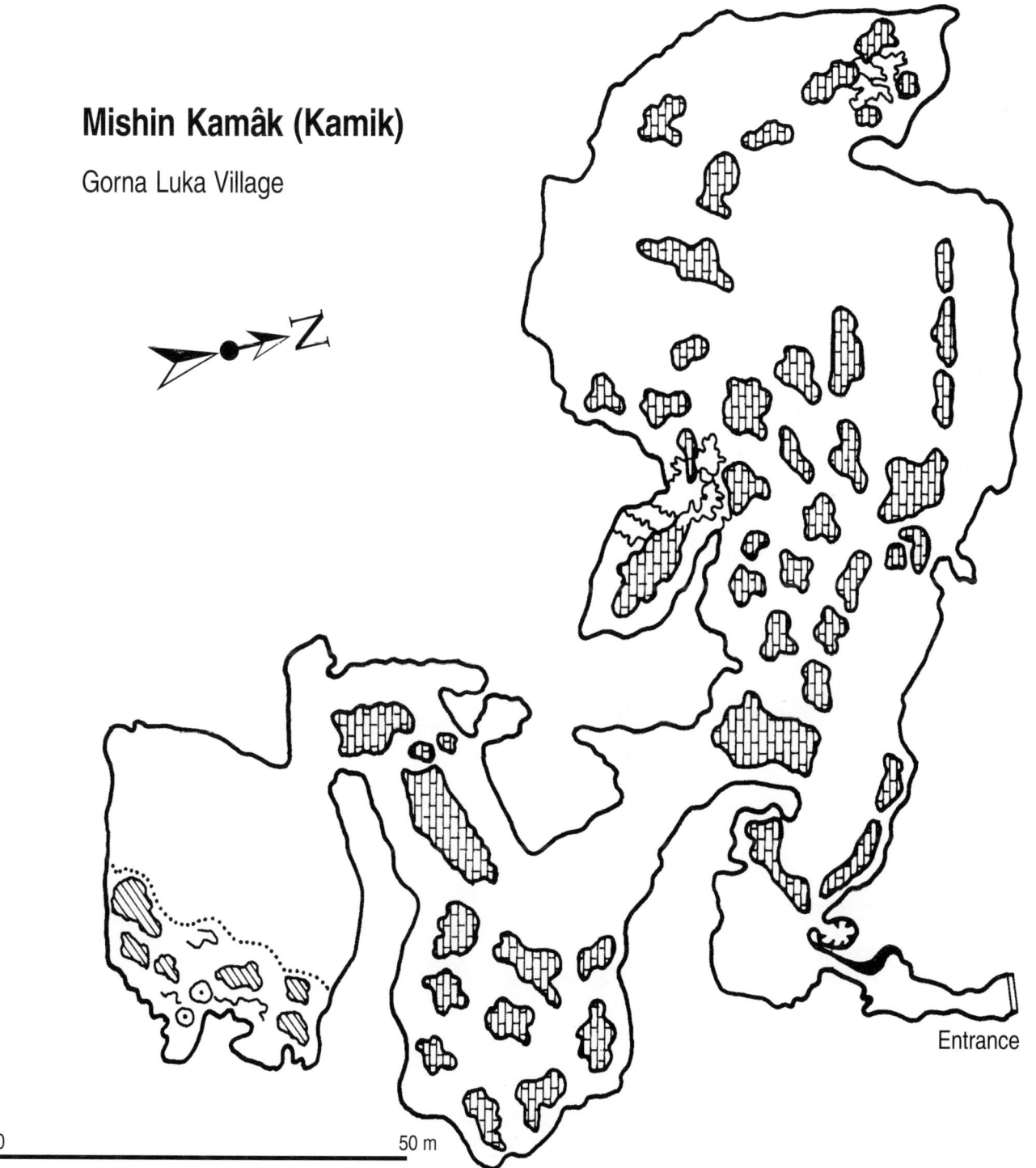

Name: **MIZHISHNITSA (40)**
Vratsa Town
Length 855 m. Denivelation -85 m
Coordinates: N 43° 12' 29.6" E 23° 28' 17.5"

A cave in Stresherski part of Vratsa Mountain, in the valley Mizhishnitsa, on the left of the footpath from Ledenika Hut to the Barkite Region. Situated in the natural reserve Vrachanski Karst.

The cave is developed in the contact zone between Middle and Upper Jurassic, in Upper Jurassic limestone, on a SE-NW fault. It collects the water from the nearby funnels (outflow 2–3 l/s). The galleries are descending and narrow, with an underground stream, starting from a side gallery, 30 m from the cave entrance. About 120 m from the entrance, on the right, there is a side gallery of the same length, with a stream joining the main river (description: T. Daaliev after M. Zlatkova).

Discovered through digging in 1988 by cavers from Dr Ivan Buresch Caving Club. Surveyed also by them (main surveyor M. Zlatkova). The exploration of this cave is not finished, continuation could be expected.

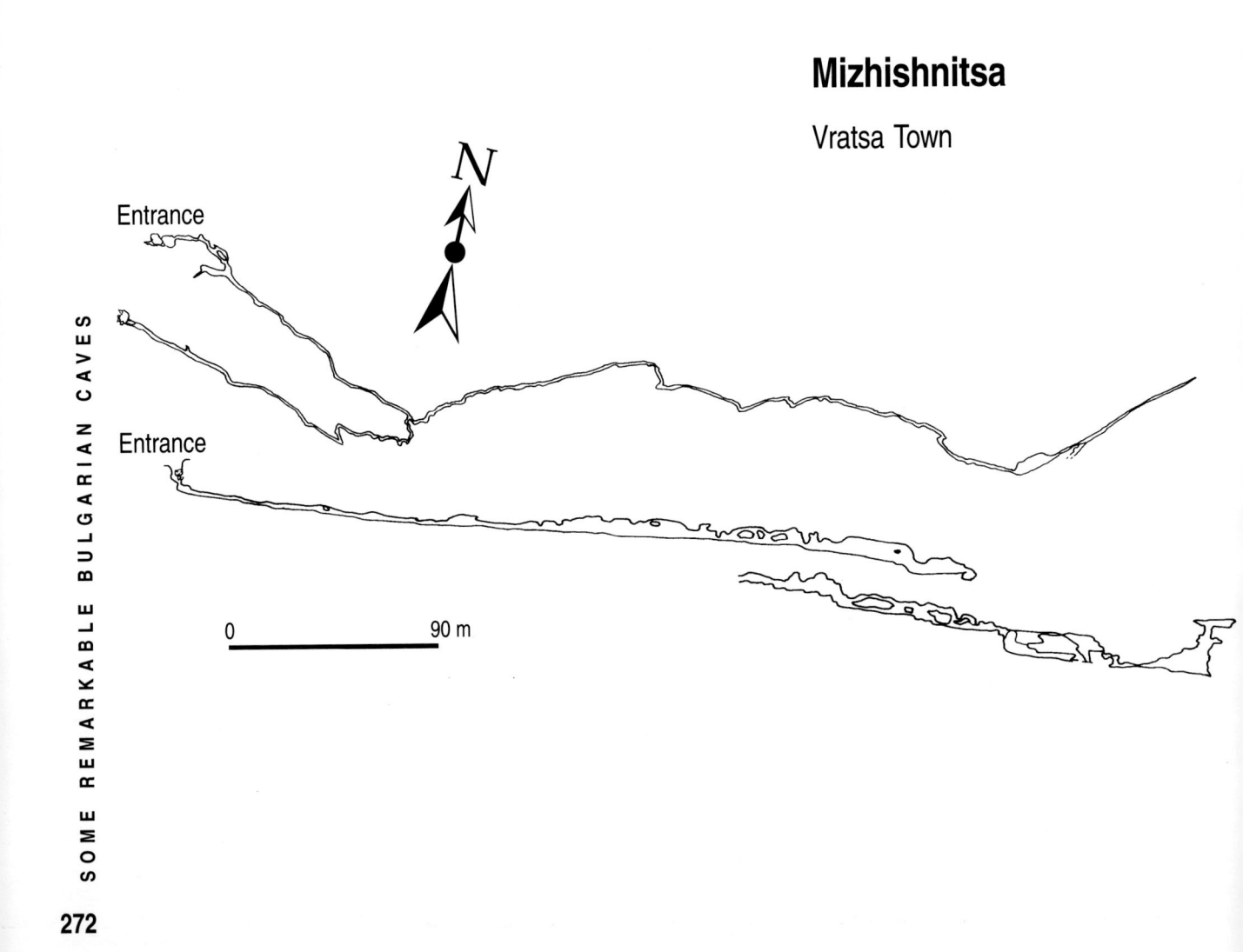

Mizhishnitsa

Vratsa Town

Name: **MLADENOVATA PESHTERA (24)**
Chiren Village Vratsa Distr.
Length 1723 m. Denivelation -20 m

A cave-sinkhole in a funnel in the locality Lipova Dramka ca. 4 km southeast of Chiren. It starts with a 15-m steep funnel, leading into a horizontal cave. From the very beginning, small, not very deep lakes start, forming a slow stream at the 150th meter. The low ceiling of the cave creates conditions for the formation of semi siphons (dangerous places when in flood). In many places there are flint concretions. Mladenovata Peshtera is a one-gallery cave with an underground stream. Few flowstone formations, many big sinter lakes.

In this cave many bones of extinct mammals have been found – elephants, reindeer, auroche, wild boar, wild horse. So many bones are rarely found in Bulgarian caves – a real paleontological museum.

The cave was discovered, explored and surveyed by Hristo Mladenov, Vladimir Fedyushkin and other cavers from Vratsa in 1962, up to 1500 m. In 1977, during the International caving expedition Ponora-77, a precise map was made by N. Landjev and Zdr. Iliev. For the first several hundred meters, P. Tranteev also participated in the surveying. In 1977 an experiment was done, which proved that the water of Mladenovata Peshtera joins the water in Ponora – the longest cave in the area. The fluorescein, released at 11.00 h on August 20, 1977 in Mladenovata Peshtera, appeared in the waterfall of Ponora at 11.00 h on August 22, 1977 (48 hours). No attempt at penetrating the siphons has ever been made.

Surveyed again in 1986 by K. Karlov, V. Peychev, N. Gergov and K. Kanev from Vratsa Caving Club. Description by T. Daaliev, according to D. Sabev, P. Tranteev and N. Landjev.

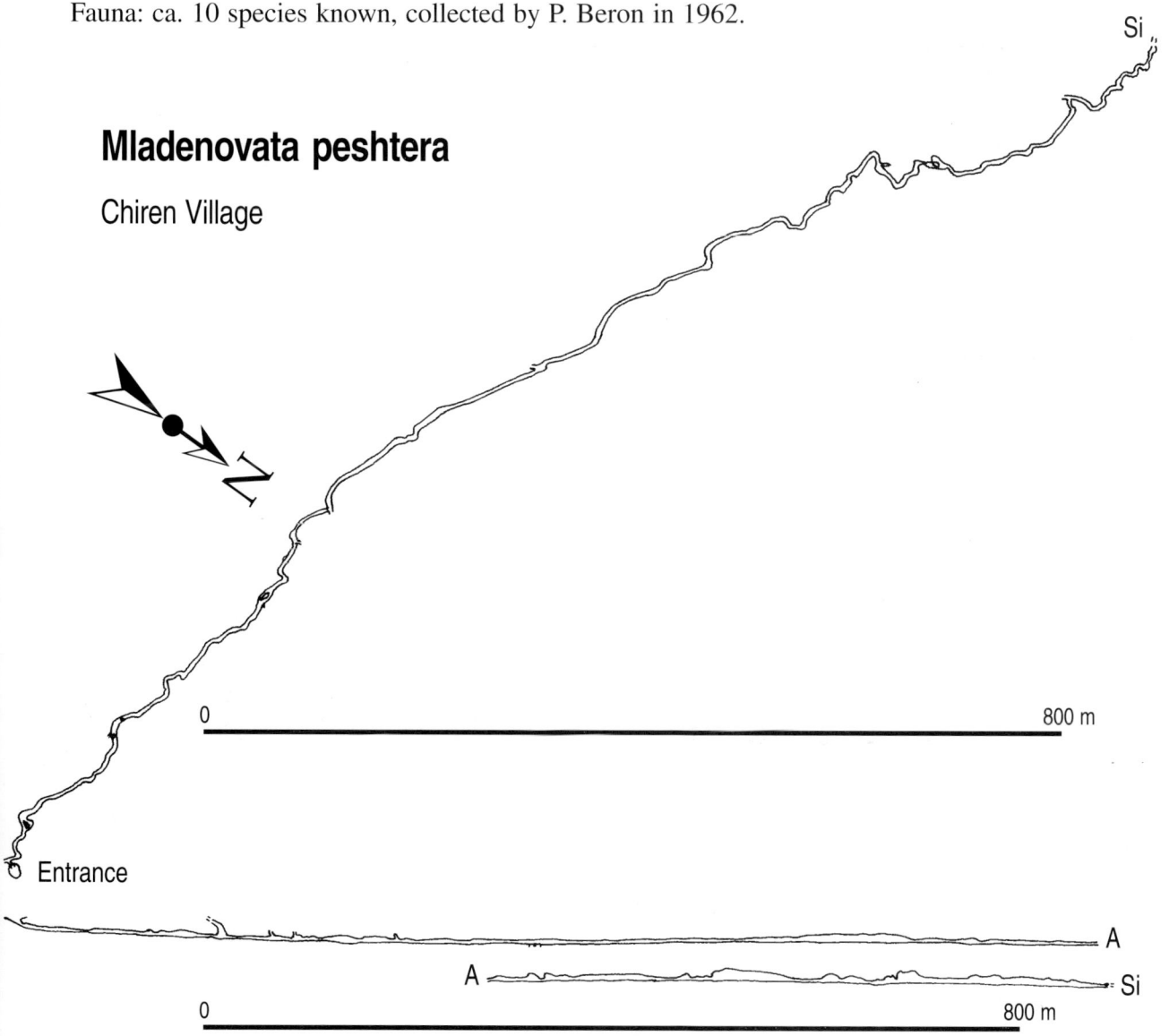

Fauna: ca. 10 species known, collected by P. Beron in 1962.

Mladenovata peshtera
Chiren Village

Name: **MOROVITSA (113)**
Glozhene Village, Lovech Distr.
Length 3200 m. Denivelation 105 m
Coordinates: N 42° 57' 55.8" E 24° 11' 16.4"

A cave SW of Glozhene, two km from the monastery St. Georgi Pobedonosets, in the locality Morov Dol. A two-storey cave in Upper Jurassic (Titonian) limestone.

The entrance faces north, 15 x 15 m. The cave starts with a large gallery, where archaeological excavations have been done. A vertical of six meters leads to the lower part of the cave. The lower part goes under the upper part in northern direction. After 130 m, a 57m deep pitch reaches the lower parts. The pitch is divided into two parts – of 27 and 30 m, and ends with a 15-m scree of bat guano. We find ourselves in a gallery bifurcating into ascending and descending galleries. The descending one is broken by a four-meter vertical. The gallery then becomes wide and is covered with boulders, bat guano and many formations, sometimes gypsum flowers.

According to the legend, in the time of the Second Bulgarian State the local people used to hide in the cave. Once, invaders walled up the entrance and those hidden in the cave died of suffocation. Since then the cave was called Morovitsa, from the word "mor" (death).

In 1912, the renown archaeologist Prof. R. Popov discovered in the cave artefacts from the Eneolith and "inscriptions" from the time of Tzar Ivan Shishman. Until 1967, only 320 m of the cave had been explored. In the same year, N. Genov, I. Rashkov, A. Bliznakov, V. Nedkov and other cavers from Planinets Sofia explored and surveyed the entire length of the cave. In 1982, the Archaeological museums in Lovech and Sofia carried out archaeological excavations, together with cavers from Iskar Sofia Caving Club. Tourists, who do not have special training and equipment, can visit only the first 320 m.

Protected by Decree 2810/10.10.1962, together with the adjacent five ha.

Fauna: *Paranemastoma radewi* (Opiliones, troglophile). Many other animals remain under study.

Morovitsa

Glozhene Village

The preparation for penetration – photo Trifon Daaliev

Group of stalagmites – photo Trifon Daaliev

Stalagmites – photo Trifon Daaliev

Name: **MUSSINSKATA PESHTERA (159)**
Mussina Village, Veliko Tarnovo Distr.
Length 532 m, Denivelation 6 m
Coordinates: N 43° 09' 02.1" E 25° 25' 41.8"

A cave by the village Mussina, at the source of Mussinska Reka. There is an old water mill at the entrance. There, we find an octagonal water catchment from where the water was piped to Nikopolis ad Istrum, a town built by Emperor Trayan to commemorate his victory over the Dacians. The cave has two grilled entrances. Through the lower one runs water. The upper part is dry and with electric light for local visits. It ends with the Blue Lake, the siphon of which was penetrated by V. Nedkov. The siphon is 30 m long and 5–6 m deep. Its end reaches a main water gallery (with a parallel dry gallery), which has a total length of 70–80 m. A new siphon is reached, 30 m long, penetrated in 1986 by Milen Dimitrov and Ilko Gunov. In the same year, Ilko Gunov and Valentin Chapanov surveyed another 150 m, bringing the cave length to 532 m. The cave is still not completely explored.

Another part of the cave contains a lake, 1.20 m deep, covered with concrete plates for preventing the water to drain. Another gallery goes parallel to the show gallery. After 20 m we descend into a 3-m vertical with a gallery, leading to the underground stream and then to a siphon.

The cave is almost without cave formations, in the show part there are fragments of ceramics and bat guano. Air T° 13°C, water T° 14.1°C. (description after A. Strezov and T. Daaliev, 1978).

The cave was explored in 1988 in order to build a trout hatchery. Participants: A. Strezov, V. Peltekov, Iv. Lichkov, L. Adamov, L. Popov, P. Evtimov, T. Daaliev.

Fauna: no troglobites known.

Mussinskata peshtera
Mussina Village

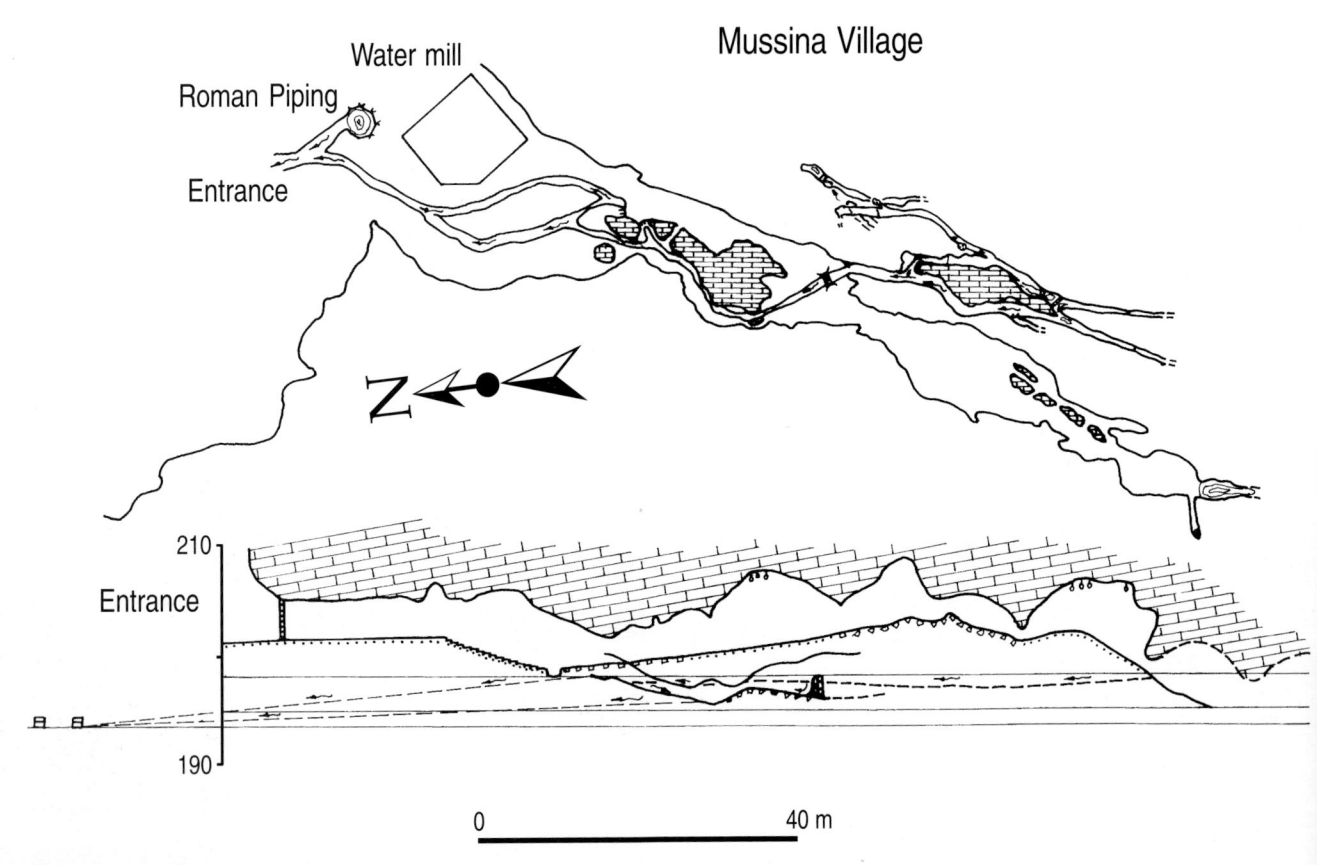

Name: **NANOVITSA (110)**
Yablanitsa Town, Quarter Nanovitsa, Lovech Distr.
Altitude 526 m. Length 200 m. Denivelation 102 m
Coordinates: N 43° 04' 31.7" E 24° 16' 46.5"

A pothole, 12 km from Yablanitsa on the asphalt road to Nanovitsa quarter. About 15 m west of the quarry is the small funnel with the entrance of the pothole (120 x 0.45 cm).

The pothole was shown to some cavers from Sofia in 1989 by local people. By that time only 8 m had been known, the way further being blocked by boulders. Until 1990 there had been several unsuccessful attempts to penetrate the obstruction. Then, a big group of cavers from the Army Caving Club managed to displace the boulders and to open the way down. Cavers from several clubs from Sofia and Pleven had surveyed the pothole by 1995 (Main surveyor Ina Barova).

After a 3-m chimney, a steep slope follows, where there is danger of falling stones. A total denivelation of 50 m (from the entrance) follows. We reach a chamber of 9 x 5 m. A series of verticals follow, the left leads to a depth of – 90 m, and the right – into ascending chambers with many flowstone formations.

The verticals have been equipped for SRT.

The main attachment in the first vertical of – 11 m is with a spit on the right. The second vertical (8 m) could be scaled with a ladder. After the verticals the gallery continues eastward, lowers and a narrow place with a pool is reached. Then the ceiling becomes higher and a chimney, 4 m wide, is reached. Narrow diaclases lead to a wide gallery with a branch on the right. Another 3m deep narrow place at the end of the gallery leads to the bottom of the pothole at – 102 m. A muddy stream disappears under the wall of the end gallery. A digging there allowed to reach the end parts of the pothole, but the entire map is still to be worked out. A continuation is to be expected.

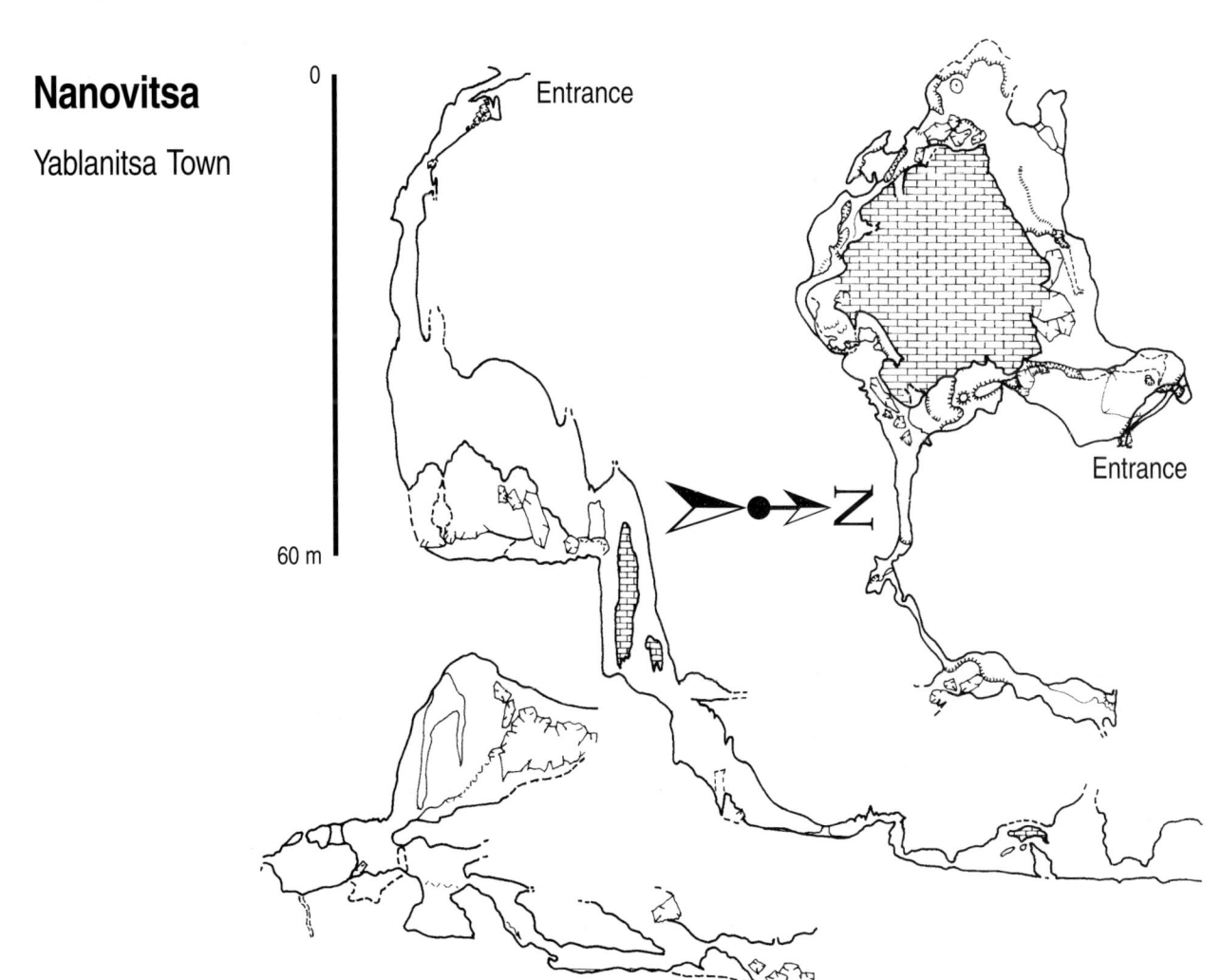

Name: **NEPRIVETLIVATA (Gornata Propast) (7)**
Belogradchik Town
Altitude 820 m. Length 158 m. Denivelation 79 m

A pothole in the area Venetsa. First descent – in February 1960 by a group, including A. Leonidov, N. Korchev, P. Beron, T. Michev and others. The map was made in 1979 by A. Leonidov, N. Gaydarov, V. Nenov, cavers from the caving club of Belogradchik. The name Neprivetlivata was given in 1965.
Fauna: the troglobites *Bulgaroniscus gueorguievi, Trichoniscus bononiensis* (Isopoda), *Beronia micevi* (Coleoptera), the bat *Myotis bechsteini* (Chiroptera).

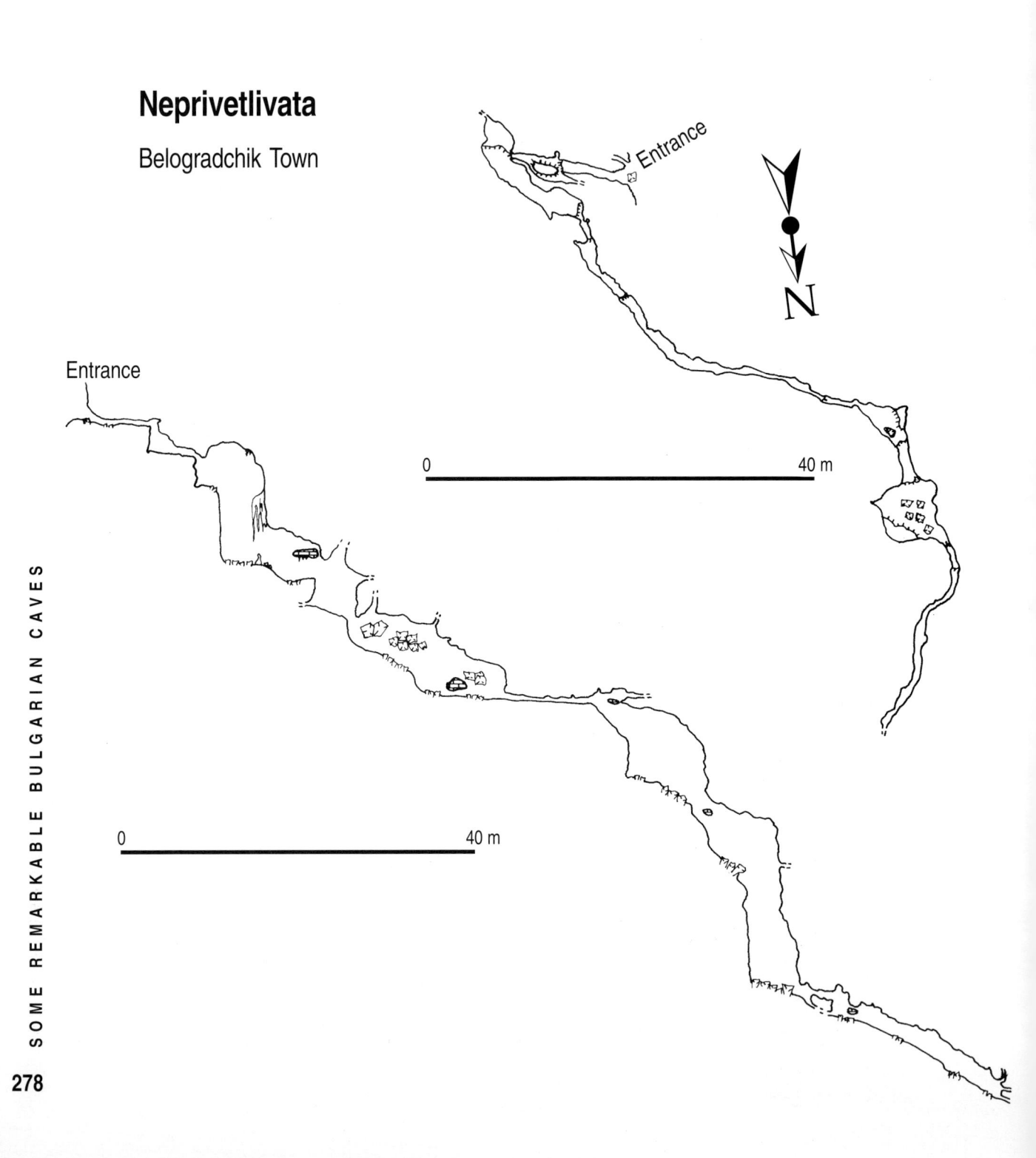

Neprivetlivata

Belogradchik Town

Name: **NEVESTINA PROPAST (41)**
Vrachanska Planina, Vratsa Distr.
Depth -76 m

A pothole ca. 3 km S–SW of Ledenika Hut, close to the locality Koloniite. The entrance has dimensions 3x2 meters. It is a large shaft with smooth walls. About 45 m from the entrance, there is a small platform, covered with leafs and branches. At about 70 m, there is a second platform, followed by a 5-m vertical. The pothole ends in a small lake. In the spring, the water from the melting snow flows in it. In the lower third, all around the year there is dripping water.

The name of the pothole comes from a legend about a young woman ("Nevesta", or bride), thrown in it, but on the bottom no remnants of humans were ever found.

Fauna: two species of troglobites – *Bulgarosoma bureschi* (Diplopoda), *Pheggomisetes globiceps stoicevi* (Coleoptera, Carabidae).

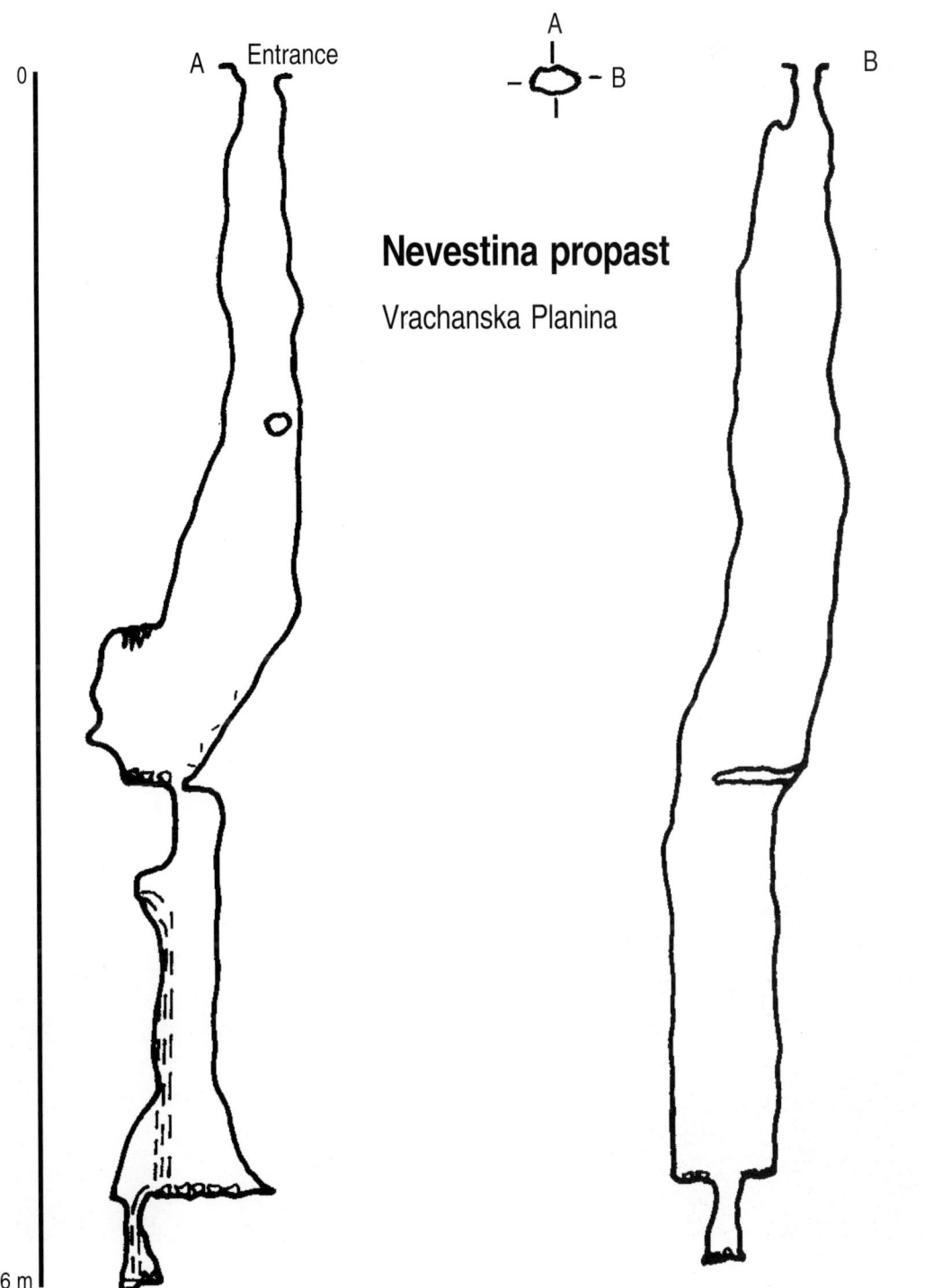

Nevestina propast

Vrachanska Planina

V. Mustakov deviate the stretcher during the cave rescue training – photo Vassil Balevski

The medows near the pothole Nevestina Propast in 1987 were the scene of tent camps – photo Trifon Daaliev

Name: **OGRADITE (S-20) (86)**
Karlukovo Village, Lovech Distr.
Altitude 235 m. Denivelation -97 m
Coordinates: N 43° 11' 25.4" E 24° 03' 12.2"

A pothole on the left (geographically) bank of Iskâr River, in the area of Karlukovo, ca. 1.5 km NW of the old Railway Station (45 min. walk on the dirt road to Resselets Village). The entrance is situated on the bottom of a steep funnel, on the left of the road, and 100 m above the LBE (Iskâr River). The pothole is branched and periodically serves as a ponor with a maximum width of 17 m and a minimum width of two meters. From the funnel a vertical of 15 m follows, then several meters of inclined platform and a 60m vertical leads to the bottom, from where through a narrowing two verticals, two meters each, are overcome. An underground river with an outflow of 2–3 l/s follows for about 6-7 m. Upstream and downstream, the pothole ends in narrow siphons. Secondary formations, represented by stalactites, stalagmites, stone "waterfalls" and sinter lakes are seen along the vertical.

Formed in Maastricht (Cretaceous) limestone.

Discovered and explored in 1983 by cavers from Studenets Pleven Caving Club. In the same year, A. Jalov and others dug out the bottom of the pothole in its upper end and discovered a prolongation, which led them to the underground river. The actual map was made in 1984 by K. Petkov, V. Dinalov and S. Gazdov from the same club. Description by S. Gazdov and K. Petkov.

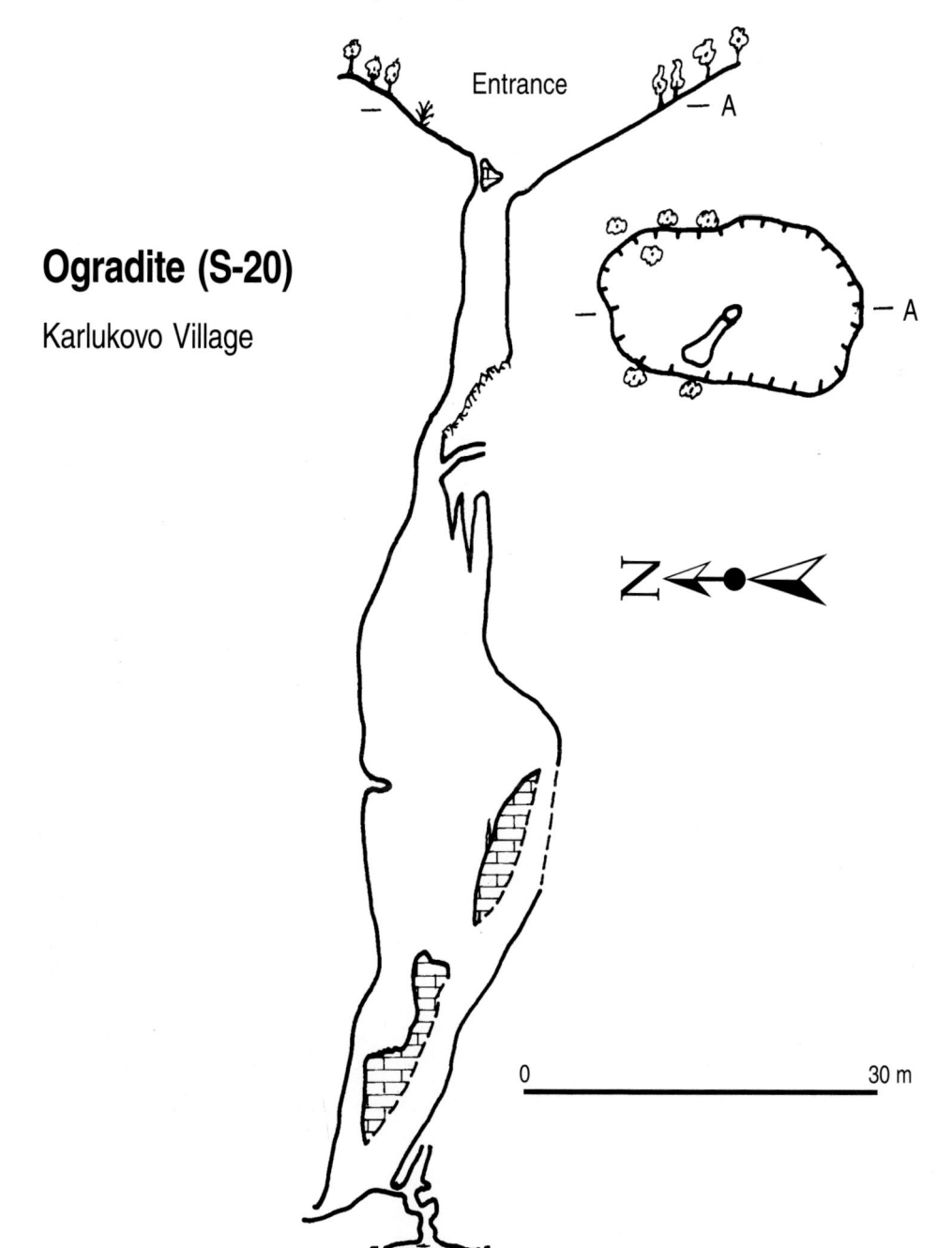

Ogradite (S-20)

Karlukovo Village

Name: **OPUSHENATA ("Smoked") (131)**
Cherni Ossam Village, Lovech Distr.
Altitude 1200 m. Length 232 m. Denivelation -32 m
Coordinates: N 42° 45' 13.2" E 24° 41' 38"

A cave with two entrances in the Biosphere Reserve Steneto. The entrances are situated at various altitudes (1187 and 1200 m) on the right (geographically) side of the valley Golyam Chaushov Dol, which ends in a 60-meter vertical in the locality Babina Pizda. Formed in Trias limestone and dolomite.

We can reach the lower entrance starting from Chaushov Dol, 20 m before the vertical, at a small valley on the right side slope. About 10 m on the left is the lower entrance. The upper entrance is about 120 m from the lower, on the top of the cliff facing the valley of Kumanitsa river (the upper part of the river Cherny Ossam). This entrance, despite its considerable dimensions, is difficult to find. A descending wet cave. The gallery is 2 m wide and 2 m high. The floor is covered at the beginning with fallen leafs, then with stone and clay. About 70 m from the entrance, a gallery branches off, leading to two "windows" in the cliff under the upper entrance. The connection between the two levels is a 30-m shaft.

The cave is poor in flowstone formations but in places stalactites, stalagmites and dendrites are seen. It is inhabited by bats.

The cave had been known to local people. Explored in 1969 by Kino Raykovski, Nayden Kereshki and Vassil Markov from Troyan Caving Club. They prepared a horizontal map of the cave and named it "Opushenata", because inside the gallery is fumigated. In the framework of the International Caving Expedition Steneto '72 the map was corrected, a vertical plan was prepared and the cave was described by K. Raykovski and A. Pencheva.

The cave is situated within the area of Central Balkan National Park.

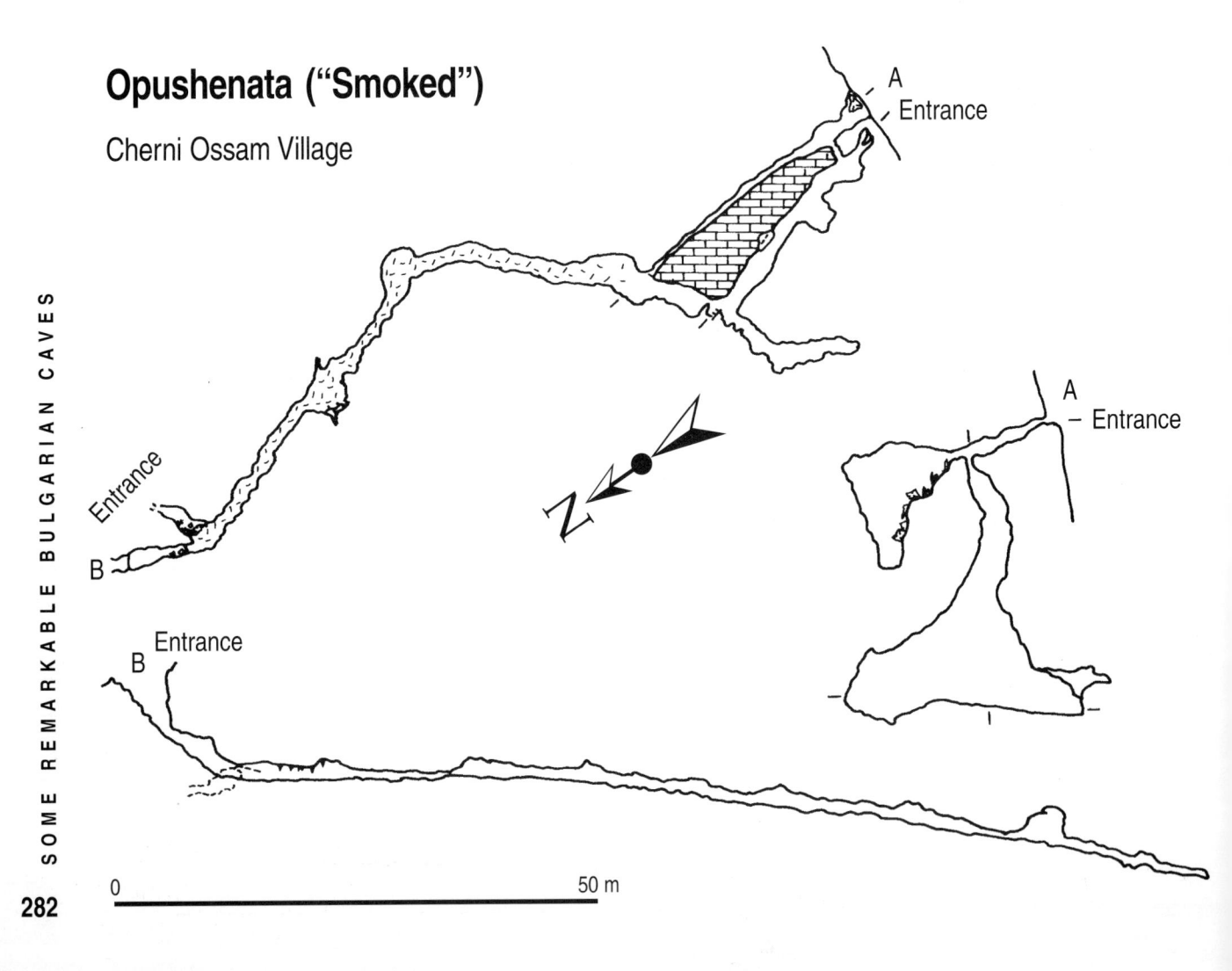

Opushenata ("Smoked")

Cherni Ossam Village

Name: **PANCHOVI GRAMADI (46)**
Zverino Village, Vratsa Distr.
Altitude 1005 m. Length 93 m. Denivelation -104 m
Coordinates: N 43° 07' 30.4" E 23° 34' 38.6"

The pothole starts with a 20-m deep pitch, lit to the bottom. A vertical of five meters follows and a platform – 3 x 3 m. One meter before the bottom, a very small opening leads to the biggest vertical of 40 m. Two ways follow until we reach a bottom gallery, 50 m long and four meters wide. The attempts to go further by digging have been unsuccessful.

The pothole was discovered in March 1968 by the cavers T. Daaliev and S. Zanev, members of Edelweiss Sofia Caving Club. After several attempts, the bottom was reached by T. Daaliev and D. Mihailov, helped by N. Gladnishki and Z. Iliev.

Panchovi gramadi

Zverino Village

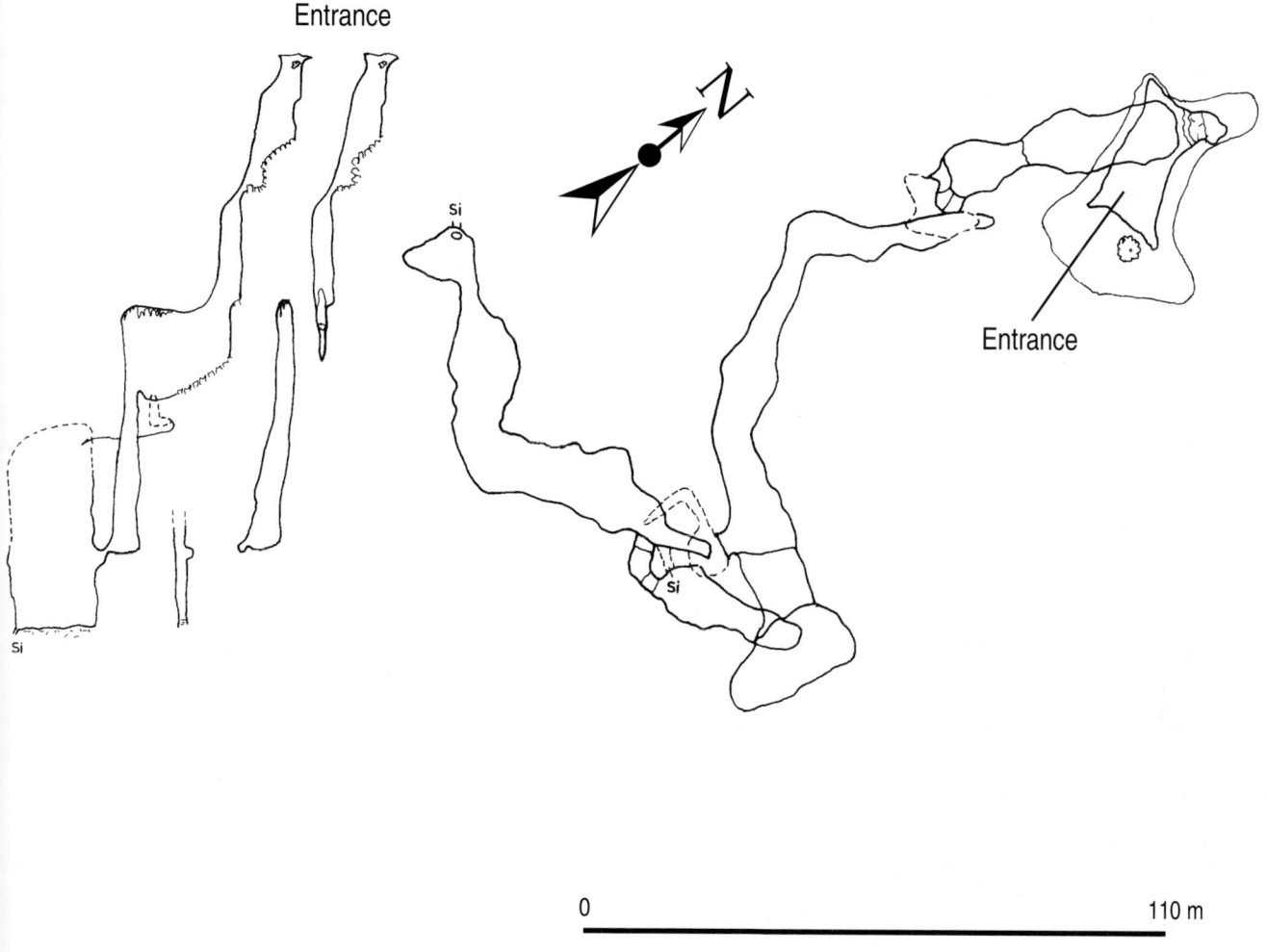

The entrance of the pit – photo Ivan Lichkov

Young researchers guided by shepherds to caves (1966).

Name: **PARASINSKATA PROPAST (18)**
Beli Mel Village, Montana Distr.
Total length 272 m. Denivelation -22 m

A cave pothole by a temporary stream, which, when there is rain, sinks to 80 m from the cave and runs through it. An entrance pitch of 13 m. The cave is branched, periodically with water and has several storeys. Sinter lakes, flowstone. It is about 300 m away from a water piping in the locality Parasinyako. Formed in limestone from the Upper Jurassic (Titonian) of Salash Syncline.

Studied for the first time by P. Tranteev, P. Beron and Z. Iliev in 1973.

Fauna: six species known, including the troglobites *Trichoniscus anophthalmus* (Isopoda) and *Serboiulus speleophilus* (Diplopoda).

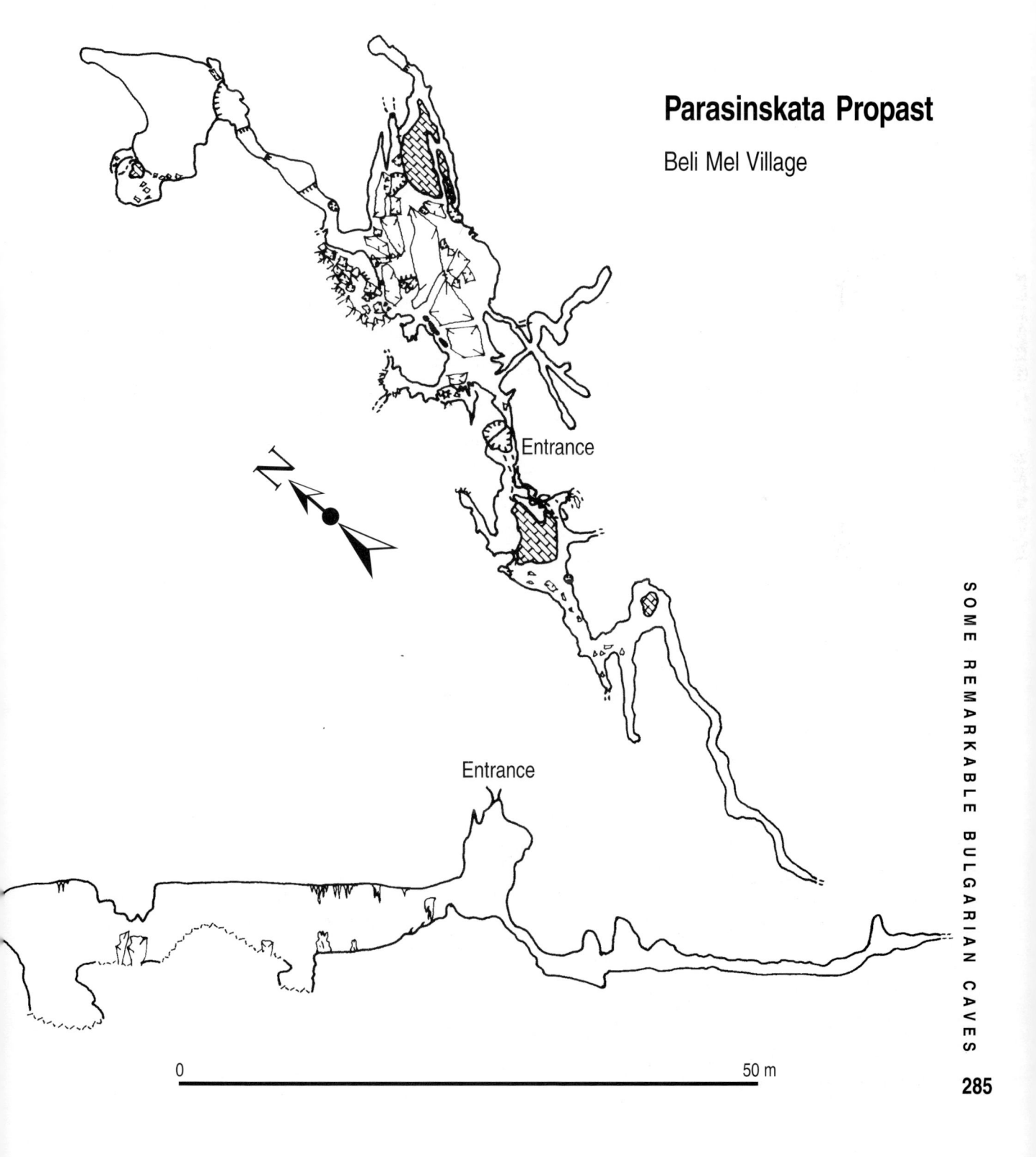

Parasinskata Propast

Beli Mel Village

Name: **PARNITSITE (103)**
Bezhanovo Village, Lovech Distr.
Length 2500 m, Denivelation -32 m
Surface 19 338 m², Volume 71 033 m³
Coordinates: N 43° 12' 02.1" E 24° 25' 58.4"

A water cave three km south of Bezhanovo, Middle Predbalkan. Formed in Upper Cretaceous limestone (Maastricht). Two entrances, the higher is called Suhiya Parnik, the lower – Vodniya Parnik. They are connected by a meandering, branched gallery. Average cross section: width seven meters, height five meters, respectively min/max 1–8 and 0.7–15 m. All genetic types of karstic accumulative formations can be observed in the cave. Sinter walls from three cm to three m, retaining the flowing water, form a descending cascade (from the upper to the lower entrances) of several lakes. The average water level in the river is 1.5 m with a span of 0.3–3 m. Deep clay and gravel deposits. At 600 m from the upper entrance, a stream of changing outflow enters the cave. The Parnishki Dol doline collects the surface water from snow melting and floods and it enters the cave through the upper entrance. One of the largest bat colonies in Bulgaria and for some species (*Myotis capaccinii*) this is a winter roost of European importance. Also important for breeding colonies in summer – sometimes up to 7–8 thousand bats.

The cave has been excavated by N. Djambazov in 1960. The first organized study of the cave took place in October 1962 by cavers from Edelweiss Sofia Caving Club (1100 m from Vodniya and 500 m from Suhiya Parnik were explored). In January 1963, the two parts were connected and the cave was declared the second longest in Bulgaria. In February 1963, for the first time the biospeleologists P. Beron, V. Beshkov and St. Andreev entered the cave through the upper entrance and left it through the lower, collecting many cave animals. After several corrections and re-surveying by the cavers from Studenets Caving Club (Pleven) in 1984 its length was fixed to 2500 m.

Both entrances and 0.1 ha of adjacent land were declared a National Monument (Decree 378/ 05.02.1964).

Fauna: more than 30 species have been identified, including the endemic troglobite woodlice *Balkanoniscus minimus* and *Beronicus capreolus* (Isopoda).

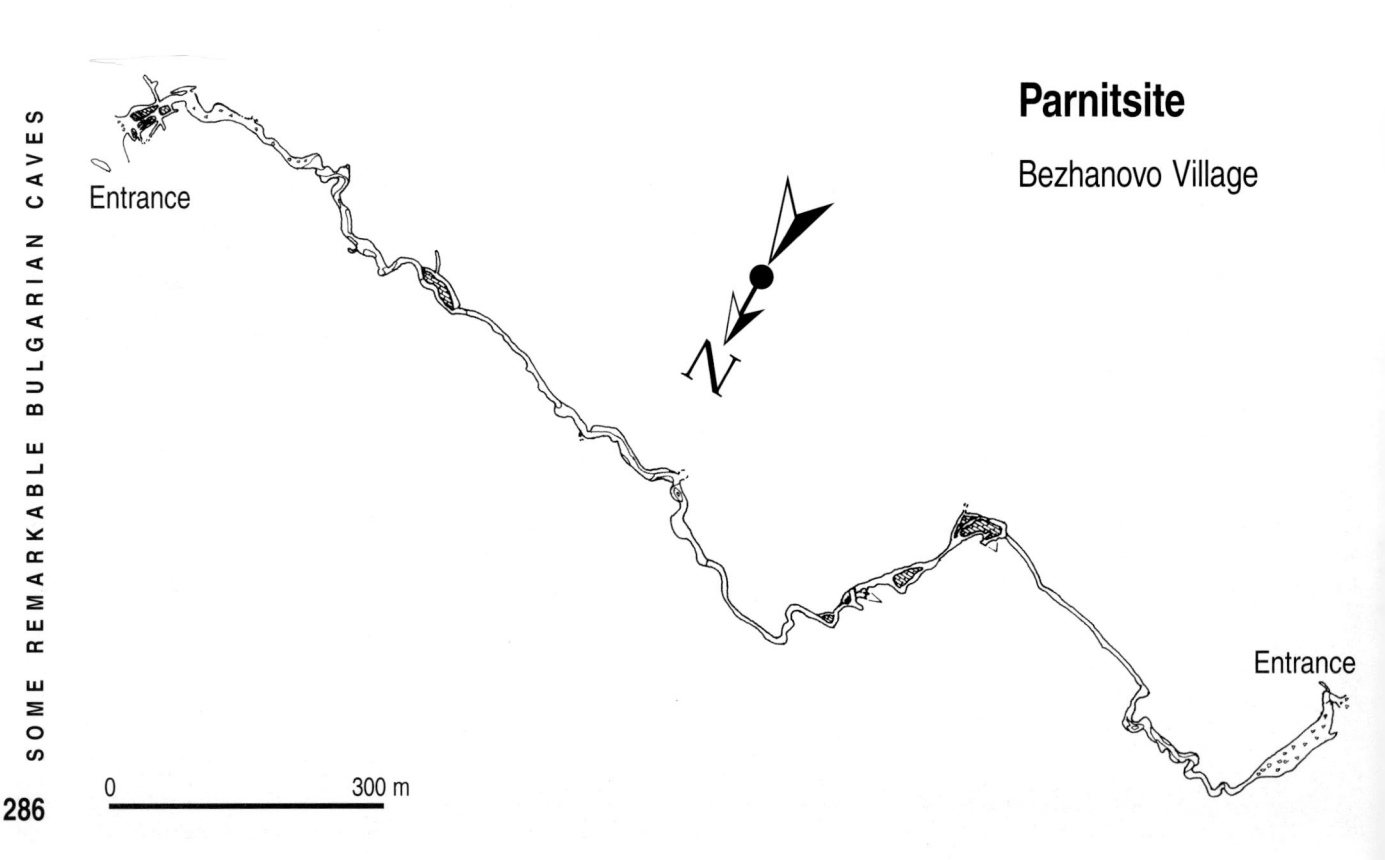

Parnitsite

Bezhanovo Village

The entrance of the Suhiya Parnik – photo Trifon Daaliev

The entrance of the Suhiya Parnik – photo Trifon Daaliev

One of the numerous sinter lakes of the cave – photo Trifon Daaliev

Name: **PARTIZANSKATA PESHTERA (114)**
Glozhene Village, Quarter Kamenna Mogila, Lovech Distr.
Altitude 305 m. Length 711 m. Denivelation -107 m

The pothole Partizanskata is situated 3.5 km south of Brestnitsa Village in Upper Jurassic limestone. Four entrances, the northern one being the best to use. The cave starts as a dry and lit gallery, leading into the first chamber. The descending floor ends at the edge of a pitch (97 m). From the 30th meter the pitch becomes bell-shaped. Walls covered with red flowstone, the floor – with blue mud.

The entrances of the pothole are situated 30 m above the level of River Vit, the bottom – 76 m below this level. It is presumed that an impermeable layer is situated between the river and the cave.

The cave had been known for a long time. During the Resistance, fighters ("partisans") used it as a shelter, hence the name Partizanskata. Surveyed first by cavers from Planinets in 1972, then in 1975, the caver Iv. Bachkov descended the pitch. The survey was completed by the Speleoclub Cherni Vrâh – Sofia.

Partizanskata peshtera

Glozhene Village

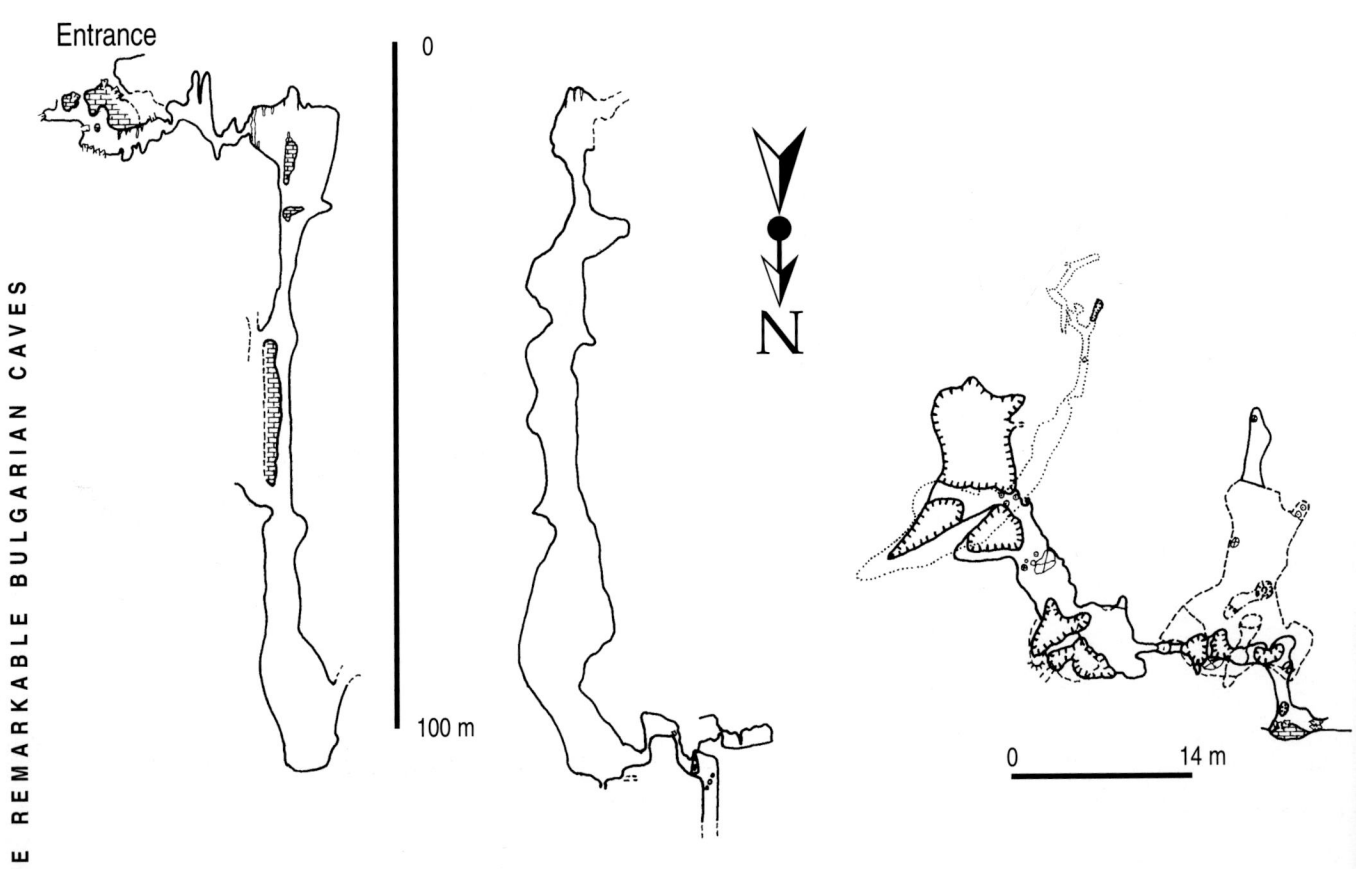

Name: **PESHKETO (27)**
Lilyache Village, Vratsa Distr.
Total length 630 m. Denivelation -27 m

Both entrances of this water cave are near the village Lilyache, on the right bank of the river. The cave is branched, diaclase, inclined, multistorey.

Peshketo has two entrances – horizontal and vertical. The horizontal entrance has dimensions 4 x 4 m. A gallery, which is 25 m long follows, leading into a chamber with dimensions 70 x 40 m. The floor of the chamber is covered with huge blocks from the crumbled ceiling. Part of the ceiling is a hole ("okno"), another part is a second entrance to the cave (35 x 15 m). A gallery starts in the northeast end of the chamber (starts with a vertical of 2.5 m). Up to this part, the cave is dry, the first 10 m are narrow. Immediately after the enlarging, we reach a cross road, which resembles the Mercedes sign. After a 1-m vertical, we come to the main gallery, which develops in west-east direction. On the left of the "mercedes", the gallery goes west for ca. 60 m. On its floor, there are blocks, clay and four lakes, the last one ending in a siphon. On the right of the "mercedes", there are two galleries. The first is circular and after 50 m it connects with the main gallery. This gallery goes east, in places it is 2–4 m wide and 1-10 m high. After 70 m, we reach a chamber of 25 x 10 m, its bottom covered with blocks. About 60 m after the chamber the gallery branches again. The left part ends after 30 m in a siphon. The right one is ascending, 50 m long. At its end, there are verticals of -3 m and -20 m. Ropes of 5 and 25 m and personal SRT equipment are needed to overcome them. The bottom of the vertical is beyond the siphon. From there the gallery is wide and high and becomes a chamber with dimensions 50 x 10 m, and a floor covered with huge boulders. At the end of the chamber, there is a siphon lake (no diving attempts have been made). When the level of the siphon is low, we can go further for another 42 m. The end of the cave is a big lake. Through the cave runs an underground stream. There is water dripping from the ceiling and beautiful flowstone formations.

The cave had been known to local people. Explored in 1965. In the big entrance hall archaeological excavations took place (conducted by B. Nikolov). The survey was done in 1983 by K. Karlov, Vl. Petrichev, G. Alipiev, V. Stoynin, I. Lilov – cavers from Veslets Vratsa Caving Club.

Peshketo

Lilyache Village

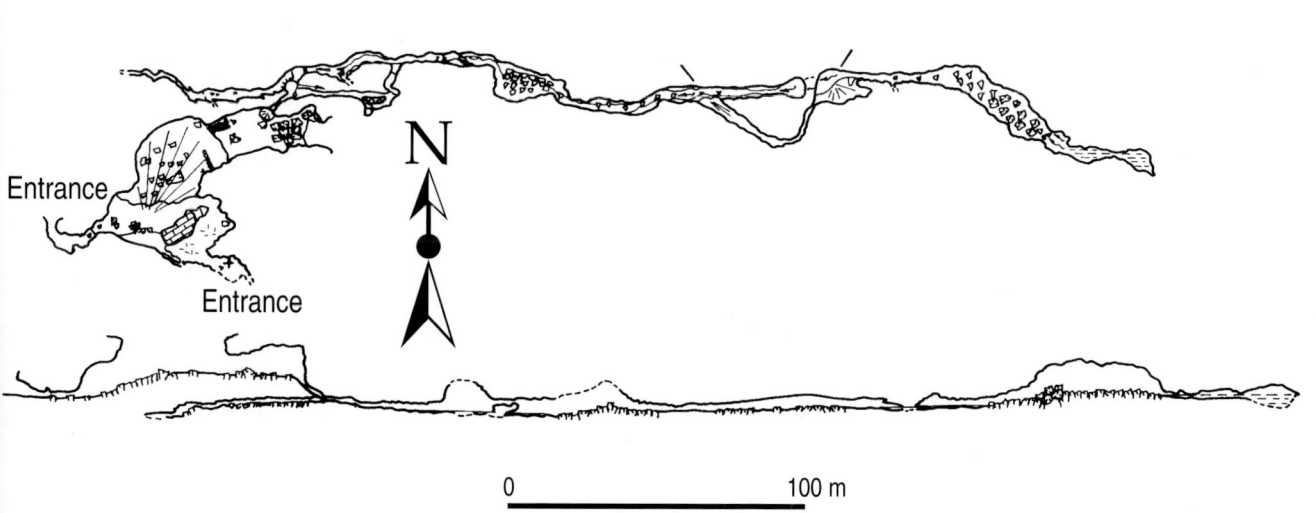

Name: **PLESHOVSKA (PLESHOVA) DUPKA (19)**
Prevala Village, Montana Distr.
Length 722 m. Denivelation -102 m

    This cave has been explored several times and is well known to local people, who can be used as guides to the caves. The exploration started in 1982 during an expedition of Akademik Sofia. In 1990, Pastrinets Montana Caving Club made an accurate map of the cave. During this expedition, attention was paid to the narrowing at the end of the cave, almost full of deposits. After their removal, the new galleries of the cave were discovered.

    After leaving the cave, the underground stream runs for at least 825 m (presumed), until reappearing in the cave Vreloto. All geological, tectonic and hydrogeological data in the area speak in favour of the hypothesis that the cave Vreloto is the most probable outlet of the underground stream of Pleshovskata Peshtera. Many caves have been discovered in the cave, most of them small and shallow. The biggest ones, deep up to 1,50 m, are the laces immediately under the highest verticals, where the strength of the falling water has been the biggest.

    Description: Pastrinets Montana Caving Club (18.10. 1992).

Pleshovska (Pleshova) dupka

Prevala Village

Name: **PONORA (23)**
Chiren Village, Vratsa Distr.
Length 3497 m. Denivelation -46 m
Coordinates: N 43° 18' 28.9" E 23° 34' 38.6"

A water cave two kilometers S/SW of Chiren, in the western foothills of Milin Kamâk ridge. Formed in Lower Cretaceous (Aptian) limestone. A branched, descending cave with its main development in W/NW direction. Prevailing shape of galleries – rectangular and triangular. Min/max dimensions of the cross sections: width 0.8–9 m, height 0.8–7 m. Many sinter walls, forming lakes, flowing into one another. The cave is a permanent sink (ponor) of the water of Lilyashka river, carrying its water to the source Zhabokrek. The cave ends with a siphon.

First studied by Hr. Mladenov and D. Ilandjiev in 1960. They penetrated to ca. 1700 m. During the First International Expedition in 1961 P. Beron, St. Andreev and others explored and surveyed 3070 m. In 1963 the group for diving at the Republican Commission of Speleology organized two attempts to connect the source Zhabokrek with the cave. After the siphon a new part of 170 m is reached. In 1965 cavers from Ivan Vazov Speleoclub penetrated another 430 m. The older surveys were corrected in 1977 and the length was established at 3172 m.

In July 1970 S. Pishtalov, B. Ivanova and K. Spassov carried out geophysical studies and a successful indicatory attempt to prove the connection between Mladenovata peshtera and Ponora. Surveyed in details by A. Siromahov, N. Gladnishki, P. Petrov, T. Daaliev, K. Kunev, Zh. Stefanov, V. Stoitsev, A. Leonidov (the map being put together by A. Leonidov). During the same expedition the participants T. Daaliev, N. Gladnishki, P. Petrov and Ya. Bozhinov discovered and mapped over 1000 m of new galleries in the cave Ponora. The connection between the caves Dushniko and Ponora was established by A. Petkova, N. Gladnishki and T. Daaliev. In 1977 again an indicatory experiment was carried out by K. Spassov and P. Delchev, proving beyond doubt that the water of Mladenovata propast reach as a side tributary the cave Ponora. The fluorescein, released in 11.00 h. on 20.08.1977 in Mladenova, appeared in the waterfall of Ponora by 11.00 on 22.08.1977 (48 hours). No attempts are known to penetrate the siphons in both caves.

In August 2001, after two divings, Franc Fasseure passed the siphons of Zhabokrek (240 m), connected them with the end parts of Ponora and brought the cave to a length of 3497 m.

Declared Natural Monument with 20 ha of adjacent land (Decree 2810/63 from 1963 of the Committee of Forests and Forestry).

Fauna: 17 species, including *Typhloiulus bureschi* (Diplopoda) and *Trichoniscus bureschi* (Isopoda). Stygobite is the Crustacean *Eucyclops s. serrulatus* (Copepoda).

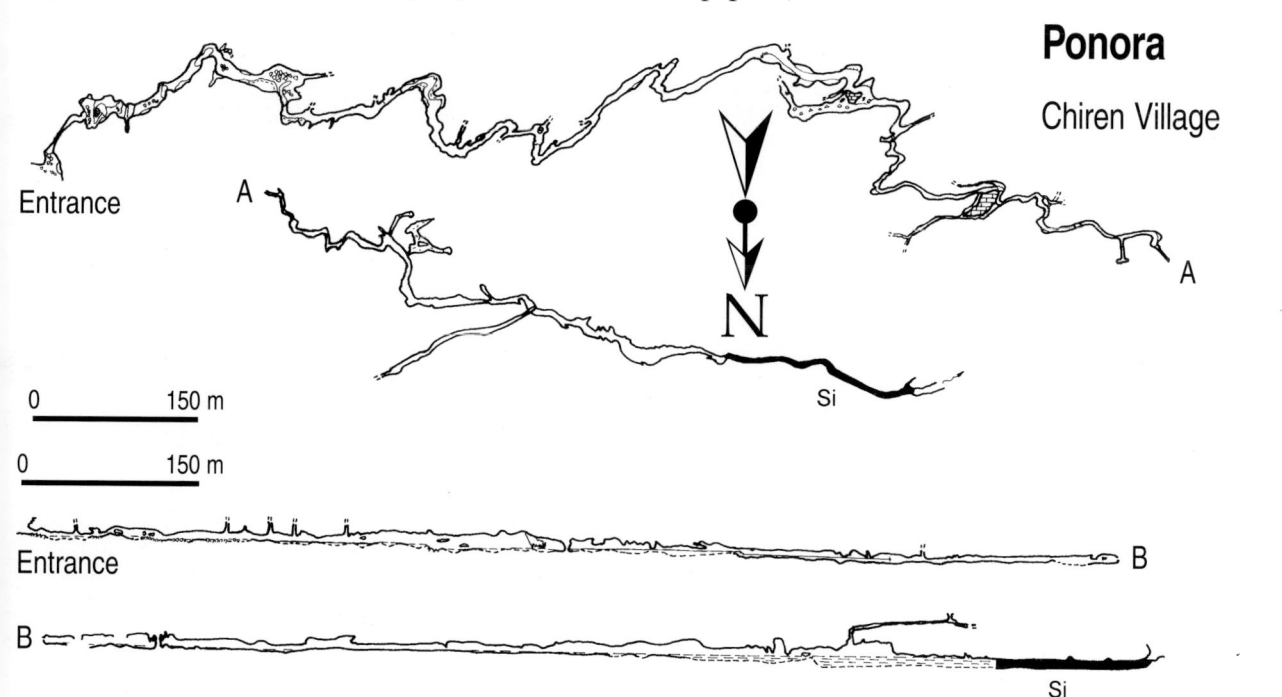

The entrance of the cave Ponora – photo Trifon Daaliev

A. Pencheva in the underground river – photo Nikolay Genov

Name: **POPOVATA PESHTERA (78)**
Gabare Village, Vratsa Distr.
Altitude 260 m. Length 530 m

A cave in the locality Vutkovskite Vartopi, SE of Gabare, in limestone from the Upper Cretaceous (Senonian – Maastrichtian), with some dolomite. Altitude above the LBE 35 m. On the surface above the cave, there are 7 funnels, situated close to one other and in a line directed at 295° (total length 105 m). In these funnels there are 4 large entrances to the cave and 4 narrow entrances. The funnels are on the right from the road Gabare – Breste. The cave is descending, labyrinthine and could be divided into 3 parts: labyrinth, southwestern and northwestern. The largest is the fourth funnel. Through it, we reach a chamber with a stony bottom. Eastwards, along a wide gallery the third entrance is reached. The east branches of the labyrinth are a complex of intersecting low galleries, 50–60 cm high. The main gallery in the southern part of the labyrinth is widening up to 4.5 m, its height is up to 1.5–2 m. In the labyrinth, there are no flowstone formations, the floor is covered with thick clay deposits. Part of the galleries are inhabited by foxes, there are many excrements. This part of the cave is also the richest in cave fauna. Near the beginning of the NW gallery, there is a pool (160 x 50 x 5 cm), its bottom covered with silt. Water T° 7.5 °C, air T° 8.5 °C. Strong air current in the labyrinth. Total length of the galleries in the labyrinth 227 m. The SW gallery is an old river bed with a groove with a small stream. Along the gallery, there are 5 lakes (the first is 1.5 m long, the fourth 2.4 m and the fifth 4.2 m). Water T° 11 °C, air T° 12 °C. In this gallery the flowstone formations of the cave are concentrated. The SW gallery is a diaclase, from its 121^{-st} meter, the height decreases to 1.5 m, the bottom is covered with clay. The length of the SW gallery (together with one branch) is 141 m, the total length of the NW gallery is 162 m. The distance from the entrance (fourth funnel) to the end of the NW gallery is 119 m. Minimal width of the galleries 40 cm. Minimum height 40 cm. Maximum width 4–8 m, maximum height 9 m. Description: Alexi Popov and Lilia Vassileva.

Fauna: including the troglobite *Typhloiulus bureschi* (Diplopoda) and the stygobite *Niphargus bureschi* (Amphipoda)

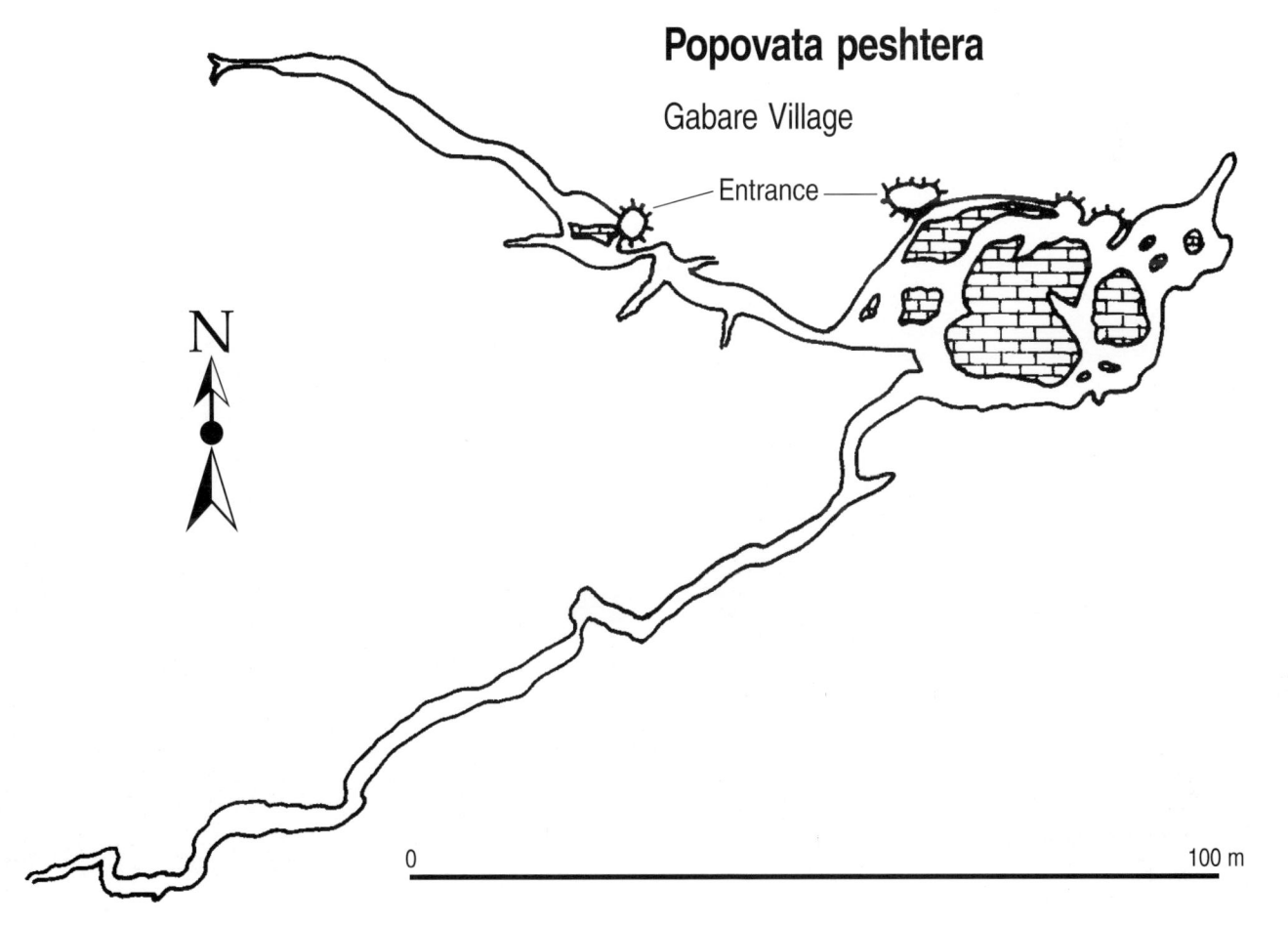

Name: **PRIKAZNA (185)**
Kotel Town, Sliven Distr.
Length 4782 m. Denivelation -37 m
Coordinates: N 42° 52' 50.8" E 26° 22' 13"

A cave complex, labyrinth-like, with asymmetric development of galleries. Formed in Cretaceous (Maastricht)limestone, on the left slope of Suhoyka river, in locality Zelenicha. In the clay deposits of the biggest chamber, called Sahara, which is 66 m long and 22 m wide, remains of cave bear (*Ursus spelaeus*) have been discovered. The cave was discovered and explored up to 700 m by the caving club of Kabile, Yambol, in April 1972. Later, cavers from Kotel and Sofia explored new passages and the cave's length reached 3200 m. In 1987, cavers from Akademik Plovdiv updated the existing map and established the latest know length of the cave.

Natural Monument (Decree 3702/29.12.1972) together with 5.2 ha of adjacent land.

The recent cave fauna includes at least four troglobites (a rare phenomenon for Kotel area): *Duvalius kotelensis* (Coleoptera, Carabidae), the collembols *Pseudosinella duodecimocellata* and *Pseudosinella bulgarica* and unidentified Isopoda.

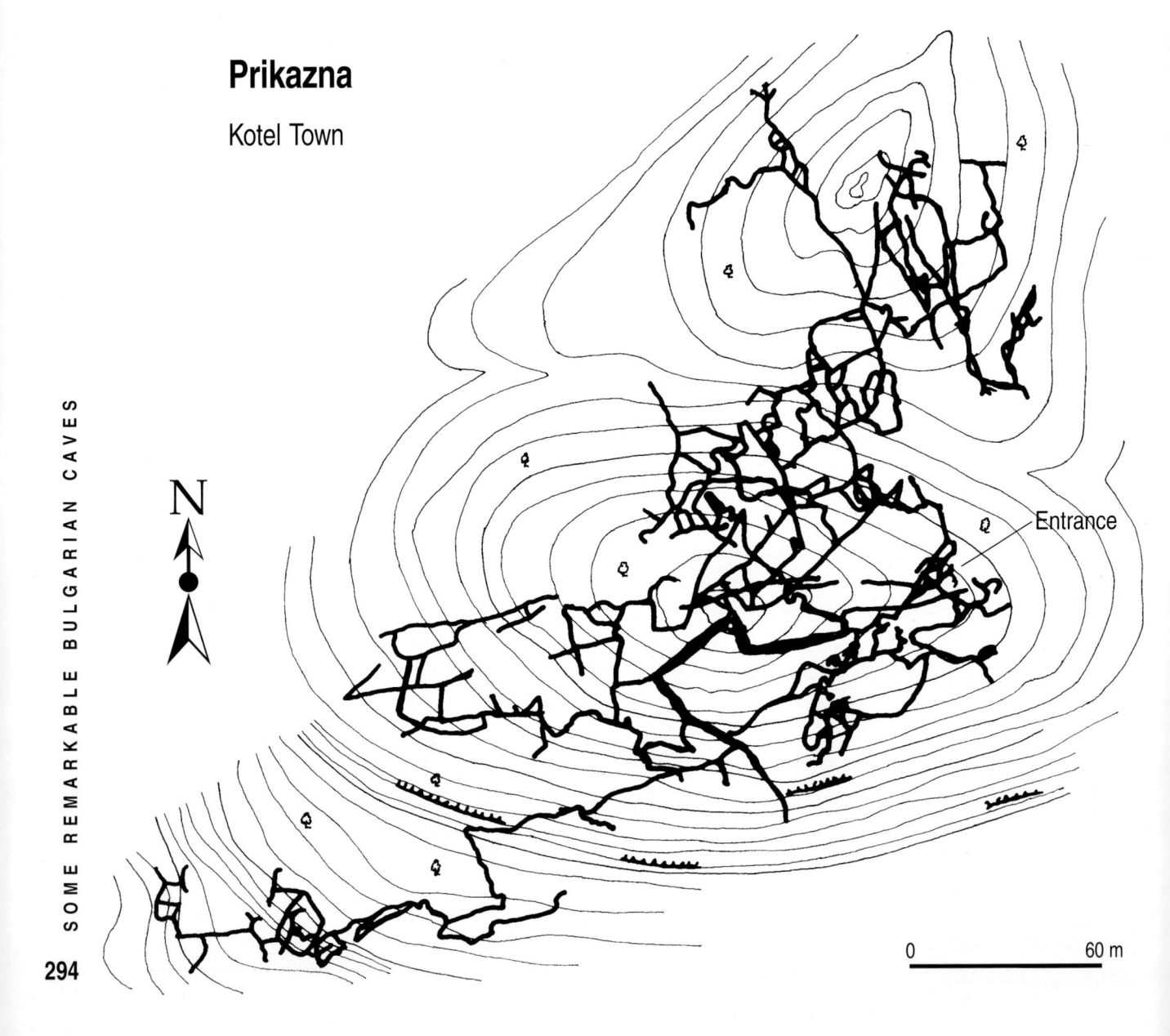

Prikazna

Kotel Town

Kotel Balkan in the area of Zelenich – photo Trifon Daaliev

The entrance of the cave Prikazna – photo Trifon Daaliev

Name: **PRILEPNATA PESHTERA (CAVE COMPLEX BOZHIYAT MOST – PRILEPNATA PESHTERA) (26)**
Lilyache Village, Vratsa Distr.
Altitude 216 m. Length of the complex 443 m. Denivelation -14.5 m (11.80; -2.70 m)
Coordinates: N 43° 18' 56.3" E 23° 33' 10.3"

A rock bridge with a cave, 2.5 km SE of Lilyache, in the valley Lilyashki Dol, SW of the hill Vrachanska Mogila (high up to 285 m), Western Predbalkan.

The cave complex is formed in reef limestone from the Urgonian, part of Mramoren Syncline.

Bozhiyat Most (God's Bridge) is a through cave, made by the water of the stream Lilyashki Potok, starting from the source Zhabokrek. It has 3 entrances – Upper, Lower and Middle. The Upper (initial) is the one through which the stream enters the cave. It is 30 m wide and 19.70 m high. Through the Lower the stream comes out and it is 39.40 m wide 11.40 m high. The Middle is in the middle of the cave and takes part of the roof and of the northern wall of the cave. The entrances are interconnected by a wide gallery, 10 m high and 128 m long, developed SE-NW. In the middle of the cave, there is a lake, which is 20 m long and up to 2 m deep. About 26 m from the NE end of the cave, on the right side of the gallery, which is 7 m high, is the entrance to the Bat's Cave (Prilepnata Peshtera). The cave is developed SE–E and has average dimensions, it is 3 m wide and 4 m high. It is 189 m long, with no flowstone formations. The entrance parts are covered with thick layers of bat guano. The cave is a roost of a large bat colony of the species *Rhinolophus ferrumequinum, Rh. hipposideros, Rh. euryale, Myotis emarginatus, M. capaccinii, M. myotis, M. blythi* and *Miniopterus schreibersi*. The complex, together with the labyrinth of narrow galleries around the Upper entrance of Bozhiya Most, has a total length of 443 m. Its denivelation from the upper entrance is 14.50 m (11.80; -2.70 m).

This natural phenomenon was first described in 1875 by Petko Slaveykov (Promenade across a country). Later (1895,1898,1900), H. and K. Škorpil published a sketch and a short description of Bozhiya Most. The first exploration was conducted by Veslets Vratsa Caving Club in 1962 and during the International Expedition in 1963. The first caver to reach Prilepnata Peshtera was Petko Nedkov, then P. Beron collected cave fauna. A detailed and accurate survey of the complex was done by A. Spassov, V. Daskalov and B. Kolev during the National Expedition Ponora'77 in 1977.

Declared a protected natural site, together with 15 ha of adjacent land (Decree 378/05.02.1964).

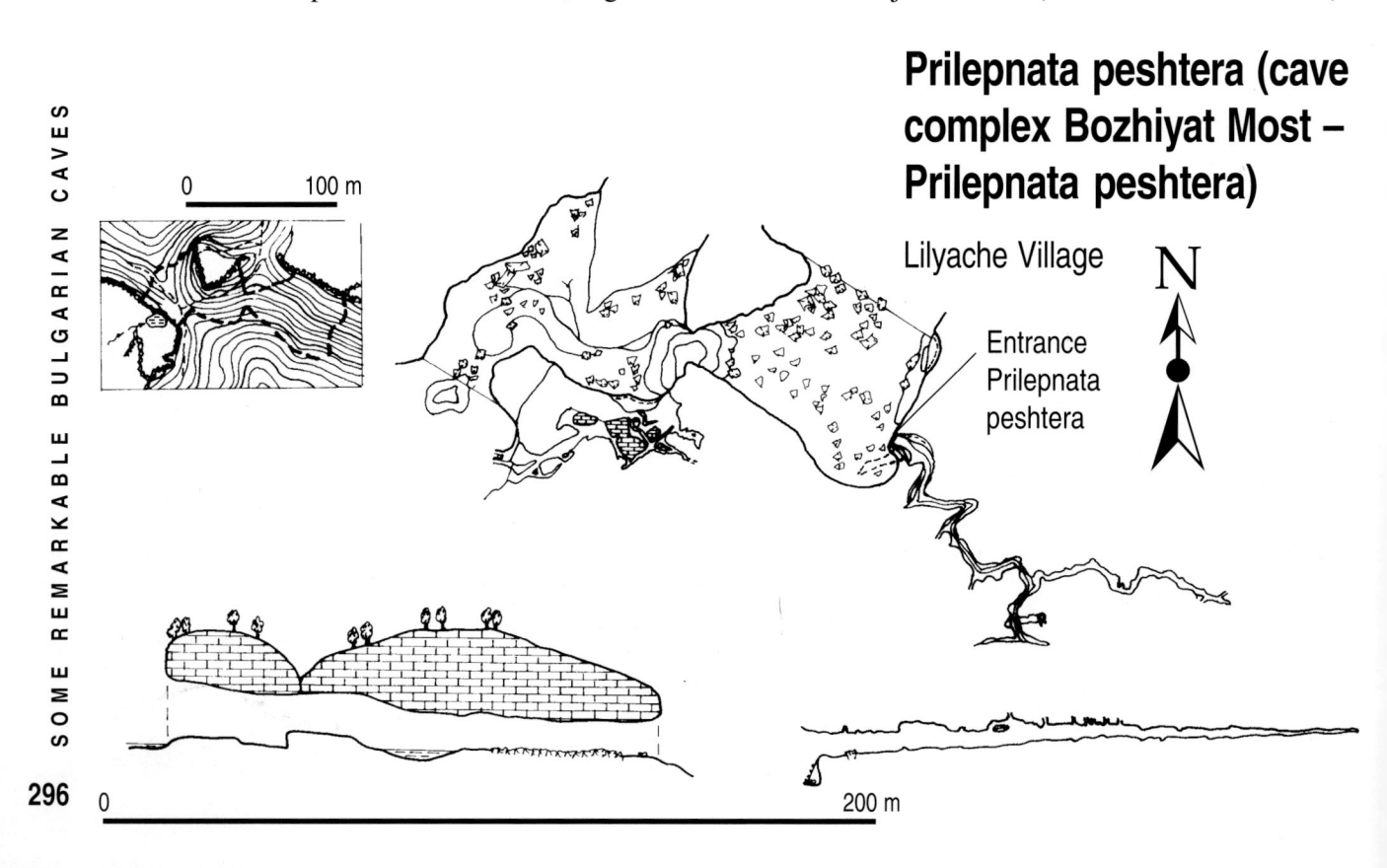

Prilepnata peshtera (cave complex Bozhiyat Most – Prilepnata peshtera)

Lilyache Village
Entrance Prilepnata peshtera

In the Temple – photo Valery Peltekov

One of the entrances of Bozhia most – photo Trifon Daaliev

Name: **PROHODNA (93)**
Karlukovo Village, Lovech Distr.
Altitude 216 m. Length 240 m. Denivelation -50 m
Coordinates: N 43° 10' 22" E 24° 04' 33"

    A natural bridge cave near Karlukovo. Formed by the river Malâk Iskâr. Two entrances. Height of the upper entrance 29 m, of the lower entrance 42.5 m. The valley after the lower entrance is the remnant of a former, much bigger cave, which is now collapsing. The trends towards disappearing of the cave could be witnessed from two large "windows" on the ceiling, called "Oknata". As early as the Neolith and the Eneolith, the primitive man appreciated the qualities of the cave and many artefacts, which are evidence of that, have been discovered during excavations.

    The cave has been declared a National Monument and Protected Area (together with 1.5 ha of adjacent land, Decree 2810 from 1963). Popov (1984) considers the cave being 1 800 000 years old.

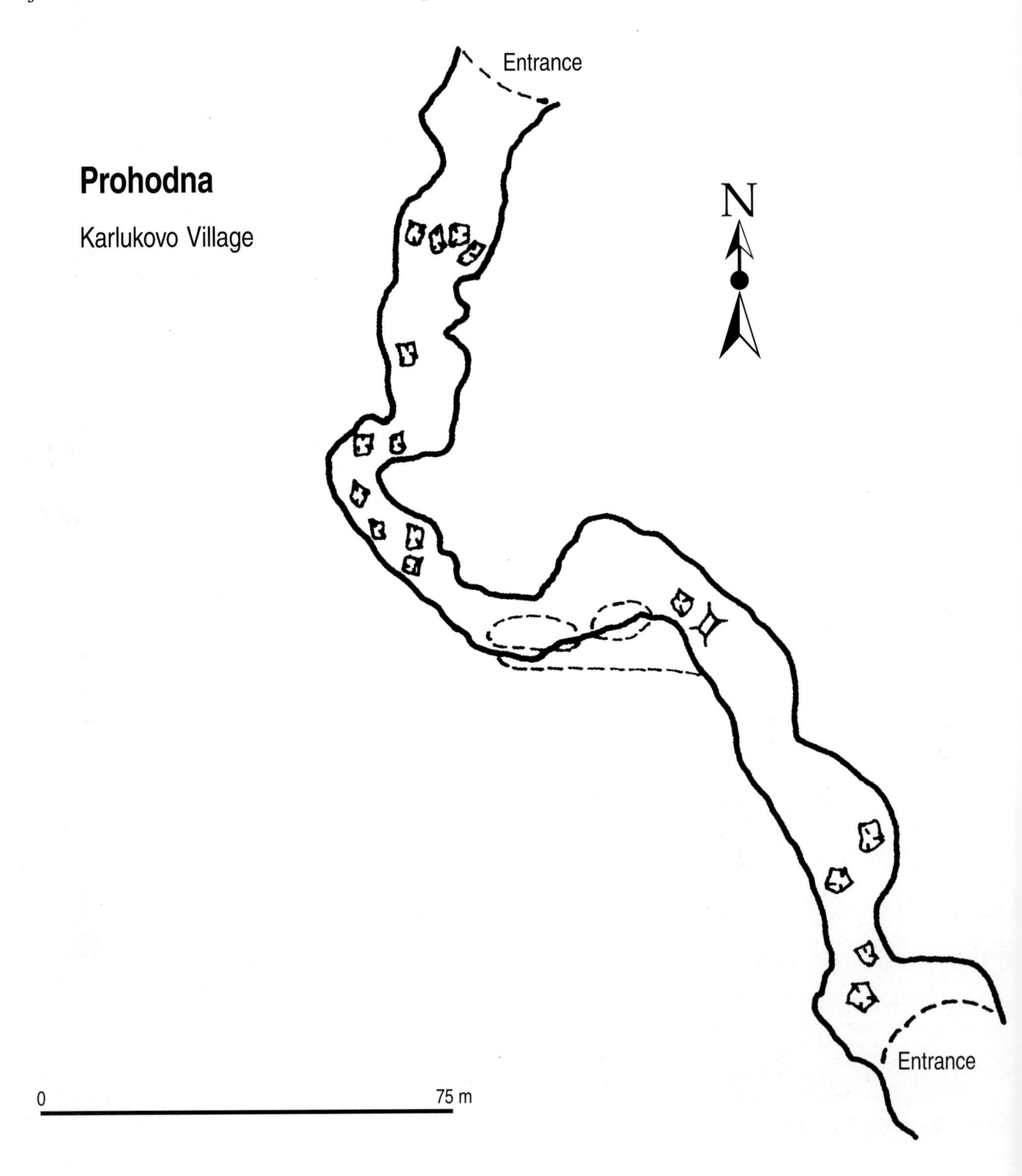

Prohodna

Karlukovo Village

The majestic 42.5 – meters high entrance of Prohodna Cave – photo Valery Peltekov

Prohodna Cave – Upper entrance – photo Valery Peltekov

The majestic 42.5 – meters high entrance of Prohodna Cave – photo Nikolay Genov

Prohodna Cave – Oknata – photo Valery Peltekov

Name: **PROLAZKATA (DERVENTSKATA) PESHTERA (176)**
Prolaz Village, Targovishte Distr.
Length 569 m. Denivelation 30 m
Coordinates: N 43° 10' 36.1" E 26° 30' 41"

A cave, about 1 – 1.5 km NE of Prolaz Village, about 150-200 m above the road Sofia – Varna, on the left (geographically) side of a small valley, about 6 m above its bottom.

The cave entrance is narrow at the beginning, subsequently enlarged by blasting works. Immediately after the entrance, the cave starts with a 5-m vertical, followed by a gallery, 60 m long, which in some places becomes narrower and lower. The entrance gallery leads into a big chamber, its floor covered with fallen blocks. From the E-NE end of the chamber starts a gallery, at the beginning of which there are two potholes in thick deposits and other materials. The first one is overcome through two verticals (8 m altogether), after which a clay wall of the same height is climbed. On the bottom of the pothole, in dry weather, there is a small lake, which in the spring and strong rain becomes several meters deep. Then, the lake takes the entire surface of the bottom. After the next hole, the gallery becomes horizontal and easy to penetrate. On the walls and the floor of this gallery, there are some flowstone formations (in some places remarkable) – helictites, anomalous stalactites, sinter dishes, columns, etc. About 150 m from the big chamber, the gallery comes to a dead end.

The cave is wet, poor in formations, especially in the initial parts. In some places, the floor is covered with bat guano.

The cave Prolazka Peshtera had been known to local people. The first exploration done by Dr K. Tuleshkov in 1946. In 1966, a map was made for the transformation of the cave into a show cave by Bozhan Marinov. The cave is declared a Natural Monument, according to the Law on Protection of Nature.

Fauna: so far 15 animal species have been recorded (no troglobites). The pseudoscorpion *Neobisium intermedium* Mahnert, 1974 was described from this cave. *Duvalius* sp. (troglobitic Carabid beetle) is still undescribed.

Prolazkata (Derventskata) peshtera

Prolaz Village

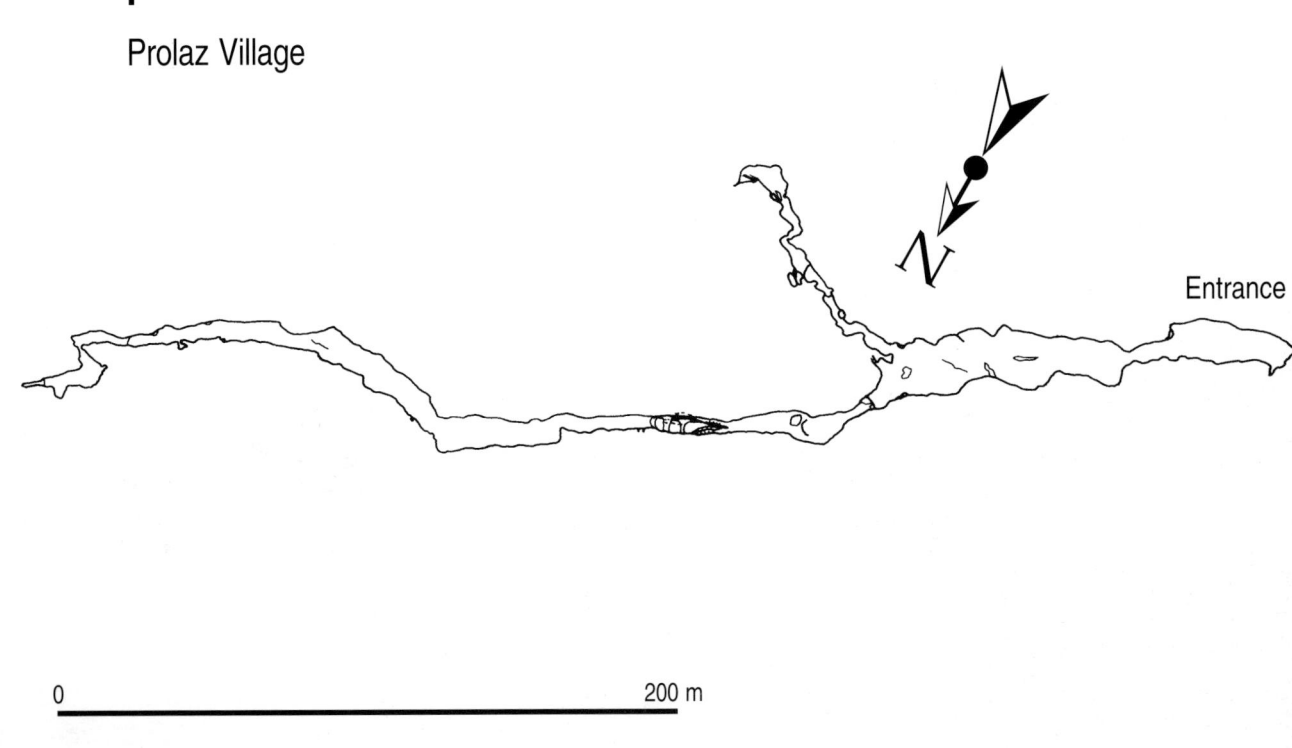

The entrance of Prolazkata Cave – photo Trifon Daaliev

The big column

Name: **PROPAST (6)**
Oreshets Railway Station, Vidin Distr.
Length 318 m. Denivelation 62 m

A cave pothole, ca. 1000 m west of the gas station of Oreshets Railway Station. Entrance – 4 x 4 m. The entrance vertical is 25 m deep. Its bottom is a cone from which a gallery starts, high up to 5–6 m and wide up to 7–8 m. Its bottom is covered mainly with clay and angular gravel. The gallery ends in a narrow siphon. It contains stalactites, stalagmites, stone "waterfalls" and sinter lakes. The cave is periodically inundated and from the end siphon, when raining and during snow melting, water goes out. The air temperature is 11 °C.

The cave was explored and surveyed in 1971 by P. Tranteev and Z. Iliev. They calculated its length at 245 m and its denivelation at – 62 m. In 1977, S. Gavrilov, E. Gesheva and P. Parvanova again surveyed the cave and established its length at 318 m and its denivelation at – 58 m.

The fauna of the cave is represented by bats and by the troglobite Isopode *Trichoniscus bononiensis* (Isopoda) and the beetle *Beronia micevi* (????), collected by P. Beron.

Propast

Oreshets Railway Station

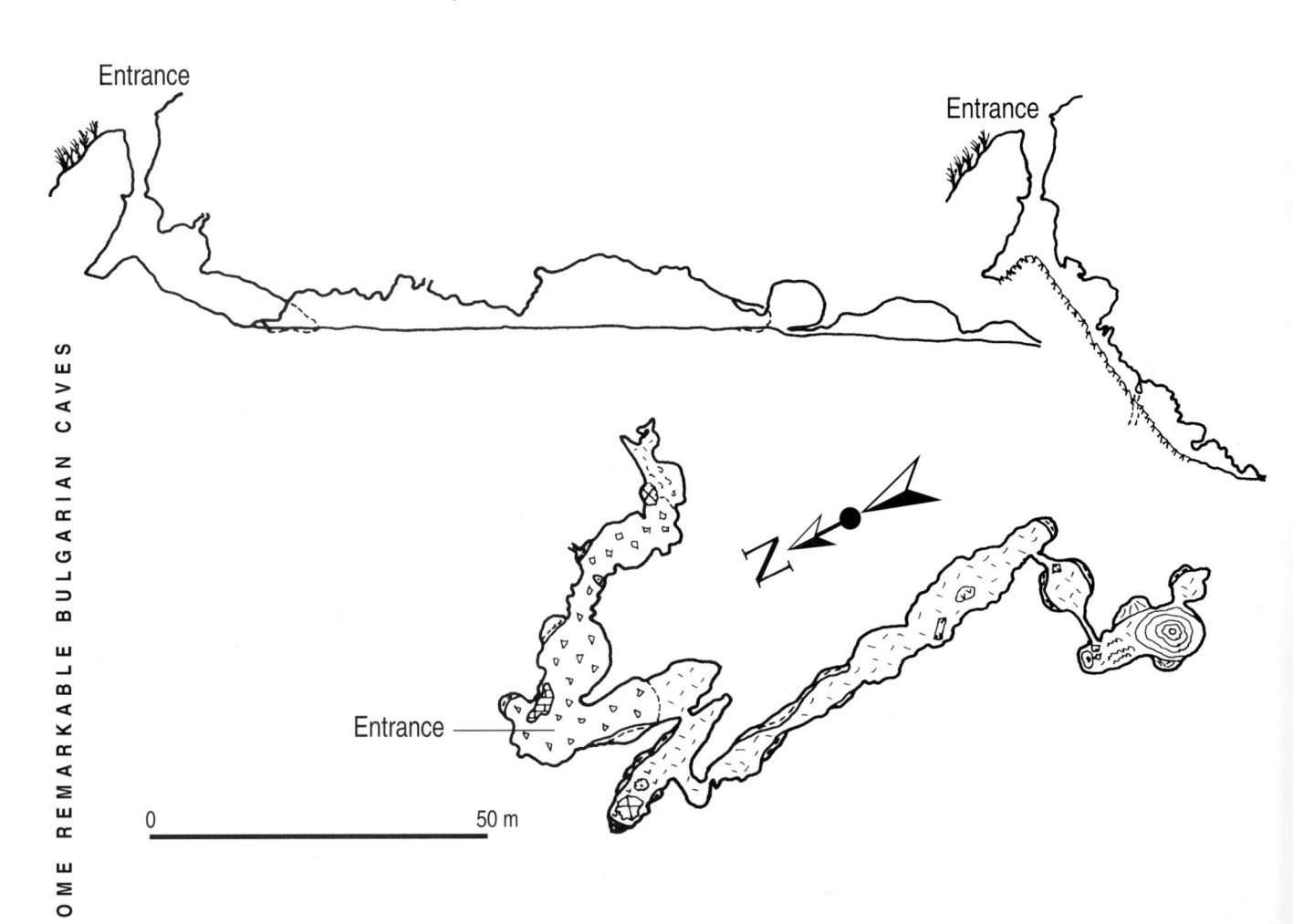

Name: **PROPASTTA NA BENYO ILYOV (84)**
Breste Village, Pleven Distr.
Total length 616 m. Denivelation -55 m

A cave pothole, about 2 km N-NW of Breste, in a thinning forest in the area of Nenchovskoto.

The entrance vertical (14 x 6 m) is 47 m deep. We enter a big chamber (30x10 m), from which three galleries start. The first one is directed to N-NW and is 135 m long. The second is directed to southwest and is 90 m long. At the end of the two galleries, there are gryphons, from which underground streams run out. They merge in a big chamber and form a river, running along the third gallery. It is directed to the east, its length is 280 m and its end is a siphon. The cave is a diaclase, with maximum height of the ceiling in the eastern gallery (in places, more than 20 m). After 10 hours of torrent rain, on April 25, 1977, the water level rose by 40 cm. There are some flowstone formations in the cave (stalactites, stalagmites, ribs, curtains, dendrites, etc.).

The name of the cave comes from the name of the land owner (Benyo Ilyov). The cave was explored during several expeditions of cavers from Sofia in 1977. The cave was surveyed in 1989 during the National Expedition „Reselets'89" by A. Velchov, aided by A. Baldjiev and D. Lyubenova.

Propastta na Benyo Ilyov

Breste Village

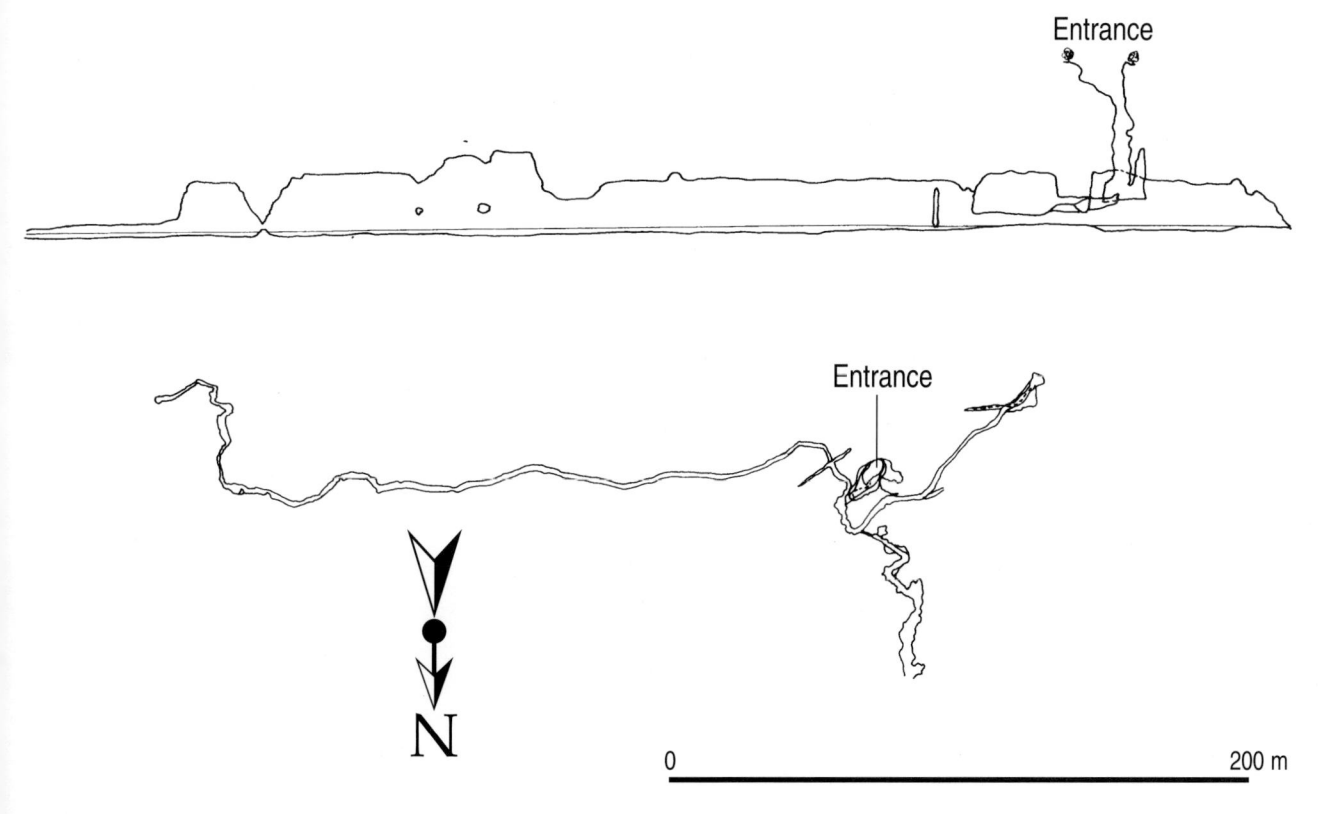

Name: **PTICHA DUPKA (132)**
Cherni Ossâm Village, Lovech Distr.
Altitude 1260 m. Length 652 m. Denivelation -108 m

A pothole in Steneto Biosphere Reserve. The entrance in Trias limestone and dolomite opens north of Rediya Trap in a beech forest. Two entrances on a steep slope. It starts with a vertical of 70 m. The bottom is oval in shape, there are blocks covered with earth from the surface. After two other verticals, the true bottom is reached, and still the light from the entrance can be seen. A small stream, two galleries, an ascending one, which goes west, and a descending one, going east. Total length of the descending gallery 302 m. In the last chamber, flowstone formations are concentrated. There is a sinter lake with blue water and yellow cave pearls and stone lilies complete the beauty of this chamber. Its total length is 100 m. Length of the ascending gallery 350 m. A variable stream at the bottom of the horizontal gallery.

First studied by Polish speleologists (T. Uminski and others, September 1957). In July 1962, the cavers Hr. Delchev, St. Andreev and P. Neykovski from Akademik Sofia Caving Club explored and surveyed the pothole. Animals were collected also by P. Beron in 1998, at the end of the 56 days stay of Emiliya Gateva alone in the pothole.

Some days before the start of the Republican Expedition Troyan -72 V. Balevski and V. Markov discovered a prolongation with the most beautiful chamber of Pticha Dupka. N. Gladnishki, V. Balevski, P. Petrov and T. Daaliev surveyed and described the pothole during the expedition Troyan'75.

Troglobites and stygobites: *Niphargus bureschi* (Amphipoda), *Anamastigona alba* (Diplopoda), *Duvalius* sp. (Coleoptera). Troglophile: *Spinophalus uminskii* (Gastropoda).

The pothole is situated within the boundaries of Central Balkan National Park.

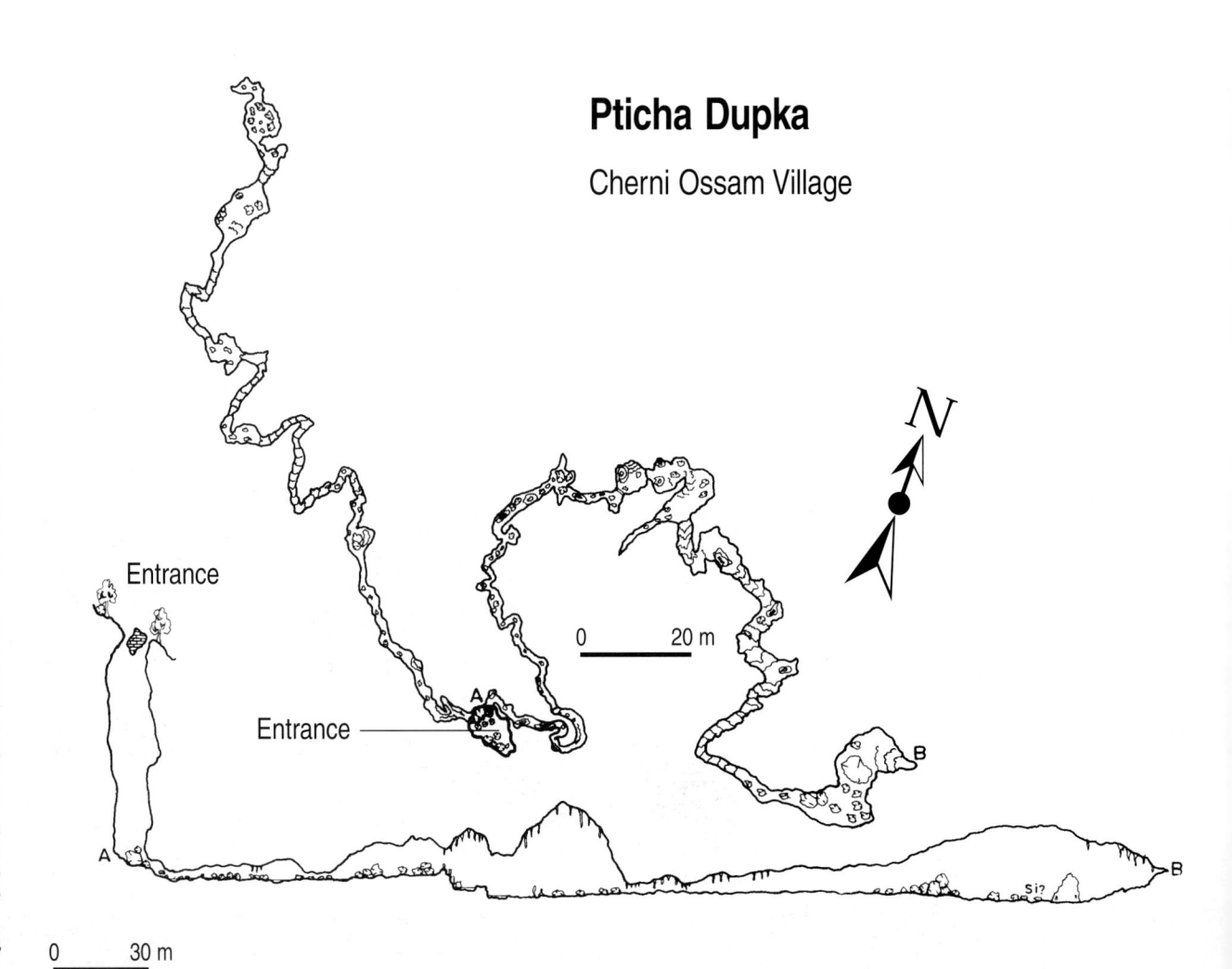

Pticha Dupka

Cherni Ossam Village

After this entrance is 70-meters pitch of Pticha Dupka – photo Vassil Balevski

The descent of the pitch – photo Vassil Balevski

Milk white column – photo Vassil Balevski

The Big Hall is one of the most beautiful in Bulgaria – photo Vassil Balevski

Name: **PUKOYA (44)**
Pavolche Village, Vratsa Distr.
Length 48 m. Denivelation -178 m

A pothole in the area of Krushovitza in Upper Jurassic – Lower Cretaceous limestone. It begins with a narrow horizontal gallery, then a shaft follows, broken by five platforms at the 7, 27, 33, 53 and 103-th meter. After several pitches (the deepest being 25 m) a depth of – 78 m is reached. An inclined (70–80°) shaft of 100 m follows with a lake at the bottom. Many speleothems. Dripping water, the lake 3 x 8 m. In dry years the lake disappears.

The pothole was discovered on June 3, 1971 by K. Kârlov (Speleoclub Veslets Vratsa).

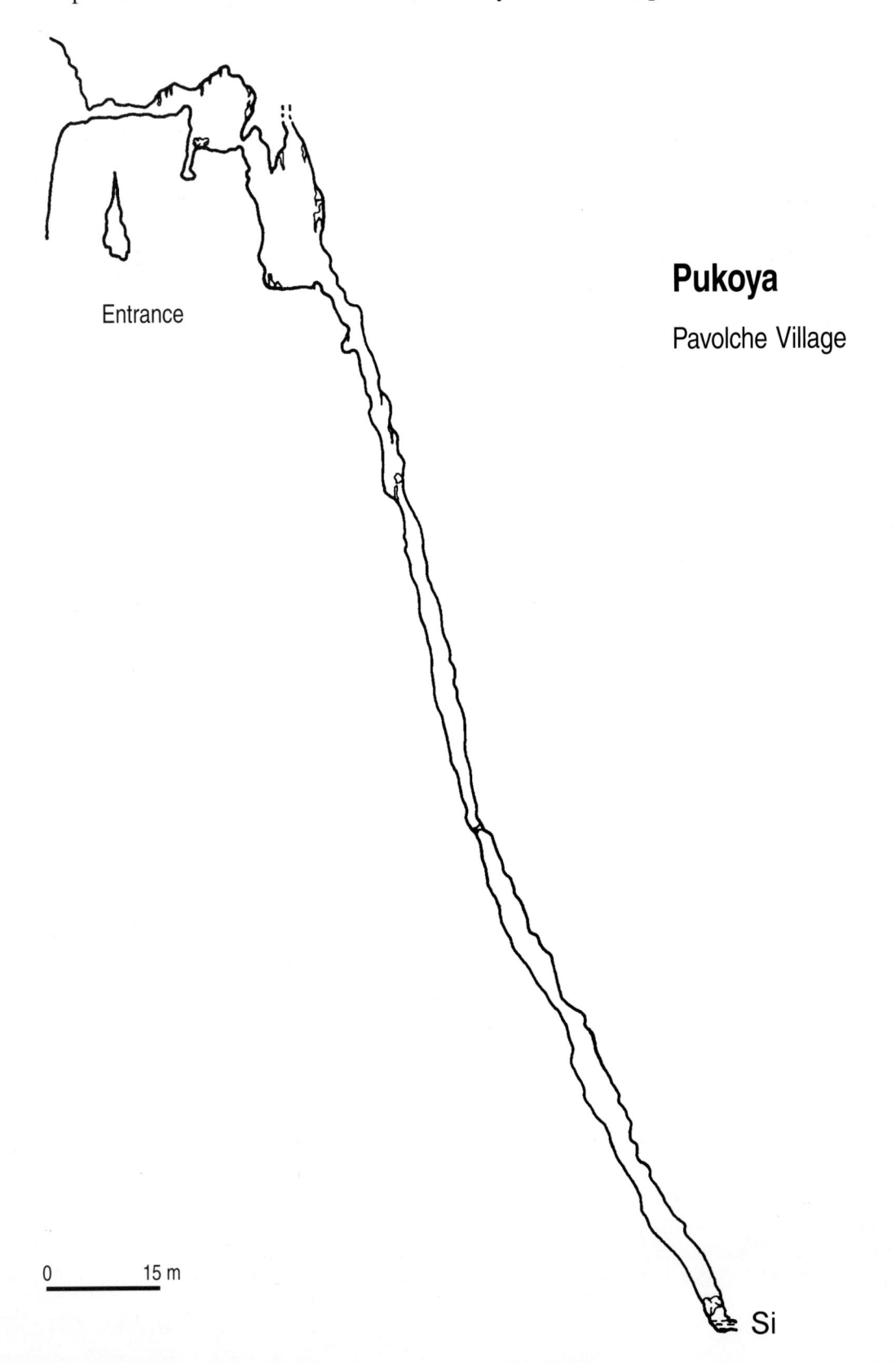

Pukoya

Pavolche Village

Name: **RADOLOVA YAMA (Toshkova dupka) (66)**
Gintsi Village, Sofia Distr.
Altitude 1150 m. Denivelation -88 m

A pothole ca. one km along the right- hand side of the upper end of Gintsi, in an old beech forest. The square entrance is at the bottom of a funnel with a diameter of 7 m, with steep slopes. The first vertical is 40 m high, a strongly inclined platform follows, then another vertical of 30 m and the bottom, which is covered with stone blocks and litter.

The name Radolova Yama comes from a local legend about a person called Radol, whose daughter fell into the pothole and he died of grief. Another person (Toshko) is believed to have given the other name Toshkova Dupka.

In December 1959 the newly created Republican Caving Commission organized an expedition for exploring the pothole (T. Michev, A. Denkov, V. Beshkov, S. Penchev). Descending on primitive ropes, they were accompanied by more than 50 local people. This could be seen from the graphic drawings made by A. Denkov to commemorate this heroic exploit. A 50-m ladder was used, then from the middle of the pothole the ladder was dropped down by rope to the bottom. A petromax lamp was hauled down along with the safety rope.

More than 15 years after the first descent, N. Gladnishki and T. Daaliev checked the map of the first group and the information that a continuation was possible. The depth was measured correctly by the first group at – 88 m, but no continuation was found.

Fauna: in this pothole the troglobites *Paranemastoma (Buresiolla) bureschi* (Opiliones) and *Pheggomisetes* sp. (Coleoptera) have been found.

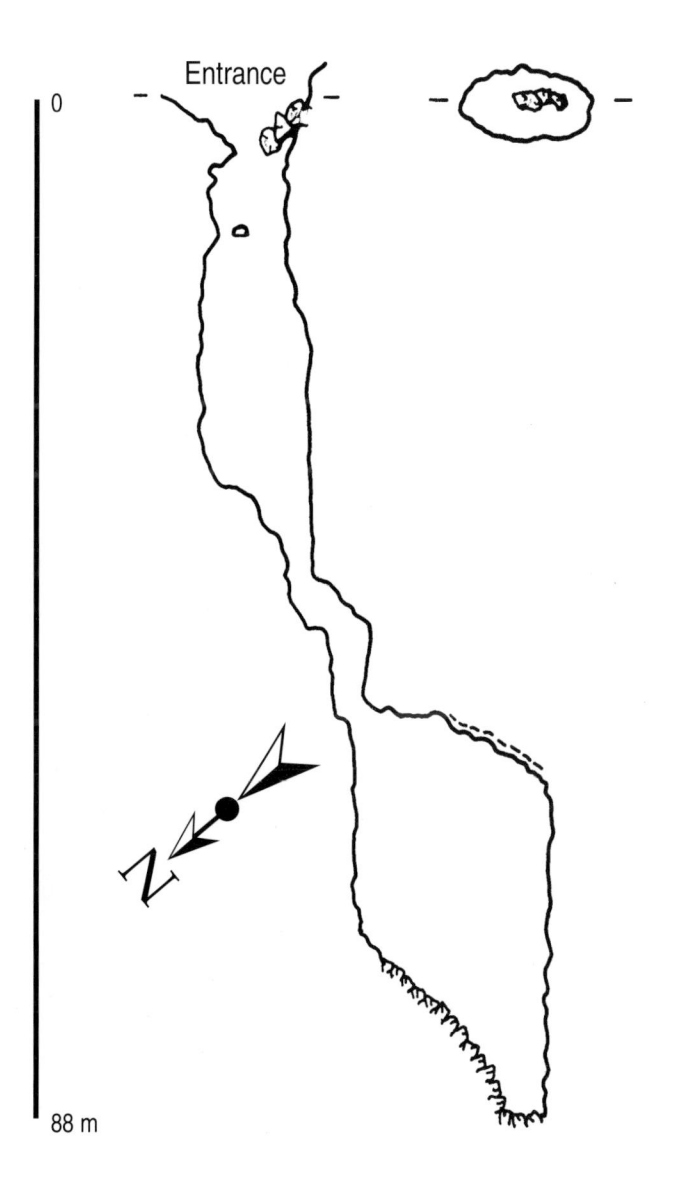

Radolova Yama

Gintsi Village

Name: **RALENA (Vodnata) (186)**
Kotel Town, Sliven Distr.
Length 450 m. Denivelation -15 m

The cave Ralena is situated in the area of Zlosten, south of Yablanovo, Eastern Stara planina, by the river Kaldere. Kotel is the starting point. Protey Sliven Caving Club should be consulted on the way to the cave. The cave entrance is marked by the sign [1] 4711.

The cave is situated southwest of Alibaba Summit (949 m), on the western slope of Chatalkaya hills, about 50 m east of the river. The entrance (50 x 50 cm) is at the foot of a small rock, facing west. After a squeeze with a gravel floor, 7 m long, we reach a small chamber. Initially, the gallery ascends, developing in NE direction, then, from point 16 a slippery descent starts to the semi siphon at point 21. In dryer years, the semi siphon has a lumen of 40–50 cm, but in wet periods the water level rises and the lumen is closed – the semi siphon becomes a siphon.

After the semi siphon, there are small ascending verticals (no rope needed). At point 40, and even more at point 45–46, there are small lakes up to 50 cm deep. In rainy weather, the second one could become a siphon.

At point 51 (4.3 m), there is a descending vertical, where a ladder is needed. About 16 m after the vertical, there is a small flooded branch to the left. At its end the lumen above the water is 10 cm. Further, a waterfall could be heard. Nobody has reached it yet.

Further along the main gallery, above point 71 on the right, there is a small ascending chamber.

Below point 71, we reach a small chamber and a descending vertical of 5 m. Then, after a steep slope, a siphon (diameter 4 m) is reached. Below the vertical, there is an ascending muddy vertical (exploration not finished, continuation possible).

The cave is formed in dark grey limestone from the Upper Cretaceous (Maastricht), in the eastern part of Kipilovo Syncline. Typical are the many irregular concretions, often in layers.

Primary forms in the cave are the evorsion domes and the many well expressed cave karren with non-dissolvable flint concretions.

Secondary (flowstone) formations are seen up to the siphon (stalactites, stalagtones, curtains and dendrites).

In the cave, graffiti have been found, left by treasure hunters. Some of the legends say, that the cave was an old Roman minting yard, or that a 120 kg gold statue of Indje Voyvoda is hidden there. In the fall of 1999, several treasure hunters tried to pump the water, using a gas pump. As a result, three of them were gased and one died.

No special biological collection has ever taken place. The main danger in the cave are the semi siphons. During floods, the water level rises and the semi siphons turn into siphons.

The exploration, observations and the description were made during expeditions of Protey Sliven Caving Club in August – September 2001 by Dimitar I. Dimitrov. The map was made by N. Dimitrov, D. Dimitrov, R. Soserova and Ch. Bankov.

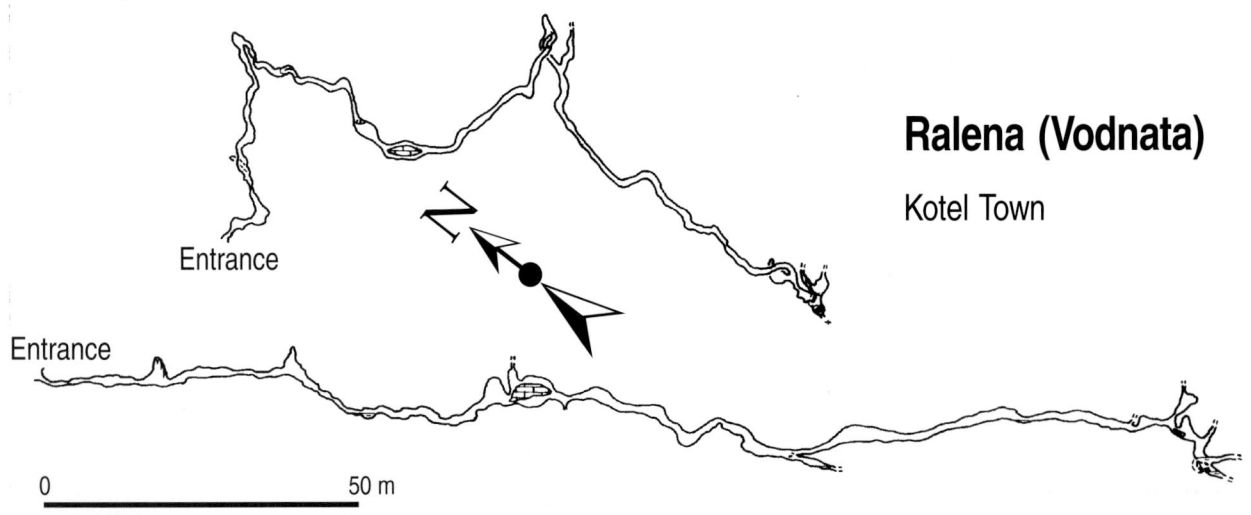

Ralena (Vodnata)

Kotel Town

Name: **RAYCHOVA DUPKA (133)**
Cherni Ossâm Village, Lovech Distr.
Altitude 1400 m. Length 3333 m. Denivelation – 377 + 10 m (the deepest in Bulgaria)
Coordinates: N 42° 46' 03.2" E 24° 41' 02"

A two-storey cave pothole in the area of Steneto near Cherni Ossam, in the upper part of the valley Malak Chaushov Dol, 8–9 km SW of Neshkovtsi.

The biggest vertical is at the beginning (-8 m), the remaining part is a descending (30 °) gallery, which is easy to penetrate. Many karst forms, sinter lakes, stone waterfalls, curtains, etc. The cave ends in a mud siphon. Formed in north-south direction, in the contact zone between the limestone and the mergels. Water temperature 7.5 °C. The main gallery is 920 m long.

Situated in the Biosphere Reserve Steneto and in the buffer zone of Central Balkan National Park.

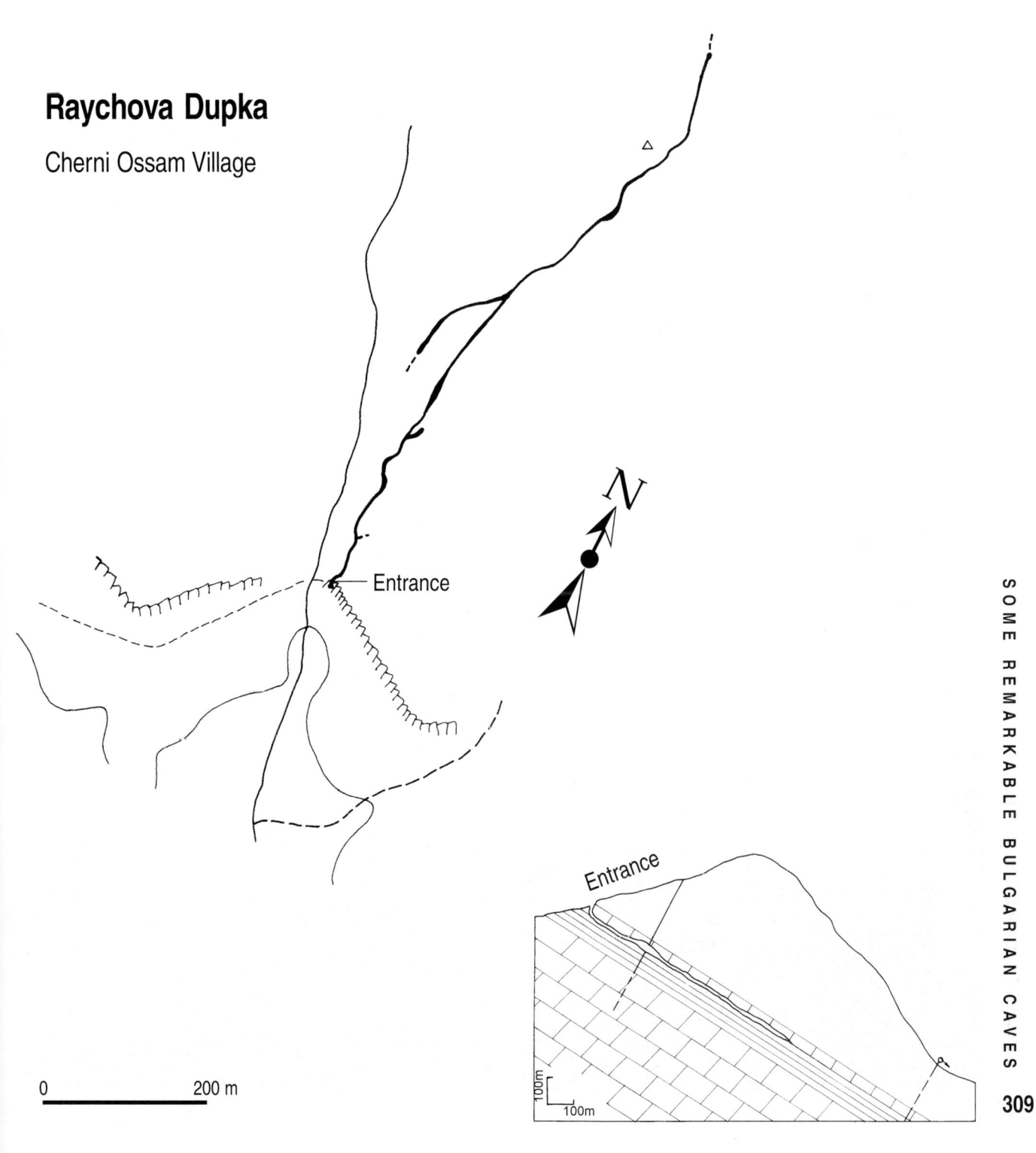

Raychova Dupka

Cherni Ossam Village

In the foots of T. Daaliev is the entrance of the deepest cave in Bulgaria – photo Trifon Daaliev

Pentcho Stanchev at the bottom of Raychova Dupka during the first cave exploration – photo Vassil Balevski

Stalactites after Mercedes Gallery – Vassil Balevski

In the Mercedes gallery – Vassil Balevski

Name: **RAZHISHKATA PESHTERA (51)**
Lakatnik Railway Station, Sofia Distr.
Length 316 m. Denivelation -22 m
Coordinates: N 43° 05' 24.2" E 23° 23' 10.4"

A cave in Lakatnik Rocks, above the cave Temnata Dupka (an older, upper storey of it). Spacious, often visited, with a well maintained footpath leading to it.

The first part of the cave is covered with dolomite "flowers" and is very dusty. Under the dust, vestiges from the Neolith lie, displaced by gold-seekers. The big trench was dug by the first caving brigade (1948). Then follow three meters of deposits, which are not yet studied.

Near the entrance, the cave formations start. The drops of water have formed some wells, which can be used for drinking. One stalagmite is often compared to an eagle. The many pieces of rocks formed by the deep frost weather give the impression of an uneven floor. Another left bent over blocks leads to big, but very damaged formations. Among blocks and destroyed formations, we reach a huge rock topple. Some sinter lakes contain a small amount of water. Dendrites and other cave formations can be seen only very high on the walls. Over the huge blocks, we reach the upper part of one of the biggest rock topples in Bulgaria, 28 m above the basement. It is clear that the cave goes on, and the new parts would be undamaged. Cavers from many generations have tried to penetrate beyond the rock topple, so far without success.

The cave has been known to local people for a long time and has been described and explored by many cavers.

Fauna: more than 50 animal species have been recorded, among which the stygobite Crustacean *Speocyclops infernus* (Copepoda) and the troglobites *Balkanoroncus hadzii* (Pseudoscorpiones), *Centromerus bulgarianus* (Araneae), *Typhloiulus bureschi* (Diplopoda), *Pseudosinella duodecimocellata* (Collembola), *Plusiocampa bureschi* (Diplura).

Razhishkata peshtera

Lakatnik Railway Station

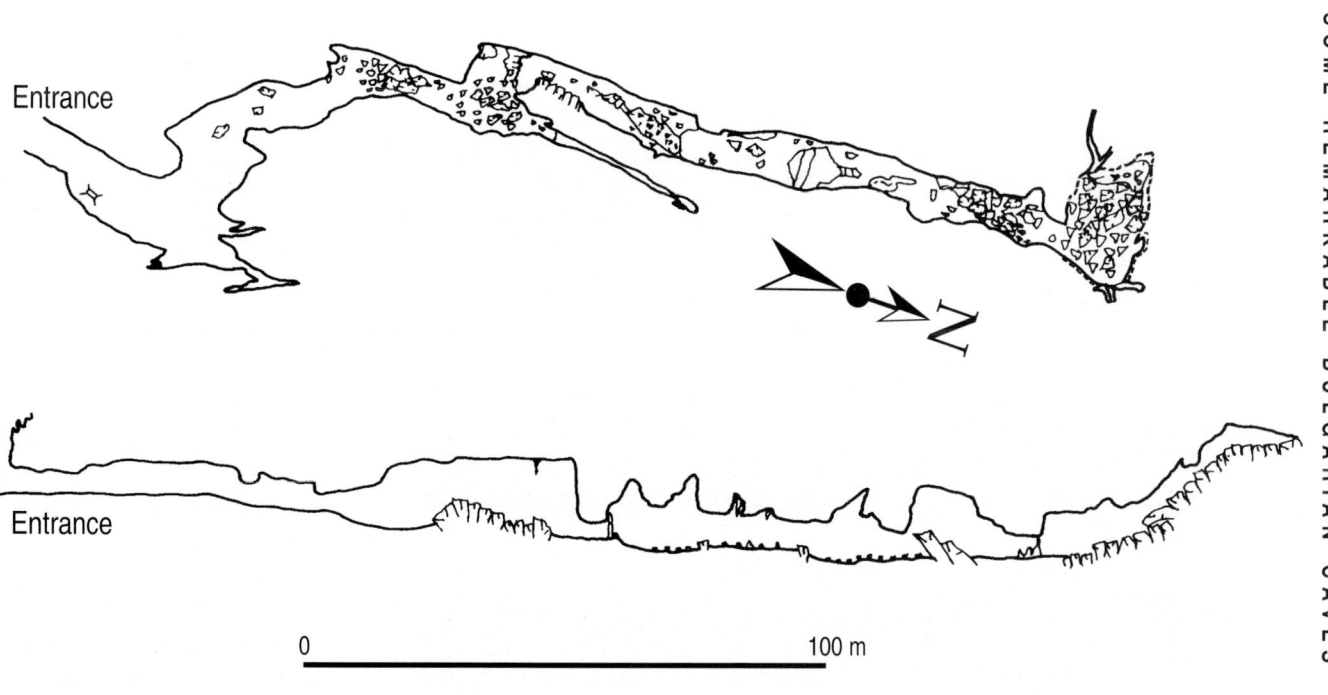

V. Peltekov with Technical University students at the entrance of Razhishkata Peshtera – photo Valery Peltekov

One of the cave's halls – photo Valery Peltekov

The first explorers of the Razhishka Peshtera – 1925

Name: **RUSHKOVITSA (Prelaz, Partizanskata) (9)**
Salash Village, Vidin Distr.
Length 422 m. Denivelation -36 m

    The cave Rushkovitsa is situated ca. 3 km NE of Salash, in the locality Rushkovitsa, in Turska Glama, on its southern side, 10 meters under the ridge. Some 3–4 m from it, its sister cave Rushkovitsa-1 (Prelaz -1) is situated.

    Rushkovitsa is a two-entrance, horizontal, dry cave. The entrance has dimensions 2 x 1 m and is facing north. About 10 m from the entrance, the gallery branches off. After 55 m, the left gallery ends in a squeeze. The floor is covered with stones and blocks (of about 1 m^3). The right gallery is 3 m wide and after 80 m it reaches the labyrinth part of the cave. Up to this place, on the floor there are many big stalactites. Here is the connection with the second entrance of the cave. For to 6–8 m the gallery winds, the height of the ceiling reaching 3–4 m. On the floor, besides the fallen blocks, there are many stalagmites. The gallery ends in a squeeze.

    Explored and surveyed during an expedition of Edelweiss, Sofia, by N. Gladnishki, Yu. Velinov, in August 1969. Surveyed as Partizanskata Peshtera (known to local people as Rushkovitsa).

Rushkovitsa (Prelaz, Partizanskata)

Salash Village

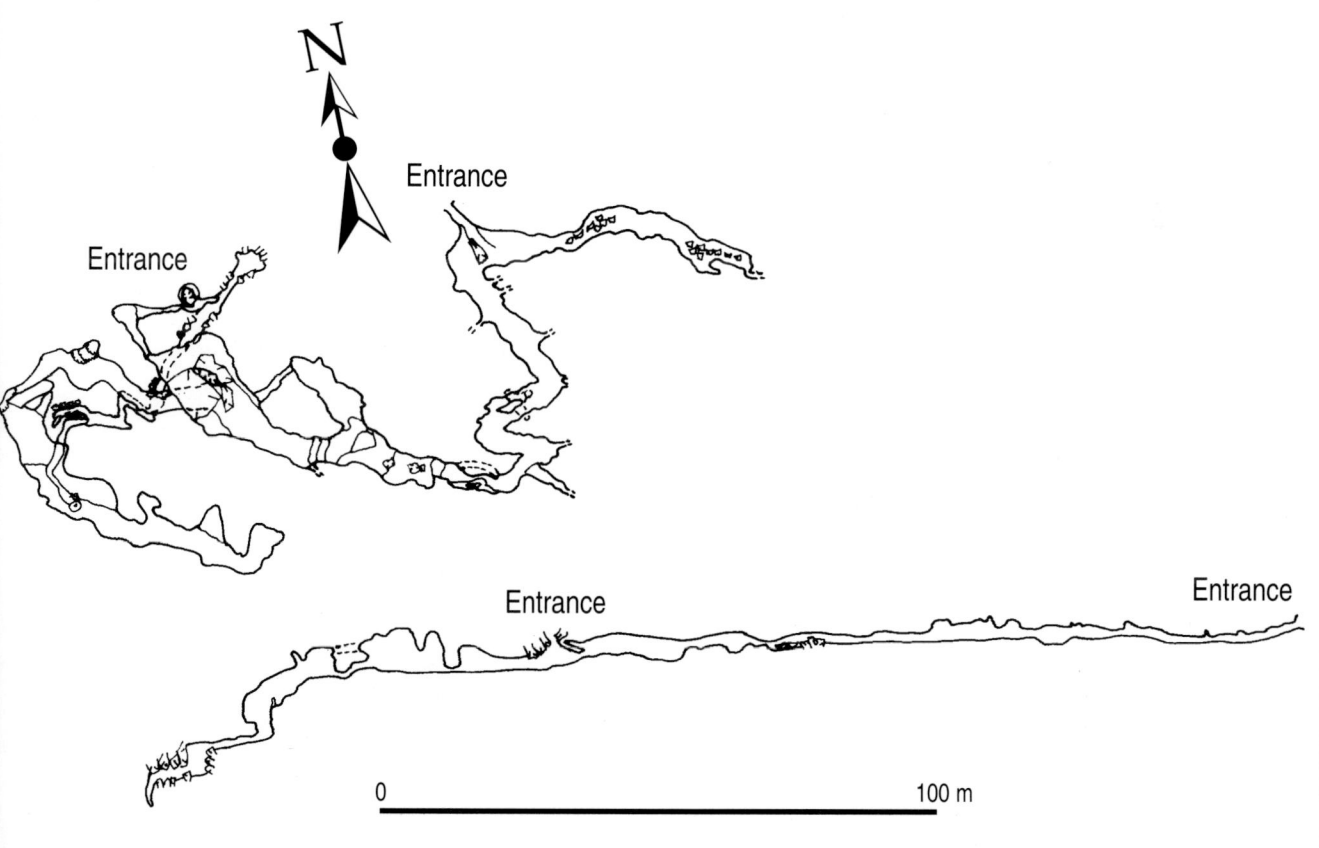

Name: **RUSSE (162)**
Emen Village, Veliko Tarnovo Distr.
Length 3306 m. Denivelation -100 m
Coordinates: N 43° 07' 19.65" E 25° 19' 53.20"

    A cave pothole in Aptian (Cretaceous) limestone, two kilometers from Emen, behind the ridge Perchema. A narrow, canyon-lake cave with several storeys and lots of clay. Many verticals, the deepest being 20 m. Crawls are followed by lakes of different depth. In the bigger chambers there are huge clay deposits, called The Clay Mountains. The most spacious chamber has dimensions 30 x 40 m.
    A water cave with several waterfalls and temperature of the water 10 °C. When in flood, visits are not advisable.
    The cave was discovered after cavers from Akademik Russe Caving Club dug a five-meter well in 1985–1986. Among the boulders, a horizontal part starts. Difficult to penetrate.

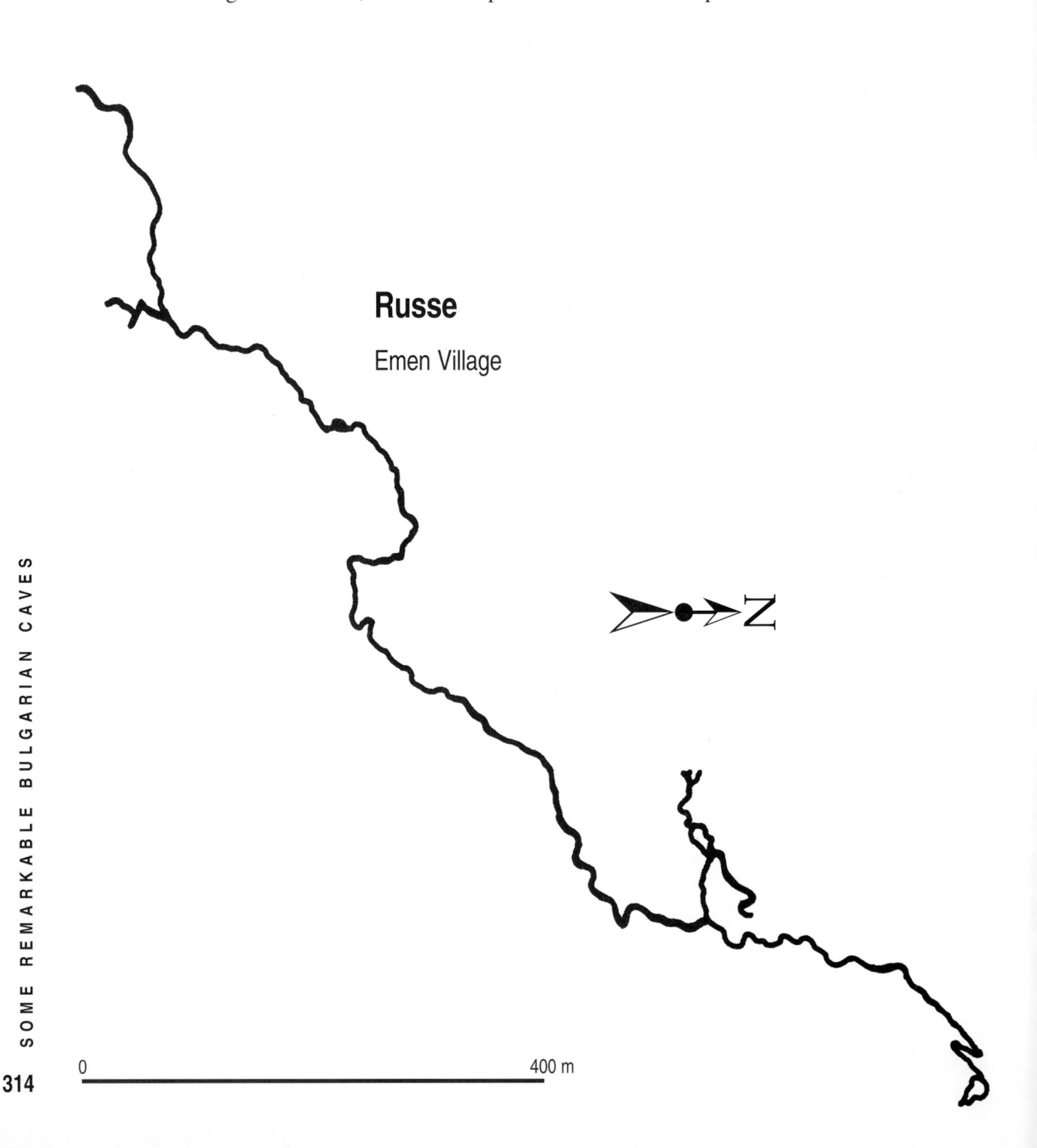

Russe

Emen Village

0 — 400 m

The entrance – photo A. Jalov

Name: **SAEVA DUPKA (108)**
Brestnitsa Village, Lovech Distr.
Length 205 m. Denivelation 22 m (-12; 10 m). Surface 15 620 m^2
Coordinates: N 43° 02' 51.3" E 24° 11' 16.4"

A cave in the Ledenishki Rid, 2.5 km from Brestnitsa, in limestone from the Upper Jurassic (Titonian).

A one-gallery, descending and ascending cave with five chambers, the biggest being Srutishteto (53 x 26 x 17 m). Karstic accumulative forms, deposits of terra rossa, boulders. Dripping water.

The first information about the cave was provided by G. Zlatarski in 1883 and Škorpil in 1900. First exploration and survey – by N. Atanassov and D. Papazov on August 20-21, 1932, and At. Stefanov and N. Atanassov on July 10-13, 1935. Again explored in 1946 and 1949. Detailed survey and scientific research done by V. Popov (geomorphologist) in 1968.

A show cave (since 1967), a Natural Monument by virtue of Decree 2180/10.10.1962), together with the adjacent 29 ha of land.

Fauna: 26 species known, including the troglobites *Neobisium bureschi* and *Balkanoroncus bureschi* (Pseudoscorpions), *Paranemastoma (Buresiolla) bureschi* (Opiliones, most probably an error) and the beetles *Netolitzkya maneki* and *Tranteeviela bulgarica*.

Saeva dupka

Brestnitsa Village

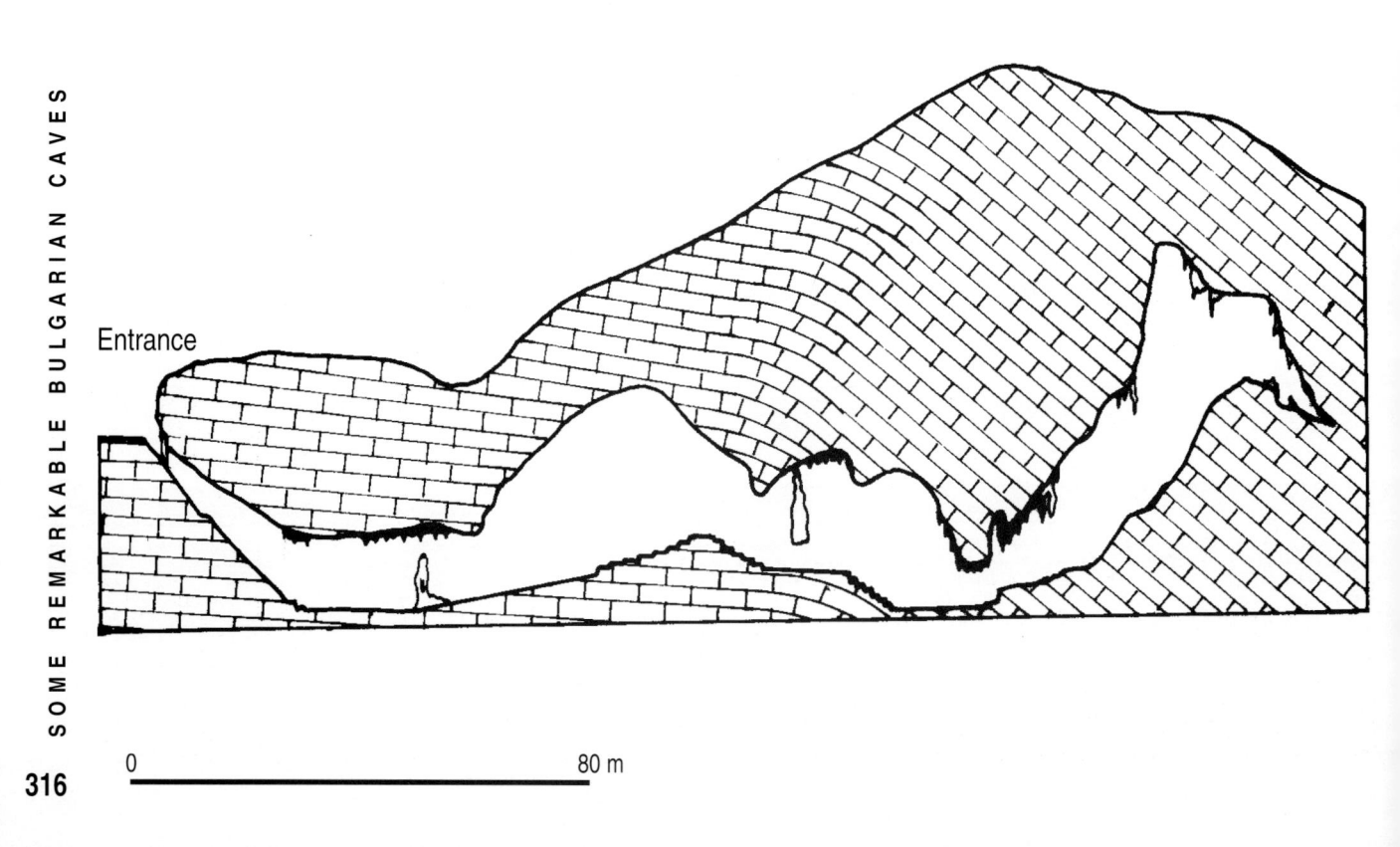

Cosmos Hall – photo Vassil Balevski

The beauty wall – photo Vassil Balevski

Name: **SAGUAROTO (GOLYAMATA PESHTERA V BILIN DOL) (67)**
Gubesh Village, Sofia Distr.
Length 2217 m. Denivelation -135 m
Coordinates: N 43° 05' 41.84" E 23° 04' 46.86"

A cave in the Trias limestone of Vidlich anticline in the doline Bilin Dol of Vuchi Baba Ridge. A Descending, two-storey, branched water cave, developed along a SW/NE fault. Elliptic profile of the cross section with dimensions: 2.5 m wide, 4 m high (except for one 7.5-m section of the end parts, where the height is up to 20 m). Many karstic forms, speleothems, stone bridges, cascade lakes and others. Along the main gallery, the lower storey and the only branch, underground streams flow, joining in a river and sinking into a boulder chock at the end of the cave.

The cave was discovered, explored and surveyed by Cherni Vrâh Sofia Caving Club in 1980–1983. The name was given by the discoverers and comes from the known Mexican cactus Saguaro (after the shape of a stalagmite).

Saguaroto (Golyamata Peshtera v Bilin Dol)

Gubesh Village

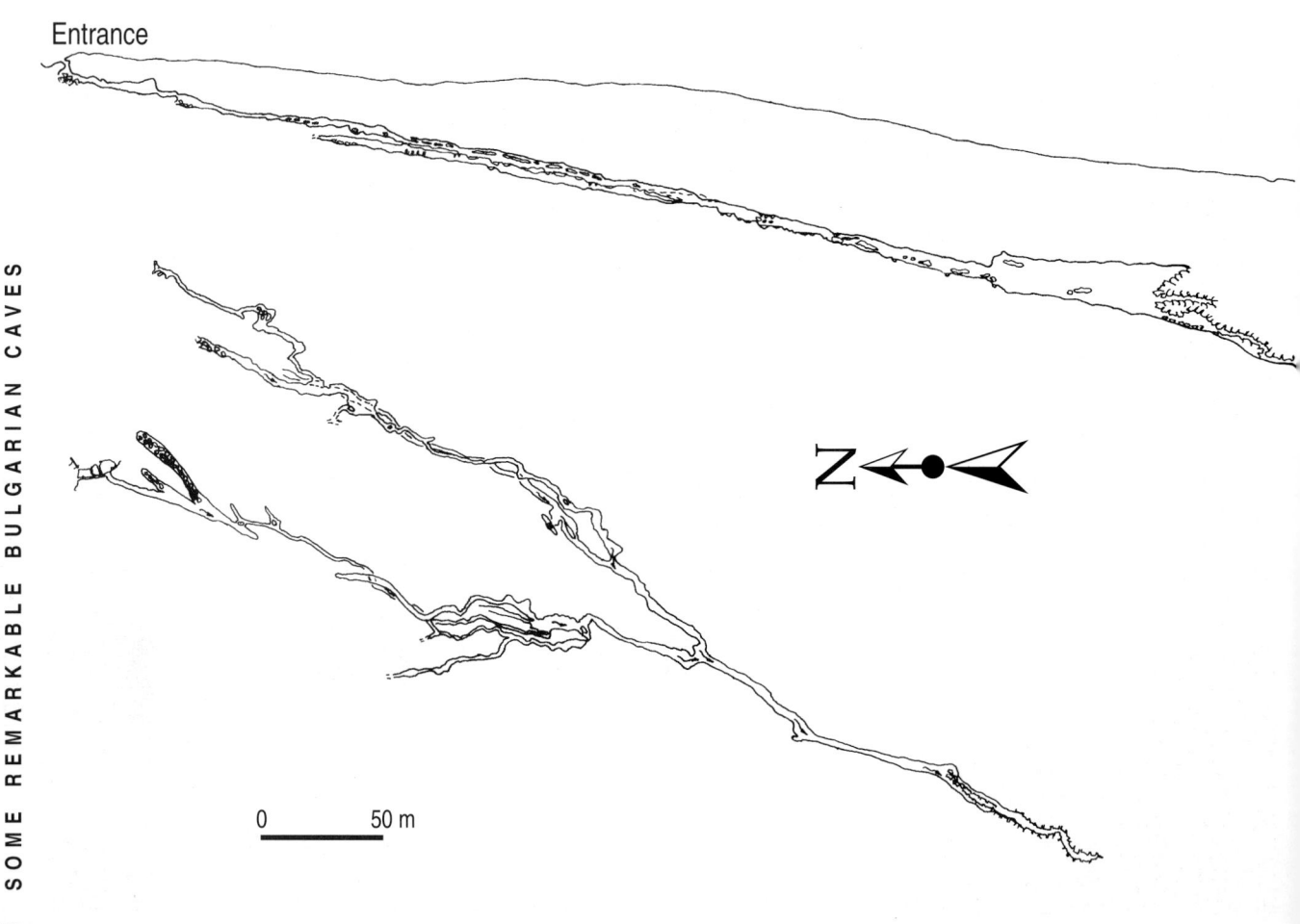

The tongue – photo BFS archive

Name: **SAMUILITSA (VASSILITSA) (85)**
Kunino Village, Vratsa Distr.
Length 190 m. Denivelation -14 m. Volume 2585 m^3
Coordinates: N 43° 11' 37.5" E 24° 00' 06.7"

A horizontal dry cave. Minimum and maximum dimensions of the cross section: width 1.8–18.2 m, height 0.5–10.5 m. The end chamber – 25 x 28 x 3.5 m. Sinter, flowstone, deep clay deposits. Dripping water. Important site for excavations (1954, N. Djambazov discovered artefacts from the Middle Palaeolith (Charantien and Mousterien)). Surveyed in 1964 by cavers from Chepelare.
Fauna: 14 species, including the Isopod *Trichoniscus bureschi*.
Protected Natural Monument, together with 3.5 ha of adjacent land (Decree 1799/30.06.1972).

Samuilitsa (Vassilitsa)

Kunino Village

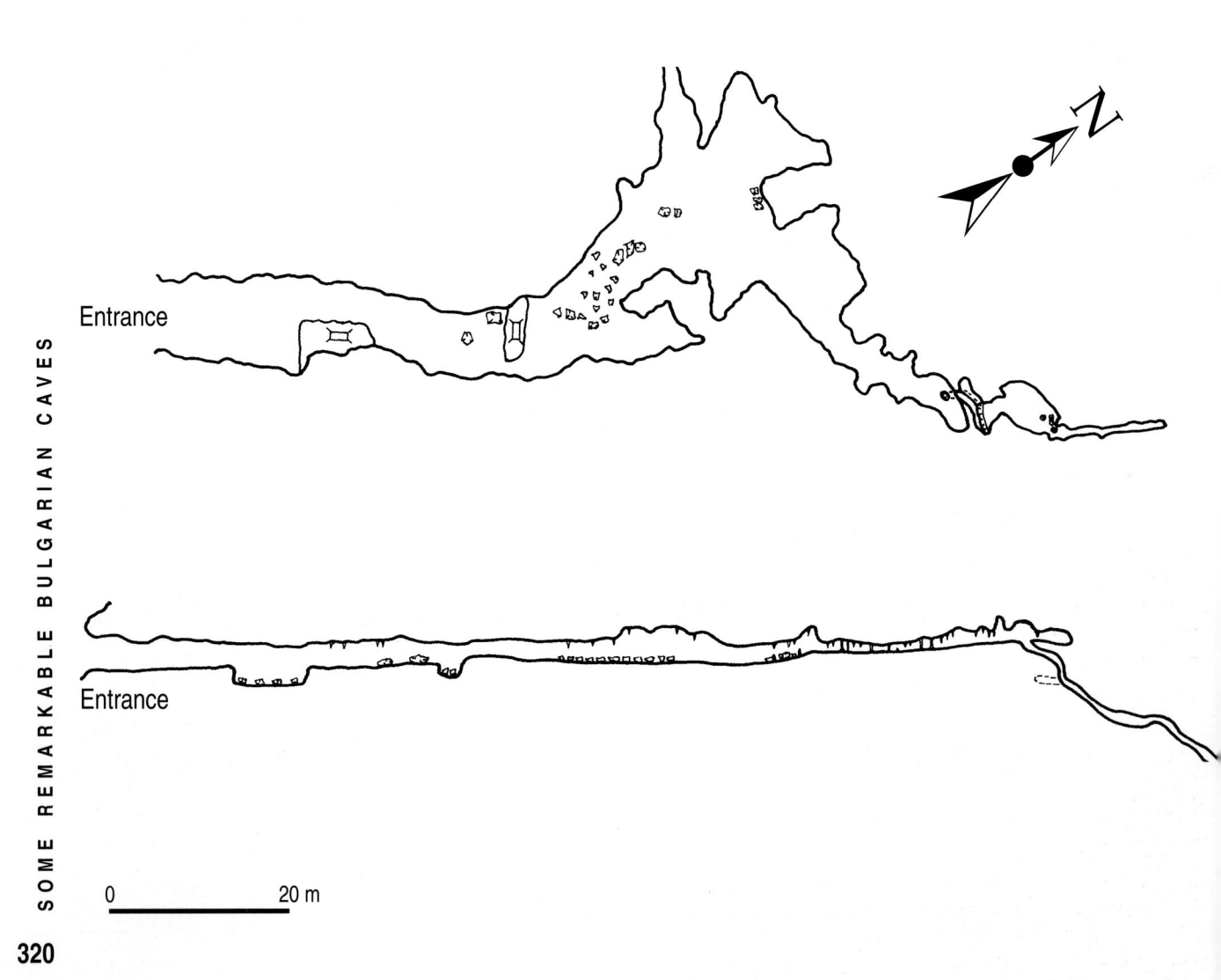

Name: **SEDLARKATA (MANDRATA) (101)**
Rakita Village, Pleven Distr.
Length 1100 m. Denivelation -13 m. Surface 3470 m². Volume 9284 m³
Coordinates: N 43° 17' 26.8" E 24° 17' 55.3"

A water cave, 3.5 km SE of Rakita, on the bank of Rakitska Reka, in the locality Ezeroto. Formed in Upper Cretacetous limestone (Maastricht).

A two-storey, branched cave, developed along two main NW/SE and SW/NE faults. Through a complex system of galleries, the upper entrance leads into the lower part, through which a river flows out of the cave. The average cross section of the galleries is: width 3.24 m, height 3.43 m. Poor in flowstone. Clay and gravel deposits. Minimum outflow on November 15, 1986–4.5 l/s.

The first study of the cave took place in 1970–1971 by cavers from Ivan Vazov Sofia Caving Club. In 1982, a team from the same club (B. Tranteev, V. Mustakov, D. Nanev, I. Yordanov) corrected the survey to its present dimensions. In 1986, cavers from R. Popov Pleven Speleoclub also explored the cave for the needs of water piping for the town of Telish. Between the two entrances is the stone bridge Sedlarkata, declared a National Monument (Decree 1799 of 1972).

Fauna: nine species known, including some troglophiles.

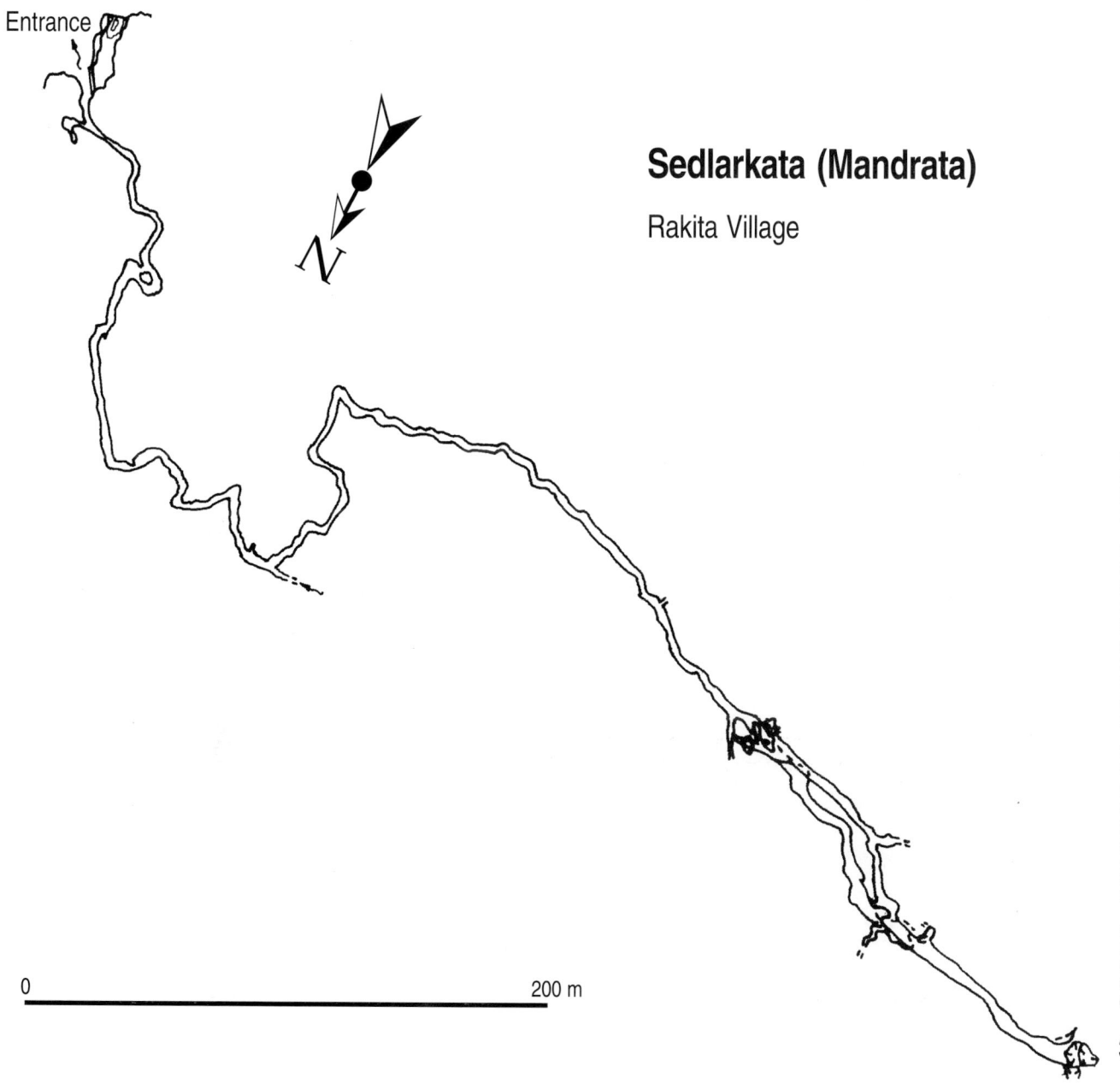

Sedlarkata (Mandrata)

Rakita Village

Name: **SERAPIONOVATA PESHTERA (50)**
Cherepish Village, Vratsa Distr.
Length 80 m. Denivelation -15 m.
Coordinates: N 43° 06' 04.5" E 23° 36' 56"

A cave at the end of the valley Klyuchni Dol, near the cave Dvuvhodovata Peshtera. It also has two entrances, after them a very inclined chamber begins. In it upper end, we find the opening of a vertical shaft with a complicated configuration.

According to a legend, monks from the Cherepish Cloister hid in the cave.

On the high, semidark ceiling of the cave, on July 6, 1960, a colony of ca. 1000 bats (*Myotis myotis* or *M. blythi*) was heard. In the lower storey, thousands of long-winged bats (*Miniopterus schreibersi*) wintered, together with several big horseshoe bats (*Rhinolophus ferrumequinum*) (observations of S. Simeonov and P. Nedkov at the end of March 1963).

The Invertebrates, which have been collected by P. Beron since July 1960, include mainly guanophiles.

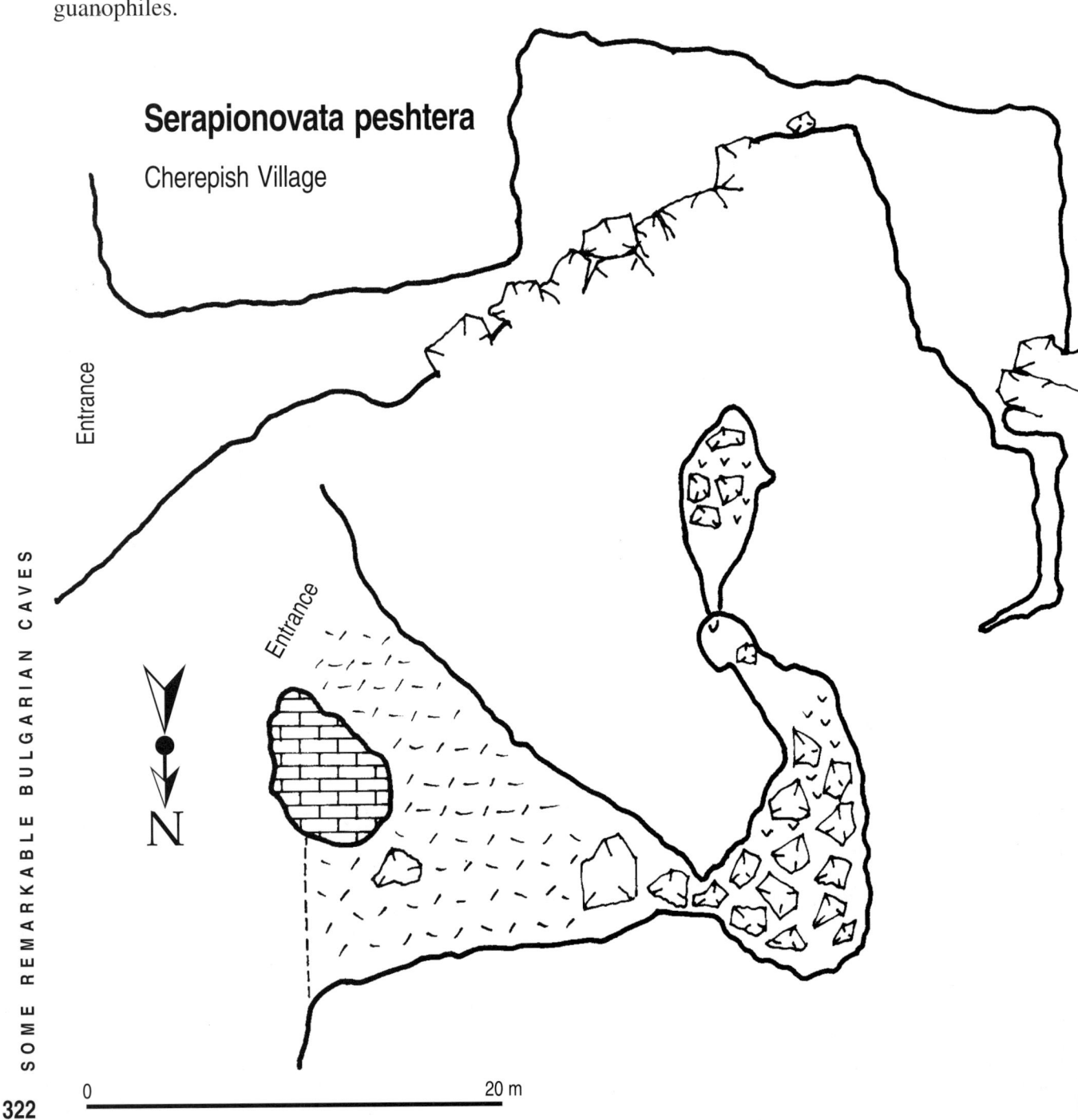

Name: **SHIPKATA (47)**
Zverino Village, Vratsa Distr.
Altitude 1005 m. Denivelation -60 m
Coordinates: N 43° 07' 29.2" E 23° 34' 36.5"

A pothole in a *Quercus – Carpinus* forest, in the area of the summit Yavorets. The entrance (6 x 4 m) is in a funnel. After several meters through a small opening, the first vertical of 30 m descends to a small platform. Here, another 30-m vertical starts, reaching the bottom of the pothole. The pothole is not equipped for SRT.

Shipkata was discovered by T. Daaliev and S. Zanev from Edelweiss Sofia Caving Club. The bottom was reached by N. Gladnishki and T. Daaliev. The survey was done in 1975 by A. Anev and V. Genchev.

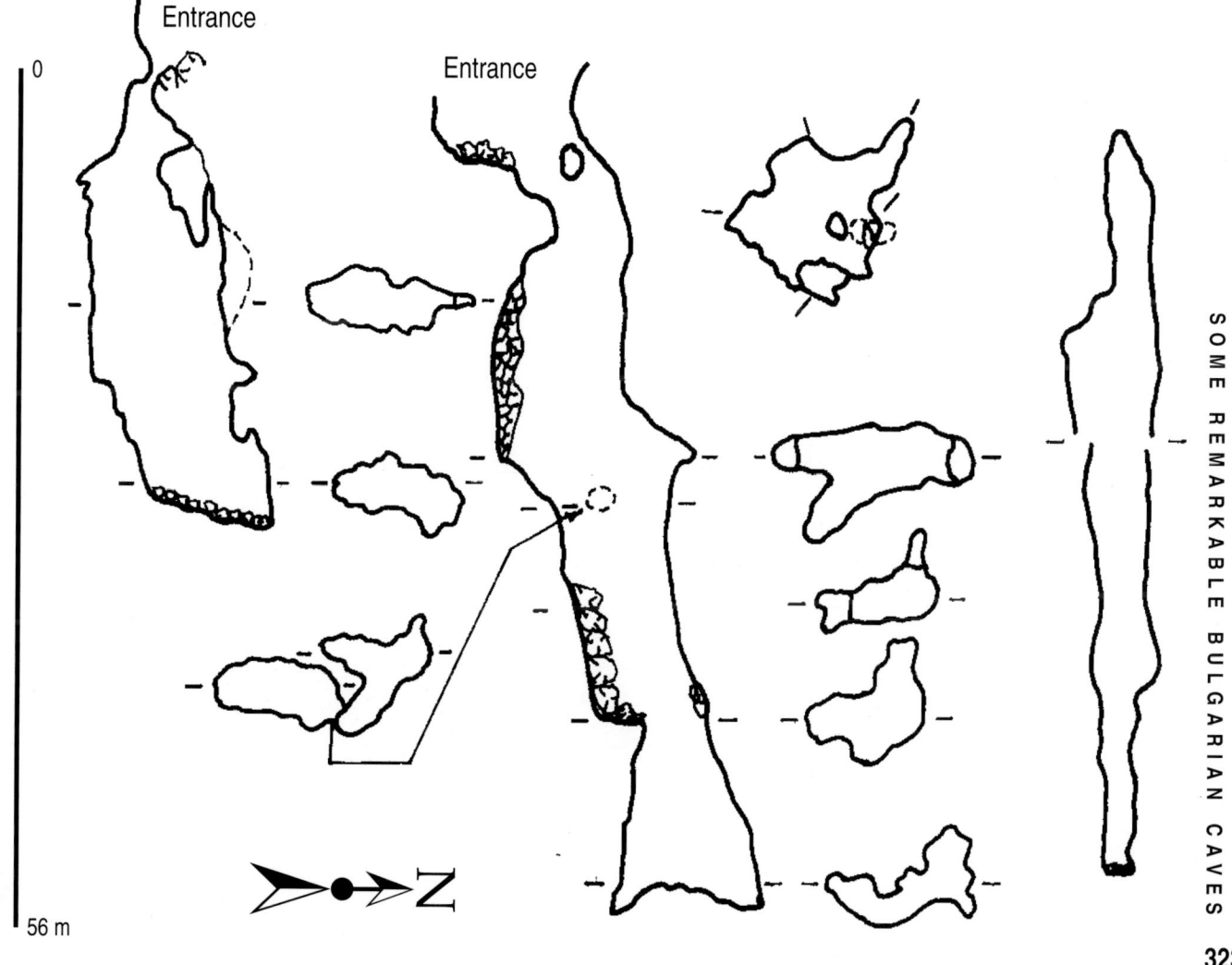

Shipkata

Zverino Village

Name: **SIFONA (NOVATA) (16)**
Dolni Lom Village, Vidin Distr.
Total length 575 m. Denivelation 43 (37.4; -5.6) m

A cave, 1135 m SE of Dolni Lom Village, Western Predbalkan, in the locality Suhata Padina.

Formed in limestone from the Upper Jurassic, inclined at 30°S, Montana Anticline. Developed along two main crevices, directed NE–SW and N–S.

The entrance is a small crevice, oriented to the south. A relatively narrow gallery follows, being at first vertical, then inclined, and with a length of 10 m. From a depth of 5.6 m, the gallery becomes ascendening. During the spring-summer season and when there are floods, in this place an impenetrable siphon is created. After 10 m, the gallery becomes up to 2 m high and wide up to 2.5 m. On its floor, an underground stream runs. From this place, the cave goes in two directions: E-SE (an upstream gallery, ca. 330 m long) and a downstream NW gallery, 120 m long. The stream appears from a siphon, runs along the galleries and disappears in a gryphon in the NW end of the cave. The height of the galleries in both directions is from 0.6 to 1.5 m., their average width being 2.5 m.

In the cave primary underground formations are observed – facets, marmites, as well as secondary flowstone formations. Anomalous formations also exist in the cave.

The cave was discovered by A. Jalov and T. Stoychev from Aleko Sofia Caving Club in 1977, when, because of the siphon, only the entrance parts were explored and surveyed. In October 1979, cavers from Bel Prilep Belogradchik Caving Club penetrated the cave and discovered the remaining parts after seeing that the siphon had retreated. The cavers N. Gaydarov, V. Nenov and Em. Velizarov surveyed the cave.

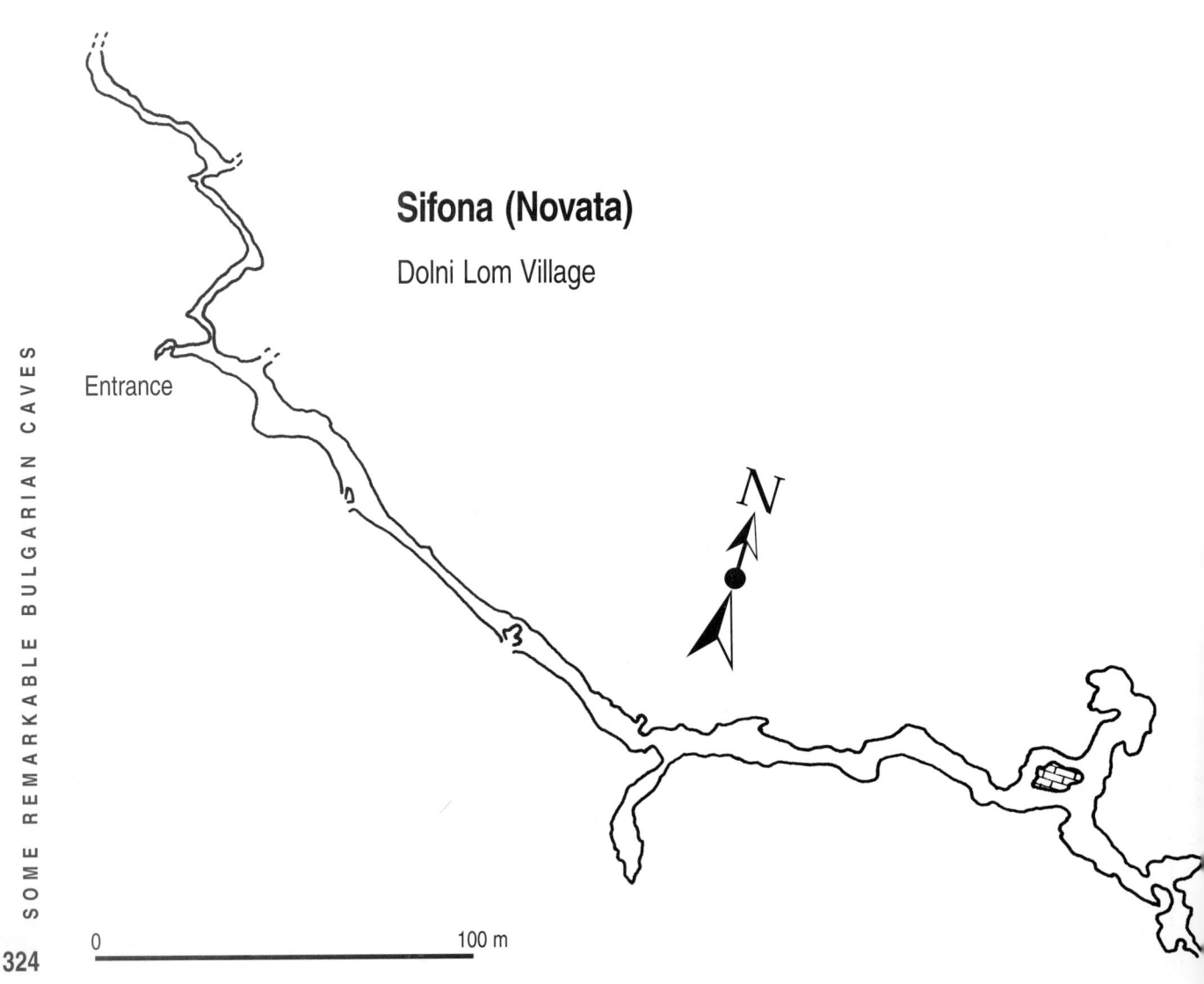

Name: **SINYOTO EZERO (105)**
Dragana Village, Lovech Distr.
Length 407 m. Denivelation -18 m. Surface 1378 m². Volume 2716 m³

We can find the cave Sinyoto Ezero in the locality Alichkovoto near Dragana Village, in a cliff on the left bank of the river. This is a permanently water cave. It is branched, horizontal, formed in limestone from the Cretaceous (Maastricht). The first 3–4 m are wide, followed by a 50-m, very low gallery. Then, a branch on the right leads to a parallel gallery (50 m long), joining the main gallery. About 80 m from the entrance, we reach an underground river, which after 30 m ends downstream in a siphon. Upstream, for ca. 200 m, the gallery is 4 m wide and 2.5–4 m high. The river forms pools, which are up to 1.5 m deep. The walls and the ceiling are covered with flowstone formations. The end is again a siphon.

In 1975, S. Nikolov, M. Karolinova and S. Savev from Studenets Pleven Caving Club explored the cave to 126 m. A complete survey of the cave was made by B. Garev, Y. Benchev and other cavers from the same club. The cave is known to local people, who gave it the name Sinyoto Ezero (The Blue Lake).

Fauna: the names of two troglobites from this cave (*Trichoniscus garevi* – Isopoda and *Duvalius garevi* – Coleoptera) commemorate Borislav Garev, an active Biospeleologist from Pleven, who passed away very early.

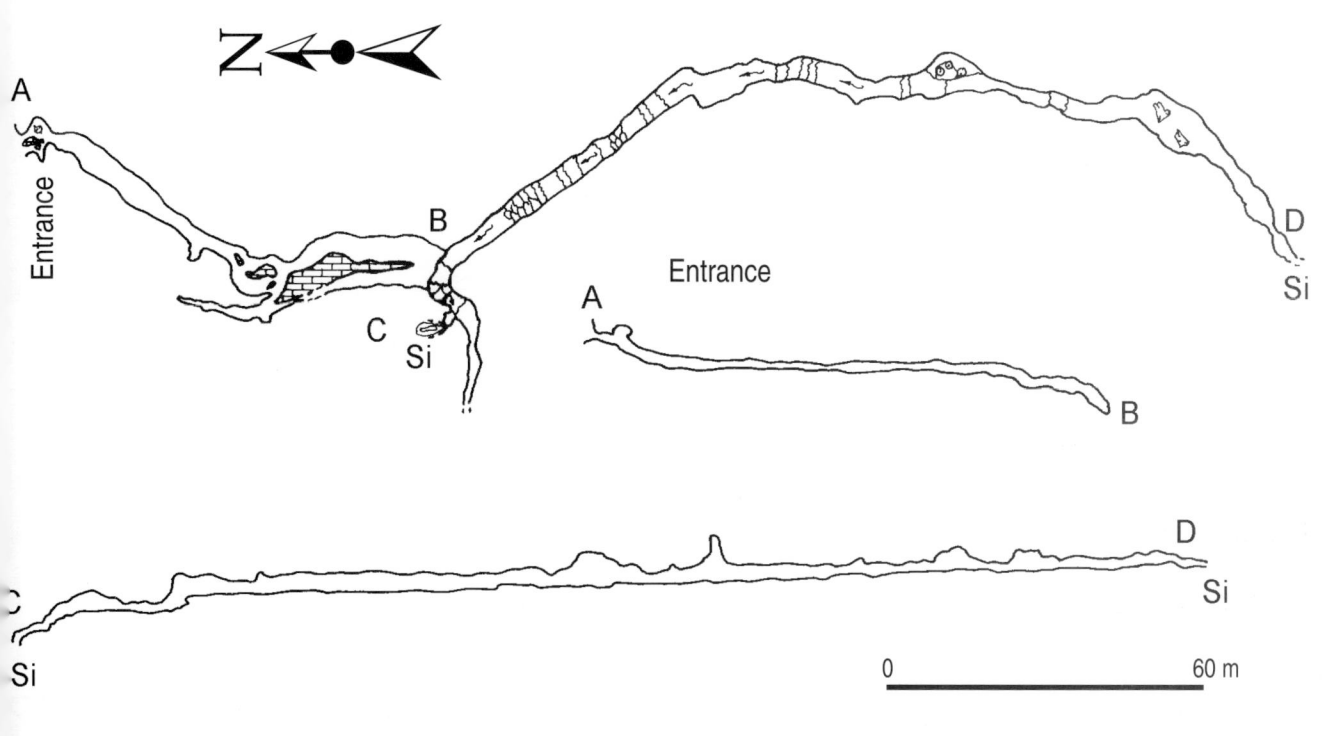

Sinyoto Ezero

Dragana Village

Name **SINYOTO KOLELO (117)**
Teteven Village, Lovech Distr.
Length 300 m

The entrance of the cave Sinyoto Kolelo is situated in the area of Vikaloto, above the steep cliffs of Teteven Anticline, south of Teteven. This entrance leads into an atrium with many stones and blocks on the floor. A gallery with very damaged curtains ends in a small vertical of 4 m, from the bottom of which starts the lower level. Here, the flowstone formations are better preserved. The cave ends in a lowering.

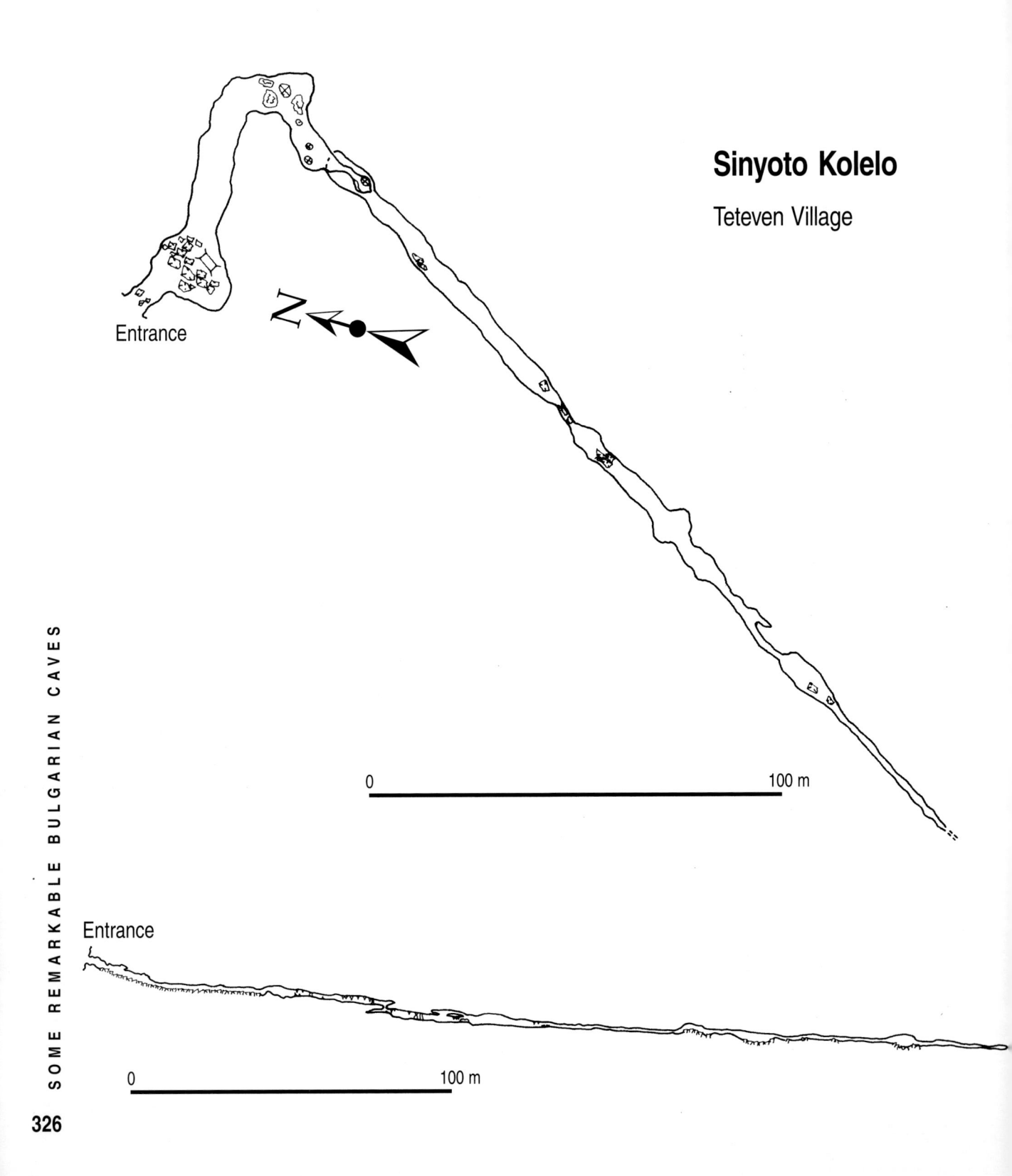

Sinyoto Kolelo

Teteven Village

Name: **SIPO I (36)**
Gorno Ozirovo Village, Vratsa Distr.
Altitude 905 m. Denivelation -80 m

The pothole Sipo I is situated in the locality of the same name, in the area of the pothole Belyar, situated ca. 5 km NW of Ledenika Hut. The entrance opens among blocks. The pothole is 78 m deep, including 76 m of full vertical. The bottom is a strongly inclined scree. The second half of the vertical is dangerous, because of falling stones. No secondary cave formations.

The pothole of Sipo is immediately close to Turloto – part of the big fault, starting from Lakatnik and ending by the Cloister of Matnitsa.

Explored by cavers from Vratsa in 1970. Surveyed in 1976 by S. Nenov and P. Tranteev. The actual map was made by L. Savchev in 1979.

Fauna (collected by P. Beron in 1963): includes the troglobite Diplopod *Typhloiulus bureschi*.

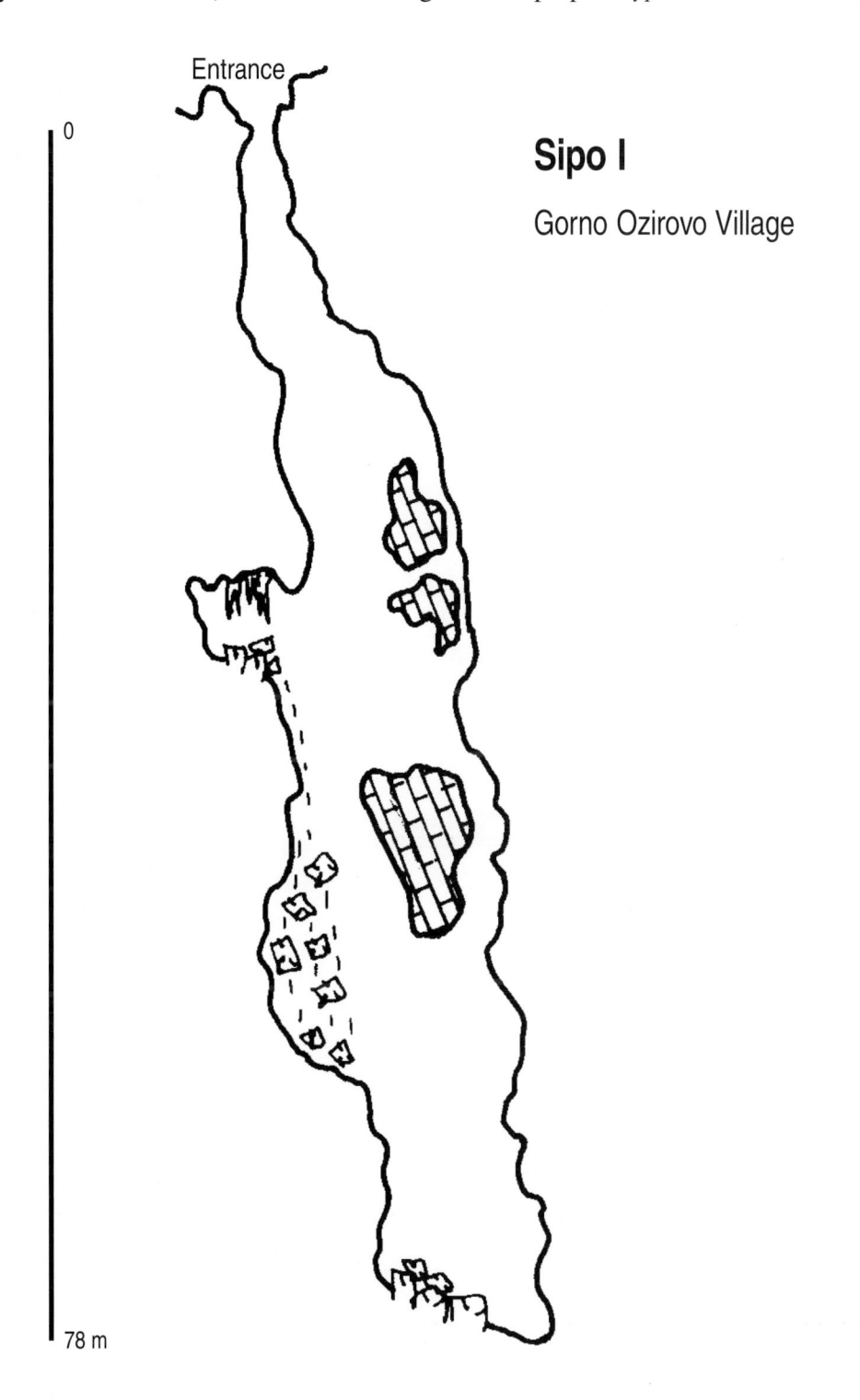

Sipo I

Gorno Ozirovo Village

Name: **SKOKA (106)**
Dragana Village, Lovech Distr.
Total length above 1000 m. Denivelation 9 m

A cave about four km from Dragana Village, upstream Kamenitsa river, in the locality Skoka. Its entrance is situated 40 m from Skoka Waterfall, on the (geographically) right bank of the river. It is a permanently water cave – through it an underground stream runs. The main gallery is 2–8 m wide and 1–6 m high. Its narrowest place is at the 230th meter of the semi siphon. The cave is horizontal, one-storey. In it there are secondary formations – stalagtones, stalactites, stone "waterfalls", curtains, ribs and peculiarly shaped stalactites.

Up to 230 m, the cave was explored by S. Savev from Studenets, Pleven, in 1976. According to him, at that point an end siphon is reached. In 1979, A. Georgiev, K. Hristov and V. Yordanov from Zlatna Panega Lukovit Caving Club made a horizontal map of the cave up to 723 m. After 1990, B. Garev and I. Gunov renewed the research in Skoka. I. Gunov overcame the semi siphon, then the siphon and discovered about 300–400 m of new galleries.

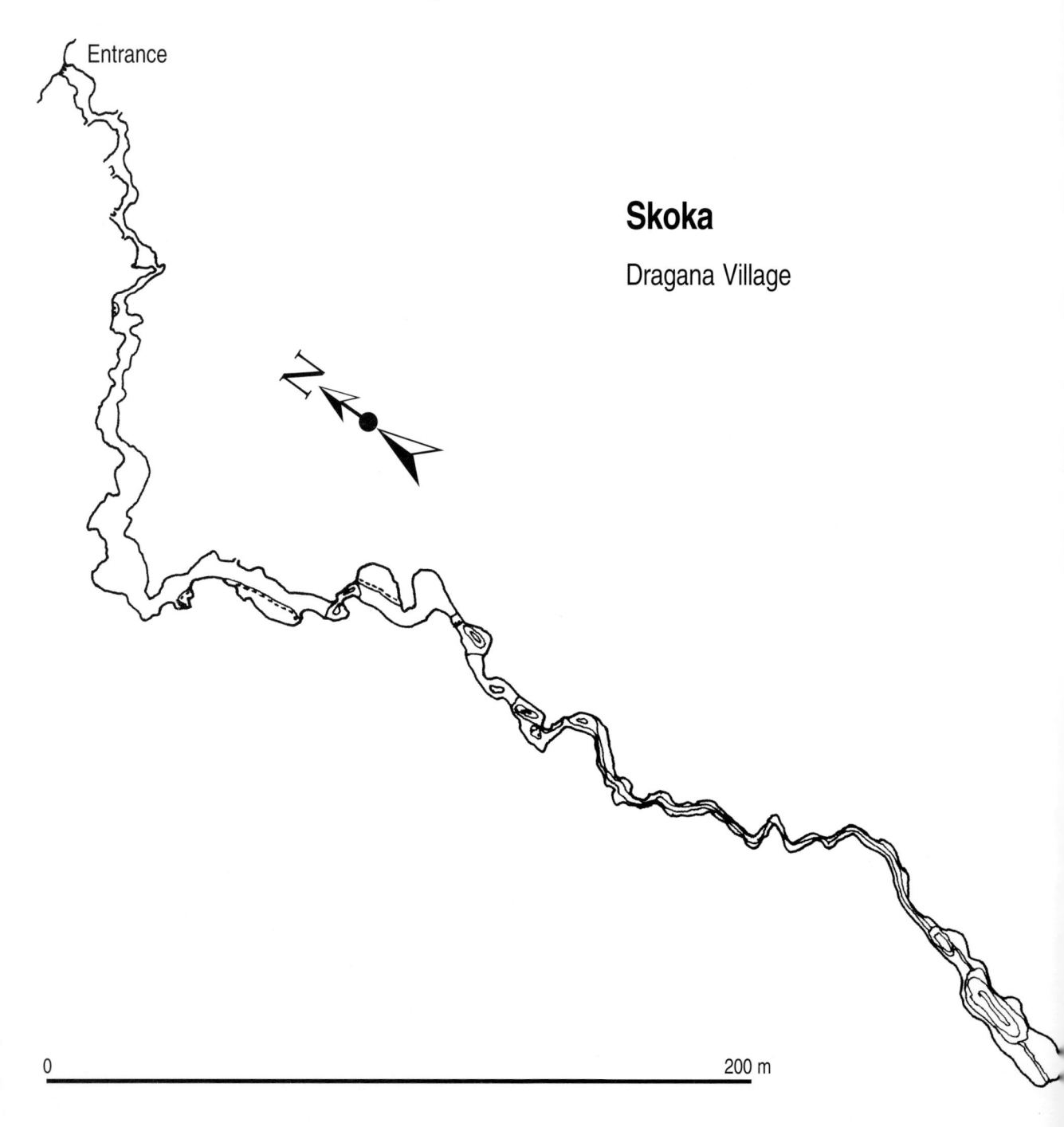

Skoka

Dragana Village

Name: **SOKOLSKATA DUPKA (PESHTERA) (38)**
Lyutadjik Village, Vratsa Distr.
Altitude 715 m. Length 815 m. Denivelation 27 m
Coordinates: N 43° 18' 46.7" E 23° 26' 17"

Following the forest road upstream Cherna River to its end, we continue further upstream to the source of the river. After ca. 170 m, we reach the entrance of the cave, which is in a small cliff and faces west. The entrance (5 x 3 m) gives access to a gallery lowering to 80 cm. After 50 m, we find the "dry siphon", where sometimes a lake is formed. This is the lowest part of the cave. After 150 m, there is a side passage to the left (50 m long)and another passage leads to a narrowing in which running water is heard. The main gallery ascends to a spacious chamber, from which an ascending passage starts, continuing for some 50 m. From the chamber the gallery goes to the left (6 m wide, 5–6 m high, the distance to the next branch is ca. 150 m). A low chamber follows with a periodical lake, 6 x 10 m. The main gallery goes straight, bifurcates and through a narrow passage we reach the main river (the end of the cave). Description according to V. Hristov from Iskar Sofia Caving Club.

Note: The water level of the cave may change in a very short time. Depending on the weather, a hydrosuite or a dingy could be used. In high waters the cave becomes impenetrable.

The fauna includes the troglobitic spider *Centromerus bulgarianus*.

Sokolskata Dupka (Peshtera)

Lyutadjik Village

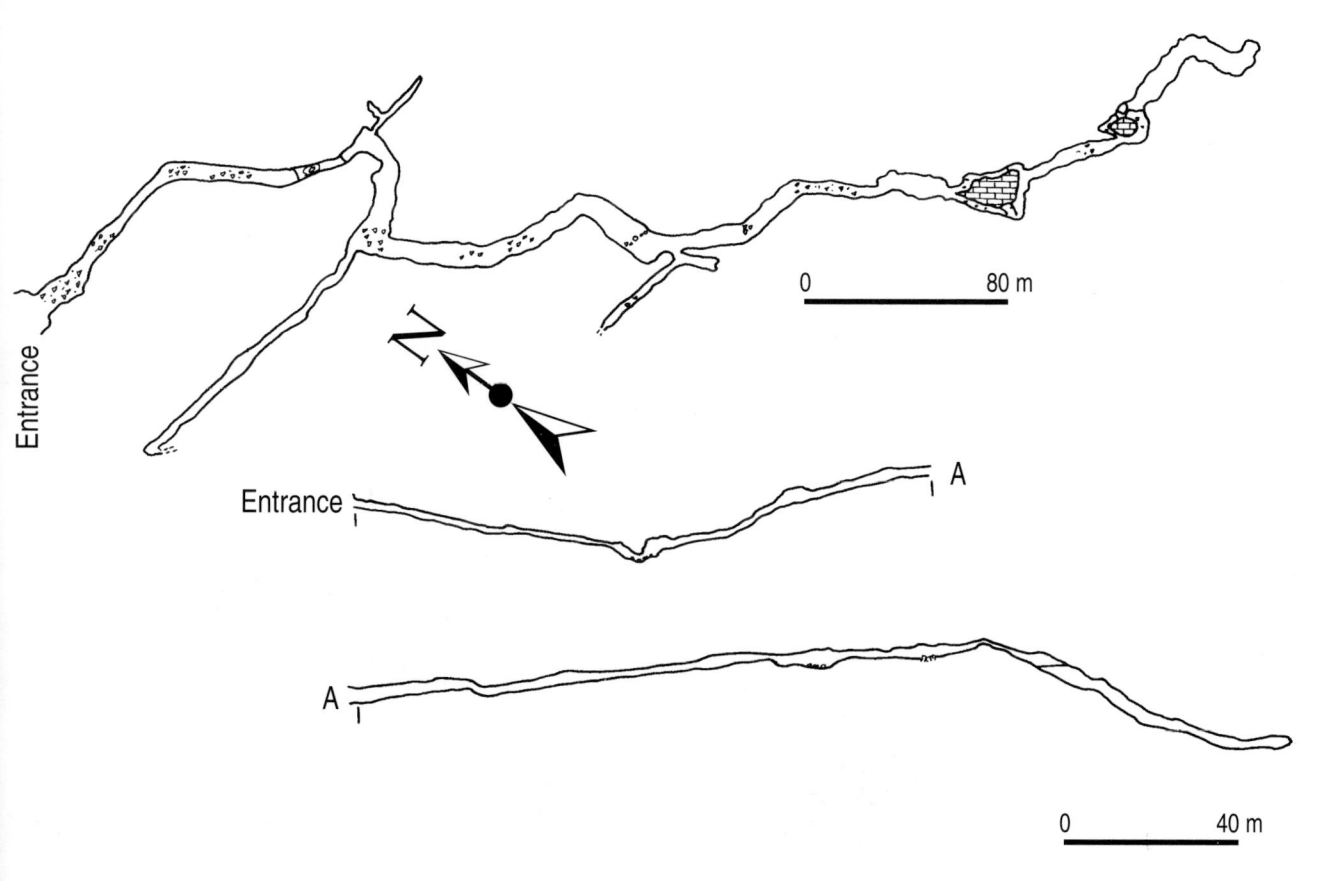

The entrance of Sokolskata Cave – photo Tsvetan Ostromski

Transport of diving equipment – photo Alexey Jalov

Name: **SOPOTSKATA PESHTERA (TÂMNATA DUPKA) (135)**
Sopot Village, Lovech Distr.
Length 1225 m. Denivelation -14 m

A cave five km SE of Sopot, in the ridge Gagayka, and on the left slope of Sopotska Reka. Formed in limestone from the Upper Cretaceous (Maastricht). One gallery. Entrance, 1.5 m high and 3 m wide. In the first 150 m the gallery is narrow and low, another part follows, which in places is four meters high and 7–8 m wide. From the 700th meter the gallery again becomes low and narrow. At the end, there is a pitch, 12 m deep and with a lake on the bottom. Poor in flowstone. Clay and gravel deposits along the final 500 m. During strong rain a stream, which runs inward, is formed.

First studied by P. Moev (geologist), who considered the cave to be 1500 m long. Explored and surveyed by the speleoclub of Planinets Sofia Tourist Society (1186 m), later corrected by the club Studenets Pleven (S. Gazdov, Ts. Hristov and P. Ignatov) to its present length.

Fauna: three species known, including the troglobite *Typhloiulus bureschi* (Diplopoda), many others are under study.

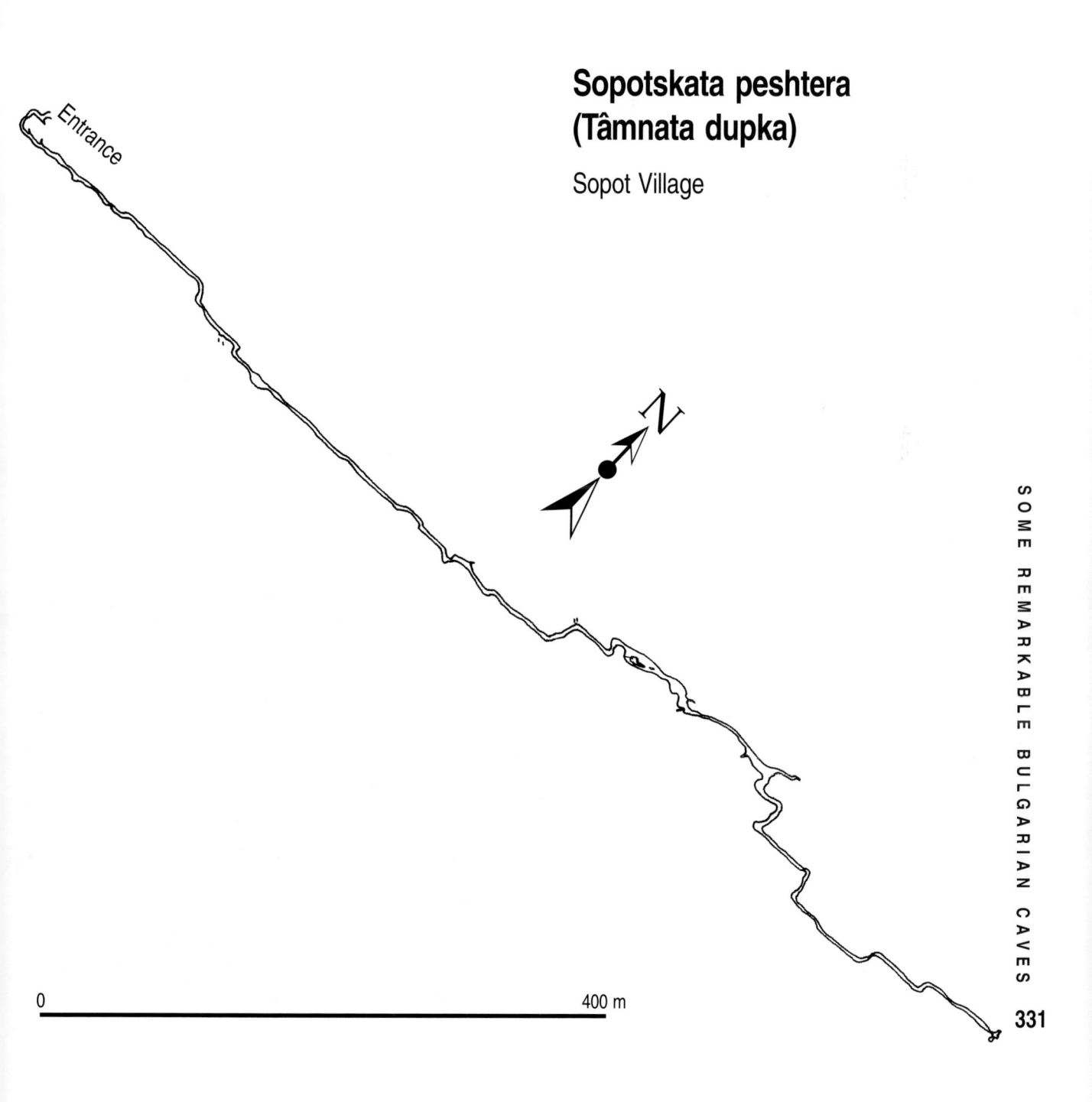

Sopotskata peshtera (Tâmnata dupka)

Sopot Village

Rest during the cave mapping – photo Nikolay Genov

Formations – photo Nikolay Genov

Name: **STARATA PRODANKA (80)**
Gabare Village, Vratsa Distr.
Altitude 250 m. Length 628 m. Denivelation -42 m
Coordinates: N 43° 17' 32.7" E 23° 55' 07.2"

A cave pothole near Gabare, about 120 m SE of the village pig farm. Known to local people also as Ezeroto (The Lake) or "Propastta s ezeroto" (The Lake Precipice). The entrance (among the rocks, in a funnel) is a vertical fissure, 5 m long, 0.5–1 m wide and 6.5 m deep. There is a descending chamber (8 x 5.5 m) on the left side. At one end is the vertical, on the left there is a small labyrinth (20 m), connected to the vertical. The pitch is 20 m deep with a stream on the bottom (The First River).

Following a low water gallery upstream, after 8 m we reach a round chamber. On its left side a stream runs out of a siphon. After crawling through muddy galleries, we again reach the river, which is closed from both sides by siphons. After 25 m upstream, we reach a passage, opening 1.5 m above the river, with many stalactites.

After another 120 m, the gallery is obstructed by cave formations. We crawl for another 180 m till we reach a two-meter pitch on the left. From here a low passage full of mud starts and after another ten meters we again reach the river. The total length of the portion, upstream the First River, is 418 m.

From the bottom of a 20-m vertical at the beginning of the cave, we reach the First Lake. Two meters above it, there is a 0.4 m passage, steeply going down to the Second Lake (elliptical, 8 m long and up to 4 m wide).

At the opposite side, the lake takes the water of the Second River. The river runs in a muddy gallery with a height, increasing from 1.5 to 3 and more meters. The depth of the river also rises, there are barriers in places. At the gallery, the river runs out of a narrow 14-m long lake at the bottom of a high diaclase. The length of this part of the cave (from the bottom of the vertical) is 99 m. A dingy is needed to cross the lake.

In 1968 cavers from Akademik – Sofia (A. Filipov, G. Antonov, R. Rahnev, H. Harizanov and V. Vekov prepared the horizontal survey of the cave and fixed its length at 553 m. The precise survey during the national expedition "Gabare' 1986", gave the actual length of the cave (628 m). K. Spassov proved by indicatory experiment for following the way of the underground water that the siphon part contains considerable bodies of water (the short way was passed for ca.10 hours). During the expedition "Kameno pole' 1987", on May 28th A. Benderev, I. Ivanov, P. Stefanov, T. Daaliev and Tz. Lichkov carried out hydrogeological observations in the cave and on the source. The river outflow on the bottom of the vertical was 17 l/sec, the temperature was 12.4°C. The oulflow of the source was 80 l/s and the water temperature 13.3 °C. These measurement show that in the siphon part of the cave there is an influx of at least 60 l/s. In 1990 divers from Pleven managed to pass 15 m in the siphon "Ezeroto" – the source draining the cave – but did not reach the cave itself.

Description after G. Antonov and A. Filipov.
Fauna: stygobite is the Crustacean *Niphargus bureschi* (Amphipoda)

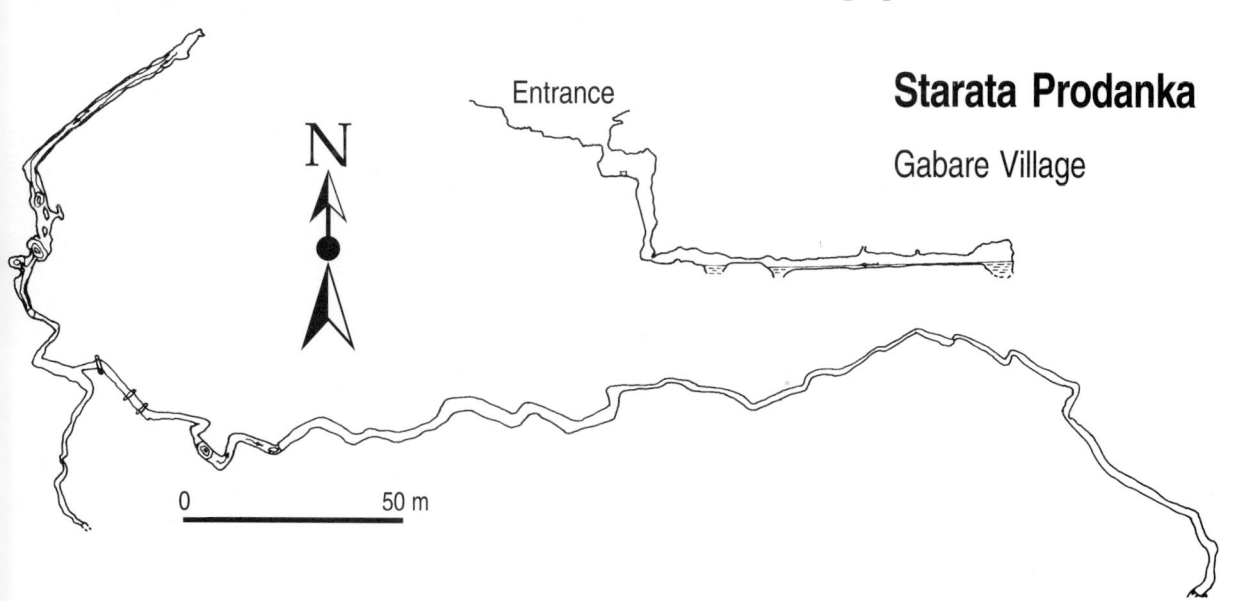

Starata Prodanka

Gabare Village

Name: **STÂLBITSA (148)**
Kârpachevo Village, Lovech Distr.
Length 145 m. Denivelation – 48 m
Coordinates: N 42° 13' 02.8" E 24° 58' 23.9"

A pothole, two km SW of Karpachevo, Lovech Distr., in the NW periphery of Devetaki Plateau, Middle Predbalkan.

Formed in limestone from the Lower Cretaceous (Aptian). Entrance (5 x 7 m), followed by a 7.5 m deep pitch, reaching the top of several boulders. The chamber is semi circular, diameter 62 m. Flowstone, stalagmites, sinter lakes. First known study in September 1926 by N. Radev. Description and survey by P. Petrov in 1933. Surveyed in details by a team led by P. Tranteev in 1965. The name comes from the wooden ladder (stalba), built by the local people and replaced later by a metal construction.

Fauna: contains the Amphipode *Niphargus ablaskiri georgievi* (stygobite).

Stâlbitsa

Karpachevo Village

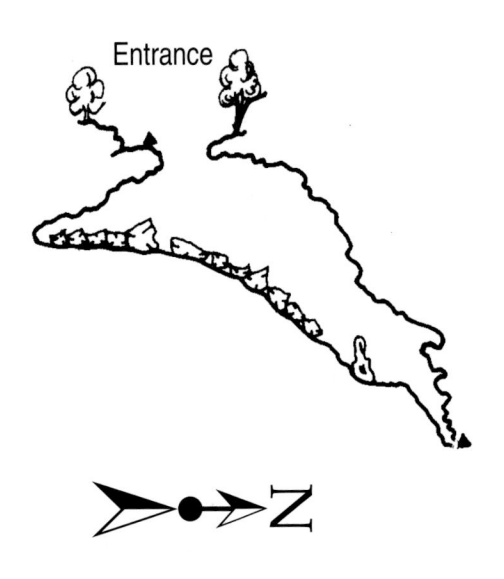

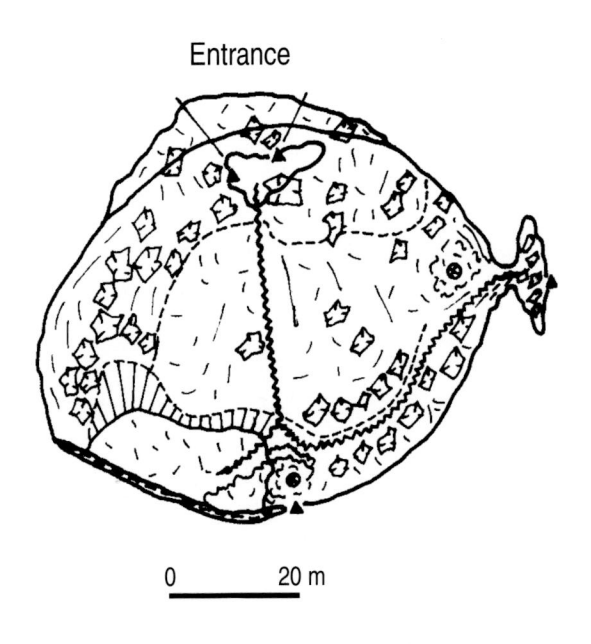

Entrance Hall of the cave Stalbitsa – photo Ivan Lichkov

Name: **STOTAKA (118)**
Cherni Vit Village, Quarter Brezovo, Lovech Distr.
Length 347 m. Denivelation -104 m

A starting point for the pothole is the road turn for Teteven-Ribaritsa and the quarter Brezovo. The entrance is situated in the lower part of Gradishka Polyana, at the beginning of a birch forest. It is a natural end of a dry valley, about 500–800 m long. Water from rain and snow melting enters the pothole, carrying with it branches, leaves, silt, etc.

The cave entrance consists of several openings of different shape and dimensions. They join the pothole at different levels.

Stotaka starts with a 7-m vertical, leading into a lighted chamber. After a squeeze, we go down for another 5 m. Up to 35 m, verticals of 5–10 m are interspersed with chambers and galleries. At about this depth a muddy gallery starts, 60 m long.

The floor is covered with clay, the rock is unstable. When water enters the pothole, it is almost impossible to avoid the verticals.

Reaching – 76 m, the spacious gallery ends. At its bottom there is a narrow place, filled with clay, sand and silt.

The pothole needs to be equipped with ladders and is not suiteable for SRT.

The pothole was explored and surveyed by cavers from Iskâr Sofia Tourist Society in 1986. A group of cavers from Sofia (Ts. Ostromski, S. Petkov and others) found a way further and the depth of the pothole became over 100 m.

Stotaka

Cherni Vit Village

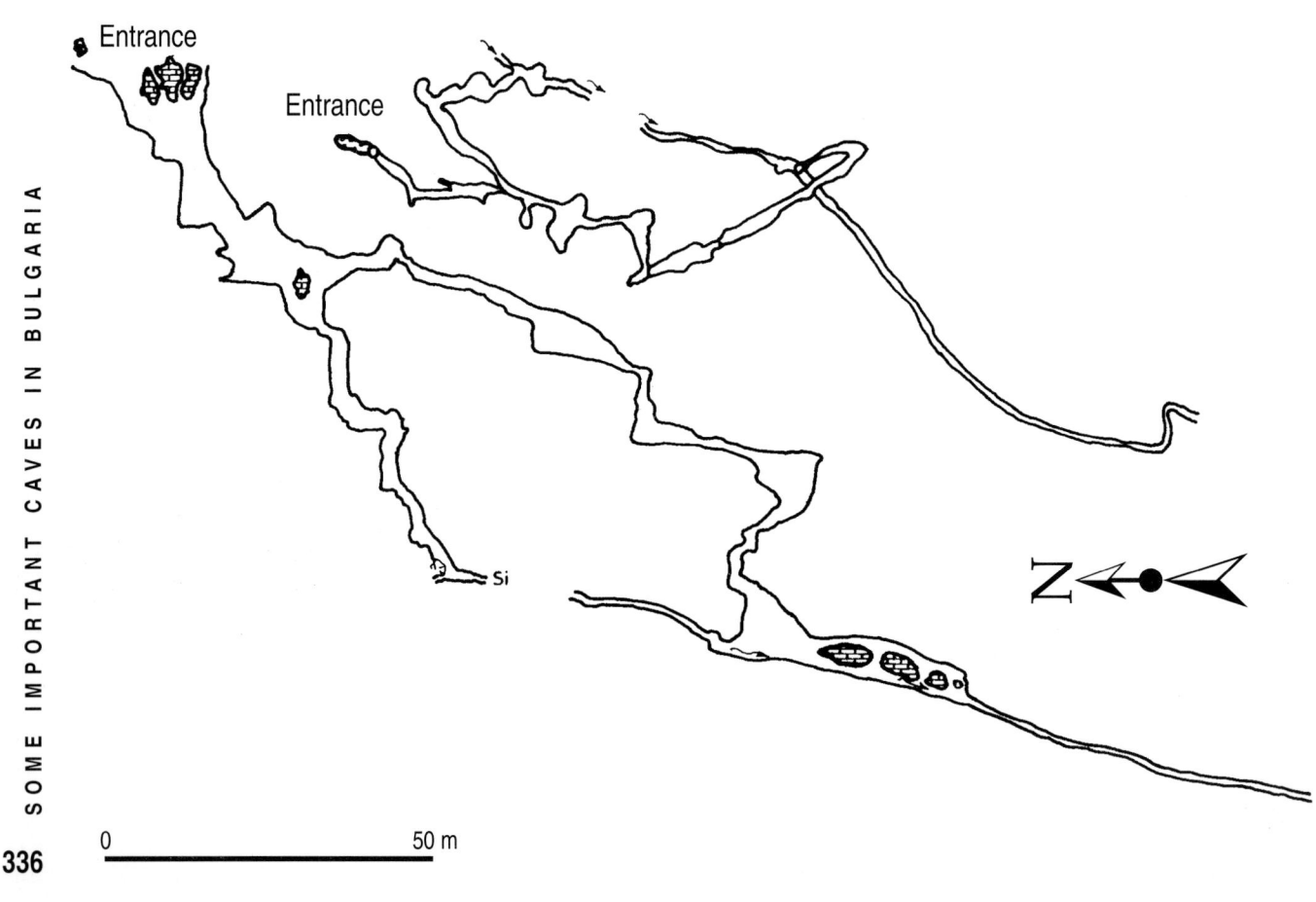

Name: **STUBLENSKA YAMA I (95)**
Karlukovo Village, Lovech District
Altitude 278 m. Length 565 m. Denivelation -72 m
Coordinates: N 43° 09' 08" E 24° 05' 07"

A cave 0.8 km SE of the southern end of Karlukovo, 100 m from the dirt road, connecting the village with Rumyantsevo, in the locality Stublenski Dol. The cave is situated in the southern periphery of the East Predbalkan, in the 206 karst region. It is formed in limestone from the Cretaceous (Maastricht), about 150 m thick, of the Karlukovo Syncline. There is an underground stream. The cave is formed along main fissures of SW-NE and NW-SE direction. The entrance is situated in the N-NE part of a stony double funnel and it is a small opening (40 x 40 cm). After the squeeze, we reach a small platform, leading to a 21-m deep bell-like vertical. About 5 m before reaching the bottom, we get to another small platform, giving access to the entrance of a horizontal gallery. After overcoming a narrow fissure and two small verticals, we arrive to a chamber, which is 10 m high. On the right, we see a +3.70 m vertical, where usually there is a rope to facilitate the climb. After several meters, we enter a bigger chamber (six meters long, 2.5 m wide and ca. seven meters high). After several narrow galleries, a small lake siphon is reached (the end of the explored parts up to 1990).

The new parts start after a 10-m vertical, ending in a shallow lake – semi siphon, ca. 3 m long. A small chamber with a stream is followed by a series of small lakes, up to 2 m deep. The gallery is ca. 2–3 m high and 2 m wide, with enlargements. After another semi siphon (2–3 m long and 1.5 m deep), another series of lakes follow. We reach a 2-m vertical with a deep lake underneath. After a difficult squeeze of 5 m, an ascending narrowing gallery meanders about 8–10 m above the water. At the end of the cave, the ceiling lowers above a shallow lake with a silt bottom. The lake becomes a siphon (the end of the cave).

The cave was discovered by A. Jalov and D. Dishovski from Helictite Sofia Caving Club on May 23, 1990, when digging in a funnel. The discoverers and L. Zhelyazkov from Strandja Burgas Caving Club penetrated the first 250 m to reach a small lake siphon. The exploration was carried out in July 1990 to discover many new parts, surveyed in 1991 by cavers from Helictite Sofia Caving Club. Other parts were discovered in 1993 and surveyed in 1994. The entire map is due to V. Mustakov, V. Mustakova, I. Ivanov, D. Lefterov, N. Sveshtarov, I. Vashev, E. Petrov, G. Slavov, S. Delchev, L. Arsov, B. Gyaurova, N. Simov and M. Stamenova.

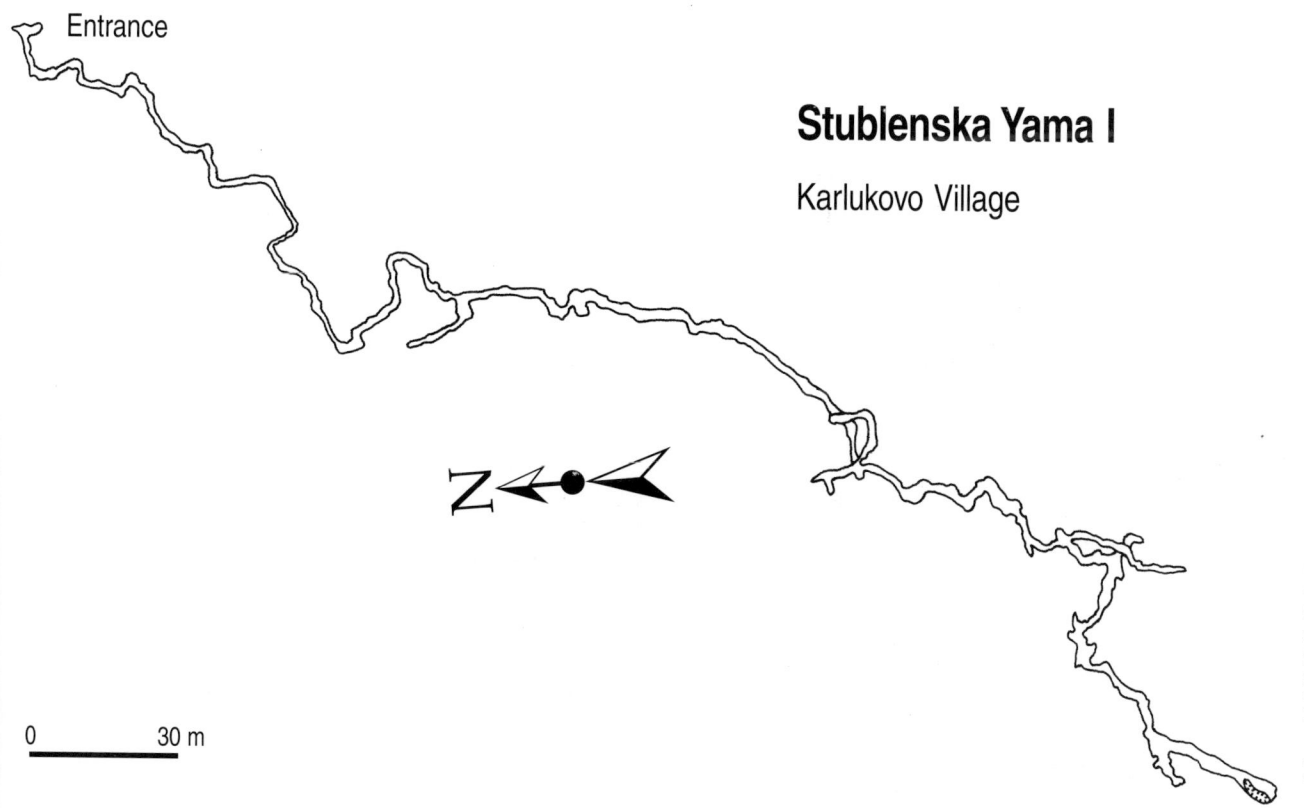

Stublenska Yama I

Karlukovo Village

Name: **STUDENATA DUPKA (CHEREPISHKATA PESHTERA) (49)**
Cherepish Railway Station, Vratsa Distr.
Altitude 209 m. Length 634 m. Denivelation: 43 m (-16, 27)
Coordinates: N 43° 05' 38.5" E 23° 37' 19.2"

Studenata Dupka (Cherepishkata Peshtera) has been formed on the right bank of Iskar River, southeast of Cherepish Railway Station, in the locality Manastirskoto. It is the longest cave known in the Cherepish area. A labyrinthine cave on three levels, the lowest being inundated. The levels and the galleries are connected by narrow places.

The cave entrance opens at the base of a small cliff, on the first river terrace of Iskar. The cave starts with a descending squeeze, leading into a small chamber. There are also much bigger chambers (The Blockage Chamber, 23 x 14 x 18 m), The Big Chamber and others.

A considerable increase of the cave length is very likely, but digging and breaking are needed. There are also intermediate levels, which are to be explored.

The cave had been known in the past, the region of Cherepish explored by cavers from Akademik Sofia Caving Club. During the road building/construction, the entrance covered by earth and gravel. The cave remained inaccessible for tens of years. In the fall of 1997, it was reopened by two cavers from Akademik. The known length of the galleries was by that time ca. 150 m, with many squeezes and clay.

In January – November 2000 many expeditions took place (cavers from Vitosha, Sofia, and other caving clubs from Sofia). As a result, a prolongation of the third level was found, as well as a lower flooded levels on the local base of erosion. Some clay obstructions were dug out and many galleries were connected. Many efforts were needed to reach the so called second level. The exploration goes slowly because of the many squeezes. At the same time, a careful survey takes place, but the complete exploration will go one for some more years. The actual map is by SC „Helictit" – Sofia.

Fauna: the new genus and species of cave beetles *Beskovia bulgarica,* collected by V. Beshkov and P. Beron in May and June 1960 from this cave. Besides this beetle, another 11 invertebrate species have been identified from this cave, including the troglobite Isopod *Trichoniscus anophthalmus*.

Studenata dupka (Cherepishkata peshtera)

Cherepish Railway Station

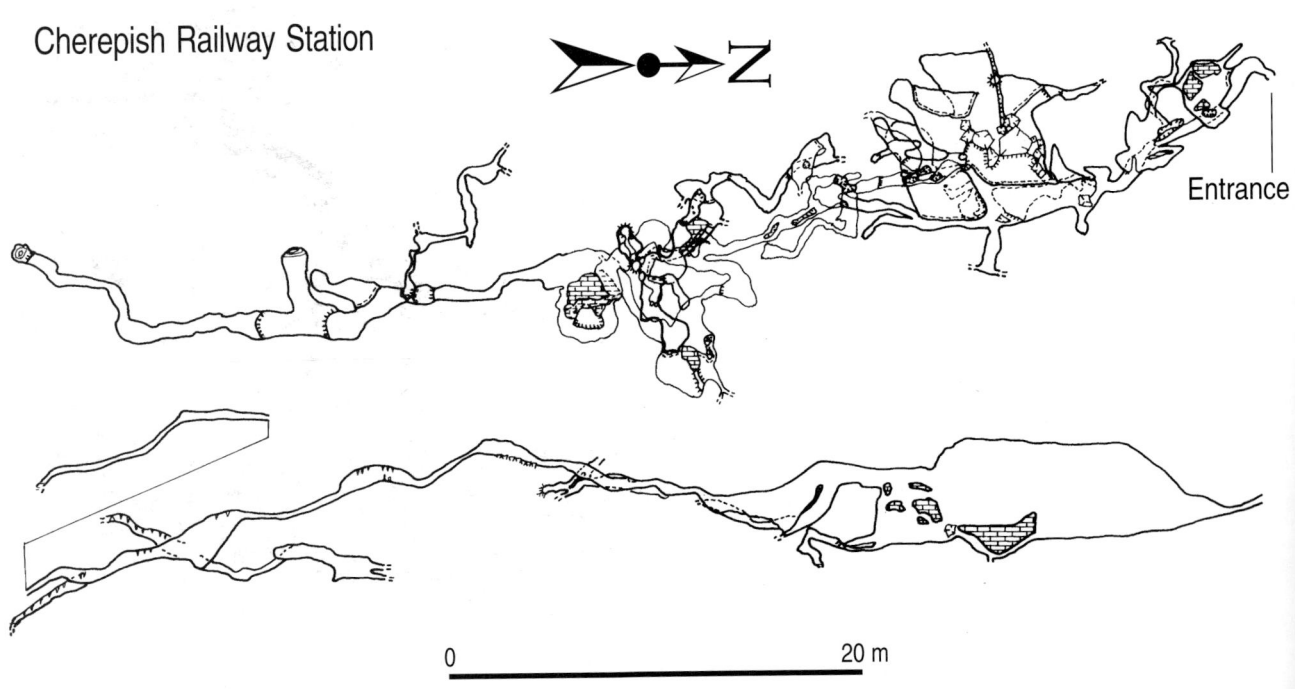

Name: **SUBATTÂ (187)**
Kotel Town, Sliven Distr.
Altitude 695 m. Length 518 m. Denivelation -55 m
Coordinates: N 42° 55' 59.1" E 26° 30' 29.8"

A cave pothole in the locality Arpalaka – Zlosten near Kotel. A two-storey, periodically watered cave, branched, rich in flowstone formations. Sink of the stream of the gully, right of Lednika.

The cave was known to the local people for a long time. Its name means "Sinking water". The entrance is six meter wide and up to eight meters high, at the bottom of a valley. In 1963, cavers from Akademik, Sofia, surveyed the cave to the first siphon. After several unsuccessful attempts to penetrate the siphon, in September 1983, a combined team from Sliven, Kotel, Yambol and Russe overcam the obstruction. In 1984, British divers participated in the exploration, studying and surveying the upper level.

The length of the second siphon is seven meters. After it, the cave branches into two galleries. The upper level is rich in flowstone formations. Lefts, after the second siphon, a horizontal gallery forms a lower storey. After many small verticals, a final siphon is reached.

The description is due to Zdravko Iliev and Mariana Petrova.

Fauna: only troglophyles and trogloxens are known.

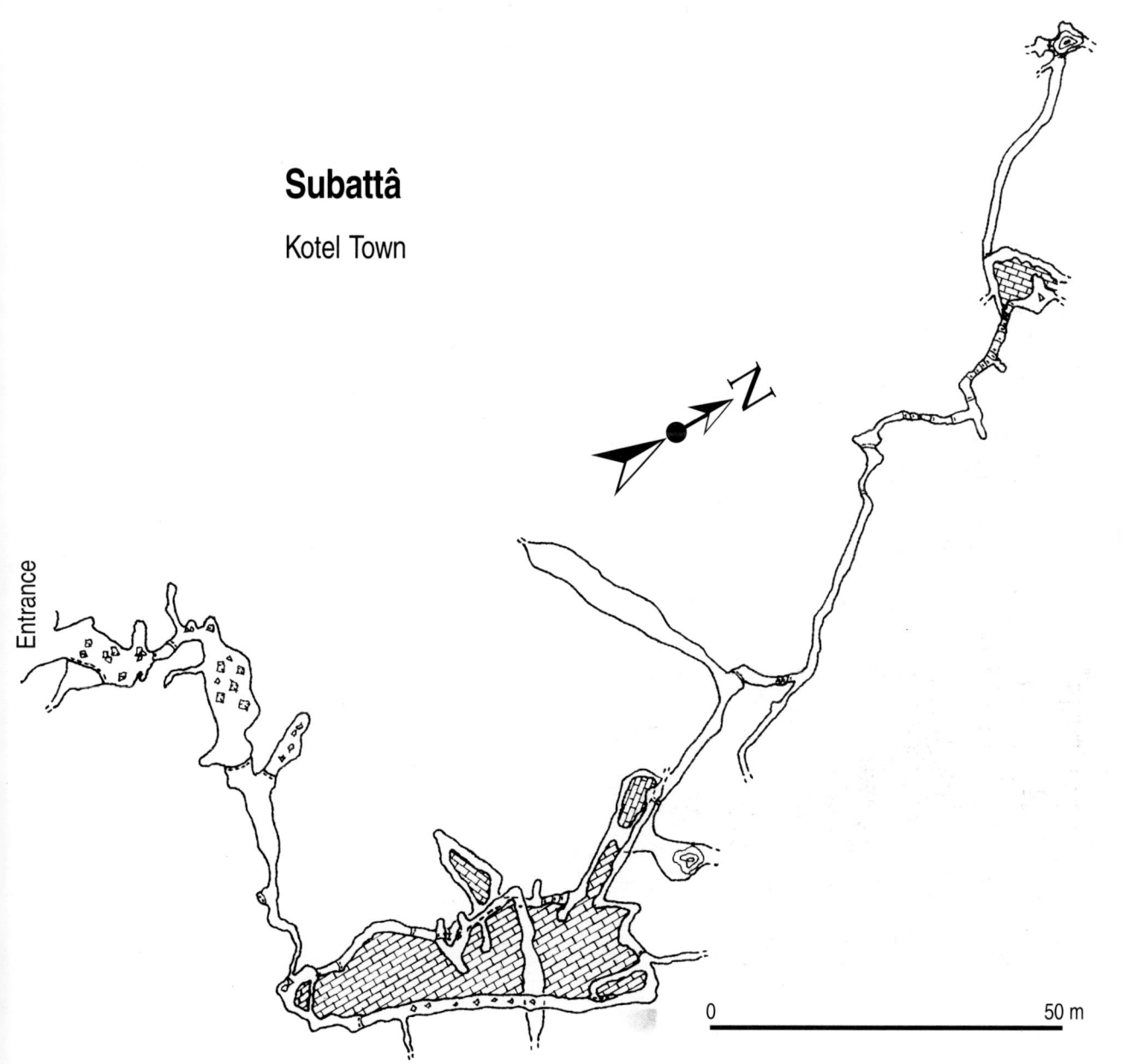

Subattâ

Kotel Town

Name: **SUHI PECH (KOZARNIKA) (3)**
Oreshets Village, Vidin Distr.
Length 218 m. Denivelation -4.5 m

A cave near Oreshets Railway Station. Entrance 6x3 m. A horizontal, wide up to 9 m and high up to 3 m, one-gallery cave. Its floor is covered with clay, guano and fallen blocks. Some groups of stalactites are seen on the ceiling in the last chamber (which is 6 m high). Air temperature 12 °C. At the end of the chamber, there is a short second floor.

Explored and surveyed in 1977 by S. Gavrilov, P. Nikolov, Ts. Tsonev and E. Donkov, cavers from Bononia Vidin Tourist Society. The archaeological remains in the cave is dated over 1.4 millions years B.C. – the first traces of human presence in Europe!

Fauna: researched by P. Beron in 1969 and by other scientists. The cave is inhabited by large bat colonies. So far, 16 animal species have been identified, including eight bat species. The Isopod *Trichoniscus bononiensis* is a troglobite. The largest bat colony in Vidin District roosts in this cave.

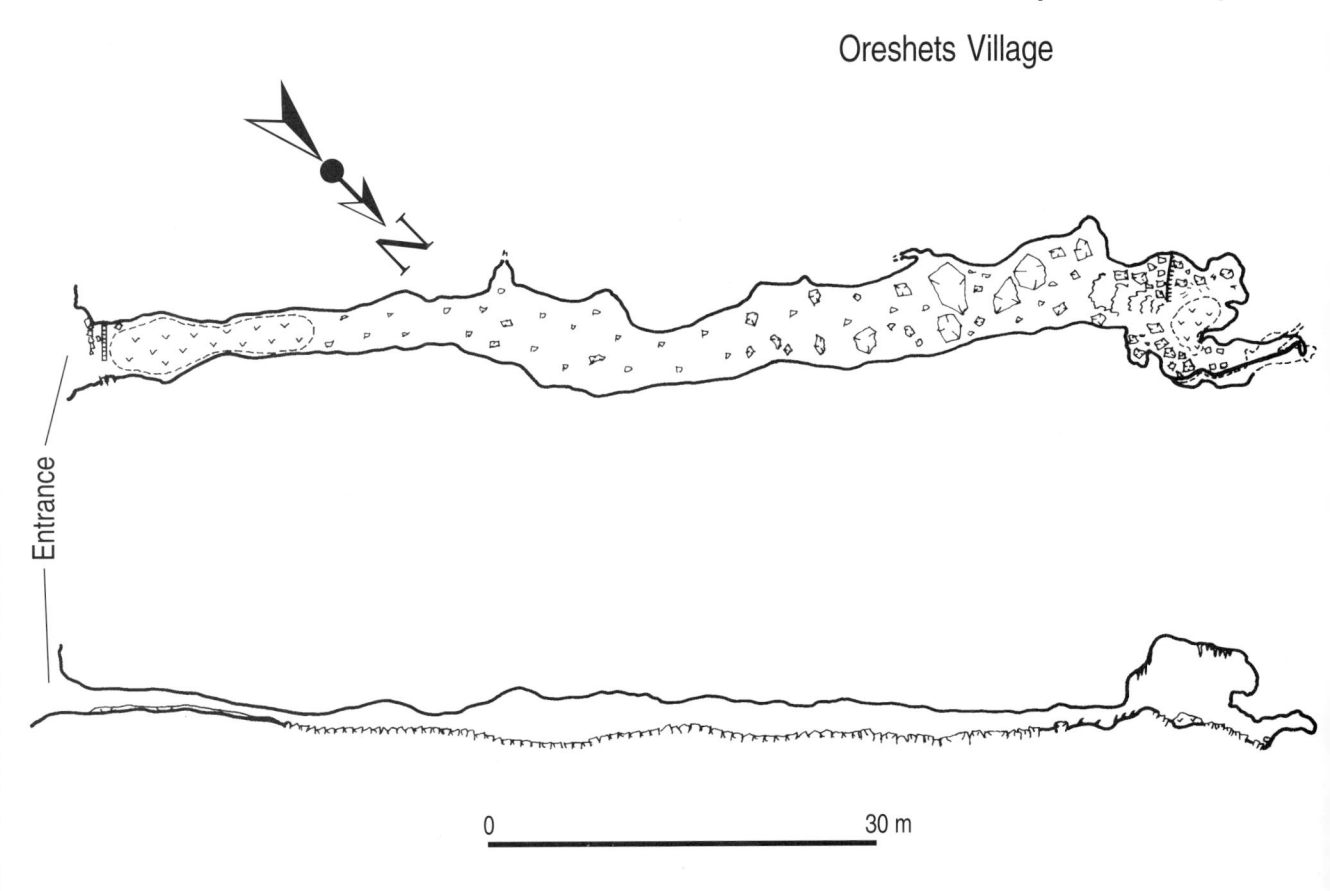

Suhi Pech (Kozarnika)

Oreshets Village

The entrance

Location of the entrance – photo Nikolay Genov

Entrance

The excavations – photo Nikolay Genov

The Lower Paleolithic remains – photo Nikolay Genov

This remain is from 1 200 000 B.C – Nikolay Genov

SOME IMPORTANT CAVES IN BULGARIA

341

Name: **SVINSKATA DUPKA (52)**
Lakatnik Railway Station
Length 240 m
Coordinates: N 43° 05' 17.02" E 23° 22' 10.64"

    A cave on the left bank of Petrenska River. About 17 m above the river bed, its triangular entrance facing West. In the wet clay of the cave floor, we can see remains of cave bears. Many of these bones have been found in the cave. The flowstone formations are very damaged by the many visitors.

    Fauna: several members of the rich cave fauna of Lakatnik karst live also in this cave, including the troglobites *Paralola buresi* (Opiliones, Laniatores), *Neobisium beroni* (Pseudoscorpiones, described from this cave and endemic to it), *Lithobius lakatnicensis* (Chilopoda), *Trachysphaera orghidani*, *Typhloiulus bureschi* (Diplopoda), *Pheggomisetes globiceps lakatnicensis* (Coleoptera, Carabidae).

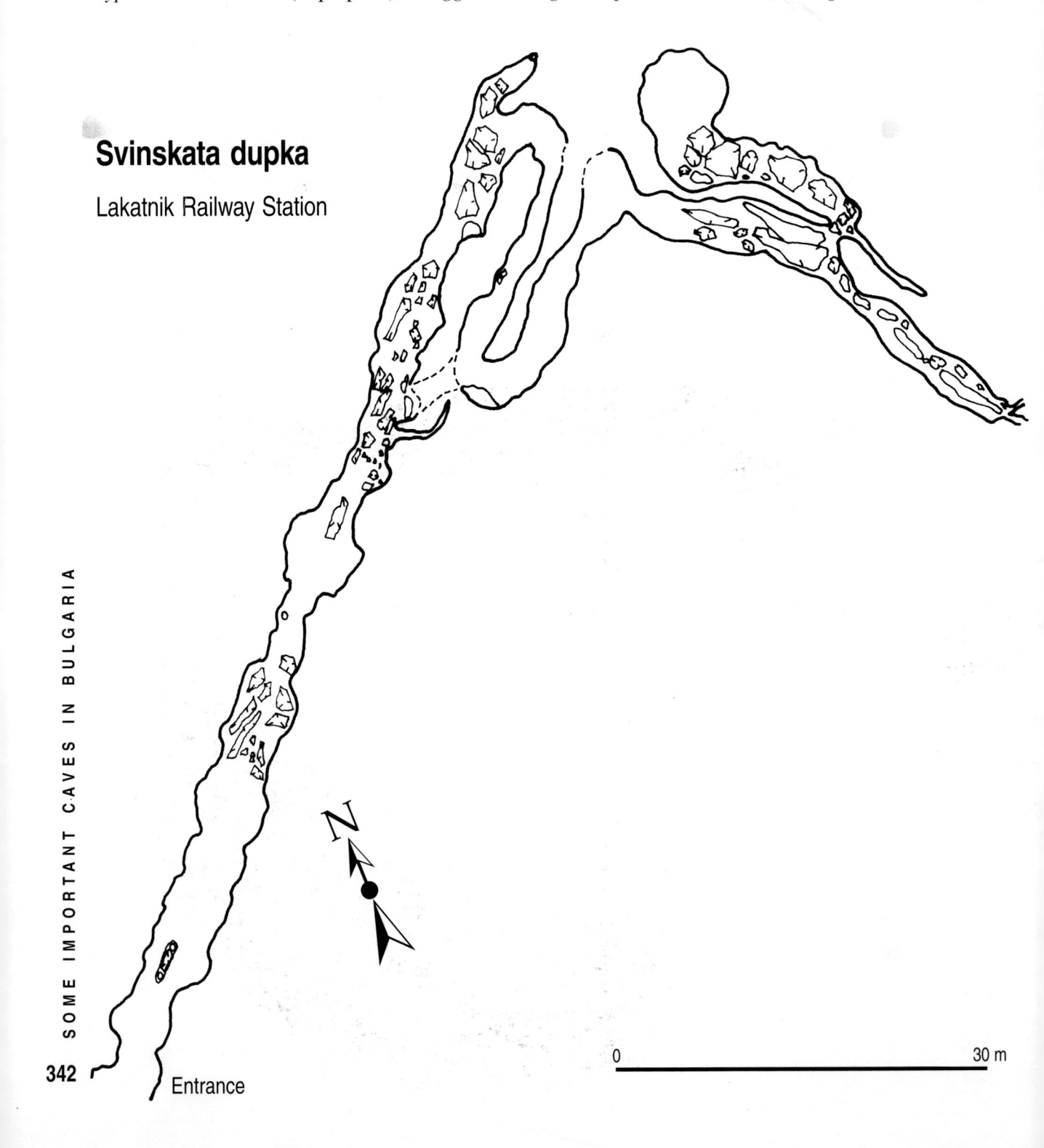

Svinskata dupka

Lakatnik Railway Station

Name: **SVIRCHOVITSA (94)**
Karlukovo Village, Lovech Distr.
Altitude 242 m. Length 231 m. Denivelation -39 m
Coordinates: N 43° 10' 15" E 24° 04' 44"

A cave, 1100 m SE of the National Caver's Hut, ca. 450 m NE of the village Karlukovo and 300 m SW of Bankovitsa, at the bottom of a funnel. Denivelation from the LBE 115.4 m. From the funnel, an underground canyon starts (about 50 m long), gradually lowering to a chamber with clay deposits. By traversing and following some old levels we can reach the upper storey. A third storey is formed in the southern (left) side, where a big cauldron (diameter 10 m, depth 5 m) is formed. In its upper part is the entrance of a 27-m vertical, which can be reached by traversing over the left side of the cauldron. No running water. There is a chamber at the bottom. The age of the cave is about 2 500 000 years (Popov, 1984).

According to a legend, the name of the cave comes from the word "svirchovina" (the local name of the plant *Sambucus nigra*), growing nearby. The local shepherds used to make whistles from it (in Bulgarian "svirki").

The cave is a National Protected Monument (Decree 2810 of 1963), together with 1.5 ha adjacent land.

In the cave 22 animal species have been recorded. Troglobite is *Typhloiulus bureschi* (Diplopoda).

Svirchovitsa

Karlukovo Village

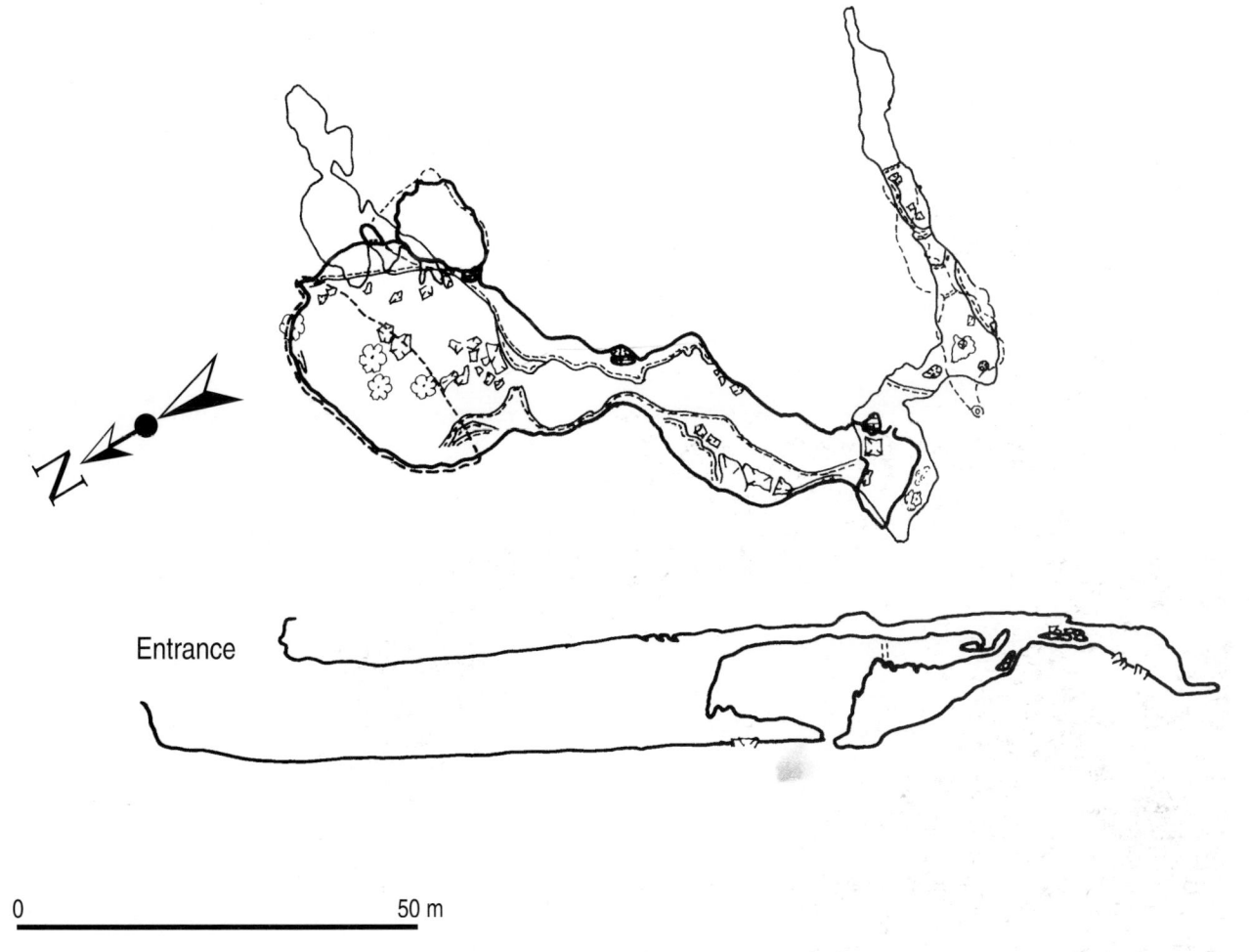

Sinter concretion – photo Petko Nedkov

Franze Habe – President of the Commission for the show caves of UIS at the entrance of Prohodna Cave – photo Trifon Daaliev

The entrance of the cave – photo Trifon Daaliev

Name: **TÂMNA DUPKA (12)**
Târgovishte Village, Vidin Distr.
Total length 276 m, length on the main axis 226 m. Denivelation -16 m

A cave in the locality Dolna Glama. Situated ca. 120 m above the local erosion basis. Descending, periodically inundated.
Fauna. Troglobites: *Trachysphaera orghidani, Serboiulus spelaeophilus* (Diplopoda).

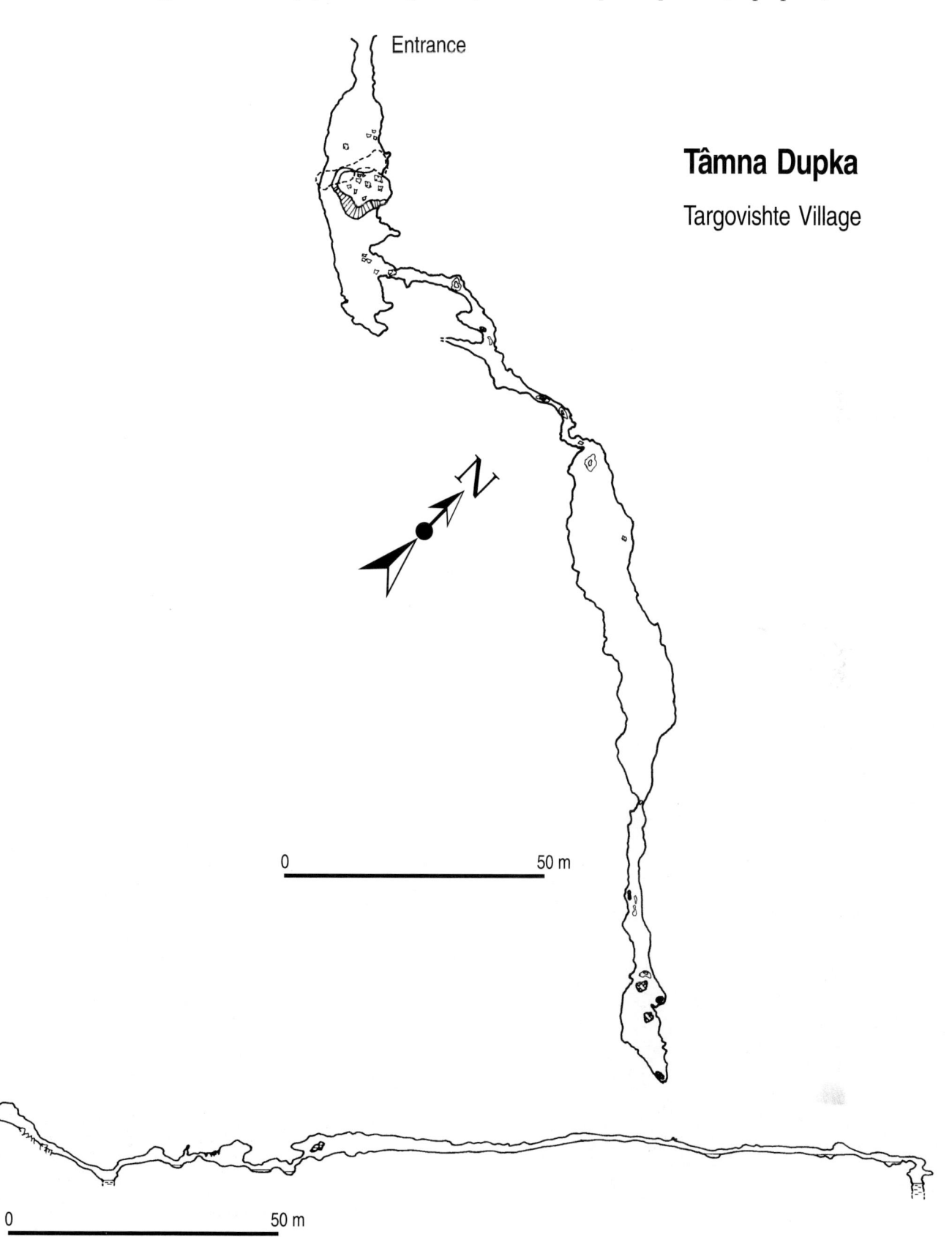

Tâmna Dupka

Targovishte Village

Name: **TEMNATA DUPKA (IZVORSKATA PESHTERA) (72)**
Kalotina/ Berende Izvor Village, Sofia Distr.
Total length 493 m. Denivelation -95 m
Coordinates: N 43° 00' 53.6" E 22° 53' 17.5"

A cave in the Trias limestone of Nishava river, 130 m high above the river and 300 m away from Berende Izvor. Length of the main gallery – 387 m. The entrance is at the bottom of a huge tunnel, under a 20-m cliff. A descending cave, broken by small verticals and lakes. Up to 70 m, there are large blocks on the floor of the gallery. Several formations (including the stalagmite The Cactus and others). T° 11 °C.

First described by Zheko Radev in 1915, with a map of the entrance part. Later, Nenko Radev explored the cave up to 112 m, collecting cave fauna. In 1962, P. Tranteev surveyed the cave and P. Beron collected more cave animals. In 1968, cavers from Akademik Sofia Caving Club made a precise map and studied the temperature and other climatic factors.

In the cave, 11 species of cave animals have been found, including three stygobites or troglobites: *Paracyclops fimbriatus, Speocyclops lindbergi* (Copepoda), *Centromerus lakatnikensis* (Araneae).

Protected Natural Monument, together with 0.2 ha of adjacent land (Decree 2810/10.10.1962).

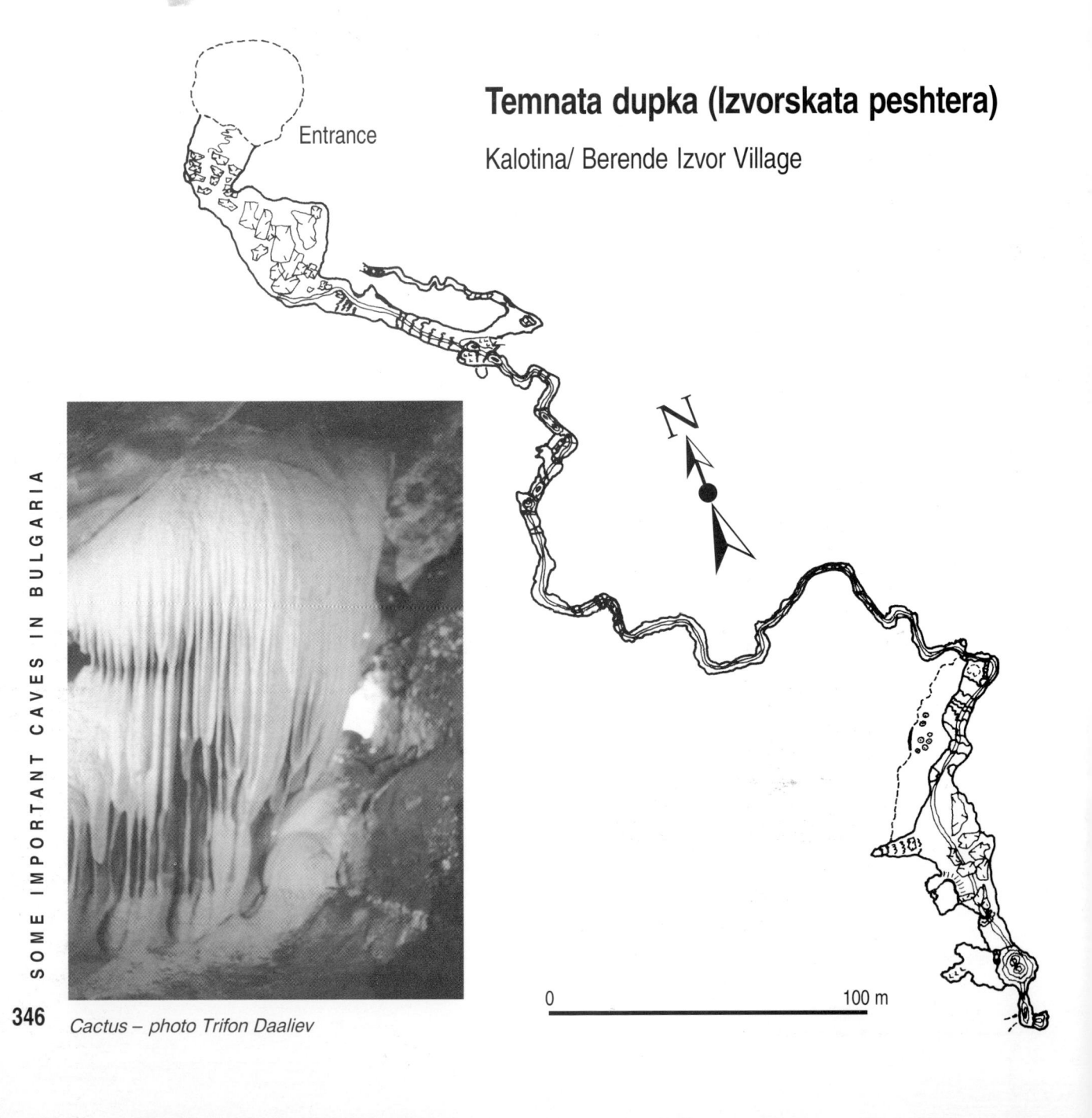

Temnata dupka (Izvorskata peshtera)

Kalotina/ Berende Izvor Village

Cactus – photo Trifon Daaliev

Name: **TEMNATA DUPKA (96)**
Karlukovo Village, Lovech Distr.
Altitude 220 m. Length 215 m. Denivelation -38 m. Surface 2054 m². Volume 6500 m³
Coordinates: N 43° 10' 30.5" E 24° 04' 32.6"

A cave, 40 m SE of the upper entrance of the cave Prohodna, at the base of a small cliff. Formed along two main faults – the entrance gallery, which is 50 meters long and directed to the west, after which the cave bents almost at 45° southwards. Here, there is a group of stalactons, surrounded radially by sinter dams, retaining the dripping water and forming a lake. On the right, we see the entrance of a 18-m vertical, ending in a chamber, which is rich in abnormal stalactites. The entrance is within an iron enclosure. The cave then gains an elliptical profile with a decreasing height, until it reaches the ceiling. Typical for the cave are the thick clay deposits. On the wall and the ceiling, the so called "leopard skin" is formed.

In 1924–1926, R. Popov made excavations in this important for archaeologists cave, discovering for the first time in Bulgaria and on the Balkan Peninsula remnants from the Late Paleolith (Orignaque). In 1984–1994, the Group for Paleolythical Studies of the Archaeological Institute in Sofia carried out international excavations and studies. The age of the artefacts found there is before 67 000 + 11 000 B.C., e.g., from the time of the Middle Paleolith.

The cave was declared a Protected National Monument (Decree 2810 from 1963), together with 1.5 ha of adjacent land.

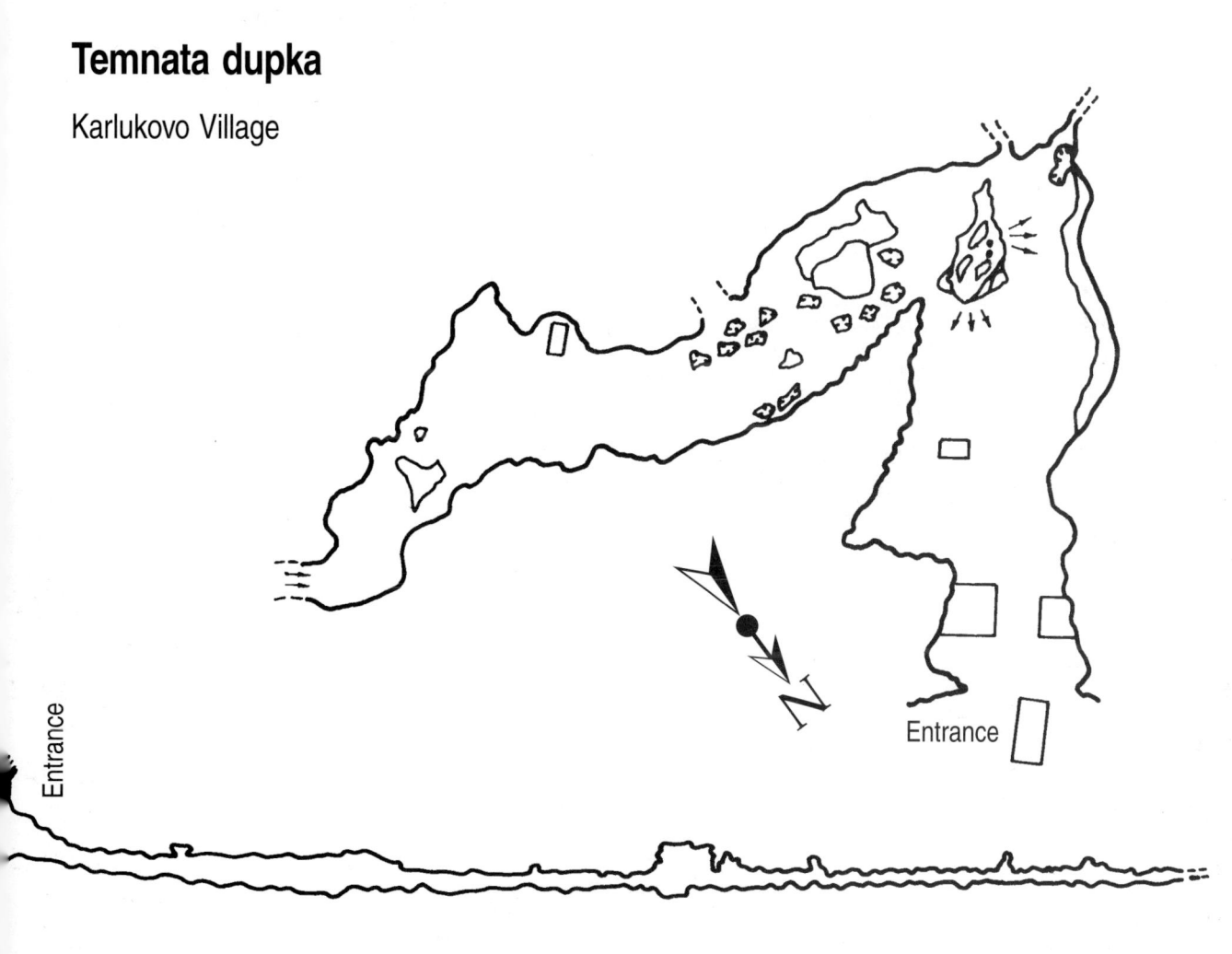

Temnata dupka

Karlukovo Village

Entrance Temnata Dupka – photo Valery Peltekov

Entrance Temnata Dupka – photo Valery Balevski

The excavations – photo Nikolay Genov

Name: **TEMNATA DUPKA (53)**
Lakatnik Railway Station Village
Altitude 400 m. Length ca. 7000 m. Denivelation +33/-21 m
Coordinates: N 43° 05' 19.9" E 23° 23' 10.6"

A river cave, 600 m NW of Lakatnik Railway Station, on the left bank of Iskar, 27 m above the river bed. Formed in dark Middle Trias limestone (Ladin) of Milanovo syncline.

It is a system of two main galleries with side parts developed in four levels. The longer NE part of the cave (800 m along the main axe) contains also the main underground river. The galleries are wide and high, with predominantly a rectangular and inverted-trapezoid cross section. The W-NW gallery is long (without the branches) 710 m. There is a river with a changing outflow, running via the underlying source Zhitolyub. During extreme rains and snow melting, the river floods through the cave entrance.

The first known visit by cavers was in 1912, when Dr Buresch studied the bat fauna. The first map and description of the first 150 m of the cave were published in 1915 by Zheko Radev. In the period 1921–1926, P. Petrov, I. Ninchovski, M. Voyniagov, P. English, I. Parvanov and other surveyed about one km of galleries. A complete and accurate description of the cave was published in 1936. In 1948 the NW parts were discovered and the cave was re-surveyed. Young speleologists, led by P. Tranteev, pushed further the study of the cave and by 1958 its known length had reached 3200 m. The cavers of Cherni Vrâh Sofia Speleoclub surveyed the cave by means of a theodolite and its dimensions were fixed at 4500 m. Since 1991, cavers from Sofia, under the guidance of Eng. J. Vuchkov from the same club, have studied and surveyed the cave in details and prepared a computerised vector map. In 2001, during a joint expedition, Belgian and French divers penetrated 240 m inside the siphon and reached a -20 m depth at the cave end. Approximately 1.5 km of new galleries was discovered in December 2004 and January 2005 from the cavers of Helictit and Vitosha Caving Clubs-Sofia.

The cave was declared a Natural Monument with an attached area of one ha (Decree 28.10/ 10.11.1962), within Vrachanski Balkan Natural Park.

The cave fauna amounts to 94 known species (the richest in Bulgaria), including 17 troglobites and stygobites: the worm *Delaya bureschi*, the water snails *Belgrandiella hessei* and *Saxurinator buresi*, the lower Crustaceans *Pseudocandona eremita*, *Speocyclops infernus* and *Diacyclops clandestinus*, the Isopods *Bureschia bulgarica* and *Protelsonia lakatnicensis*, the Amphipod *Niphargus bureschi*, the centipede *Lithobius lakatnicensis*, the millipede *Typhloiulus bureschi*, the harvestmen *Paranemastoma (Buresiolla) bureschi* and *Paralola buresi*, the Collembolan *Pseudosinella duodecimocellata*, the Diplura *Plusiocampa bureschi* (= *P. rauseri*), the beetles *Pheggomisetes globiceps lakatnicensis* and *Duvalius papasoffi*.

Temnata dupka

Lakatnik Railway Station Village

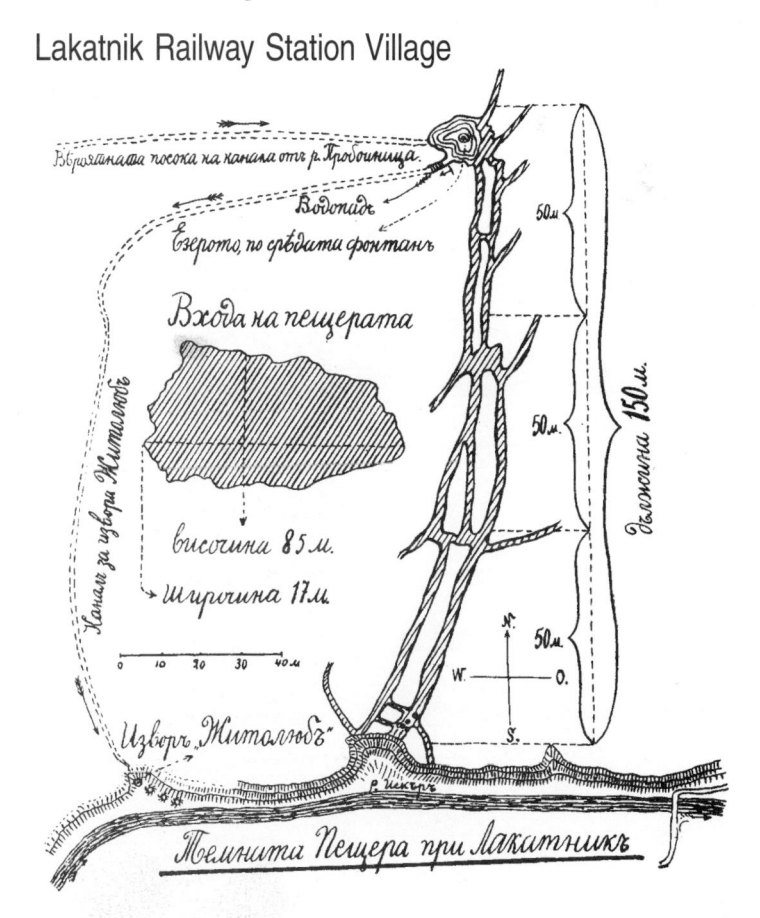

Map of the cave by Zh. Radev – modern map in the annex

The entrance of Temnata Dupka Cave – photo Valery Peltekov

The Waterfall – photo Valery Peltekov

The River – cave rescue training – photo Trifon Daaliev

Lake in the cave – photo Trifon Daaliev

Helictites in the new found passages of the cave – photo TsvetanOstromski

Chatala – the passing of the passage in 1925 – photo BFS archive

Name: **TIGANCHETO (25)**
Chiren Village, Vratsa Distr.
Altitude 132 m. Length 466 m. Denivelation -9 m
Coordinates: N 43° 18' 52.5" E 23° 33' 12.2"

The water cave Tigancheto is in the locality of the same name, one km from the station Ponora, on the left of the road Vratsa – Chiren. It is formed in limestone from the Cretaceous (Aptian). The cave is horizontal, one-storey, branched.

The cave entrance is furnace-like, 16 meters wide, with a maximum height of 1.40 m. About 30 m from the entrance, on the left there is a 100-m gallery, ending in a gryphon, from which an underground stream starts. The stream runs through the cave, forming some sinter barriers, the biggest being 1.20 m deep. The cave is a horizontal one, but before the end siphon there is an 8-meter inclined threshold, which can be overcome without a rope. On the floor, there are sinter lakes and gravel. There are many flowstone formations in the cave, also cavern pearls.

The cave is known to local people. First explored in 1963 by the Third International Expedition. The survey was done by cavers from Veslets Vratsa Caving Club (K. Karlov, K. Dobrinov, M. Baleva, V. Kissimova). Description: T. Daaliev.

Fauna: bats, harvestmen.

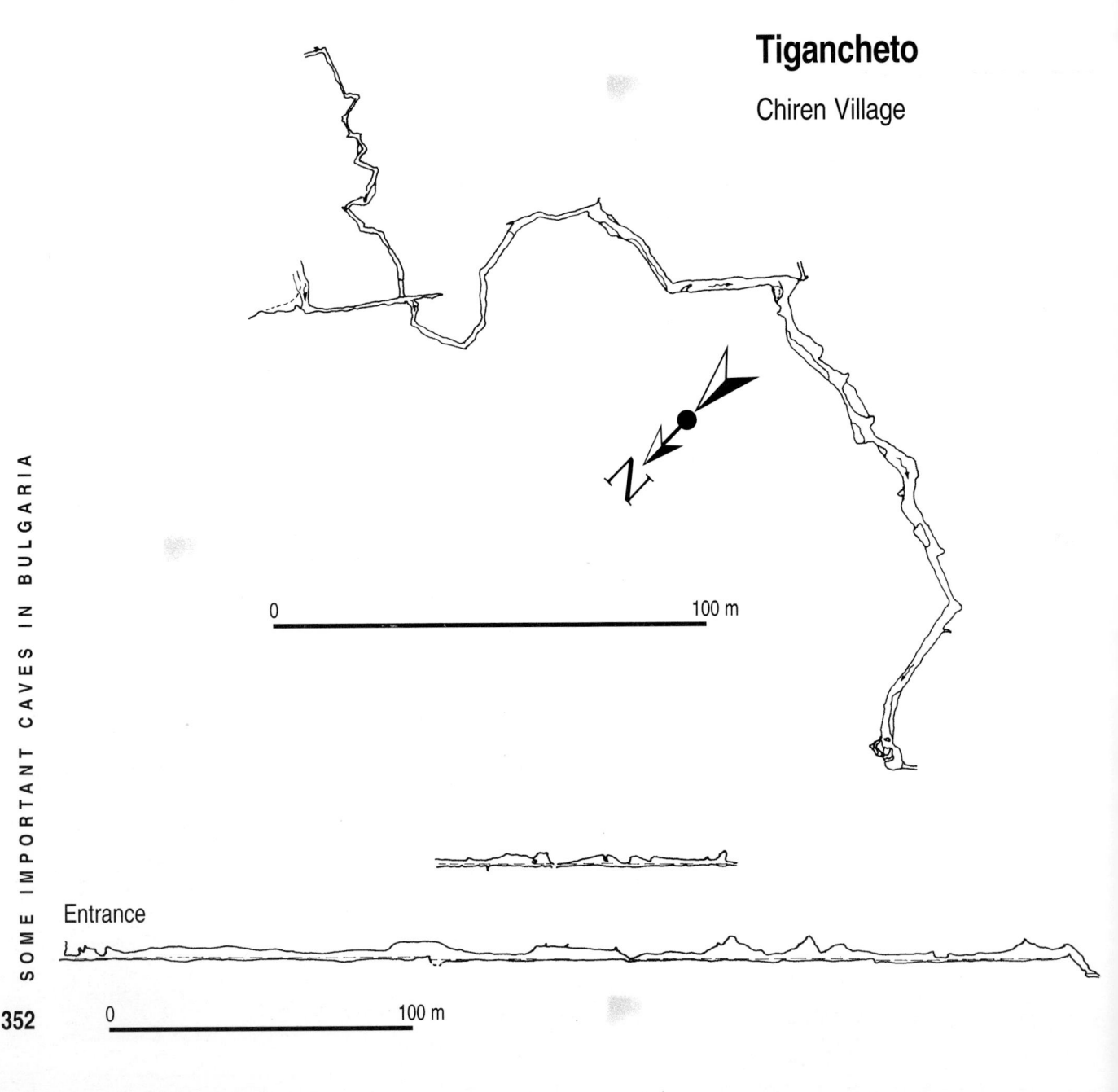

Tigancheto

Chiren Village

▲ The underground drop – sculptor of beauty

▲ The pothole Raychova dupka

▼ Stalactites in Raychova dupka – the deepest pothole in Bulgaria

▲ Magura cave – groups of paintings representing mythical dances, astronomic events, hunting scenes, etc.

▼ The Giant (Magura)

▼ The Big Hall in the cave Magura

▲ The entrance gallery of the cave Magura

▼ Formations

▼ Giant stalactite

▲ Water gallery in the cave Temnata dupka (Lakatnik, Bulgaria)

▲ Crystals in the cave Temnata dupka (Lakatnik) (Tz. Ostromski)

▼ Crystals in the cave Temnata dupka (Lakatnik) (Tz. Ostromski)

▲ The cave Golyamata Temnota (Bulgaria, Tz. Ostromski)

▼ Ice formations in the Atrium (Ledenika cave)

▼ Sinter lake in the highest part of the Concert Hall (Ledenika)

▲ Cave Lepenitsa (Bulgaria)

▼ Cave Lepenitsa

▲ Sun shines on the bottom of the pothole Stalbitsa (Iv. Lichkov)

▶ "Poplars" in the Pothole 13 (Chelopek)

▼ The Waterfall (Krushuna)

▲ Thin stalactites (Snezhanka)

▼ "The Poplars" (Snezhanka)

▲ One of the chambers in the cave Eminova dupka (Photo Tz. Ostromski)

▼ Three meter long needles in the cave Izvora na Kastrakli
(Photo Tz. Ostromski)

▼ Helictite

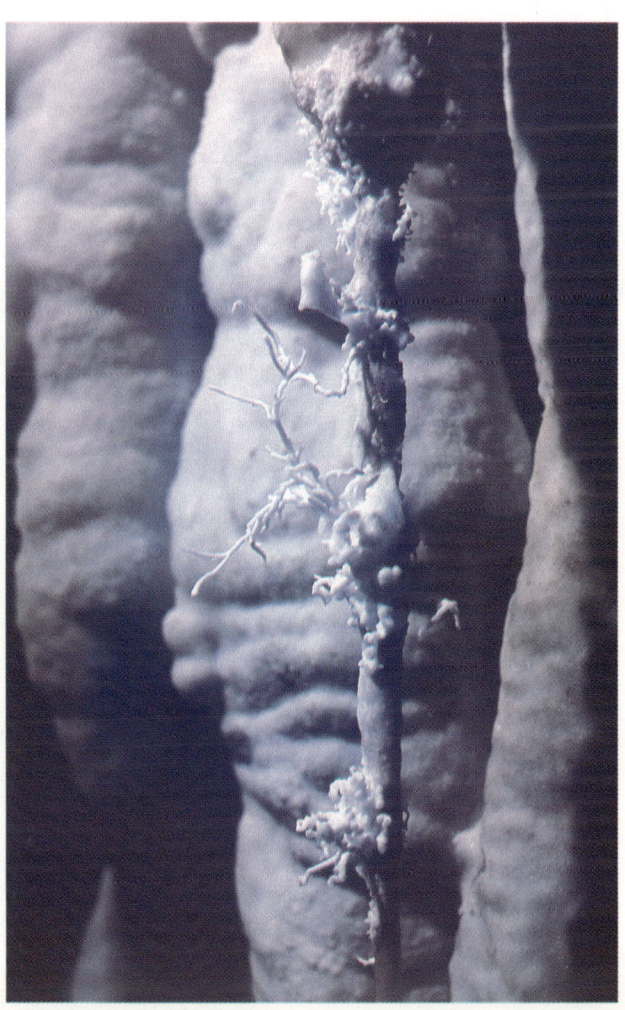

▲ Sinter floor

▼ V. Peltekov in the middle of a beautiful chamber

▼ Poplar-like stalagmites

▲ Photographer among beautiful places in the cave Yagodinskata peshtera

▼ Poplar-like stalagmites Milkwhite stalagmites and stalactites in the cave Yagodinskata peshtera

▲ The cave Yubileyna (Rhodopes, Bulgaria)

▼ The cave Yubileyna (Rhodopes, Bulgaria)   ▼ The cave Yubileyna (Rhodopes, Bulgaria)

▲ Stone waterfall in the cave Andaka (Bulgaria)   ▲ Cave lake

▼ Alexey Benderev shows one of the methods for tracing the underground water in the source Zhitolyub (Lakatnik)

◀ In the Temple

▼ Brief relax

▶ Rope contest on the Big Vertical in the cave Prohodna (Bulgaria)

◀ The waterfall in front of the cave Andaka

◀ Contest "Nedkov – Chapanov" on the waterfalls in front of the cave The Waterfall, Krushuna

▼ Ivo Tachev (kneeled), Tz. Ostromski and G. Georgiev after penetration in a cave in Ponor

▲ Sun shines over the cave river in the cave Dushnika near Iskrets

▼ The karstic source near Iskrets is the biggest dynamic source in Bulgaria. Most of it springs from the lower entrance of the cave Dushnika (up to 30 cu/sec).

Most color pictures are due to Vassil Balevski, Valery Peltekov and Trifon Daaliev

Name: **TIPCHENITSA** (97)
Karlukovo Village, Lovech Distr.
Altitude: 284 m. Length 190 m. Denivelation -78 m
Coordinates: N 43° 11' 55.6" E 24° 01' 56.4"

The cave pothole Tipchenitsa is situated on the left (geographically) bank of Iskar River, 4.5 km NW of Karlukovo. It is formed in limestone from the Cretaceous (Maastricht), in the Kameno Pole – Karlukovo Syncline, Middle Predbalkan.

An elliptic entrance with dimensions 22 x 14 m. A 35-meter vertical follows, from which a muddy descending gallery starts. About 15 m along the gallery, there is another 12-meter vertical, followed by a wide inclined gallery, high up to 15 m and 55 m long. The gallery goes further up and ends at a 17-meter vertical with a muddy siphon at the bottom. There are some stalactites and curtains in the cave.

First exploration in 1966 by V. Nedkov, N. Genov, T. Daaliev. Surveyed in 1983 by G. Gospodinova, L. Velichkova and S. Kolarova.

Tipchenitsa

Karlukovo Village

Name: **TIZOIN (68)**
Gubesh Village, Sofia Distr.
Total length 3599 m. Denivelation -320 m (fourth deepest in Bulgaria)
Coordinates: N 43° 05' 20.1" E 23° 04' 29"

A cave in limestone from the Upper Jurassic, 1.5 hours walk from Malina Hut. The entrance was dug out five meters above the sink, at the base of a cliff. The first 100 m are river meanders. Further on, the cave descends with only one branch, 80 m long, 100 m before it reaches the bottom. The two biggest verticals are of ten and eight meters. Under them, evorsion cauldrons are formed, full of water. The gallery follows further, 4–5 m wide and at a height of 6–7 m. The cave ends in a siphon. In flood, the siphon cannot take the whole quantity of water and the water level in the last 100–200 m rises very fast, which is seen from the marks on the walls. The outflow is rather variable. In dry periods it is ca. 1–2 l/s.

Discovered and explored by cavers from Cherni Vrâh Sofia Caving Club in 1981–1982.

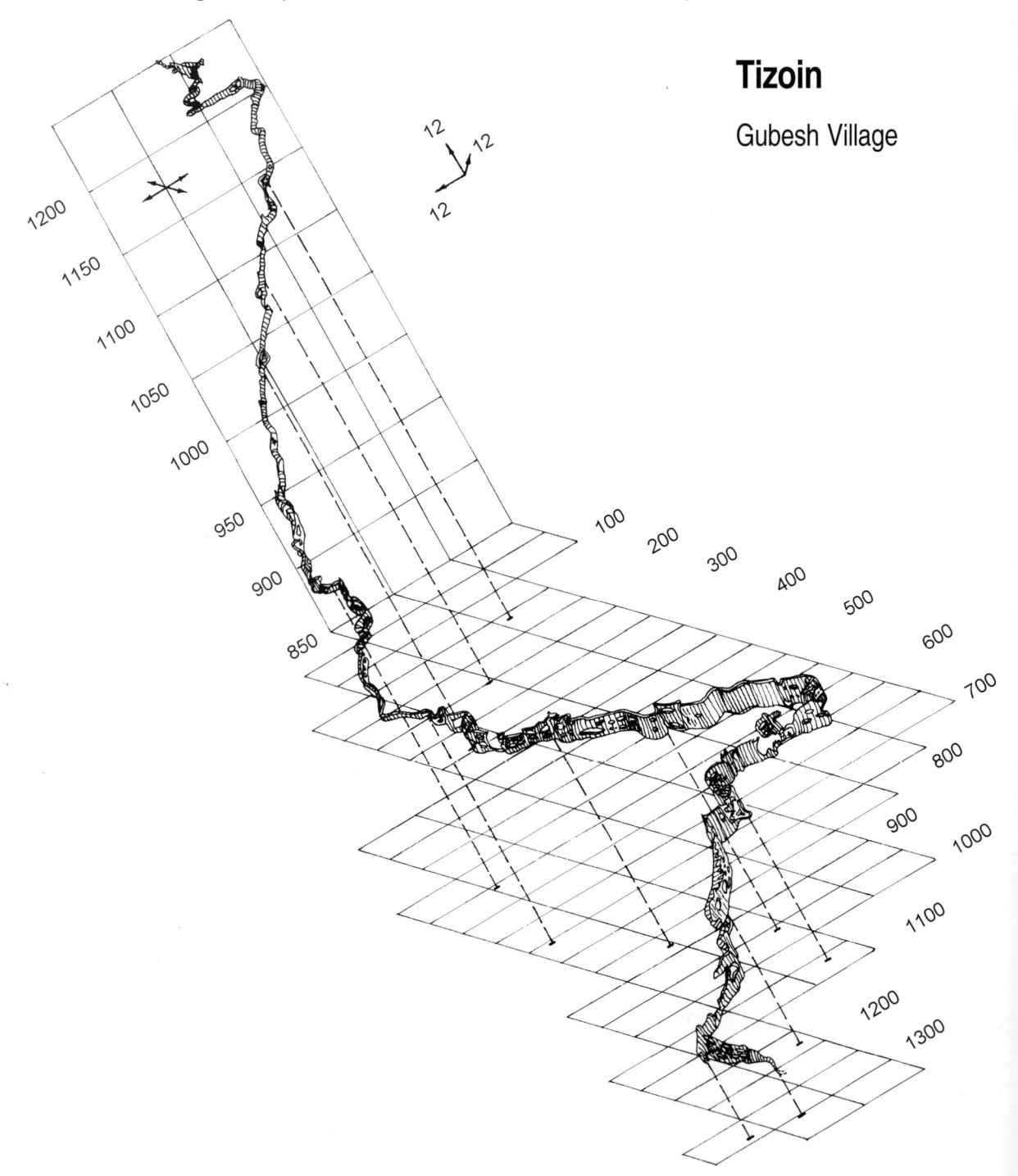

Tizoin

Gubesh Village

Z. Iliev, T. Daaliev, P. Beron and B. Marinov in front of the entrance – photo Trifon Daaliev

The karstic cliffs above Gintsi Village, with the entrances of the caves Dineva pesht and Krivata pesht – photo Trifon Daaliev

Name: **TOPLYA (134)**
Golyama Zhelyazna Village, Lovech Distr.
Length 462 m. Denivelation 5 m
Coordinates: N 42° 56' 53.8" E 24° 29' 15"

A cave, five km SE of Golyama Zhelyazna, in the western outskirts of the hill Yalovitsa, on the northern slope of Vassilyovska Planina, Middle Predbalkan. Formed in limestone from the Middle Trias. A horizontal, slightly descending, branched cave with two entrances. Shape of the cross section – mainly rectangular in the front parts of the gallery, and triangular in the back parts. Average width – 3.7 m; av. height 1.8 m. Max.-min. dimensions: width 0.8–7 m; height 0.8–3.8 m. No flowstone. Thick clay deposits along the galleries. Twenty meters SE and 12 m below the main entrance of the dry cave, there is another (source) cave, giving birth to Toplya River and it is the lowest level of the karstic water of the whole system. Entrance 5.6 x 5 m. The source drains the surface and rain waters on the N slope of Vassiljovska Planina. The average annual outflow is 309 l/sec, average min/max values are 24.5–1609 l/s.

The first studies included archaeological excavations (in 1898 by M. Koychev and in 1899 by G. Bonchev). A complex study done in June – July 1900 by Il. Stoyanov (results published in 1904) established artefacts from the Eneolith. Fossil remains from Quaternary mammals, birds and molluscs, also human bones were found. The cave was biospeleologically studied first by N. Radev in October 1925, then by other researchers. Speleologically explored by cavers from Planinets, Sofia, in 1970 and by Akademik, Sofia, in 1975. In July 1987, divers from Studenets, Pleven, conducted three consecutive dives in the source of the cave. During the first diving, I. Gunov, V. Chapanov and M. Dimitrov entered a spacious underwater chamber, 70 m from the entrance. The second diving was done by two pairs of divers. I. Gunov and Ya. Ivanov entered 25 m further, while the second pair (V. Chapanov and M. Dimitrov) surveyed the known part and penetrated another 60 m, without reaching a dry passage. The total length of the known parts is 178 m, with a maximum depth of -18 m.

Fauna: 13 species, including the troglobites *Typhloiulus bureschi* (Diplopoda), *Neobisium bulgaricum* (= *Obisium subterraneum*)(Pseudoscorpiones), *Tranteeva paradoxa* (Opiliones, Cyphophthalmi), *Plusiocampa gueorguievi* (Diplura), *Genestiellina gueorguievi* (Coleoptera) and the stygobite *Speocyclops infernus* (Copepoda).

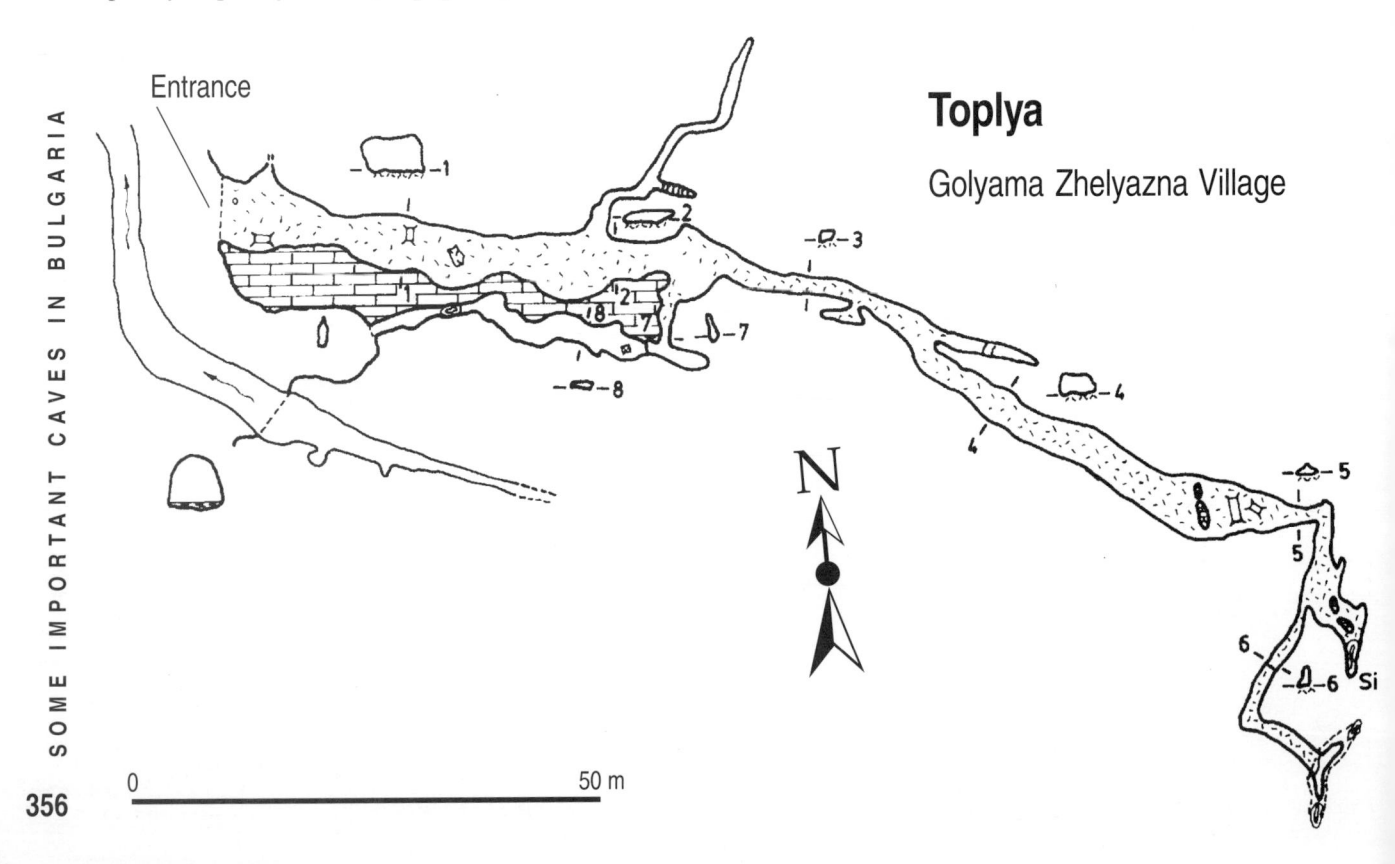

The entrance of Toplia Cave – photo Nikolay Genov

V. Nedkov is mapping the cave – photo Nikolay Genov

The first map of the cave from I.Stoyanov – 1904

SOME IMPORTANT CAVES IN BULGARIA

357

Name: **TOSHOVA DUPKA (KALNA MÂTNITSA, IZVORNA) (30)**
Glavatsi Village, Vratsa Distr.
Length 1302 m. Denivelation -63 m
Coordinates: N 43° 14' 33.2" E 25° 02' 09.2"

A cave at the base of the NE slope of Vrachanska Planina. Three entrances, three storeys. A branched cave with permanent water. Sinter formations in the two lower storeys. Thick clay deposits in the higher galleries. A stream on the second floor brings water to the principal river, coming out from the source entrance. Outflow 78–390 l/s.

First known exploration on February 16-18, 1968, by 13 cavers from Akademik, Sofia Caving Club, led by St. Andreev (surveyed 500 m, climatic and biospeleological studies). In 1971, Planinets Sofia Caving Club surveyed 963 m of the upper storeys and 130 m of the source part. Further studies by Veslets Vratsa Caving Club, also by P. Beron and other biospeleologists. A more recent study and survey took place from 9 to 16 May 1987 by a joint expedition of Pâstrinets, Montana, and Akademik Sofia Caving Clubs.

Fauna: 17 species of invertebrates, including six troglobites: *Sphaeromides bureschi, Trichoniscus anophthalmus, Vandeloniscellus bulgaricus* (Isopoda), the millipede *Typhloiulus bureschi*, the centipede *Lithobius lakatnicensis* and the beetle *Duvalius (Paraduvalius) beroni.*

Protected Natural Monument, together with 0,9 ha of adjacent land (Decree 1799/30.06.1972).

Toshova dupka
(Kalna Mâtnitsa, Izvorna)

Glavatsi Village

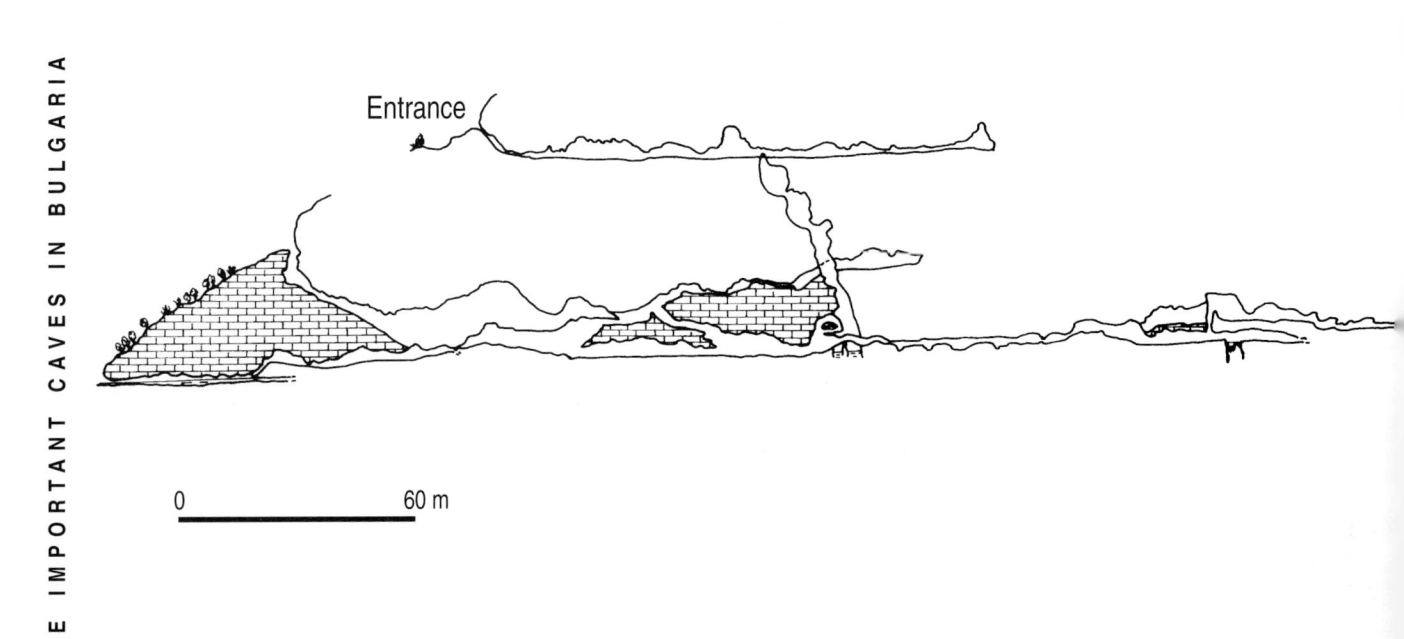

Name: **TROANA (163)**
Emen Village, Veliko Tarnovo Distr.
Length 2750 m. Denivelation -45 m
Coordinates: N 43° 07' 05.1" E 25° 18' 49.3"

A cave in the area of Emen.
Discovered, explored and surveyed up to 1743 m by Kamen Dimchev and Angel Georgiev, cavers from Akademik, Russe, in 1989. Later, Y. Yordanov and K. Stefanov from Prista Russe Caving Club reached 2750 m.
Fauna: troglobite is the beetle *Netolitzkya maneki iltschewi*.

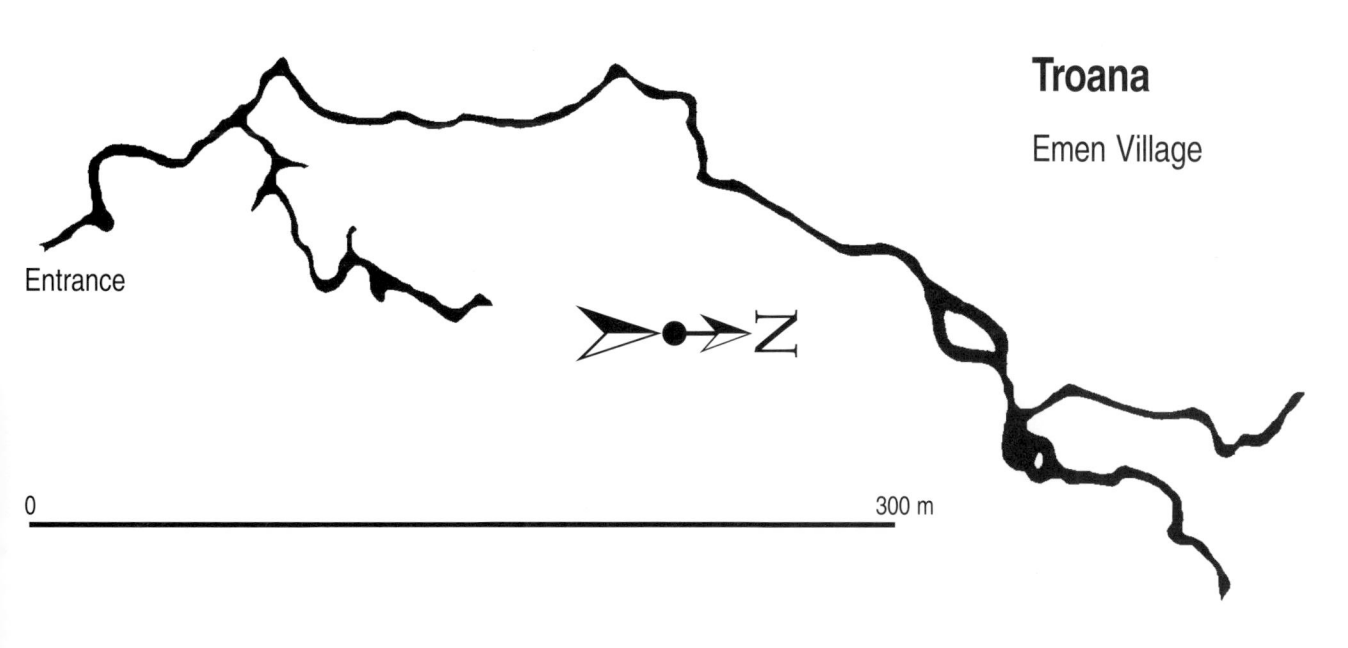

Formations – photo Constantin Kostov

Inclined stalagmite – photo Constantin Kostov

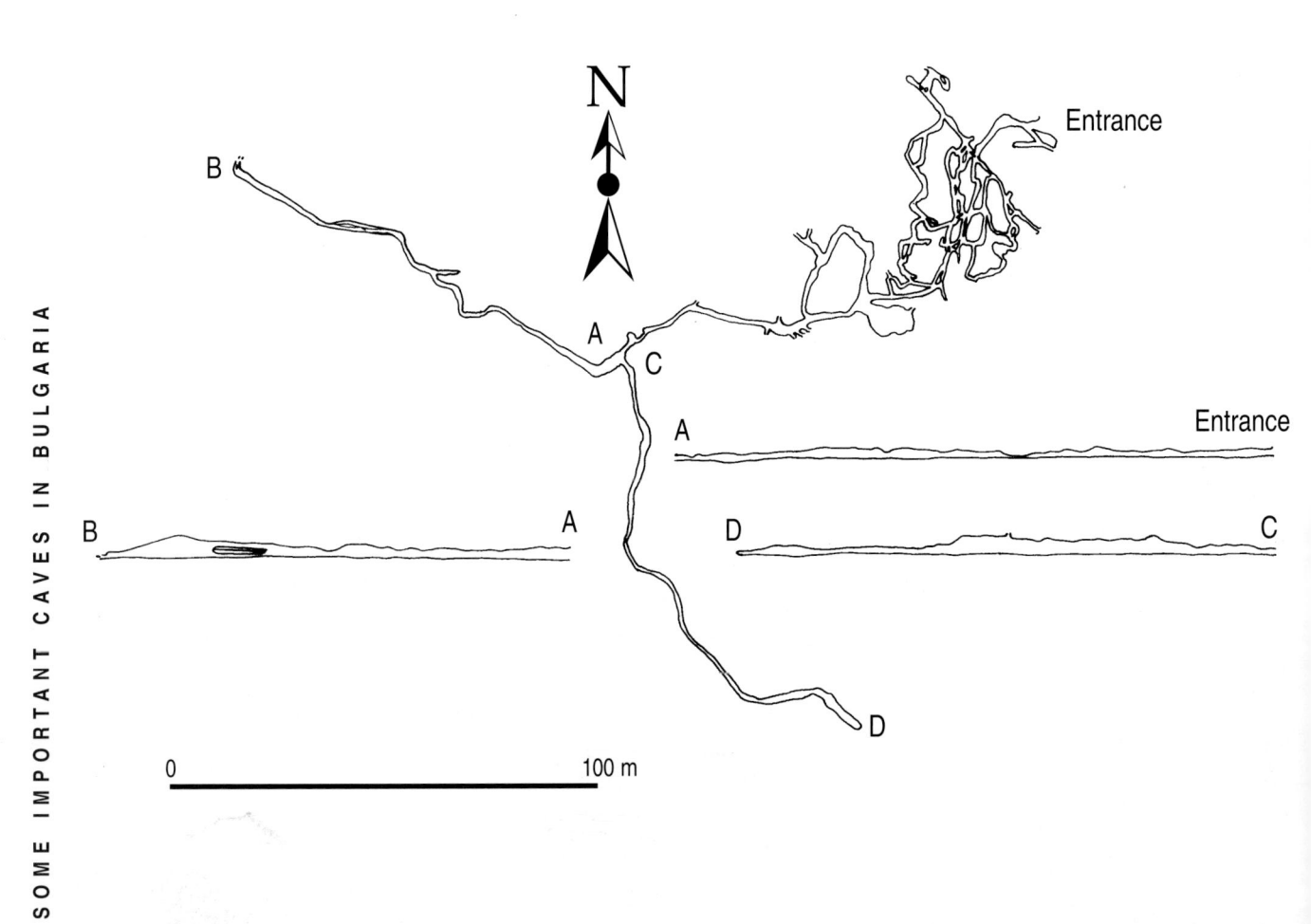

Name: **TRONA (DUHALOTO) ('123)**
Apriltsi Village, Lovech Distr.
Altitude 1364 m. Length 1040 m. Denivelation -12 m.

A water cave south of Apriltsi, in the locality Polenitsi, in the southern outskirts of the summit Novoselski Polenitsi, Troyan Balkan. Formed in Jurassic limestone. A horizontal, branched cave. Entrance – 2 m wide and 1.6 m high. Dimensions of the galleries: 0.40–2 m wide, 0.9–16 m high. The biggest chamber is 12 m long, 4.5 m wide and 13 m high. Poor in flowstone. About 80 m from the entrance, water is running out of a fissure on the ceiling and forms an inward stream. Ends in a siphon at the NE end.

The cave was discovered by M. Michev from Cherni Ossam. Explored first V. Markov by Balkansko Eho Oreshaka Caving Club in 1963, up to 700 m length. Explored and surveyed to its present length by M. Dimitrov, S. Gazdov, D. Toshkov and B. Garev, cavers from Studenets Pleven Caving Club.

Fauna: troglobite is the beetle *Hexaurus beroni* (Cholevidae).

Trona (Duhaloto)

Apriltsi Village

Name: **TSAKOVSKA PESHT (TSAKONICHKI PECH) (76)**
Tsakonitsa Village, Vratsa Distr.
Length 485 m

    A through cave near Tsakonitsa Village, used for cheese production. Both entrances have almost the same height (8 m). A horizontal water cave, 345 m long on the main axis. The bottom of the gallery is covered with thick layers and fallen blocks, through which a small underground stream runs. About 160 m from the entrance, on the left towards SW, a higher positioned branch starts, which is 140 m long. Wet and relatively narrow, but after the first third it becomes extremely beautiful.
    In the cave artefacts from the Iron Age have been found.
    Explored in 1949 and 1959 by P. Tranteev.

Transparent stalagtons from the pothole Pticha Dupka – photo Vassil Balevski

Name: **TSARKVISHTE (TSARKVETO) (62)**
Breze Village, Sofia Distr.
Altitude 900 m. Length 106 m. Denivelation 9 m

A cave on the right side of Brezenska Reka. Descending, one-gallery. At the beginning it is very low and narrow, then it becomes wider, In old times it was full of aragonite formations, which are now broken. A small lake is believed to be healing the sick who have left spoons and coins. Air T° 11 °C, clay T° 10.4 °C. The cave was surveyed and explored by Akademik Sofia Caving Club.

Fauna: includes the centipede *Eupolybothrus gloriastygis* (Chilopoda), a troglobite, and the stygobites *Stygoelaphoidella elegans* and *Speocyclops lindbergi* (Crustacea, Copepoda).

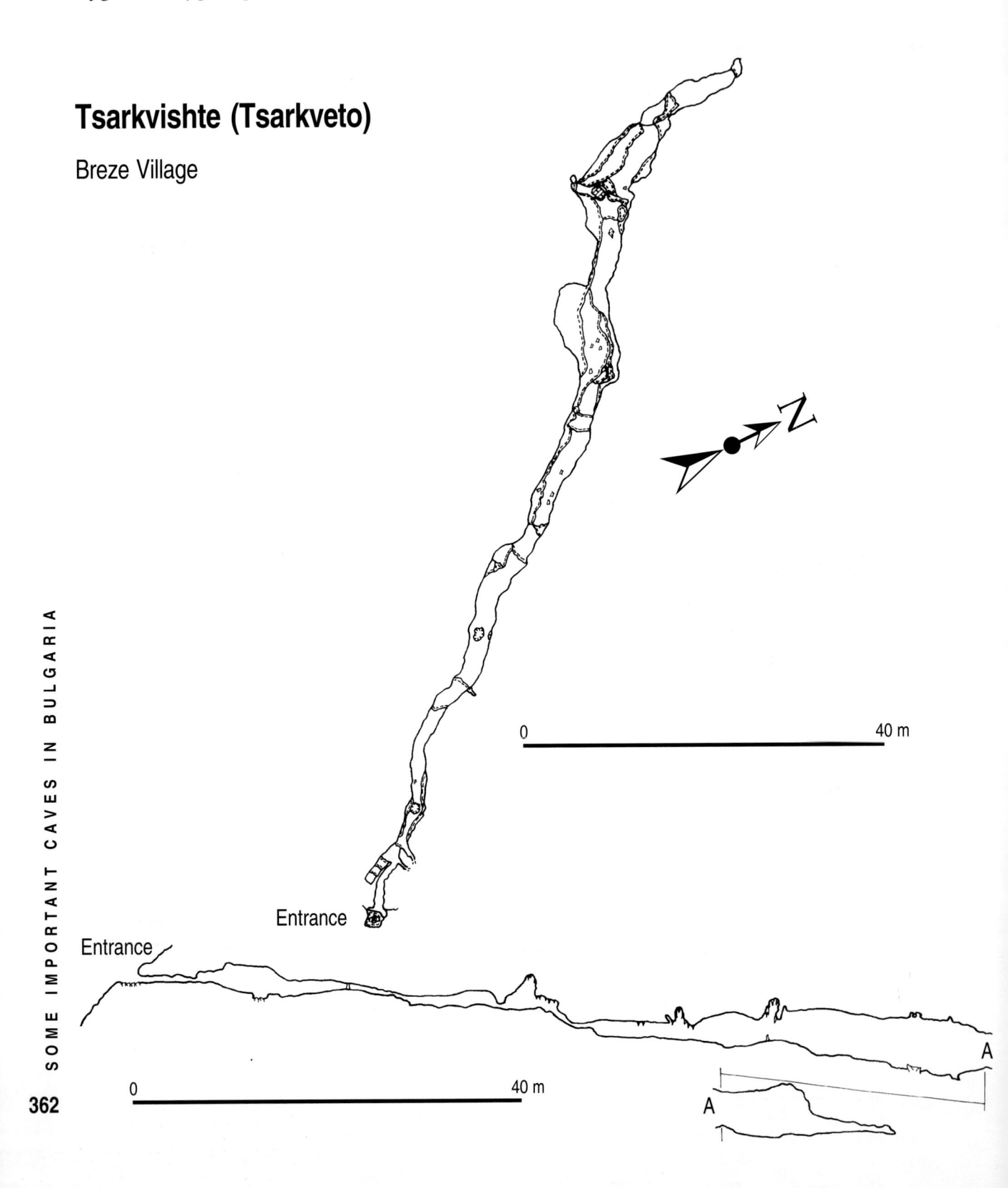

Tsarkvishte (Tsarkveto)

Breze Village

Name: **URUSHKA MAARA (PROYNOVATA) (157)**
Krushuna Village, Lovech Distr.
Altitude: 191 m. Length 1600 m
Coordinates: N 43° 14' 43.3" E 25° 01' 51.3"

A source cave, two km SW of Krushuna, in the periphery of Devetaki Plateau, Middle Predbalkan. Formed in limestone of the Lower Cretaceous (Aptian). A branched, two-storey cave, developing S-SW. Entrance, 10 m wide and 4 m high. Upstream, there is a siphon, circumvented by a dry gallery on the east. A gallery follows, which is ca. 300 m long, first developing to SE, then to SW, with a rectangular cross section (2 x 2 m). Many thick sinter walls, forming about 12 lakes, 1–2 m deep. The development of the cave is about 200 m to the south. In this sector, the profile is elongated, 1–1.5 m long and 8–9 m high. There are remaining parts of an older upper storey. The gallery is meandering and slightly ascending E-SE, retaining its shape. The lakes are transformed in a stream, starting from a siphon at the end of the cave. Sinter walls along the cave, many flowstone formations. The source is with a variable outflow. First explored by Nenko Radev in 1924. In 1955, the archaeologists N. Djambazov and V. Mikov carried out some research. Up to 1965, some 200 m of the cave had been known. In the same year, a Bulgarian – Hungarian expedition proceeded further in the cave and the research was completed in 1968, during a National Expedition. In May 1989, the divers V. Nedkov and V. Chapanov penetrated the 30-m siphon at the end of the cave, but found their death in the following gallery (with a toxic gas).

Fauna: 12 species, including the stygobite *Niphargus ablaskiri georgievi* (Amphipoda). The cave is inhabited by a large bat colony.

Urushka Maara (Proynovata)

Krushuna Village

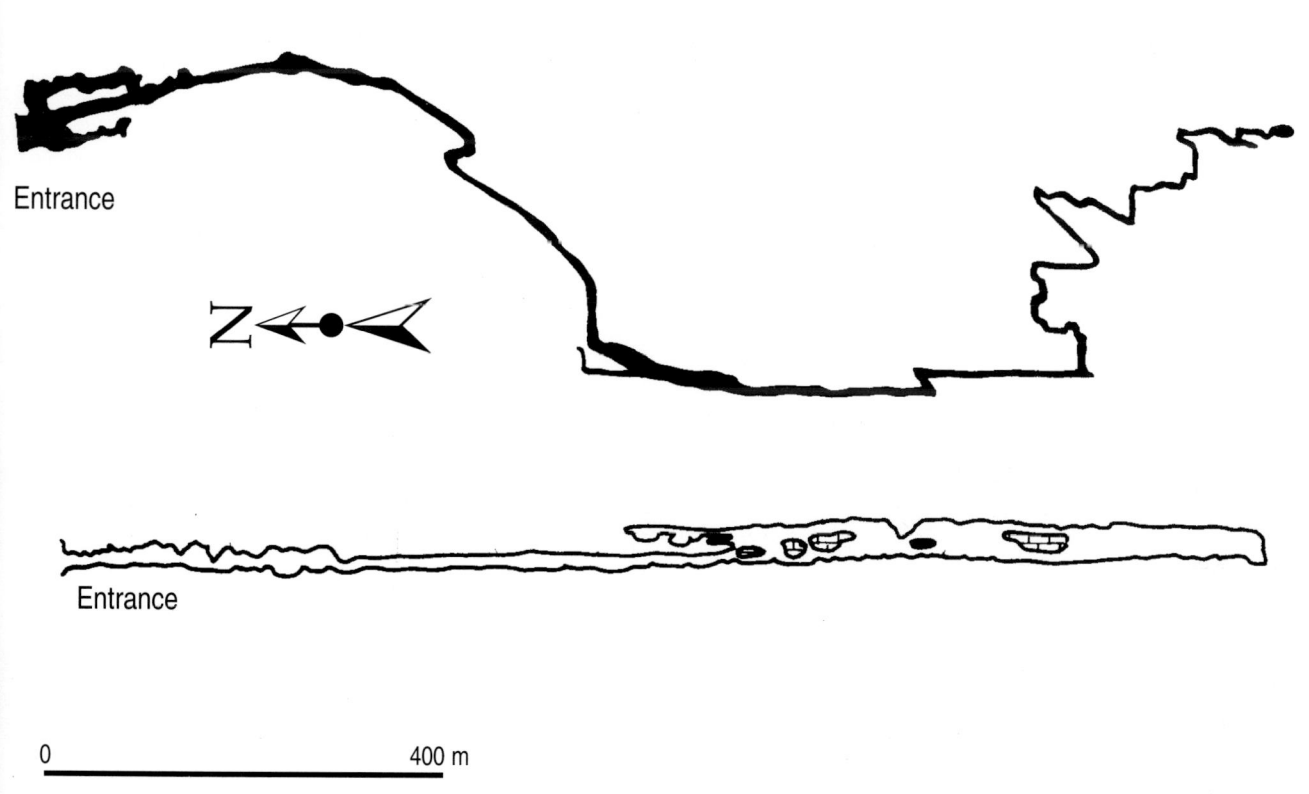

The entrance of the cave Urushka Maara – photo Trifon Daaliev

Travertine waterfall – photo Vassil Balevski and Trifon Daaliev

Name: **UZHASA NA IMANIYARITE (188)**
Kotel Town, Sliven Distr.
Altitude 710 m. Length 440 m. Denivelation -160 m
Coordinates: N 42° 56' 01.8" E 26° 30' 10.5"

A cave pothole in the Senonian limestone of Zlosten karstic area, 3.3 km NE of Kotel. A cascade cave, broken by verticals of different height, the deepest being 45 m. Through the horizontal gallery, an underground stream runs with a small outflow. Siphon at the bottom. T° 8.5 °C.

Known to local people and gold seekers for a long time. Surveyed in 1964 by members of Akademik Sofia Caving Club and in 1979 by Protey Sliven Caving Club.

Uzhasa na Imaniyarite

Kotel Town

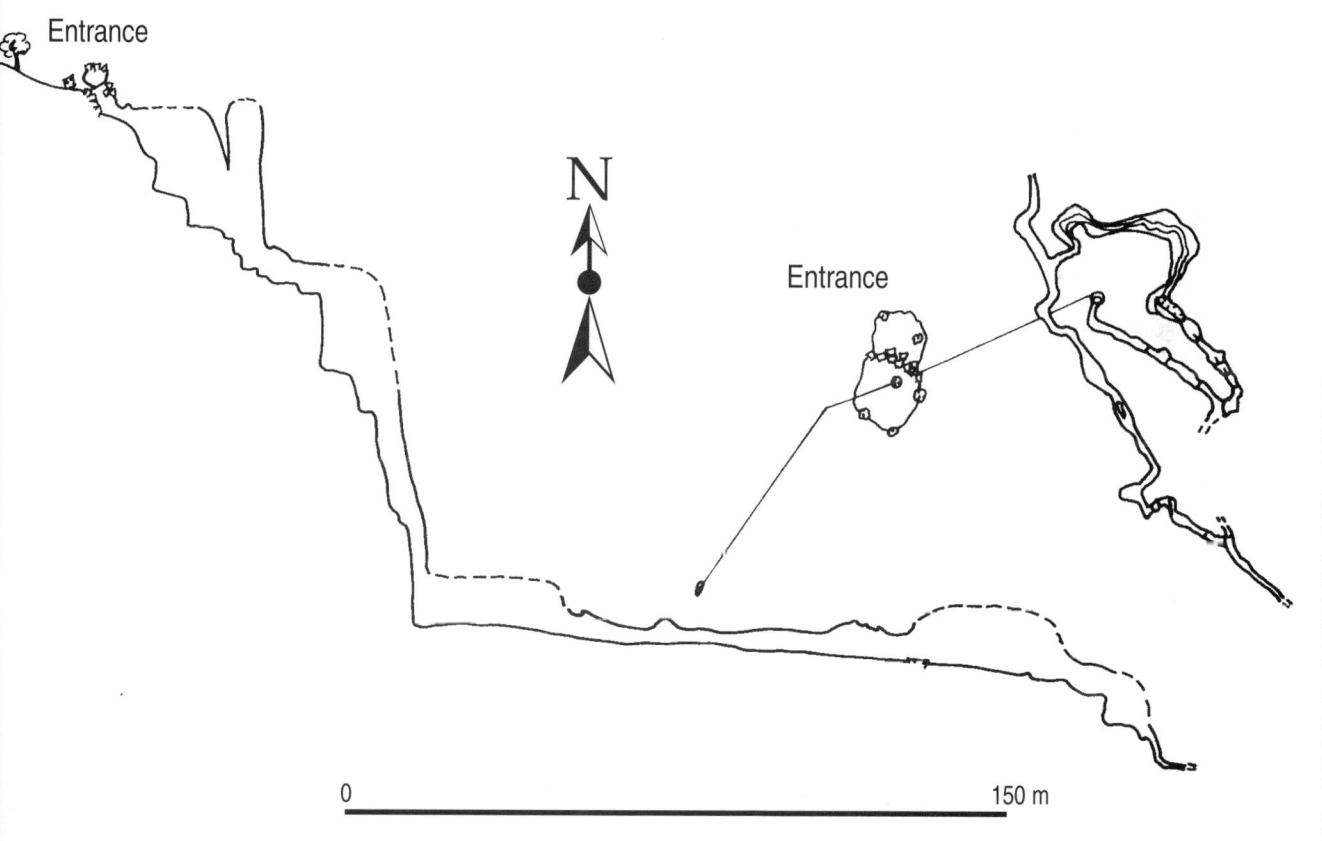

Name: **VÂRLATA (126)**
Cherni Ossam Village, Lovech Distr.
Altitude ca. 1300 m. Length 1110 m. Denivelation -41 m

A cave in the eastern part of the karstic canyon Steneto in the area of Cherni Ossam, within the protected area on the left bank of Kumanitsa river (the upper part of the river Cherny Ossam, from the source to the sink), above the nice waterfall Golemiya Kazan. There are several ways to reach the cave, which is 500 m far (20 min walk) from the locality Kozi Brod, downstream Kumanitsa river. Entering the cave in the spring or during snow melting or heavy rain is dangerous, sometimes even impossible, as part of the flood water in the canyon enters through the entrance or through smaller sinks.

The cave entrance (1.5 x 2.5 m) is in a funnel under a small cliff at the forest limit. A descending, one-storey, branched and permanently watered cave. Bifurcates about 20 m. from the entrance. The right side gallery (80 m long) goes SE. Many stalactites on the ceiling.

The left side (the main) gallery goes NE. It is 2–6 m wide and 1-8 m high. On the floor, sinter lakes and fallen boulders are seen. At the beginning of the gallery, from the left, the first water comes. After 100 m, the main gallery branches off again.

The right branch is ascending, 150 m long, ending in a narrowing, developing NW–W. It is wide and high, with a floor covered with stones.

The right side goes NE for about 450 m and ends in a blockage. The floor is covered with blocks and sinter lakes. There is a way through the blockage, leading into the end gallery (250 m long). The cave ends in a narrow and, according to the explorers, impenetrable water siphon.

Rich in flowstone formations. In the cave beautiful sinter lakes, fantastic stalactites and stalagmites, stalagtones and stone waterfalls have been observed.

The cave Varlata is easily penetrated, but there is a danger of crumbling of the entrance parts and of a blockage in the Big Chamber. During floods, it is possible that the low end parts sink under the water.

The cave was discovered when in 1988 Troyan cavers dug out the entrance funnel. In the following 1989, cavers from Steneto organized an expedition to survey the cave. The map was prepared by Detelina Petkova, Aleksander Radulov, Krasen Radulov and Emilia Gateva.

The cave falls within the buffer zone of the Reserve Steneto and within the limits of Central Balkan National Park.

Description by Detelina Marinova.

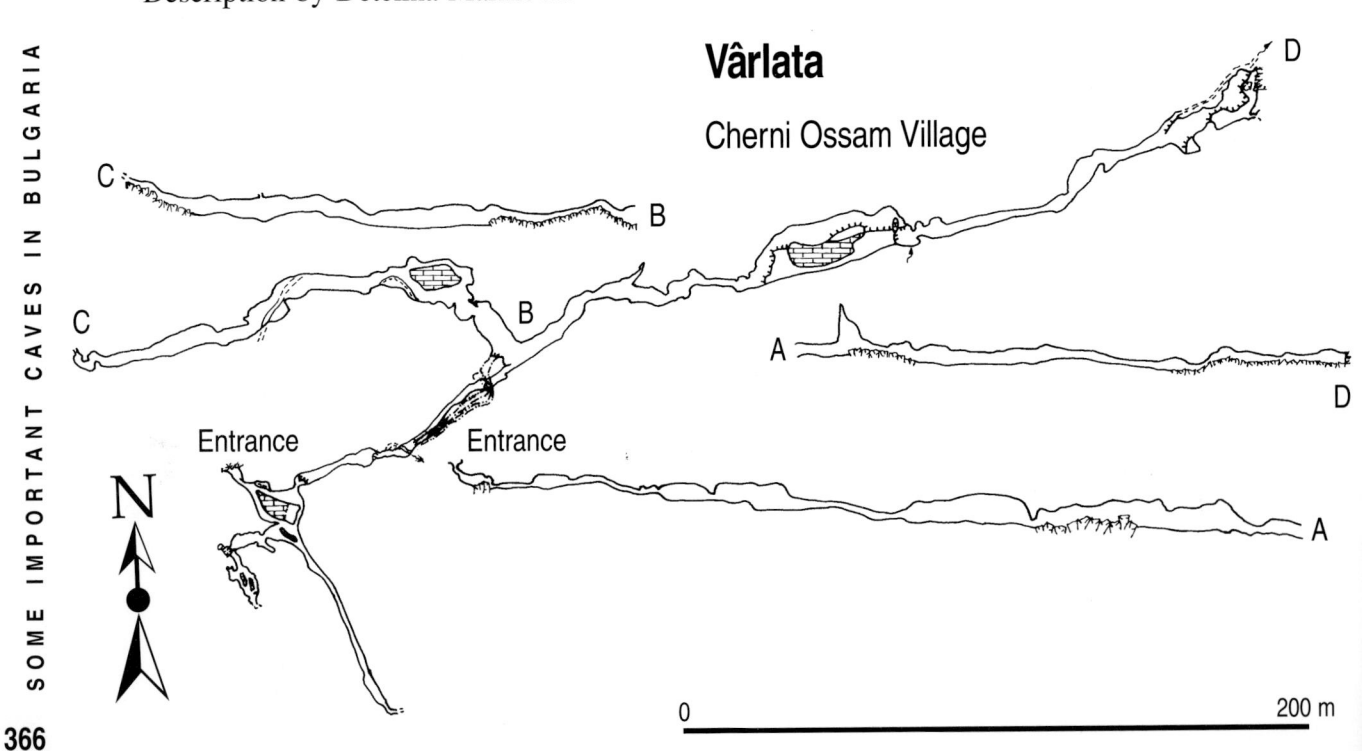

Name: **VÂRTESHKATA (45)**
Zverino Village, Vratsa Distr.
Altitude 1160 m. Length 158 m. Denivelation -74 m
Coordinates: N 43° 08' 27,6" E 23° 33' 27,6"

A pothole near Zverino. Entrance 8 x 8 m, situated in an old beech forest. The first vertical (25 m) leads to a flat bottom with fallen leaves. On the left of the vertical, there is another vertical of 10 m, leading into a 25-m long gallery.

The main gallery goes to boulders, behind which a 4-m vertical follows, leading into a hall, which is 50 m long, 8 m large and the height is more than 10 m. In the gallery, there are beautiful poplar-like stalagmites, about 5 m high. At the end, we reach the most beautiful chamber of the cave, covered with dendrites, stalagmites, stalactites, sinter lakes and other secondary formations. The cave is humid, but there is no running water. T° 13 °C.

The pothole has been known to the local people for a long time. First penetrated by T. Daaliev, V. Gyaurov, cavers from Edelweiss – Sofia Caving Club in 1968.

Fauna: includes the troglobites *Radevia hanusi, Pheggomisetes* sp. (Coleoptera), *Bulgarosoma bureschi* (Diplopoda).

Vârteshkata

Zverino Village

Pit entrance of Varteshkata cave – photo Valery Peltekov

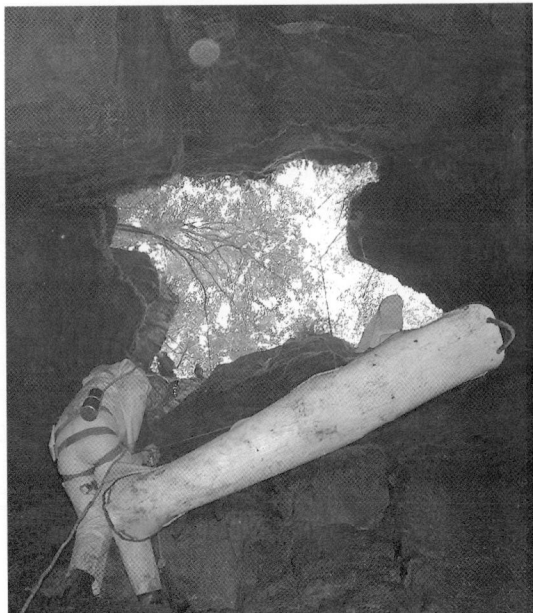

Extraction of a casualty – photo Valery Peltekov

The stalagmites are the cave's beauty – photo Valery Peltekov

Name: **VENETSA (4)**
Oreshets Railway Station, Vidin Distr.
Total length 220 m. Denivelation 27.5 m
Coordinates: N 43° 37' 47.2" E 22° 44' 24"

A cave one km southwest of Oreshets Railway Station. Formed in Jurassic limestone, forming the ridge Belogradchishki Venets, Western Predbalkan. Entrance – rectangular (2 x 1.5 m). Five interconnected chambers follow, developed in N-S direction.

About 3 m below, we reach the first chamber. Through a narrow squeeze, the second chamber is reached. Through an opening in its northern wall, a third chamber can be entered, which is 20 m long, 15 m wide and between 3 and 6 m high. Here, the whole splendour of the cave is seen. There are all kinds of flowstone formations – stalactites, stalagmites, stalagtones, corallites, sinter lakes – colored in yellow, brown and red. A fourth chamber follows – a labyrinth of blocks and stalagtones. In places, the floor is covered with clay, in others – with thick sinter crust. In the lowest part of the chamber, which is the bottom of a former deep lake, there are peculiar club-like formations. Among the stalactites there are some helictites. The fifth and last chamber is relatively poor in formations. Here, there are a series of 6 potholes, the deepest being 12 m. The ceiling of the chamber is cracked and many vertical crevices are seen.

The cave temperature is 9.6 °C, the humidity – 100 %.

The cave was found during blasting works in the quarry near Oreshets Railway Station in 1970. It was explored and surveyed in 1971 by Bel Prilep Belogradchik Caving Club. The map was checked and completed in 1973 by cavers from Akademik Sofia Caving Club, together with some meteorological, geomorphological and biospeleological observations. After P. Neykovski and other sources.

Fauna (collected by P. Beron on October 17.1971): includes the beetle *Beronia micevi* (Cholevidae) the troglobitic Collembola *Protaphorura beroni* has been described from this cave.

Venetsa

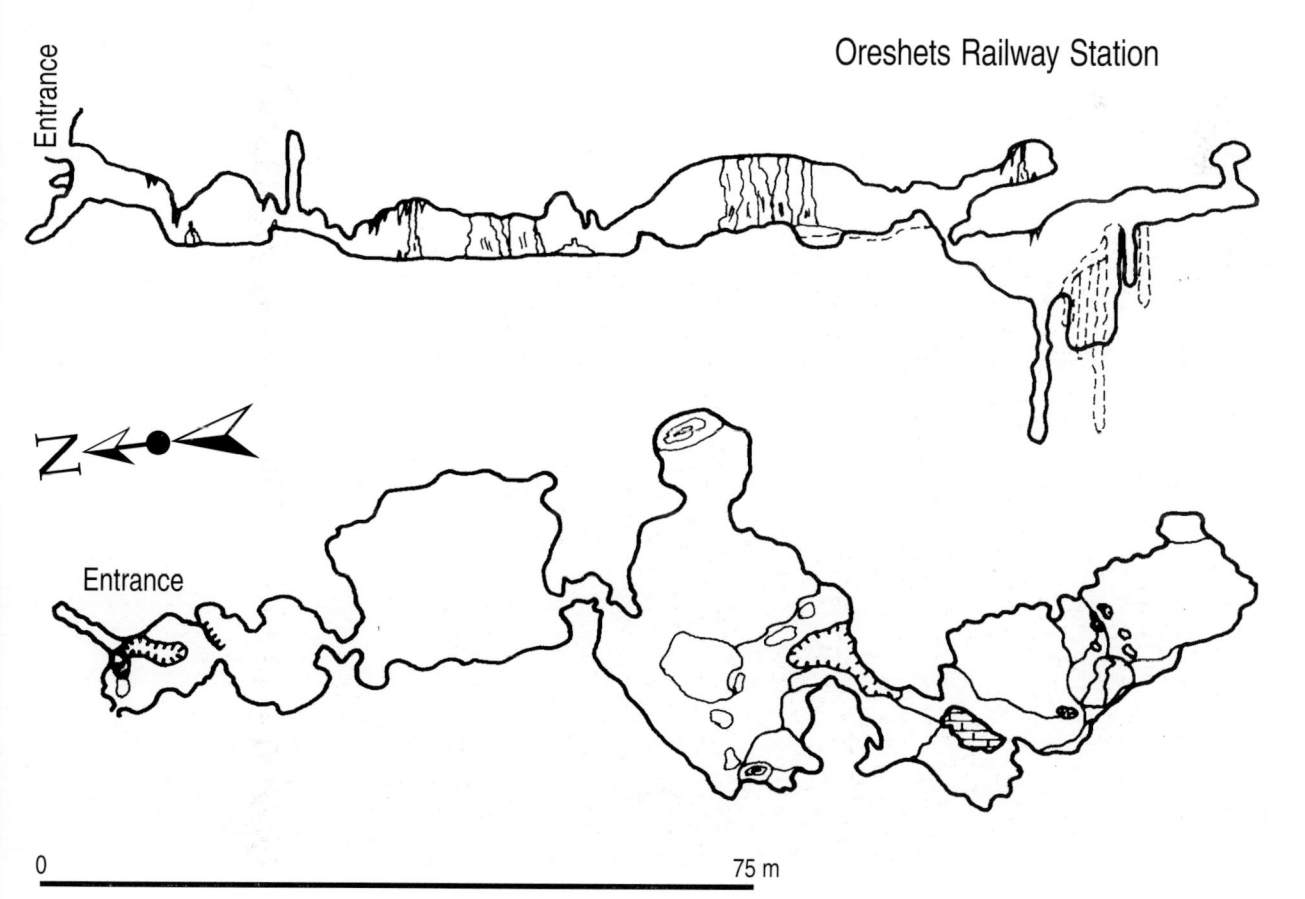

Oreshets Railway Station

Crystals are growing on stalagtites.

Name: **VODNATA PESHT (74)**
Lipnitsa Village, Sofia Distr.
Length 1100 m. Denivelation 42 m
Coordinates: N 43° 00' 42.8" E 23° 44' 55.3"

A water cave 4 km S of Lipnitsa. The entrance is situated at the base of a cliff, in the locality Rugyova Padina in the north, on the right of the road to Novachene Village.

The entrance is triangular, 3 m wide and 6 m high, facing south. A stream of changing outflow comes out of the entrance, running along the first level of the cave.

An ascending horizontal water cave with three levels. About 200 m from the entrance, the cave bifurcates. The right-side gallery is dry, on three levels. On the left, the main water gallery follows, ending at the 340th meter in a siphon. The first siphon was dug out and overcome in 1975. After the siphon, 342 m have been surveyed. The gallery ends in a second siphon.

The cave is rich in flowstone formations (many stalactites, stalagmites, stalagtones, stone waterfalls, in different colors).

In the entrance parts of the cave there are small bat colonies.

The cave had been known to local people for a long time. Explored in April 1963 by A. Petkova, V. Gyaurov and R. Zahariev from Edelweiss Sofia Caving Club. They surveyed and explored the cave. Later, the exploration was further carried out by cavers from Botevgrad. They penetrated the siphon and brought the cave to its present length. Tsvetko Tsvetkov from Biser Botevgrad Caving Club made a new survey of all galleries. In 1991–1995, Iv. Aleksiev from Edelweiss Sofia Caving Club made a 3D model of the cave. So far, 570 m of galleries have been surveyed up to the siphon and 342 m beyond it, altogether 912 m. The total length of the explored galleries exceeds 1000 m, but some of them are not mapped. Description after A. Petkova.

Fauna: collected since 1960 by V. Beshkov, P. Beron, Iv. Pandurski and other biospeleologists. So far, 16 species have been identified, including the stygobites *Diacyclops clandestinus* and *Stygoelaphoidella stygia* (Copepoda) and the troglobites *Typhloiulus bureschi* (Diplopoda) and *Tricyphoniscus bureschi* (Isopoda).

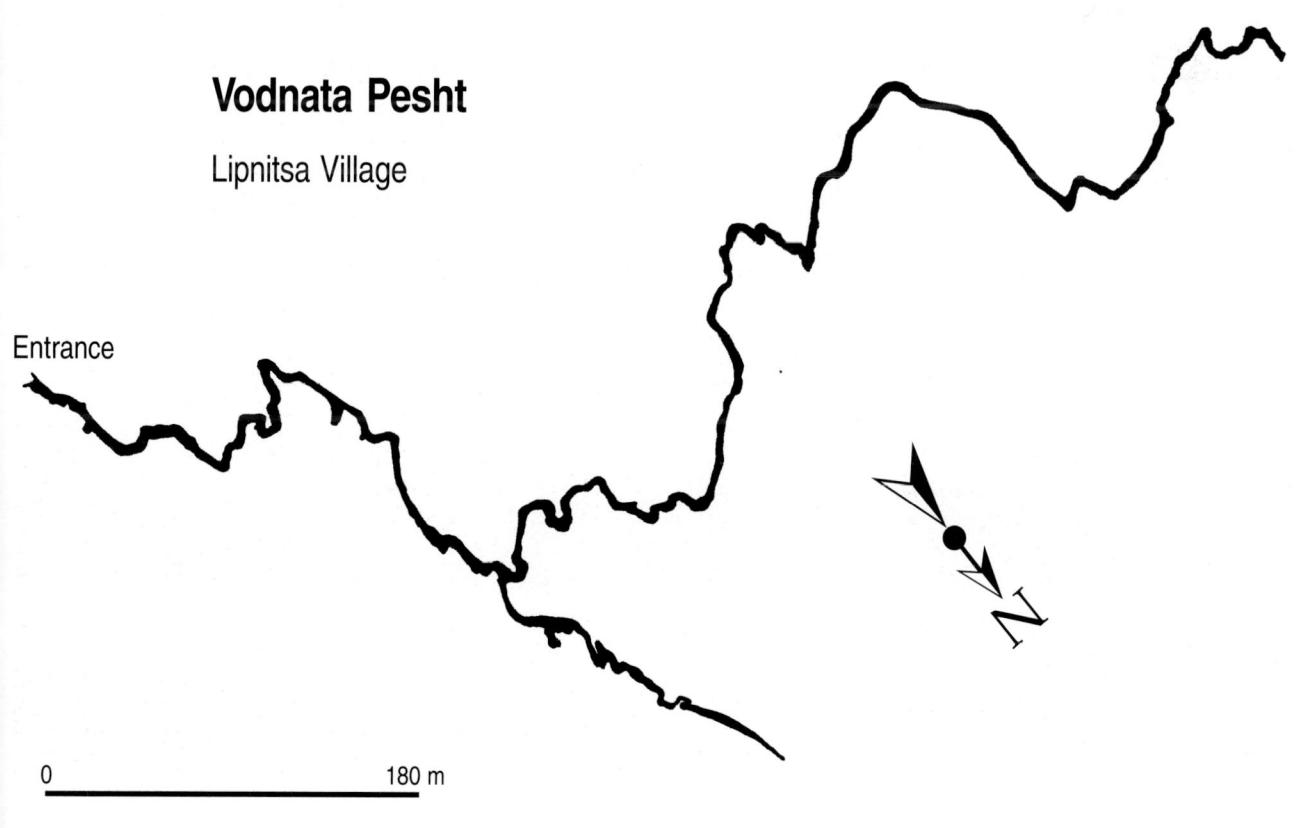

Water Cave entrance – photo archive speleo club „Edelweiss" Sofia – Angelina Petkova

Name: **VODNATA PESHTERA (or TSEROVSKATA PESHTERA) (59)**
Tserovo Village, Sofia Distr.
Length 3264 m. Denivelation +85 m.
Coordinates: N 43° 00' 25.4" E 23° 20' 39.9"

A water cave at the base of the cliff Kamiko, on the E slope of Ponor Planina. Formed in limestone from the Lower Trias. Two superposed entrances, in the lower the water is piped. A one-gallery cave, developed along three main faults. A one-side gallery near the entrance and another one 220 m before the end of the cave. Average width – 2.50 m, height by the 180th m – 15 m, and 250 m from the entrance, the gallery becomes lower (1.4–1.5 m). Boulder chocks at 650, 850, 1050 and 1220 m from the entrance, the last one being the biggest. Flowstone formations in the deeper parts of the cave. The cave drains a complex karstic system, the source having a changing outflow of 36–200 dm^3/s. Used for drinking water. The cave was first recorded in 1915 by Zh. Radev, who surveyed the first 720 m. In 1924–1925, Dr Buresch and his group explored the first 300 m, searching for cave animals. Surveyed in 1963 by Akademik Sofia Speleoclub up to 2140 m and in 1979 by Edelweiss Sofia Speleoclub up its present dimensions. In 1990, a group of hydrogeologists led by K. Spassov explored the hydrology of the cave and the influence of the nearby quarries. The cave could be visited only with a permit.

Fauna: so far, 47 species recorded including 7 troglobites and stygobites: *Sphaeromides bureschi* and *Bureschia bulgarica* (Isopoda), *Typhloiulus bureschi* (Diplopoda), *Eupolybothrus andreevi* (Chilopoda), *Paranemastoma (Buresiolla) bureschi* (Opiliones), *Acanthocyclops radevi* (Copepoda) and *Plusiocampa bureschi* (Diplura).

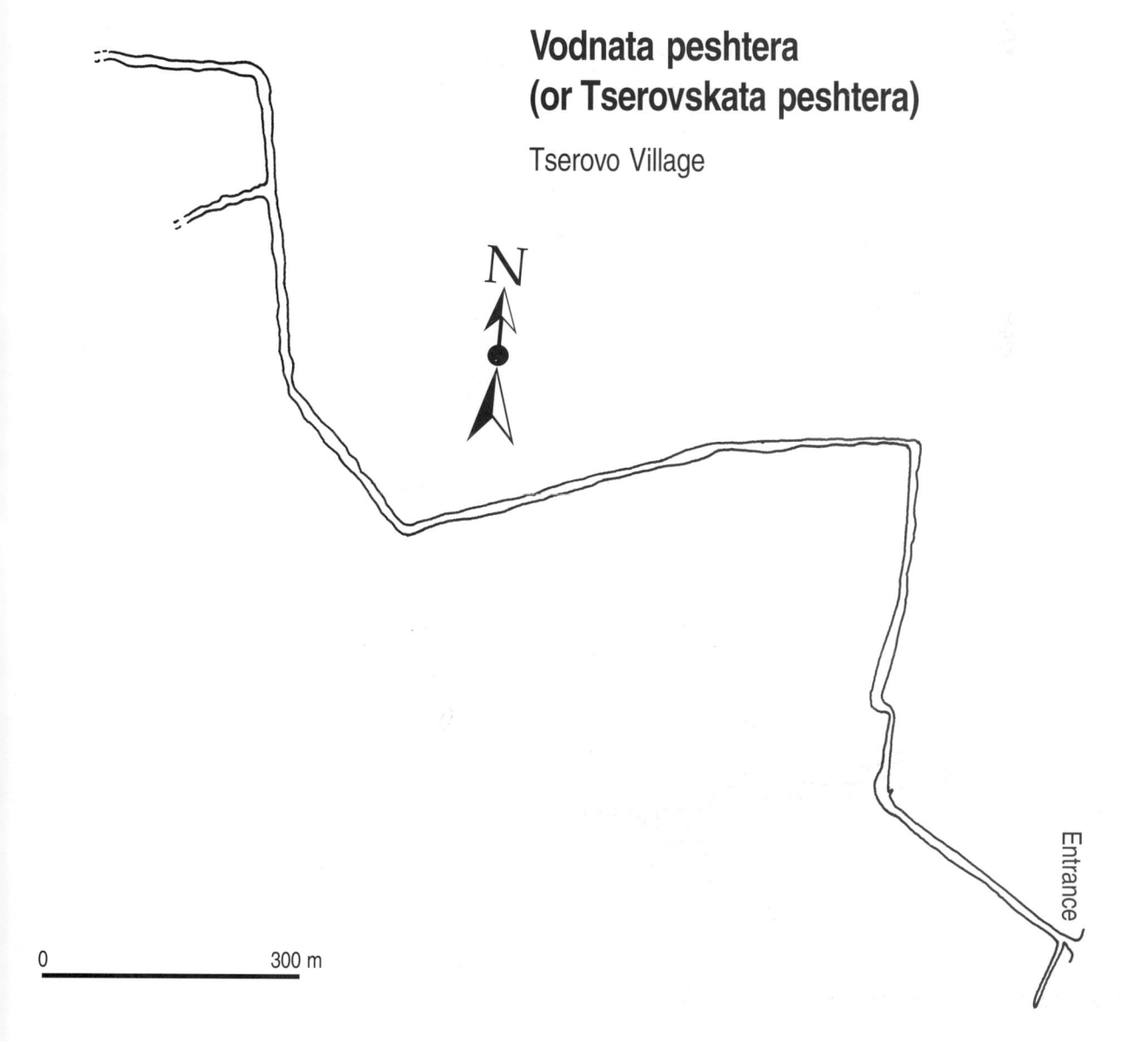

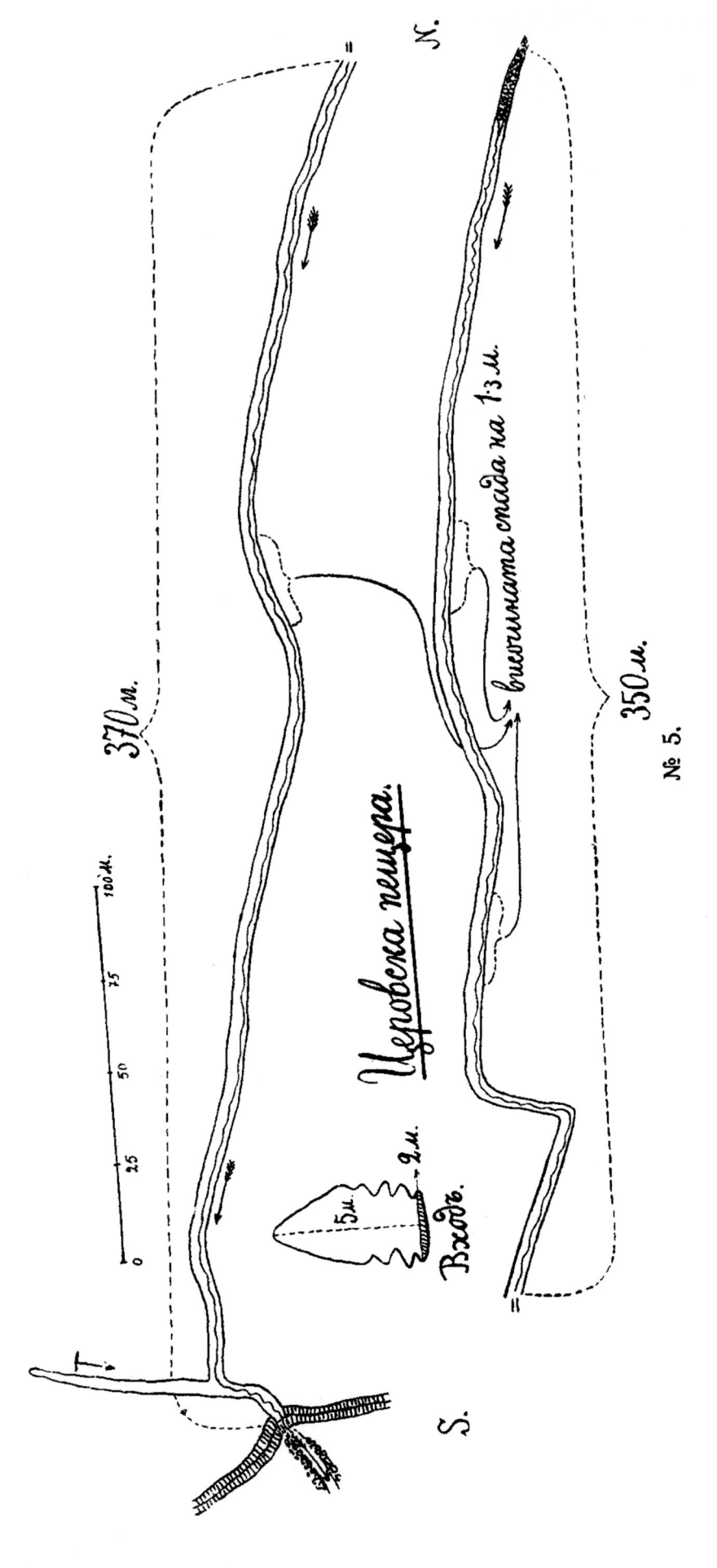

The first map of the cave is made by prof. Zh. Radev in 1915

Name: **VODNATA PROPAST (Propust) (81)**
Kameno Pole Village, Vratsa Distr.
Altitude 433 m. Total length 695 m

The entrance of Vodnata Propast is situated W-NW of Drashan Village, ca. three km in a straight line. It forms a part of the system Vodna Propast – Bayov Komin – Drashanska Peshtera.

A one-gallery horizontal cave, developing basically on the east. In the first 500 m the gallery is 1–2 m wide and high and there are several shallow lakes. The remaining part of the bottom is covered with mud and fallen stones. There are in many places stalactites and stalagmites. Five hundred meters from the big lake an underground stream starts, running at the bottom of the gallery up to its end (lake siphon). In the last 200 meters, the gallery is higher and wider, also with stalactites and stalagmites.

Explored and surveyed in 1970–1975 by cavers from Veslets Vratsa Caving Club, led by K. Karlov. The surface survey in 1975 proved theoretically the connection between the three caves. The water runs through Vodnata Propast (695 m), enters the cave Bayov Komin, runs through the entire cave (2196 m) and surfaces through Drashanskata Peshtera (578 m). After the eventual connecting of the three caves the length of the system could surpass 4500 m.

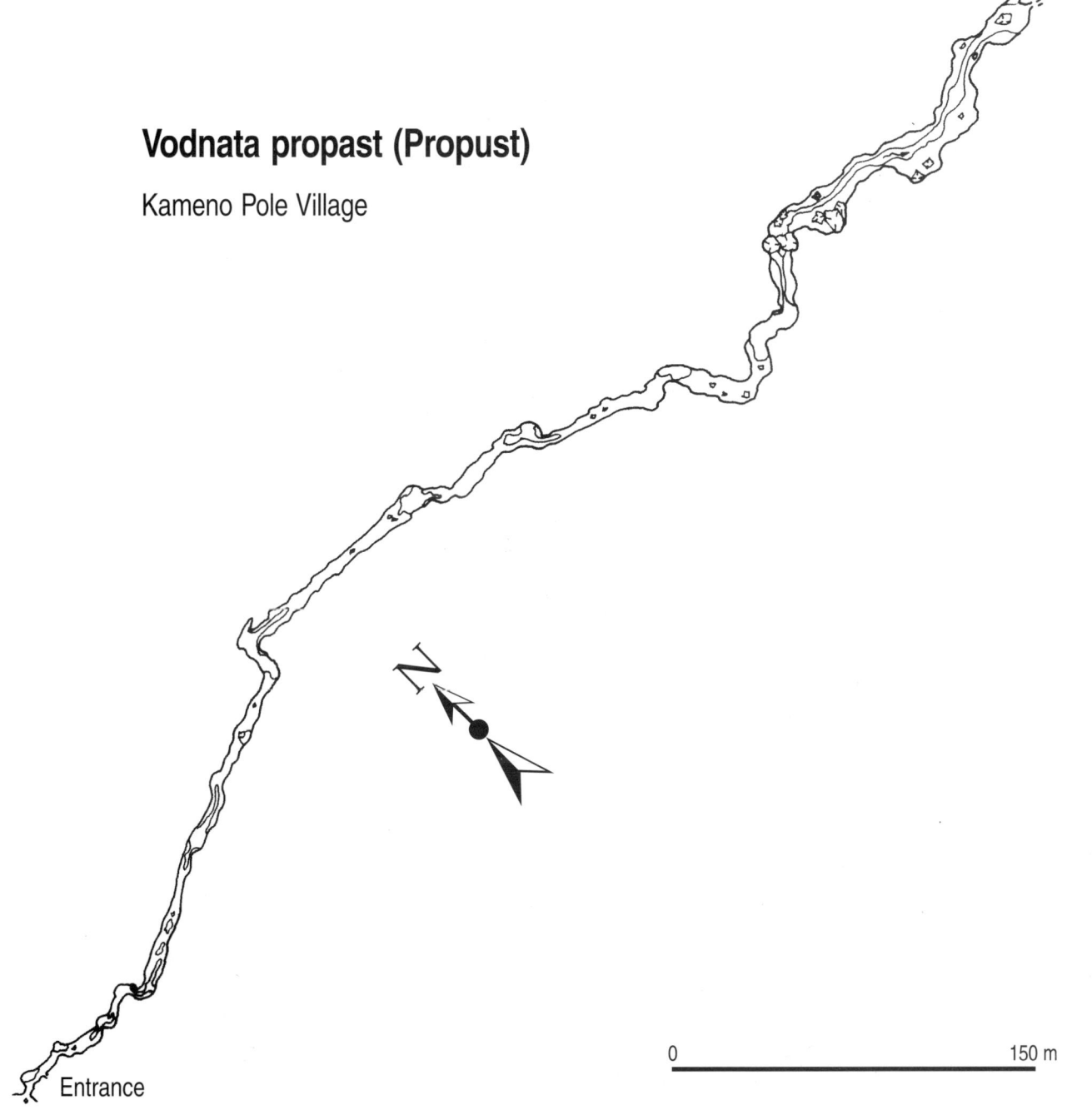

Vodnata propast (Propust)

Kameno Pole Village

Name: **VODNI PECH (15)**
Dolni Lom Village, Vidin Distr.
Altitude 450 m. Length 1300 m

A water cave on the right bank of Lom River. Formed in limestone from the Upper Jurassic – Lower Cretaceous. A horizontal, labyrinth, branched cave with a semi elliptical cross section of galleries. Poor in speleothems. Lateral clay deposits in the water gallery. A source cave with an outflow of 13–377 l/s. It was discovered in 1981 during a National Expedition by R. Metodiev from Rudnichar Pernik Caving Club. The actual map of the cave was made later by members of Edelweiss Sofia Caving Club. The underground river is piped for drinking water. Visits with a permit only.

Fauna: 29 known species, including the stygobites *Niphargus pecarensis* and *N. bureschi* (Amphipoda), the troglobites *Bulgaroniscus gueorguievi* (Isopoda), *Typhloiulus bureschi* and *Serboiulus spelaeophilus* (Diplopoda), *Plusiocampa vodniensis* (Diplura), *Beronia andreevi* (Cholevidae) and a non described Pseudoscorpion.

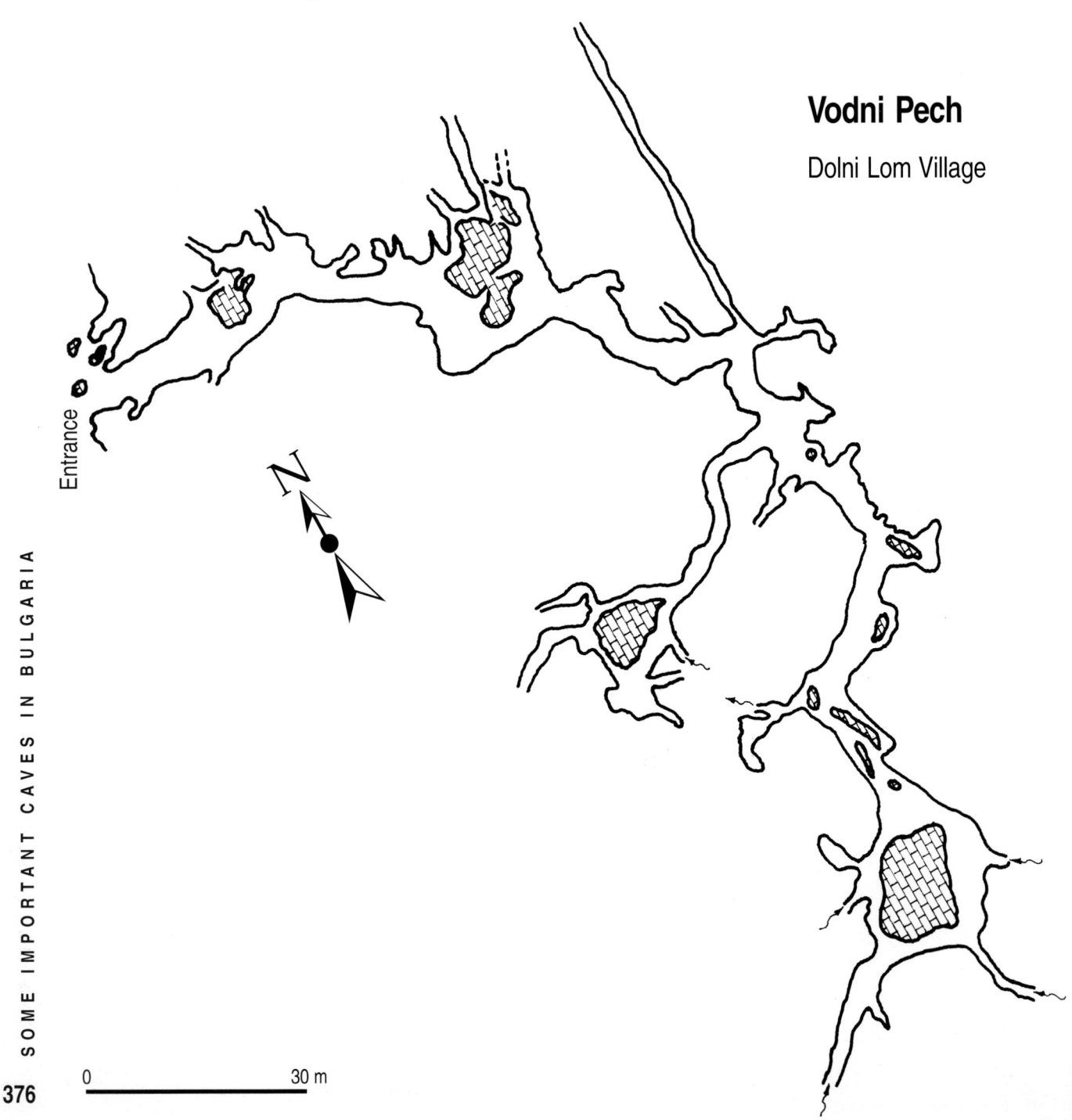

Vodni Pech

Dolni Lom Village

Name: **VODNITE DUPKI (124)**
Qu. Vidima, Apriltsi Town, Lovech Distr.
Length 813 m Denivelation 1.8 m
Coordinates: N 42° 44' 014" E 24° 53' 983"

Cave 2.5 km south-southeast from Pleven Hut, in the area of Severen Djendem, on the north slope of Botev Peak, Central Balkan. Formed in massif Trias limestone along two main fault systems with direction N-NE – S – SW and NW-SE. There are two superposed entrances. The upper is dry, triangular in shape and with dimensions. 8 m wide and 3 m high. The lower entrance gives birth to a stream flowing out. It is semielliptic, with dimensions 9.4 x 8 m. From the lower entrance starts a stony gallery, wide up to 6 and high up to 5 m, along which runs the underground stream. First it goes east, but after 86 m sharply bends to the nord and after 40 m joins the middle of the gallery, starting from the upper entrance. Immediately prior to the place of junction is developed is developed a labyrinth of narrow and low dry galleries. The gallery starting from the upper entrance is 135 m long. It is stony, triangular in cross section, up to 14 m wide and up to 3 m high. The water gallery is ca. 250 m long and is going NE. It is with triangular profile, wide up to 1 to 7 m and high 0.8–1 m. The cave is still active and only primary karstic forms are observed – fissures, levels of denudation, blocages and isolated gravitation blocks. The secondary formations are represented by dendrites only. The temperature of the underground river is 5 °C.

N. Korchev, S. Paskov, B. Marinov were the first to study this cave (1961), the first two having mapped the cave (N. Korchev – "Otechestven Front", 18.11.1961). In 1971, during the national expedition "Djendema – 1971" have been mapped 500 m of the cave.

Next exploration was done in 1975–76 by Stratesh Caving Club from the tourist society in Lovech. The cave was studied in detail and surveyed by cavers from Studenec Caving Club from Kaylashka Dolina-Pleven Tourist Society in 1982. The cave is on the territory of Central Balkan National Park.

Fauna: among the animals, collected by P. Beron, there are troglobitic beetles and pseudoscorpions, still under study.

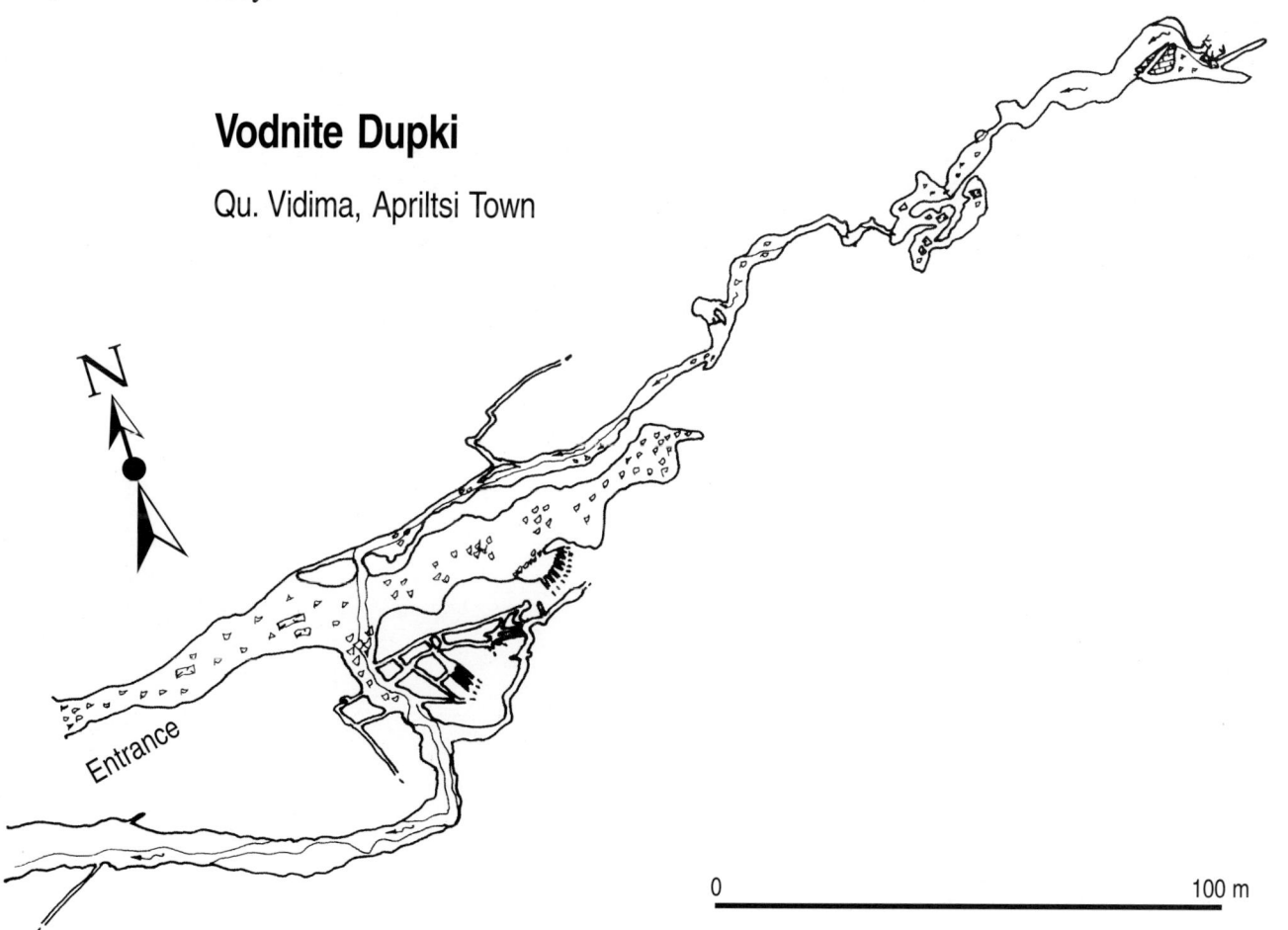

Vodnite Dupki

Qu. Vidima, Apriltsi Town

Ice concretions in the entrance of the cave Vodnite Dupki – photo Valery Peltekov

Name: **VODNITSATA (142)**
Âglen Village, Lovech Distr.
Length 249 m. Denivelation -28 m. Volume 3199 m³

A cave on the plateau between the rivers Vit and Kamenitsa, in one of the many funnels. Starts with a 7–8 m vertical. Two galleries, merging in places. Many sinter formations and stalactites, a one meter deep lake.

The initial exploration of the cave was done by Edelweiss Sofia Caving Club, together with the first survey. Surveyed again by Z. Iliev, D. Garev, V. Pashovski, I. Benchev in 1983 during the expedition Chernelka-Kamenitsa '83

Fauna: contains the troglobite *Trichoniscus garevi* (Isopoda)

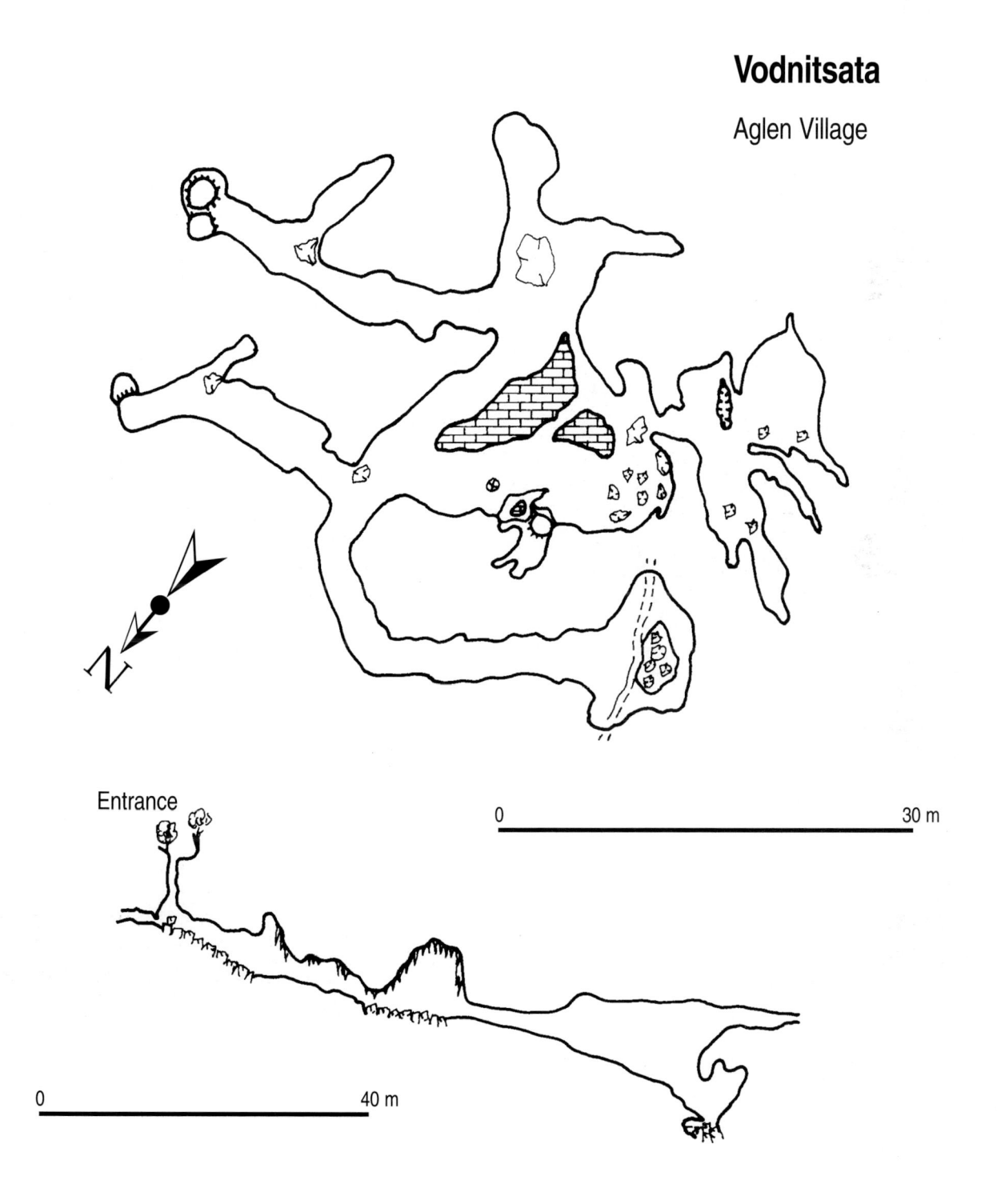

Vodnitsata

Aglen Village

Name: **VODOPADA (MAARATA) (153)**
Krushuna Village, Lovech Distr.
Length 1995 m
Coordinates: N 43° 14' 33.2" E 25° 02' 09.2"

A source cave, 1.5 km S of Krushuna, in the N periphery of Devetaki Plateau, Middle Predbalkan. Formed in Lower Cretaceous (Aptian) limestone. A branched, ascending and permanently water cave. Up to 230 m, the entrance gallery is directed N, then it turns W and goes for 100 m in this direction, being up to 7 m wide. The main gallery receives three galleries one after the other, one branch, wide up to 6 m, abducts the cave water inward and ends in a siphon. About 90 m before the end of the western sector, the main gallery takes a N direction and after 80 m it bends NW and ends in a siphon. Sinter walls, which dam the big lakes, are typical. Vodopada and the connected Boninska Peshtera drain the northern part of Devetashko Plateau. Average annual outflow of the source 100 l/s. Up to 1965, only 160 m (before the siphon) were known. During the Bulgarian – Hungarian Expedition in 1965, P. Moczari penetrated the siphon and discovered a continuation, thus bringing the cave's length to 1775 m. In 1971, a Bulgarian National Expedition surveyed another 200 m and the known length became 1995 m. In 1995, the caver and diver from Pleven Kr. Petkov (Studenets Pleven Caving Club) penetrated another 30 meters of siphon in one of the branches and, after 200 m, reached another siphon.

Fauna: 18 species known, the only troglobite being *Pseudosinella duodecimocellata* (Collembola)

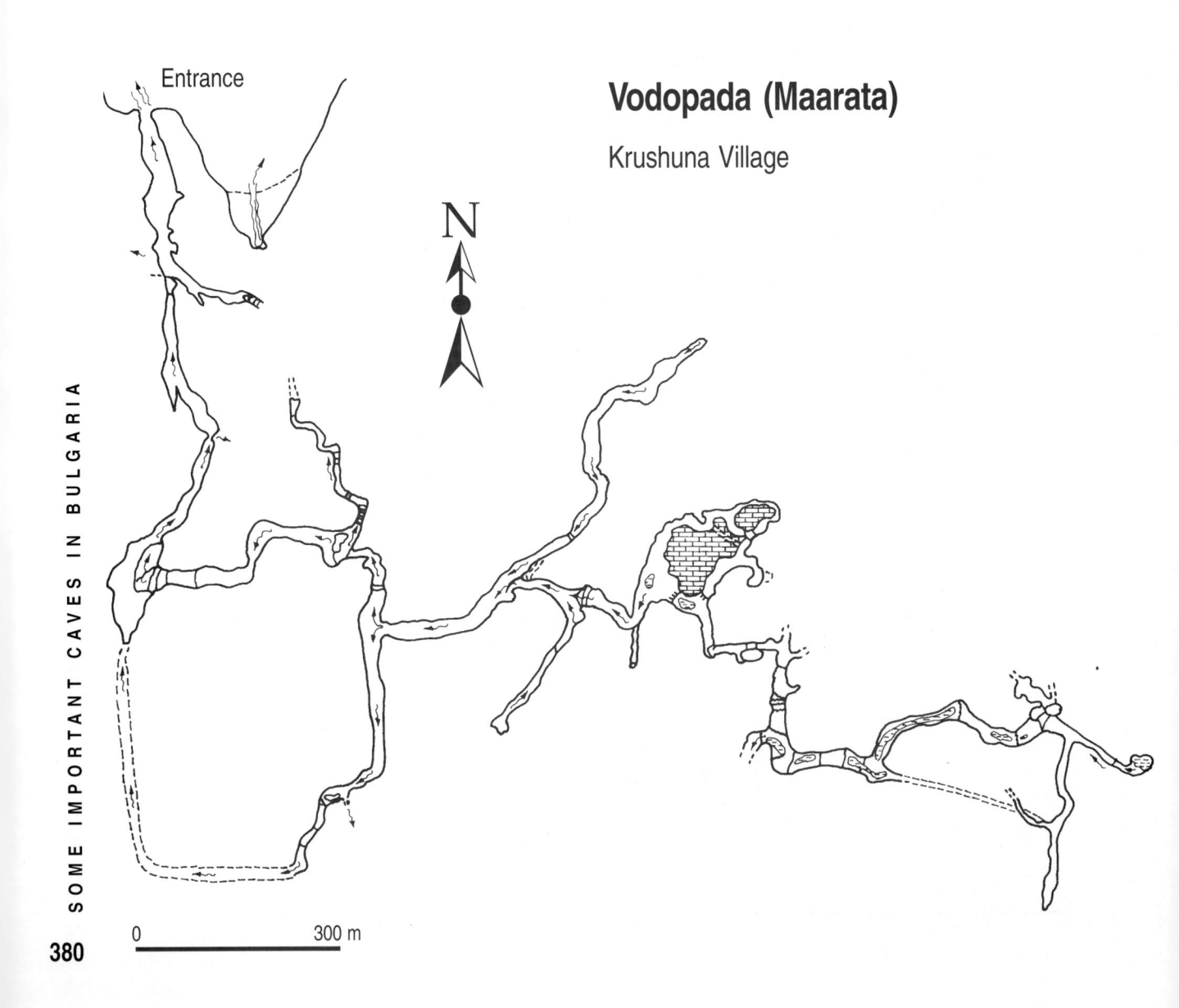

Vodopada (Maarata)

Krushuna Village

V. Peltekov, P.Yotov and T. Daaliev in front of the entrance of the cave Vodopada – photo Vassil Balevski

Penetration in the cave – photo Vassil Balevski

Name: **YAME – 3 (SREDNITSA) (13)**
Targovishte Village, Vidin Distr.
Altitude 810 m. Length 170 m. Denivelation -58 m

A pothole in the locality Klyun, ca. 25 m from Yame – 2. Length on the main axis 137 m. The entrance vertical, 25 m deep, is in the middle of a horizontal gallery, 2 to 8 m wide and 3 to 12 m high. At the end of the gallery there is another 25-m deep vertical. About 2 m above its bottom, there are two short opposite galleries. Both verticals are equipped for SRT (use of natural points of attachment). The pothole is rich in flowstone formations, which start several meters before the beginning of the vertical. In the horizontal gallery there are stone waterfalls, drums, stalactites and stalagmites.

Known to local people. First descended by D. Ilandjiev from Ivan Vazov Tourist Society in 1964. The first map is due to P. Traykov, A. Ivanova and A. Dimitrov during the Caving Expedition in 1969. The actual map was made in 1981 by a team within the Caving Expedition Varbovo'81 (T. Daaliev, N. Landjev, D. Nanev).

Fauna (studied by Hr. Deltshev in 1967 and by P. Beron in 1969): 6 species known, including the troglobites *Trichoniscus bononiensis* (Isopoda) and *Serboiulus speleophilus* (Diplopoda).

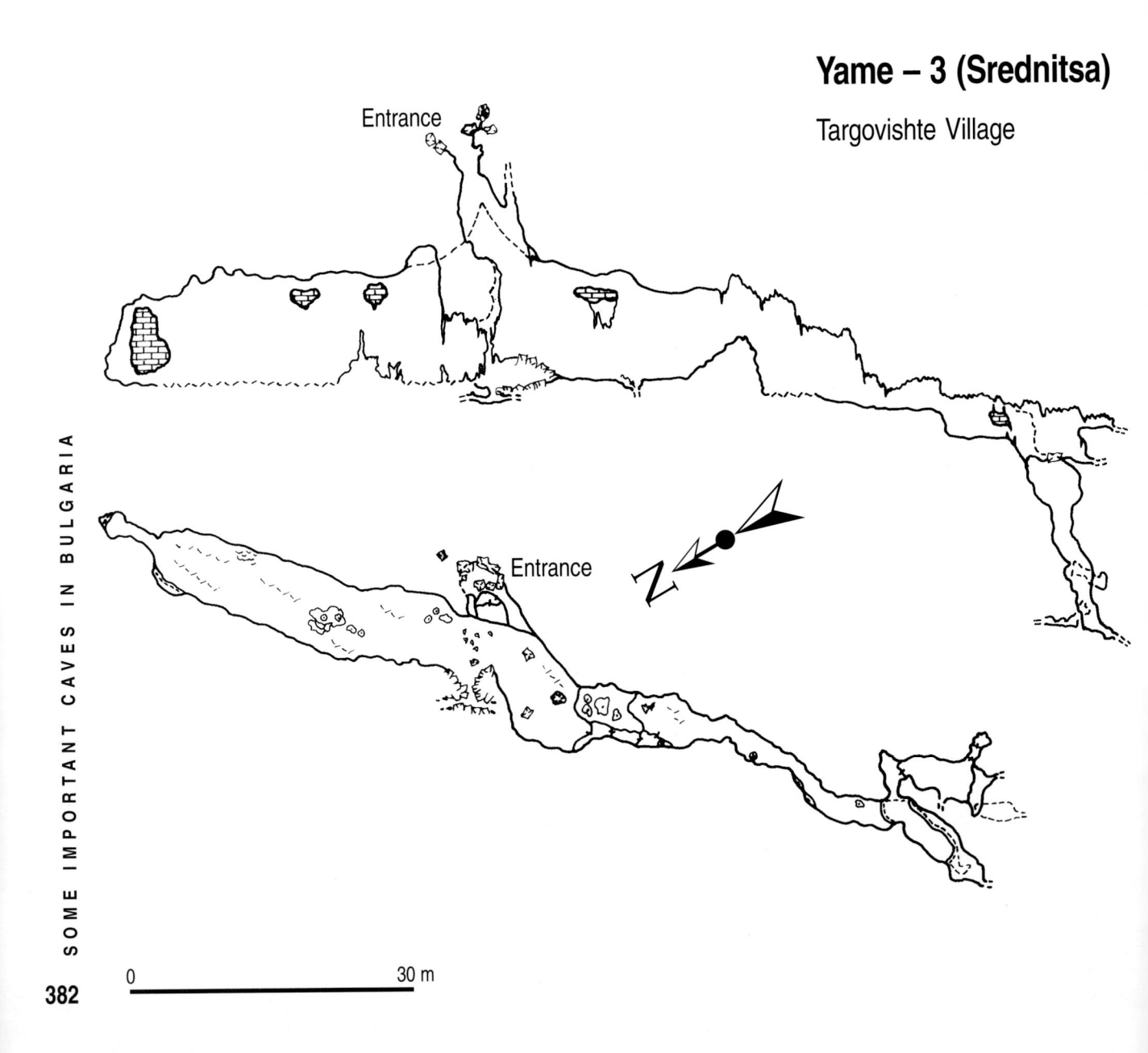

Yame – 3 (Srednitsa)

Targovishte Village

Name: **YASENSKI OBLIK 2 (111)**
Yablanitsa Town, Lovech Distr.
Denivelation -108 m

A pothole on the hill Yablanishko Bârdo (Nanovitsa Karst Plateau), in the area of the pothole Bezdânniya Pchelin. The name comes from the valley 400 m east of the sump near Bezdanniya Pchelin. The entrance is small, situated 8–10 m on the right-hand side of the road. About 10 m further, almost at the same height, is the entrance of the 7 meter deep Yasensky Oblik 1.

The pothole was explored in the period 1993–1995 by the cavers St. Petkov, Tsv. Ostromski, M. Zlatkova, etc. The entrance of the pothole was opened by digging and several parts have been enlarged in order to penetrate further. Nevertheless, the pothole remains very narrow and is not for bigger cavers. There are 8 relatively small and narrow verticals (up to 17 m).

The pothole was surveyed by St. Petkov, the description belongs to Tsvetan Ostromski.

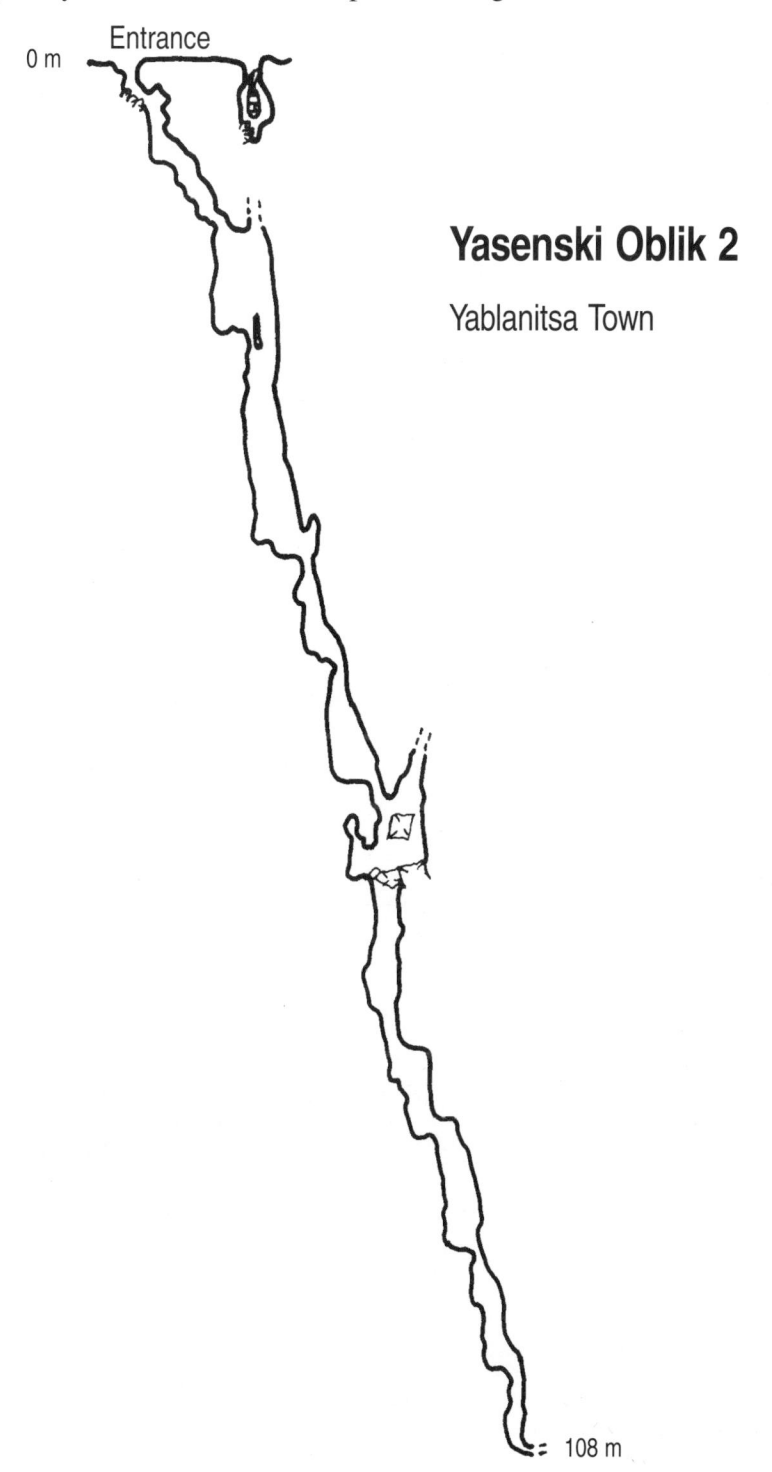

Yasenski Oblik 2

Yablanitsa Town

Name: **YAVORETS (48)**
Zverino Village, Vratsa Distr.
Altitude 1228 m. Length 50 m. Denivelation -147 m
Coordinates: N 43° 07' 41.9" E 23° 34' 12.6"

A pothole in limestone from the Upper Jurassic – Lower Cretacetous. The entrances are situated 60–70 m south of the ridge between Yavorets and Markova Mogila. One of them is triangular (0.80 x 0.80 m), the other is 10 m higher on the diaclase and is a bit bigger. The pothole starts with a 73m shaft with dimensions 0.8–2 m x 8–13 m. The bottom is at 45° and covered with pebbles. In a chamber with sinters, the second pitch (19 m deep, 2–3 x 1.5–2 m) starts. The last shaft is elliptic and 64 m deep.

The pothole was discovered in June 1966 by T. Daaliev, S. Zanev and Eng. Velinov, members of the Caving club Edelweiss Sofia. Named by the discoverers after the nearby hill. The bottom was reached on September 10, 1966, by T. Daaliev, I. Zanev (Edelweiss) and Georgi Makedonetsa (Vitosha, Sofia).

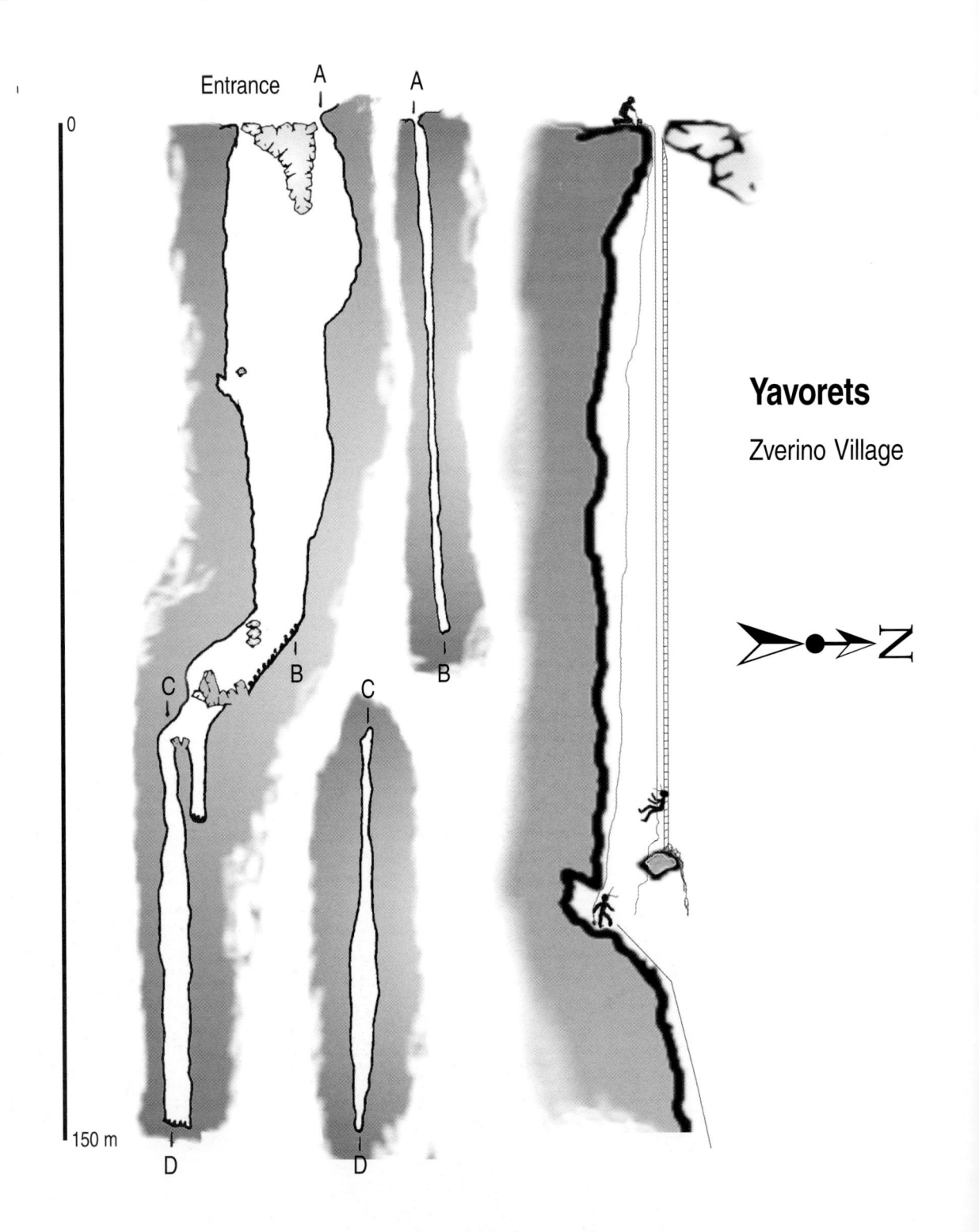

Yavorets

Zverino Village

T. Daaliev in front of the entrance of the pothole Yavorets

Name: **YULEN ERE (120)**
Hristo Danovo Village, Plovdiv Distr.
Length 227 m. Denivelation 17 m. Surface 487 m². Volume 3687 m³
Coordinates: N 42° 44' 26.5" E 24° 36' 16.9"

A cave in the locality Yulen Ere, in the valley of Kodja Dere. A mighty stream runs out of the cave, ca. 250–300 above the river. The underground stream runs along the whole length of the cave, so the cavers have to move over the water, as deep pools are formed in some places. Explored by cavers from Akademik Plovdiv Caving Club.

Fauna (studied by P. Beron and T. Ivanova): includes the troglobite *Trichoniscus bulgaricus* (Isopoda).

Protected Natural Monument, together with 0.8 ha (Decree 206/23.03.1981).

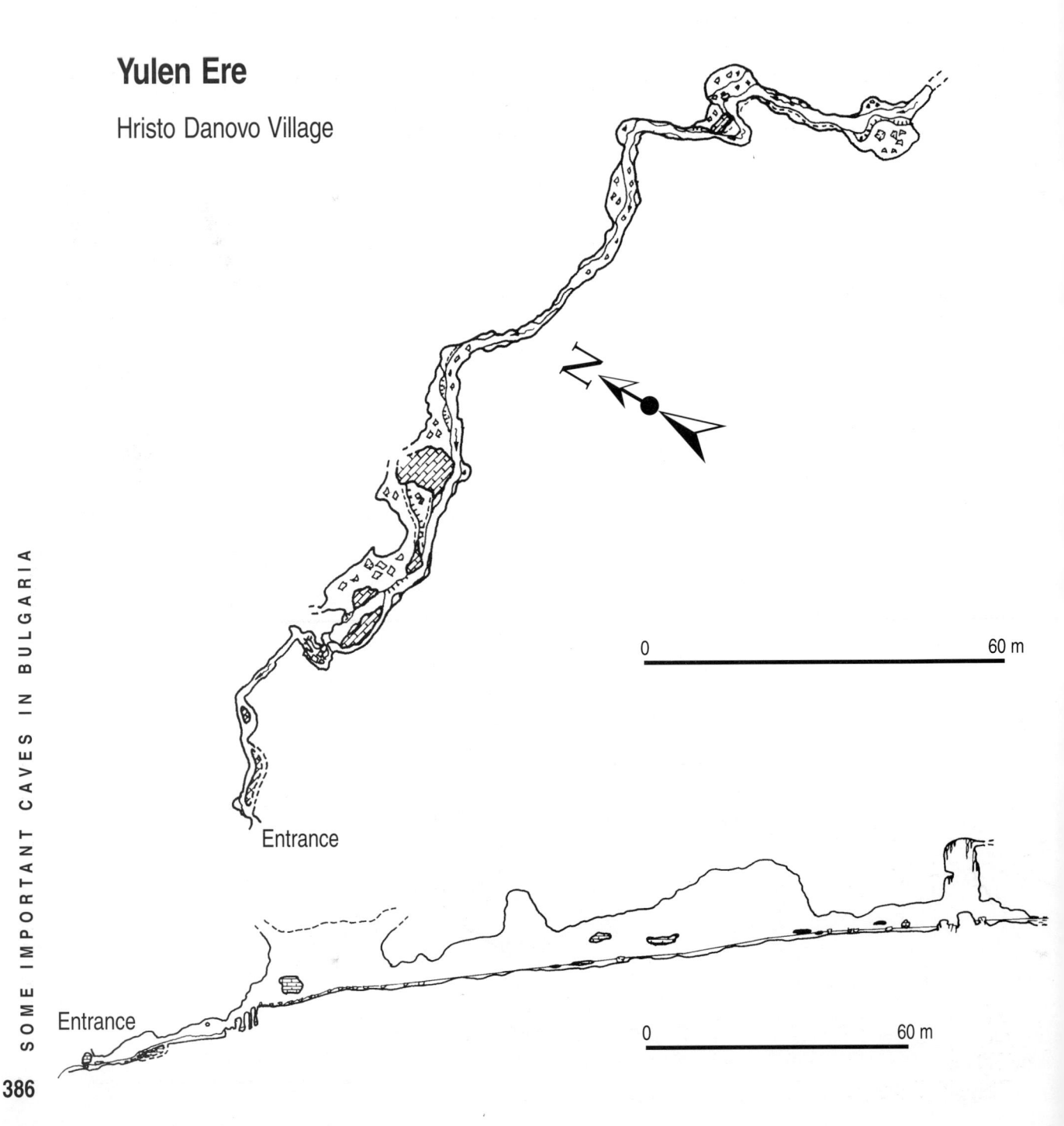

Yulen Ere
Hristo Danovo Village

Name: **ZADÂNENKA (90)**
Karlukovo Village, Lovech Distr.
Altitude 170 m. Length 1150 m. Denivelation 26 (-22; 4) m
Coordinates: N 43° 10' 35.7" E 24° 04' 14.1"

A cave in the locality Zadânen Dol, in Maastrichtian limestone. Altitude above the LBE 43 m. A horizontal, slightly branched, with many storeys. It follows eastwards, 3 m tall and 4–4.5 m wide. In this part there are clearly visible old morphological levels. About 50 meters on the left a 24meter long gallery branches out. After another 60 meters, a sill (+1) follows, where the gallery lowers to 1–0.8 m and keeps this height almost to the end. About 160–180 m from the entrance, a 24-meter pitch gives access to a lower part with a stream, starting from one end of the gallery and sinking into a siphon at its end. ATTENTION! At the bottom there is CO_2! In the main gallery, there are a series of muddy pools. After 150 m from the branching, we reach the only bigger chamber in the cave. After overcoming a 3m vertical, we enter a higher situated gallery. After another ca. 100 m, we come to a big boulder blocking the way. The main gallery goes south, then southwest and after ca. 80 m it ends in a siphon, which is 2 meters long and was penetrated by K. Georgiev in 1988.

The main gallery is periodically wet, the lower part is with water. The cave is poor in flowstone formations. Explored in 1982 by a group led by V. Mustakov.

Fauna: the stygobite *Niphargus bureschi* (Amphipoda) and the troglobite *Typhloiulus bureschi* (Diplopoda).

Zadânenka

Karlukovo Village

Cave rescue training in front of the cave entrance (Zadànenka) – photo Trifon Daaliev

Name: **ZHABATA (21)**
Portitovtsi Village, Montana Distr.
Total length 609 m. Denivelation 1.5 m. Surface 1696 m^2

The cave Zhabata is in the area of Portitovtsi, in the locality Zli Dol. The exit point is the village Beli Breg. The distance from Beli Breg to the cave is 2 km, azimuth of 294°, from Portitovtsi – ca. 4 km, azimuth 20°.

The entrance is situated at the bottom of a funnel, 9 meter deep, overgrown with bushes and trees. The entrance faces southeast. It is elliptic (0.40 x 0.50 m). The distance above the LBE (Ogosta River) is 20 m.

The cave Zhabata is horizontal, brachyclasic, labyrinthine, one-storey and dry, in yellow grey Sarmatian shell limestone. The average height in the cave galleries is 1 m, rarely more than 1.40 m. No connection with surface water. The cross section is elliptic, the floor follows the horizontal layers. No flowstone formations.

The cave is well known to local people, mainly from treasure hunting activities. In most galleries there are pits from digging.

The cave was explored, surveyed and described in March 1985 by cavers from Pastrinets Montana Caving Club (V. Georgiev, A. Georgiev, G. Georgiev, Iv. Ivanov, I. Todorov).

The name "Zhabata" (The Frog) comes from a legend about a frog-like statue, connected with a hidden treasure (after V. Georgiev).

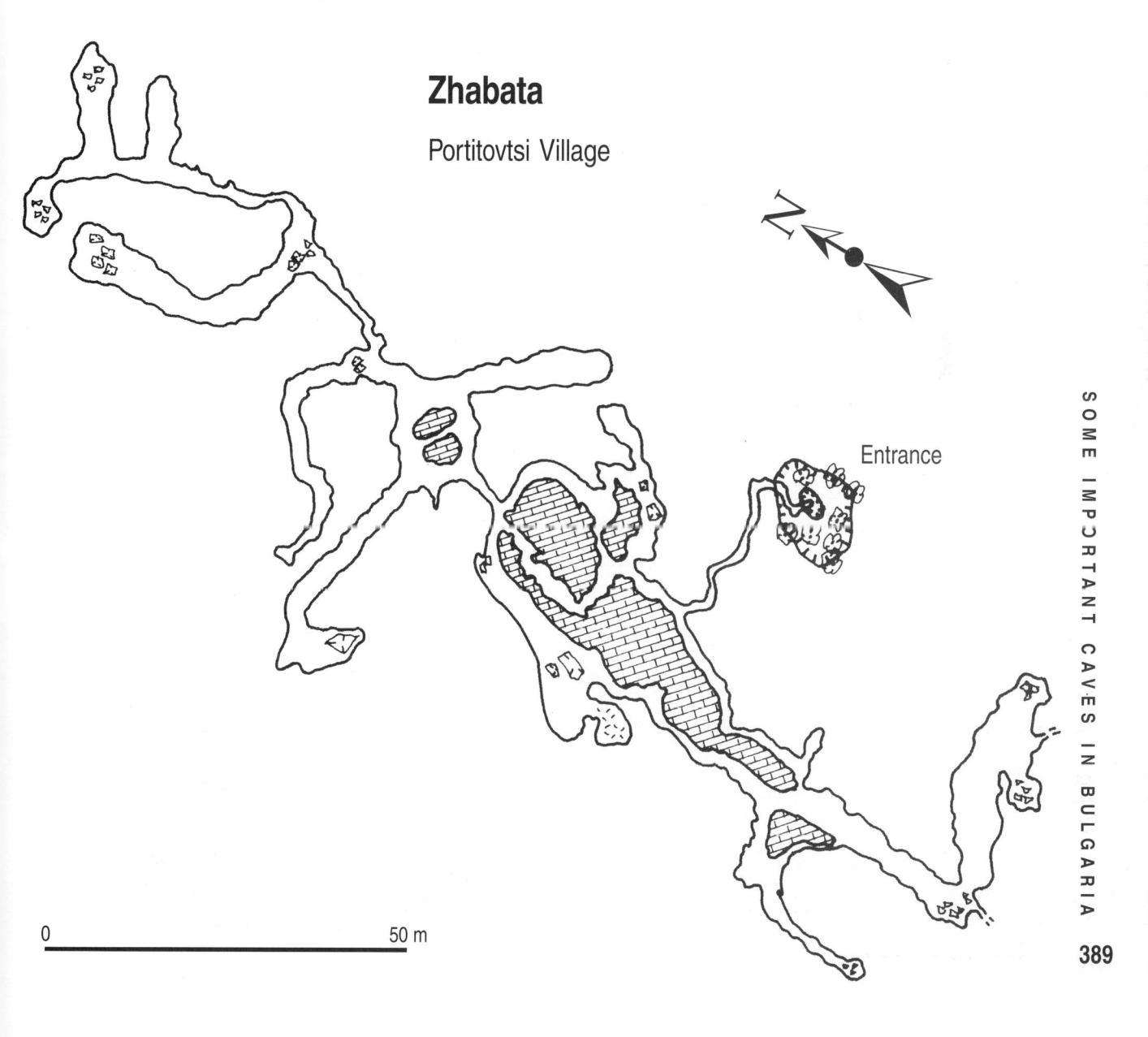

Name: **ZIDANKA (91)**
Karlukovo Village, Lovech Distr.
Altitude 170 m. Length 435 m
Coordinates (at the entrance of Zadanen Dol): N 43° 10' 35.7" E 24° 04' 14.1"

A two-entrance cave near Karlukovo. Distance above the LBE 42 m. Its galleries cross the end part of the massif, formed between the western vertical slope of Zadanen Dol and the cliff on the right bank of the Iskar River. The upper entrance is situated on the left of the Waterfall, reached by a traverse. The wide and clear atrium is followed by a gallery, first going west, then north, until coming out above the Iskar River. On the left of the main gallery, there is a branch with secondary formations. Here, the gallery lowers and is filled with clay formations. Almost at its end, the gallery is intersected by a crevice, 2 m wide (perhaps an old sink). The main gallery is covered with angular stones, at its northern entrance there is also clay dust. As a result of frost weathering, the walls are with horizontal and vertical lines, which remind of masonry, hence the name Zidanka. The northern entrance is elliptic and could be reached following a steep path, which starts on the left of the mouth of Zadanen Dol.

This cave has been explored and surveyed many times. Its age has been assessed at 2 500 000 years (Popov, 1984).

Fauna: 5 species have been identified, including the millipede *Typhloiulus bureschi* (Diplopoda, troglobite) and the Crustacean *Elaphoidella balcanica* (Harpacticoida) – stygobite.

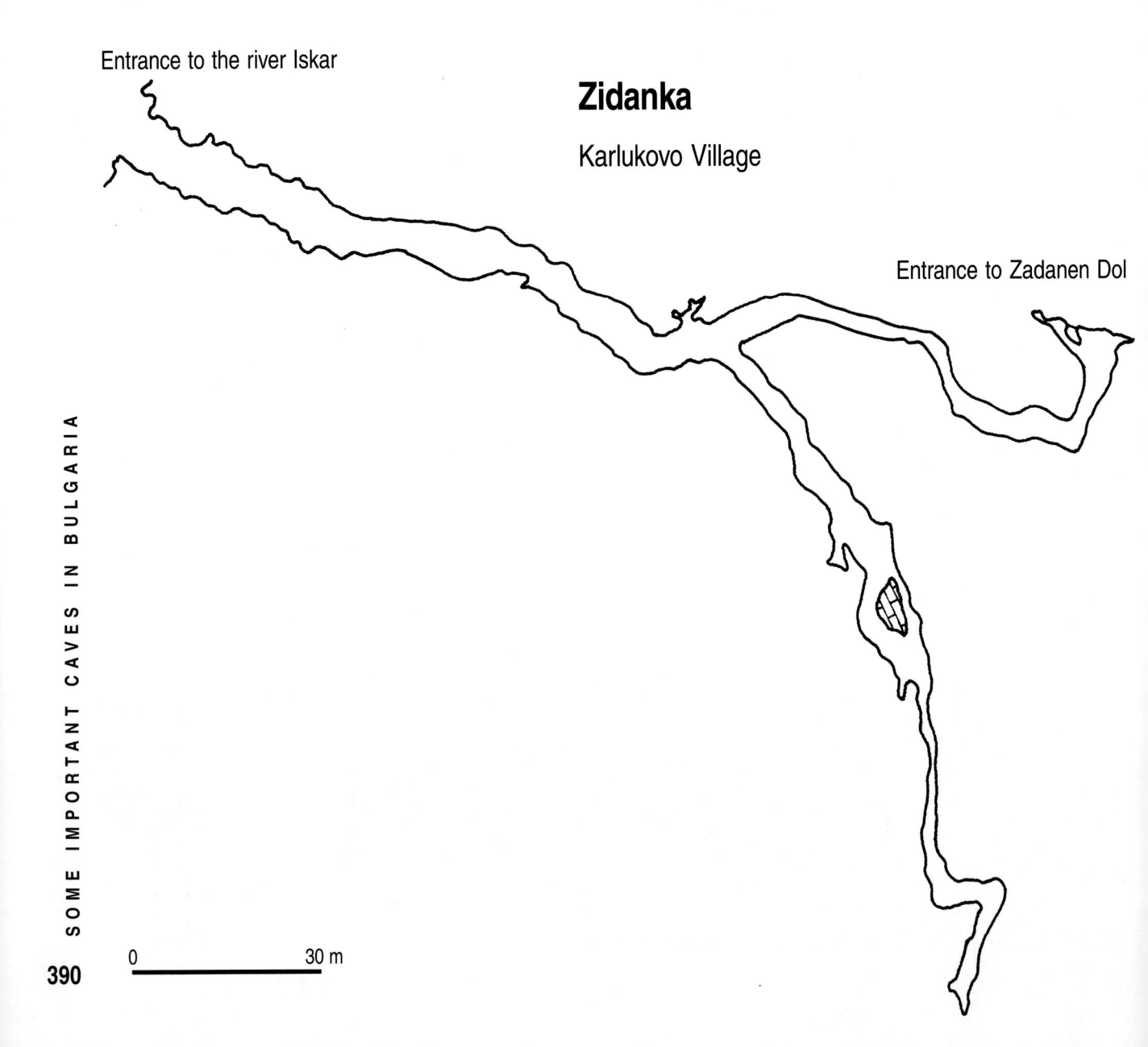

The entrance – view from Zadanen Dol – photo Trifon Daaliev

The National Caving Center in Karlukovo – photo Trifon Daaliev)

The second entrance of Zidanka is looking at Iskar River – photo Trifon Daaliev

Name: **ZIDANKA (55)**
Lakatnik Railway Station
Altitude 390 m. Total length ca. 400 m
Coordinates: N 43° 05' 19.9" E 23° 23' 10.6"

A cave left of Temnata Dupka entrance, with a semi-destroyed wall at the entrance. The formations are similar to the ones in Temnata Dupka, but without running water. Zidanka is somewhat older than Temnata Dupka and is connected with it. Even in the initial parts, a narrow gallery leads to Razliva and the map shows that the distance between the end points of Zidanka and Trapezariyata in Temnata Dupka is only approximate 120 m.

While exploring this possible connection, the cavers from Edelweiss Sofia Caving Club discovered the Pearl Chamber (Bisernata Zala), but the cave pearls were soon after that stolen by other visitors. We can reach this chamber crawling among the blocks before the cave's end. It is a pity that the chamber was not closed in time to save the rare cave pearls.

Map and description by P. Tranteev.

Fauna: 19 species are known, including 8 troglobites: *Paranemastoma (Buresiolla) bureschi, Paralola buresi* (Opiliones), *Centromerus bulgarianus* (Araneae), *Lithobius lakatnicensis* (Chilopoda), *Trachysphaera orghidani, Typhloiulus bureschi* (also var. *obscurus*)(Diplopoda), *Pheggomisetes globiceps lakatnicensis, Duvalius papasoffi* (Coleoptera).

Zidanka

Lakatnik Railway Station

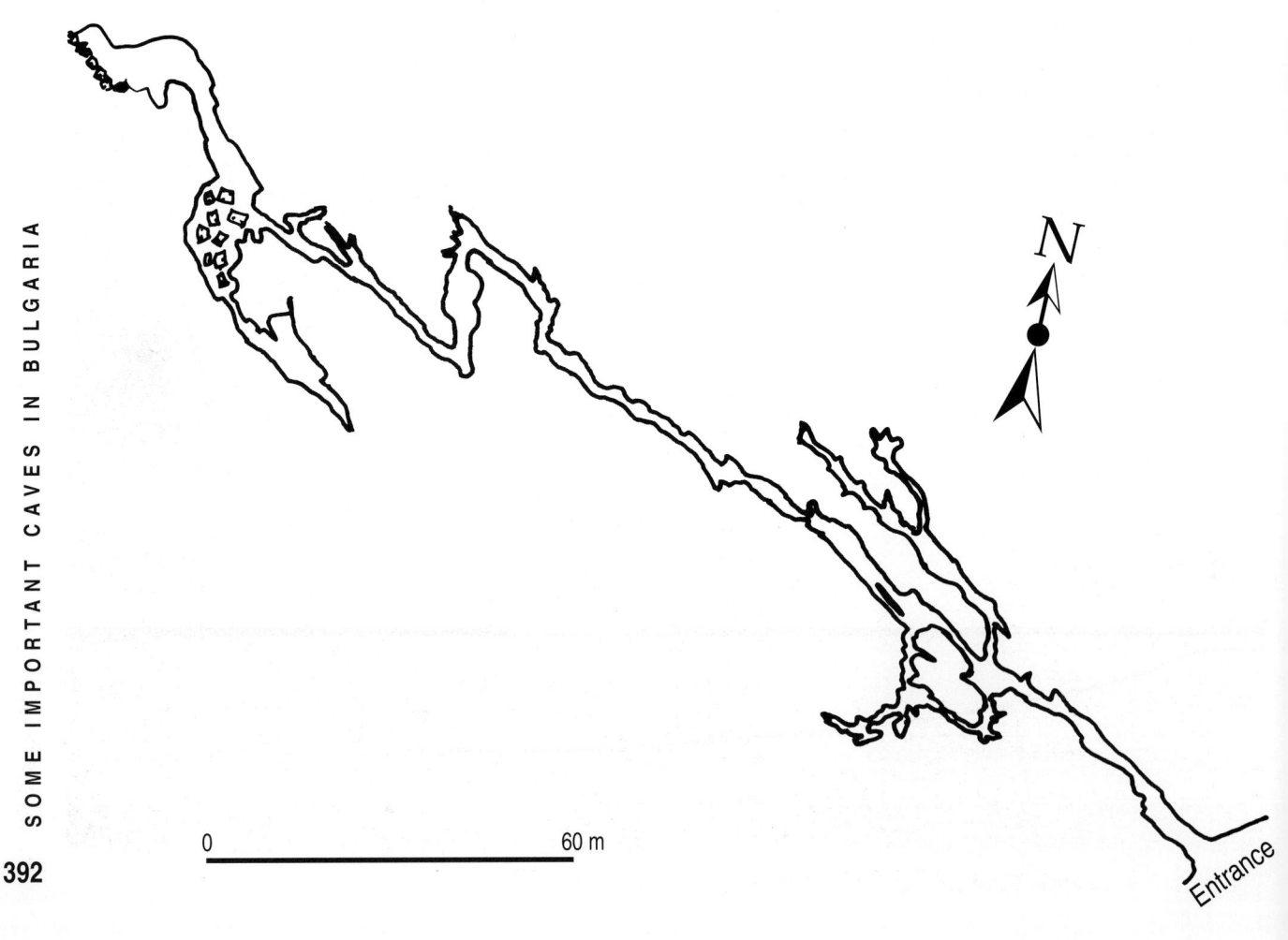

The Pearls Room (Bisernata Zala) – archive Angelina Petkova

Name: **ZMEYOVA DUPKA (28)**
Bistrets Village, Vratsa Distr.
Length 122 m. Denivelation -52 m
Coordinates: N 43° 14' 08.5" E 23° 27' 41"

    A pothole, not far from Haydushkata Propast. The chamber has a cross section of 30 – 35 m and is a prolongation of the pothole. The end is a chamber with very beautiful flowstone formations, which were later heavily damaged.
    The pothole was discovered in 1960 by Dimitar Ilandjiev.
    Fauna (collected by P. Beron in 1960): 7 sp., including the troglobites *Paranemastoma (Buresiolla) bureschi* (Opiliones), *Pheggomisetes buresi* (Col., Carabidae) and *Radevia hanusi* (Col., Cholevidae).

Zmeyova dupka

Bistrets Village

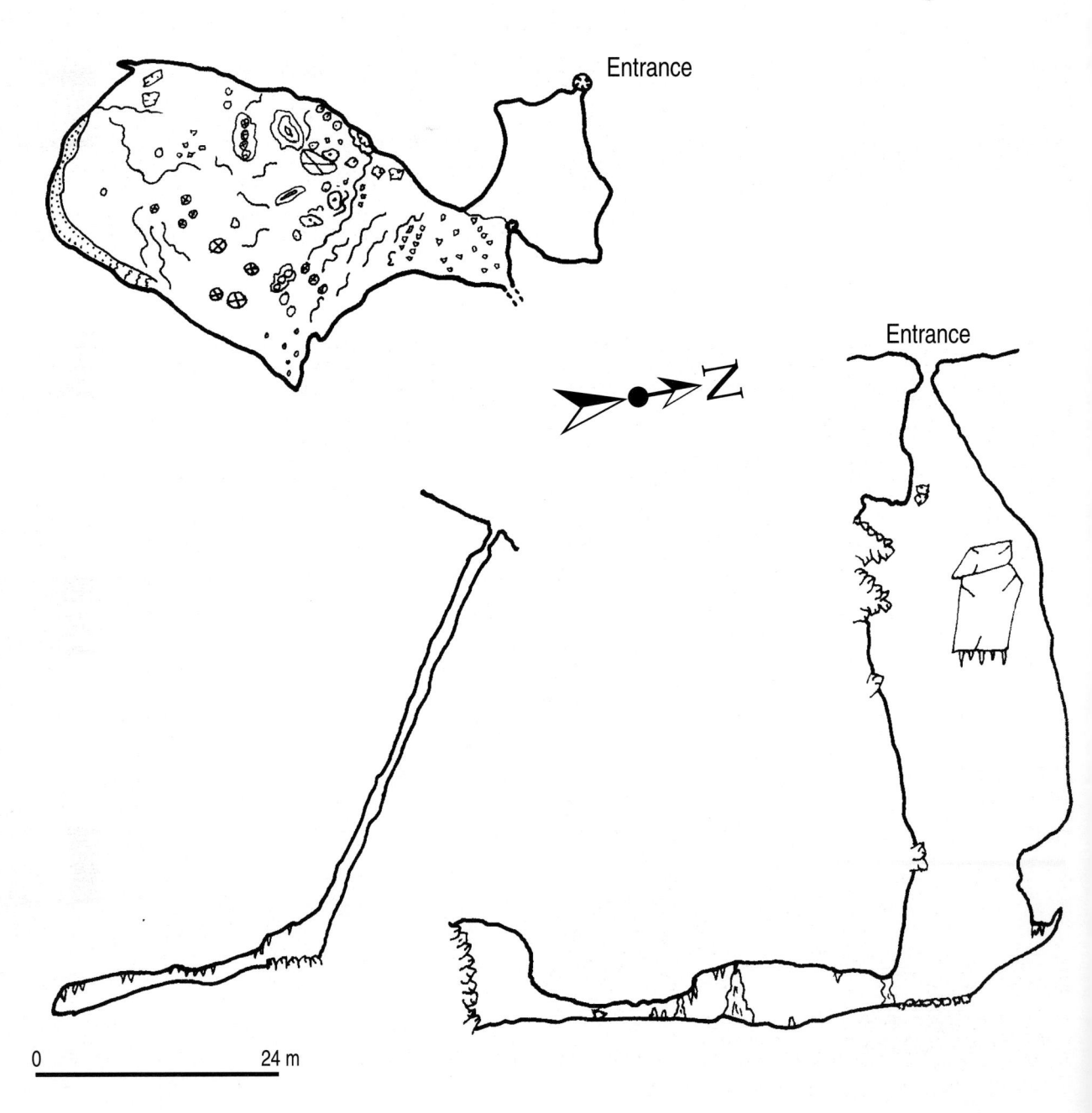

Amazing underground beauty of the bottom hall – photo Mihail Kvartirnikov

Tha hall at the bottom of the pothole – photo Mihail Kvartirnikov

CAVES IN THE TRANSITIONAL REGION

Name: **AKADEMIK (198)**
Bosnek Village, Pernik Distr.
Total length 350 m. Denivelation -15 m
Coordinates: N 42° 29' 30" E 23° 11' 24"

A cave in Bosnek, the entrance is immediately near the road to Chuypetlovo, within the confines of the village.

The first ca. 100 meters represent low and narrow galleries, where in some places one has to crawl and to pass through squeezes. After that, the gallery becomes wide and high, leading into a big chamber (100 x 30 m). Through a part of the cave, an underground stream with a changing outflow (60–100 l/s on 1.04. 1978) runs. The fluorescein test proved the connection between the underground stream and the source in Bosnek Village. At an outflow of ca. 800 l/s. the water passes the distance of ca. 900 m in ~ 5 h. The gallery and the chamber are rich in flowstone formations – stalactites, stalagmites, helictites, stone "waterfalls", sinter pockets, leopard skin, aragonite. In the chamber there is one lake and pools.

The entrance parts were discovered in November 1977, when the movie "Duhlata" was shot. Explored in 1978 by cavers from Student's Caving Club Akademik Sofia. In the same year, G. Markov discovered another 50 m of galleries. The map was made by P. Georgiev, O. Stoyanov, R. Kancheva and Kr. Kanchev in 1978.

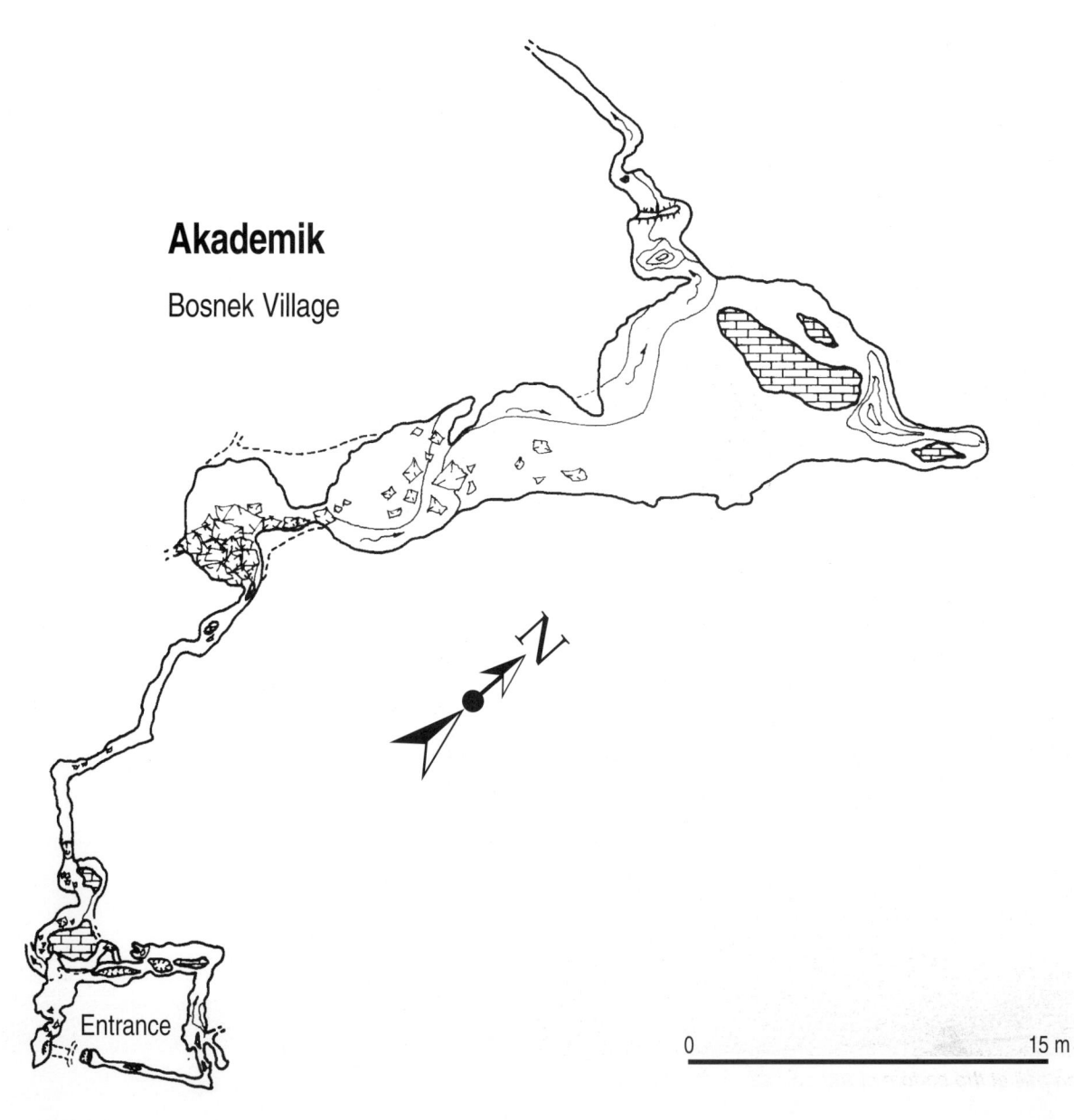

D.Daalieva and V. Peltekov – Junior in front of the cave entrance – photo Trifon Daaliev

Crystals in the cave Duhlata – photo Mihail Kvartirnikov

Cavers in Akademik Cave – photo Trifon Daaliev

Name: **BOYCHOVATA PESHTERA** (Boychova dupka) (201)
Logodash Village, Sofia Distr.
Length 200 m. Denivelation 27 m (+22, -5)
Coordinates: N 41° 58' 08.4" E 22° 56' 22.2"

A cave in the quarry in the locality Boychova Skala (Boychovtsi), ca. 2.5 km south of Logodash Village, west of Blagoevgrad.

Formed in Middle Trias limestone. A horizontal, not branched, dry cave. Developed along a basic tectonic fault, directed E/W, and an interlayer fissure, sinking E/SE at 35 degrees inclination. In the central part, there is a chamber, which is 30 m long, 20 m wide and three meters high. Rich in flowstone formations. Deposits of terra rossa. Infiltration water and drops periodically form a lake.

Opened during quarrying works in 1979. Explored and surveyed by the caving club of Aygidik Blagoevgrad Tourist Society (Iliya Iliev and others) and in 1981 by specialists from the Institute of Geography (BAS) and BFS (P. Beron, P. Stefanov, T. Daaliev).

Fauna: collected by P. Beron since 1981, later also by B. Petrov and other biologists and includes the troglobitic Isopod *Cordioniscus bulgaricus*, the northernmost representative of Styloniscidae (Isopoda) family.

Protected Natural Site with 4 ha adjacent area (Decree 542/23.05.1984).

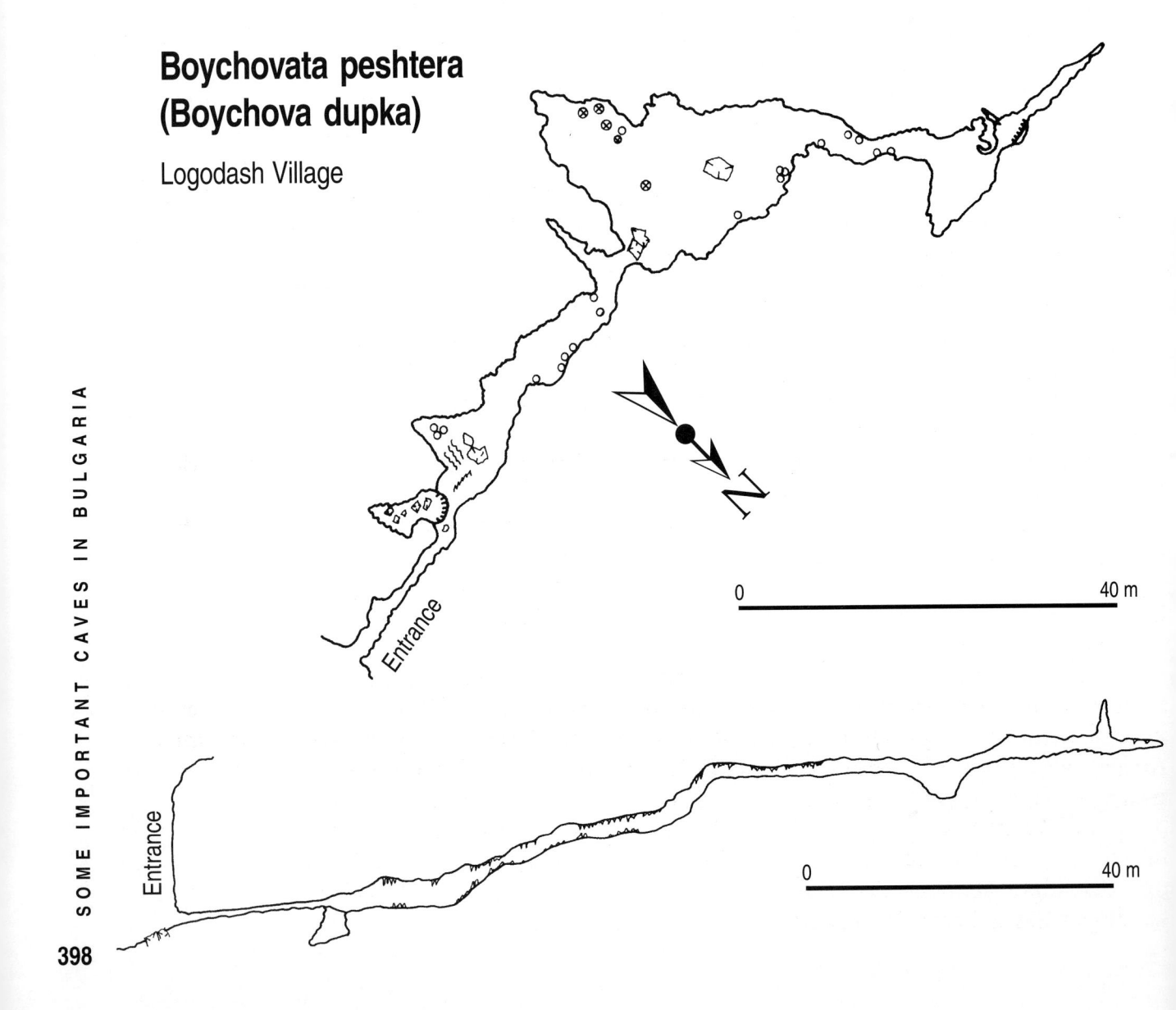

Boychovata peshtera (Boychova dupka)

Logodash Village

Name : **DUHLATA (199)**
Bosnek Village, Sofia Distr.
Altitude of entrance 972 m. Length more than 18 000 m (2006). Denivelation -70 m
Coordinates: N 42° 29' 41.6" E 23° 11' 46.3"

On the southern slope of Vitosha, SW of Cherni Vrah, 35 km from Sofia, on the Struma River, in one of the most important karst regions in Bulgaria – the Bosnek Region. The cave Duhlata is central to this region, the longest and most complicated cave labyrinth in Bulgaria. It includes a six level system of interconnected underground galleries, chambers, pitches and rivers. Seven permanent, one periodical and one adjacent rivers run underground. "The Big River" is with the biggest outflow and is the central river of the cave. All other seven rivers join it, except for the adjacent one. The end point of the waters in the system is the fountain Duhlata in Bosnek Village (918 m altitude). The cave is situated on the right bank of Struma River, 1.5 km upstream of Bosnek Village, on the southeastern slope of the ridge Golema Mogila.

The cave was formed by the water of Struma River in the limestone and the dolomite of Middle Trias. The old entrance is situated at 972 m altitude, 6 m above the level of Struma, in a small cliff. From the entrance hall (The Atrium) we follow a narrow passage on the right. A small lake (Ayazmoto) is reached, where different objects (buttons, coins, pencils, etc.) were left by visitors in the hope to be healed. After crawling, we reach the largest chamber of the cave – Stalagton, then the labyrinth of Prashnite Zali – several small chambers interconnected by narrow passages. After the hall called Urinarnik, we descend a 10 m steep gallery – the First Mercedes (3-direction point). Malkata Duhla follows, from where we reach The Second Mercedes, then after several chambers, we reach the end of Malkata Reka (The Small River).

The right gallery leads to the so called Suh Kray (Dry End) with a periodical stream in rainy weather. The right gallery leads to the chamber Nosorog (Rhinoceros). Over a steep plate, the Big River is reached. On the left, the labyrinth of Novata Duhla is reached.

From Ribkata, we can follow the Big River downstream for ca. 600 m, circumventing the obstructed parts, using a level which is 4–5 m higher than the river bed. Narrow chimneys connect this part with Novata Duhla. After two semi siphons, the river ends in a siphon. On the way left of the underground river, there are multilevel galleries over the Deep River and the Sixth River.

From Ribkata, after 50–60 m of narrow passages upstream (The Termopyles), a wide canyon is reached. Nearby is the junction with Malkata Reka (The Small River). After 100 m upstream along the Big River, we reach the junction with the Rain River, connected with the labyrinth of Malkata Duhla.

The last two rivers are with waterfalls, the entrance of the "Akademik parts" is 10 m far from the old entrance of the cave.

The cave had been known for a long time. It was mentioned in the book "Kražski Yavleniya" (Karst Phenomena)(1900) by K. and H. Škorpil. In 1926, P. Deliradev wrote in his monograph "Vitosha": "The cave is with several inner entrances. It is possible to penetrate for some 100 m. Further the galleries become narrow and are not accessible to tourists".

The exploration of the inner parts of the cave started in the late 50s. Cavers lake T. Michev, V. Beshkov, A. Denkov, P. Nedkov, L. Popov, P. Beron and others managed to discover the way to the Big River.

The organized exploration started at the end of 1965. A group of five cavers: M. Kvartirnikov and S. Penchev from the Republican Caving Commission, D. Konstantinov, A. Filipov and A. Strezov from Akademik Sofia Caving Club surveyed until 1970 more than 6000 m of cave galleries. Kambanariyata, Aragonitovata galeriya, Labirinta na Tronnite Zali, Malkata Duhla, Gubilishteto, the upstream parts of the Big River, the upper level and the connections between the various levels were discovered.

After 1970, the participation of Akademik Sofia Caving Club in the study of the cave became more intensive. Successful work was done by the mineralogists M. Maleev, A. Filipov, M. Raynova and others. The level above the Big River was surveyed and in March 1970 the "Akademik parts" – over 1.5 km of galleries and chambers – were discovered and surveyed.

In October, an expedition of the Sofia based caving clubs (Aleko, Planinets, Cherni Vrah, Akademik, Pernik, Ivan Vazov) surveyed another 350 m.

In 1971, explorations and surveys were carried out mainly by A. Popov, G. Markov, A. Strezov and P. Veselinov from Akademik Sofia Caving Club. In September, the film "Ways to the Abysses" was shot in the Aragonite Gallery.

In October 1971, a Republican expedition was carried out (cavers from Russe, Haskovo, Shumen, Aleko, Akademik, Ivan Vazov, Planinets, Botevgrad), and as a result 200 m of new galleries were surveyed. In 1972, road works opened a second entrance to the cave, a short cut to Tronnite Zali.

In May 1973, another such expedition with cavers from Yambol, Russe, Kotel, Botevgrad, Peshtera and Sofia surveyed another 630 meters of new parts. In 1974, the work to open the cave as a show cave started. A tunnel was dug out and then walled.

In October and November 1974, parts of Novata Duhla, situated above the Deep River, were discovered. Up to September 1975, 3163 m from the newly discovered parts were surveyed. In October 1976, the labyrinth of galleries between the Second and the Third Semi Siphon was discovered. Over 700 meters of galleries were surveyed with the active participation of Rumyana Panayotova, Lydia Popova, Georgi Markov, Alexander Strezov, Ognyan Stoyanov, Petko Saynov, Maria Zlatkova, A. Ivanchev (Aleko), Ivo Pozharevski, Bozhidar Nikolov and others.

In October 1977, while filming the movie Duhlata, the caves Pepelyankata and Akademik were discovered, which are separate sections of Duhlata system – the river in the cave Akademik joins the Big River of Duhlata after the siphon and surfaces at the sources in Bosnek Village.

In June 1979, a Republican expedition (P. Neykovski, P. Tranteev, A. Strezov and 34 cavers from the whole country worked for detailed answer to many questions. A third entrance, leading to the fabulously beautiful chambers of Novite chasti (The New Parts) was discovered and later walled. In 1980 cavers from Akademik tried to go through the siphon of the Big River (120 x 80 cm).

September 7 to September 10, 1980 expedition of Akademik Sofia Caving Club with 10 participants surveyed 600 m of new galleries.

In 1982 in Duhlata started the permanent expedition Chisti Peshteri (Neat Caves) – about 70 cavers participated, collecting large quantities of garbage.

On May 9–10, 1987, Atanas Russev and his team discovered and surveyed the "Edelweiss parts" – more than one km of new galleries.

In January 1989, a team of cavers from Akademik Sofia Caving Club (K. Krachunov and others), after negotiating a chimney above the 6th river, reached a series of large chambers and galleries – another 290 m were surveyed.

In November 1993, after climbing the canyon at the siphon of the Big River, a group including Aglika Gyaurova and Ivo Kalushev reached a labyrinth of large chambers, about 20 m above the river, ca. 600 m, which were not explored to the end.

The exploration of Duhlata has a long history and is still going on with the efforts of cavers from Akademik, Edelweiss and Extreme clubs from Sofia.

The cave was declared a Natural Monument (Decree 2810/ 10.10.1962, with 5 ha adjacent land). In this cave almost all possible flowstone formations are found.

Known troglobites and stygobites: *Diacyclops chappuisi, D. strimonis, Elaphoidella pandurskyi, Parastenocaris* sp. (Copepoda), *Neobisium kwartirnikovi* (Pseudoscorpiones).

Description: Alexander Strezov (Akademik Sofia Caving Club).

The first explorers of Duhlata Cave – photo Mihail Kvartirnikov

Duhlata

Bosnek Village

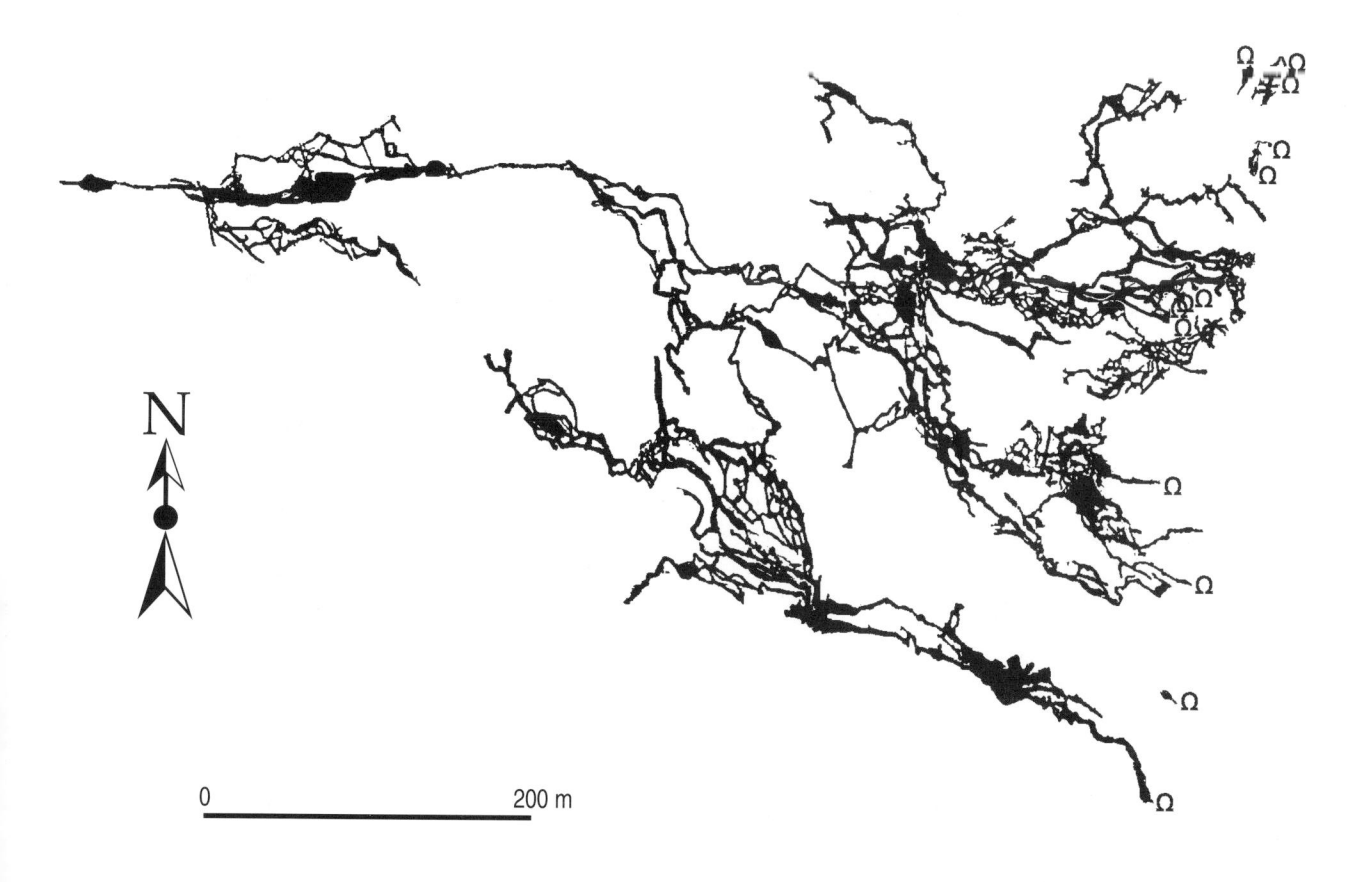

The helictites – photo Mihail Kvartirnikov

The Sinter Chairs – photo Mihail Kvartirnikov

Name: **PPD ("Introduction to Speleology") (197)**
Bosnek Village, Pernik Distr.
Altitude 1050 m. Length 1020 m. Denivelation 125 m
Coordinates: N 42° 29' 49" E 23° 12' 41"

A cave in Trias dolomite and limestone, about 2,5 km from Bosnek, on the way to Chuypetlovo, on the right bank of Struma. The cave has four small verticals (up to ten meters). Two underground streams: Golyamata Reka (sinking at 125 m) and Malkata Reka (sinking at 100 m).

Discovered in 1981 by members of Akademik Sofia Caving Club as a small hole, steaming in winter. Digging and enlarging the first ten meters of the hole took two years. The map was made by G. Markov, M. Zlatkova, Y. Atanasov, I. Barova, Tz. Ostromski and D. Angelov.

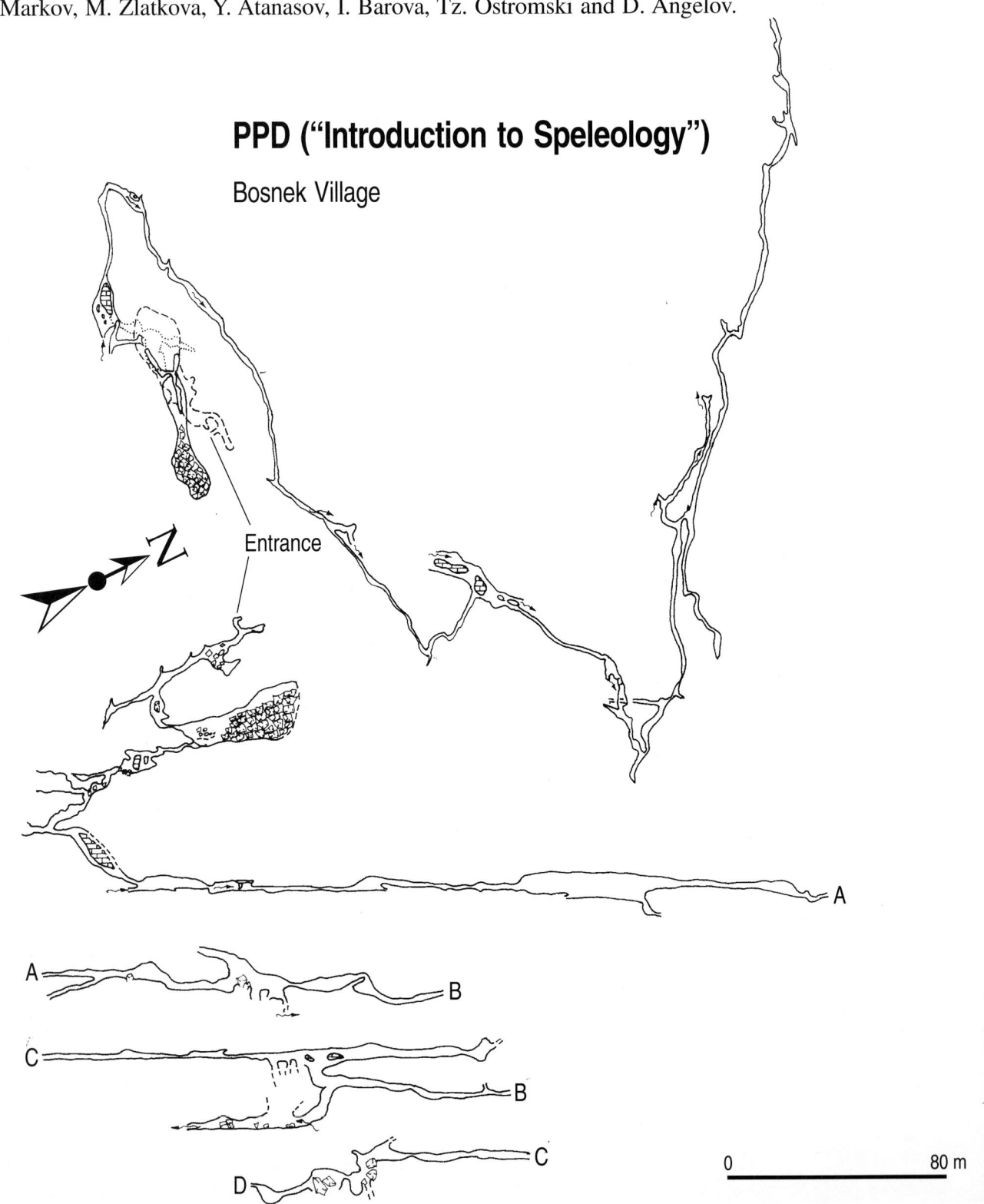

Name: **PROPADA (PARAMUNSKA YAMA) (194)**
Paramun Village, Pernik Distr.
Altitude 1068 m. Denivelation -63 m
Coordinates: N 42° 48' 13.7" E 22° 44' 16.8"

A pothole 1.5 km NW of Paramun Village, in the Eastern slope of Strazha (Paramuska Planina Mt.). Formed in organogenic Upper Jurassic – Lower Cretaceous limestone from the eastern periphery of Lyubash Horst – Monocline.

The pothole opens in a funnel with steep slopes. There is only one vertical, slightly narrowing 15 m from the entrance. Here, the only intermediate transfer could be organized. The vertical widens into an ideal bell. The bottom of the pothole is a small inclined chamber with a cone deposit of mud under the vertical.

The pothole is rarely visited. In the upper part of the vertical some hanging stones are to be taken into account. Description: A. Jalov.

First explored in 1960 by a team of the Republican Commission for Speleology, the map being responsibility of P. Tranteev. In November 1979, during an expedition of Rudnichar Pernik Caving Club a more precise survey took place (M. Metodiev, R. Metodiev, Z. Velinova).

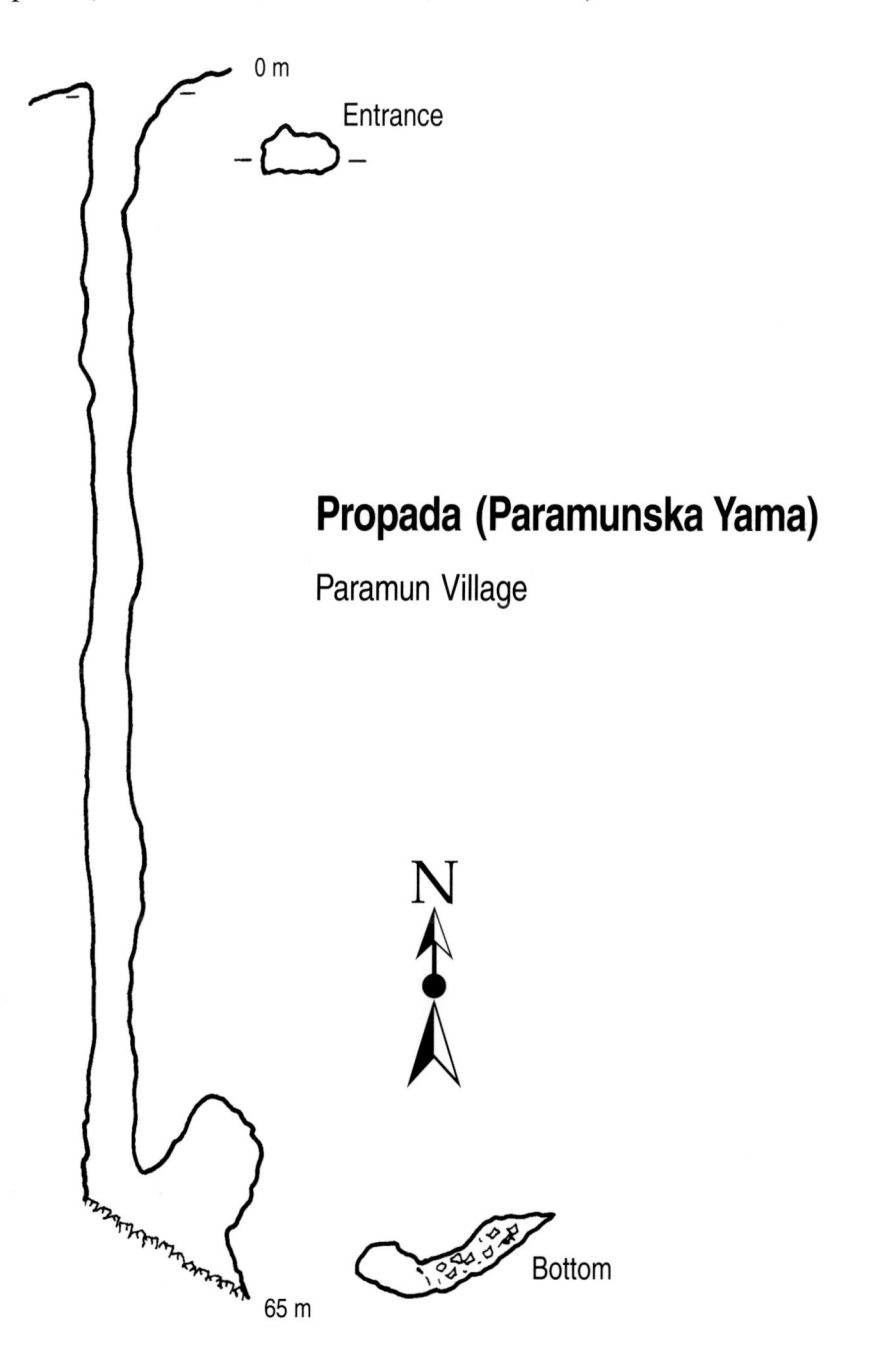

Propada (Paramunska Yama)
Paramun Village

Name: **VRELOTO (196)**
Bosnek Village, Pernik Distr.
Length 5300 m. Denivelation +90 m

A water cave in Trias limestone and dolomite. The underground river explored up to 1600 m. A one-channel cave, with several storeys. The galleries are very spacious and are rich in Mondmilch, crystals and speleothems. There are gypsum crystals, long up to 23 cm. Big deposits of fossil bones.

The source Vreloto is one of the biggest in Bosnek area (350 l/s) and takes its water from the sinks of Struma in the locality Djeranitsa.

On November 15, 1980, three cavers from Akademik Sofia Caving Club first entered the cave and reached a siphon, 820 m from the entrance. The other galleries have been explored by members of Extreme Speleoclub – Sofia directed by A. Russev.

The penetration of the cave is difficult because of the many unstable boulders and it requires a long stay underground.

Vreloto

Bosnek Village

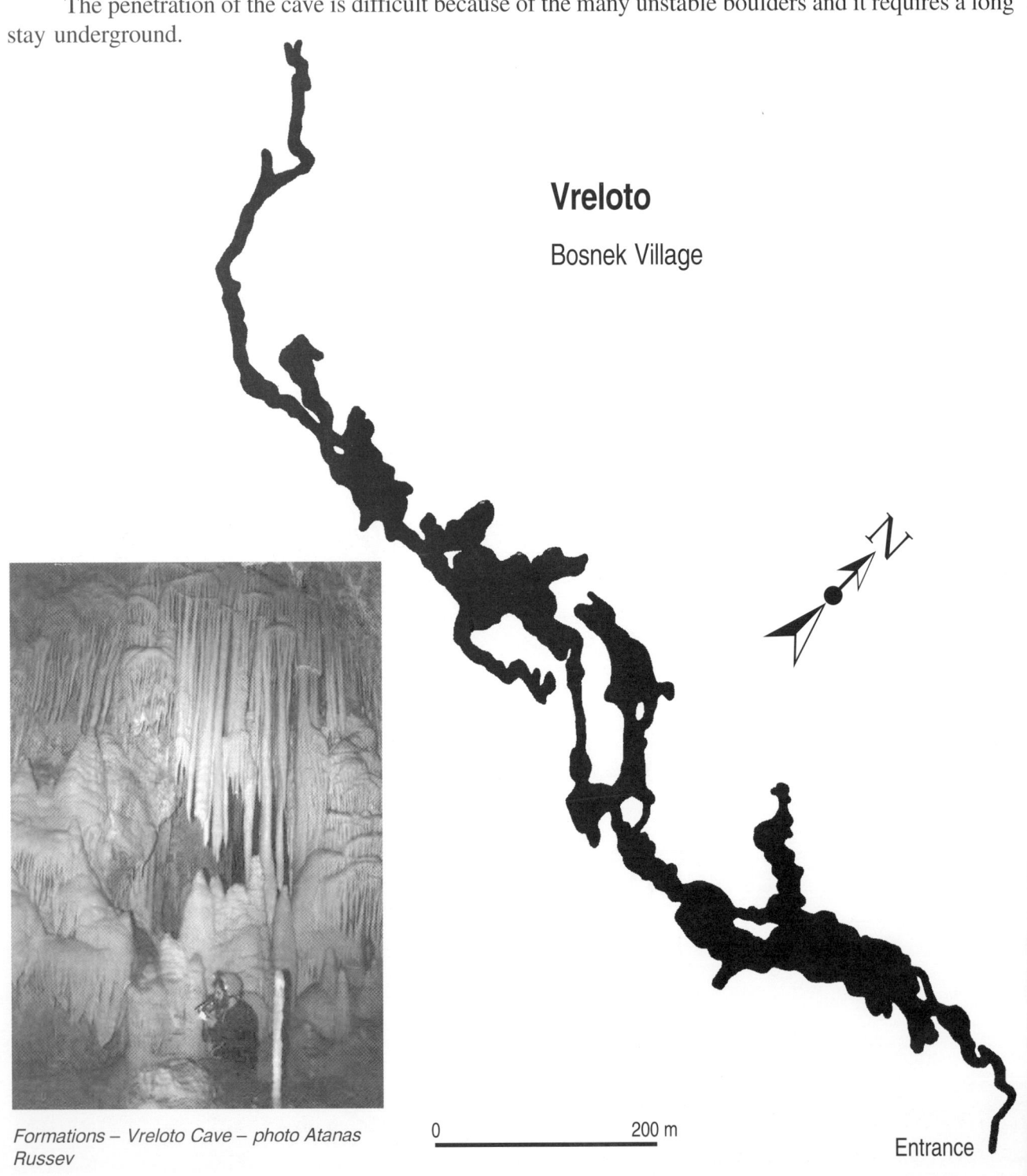

Formations – Vreloto Cave – photo Atanas Russev

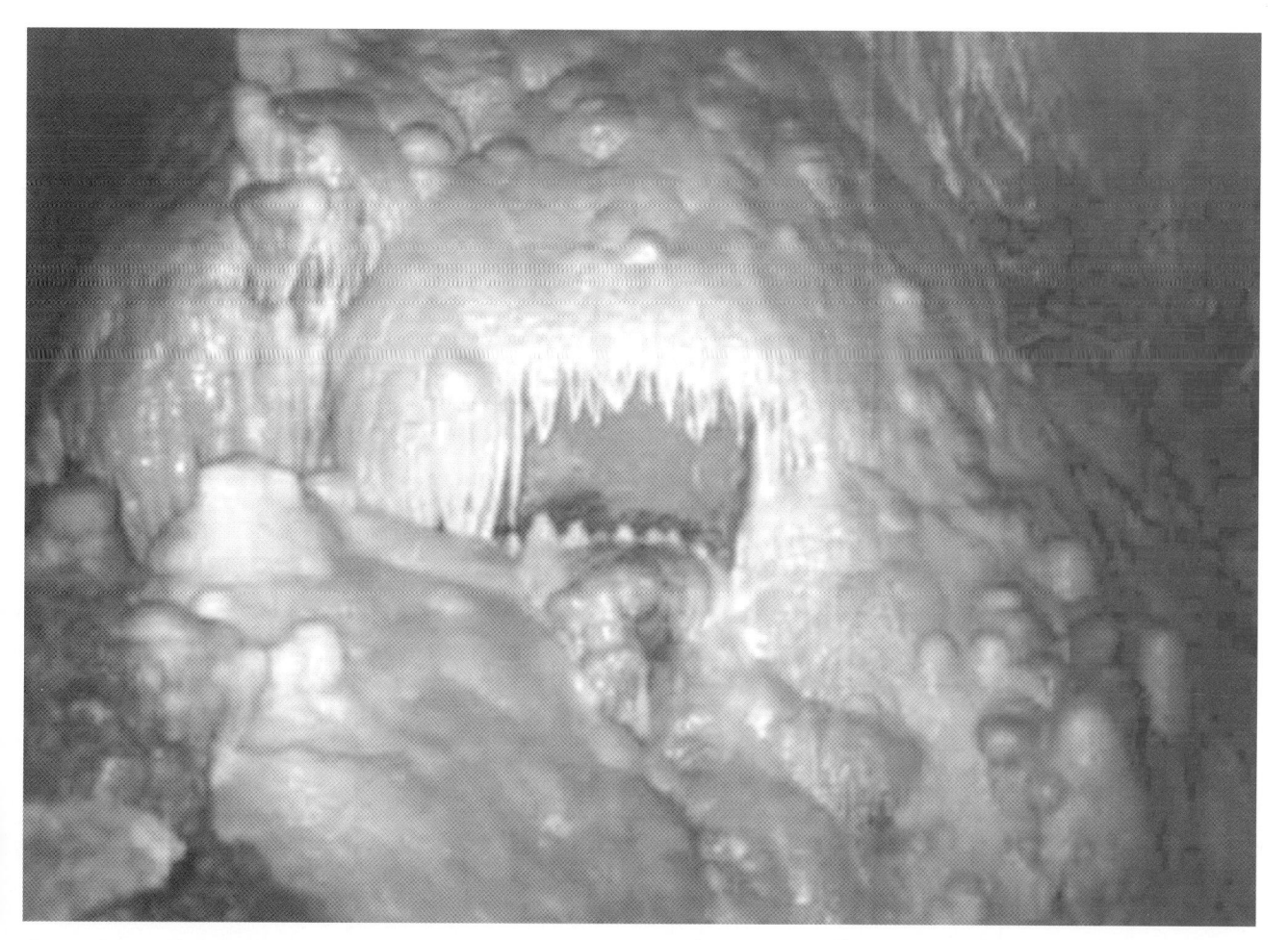

Formations – Vreloto Cave – photo Atanas Russev

Crystal – Vreloto Cave – photo Atanas Russev

Name: **YAMKATA (YAMKITE) (195)**
Rayantsi Village, Pernik Distr.
Length 358 m. Denivelation -66 m

A cave pothole, 2 km NW of Rayantsi, near the quarter Golema Livada on the southern slope of Konski Vrah, the area of Kraishte. Formed in dolomitized limestone from the Middle Trias.

The pothole has two entrances, opened at the bottom of funnels. The bigger one has a diameter of 15 m and depth of 16 m, the smaller (SW) has a diameter of 3.4 m and the same depth.

After descending the entrance verticals, we reach a wide stone-covered bottom. From its southeastern part a labyrinth of horizontal and inclined galleries, verticals and superposed chambers follow, the biggest being with dimensions 28 x 8 x 9 m. The floor is covered with gravel and clay, in some places there are stone blocks. The flowstone formations (stalactites, stalagmites, stalagtons, curtains, dendrites, coralites, etc.) are mostly dry.

The cave was studied first in 1976 by the Pernik Caving Club. In February 1981, it was surveyed by M. Metodiev, R. Metodiev and V. Mirchev from Rudnichar Pernik Caving Club, and again in 1994 by Helictite Sofia Caving Club. In 1999 cavers from Iskar Sofia Caving Club described the setting of the pothole (description: A. Jalov after V. Mustakov and I. Tachev).

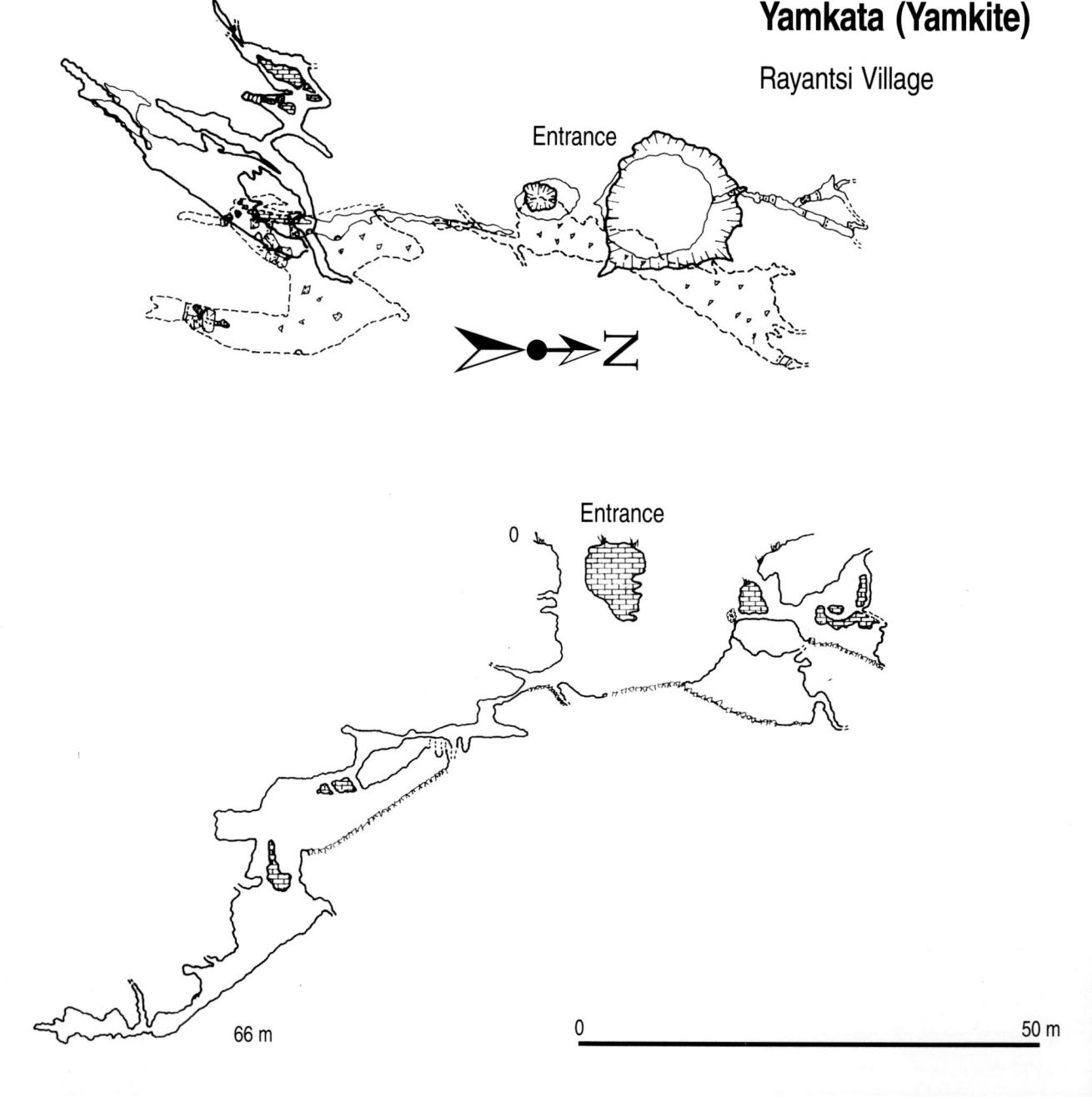

Yamkata (Yamkite)

Rayantsi Village

Name **ZHIVATA VODA (Ayazmoto) (200)**
Bosnek Village, Pernik Distr.
Length 162 m. Denivelation -7 m
Coordinates: N 42° 31' 30" E 23° 11' 58"

A cave NE of the village Bosnek, on the SW slope of Vitosha. Named after the nearby source Zhivata Voda (The Live Water). After the entrance hall, which is 14 m long, up to 11 m wide and 2 m high, a gallery starts, heading E for some 130 m. Here, the width is 1.5–2.5 m and the height 1–4.40 m. In some places the floor is covered with clay and sinter deposits with small sinter lakes and a dozen of stalagmites. A descending threshold of 2.5 m follows and a chamber is reached, which is 11 m long, 12 m wide and ca. 8 m high, with boulders at the bottom. About 2.5 m below there is a narrow gallery through which an underground stream flows coming from the neighbouring cave, called Suha Peshtera. After the chamber, the gallery bends to the north and after 20 m ends in a small lake. Flowstone formations in the end parts.

Recorded first in 1875, then in 1895 by K. and H. Škorpil.

Surveyed in 1975 by L. Popov, St. Tsonev and S. Badeva, cavers of Aleko Konstantinov Sofia Caving Club. In 1989, another map was made by M. Zlatkova, D. Kozhuharov, Ya. Bozhinov and I. Naneva from Zdravets Sofia Caving Club.

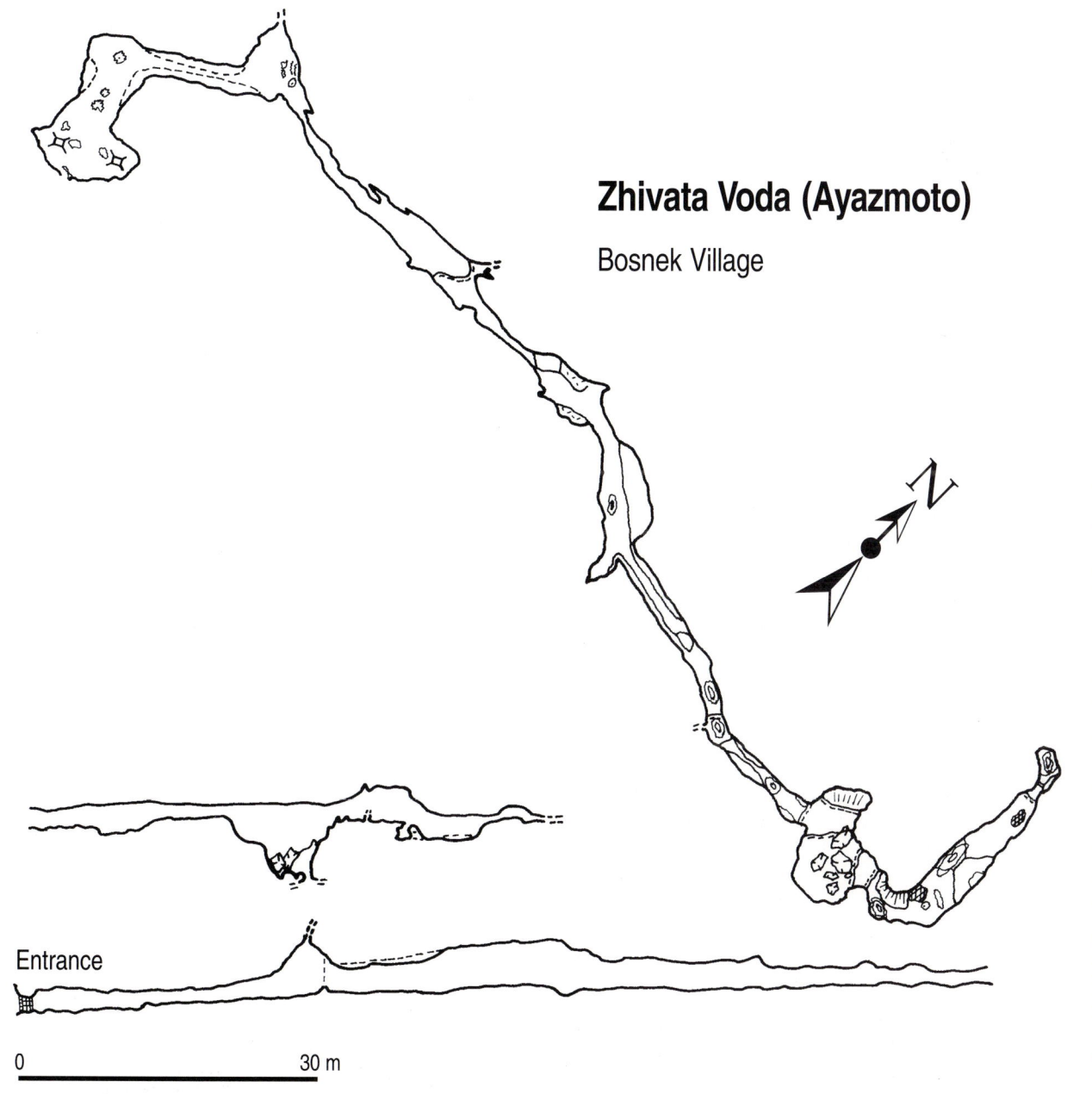

Zhivata Voda (Ayazmoto)
Bosnek Village

CAVES IN RILA – RHODOPEAN REGION
CAVES IN PIRIN

Name: **20 YEARS OF AKADEMIK (211)**
Bansko Town, Blagoevgrad Distr.
Altitude 2450 m. Lenght 80 m. Denivelation -118 m

In the northern part of Pirin, in the circus Banski Suhodol, 300 m before the joint of Koteshki Chal and the main karstic ridge Koncheto. Snow is accumulated in the entrance parts. One big shaft of – 55 m. The two entrances of the shaft are joint in a gallery, the bottom of which is covered with stones. T° 1.5 °C. Discovered and surveyed in 1978 m.

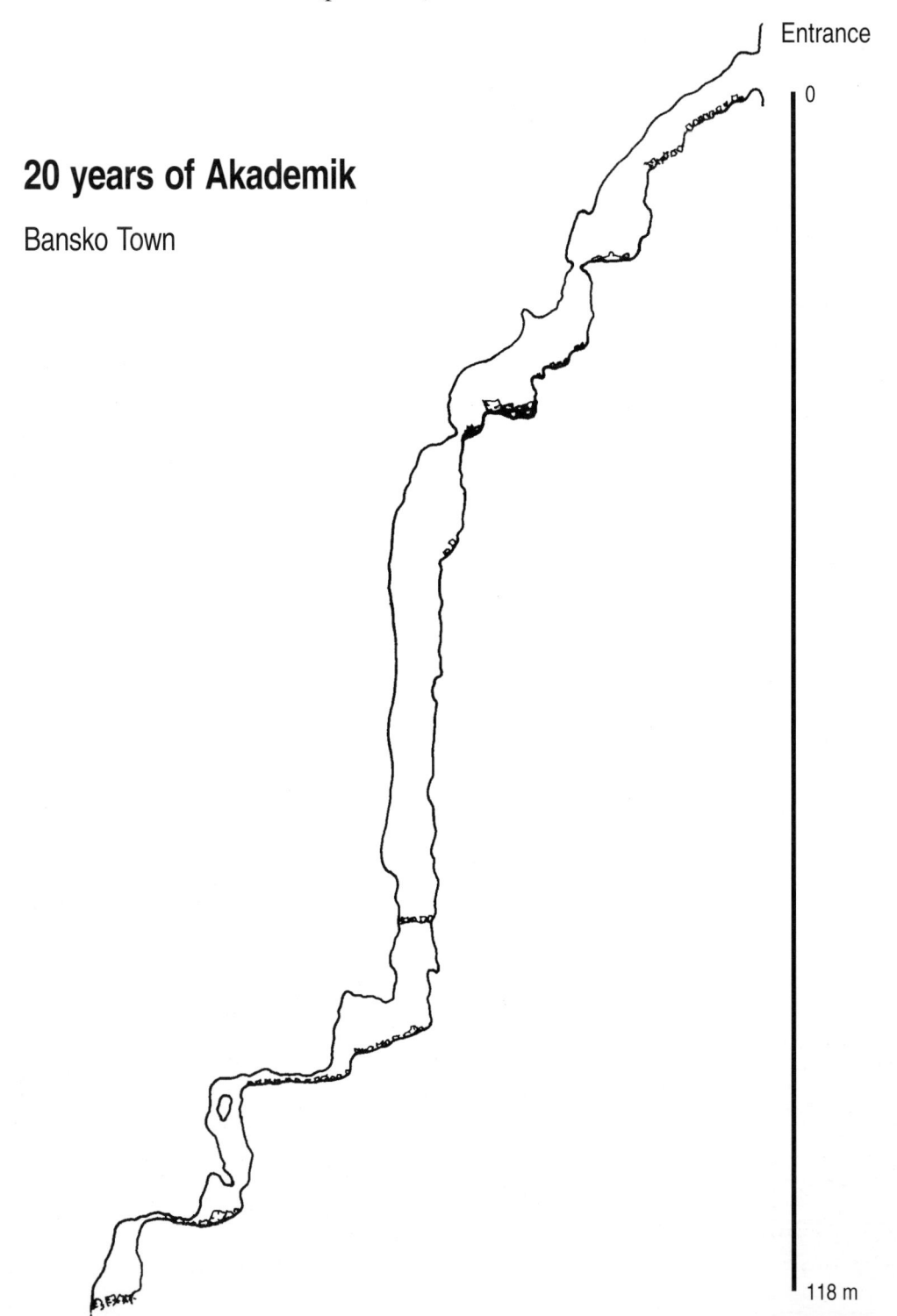

Situated in Pirin National Park (protected).

20 years of Akademik

Bansko Town

The entrance of the pit – photo Ivan Lichkov

Name: **ALEKO (203)**
Ilindentsi Village, Blagoevgrad Distr.
Altitude ca. 1700 m. Length 547 m. Denivelation -132 m

A descending branched pothole on the grounds of Pirin National Park, in the Sinanitsa part, 300 m under the summit Sharaliya (2172 m), on the NW slope. It has two entrances, 11 m difference in their levels, on the left bank of Razkolska river. Some of the galleries are due to mining. Explored by the Speleoclub of Aleko Konstantinov, Sofia Tourist Society, in February 1978–May 1985. The bottom was reached on March 29, 1978 by A. Jalov, Y. Bozhinov, S. Tsonev, Hr. Iliev. Map: K. Bonev, D. Vangelov, S. Ivanov during the club expedition Pirin-85. Perhaps in XVI Century Saxons were mining ore in these parts.

Aleko

Ilindentsi Village

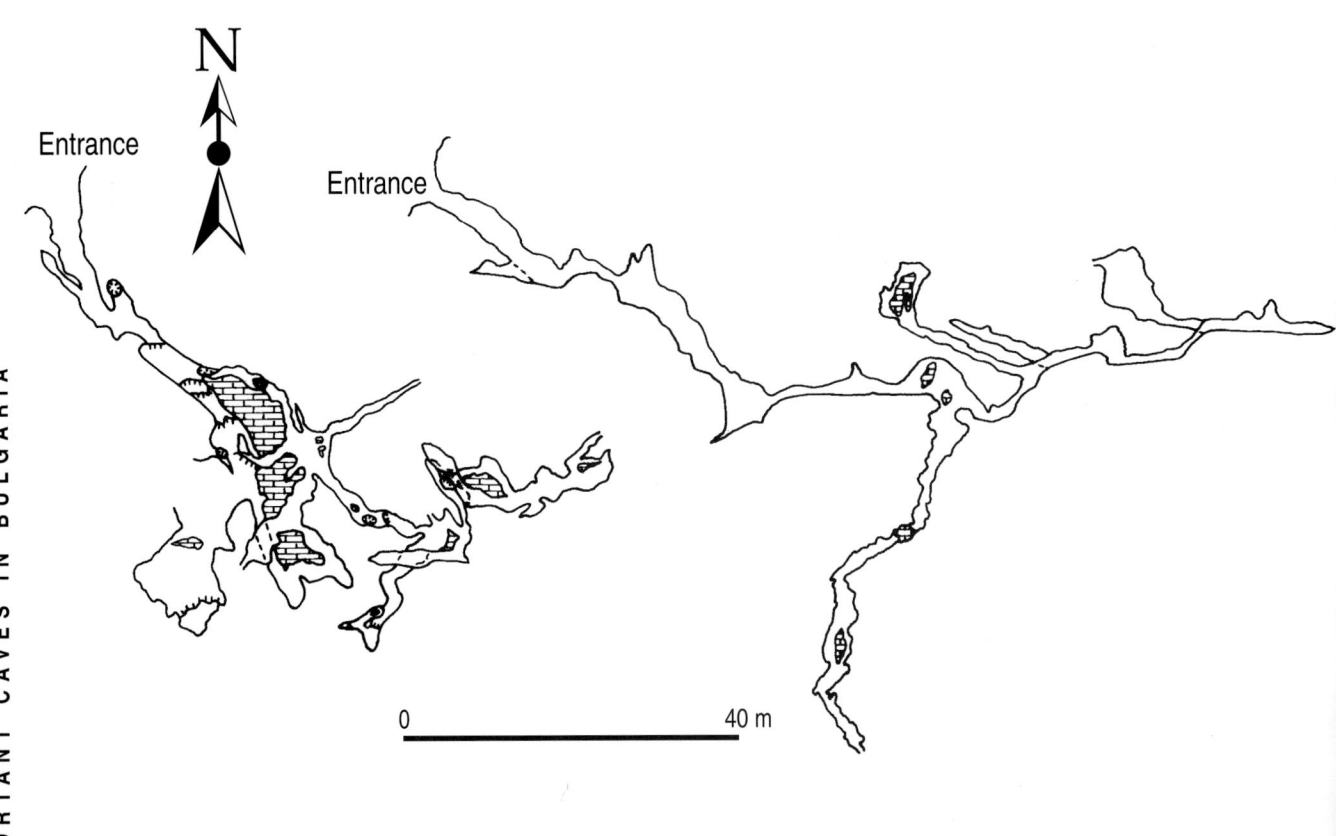

Name: **BANDERITSA (212)**
Banderitsa Hut (North Pirin), Blagoevgrad Distr.
Altitude: 1778 m. Length 243 m. Denivelation -125 m
Coordinates: N 41° 46' 04.4" E 23° 25' 57"

A pothole in Proterozoan marble, 10 minutes from the hut Banderitsa, on the left bank of the river near the contact zone. The water in the pothole re-appears 900 m lower, near Bansko. A narrowing at the end. Difficult access because of low air temperature (4 °C) and water temperatures (2.8 °C). The water capacity at the fifth waterfall is 20 l/s, and cannot be avoided. Discovered in November 1970 by five cavers from the Speleoclub Akademik, Sofia. In December 1970, the bottom was reached by P. Delchev, K. Spassov, A. Filipov and I. Nelchinov. The pothole is included within Pirin National Park and is protected as a Natural Monument (Decree 1799/ 30.06.1972).

Banderitsa

Banderitsa Hut (North Pirin)

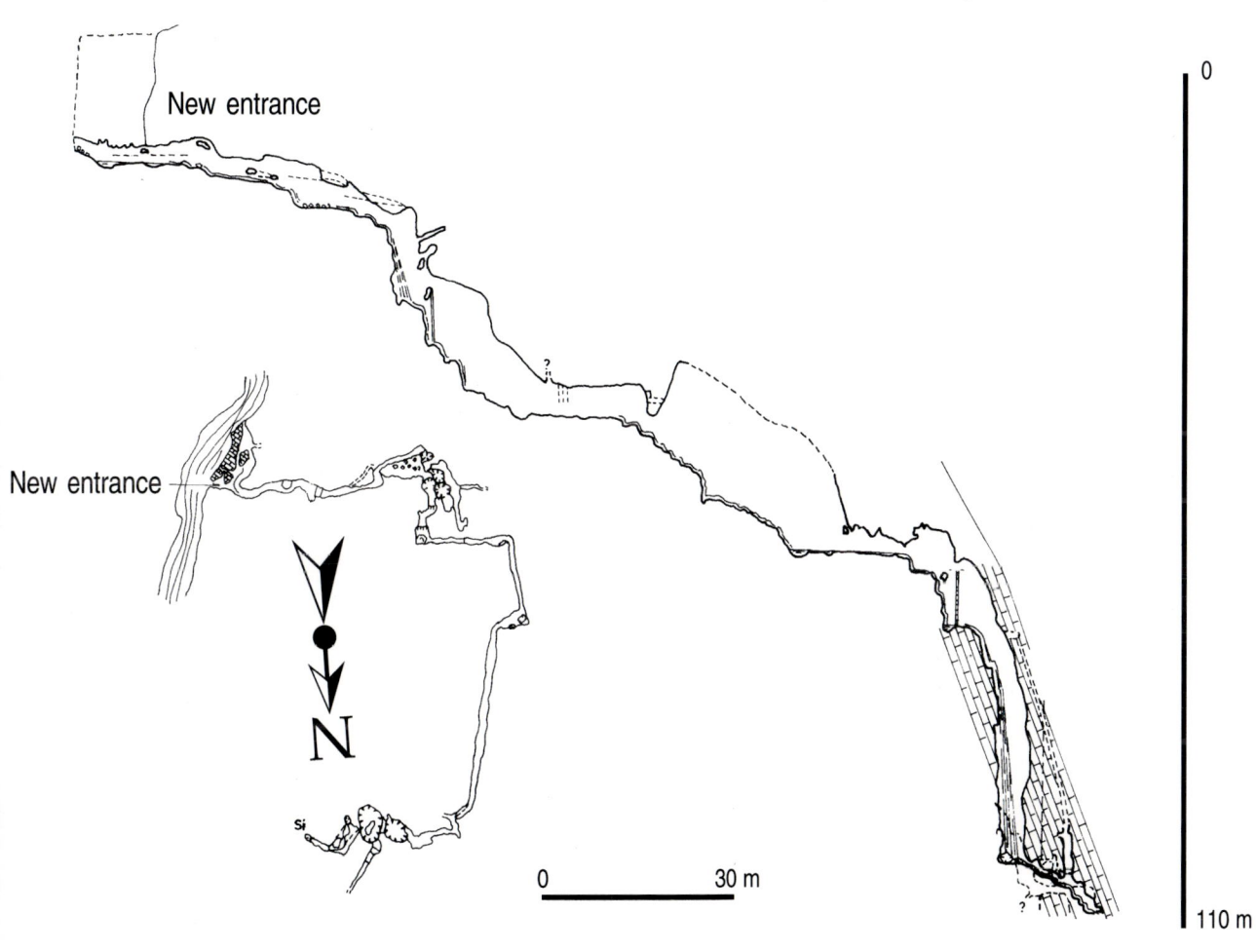

Name: **CHELYUSTNITSA (209)**
Razlog Town, Bayuvi Dupki Circus
Altitude 2300 m. Denivelation -103 m
Coordinates: N 41° 48' 05" E 23° 23' 06"

A pothole in the middle part of Bayuvi Dupki Circus. On the surface, a crevice is clearly seen, about 50 m long, in the middle of which the entrance is situated. Discovered in 1985 by I. Barova, A. Drazhev, Tz. Ostromski, V. Shekerdzhieva, D. Angelov, cavers of Akademik – Sofia Caving Club during the expedition "Pirin'85". Name given by the expedition.

The pothole consists of one vertical and is rather narrow. Rope is not needed for penetration. One difficulty is the narrow place at – 30 m, which stopped the first explorers, and the boulders at – 98 m. The pothole ends with an impenetrable narrowing, under which at least another 25–30 m are seen. Danger of falling stones. The boulders at – 98 m are rather stable (Dimitar Angelov informs). Consists of 7–8 big blocks, well jammed.

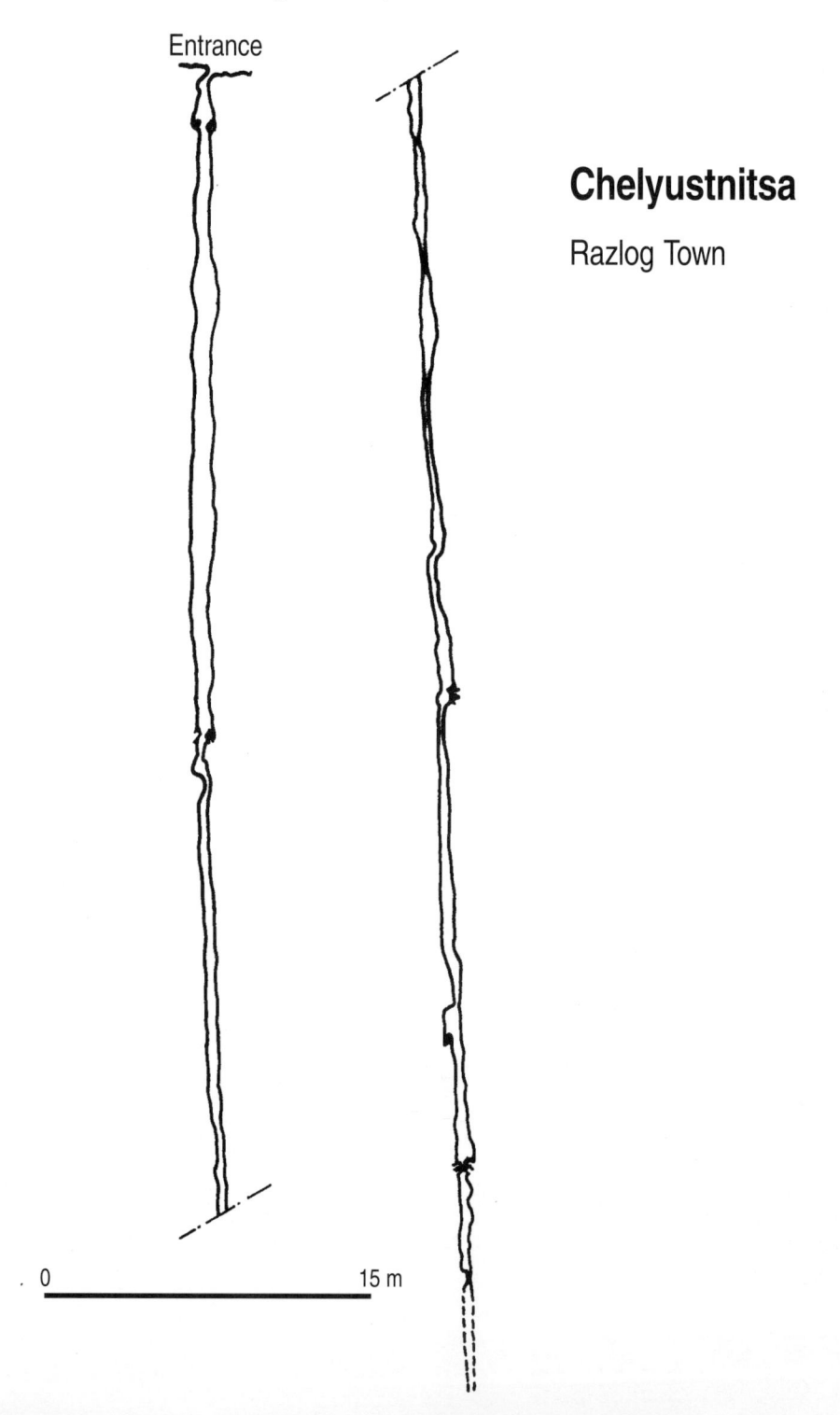

Situated within Pirin National Park (protected).

Chelyustnitsa

Razlog Town

The entrance of the cave Chelyustnitsa – photo Ivan Lichkov

The circus Kamenititsa and the ridge Sredonosa, behind which is the circus Bayuvi dupki

Name: **KAMENITITSA No 14 (208)**
Razlog Town, Kamenititsa Circus
Length 22 m. Denivelation -103 m
Coordinates: N 41° 48' 00" E 23° 22' 36"

From August 10 to 22, 1972, the second expedition of Akademik to Kamenititsa Circus took place. In pothole No 14 the cavers reached – 32 m, when a huge boulder stopped them. During the third expedition (October 1–8, 1973), four cavers managed to remove the boulder and reach – 103 m.

The entrance is situated in the upper part of the circus. Two entrances (5 x 1.5 m), joint at – 4 m, a steep gallery up to – 26 m follows. There, a small chamber is formed, followed by an almost vertical gallery, up to – 56 m. The next vertical is 40 m deep, ending with blocks, allowing some more meters.

This is the deepest pothole in the circus. It is not equipped for SRT.

Situated within Pirin National Park (protected).

Kamenititsa No 14

Razlog Town

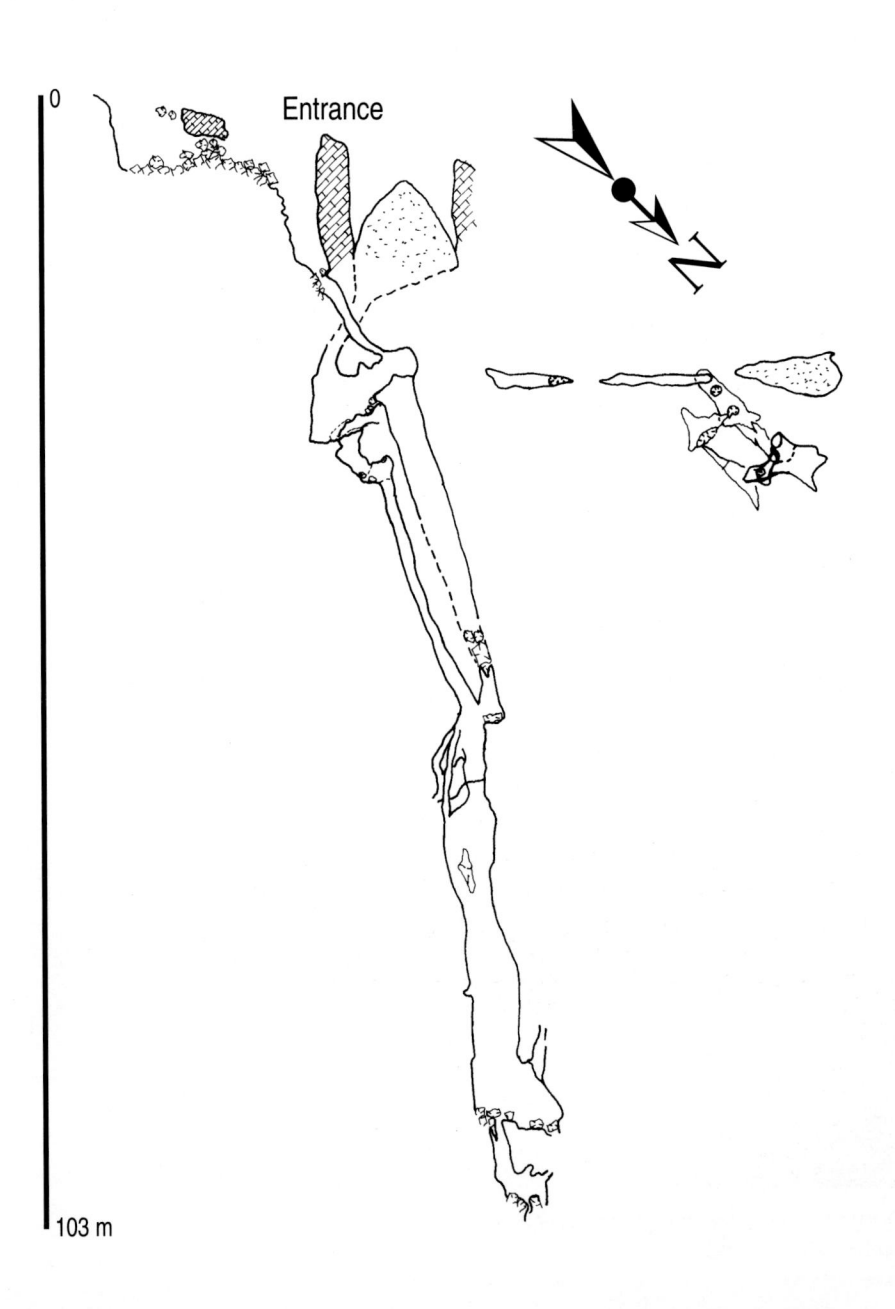

G.Georgiev in front of the entrance of the cave Kamenititsa No 14 – photo Ivan Lichkov

Name: **PROPAST (Pothole) No 9 (Devyatkata, System 9–11) (214)**
Bansko Town, Banski Suhodol Circus
Altitude 2525 m. Length 311 m. Reached denivelation -225 m

The pothole was discovered in 1977 during an expedition of Akademik Sofia Caving Club. The map published here was made also in 1977 by V. Vassilev and I. Lichkov. During another expedition in 1981 Pothole N 9 was connected to Pothole N 11 and a depth of 100 m was reached.

In the period August 25 – September 2, 2001, an expedition was organized in the circus Banski Suhodol in Northern Pirin. Organizer of the expedition was Vitosha Sofia Caving Club, with the participation of other cavers. The idea was to break down the boulder at the bottom of Pothole 9 (-138 m) and find a continuation after it.

After several hours of work the boulder was broken down and raised to the surface by means of polyspasts. A. Drazhev, Y. Tsvetanov and N. Orlov were thin enough to penetrate behind it and remove the remaining stones. Then, they reached a vertical leading to a big chamber. The next day, a group, formed by V. Stefanov, K. Bogacheva and P. Bakalov, descended the vertical (15–20 m) and reached a big chamber covered with boulders. Several new verticals were detected and one was descended to the bottom. There was no time to finish the exploration, but there are still good prospects to do that. There is not a complete map of the pothole, but its depth up to the stone is 138 m. From there to the bottom of the big chamber there are another 35–40 m, which probably makes Pothole No 9 the deepest in Pirin. The pothole Vihrenska Propast is (according to the survey of 1999) – 170 m (K. Stoichkov reports).

The actual map was made by Helictit Caving Club – Sofia.
Situated on the territory of Pirin National Park (protected).

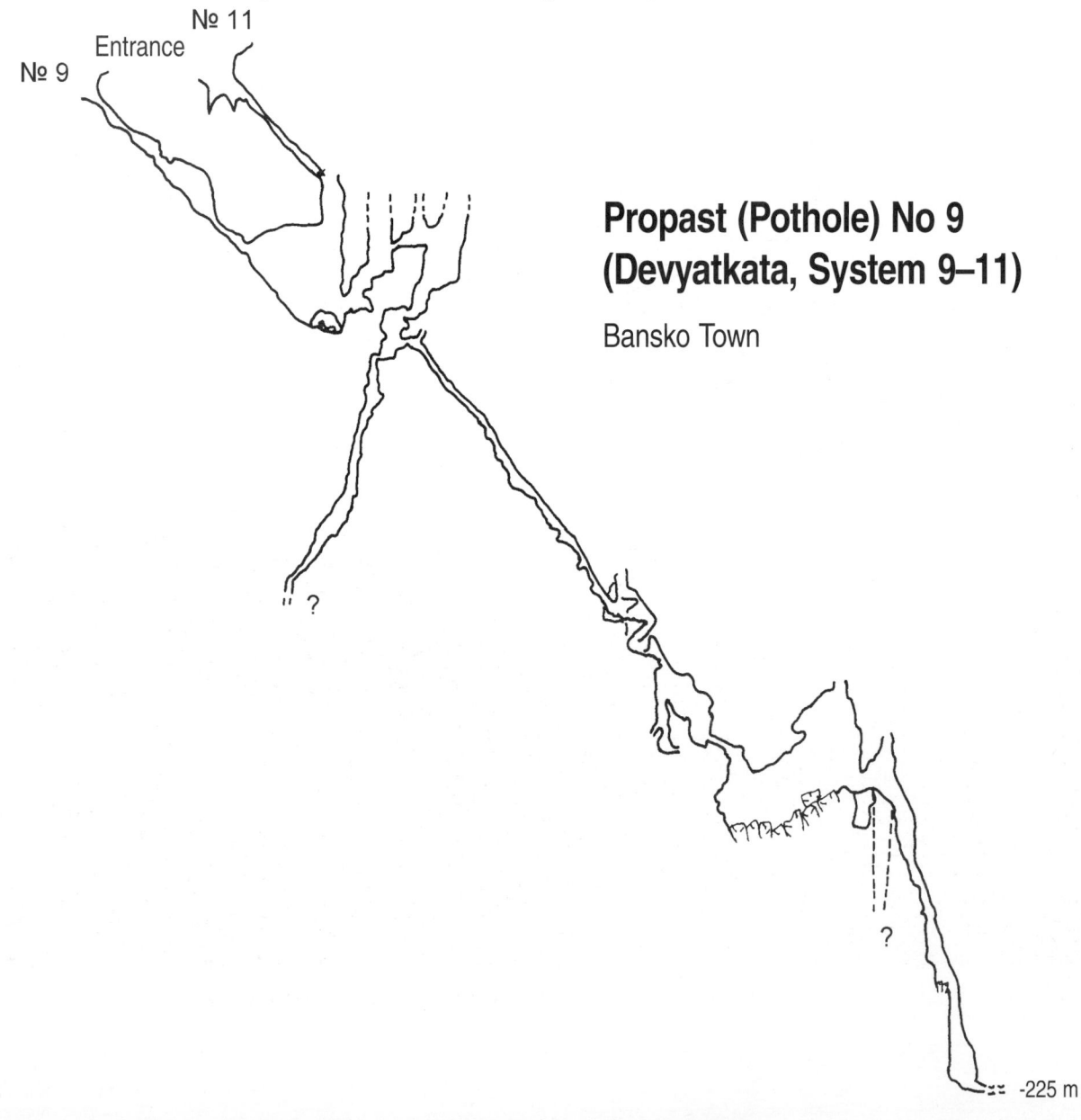

Propast (Pothole) No 9 (Devyatkata, System 9–11)
Bansko Town

V. Oshanov (Vilicata) – one of the first explorers of the cave – photo Ivan Lichkov

The carrying of luggage to the camp in Banski Suhodol Circus is done with mules – photo Trifon Daaliev, 1981

Name: **PROPAST K -18 (LEDENATA) (206)**
Razlog, Blagoevgrad Distr.
Circus Kamenititsa, Pirin
Denivelation -126 m
Coordinates: N 41° 48' 05" E 23° 22' 42"

The entrance is oval with dimensions 8 x 3 m.
It is a cascade pot hole. Starts with a 20 m vertical, followed by verticals of 10 and again 20 m. The bottom is covered with ice. Than starts in inclined gallery, ending in another 20 m vertical with ice covered bottom. The pot hole ends in a squeeze with strong draft – hammering could give way further.
Discovered and explored by cavers from the international expedition "Pirin'2002". Surveyed by André Dawagne and Sebastien Dujardin from Belgium.
Within Pirin National Park.

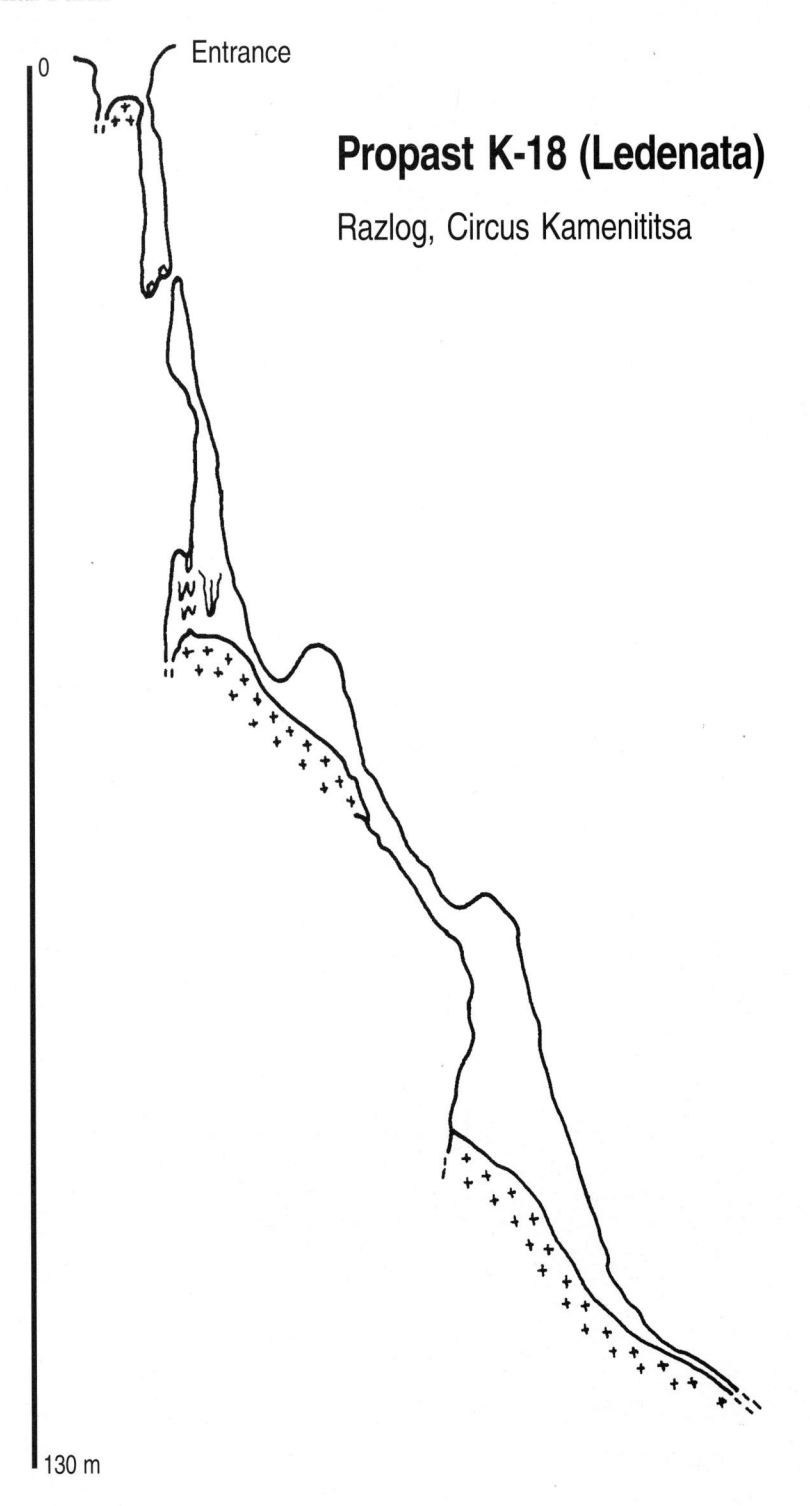

Propast K-18 (Ledenata)

Razlog, Circus Kamenititsa

The entrance is covered by snow – photo Ivan Lichkov

Exploration of the pothole K-18 in 2002

Name: **PROPASTNA PESHTERA** (Cave – Pot hole) N 29 (205)
Bayuvi Dupki Circus
Razlog Town, Blagoevgrad Distr.
Length 111 m. Denivelation -82 m
Coordinates: N 41° 48' 07" E 23° 23' 15"

A cave pothole above the first rigel in the valley of the lower couloir. In the pothole there is always some snow, so one penetrates the first 10 meters between the snow and the rock wall. After the first inclined shaft, there are ca. 50 m of a horizontal gallery, at the end of which a 41-m deep vertical starts. Right above it, there are several big blocks, by the 14th meter a huge boulder is jammed resembling a balcony. At the bottom, there is a blockage, through which one falls into a narrow crevice, 7 m long, at the beginning of which there are dendrites. The crevice is further transformed into a second pitch (30 m). At the 19th meter, there is a platform where one can rest.

The walls are moist. In the first pitch there is dripping water, which further down forms a small stream. No flowstone formations, the temperature in August 1998 was 1.6 °C, at the entrance 0.8 °C, average temperature 1.5 °C.

Cave Pothole No 29 was explored first on the July 9–11, 1968 by the "Pirin'68" expedition of Akademik Sofia Caving Club (Valeri Vassilev, Igor Gachkovski, Zhelyo Marinov, Radoslav Rahnev, Valeri Nazmovich, Tsvetan Lichkov, Petar Delchev, Proyno Somov, Assya Ermalova).

The map was prepared by the participants in that expedition and the present description is due to P. Delchev.

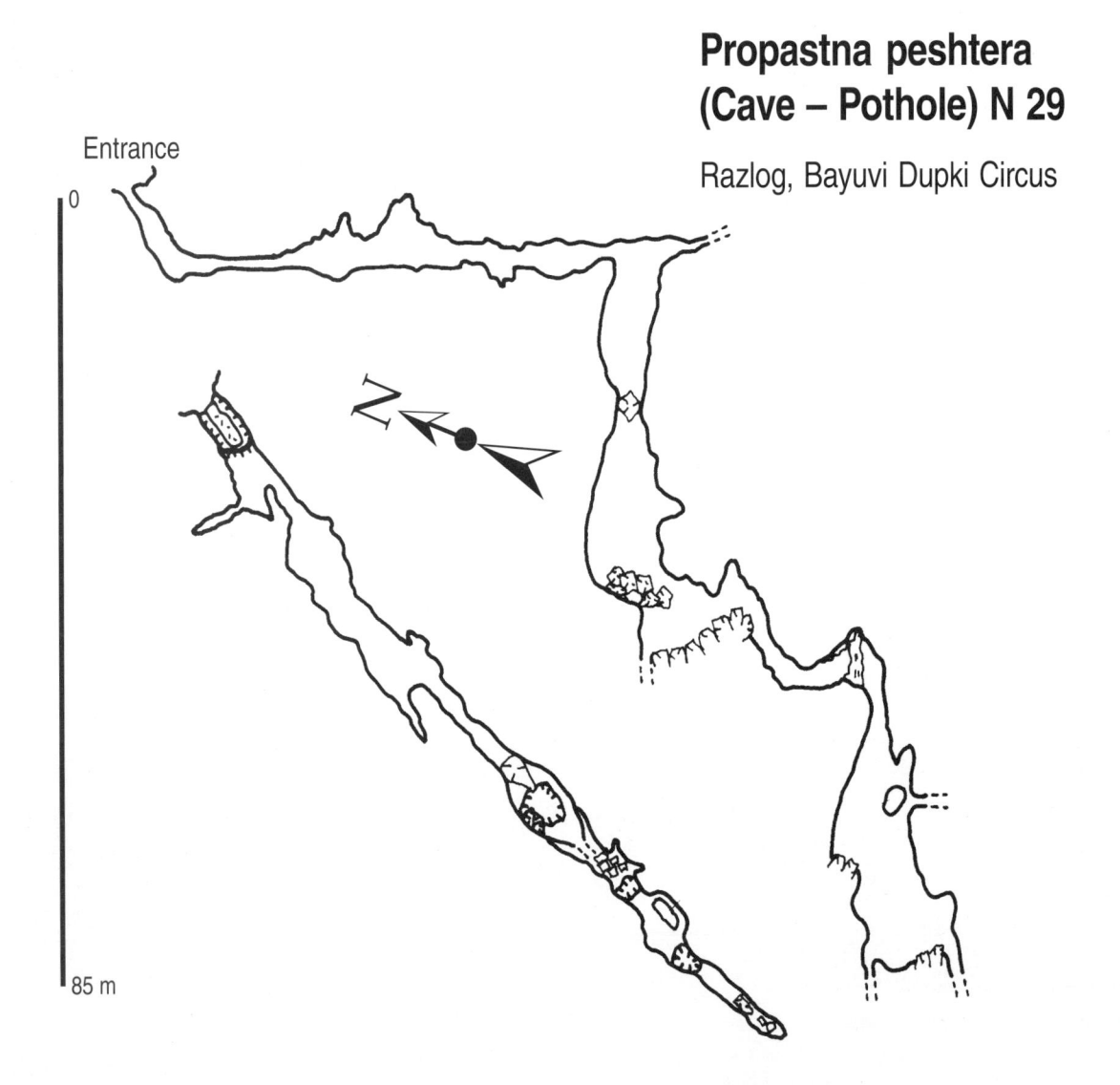

Situated within Pirin National Park (protected).

Propastna peshtera (Cave – Pothole) N 29

Razlog, Bayuvi Dupki Circus

In front of the entrance – photo Ivan Lichkov

Name: **PROPAST K-19 (KAMENITITSA – 19, Bulgaria – France – Belgium) (204)**
Razlog Town, Blagoevgrad District
Kamenititsa Circus
Altitude 2450 m. Denivelation -136 m
Coordinates: N 41° 48' 02" E 23° 22' 46"

Propast K-19 is situated in the upper part of Kamenititsa Circus, 4 hours from Yavorov Chalet and 1 h from the shelter Koncheto. It is a cascade pothole. It starts with a funnel of 10x3 m and a short descending gallery with a squeeze at the end. After the squeeze, we find four smaller verticals and two bigger – one of 36 m and one of 22 m. The end squeeze has not been followed, but a strong draft can be noted – maybe there is a way further down.

The entrance of K 19 – BFB was discovered on August 23, 2002, by Nikolay Orlov, together with Nancy Rosseti and Marilyne Hanin from France and was marked as NMN (during the international caving expedition "Pirin'2002"). At that time, part of the first vertical (to the second platform) was descended. On the next day, all three reached a chamber, after which there was a very narrow passage. After several hours of hammering, they opened the way to the second and the third verticals. On August 24–25, the Frenchmen Damyan and Emanuel reached the bottom at -112 m, equipping the four verticals of the pothole.

Some days later, Mathieu Berger and Oliver Peron from France surveyed the pothole, leaving several (especially the squeeze at the bottom with a strong air current, suggesting a very deep pothole).

In 2005 the pothole was explored to -136 m by M. Tranteev and other cavers.

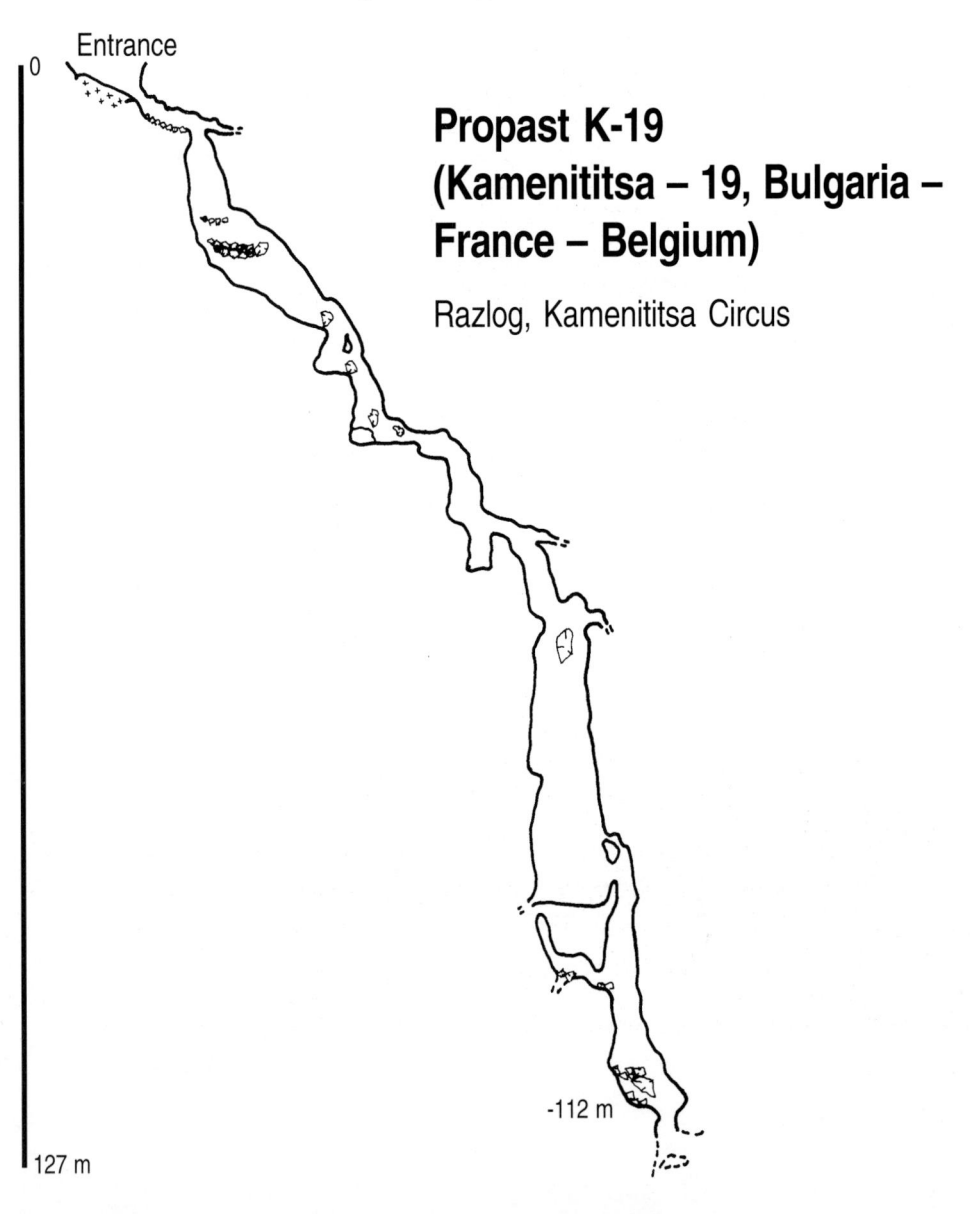

Situated within Pirin National Park (protected).

**Propast K-19
(Kamenititsa – 19, Bulgaria –
France – Belgium)**

Razlog, Kamenititsa Circus

The entrance – photo Ivan Lichkov

Name: **PROPAST 35 (CYCLOPE) (207)**
Kamenititsa Circus
Razlog Town, Blagoevgrad Distr.
Length 158 m. Denivelation: -87 m
Coordinates: N 41° 48' 19" E 23° 23' 04"

A pothole in the circus Kamenititsa, Pirin, on the slope above the Second Rigel (on the right of the avalanche couloirs, above a small stone vertical). A one-level, non-branched cave pothole. Formed in white marble with grey stripe coloration. The autochthonous formations are the result of the frost transformations.

Near the entrance there is a dangerous blockage. East of the second vertical, a stream appears from the ceiling, runs under the blockage and reappears at the vertical before the end chamber.

The deposits are ochre–whitish in colour. The ceiling is usually smooth and horizontal. Secondary formations are missing, except for scarce dendrites on the inclined walls, result of the dripping water.

In the first well, bones and the scull of a boar have been found.

Explored first by a team of the cavers Kliment Burin, Miron Savchin, Klava Stelmolchuk, Zhclyo Marinov, Ivan Mateev, Georgi Antonov, Ala Ermolova, Lilo Lilov and Vassil Gruev. The present description and map were made by Vassil Gruev in 1968.

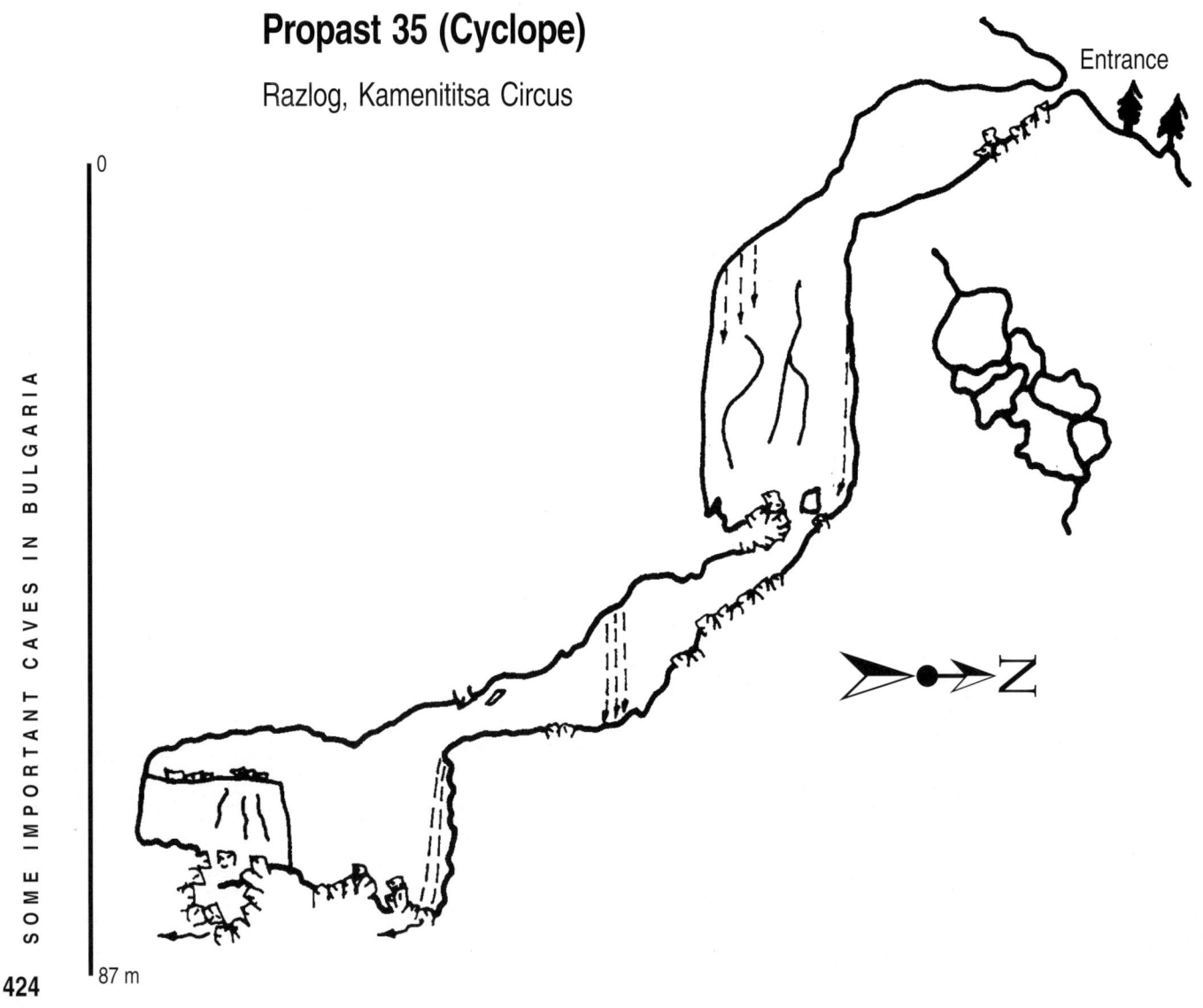

Situated within Pirin National Park (protected).

Propast 35 (Cyclope)

Razlog, Kamenititsa Circus

The entrance of Cyclop Cave – photo Ivan Lichkov

Name: **SHARALIYSKATA PESHTERA (202)**
Village Ilindentsi, Blagoevgrad Distr.
Length 470 m. Denivelation 52 m (-10, +40). Surface 1815 m
Coordinates: N 41° 42' 47" E 23° 18' 51"

A cave 10 km NE of Ilindentsi. The cave is developed in the SE slope of Kurtov Rid, reaching the summit of Sharaliya (2172 m). Formed in Proterozoan calcite-dolomite marbles. A cavern without a natural exit, an ascending-descending branched cavern. Galleries connect four chambers at different levels, the largest being the entrance chamber with a surface of 902 m^2. Sinter formations on the walls and the floor everywhere in the cave. Calcite is widespread, aragonite is present everywhere as needle-shaped and needle-plate transparent crystals. Pearls, moon milk. The infiltrating water forms five stagnant lakes.

The cave was discovered in the 50s during mining works. In 1962, cavers from Sandanski (Edelweiss Tourist Society) and from Sofia (Planinets Sofia Caving Club) studied the cave and collected cave animals. In 1975, 1977 and after 1983 new galleries and chambers were discovered, including by students of Sofia University. Surveyed in 1979 by L. Popov and V. Vassilev during an expedition of Aleko Sofia Caving Club.

Fauna. Troglobites: *Pseudosinella bulgarica* (Collembola), *Duvalius pirinicus* (Coleoptera).

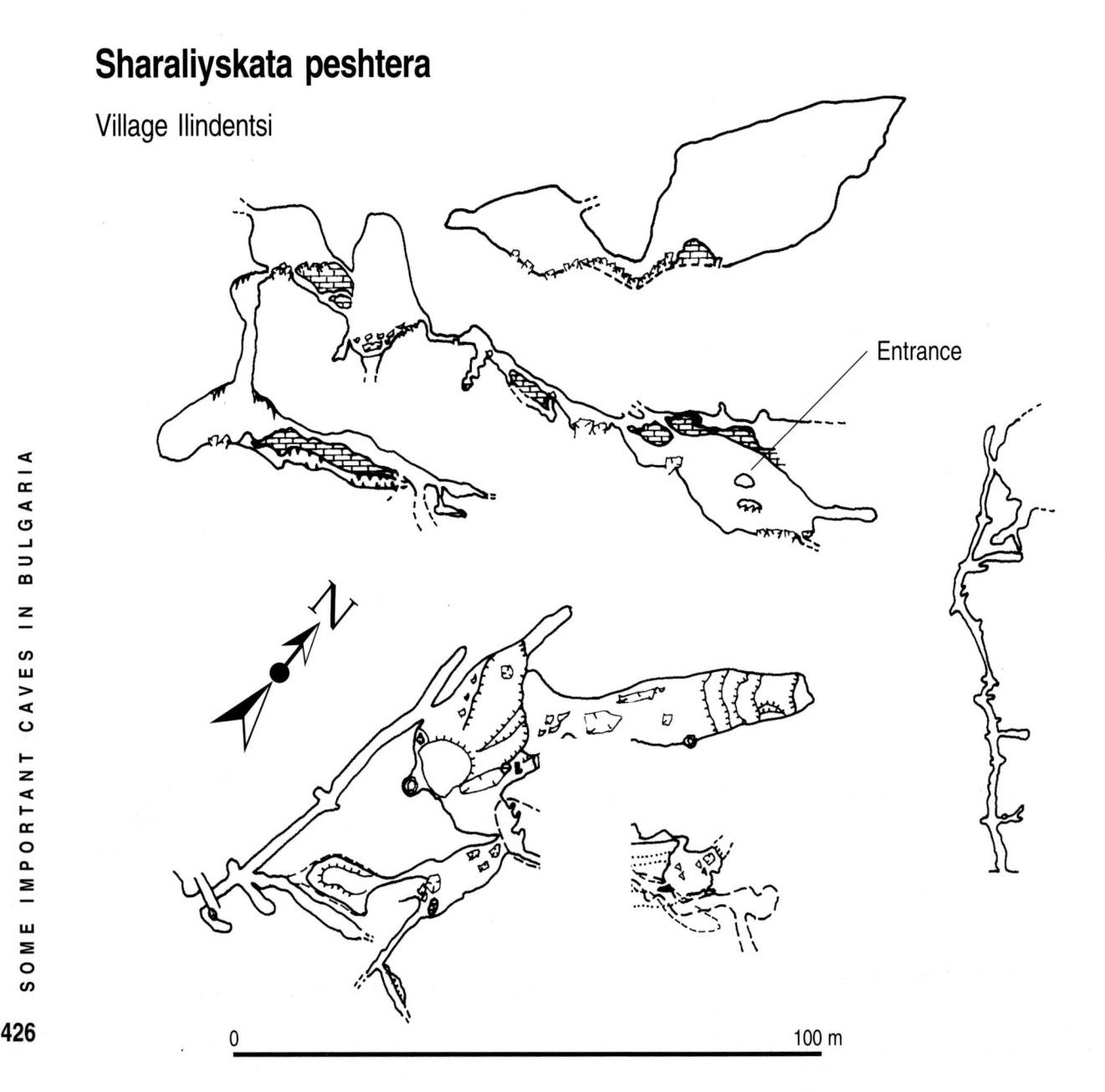

Sharaliyskata peshtera

Village Ilindentsi

Name: **SPROPADNALOTO (PROPADNALOTO) (210)**
Razlog Town
Length 605 m. Denivelation – 8 m.
Coordinates: N 41° 51' 04.3" E 23° 25' 57.7"

The cave entrance is situated 4 km SW of Razlog, in a military area. It opened during the earthquake in 1904. By 1965 the known length was ca. 250 m. In 1967 new parts were discovered and since 1971 cavers from Akademik – Sofia Caving Club have surveyed the cave up to the 605 m known till now.

One of the few Bulgarian caves in a conglomerate. About 60 m from the entrance, an underground stream runs. Water temperature about 6 °C. The cave has few flowstone formations.

The water of Spropadnaloto emerges 300 m from the entrance and from there Rakovitsa River starts.

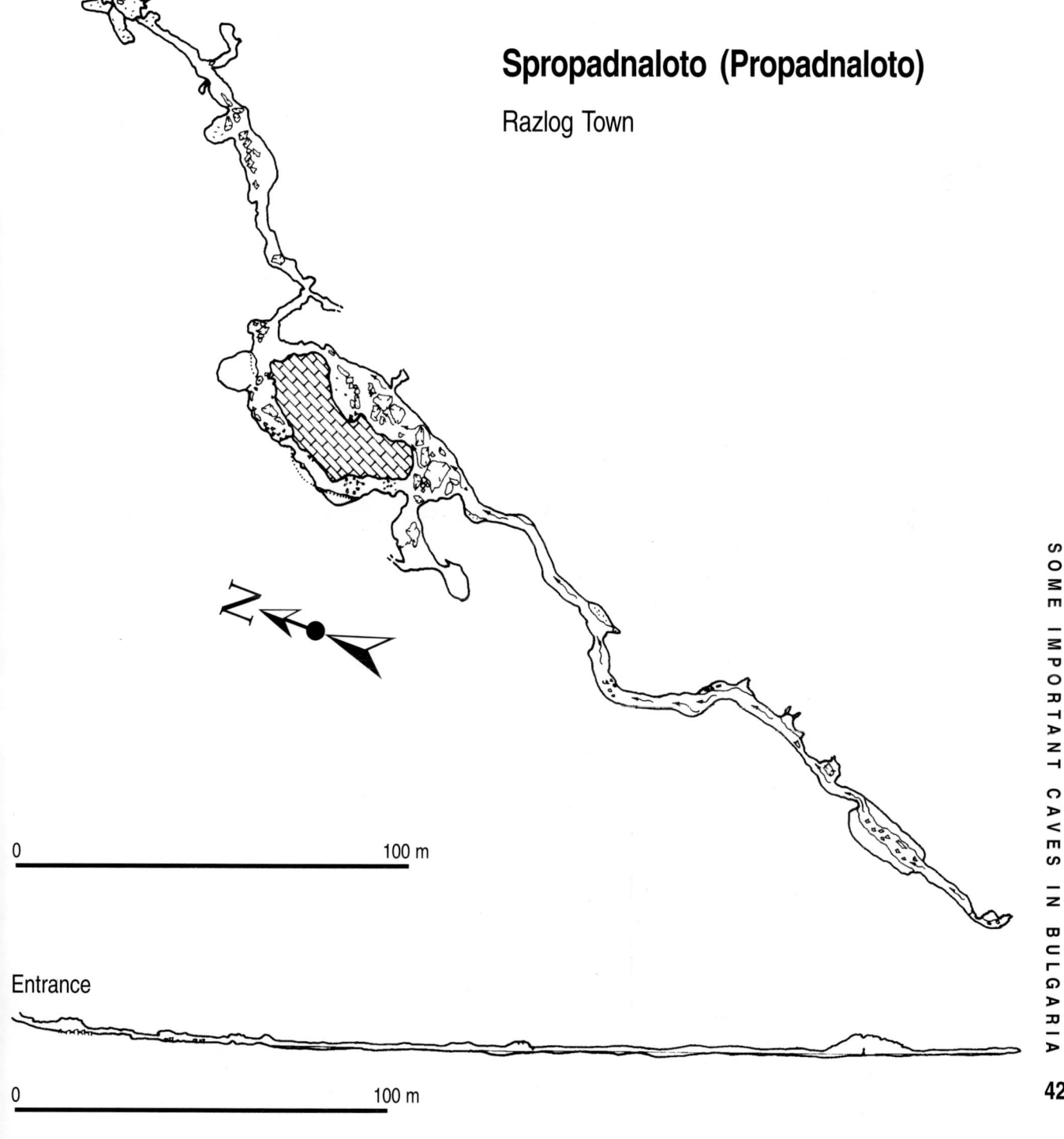

Spropadnaloto (Propadnaloto)

Razlog Town

Serge Delaby - President of Belgian Federation of Speleology in front of the entrance of the Spropadnaloto – photo Trifon Daaliev

Name: **VIHRENSKA PROPAST (213)**
Vihren Summit, Pirin, Blagoevgrad Distr.
Altitude 2650 m. Length 396 m. Denivelation -170 m
Coordinates: N 43° 43' 41.1" E 23° 24' 52.2"

On the eastern edge near Vihren summit, formed in strongly inclined Proterozoan marble. This is the deepest pothole in Pirin and the highest cavern between the Alps and the Caucasus. Exploration difficult. Temperature in the first pitch is 1 °C. Four pitches (45, 35, 41 and 16 m), in the first pitch there is icefall.

The first information about the cave was supplied by Polish cavers. Explored in 1976–79 by Speleoclub Akademik, Sofia, up to -132 m. In 1995–1996 cavers from the Speleoclub Helictite, Sofia, discovered new sectors, reaching a depth of 200 m. P. Beron collected cave fauna in August 1972.

Fauna: Opiliones, Chiroptera and the interesting caddis – fly *Micropterna caesareica* (Trichoptera).

A Protected Natural Monument, together with 0.3 ha of adjacent land (Decree 1187/19.04.1976). Situated on the territory of Pirin National Park.

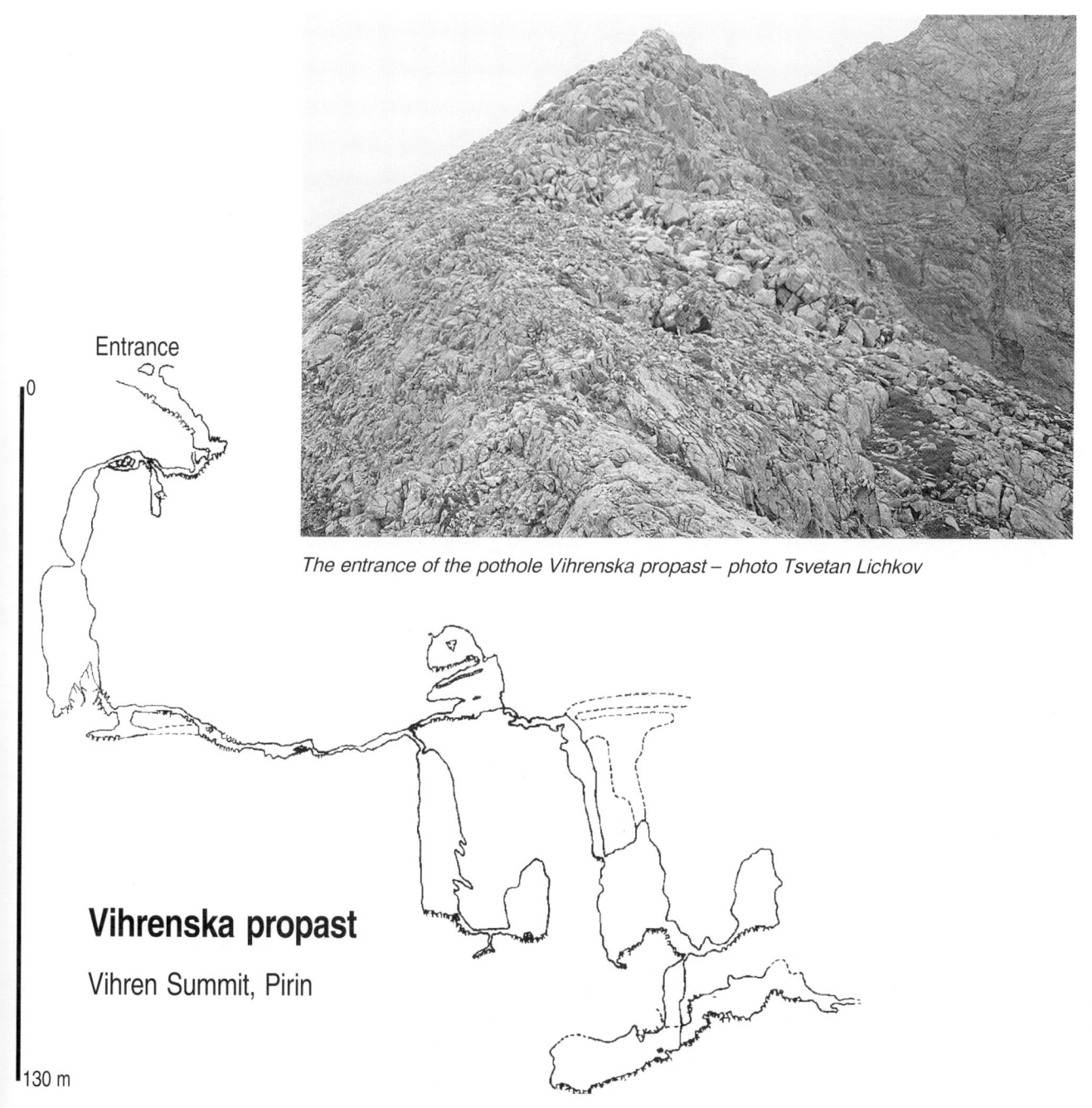

The entrance of the pothole Vihrenska propast – photo Tsvetan Lichkov

Vihrenska propast
Vihren Summit, Pirin

CAVES IN THE RHODOPES

Name: **AHMETYOVA DUPKA (PROKLETATA, DOBROSTANSKI BISER) (240)**
Dobrostan Village, Plovdiv Distr.
Length 43.7 m. Denivelation -14 m

A pothole, 5 km SW of Dobrostan, 0.7 km W of Martsiganitsa Hut, Gogova Padina, Dobrostanski Rid. Formed in Rifean marbles. Surface 704 m². Only one chamber, depth of the entrance hall 12 m. The chamber is rectangular with max. dimensions – 20 m long, 11 m wide, four meters high. Big boulders in the center of the SW part of the chamber. All kinds of sinter pools. Explored for the first time in 1963. Surveyed in 1975 by P. Tranteev, V. Stoitsev and Al. Leonidov. A show cave since 1990. Named after the Turkish oppressor Ahmed Aga.

Ahmetyova dupka
(Prokletata, Dobrostanski biser)
Dobrostan Village

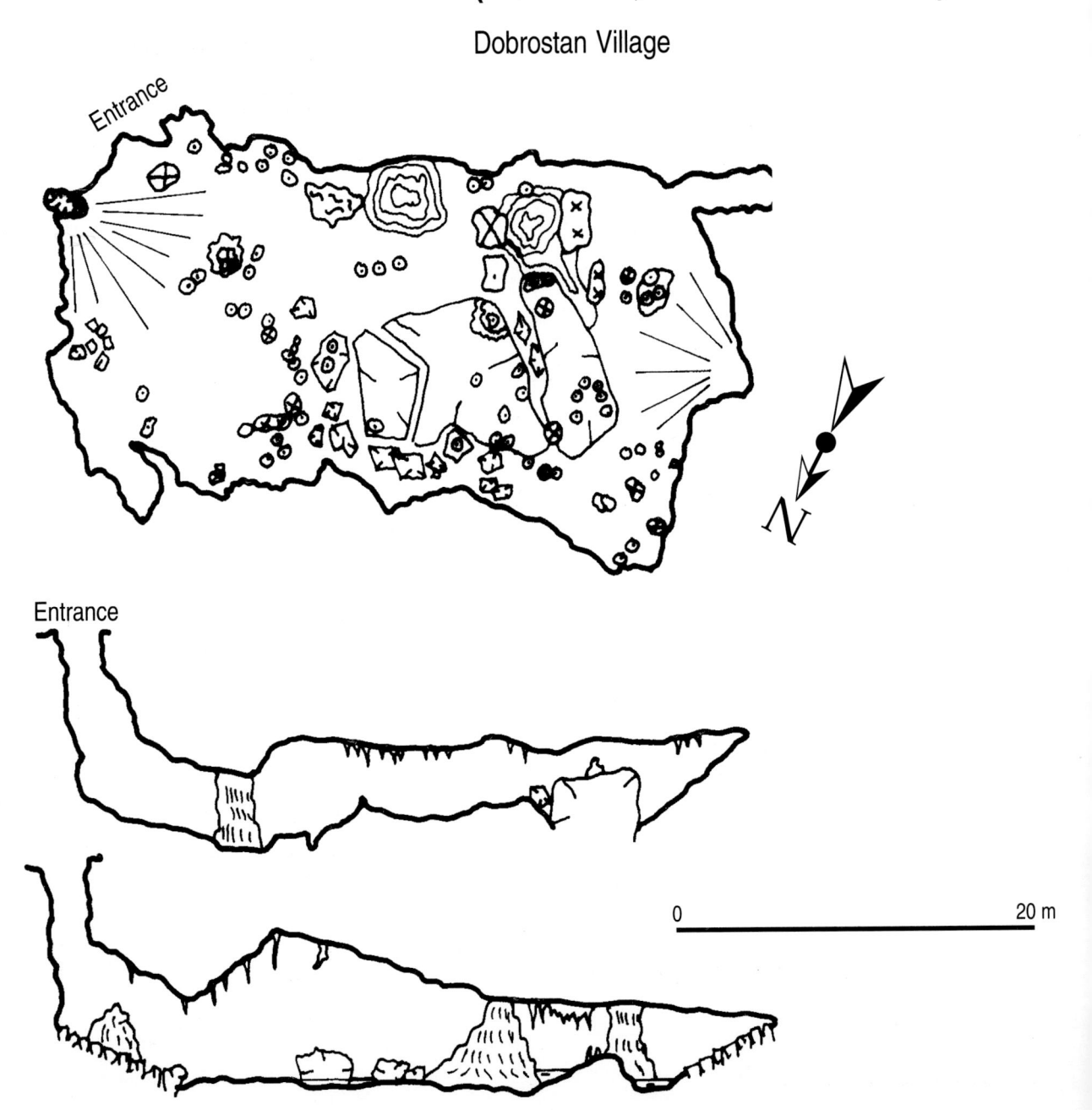

Name: **BORIKOVSKATA PESHTERA (Chervena dupka) (247)**
Borikovo Village, Smolyan Distr.
Length 470 m. Denivelation 10 m
Coordinates: N 41° 29' 09.6" E 24° 36' 06.8"

A cave 1, 5 km S of Borikovo, upstream the river. Height above the LBE – 130 m. The cave entrance, 150 m above the river bed at the foot of a cliff, is a rhomboid, 4 x 1.5 m. The cave in a monotonous one, a one gallery channel, 4–5 m wide with an average height of 3–4 m, 7 m maximum and a total length reaching 470 m. Formed in Proterozoan marble. Up to 270 m, the floor is covered mainly with clay, from 270 to 380 m – mainly with calcite deposits. Only the end of the cave is rich in flowstone formations, the remaining part of the cave is poor in such formations.

The cave is rich in paleontological material.

Borikovskata Peshtera was explored, surveyed and described in 1967 by D. Raychev, G. Raychev, T. Cholakov and other cavers from Chepelare (expeditions of Studenets Caving Club (Rodopski Peshternyak, 34, 1967).

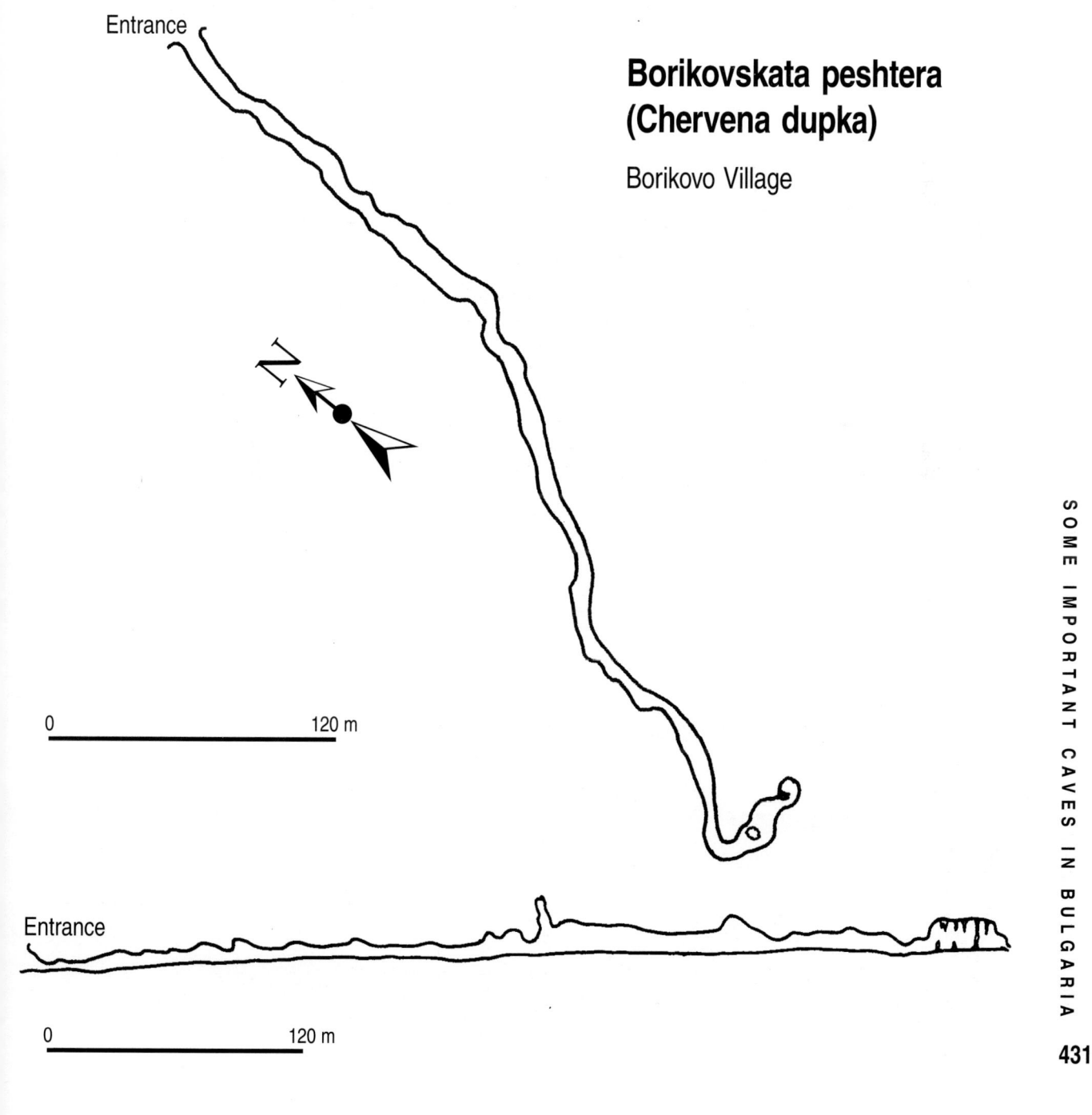

Borikovskata peshtera (Chervena dupka)

Borikovo Village

The explorers in front of the entrance – photo Trifon Daaliev

The river Arda near Mogilitsa village – photo Trifon Daaliev

V. Popov is washing in Arda River the samples from Borikovskata peshtera Cave – photo Trifon Daaliev

Name: **CHELEVESHNITSA (CHELEVESHKATA, CHOVESHKATA, CHILYASHKATA) (233)**
Orehovo Village, Smolyan Distr.
Length 305 m. Denivelation -12 m. Volume 3456 m^3

A cave 0.7 km NW from Orehovo, in the locality Dupleski Kamak, on the left bank of Oreshitsa River, Chernatitsa Ridge. Formed in Proterozoan marble. A descending, three-storey, branched cave. The axe of the main storey is 185 m long. Upper storey 65 m long. Lower storey 33 m. Min-max dimensions of the cross-section of galleries: 0.6–11.6 m wide, 0.3–9 m high. All kinds of flowstone. Dripping infiltration water. The first written information about the cave dates back to 1890, (the school teacher A. Cherpokov). In 1902, studied by G. Bonchev. In July 1924, biological exploration by D. Ilchev, P. Drenski and N. Radev. Studied and surveyed in detail by Chepelare Caving Club 1964, completed by the National Expedition Trite Cheleveshnitsi in 1978. Known from the stories about local inhabitants who hid in the cave and suffocated in a fire somewhere in the 18 or the beginning of 19 century, which is confirmed by the many human bones found in the cave.

Fauna: five species known, including the troglobites *Troglodicus tridentifer* (Diplopoda) and *Lithobius lakatnicensis* (Chilopoda).

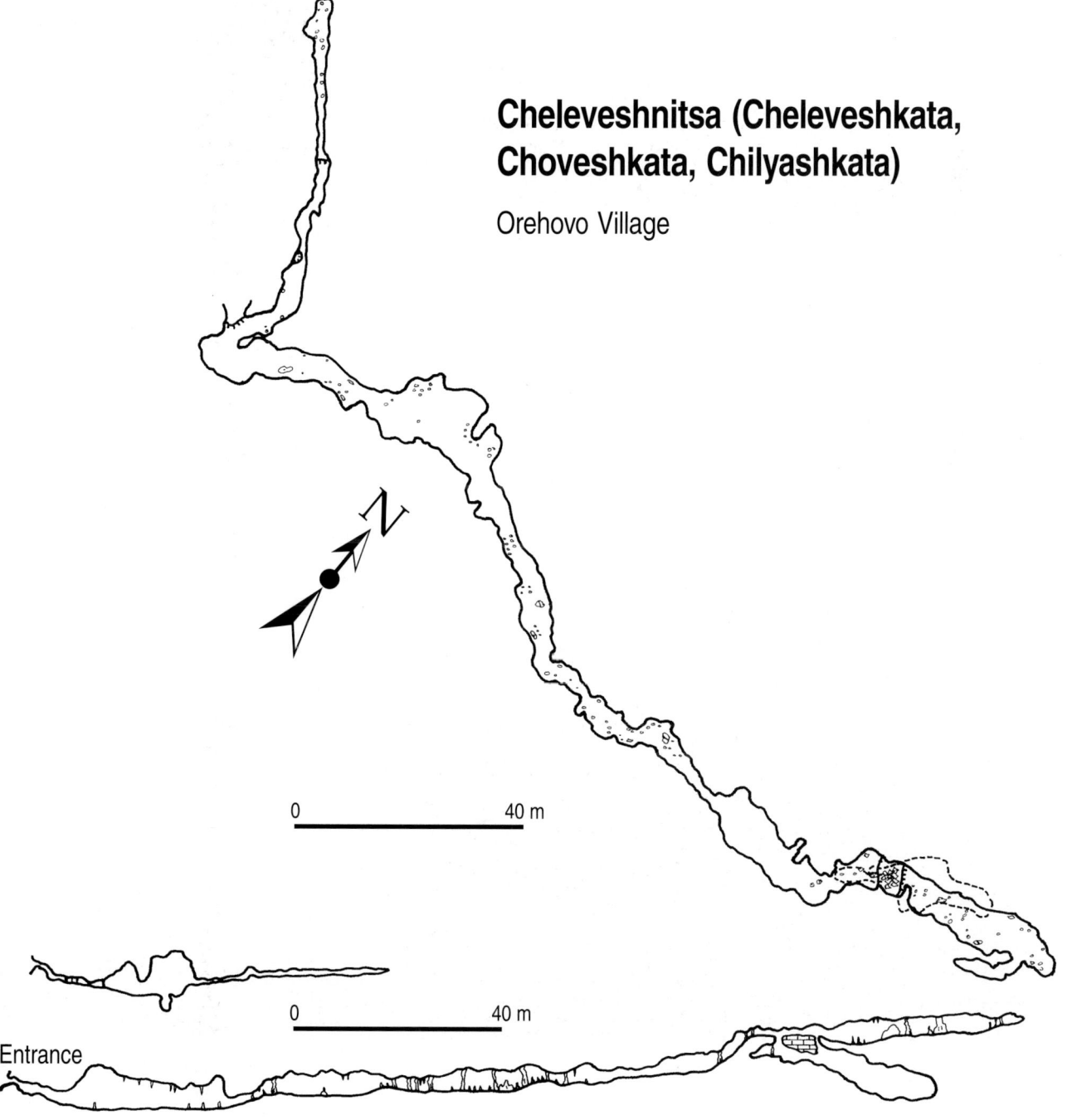

Cheleveshnitsa (Cheleveshkata, Choveshkata, Chilyashkata)
Orehovo Village

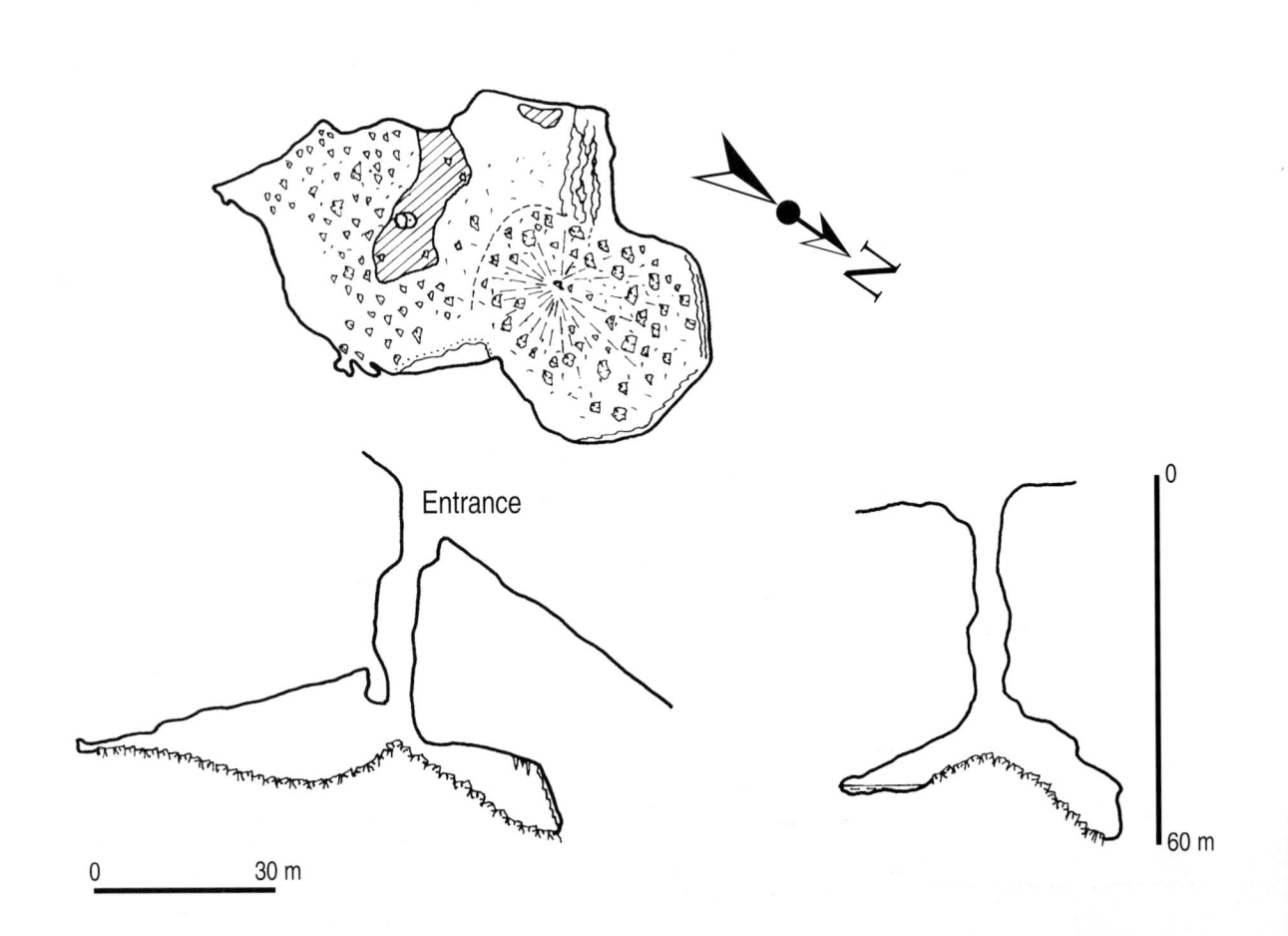

Name: **CHELEVESHNITSA (234)**
Pavelsko Village, Smolyan Distr.
Length 85 m. Denivelation -60 m

The cave – pot hole Cheleveshnitsa is situated 2.4 km south of Pavelsko, on the left bank of a tributary of Chaya River. The entrance has dimensions 7 x 4 m.

The vertical 33 m deep entrance pitch leads straight to the huge only chamber of the cave. It is 85 m long, 40 m wide and 52 m high. It is a strongly faulted zone. Behind a huge cone of crumbled blocks and gravel one can see many shining sinter walls. Some are dry, full of millet-size cave pearls. Others are full of water. Not far is a drained lake, longtime ago being 12 m deep. Behind it there is a stone forest – colonade of stalaktons. The legend says that this is the very place of the descent of Orphaeus underground to look for his beloved Eurydice. Indirect prove for the tragic events in the past here could be the human skeleton discovered by Chepelare cavers during the first exploration of the cave.

In the exploration of the cave took part also the Russian cavers Igor Efremov and Viktor Dublyanski, members of the VI th International caving expedition in 1966. The description was done by Dimitar Sabev and the map by A. Dimitrov and V. Dublyanski. The survey was redone in 1977 by Akademik – Plovdiv Caving Club.

In 1978 F. Filchev noticed in the cave 26 alpine choughs (*Pyrrhocorax graculus*).

Cheleveshnitsa

Pavelsko Village

Name: **CHELEVESHNITSA (CHELOVESHNITSA) (231)**
Zabardo Village, Smolyan Distr.
Length 65 m. Depth 51 m

Pot hole above Zabardo Village. The small entrance is hidden among jinever and rocks in the place called Aydarski Kamak. The sagand says that not far from the pot hole was the castle Zagrad of Gordyu Voyvoda. After the taking of the castle the turks have thrown in the pot hole all defenders still alive, together with the dead and of their horses and weapons. The walls of the castle still stay, and the legend found confirmation after the exploration of the pot hole by cavers from the Second (1962) and the Fourth International Expeditions which took to the surface many human bones and 9 roasted old weapons. The bones were relatively well preserved and show that tall adult males have been thrown there. The type and manufacture of the weapons indicate approximately the end of the Second Bulgarian Kingdom (end of 14th Century).

A 36 m vertical leads to the top of a huge cone of stones and clay, kept in places by loose stone blocks. At the base of this cone have been found the iron weapons. After an incident which almost buried four cavers entry to the pot hole was forbidden.

The map was prepared by P. Tranteev and Hr. Delchev in 1962 during the Second International Expedition and the description – by Dimitar Sabev in his booklet "Rupchoskite peshteri" (1967). The actual map is responsibility of P. Petrov, Zh. Stefanov, V. Gogov, I. Hristova and L. Petrova in 1978 during the expedition "Trite Cheleveshnitsi" ("Orehovo'78").

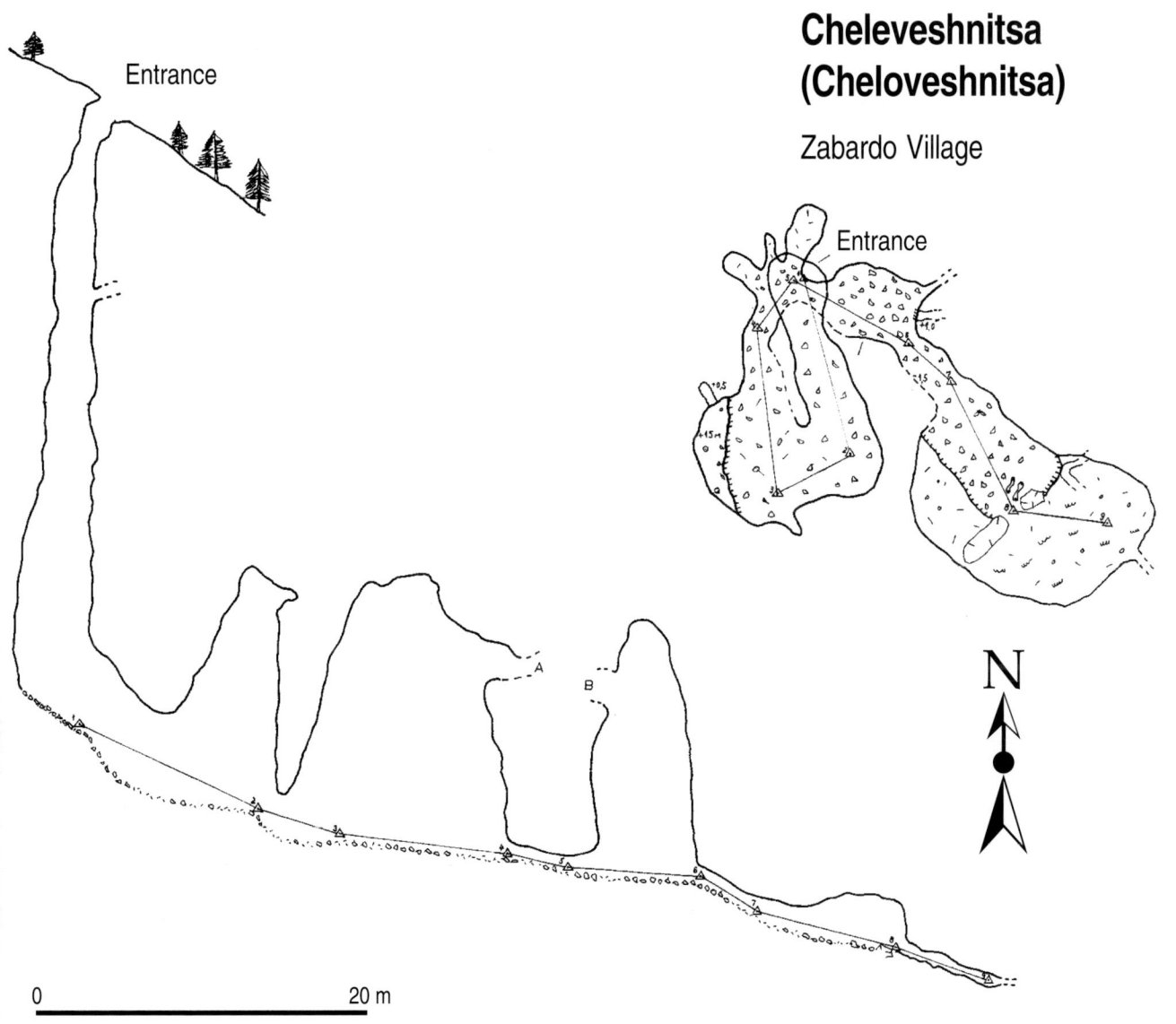

Cheleveshnitsa (Cheloveshnitsa)

Zabardo Village

Name: **CHUDNITE MOSTOVE (Erkyupriya) (232)**
Village Zabardo, Smolyan Distr.
Altitude 1490 m. Length 107 m. Denivelation -16.5 m
Coordinates: N 41° 49' 08.7" E 24° 34' 58"

The natural phenomenon Chudnite Mostove is situated 3.71 km N-NW of Zabardo village, in the middle part of the valley Aydarski Dol, mountain massif Chernatitsa.

Formed in Proterozoan marbles. A system of two rock bridges, 70 m from each other – Upper (Goren) and Lower (Dolen). The upper one is 51 m long and is developed westwards. The entrance faces E and has the following dimensions – width 11 and height 27 m. The other entrance is 52 m wide and 51 m high (the highest cave entrance in Bulgaria). The entrance of the lower rock bridge (Dolnata Peshtera) is triangular and has dimensions as follows: width 5 m and height 26 m. A gallery follows, which is 56 m long and with approximately the same dimensions. First, it goes to the west, then to the north. Along the gallery there are 8 verticals, it's the gallery denivelation reaching 16.5 m. The dimensions of the exit opening are: 5 m wide and 41 m high. Through the upper bridge a stream runs, coming from the upper (E) part of Aydarsko Dere. When the outflow of the stream is small, its water disappears in a ponor, situated 25 m below the exit of the bridge. When in flood, the stream runs further in the valley between the two bridges and passes through the lower bridge (cave) from where it runs further into the canyon downstream.

The famous stone bridge was first recorded and described in 1885 by Stoyu Shishkov (Journal Balgarski Pregled, 9 and 10), then by K. and H. Škorpil in their book "Krazhski Yavleniya" (1900 c). Later, studied many times. The actual map was made in September 1990 by A. Jalov from Heliktit – Sofia and G. Ilchevski from Studenets – Chepelare Caving Clubs.

Declared a Natural Monument, together with 39.7 ha of adjacent land (Decree 2813/08.11.1961).

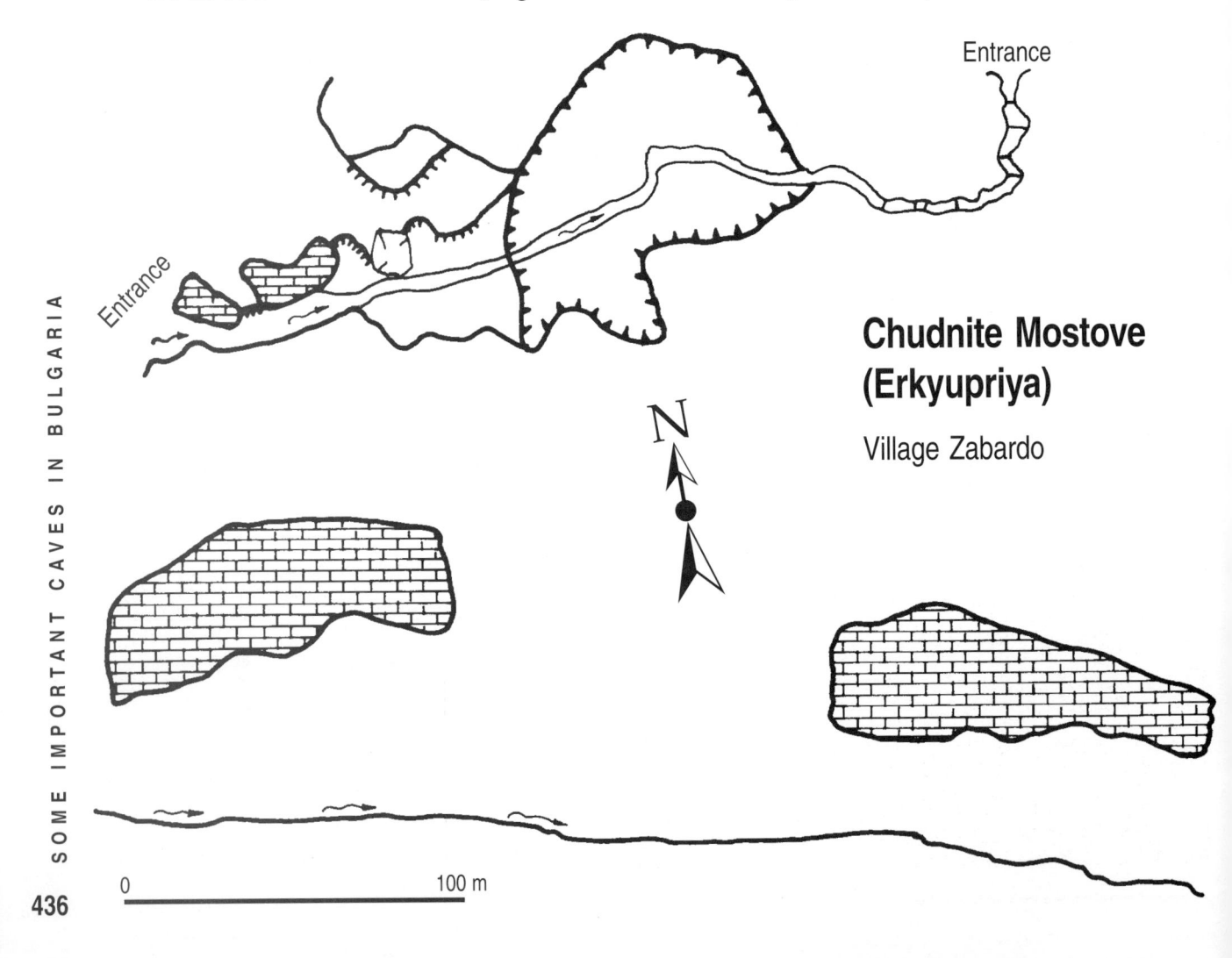

Chudnite Mostove (Erkyupriya)

Village Zabardo

Chudnite Mostove – the highest entrance of a stone bridge in Bulgaria – photo Alexey Jalov

Name: **DRANGALESHKA (DANGALASHKA) DUPKA (228)**
Mugla Village, Smolyan Distr.
Length 1142 m. Denivelation – 255 m
Coordinates: N 41° 38' 50.7" E 24° 28' 12.6"

A cave entrance situated 80 meters above the river, 20 meters from the road. Dimensions 1.5 x 2 m. The entrance vertical is a 45m pitch, expanding bell-shaped. Several other verticals lead to the "old bottom" at – 165 m.

The new sectors, discovered in 1991–1994, are to the side in the middle of the third pitch at – 110 m. These parts are muddy, watery and narrow. The cavers penetrating there should be equipped with water suites, with reliable light and have to be well trained. The river outflow is 30–50 l/s. The underground river of Drangaleshka Dupka is 80 m under the level of Muglenska reka.

The cave was explored and surveyed first on May 15, 1965. First penetrated by Ts. Lichkov, V. Gyaurov and in 1991 cavers from Sofia (Ts. Ostromski and M. Zlatkova) discovered a descent -111 m from the entrance. The bottom of the system is 175 metres below the river level. Fauna collected by P. Beron in 1998.

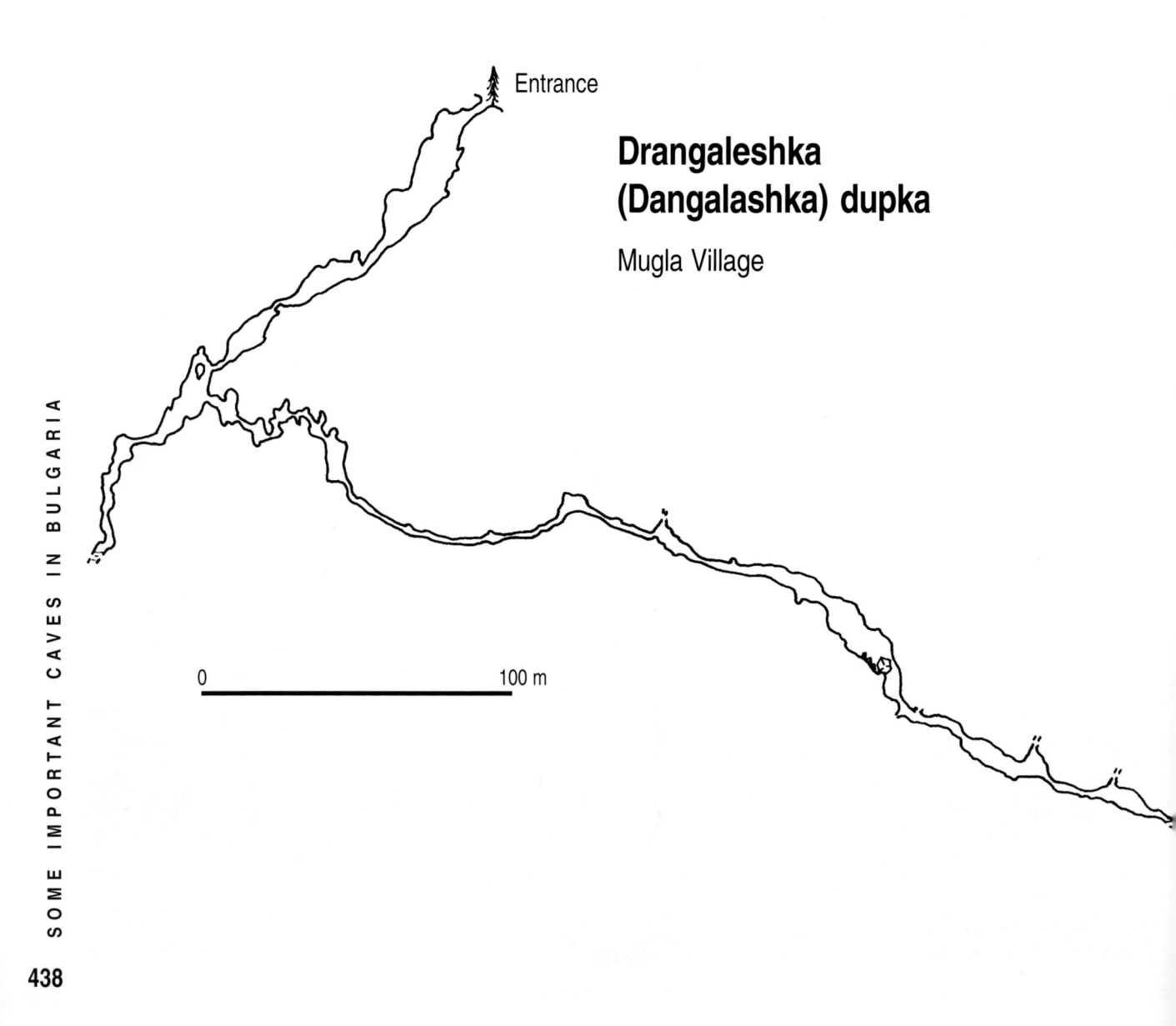

Drangaleshka (Dangalashka) dupka

Mugla Village

The entrance of the Drangaleshkata cave – photo Trifon Daaliev

The pothole Drangaleshka dupka

Trigrad Gorge – photo Trifon Daaliev

Name: **DRUZHBA (SVETI HRIS) (241)**
Dobrostan Village, Plovdiv Distr.
Altitude 1305 m. Denivelation -130 m

A pothole in Proterozoan marble, about 30 minutes SW of Martsiganitsa Hut, in the locality Sveti Hris. A round entrance, diameter ca. 3 m. This shape is preserved up to the 25th meter, where a platform of 2–3 m^2 is reached. From the platform a vertical of 95 m starts (according to the survey). Dimensions of the bottom: 10 x 26 m, many fallen rocks. Its NW side could be eventually penetrated, and a new part is to be expected, moreover that the local base of erosion is at least 200–300 m lower. Very few cave formations. T° 7.3 °C.

The pothole had been known to local people for a long time. In 1962, at the end of the II International Expedition, the first attempt to descend it took place (P. Beron and K. Kowalski got up to 25 m, because of the lack of time and climbing gear). Named Druzhba (Friendship), in honour of the international expedition. In the same year, the caver from Plovdiv A. Siderov descended to the bottom and declared the pothole depth at 211 m. On November 10,1968, N. Genov and I. Rashkov surveyed the pothole, recording a depth of 130 m.

Fauna: among the few cave animals are the large Diplopod troglophiles *Balkanopetalum beskovi*.

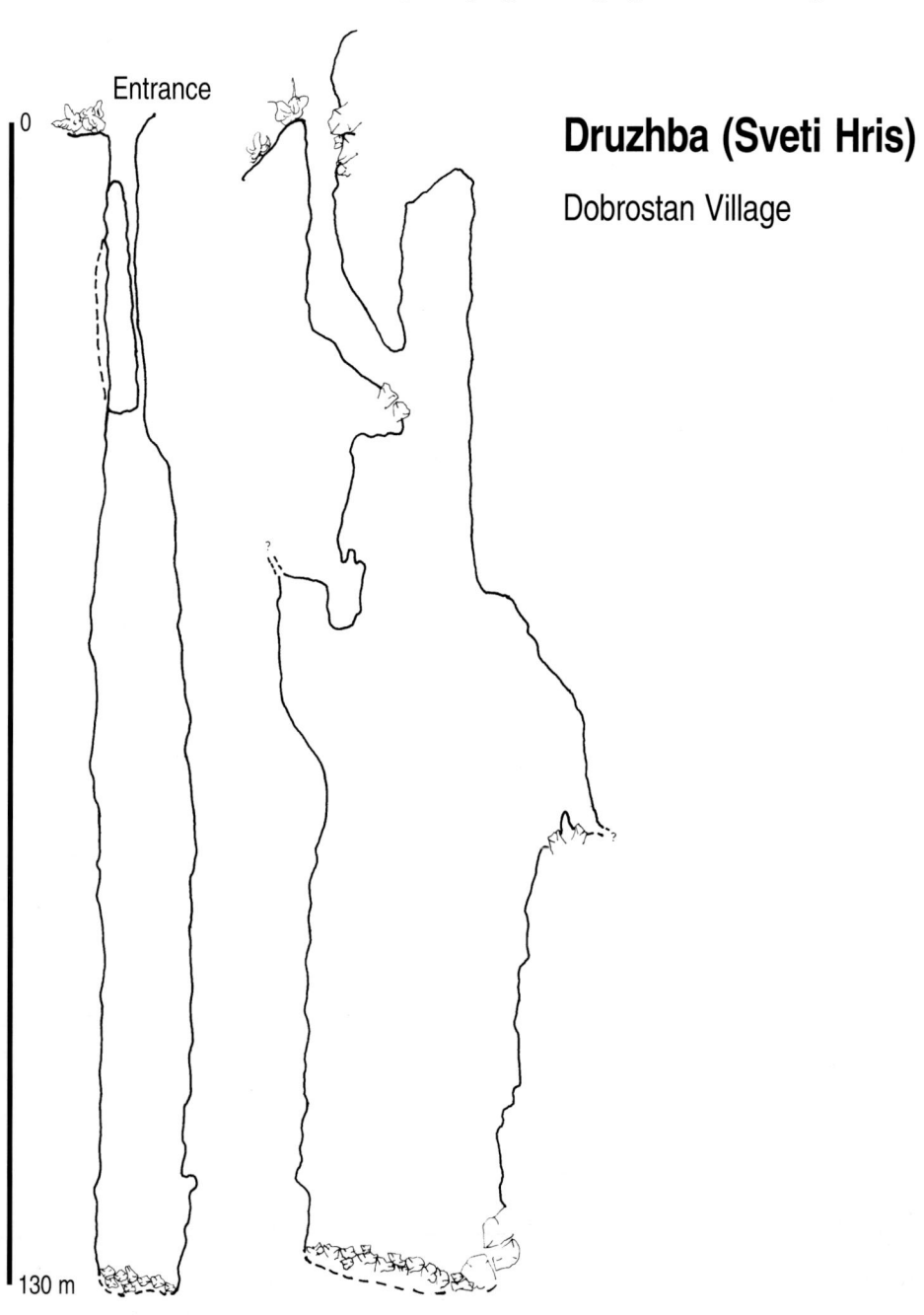

Druzhba (Sveti Hris)

Dobrostan Village

Name: **DYAVOLSKOTO GÂRLO (HARLOGA) (226)**
Trigrad Village, Smolyan Distr.
Altitude 1150 m Lenght 480 m Denivelation 89 m
Coordinates: N 41° 36' 46" E 24° 22' 53"

    A pothole in Proterozoan marble, at the beginning of the spectacular Trigrad Gorge, 1.5 km from Trigrad Village. A large entrance leads into the first shaft (diaclase of 51 m). The entrance waterfall is 34 m high, now avoided by using an iron ladder, leads to a big chamber, called "Roaming" (100 x 40 m). In the pothole, there are 15 waterfalls, all of them terminated by evorsion cauldrons. From the end of the Big Chamber to the end of Dyavolskoto Garlo, 272 m of horizontal galleries have been surveyed, intersected by 13 cataracts, 0.60 to 8 m high, with a total denivelation of 38 meters. In this part a dingy is needed. When in flood, the amount of water entering the pothole is over 2300 l/s. In the dry period, this quantity is about 300–320 l/s.

    In the summer of 1961, N. Korchev and his group undertook the first attempt to penetrate Djavolskoto Gârlo, but the difficulties were too many for the technique of that time. In 1969, 7 cavers from Cherni Vrâh Caving Club (Sofia), led by A. Petkova, reached the bottom (V. Gogov and V. Kitov). Two divers drowned in the end siphon in 1970 A dangerous system, requiring good equipment and skilled cavers. A show cave, which is open all year round.

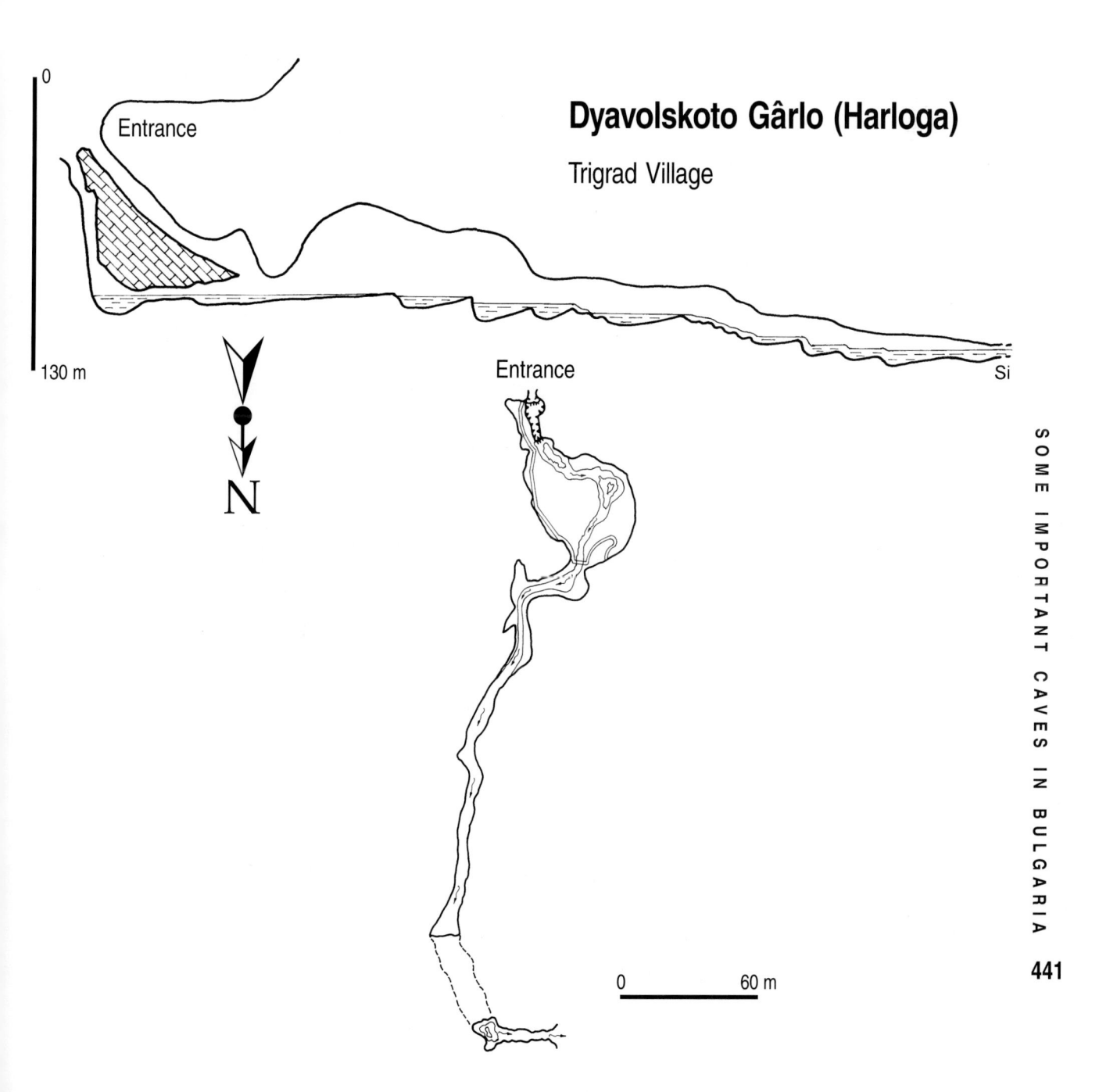

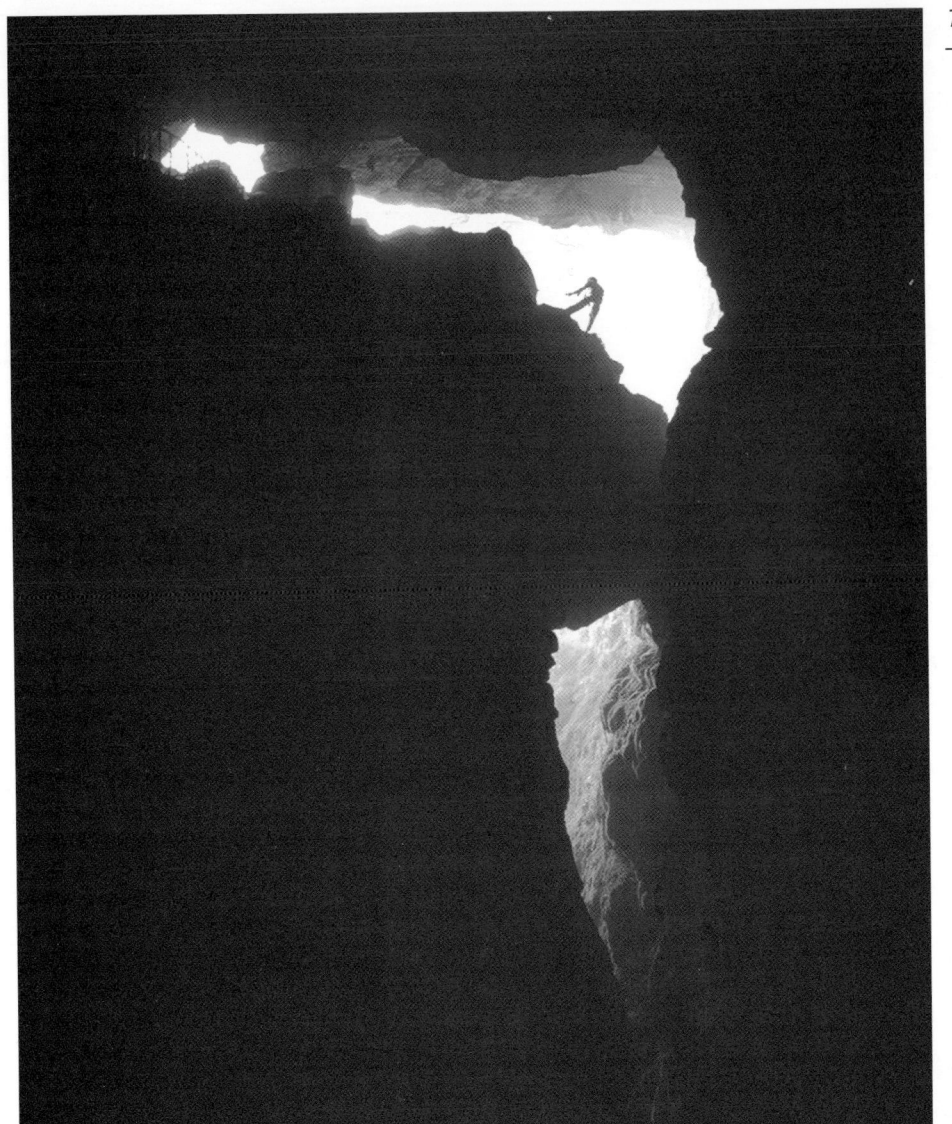

The rappelling down to the cave – photo Vassil Balevski

P.Tranteev and V.Markov are mapping the cave – 1975 – photo Vassil Balevski

SOME IMPORTANT CAVES IN BULGARIA

442

Name: **EMINOVA DUPKA (222)**
Orpheus Hut, Smolyan Distr.
Total length 635 m. Denivelation -35 m

A two-entrance cave about 3 km SE of Orpheus Hut, on the left bank of Kastrakli Gorge, in the karstic cliff above the road to Teshel, immediately after the water joint with Eminov Dol valley. The bigger entrance is crescent-shaped and faces west, towards Eminov Dol. It can be reached only with a rope. The upper entrance (easier to enter) is on the edge of a 20m cliff above the Arrow Cave. To find it one needs a guide.

The cave begins with a vertical, for which a 20-m rope is needed. After a 2 m obstacle, a wide and lit gallery is reached. About 10 m on the left is the main entrance. About 10 m on the right, the gallery ends with flowstone formations. Before that, a narrow branch on the left leads to a wide and high gallery, parallel to the previous one. After 12 m on the left a sinter waterfall is reached. We can crawl for another 7 m in a squeeze, after which the continuation has not been explored. A strong draft is an indication for the existence of a third entrance.

In the other direction, the gallery gradually lowers. Its most narrow part was dug out by Ts. Ostromski and S. Petkov on May 1, 1992.

Then the gallery widens and after 20 m we reach a wide diaclasic gallery, high up to 10–12 m and with many flowstone formations. On the right, climbing on dry sinters, we can enter a gallery with a level floor with dendrites. After 150 m, there is a big boulder mass and the gallery leads to the left. Before the bent there is a 50m branch on the left.

Following the main gallery, after 50 m we reach another crossroad. Straight on after 7–8 m there is a 18m vertical. In its upper end there is an ascending 20 m gallery.

After climbing to the upper level and then descending, the right branch (the main gallery), leads us to the most beautiful part of the cave. The ceiling lowers and, after a nice lake, the way further is blocked. The tree roots suggest that the surface may be near.

Another big gallery goes NE. Seventy (70) m after the chamber, on the left there is a low and narrow branch, 55 m long, ending in beautiful flowstone formations. After another 40 m along the main gallery, a branch on the left follows, with a 15m vertical and a horizontal gallery in the middle of the vertical.

Hydrogeologically, this subregion depends on the water from the source near Teshel. The river Krichim forms its hydrogeological boundary on the south. The marble on both sides of Kastrakliysko Dere border forms a picturesque gorge up to 300 m deep. Karst water sinks and re-appears in different places, which is a sign of deeper karstification and of a chance to find also other caves.

The cave pothole Eminova Dupka was discovered by St. Petkov and Ts. Ostromski through digging in 1992. The cave was explored and surveyed in the same year by Ts. Ostromski, S. Petkov, M. Zlatkova, Zdr. Iliev, G. Raychev, N. Milev and other cavers.

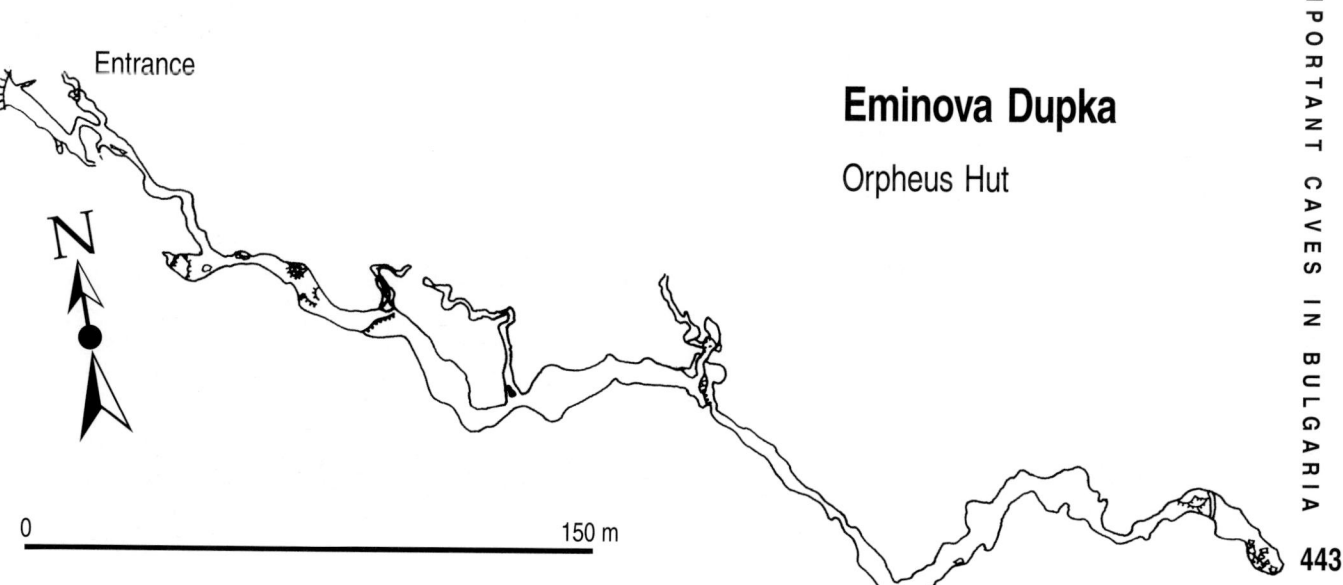

Eminova Dupka

Orpheus Hut

Ts. Ostromski in front of the cave entrance – photo Tsvetan Ostromski

Concretions – photo Tsvetan Ostromski

Name: **GARGINA DUPKA (Garvanitsa) (239)**
Mostovo Village, Plovdiv Distr.
Altitude 905 m. Length 524 m. Denivelation 38 m
Coordinates: N 41° 51' 01.4" E 24° 56' 02.2"

A cave by the village Mostovo. The village is situated on a rock terrace of white marble, terminating in an abrupt cliff. Underneath, the two entrances of Gargina Dupka are situated. The cave name comes from the alpine choughs (*Pyrrhocorax graculus,* in Bulgarian "Haydushki gargi"), nesting near the entrance.

The cave has two entrances. A stream runs out the lower (A) entrance, the upper entrance is overgrown with *Hedera helix*. After the entrance, the gallery goes up until reaching the first chamber. It is high and lit by the light which comes from the second entrance. In spring and late in the autumn, as well as after long rains, in this place a big shallow lake is formed. The way inside is through a 2.5m waterfall. After a meandering gallery and big boulders, we reach the Bat Chamber, with guano. The second gallery goes right. The floor is muddy, the walls are covered with dendrites and secondary formations. Through two shallow wells, it is possible to reach the underground stream. The way further is hindered by the low ceiling (15 cm). The siphon upstream was penetrated first in 1980. Some 20 m of new galleries have been were discovered, together with a new impenetrable siphon. After the wells, the cave becomes lower and ends. After a steep climb on slippery guano in the Bat Chamber, we reach a narrow ascending gallery. After a five-meter vertical, we enter a chamber. A narrow blind gallery follows.

The first survey of the cave was done in 1970 by Vl. Popov and B. Ilyuhin. Another, more complete, map was made in 1983 by members of Akademik Plovdiv Caving Club.

Fauna: many bats, choughs, guano fauna (15 species). No troglobites.

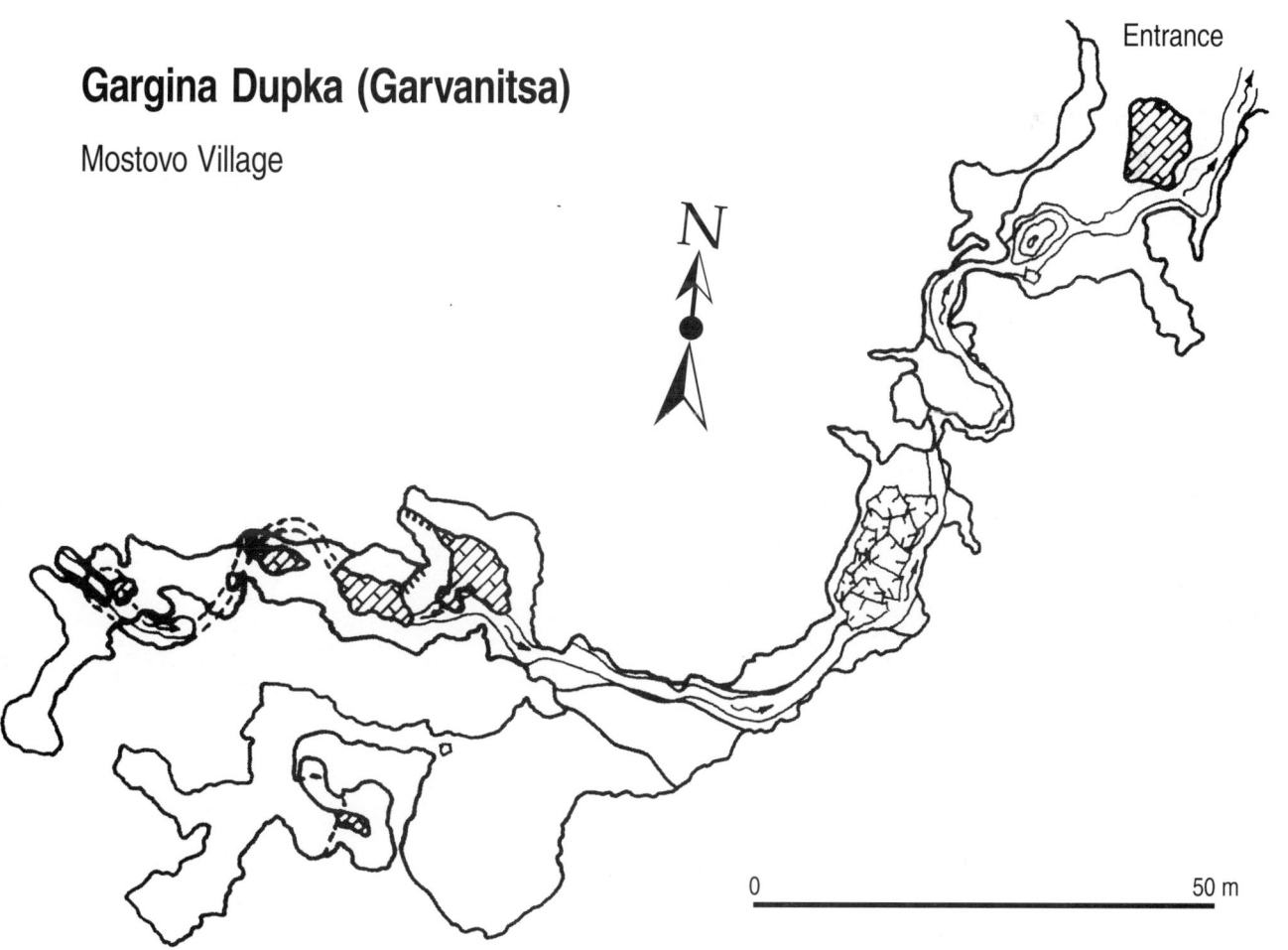

Gargina Dupka (Garvanitsa)

Mostovo Village

Name : **GARVANITSA (KOSOVSKATA PESHTERA) (235)**
Kosovo Village, Smolyan Distr.
Length 897 m. Denivelation 30 m

An ascending cave in marble near Kosovo Village in the locality Dalbokoto Dere. Mentioned for the first time in writing as early as 1895, 1898 and 1900 by K. and H. Škorpil in their books "Sources and Sinks" and "Karst phenomena in Bulgaria". The first recorded visit was made by cavers from Plovdiv, led by At. Siderov, in 1961. The cave was explored and surveyed first in 1962 by cavers from the Second International Caving Expedition and by Polish cavers. In 1978, during the Republican Expedition Trite Cheleveshnitsi another attempt (up to 180 m) was made by Z. Iliev, A. Jalov and L. Adamov. A complete map of the cave was prepared in the period August 9–19, 1983 by cavers from Akademik Plovdiv Caving Club, led by Iv. Petrov.

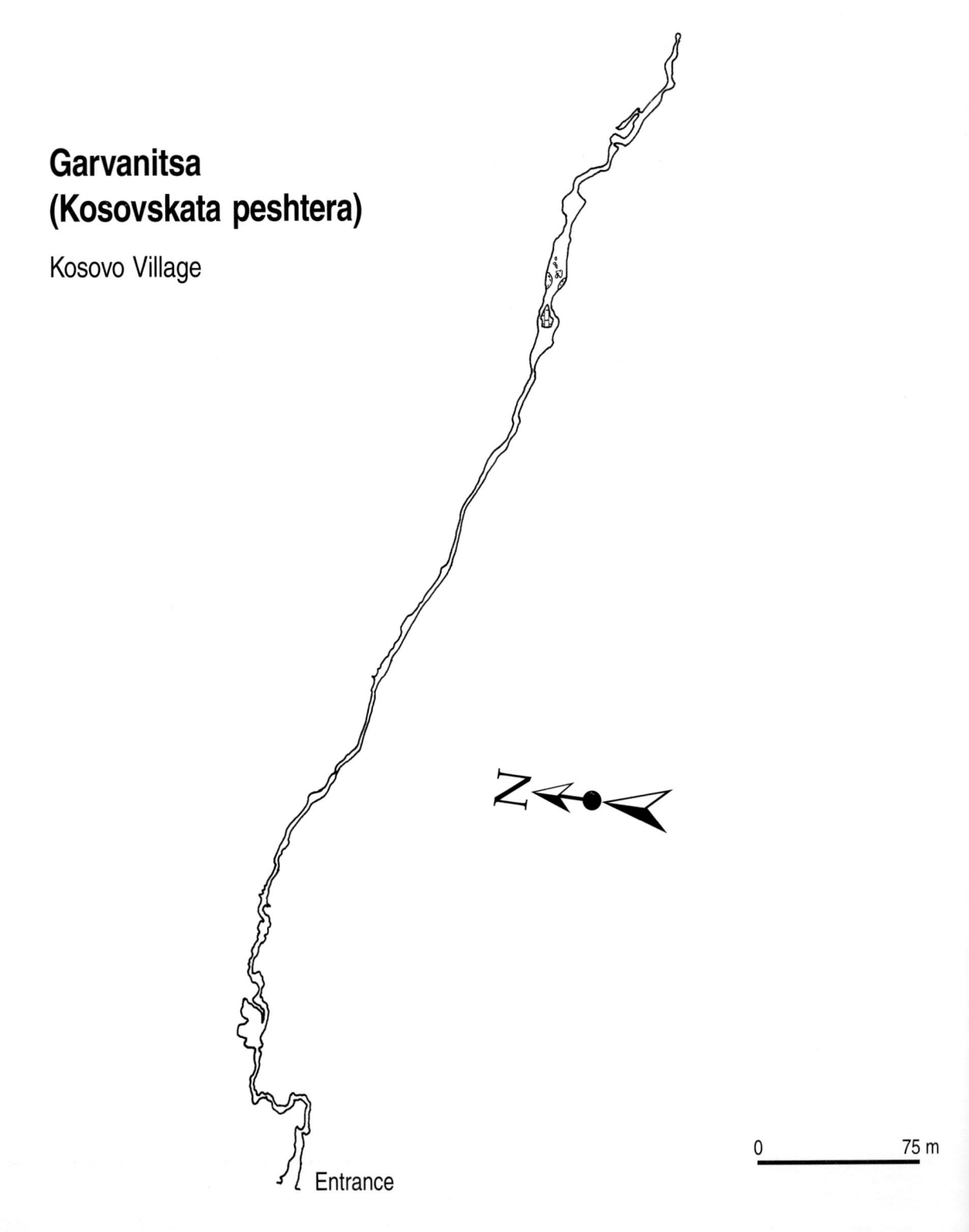

**Garvanitsa
(Kosovskata peshtera)**

Kosovo Village

Name: **GOLOBOITSA 1 and 2 (248)**
Koshnitsa Village, Smolyan Distr.
Total length 500 m. Denivelation 20 m

Caves about 3 km from the village Koshnitsa, in Garga Dere. These are two separate caves and a source, interconnected and formed in Proterozoan marble. A river with a non-constant outflow of 30–3000 l/s runs out of the lowest situated cave gallery, which is 16 m long. The connection between the source and the underground stream in the cave Goloboitsa 1 has been proved by an indicatory experiment, carried out by D. Raytshev in 1967. The second entrance of the cave Goloboitsa 1 is situated 57 m above the karstic source. Only the first 20 m of the caves are flooded. About 30–40 m before the entrance, the underground stream disappears in a siphon and reappears as the karstic source Goloboitsa. About 100 m from the entrance, the ceiling lowers and the lumen between it and the water table is small. In flood, this lumen is filled and in this place a siphon is formed. Later, in some places the gallery becomes higher than 10 m and wide up to 5–6 m. The depth of the underground stream is ca. 1 m, and its width is 1–2 m. It ends in a siphon. The bottom is covered with sand and the water is transparently clear. No attempt to overcome it has ever been made. Water T° is 8 °C.

The third entrance, situated 10 m above Goloboitsa 1, gives access to Goloboitsa 2. Both caves are connected at the beginning by a 12m vertical. The total length of this cave is 140 m. The first 70 m of the gallery are horizontal, the remaining part to the end is inclined. Besides several narrow places after the entrance, the gallery is 5–6 m wide and 6–7 m high, and it ends in a narrowing. After the first third of the cave, stalactites are observed, together with other flowstone formations.

The caves were explored, surveyed and described in 1967 by D. Raychev, G. Raychev, T. Cholakov and other cavers during the expeditions of Studenets Chepelare Caving Club (the Journal Rodopski Peshternyak, 34, 1967).

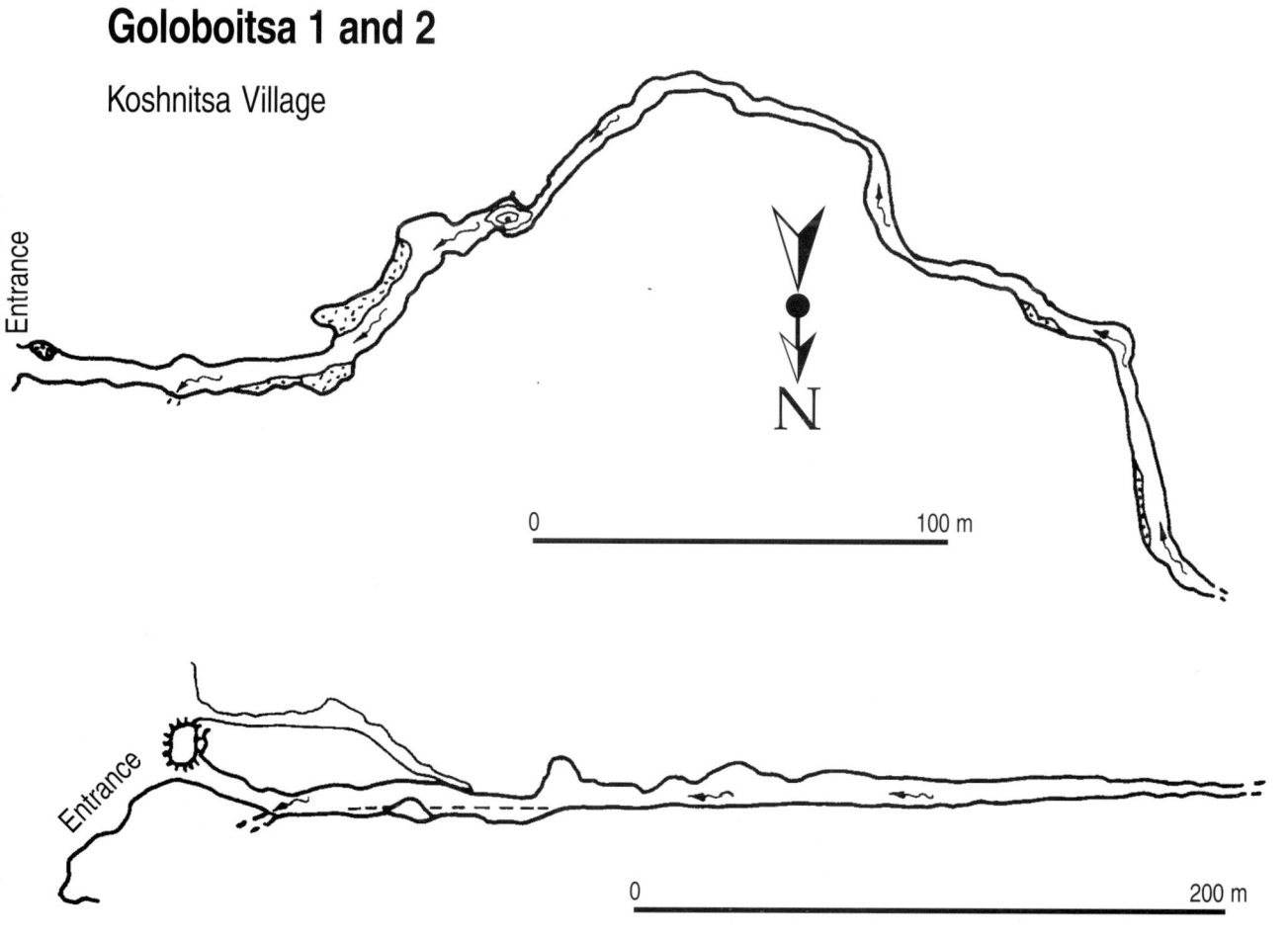

Goloboitsa 1 and 2

Koshnitsa Village

Name: **HARAMIYSKA DUPKA (227)**
Trigrad Village, Smolyan Distr.
Length 510 m. Denivelation -42 m

A cave one kilometer N of Trigrad, on the right bank of Trigradska river. Formed in Proterozoan marble. A two entrance (tunnel), two-storey cave. Until 1977 known as two separate caves: Haramiyska I and Haramiyska II. The entrance of the first one is 20 m above the base of a cliff. A horizontal gallery follows at the end of which a 34 m high wall opens, connected with an ascending gallery, reaching Haramiyska II. Poor in flowstone. Clay deposits up to one meter thickness. Dripping water. In some seasons there are mighty ice formations. First studied in 1924 by D. Ilchev, P. Drensky and N. Radev. In 1965, surveyed by L. Popov (National Expedition). In 1967, cavers from Studenets Caving Club again explored the cave and in 1977 they connected the two parts. Artefacts dated from the Eneolith to the Bronze Age were found during archaeological excavations in 1981–1983 by Hr. Valchanova.

Fauna: four species known, including the Dipluran *Plusiocampa bulgarica*.

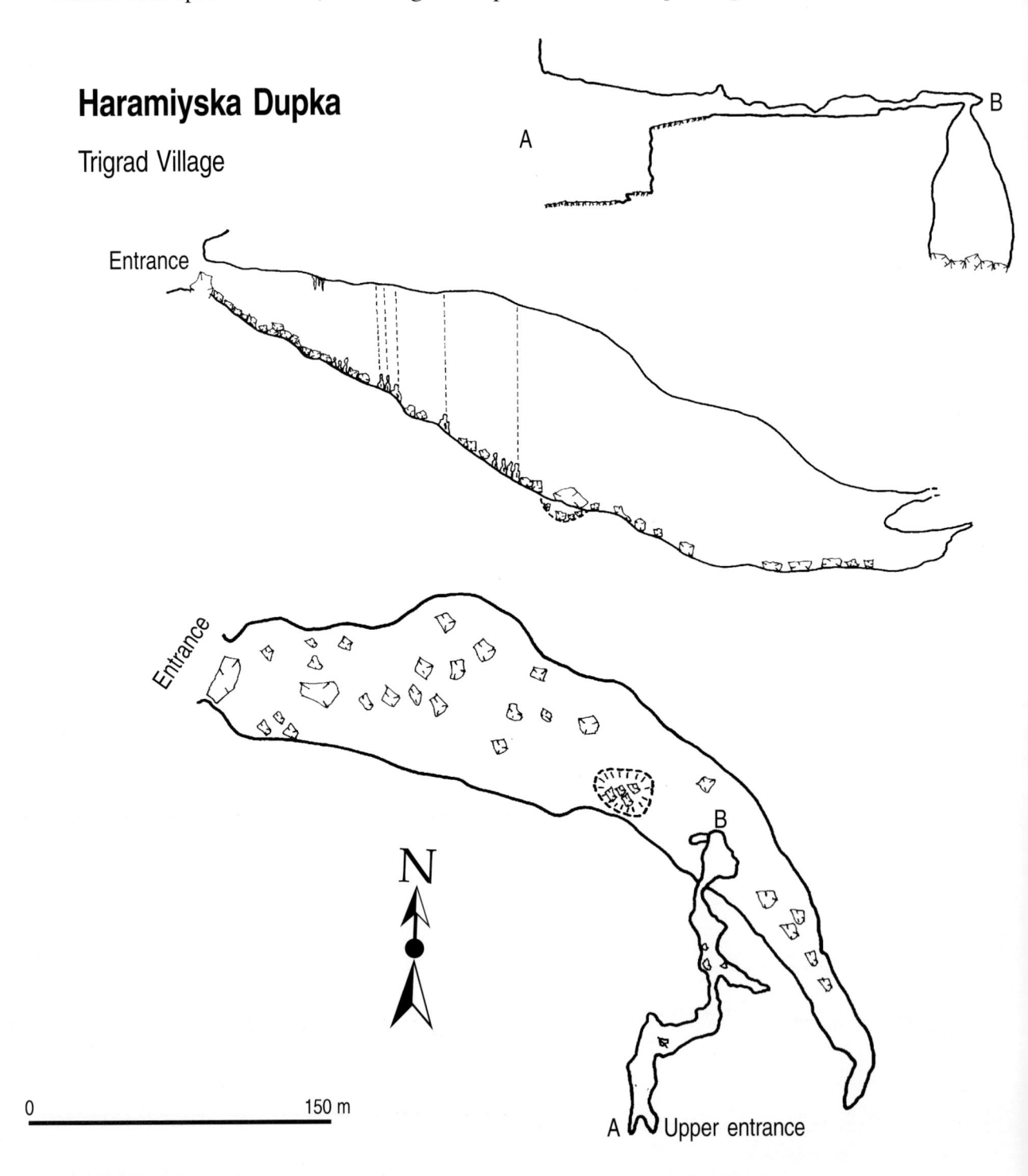

Name: **HRALUPA (Hralup) (245)**
Dobrostan Village, Plovdiv Distr.
Altitude 860 m. Length 311 m. Denivelation 10 m

A cave in marbles from the Precambrian, 4.5 km SW of Dobrostan, in the locality Kozi Rog, high above the left bank (geographically) of the river Sushitsa, Dobrostan Karst Massif, in the east end of the North-Rhodopean Anticline.

The cave has two entrances. The bigger one faces west, the smaller – south. A horizontal gallery without branches follows and ends with blocks. The cave is rich in secondary flowstone formations.

There is a legend that the corpse of a dead man from the village had been thrown in the cave, which was proved when human bones and a knife were found (description of A. Jalov after Radmil Pandev).

The cave was explored by the Second International Expedition in July 1962. Surveyed also in 1970 by B. Kolev from Aida Haskovo Caving Club. So far 8 animal species have been identified from this cave, including the troglobite Isopod *Rhodopioniscus beroni*.

Hralupa (Hralup)

Dobrostan Village

Name: **IVANOVA VODA (243)**
Dobrostan Village, Plovdiv Distr.
Length 695 m. Denivelation 131 m

A cave pothole in the locality Vodite, near Martsiganitsa Hut, in the lowest part of a huge doline. Formed in Proterozoan Marble. It starts with a descending gallery up to 40 m long, ending at the edge of a 45-m shaft. From the bottom of the shaft a winding horizontal gallery starts, reaching a 5–6 m vertical, leading to an underground lake. In rare cases there is no water in it. The cave is rich in flowstone.

In 1962, during the Second International Expedition Hr. Deltchev and St. Andreev reached the end of the cave and collected (together with P. Beron) cave animals. They surveyed the cave. Animals were collected also by British cavers in 1967 (published by M. Hazelton in 1970). Stygobite is *Niphargus* sp.

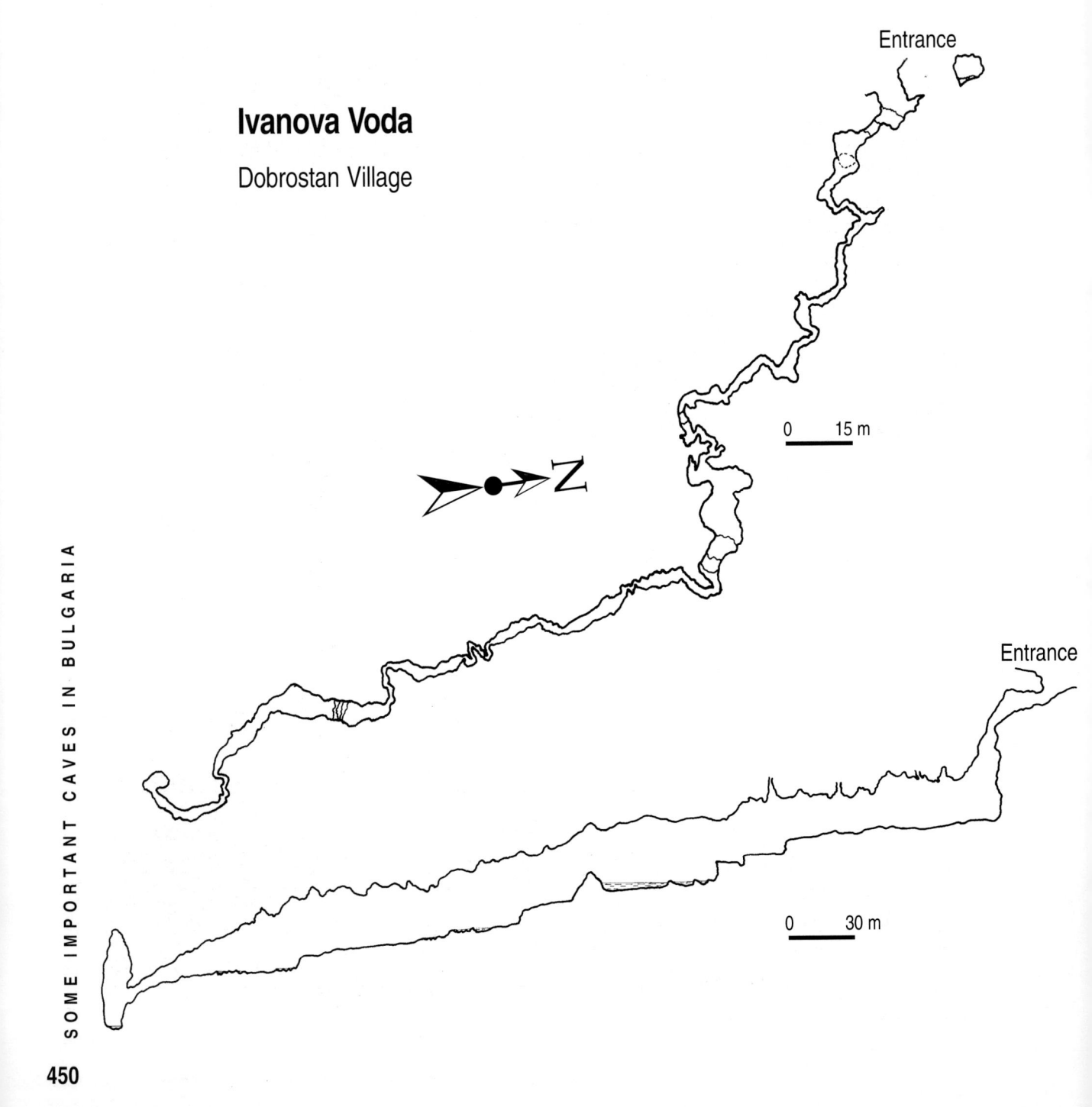

Ivanova Voda

Dobrostan Village

V. Nedkov and N.Gladnishki before the penetration – photo Trifon Daaliev

Name: **IZVORA (IZVORA NA KASTRAKLI) (223)**
Borino Village, Smolyan Distr.
Length 2480 m. Denivelation 25 m

A cave near Orfey Hut, in the Kastrakli Gorge. D. Raychev penetrated the first 150-200 m of the cave. In 1992, cavers from Sofia (St. Petkov, Zdr. Iliev, M. Zlatkova, T. and S. Medarovi, Ts. Ostromski, T. Ivanova and others) started their research.

It is possible to reach the cave via the old road from Teshel to Orphey Hut (up to 2 km from Orphey Hut still good for going by car). The entrance is a periodical source on the right bank of Izvorska River, one meter above the river (dry in summer), under the road ca. 200 m before the junction with Svinski Dol Valley.

The first 180 m of the cave are entirely below the level of the entrance. When in flood a long siphon is formed and the penetration is impossible. The observations in 1992–2002 have shown that the cave is dry in the second half of summer, in autumn and in winter. Penetration during rain or abrupt warming in winter is not recommended, because of flood hazard. It has been shown that the cave gets water from the sink of Izvorska River below Orphey Hut.

The cave is barely branching, horizontal on the main axis, with several ascending branches to the right. In many places on the main axis when there is no water several sand siphons are formed. They have to be dug out every year. So far 8 have been penetrated, the second one (from the entrance) being the longest (16 m), the sixth is the deepest (14 m). So far 2480 m have been surveyed. Another 300 m (non surveyed) have been explored, without reaching the end.

Until 1988 Izvornata Peshtera Cave was a permanent karstic source in Kastrakliysko Dere Valley. After the obstruction of the sinks above the valley the source became periodical and stays dry for 6–7 months every year. Near the entrance remains a small lake. There is a second source, which drains the ca. 200 m long siphon parts in the first part of the cave. May be the running water will be reached after several more sand siphons, so the most interesting parts of the cave are to be discovered.

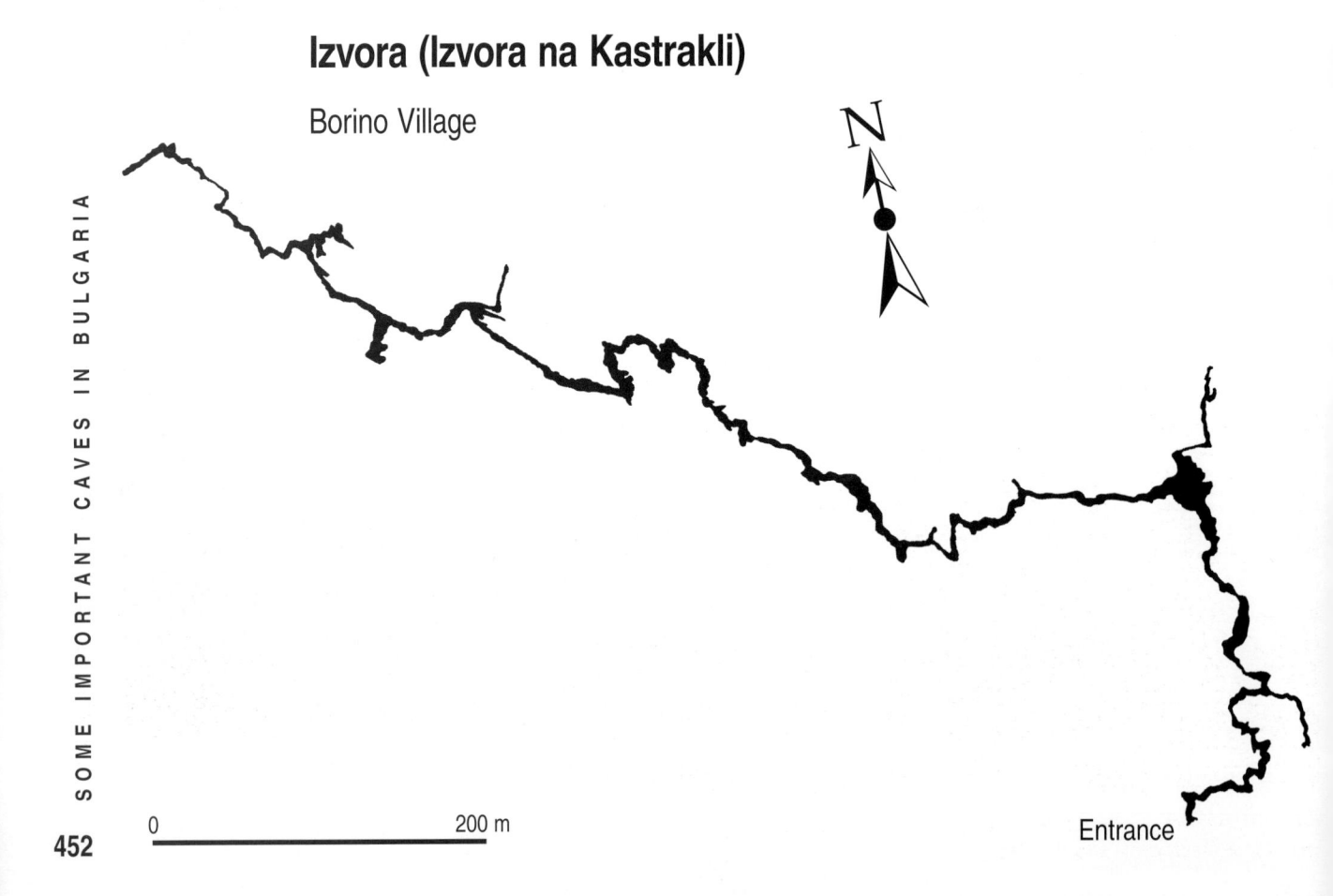

Izvora (Izvora na Kastrakli)

Borino Village

I. Tachev and N. Milev before exploration – photo Tsvetan Ostromski

Needle-shaped stalactites are located in some galleries and rooms – photo Tsvetan Ostromski

Yagodinskata peshtera – photo Vassil Balevski

Name: **KAMBANKITE (PROPAST M-4) (229)**
Mugla Village, Smolyan Distr.
Altitude 1625 m. Length of the main gallery 456 m. Denivelation -158 m

A pothole in Chamla area of Trigrad karstic region, between the villages Mugla and Trigrad, in Domus Gyol locality. Formed in Proterozoan marble, in a fault area, separating the marble from the sediments from the Paleogene. The entrance is in a karst funnel with a diam. of 15 m and height ca. 8 m, from which a gallery starts. There are five shafts in this pothole. The entrance vertical is 10 m, the second is 10 m, the third is 22 m, the fourth is 18 m and the fifth is 25 m deep. The verticals are interconnected by horizontal galleries.

The pothole was discovered during a club expedition of Sredets Sofia in the summer of 1987 by K. Stoyanov, K. Garbev, Y. Georgiev and others.

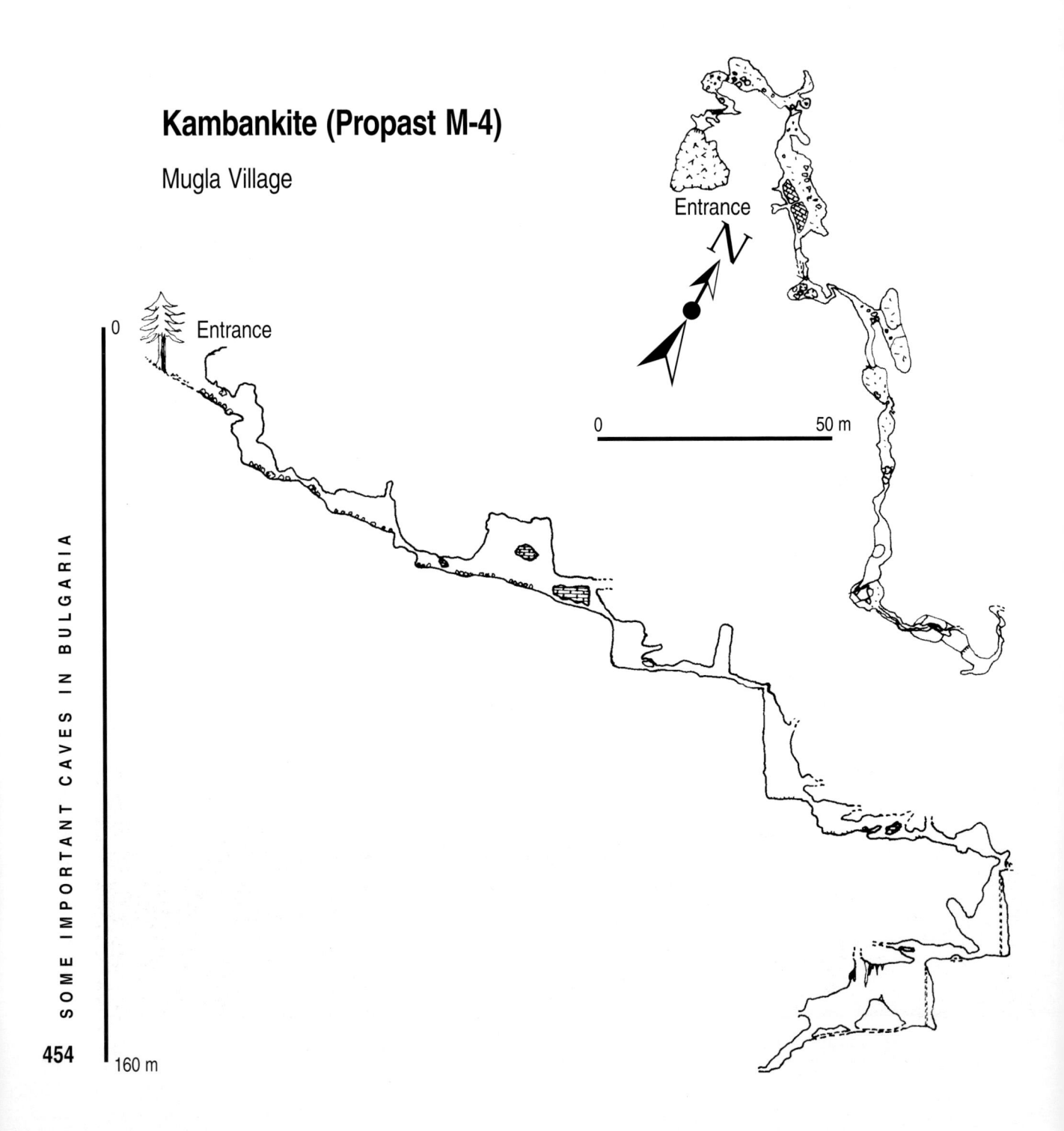

Kambankite (Propast M-4)

Mugla Village

Name: **KARANGIL (251)**
Shiroko Pole Village, Kardjali Distr.
Length 490 m

A three-storey cave near Shiroko Pole, below the medieval castle Hissar. The big chamber is 120 m away from the entrance. It is the former bottom of an underground lake, but after the building of the nearby fountain the water drained and disappeared. On the floor of the gallery of the first floor, there are rocks blocking the advancement of cavers. We reach the second floor after overcoming a 10meter chimney by rope or ladder. On the floor, there are stones and bat guano, the air is dry and warm. There is a danger of stone falls. Via a five-meter narrow passage we pass from the second to the third floor, which is wet and with a permanent temperature. There are flowstone formations on the ceiling and walls.

On the floor of the gallery there is ceramics from the Early Bronze Age. There is a hypothesis that the third floor was used in this period by the local tribe to preserve dairy and other products. As it is difficult to believe that the jars were carried from the first floor, presumably another entrance existed. Through a careful search, an entrance leading directly to the third floor has been discovered. This entrance is the one, used by the Thracians.

The first floor was surveyed by B. Kolev. All three levels were surveyed by M. Gumarov, G. Siderov, Y. Hristov and R. Popov during the expedition "Shiroko Pole'86".

Description after M. Gumarov.

Fauna: 10 species identified. The spider species *Centromerus milleri* has been described from this cave. Troglobite is the beetle *Bureschiana drenskii* (Col., Cholevidae).

Karangil
Shiroko Pole Village

Name: **KLADETO (249)**
Rudozem, Smolyan Distr.
Length 263 m. Denivelation -147 m

A pothole on the eastern slope of Kavgadjika Ridge, in the locality Kladeto. Formed in sediments from the Paleogene (breccia – a conglomerate of marble pieces with clay-carbonate matrix, among which metamorphytes are observed. Sinter formations, red clay, small gravel and sand, tectonic and corrosion-gravitation boulders.

Discovered and explored by Studenets Chepelare Caving Club in 1988 and surveyed by A. Baldjiev, M. Tranteev and G. Raychev.

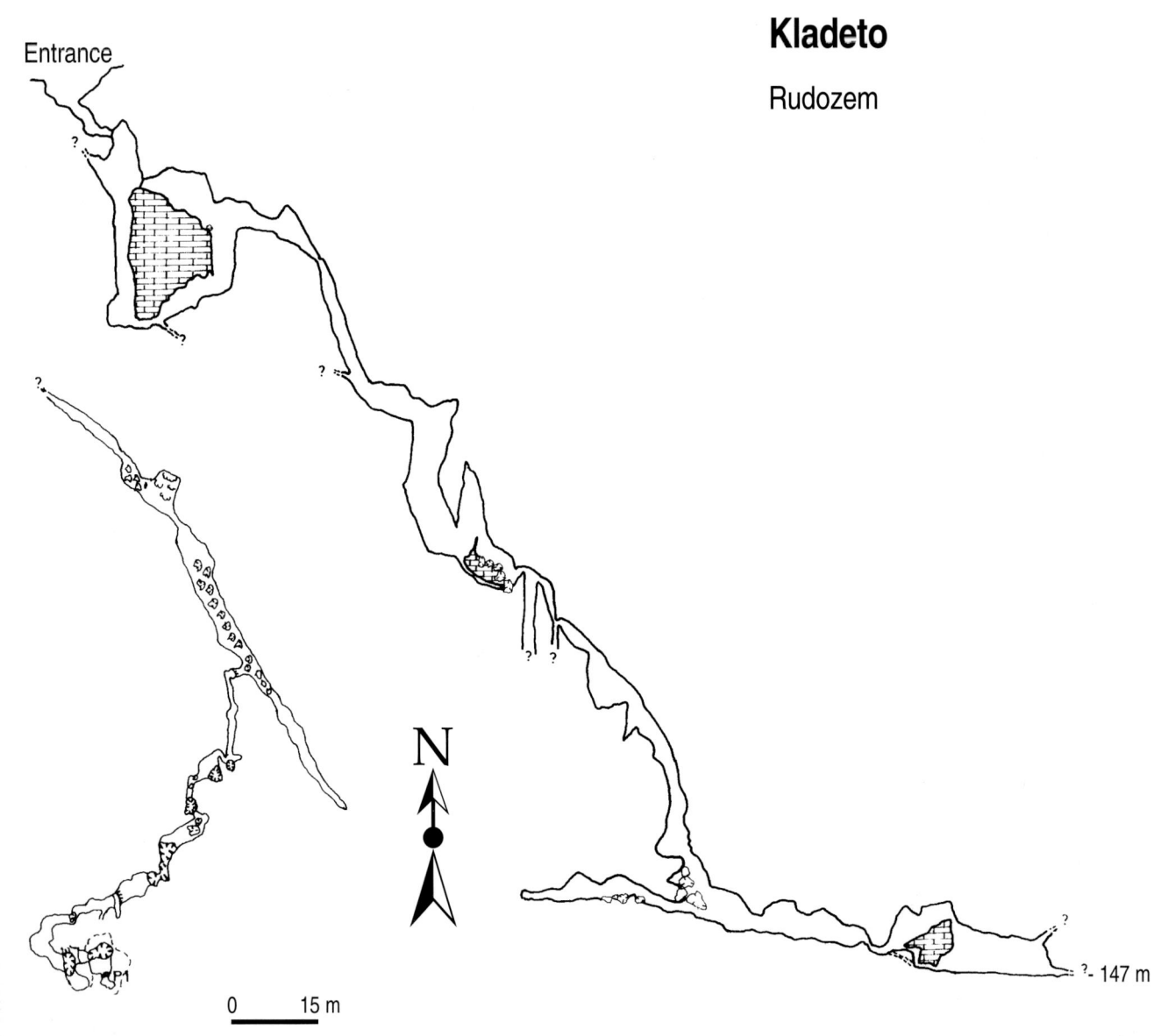

Kladeto

Rudozem

Name: **LEDNITSATA (230)**
Gela Village, Smolyan Distr.
Altitude 1538 m. Length 1419 m. Denivelation 108 m
Coordinates: N 41° 38' 52.8" E 24° 31' 27'1"

Cave in Proterozoan marble, 3.9 km SW of Gela, in the locality Kekeva, on the right bank of Krushov Dol.

From the entrance, a large descending gallery leads into a spacious chamber, 65 m long and 25 m wide, high up to 8 m. In the chamber, almost throughout the year ice formations can be found. Then, the gallery follows the underground stream in northeast direction. All galleries have rich decoration of aragonite and other cave formations.

After more than one kilometer, the stream joins another stream with a much bigger outflow, coming from a siphon (360 l/s). Another bigger underground river flows first SW, then NW and reaches another siphon, which has not been penetrated yet.

The first know description of the cave was due to an anonymous author in 1902. The first map of the entrance part was published by the zoologist P. Drensky, followed by the article of Ivan Popnikolov "Attempt for a morphological and climatic description of Lednitsata" (Bâlgarski Turist, 1929). In 1954 geologists led by Prof. D. Jaranov penetrated deeper into the cave, claiming a length of 860 m. Actually, the bottom of the cave was reached on September 3, 1964, by D. Sabev, D. Raychev, M. Raycheva, Hr. Delchev, T. Sarâmov and others. It took them 13 days. In 1965, D. Raychev and D. Sabev published a special issue of the journal Rodopski Peshternyak (Lednitsata, 99 p.), with a detailed description of the cave.

Protected by virtue of Decree 1799/02.07.1968 (together with 4.6 ha of adjacent area).
Fauna: Known stygobite *Speocyclops rhodopensis* (Copepoda).

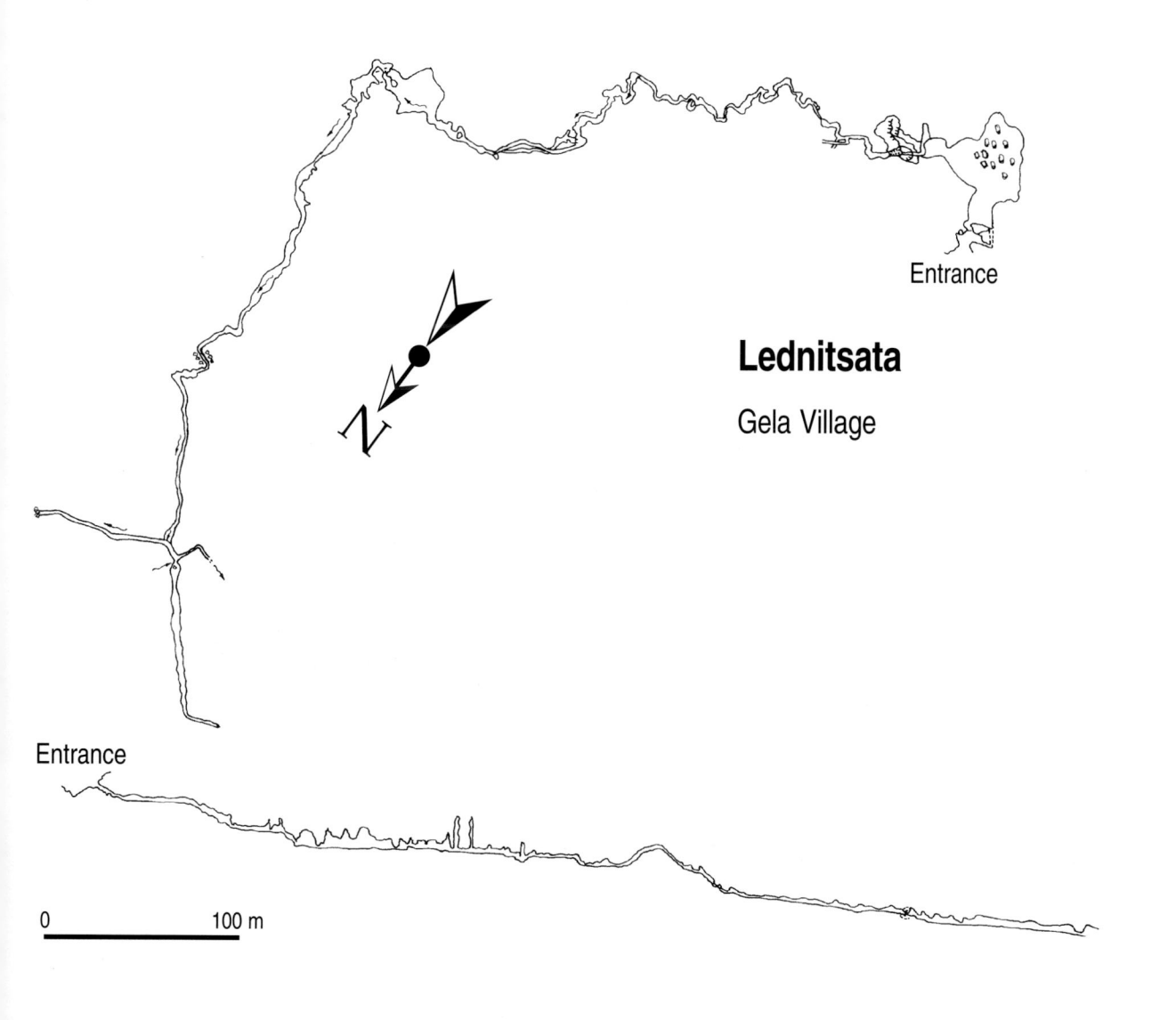

Lednitsata
Gela Village

Name: **LEPENITSA (VODNATA, MOKRATA, IZVORA) (216)**
Velingrad Town, Pazardjik Distr.
Length 1525 m
Coordinates: N 41° 57' 14" E 24° 00' 43.2"

A water cave in Protozoan marble, 13 km SE of Velingrad, in the area of Rakitovo, on the left slope of Chukura River. The first known research was done by people from Rakitovo. In 1921 the founders of Rakitovo Caving Society again explored the cave. In the period 1925-1927, P. Drenski and Iv. Buresch carried out the first biospeleological research. The first 400 m of the cave were surveyed by Eng. Pavel Petrov, A. Popov and Iv. Andonov in 1931. Another map of 1100 m of the cave was prepared by the National Expedition in 1960. The cave was explored and surveyed in details by P. and B. Tranteev in 1971-73.

Declared a National Monument (together with 5 ha of adjacent area by Decree of 1963).

Fauna: 33 species known, including the troglobites *Niphargus bureschi* (Amphipoda), the beetle *Duvalius* (*Paraduvalius*) *bureschi* (Coleoptera, Carabidae) and the Dipluran *Plusiocampa bulgarica*.

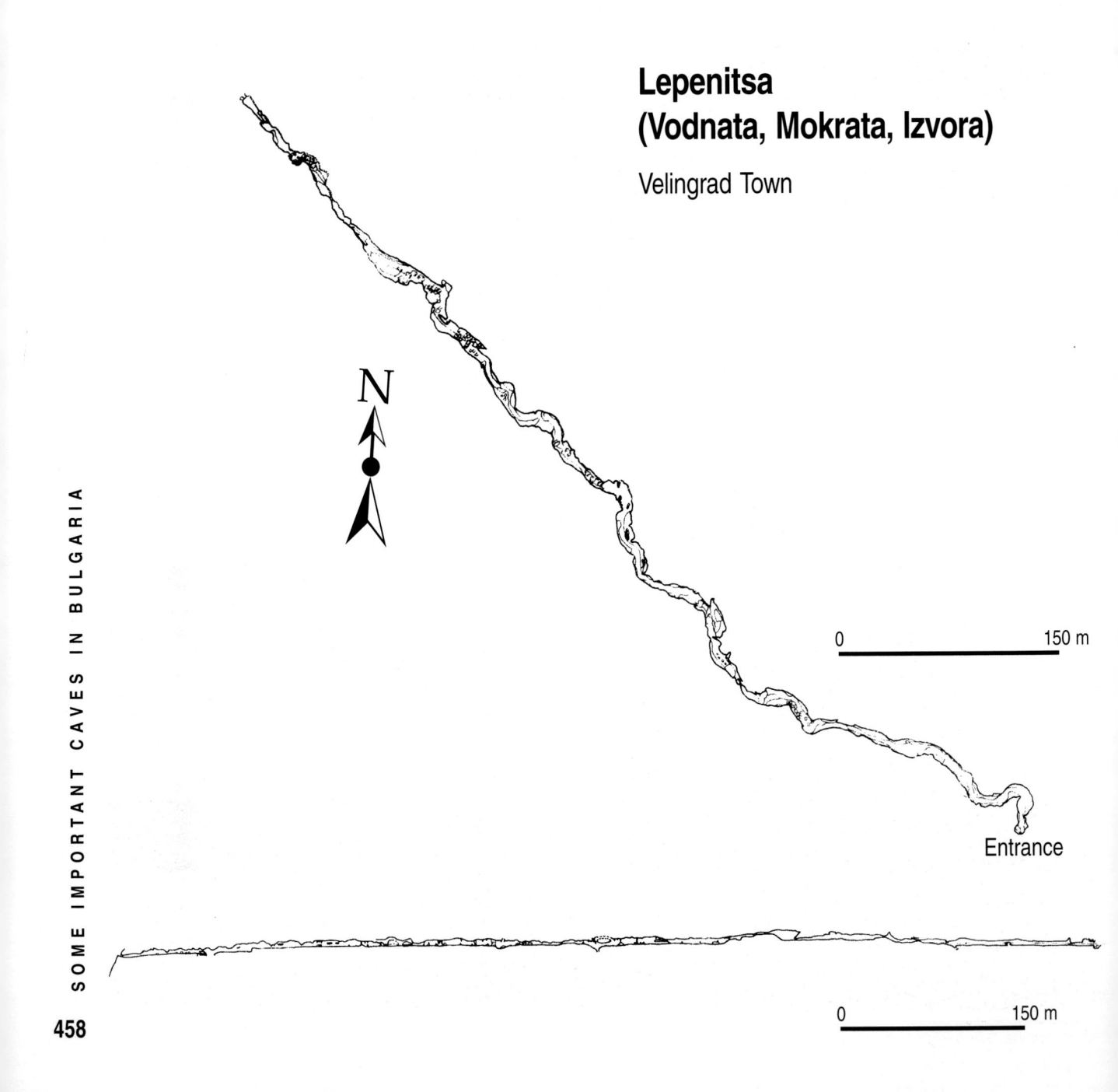

**Lepenitsa
(Vodnata, Mokrata, Izvora)**

Velingrad Town

Valeri Peltekov climbs in a chimney in the cave Lepenitsa – photo Vassil Balevski

The river passage – photo Trifon Daaliev

Name: **LISEK (PANTYOLOVA) (236)**
Dryanovo Village, Plovdiv Distr.
Length 39 m. Denivelation -164 m

A pothole in the area of the village Dryanovo, seven km from the town Laki. Formed in Proterozoan marble. There are two verticals of more than 80 m each. Flowstone and big stalactites in both. On the bottom of the second pitch many flowstone forms, dendrites and other formations.

Discovered by S. Nenov and other cavers from the expedition in 1983 on information by a local shepherd. Ten years later, cavers from Iskar Sofia Caving Club discovered a parallel shaft.

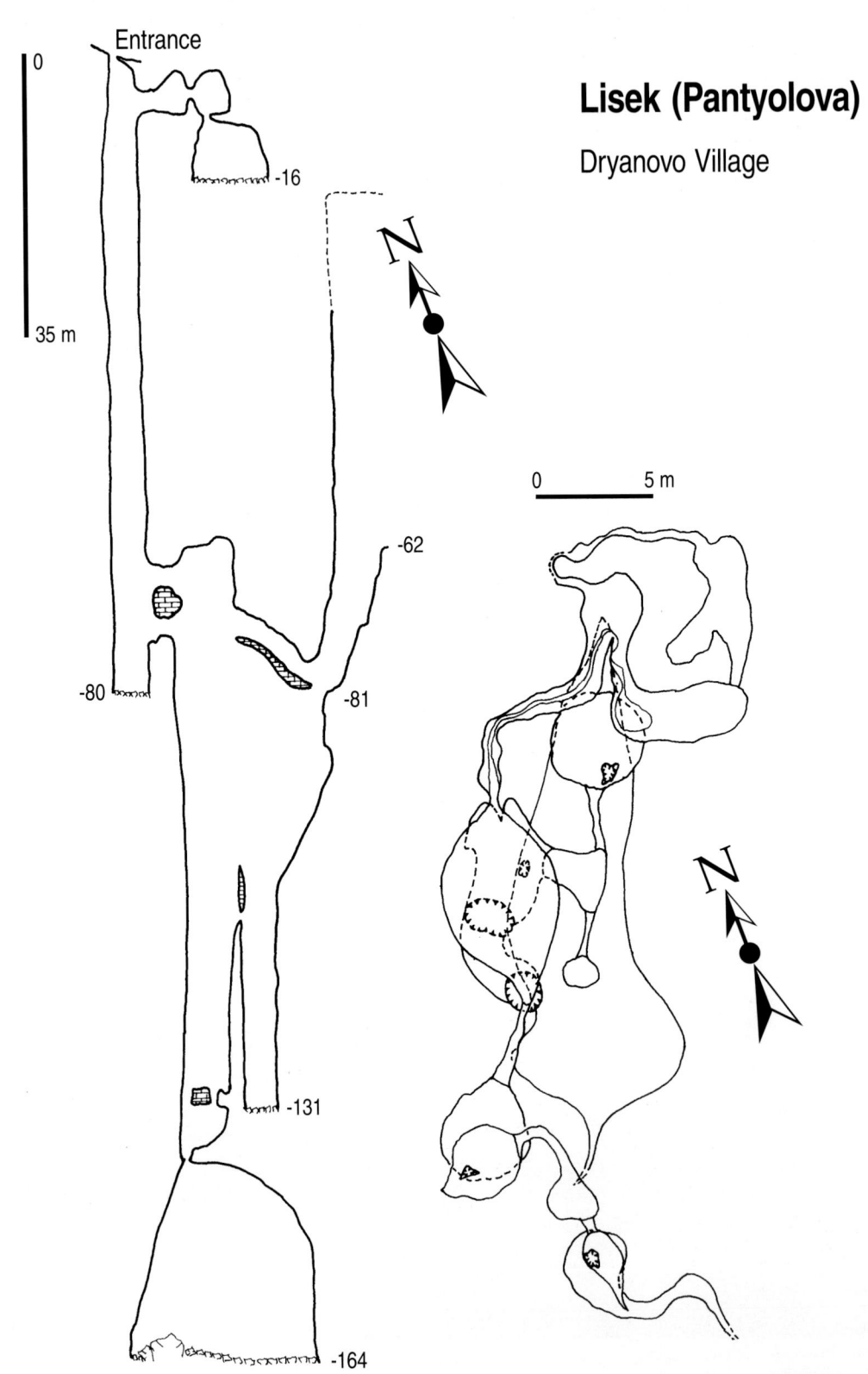

Lisek (Pantyolova)

Dryanovo Village

Name: **MANUILOVATA DUPKA (MANAILOVATA PESHTERA) (215)**
Ribnovo Village, Blagoevgrad Distr.
Altitude 1150 m. Length 2119 m. Denivelation -115 m

A cave between Ossikovo and Ribnovo. The entrance is on the steep left slope of Manailovsko Dere, in Proterozoan marble. Manuilovata Peshtera is a descending water cascade cave. After the first 350 meters, it branches into two main galleries with short secondary branches. The lower gallery is 605 m long (together with the branches 845 m), having a 105-m denivelation from the entrance and an average inclination of 17.36%. The upper gallery is 1150 m long (1330 m together with the branches), having a 115-m denivelation from the entrance. The supposed outlet of the underground river is the karstic source at the water junction of Ossikovsko Dere and Manuilovsko Dere.

The cave had been known to the local people for a long time. The first speleological exploration was carried out by the Committee for Cave Tourism in 1960 by P. Beron and S. Penchev. They surveyed and described the lower part (surveyed 900 m of galleries) and collected cave fauna (Beron, Penchev, 1961). In 1971 another expedition of the Bulgarian Caving Federation again surveyed the lower part and some distance of the upper gallery, altogether 1026 m. In 1975–1978, several scientific expeditions led by P. Stefanov explored the cave, discovering and mapping another 770 m of galleries. The map of the whole cave was completed and experiments were carried out to determine the way and the parameters of the underground river.

Troglobites and stygobites: *Eucyclops subterraneus* (Copepoda), *Stygiosoma beroni* (Diplopoda).

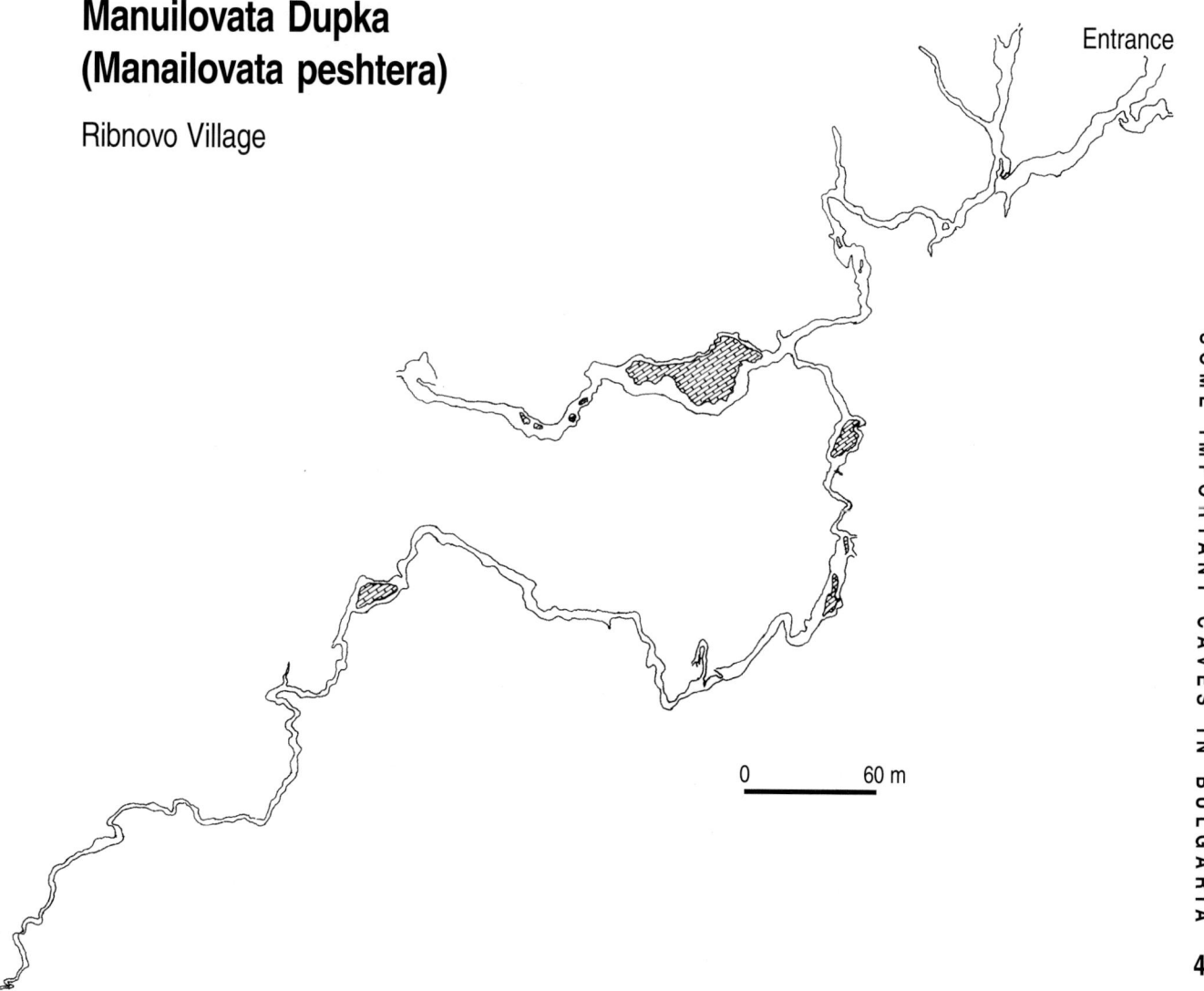

**Manuilovata Dupka
(Manailovata peshtera)**

Ribnovo Village

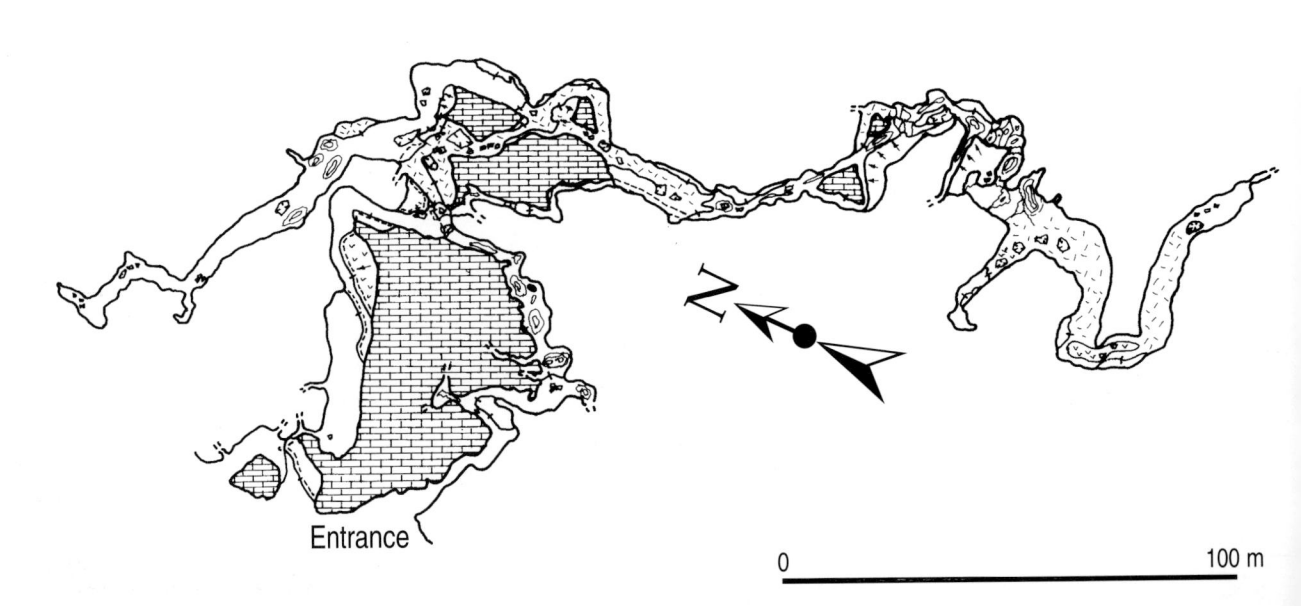

Name: **NOVATA PESHTERA (218)**
Peshtera Town, Pazardjik Distr.
Length 846 m

A cave two km SW of Peshtera Town, on the road to Batak, 400 m upstream the river Novomahlenska Reka, in a cliff, 100 m from a water catchmentand 52 m above the river bed.

Novata Peshtera is a two entrance cave. The galleries starting from the two entrances join in a large chamber after ca. 100 m. On the floor there are stalagmites, on the walls and the ceiling – stalactites. From the chamber, the cave develops in two main directions – NE and S. The northeastern gallery is 100 m long, 3 to 6 m wide and 1 to 4 meters high. The floor is covered with stalagmites. South of the chamber, three galleries start, which later merge and go further as a gallery, which is 7–8 m wide, on the floor of which there are several lakes. This gallery is followed by another one, with a floor covered by clay. This gallery is situated on the lowest level (compared with the others), that is why the water from the entire cave collects there. Throughout most of the year, this gallery is inundated and bears the name The Water Branch. This part of the cave is the richest in flowstone formations, and also the longest. There are several chambers in it. The first one is called Suhite Ezertsa (The Dry Small Lakes). The next chamber is the biggest and on its ceiling many bats roost, hence the name The Fairy Kingdom of Bats. A smaller chamber with tender stalactites and huge stalagmites follows. Further, the gallery becomes narrower and ends in a squeeze.

Discovered in 1931 by the local tourists K. Turtov, G. Gadzhev and one Italian speleologist. The name Novata (The New) was given by the discoverers. After cleaning the small entrance, they entered the cave. To preserve the stalactite decoration, tourists from Peshtera mounted an iron gate (80 x 80 cm) with a padlock.

The cave was described and surveyed for the first time by the Zoologist Pencho Drenski from the Royal Museum of Natural History in Sofia. Surveyed again in 1975 by F. Filchev, A. Andreev, G. Pehlyov, K. Kostov, I. Kozarev and P. Turtev, cavers from Kupena Peshtera Caving Club. During that time, the second entrance of the cave was discovered.

Fauna (studied by N. Atanassov and P. Drenski in the 30s, by P. Beron on September 12, 1962 and later by other cavers): includes the troglobites *Bulgaronethes haplophthalmoides* (Isopoda), *Lithobius lakatnicensis* and *L. stygius* (Chilopoda), *Plusiocampa* cf. *beroni* (Diplura), the troglophiles *Pseudacherontides spelaea* (Collembola), *Balkanopetalum rhodopinum* (Diplopoda), and others, 22 sp. in total.

Novata peshtera

Peshtera Town

Name: **PAVLA (VODNATA PESHTERA) (221)**
Ravnogor Village, Pazardjik Distr.
Total length 225 m. Denivelation +25 m

A cave in the area called Pavla, east of Ravnogor. It can be reached following a dirt road, on the left of the paved road Bratsigovo – Ravnogor. Entrance: 2 m wide, 5 m high. It is horizontal, slightly ascending, with a temporary stream on the floor. Sinters with water. About 45 m from the entrance, a 6-m chimney leads to a gallery, which is 15 m long. About 187 m from the entrance, the main gallery branches off.
Fauna: five species identified.

Name: **SAMARA (SAMARSKATA PESHTERA) (250)**
Ribino Village, Kardjali Distr.
Length 327 m. Denivelation +10 m

The cave Samara (Semer-ini in Turkish) is situated in Krumovgrad Karstic Region, near Ribino Village, in the locality Talashman Dere. The name is due to the shape of the entrance ("Samar" means donkey saddle).

The cave is formed in grey organogenic, very fissured limestone from the Paleogene. It consists of one main gallery, which is easy to walk. In places, in gets higher and wider, forming several chambers and one side branch, which is on the right of the 55th meter and 80 m long.

At the 50th meter along the main gallery, a wall without joining plaster has been built. Here, there is a low and wide chamber. There are ceramic pieces on the floor. In the side gallery a stream runs the gallery gets wider and ends in a big chamber. The main gallery is 200 m long. On the walls, there are curtains and other flowstone formations. In sinter bowls on the floor, there are cave pearls, 1–3 cm in size. The air temperature is 10 °C.

The cave was explored during a joint caving expedition of Rhodopi Kardjali and Aida Haskovo Caving Clubs in July 1975 (description after B. Kolev and M. Gumarov).

Fauna: collected by P. Beron, B. Petrov, P. Stoev, T. Ivanova and others during different visits. It contains 23 identified species, including the troglobite *Trichoniscus rhodopiense* (Isopoda), as well as the interesting troglophiles *Balkanopetalum petrovi* (Diplopoda), *Balkanodiscus frivaldskyanus* (Gastropoda), the fly *Brachytarsina flavipennis*, the only freshwater crab, found in Bulgarian caves, etc.

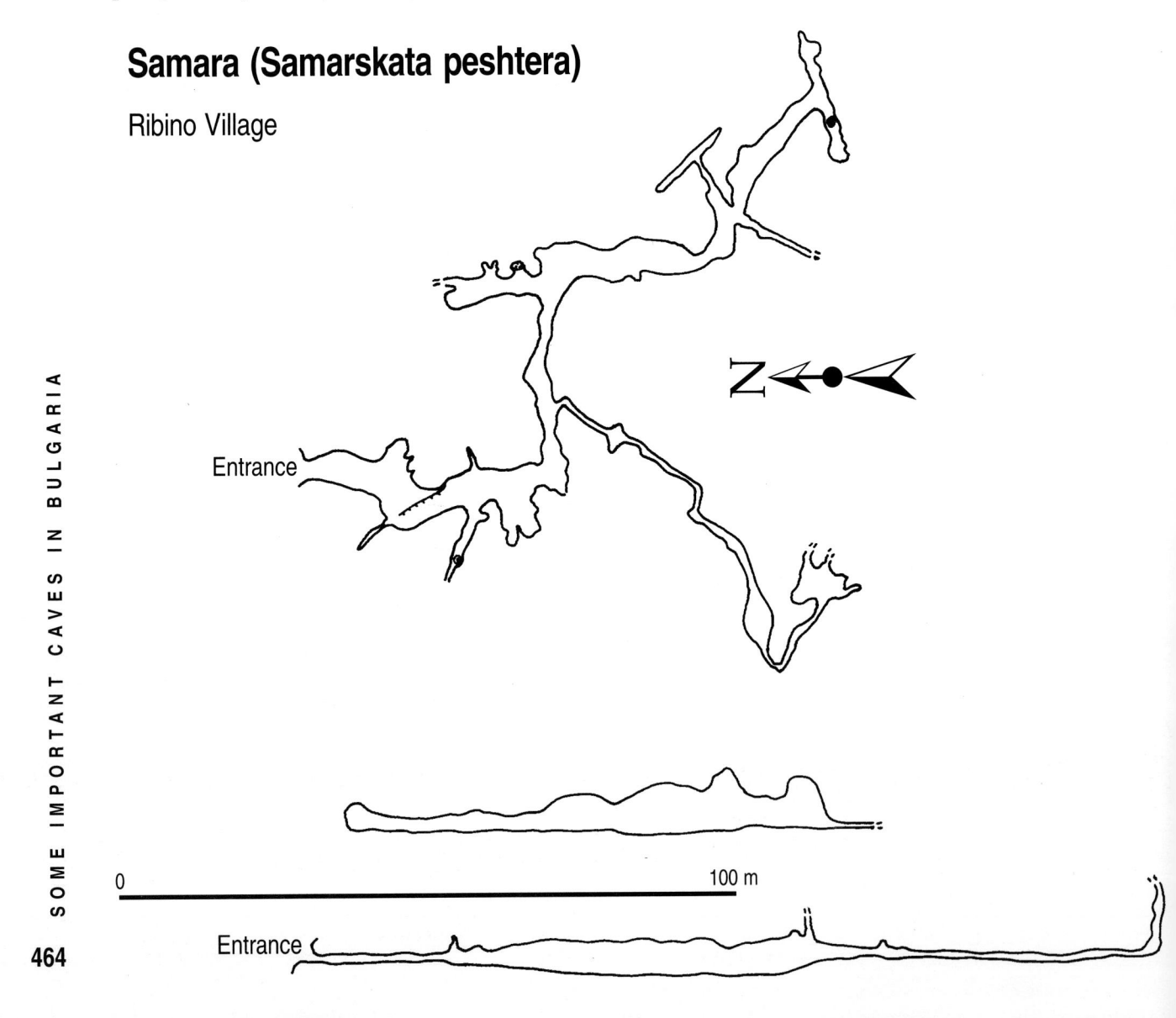

Samara (Samarskata peshtera)

Ribino Village

Name **SANCHOVA DUPKA (224)**
Yagodina Village, Smolyan Distr.
Length 888 m, on the main axis 455 m. Denivelation +18 m
Coordinates: N 41° 38' 0.51" E 24° 20.07', 33"

A cave near Yagodina, the youngest in this region (the same age as the middle storey of Yagodinskata Peshtera, or the early Pleistocene). From the central chamber (10 x 12 m), two galleries start – northern and southern. In the two parts there are bigger chambers (20 x 30 m). The gallery, starting from the northwest end, leads to the northern part of the cave. This gallery connects three huge chambers with floors, covered with boulders. The southernmost chamber is with dimensions 27 x 32 m. From its southeast end a narrow 150-meter gallery starts, adorned with secondary karstic formations. Several water levels are seen.

There is a hypothesis, that the cave Dolna Karanska Dupka is an upper storey of Sanchova Dupka. The air temperature in different parts of the cave is between 6.8 and 9.8 °C.

Sanchova Dupka was discovered and explored by members of the Chepelare Caving Club as early as 1965, but during the International caving expedition in 1966 it was studied by V. Dublyanski (Russia), Kurt Brendel and Friedrich Schuster (Germany), Dimitar Sabev and Dimitar Raychev.

Fauna: 12 species of animals have been recorded, but so far no troglobites.

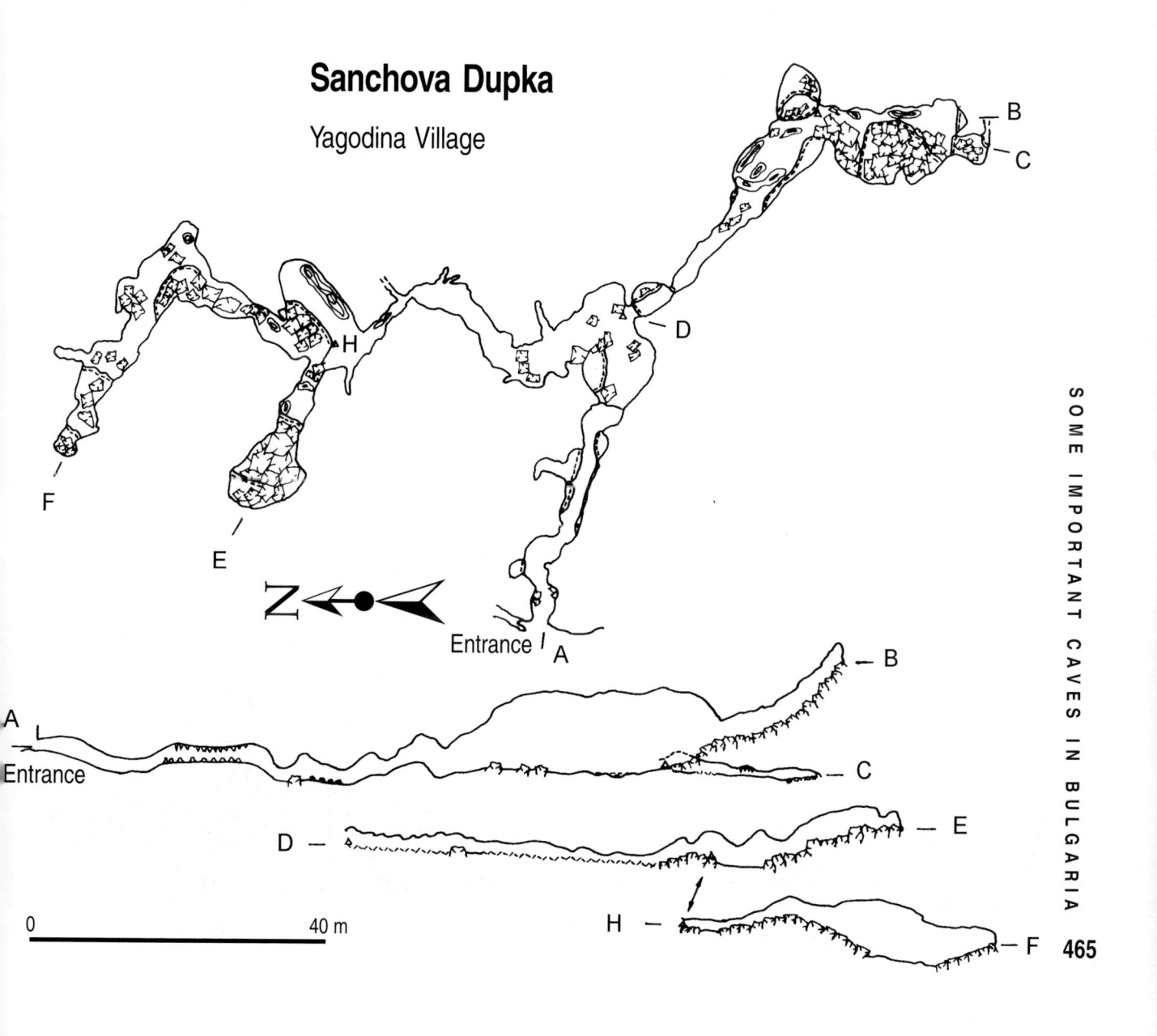

K. Hadjiiski in front of the entrance of the cave Sanchova Dupka – photo Alexey Jalov

The Wolf's Jump in the gorge Buynovskoto zhdrelo – the narrowest place Yagodinska – Leopard Skin – photo Vassil Balevski

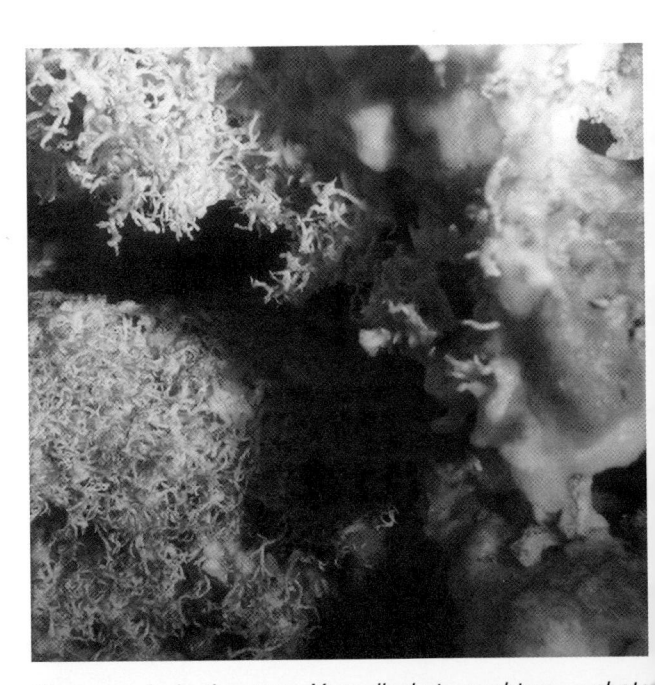

Helictites in Yagodinskata peshtera – photo Vassil Balevski

Sinter floor in the cave Yagodinskata peshtera

Cave pearls in the cave Yagodinskata peshtera – photo Vassil Balevski

Name: **SHEPRAN (SHEPRA) DUPKA (237)**
Belitsa Village, Smolyan Distr.
Length 260 m. Denivelation 20 m. Volume 3011 m^3

A cave 800 m NW of Belitsa, one of the most beautiful caves in Rupchos. A spacious, almost horizontal gallery leads to a "stone forest" of stalactites, stalagmites and other flowstone formations. We pass by seven short side branches of the cave, ending in boulders or stalagtones. A 100 m from the cave entrance, 15 m of the east wall of the gallery are overgrown with snow-white dendrites and helictites (The Gallery of the Stone Flower). A bit further, the gallery bifurcates and reaches the 18-meter shaft Imanyarska Propast, situated almost at the end of the cave. After the pothole, a stalactite chamber is reached at the accessible end of the cave. A narrow crevice is reached, where running water is heard.

Explored and surveyed by Dimitar Sabev, who described it in his book "Rupchoskite Peshteri" [The caves of Rupchos] (1967). The map was made by Nikolay Gladnishki and T. Daaliev in 1975.

Fauna: from this cave the troglobite Isopods *Cordioniscus schmalfussi* and *Trichoniscus petrovi* have been described, as well as a third troglobitic Isopod, *Rhodopioniscus beroni* and many white Amphipods.

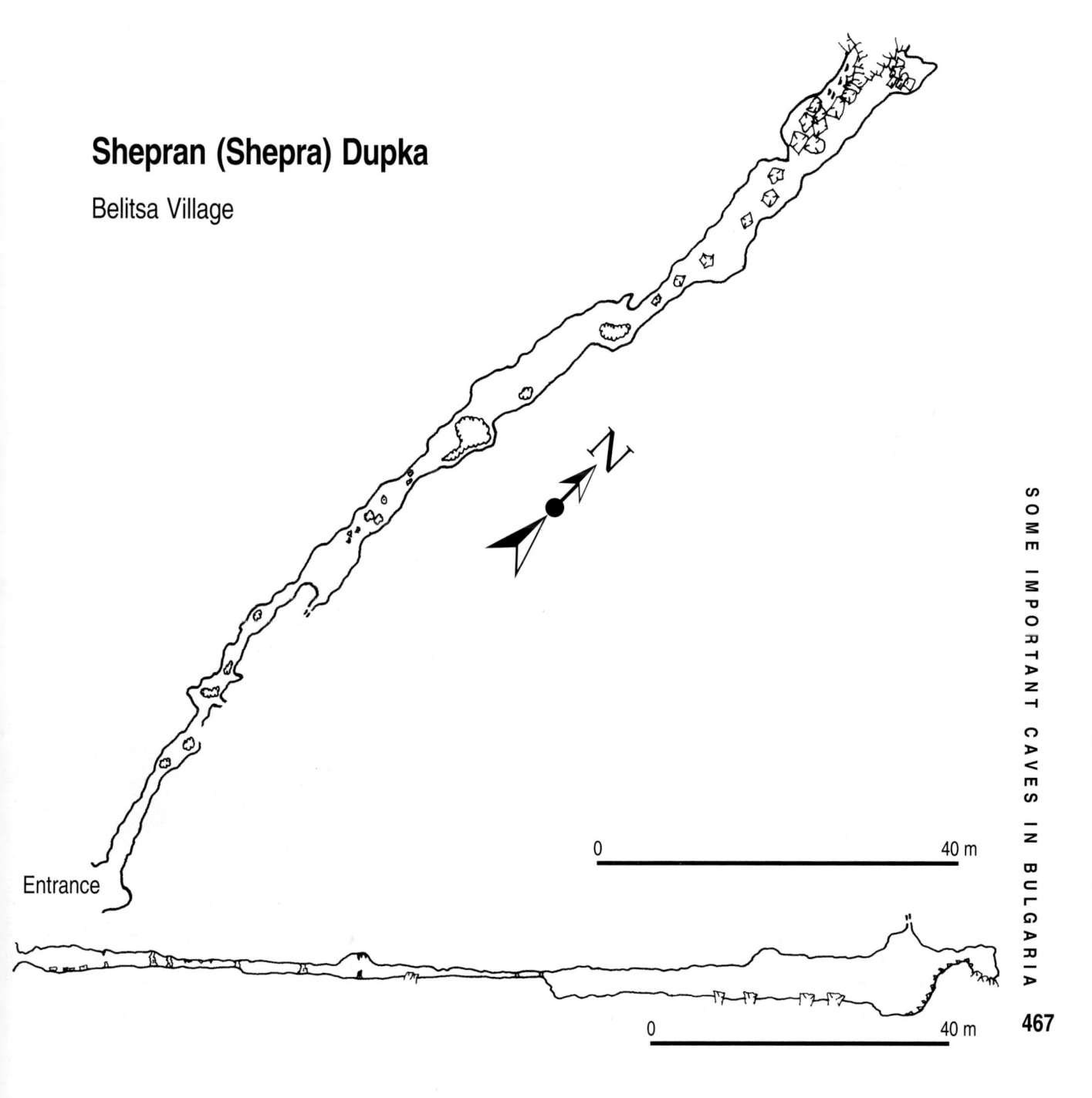

Shepran (Shepra) Dupka
Belitsa Village

Entrance

Name: **SNEZHANKA** (Snow White, called so by A. Petrov because of the white stalactites) **(219)**
Peshtera Town
Length 230 m. Denivelation -18 m. Surface 3150 m^2
Coordinates: N 42° 00' 40.6" E 24° 16' 30.7"

A show cave SW of the town Peshtera, on the left slope of Novomahlenska River, near the base of Lilova Skala, northern part of Batashka Mountain. A one gallery, descending cave, developed along two basic systems of faults directed SE/NW and S/N. Six chambers, the biggest being 60 x 40 x 12 m. Variety of sinter forms everywhere. Sinter lakes formed by infiltration water. Discovered on January 2, 1961 by G. Kotsev, G. Zlatarev and B. Evtimov (members of Speleoclub Kupena Peshtera). In the same year, it was studied by the scientist G. Markov, N. Djambazov, Vl. Popov and A. Petrov. Material from the early Iron Age and Roman-Thracian time has been discovered. In 1968 a precise topographic and geomorphological survey was done, then in the 80s geophysical studies were carried out.

It was equipped as a show cave in 1968, managed by the Tourist Society Kupena. Declared a Natural Monument (together with 25.1 ha), by Decree 504/11.07.1979. Since 1983 included in the buffer zone of Kupena Natural Reserve.

Fauna (studied by P. Beron, P. Stoev and B. Petrov): Several species under study, troglobites are *Lithobius lakatnicensis* (Chilopoda) and *Trechini* gen. sp. indet. (Col., Carabidae).

Snezhanka

Peshtera Town

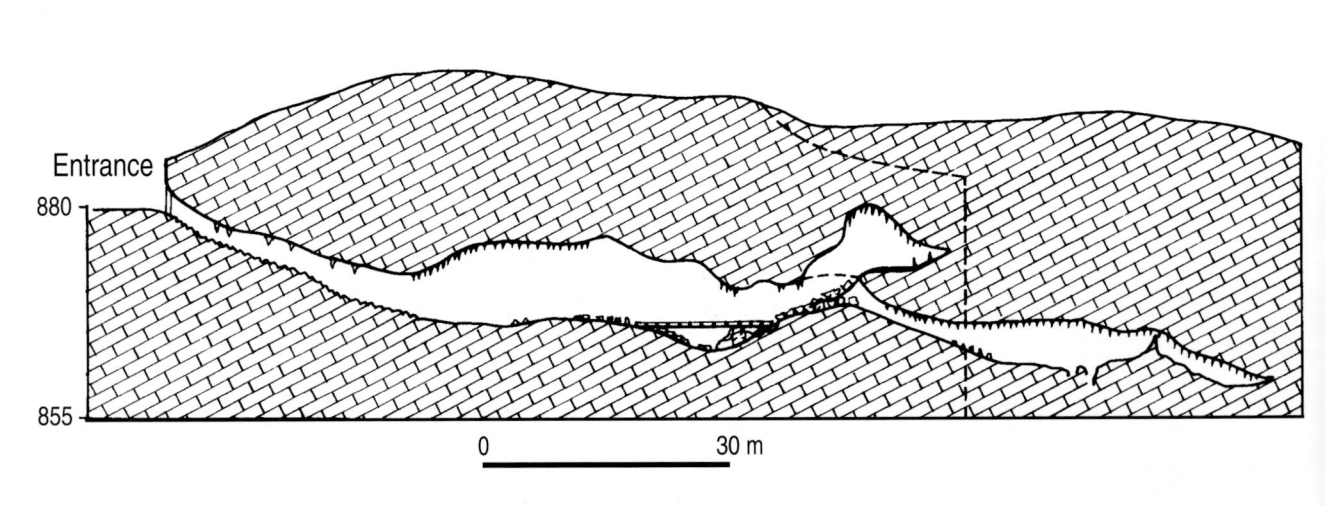

Snezhanka – T. Daaliev in front of the entrance of Snezhanka – photo Valery Peltekov

The beauty of the Big Hall – photo Valery Peltekov

The Udder – photo Valery Peltekov

SOME IMPORTANT CAVES IN BULGARIA

Name: **TOPCHIKA (244)**
Dobrostan Village, Plovdiv Distr.
Length 727 m. Denivelation -61 m
Coordinates: N 41° 52' 43.3" E 24° 53' 51.7"

A cave SW of Dobrostan, in the area of Pastyolov Chuchul, high on the left slope of Sushitsa River, Dobrostan Ridge. Formed in marbles from the Upper-Middle Rifey. A pothole like, two-storey, branched cave with main development to NW. The upper dry storey is 110 m long (without the branches). Fifty (50) m from the entrance, a 45meter deep shaft opens, giving access to the lower section. Semielliptic shape of the cross section, 12 m wide and 3.5 m high. Flowstone formations mainly in the lower part. Clay deposits in both parts. Dripping water.

The cave was explored first in August 1958 by the alpinist S. Zadgorski, who descended into the shaft. In the autumn of the same year, 7 cavers from Plovdiv (N. Atanasov, S. Stoychev, P. Kostov, D. Beslimov, D. Russev and S. Paskov) studied in details and surveyed the cave. In 1967, the club Young Traveller in Plovdiv, led by S. Stoychev, discovered rock engravings at the cave entrance. The prehistoric engravings were studied by specialists – archaeologists. The cave was surveyed again in 1970 by Russian speleologists during an International Expedition and later by Akademik Plovdiv Caving Club.

Fauna: studied by P. Beron 1962 and includes the troglobite Isopod *Rhodopioniscus beroni,* the troglophile Diplopod *Balkanopetalum beskovi* and other interesting species.

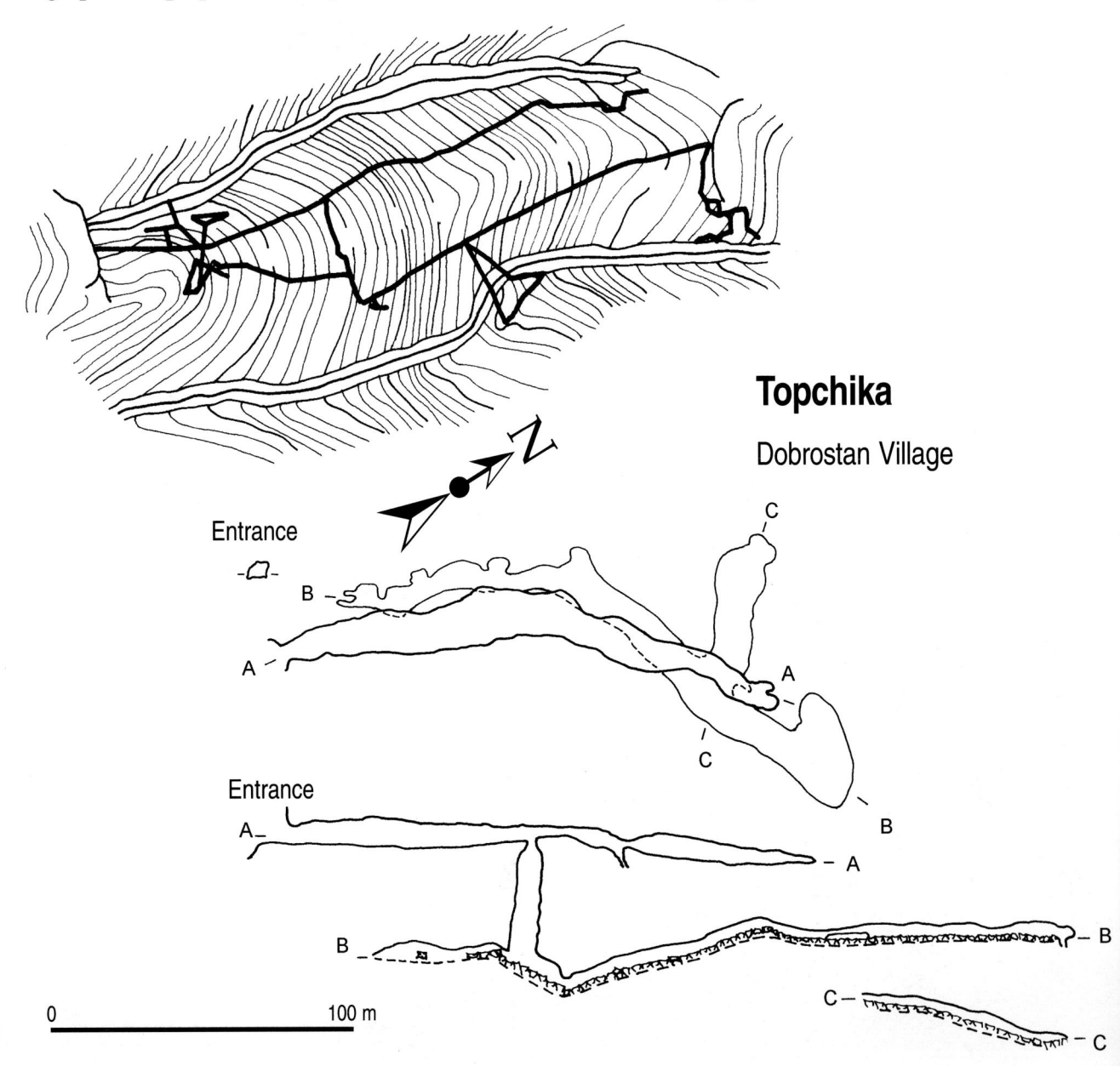

Topchika

Dobrostan Village

Name: **UHLOVITSA (ULTSATA) (246)**
Mogilitsa Village, Smolyan Distr.
Length 330 m. Denivelation -25 m
Coordinates: N 41° 30' 52.2" E 24° 39' 43.2"

    A show cave 2.3 km NE of Mogilitsa, on the left bank of Arda and the S slope of the Kaynadinski Rid, Central part of the Western Rhodopes. Formed in low-crystalline massive Proterozoan marbles in the southern periclinal (Borikovska) part of the South Rhodopean anticline. A two-storey pothole, developed along the basic fissure, directed N/S. Rich decoration of all kinds of sinter forms.
    Discovered and explored in 1968–1969 by the Speleoclub Studenets Chepelare. Inaugurated as a show cave in 1983.
    A Natural Monument with 1 ha adherent land (Decree 238/04.05.1979).

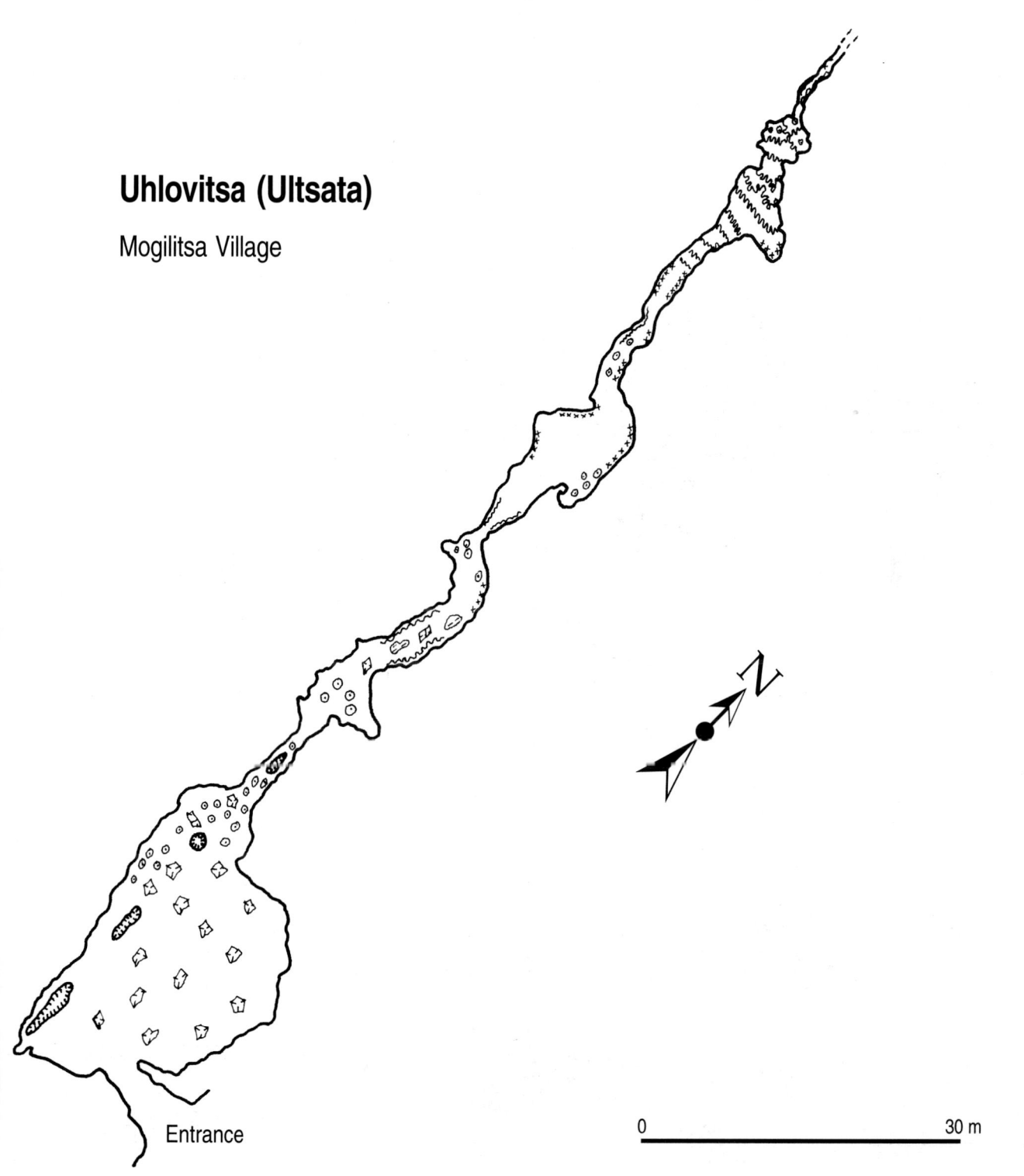

Uhlovitsa (Ultsata)

Mogilitsa Village

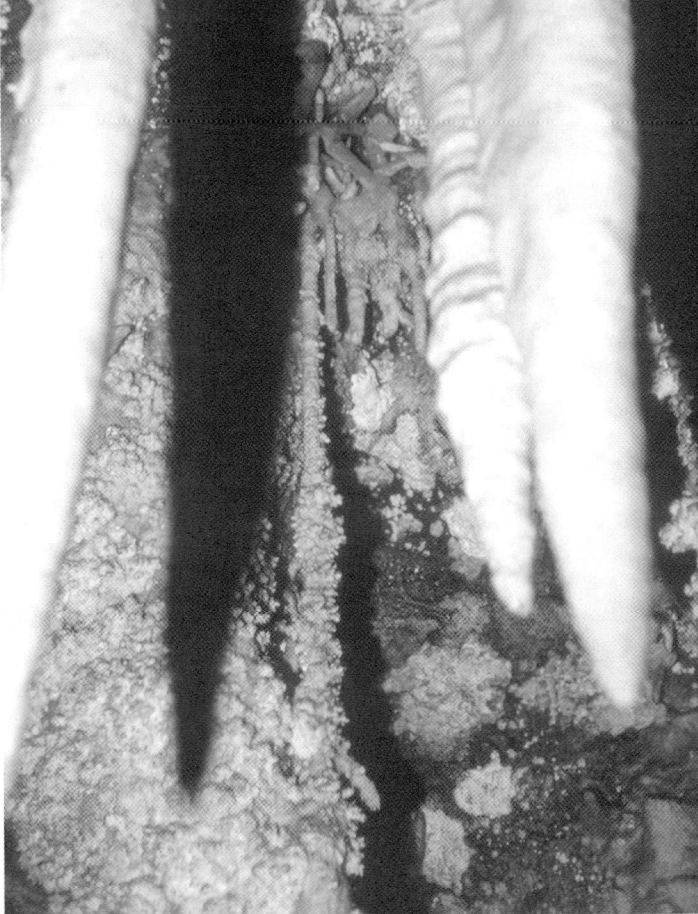

The decoration of the walls – photo Trifon Daaliev

Stalactites – photo Trifon Daaliev

The stone waterfall – photo Vassil Balevski

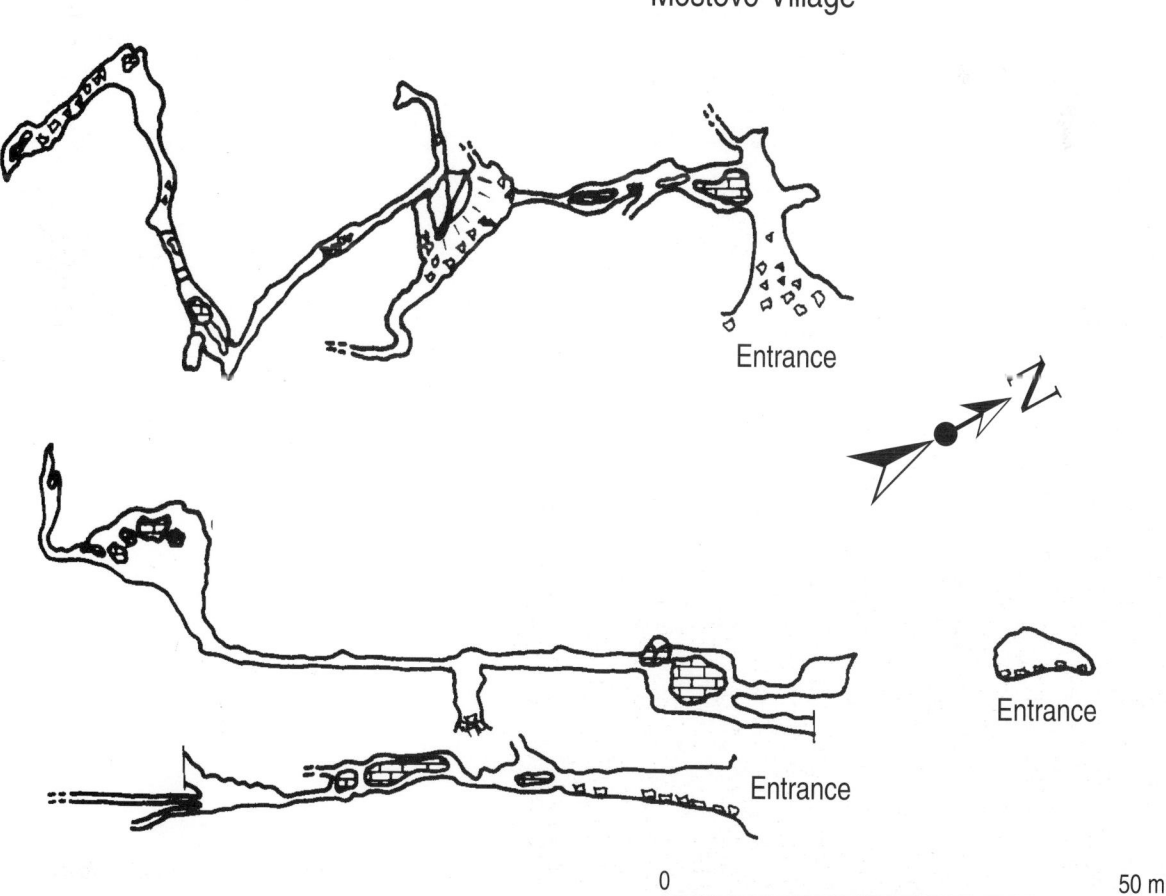

Name: **VODNATA PESHTERA (BOROVSKATA VODNA PESHTERA) (238)**
Mostovo Village, Plovdiv Distr.
Altitude 750 m. Total length 455 m. Denivelation 53 m (-10.45, +42.51 m)

The cave is known as Vodnata (The Water Cave), because of the underground stream, coming out of the cave and joining Sushitsa River. It is called also Borovska, as it is between Mostovo and Borovo.

The cave entrance is on the left bank (geographically) of Sushitsa River, about 1.5 hours of walk from the natural bridge Erkyupriya. The entrance faces east and is rather big (7 m high, 22 m wide). It is situated about 3 m above the river bed. When the river floods, there is a danger of water entering the cave.

A big boulder splits the gallery into two parts, later merging again. The passage is descending and narrowing. After circumventing two sand siphons, we reach a chamber. A gallery leads to a bigger hall, which is blocked in front. A branch on the right leads to a long and narrow gallery, parallel to the big chamber. Then, another passage widens as a gallery, leading to another chamber. We circumvent a deep well without a way further, then another gallery leading to the last chamber with a high chimney.

The cave is with several floors, branched and periodically inundated. It is formed in Proterozoan marble. In the past it used to be a river bed, hence the sand and gravel on its floor. Many flowstone formations are seen everywhere in the cave – stalactites of different shape, curtains, dendrites.

This cave was explored first by a Bulgarian-Russian team during the International Expedition "Dobrostan'70". The survey was done by A. Petkova, L. Popov, V. Volkov and M. Eygel, and again in 1981–1983 by T. Todorov, I. Petrov, V. Grancharov and G. Mindev from the Student's Caving Club Akademik, Plovdiv. After digging in several sand siphons, they added another 50 m to the cave's length.

**Vodnata peshtera
(Borovskata vodna peshtera)**

Mostovo Village

Name: **VODNATA PESHTERA (217)**
Peshtera Town, Pazardjik Distr.
Altitude 900 m. Length 1114 m. Denivelation +50 m

A cave about one and half hours walk from Peshtera Town. After 2.5 km on the road from Peshtera Town to Batak, there is a deviation, where the road to Snezhanka starts. After another 2.5 km, we reach the area of Angelovi Nivi (on the left). From here we can see the majestic peak Kupena – rocky in its upper part. The cave is found at the base of a cliff on its Eastern side. From the bed of Novomahlenska Reka River, a footpath to the summit starts. Here we must look for the old excavations for the pipeline to Snejanka. Following the pipeline, we reach the cave entrance. It is important to not disturb the cave water very much.

It is an inclined, branched and diaclasic water cave. In some places is very narrow. In the first 200 m there are no flowstone formations.

The cave was described for the first time by F. Filchev in 1976. During the District Caving Expedition Peshtera-76, members of the caving clubs in Dryanovo and Peshtera did a horizontal survey up to 200 m. During the next years, over 800 m of galleries have been explored.

Vodnata peshtera

Peshtera Town

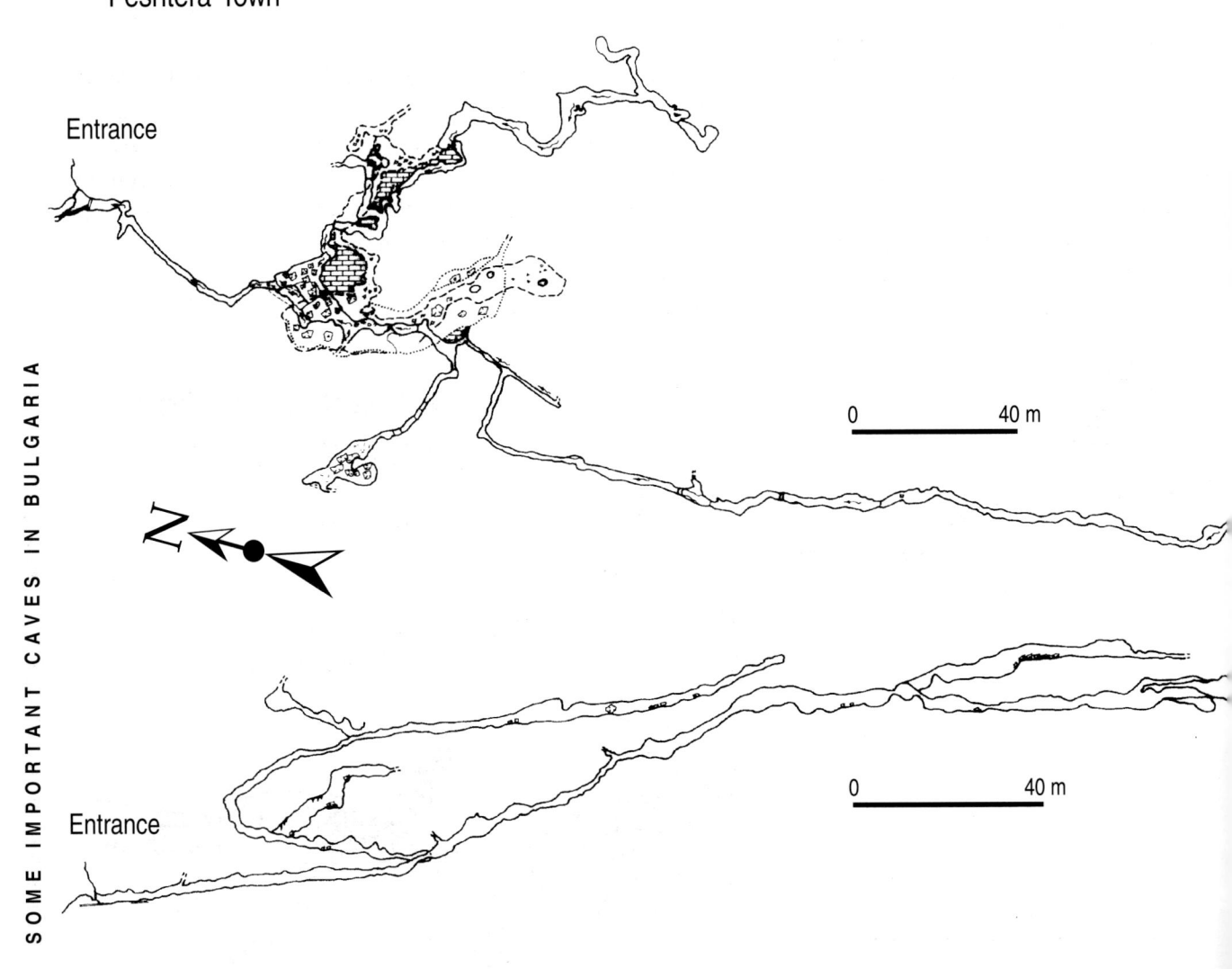

Name: **YAGODINSKA PESHTERA (IMAMOVA DUPKA) (225)**
Yagodina Village, Smolyan Distr.
Length 8501 m. Denivelation -36 m
Coordinates: N 41 37' 27.9" E 24 19' 30"

The cave is the longest in the Rhodopes and the third in length in Bulgaria. It is formed in Proterozoan marble, three kilometers SW of Jagodina, on the right bank of Buyanovska river. The natural entrance of the cave is 30 m higher than the river bed. The artificial entrance of the cave was dug out one kilometer upstream. A twostorey, labyrinthine cave system. There are large and small speleothems of different shapes in the galleries and chambers.

The entrance gallery leads into the atrium with a depth of 15 m and denivelation of 27 m. This is the connection with the lower part. After descending the pitch, we reach the 600 m gallery, called D. Sabev, connecting with the Tourist gallery and the Northern Sector. As a table, the sectors have the following dimensions:

First storey – 1176 m, 3939 m², 8470 m³
Northern sector – 1343 m, 3984 m², 18 492 m³
Southwestern Labyrinth – 2720 m, 10 662 m², 20 365 m³
South Labyrinth – 3061 m, 6948 m², 10 123 m³
Total – 8501 m, 23 535 m², 57 450 m³

When the water of Buynovska River rises, part of them sink into the lower part of the cave (last case in 1990) T° 8 °C.

The cave has been known to local people forever, sheltering people from the Eneolith to recent time. In 1928, the first excavations were attempted by V. Mikov. Later, the cave was visited by another Bulgarian archaeologist (Petar Detev) who excavated the entrance chamber.

Speleological research of the cave started in 1964 by cavers from Chepelare. The total length, mapped in 1964–1965 by D. Raichev and D. Sabev, was 6455 m. In 1965–1966, the Smolyan Museum (under M. Deyanova) organized archaeological excavations. They discovered artefacts from the Eneolith time. After a special survey in 1969–1973, the cave was opened as a show cave in 1982. The tourist circuit (through two artificial tunnels) is two kilometers long. There are beautiful speleothems all the way (stalactites, stalagmites, sinter lakes, cave pearls, leopard skin and others). Pottery paint in yellow, red and black.

The fauna includes 30 known species, among them the troglobite *Troglodicus meridionale* (Diplopoda), described from this cave, and *Plusiocampa bulgarica* (Diplura).

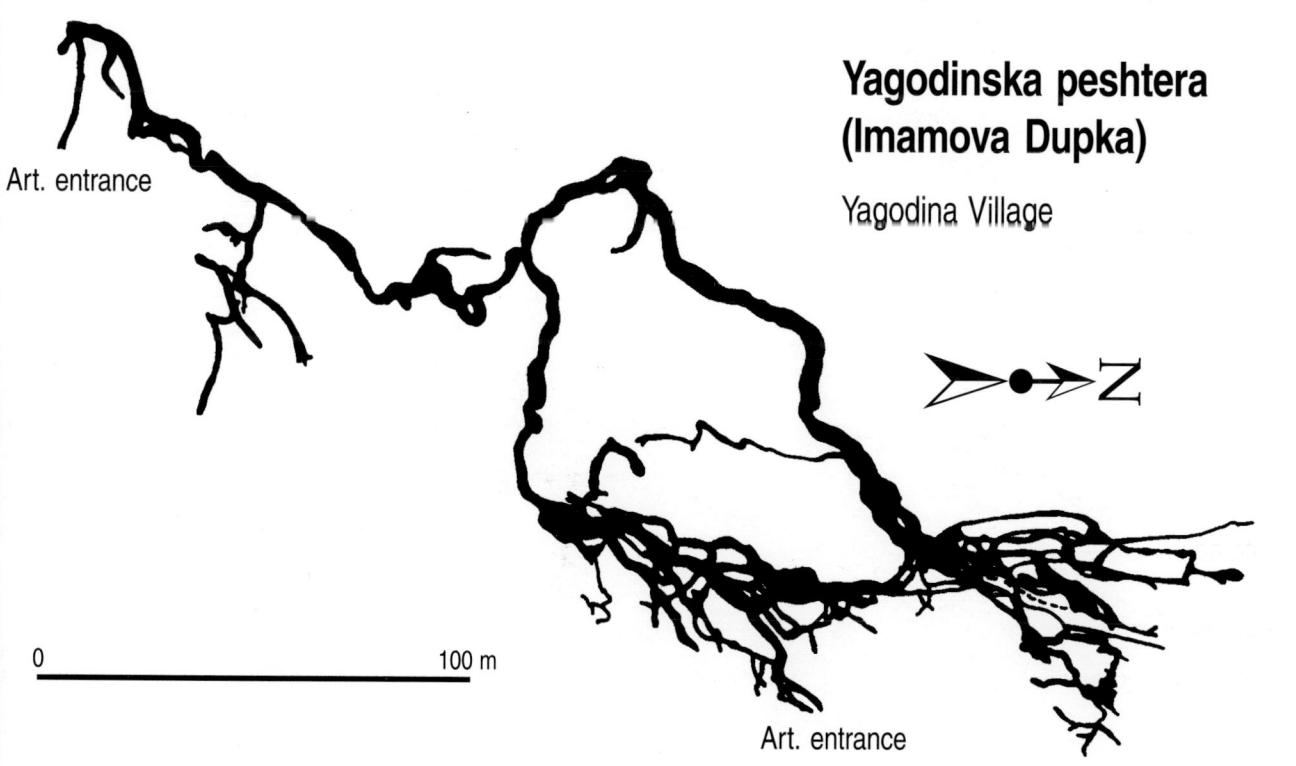

Yagodinska peshtera (Imamova Dupka)
Yagodina Village

A. Jalov in front of the natural entrance of the cave Yagodinska – photo Alexey Jalov

The pearls – photo Vassil Balevski, Valery Pelteokov, Trifon Daaliev

The decoration is fantastic – photo Vassil Balevski, Valery Pelteokov, Trifon Daaliev

Name: **YUBILEYNA (220)**
Peshtera Town
Length 814 m. Denivelation 18 m

A two-storey, branched and (initially) very beautiful cave, about 2 km SW of Peshtera Town, on the road to Batak Town, 1000 m upstream Novomahlenska River, on the right (geographically) bank, about 20 m above the river.

The entrance has dimensions 2 x 2 m. A heavily inclined entrance gallery leads to the first area of blocks. After this area is the biggest chamber in the cave. It is 40–50 m long and 17–18 m wide. There are many blocks and secondary flowstone formations. The chamber called Topolkite (The Poplars) follows with many stalactites and stalagmites.

From this chamber, the main gallery goes straight. On the right, there is a branch, called Terasata, where there are several openings of different depth (from 6 to 11 m), connected to the lower (water) level of the cave after the siphon and leading to a gallery of ca. 60 m. The main gallery is 4-5 m wide and at 40–50 m it narrows to 0.50–1 m. This narrowing is 7–8 m long. After the squeeze the gallery gets wider and ends in a vertical of 7–8 m (no need of rope), falling in the water gallery. Downstream, this gallery is 100 m long and ends in a siphon. After the siphon, the water reappears in the gallery bellow the terrace.

Upstream, the gallery bifurcates. The right gallery crosses the block area and again we reach wide galleries, full of flowstone formations, sinter lakes, stalactites and stalagmites. The left one is poor in flowstone formations. After 50–60 m, both galleries merge together, then bifurcate again and join the main gallery, with many branches at the end. The end of the cave is a narrowing, from which an underground stream runs with a non-constant outflow of 10–20 l/s and T° 8–9 °C.

The cave was adapted partially for visitors by F. Filchev.

Yubileyna was discovered during an exploration of Novomahlenska River in 1974 by F. Filchev, A. Andreev and other cavers from Peshtera Town. The name was given by the discoverers. In 1979, during the gathering "Peshtera'79" the cavers A. Anev and T. Daaliev surveyed ca. 300 m of the cave. A detailed survey was done by F. Filchev, A. Andreev, K. Kiev, Sht. Vassilev and other cavers from Peshtera Town during the next years, together with the elaboration of a project on arranging it a show cave.

Fauna (studied by P. Beron in 1973 and 2005) includes the troglobite centipedes *Lithobius lakatnicensis* (Chilopoda) and *L. stygius*.

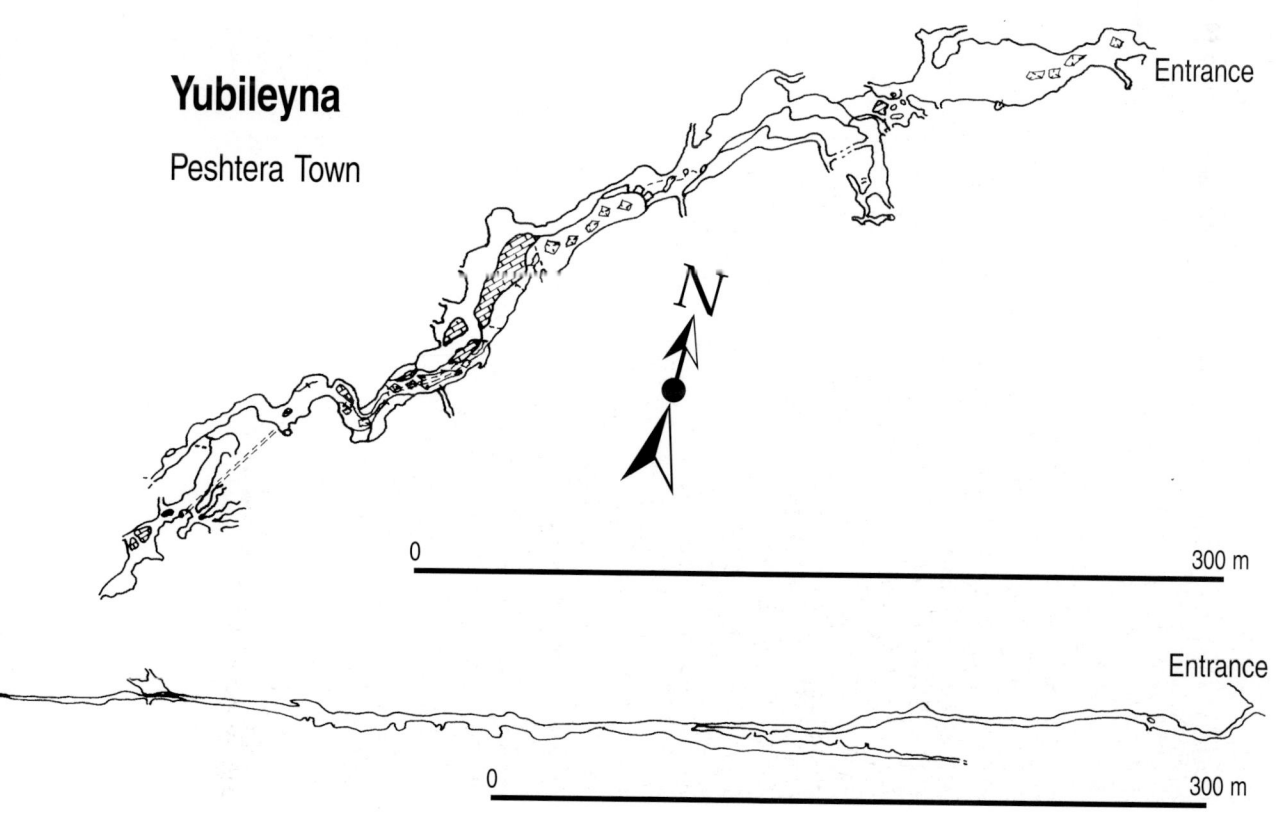

V. Peltekov and T. Daaliev in front of the entrance of the cave Yubileyna – photo Valery Peltekov

The entrance from inside – photo Valery Peltekov

The beauty of the cave – photo Valery Peltekov

Name: **ZMIYN BURUN (242)**
Mostovo Village, Plovdiv Distr.
Altitude: 930 m. Total length 155.55 m. Denivelation +22; -86 m

A pothole on the slope of Mostovska Sushitsa, difficult to find because of its small entrance and lack of landmarks. The entrance is near a big tree with a triangular metal mark with a registration number. The difference between the altitudes of the ridge and the cave entrance is ca. 150–200 m. Some big limestone slabs just under the entrance level could orient us.

Cavers from Paldin Plovdiv Caving Club could be contacted for guiding.

The cave starts with a 30° slope ending with a siphon. The floor is covered with clay. There are some flowstone formations. The gallery gets narrow and reaches a vertical. After ca. 80 m is the bottom, often with filtration water.

The cave was discovered by cavers from Plovdiv in 1986 with the help of local shepherds. Attempts have been made by cavers of Akademik to dig the bottom, but with no success. Surveyed twice – in 1970 and in 1983.

Survey: Akademik Sofia Caving Club, 1983. Chief surveyor: T. Todorov.

Fauna: five species known, no troglobites.

Zmiyn Burun

Mostovo Village

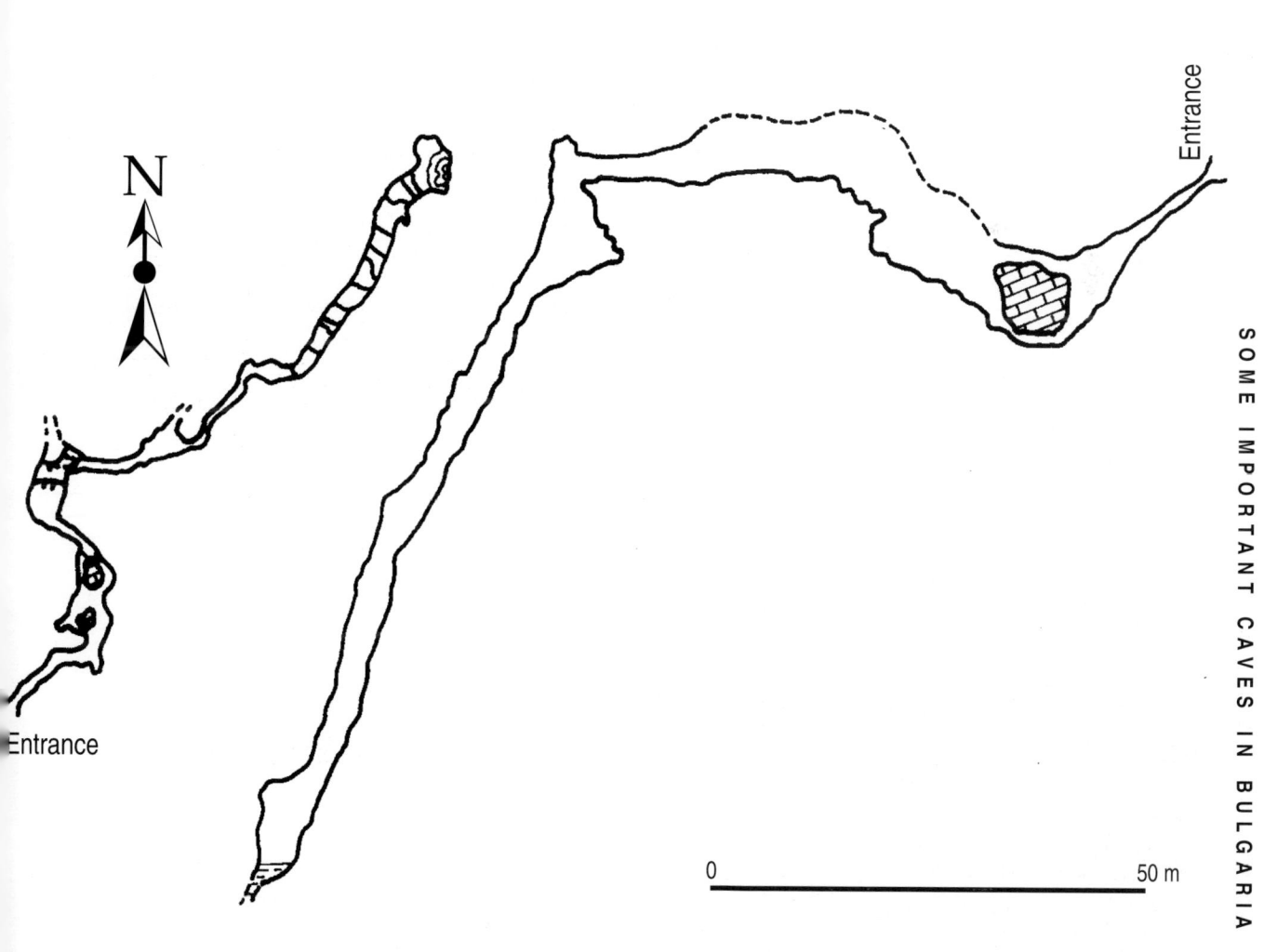

CAVES IN STRANDJA AND SAKAR

Name: **BOZKITE (Babini Bozki) (252)**
Mramor Village, Yambol Distr.
Total length 324 m. Denivelation -8 m

A cave in the valley of Manastirska river, three kilometres from Mramor. Entrance 4 m high, 3 m wide. Length of the main gallery 60 m. Wet, with a non-permanent stream, disappearing in a siphon at 120 m. The name of the cave comes from the shape of the cave formations. Many bats (up to 2–3000) and guano. No troglobites found so far. The cave was explored and surveyed by Zl. Kachanov and other cavers from the caving club of Kabile Yambol, and before them by Nenko Radev in 1926 and by P. Beron since 1959.

Fauna: more than 20 species known, but no troglobites. Big bat colonies (2-3000 ex.).

Bozkite (Babini Bozki)

Mramor Village

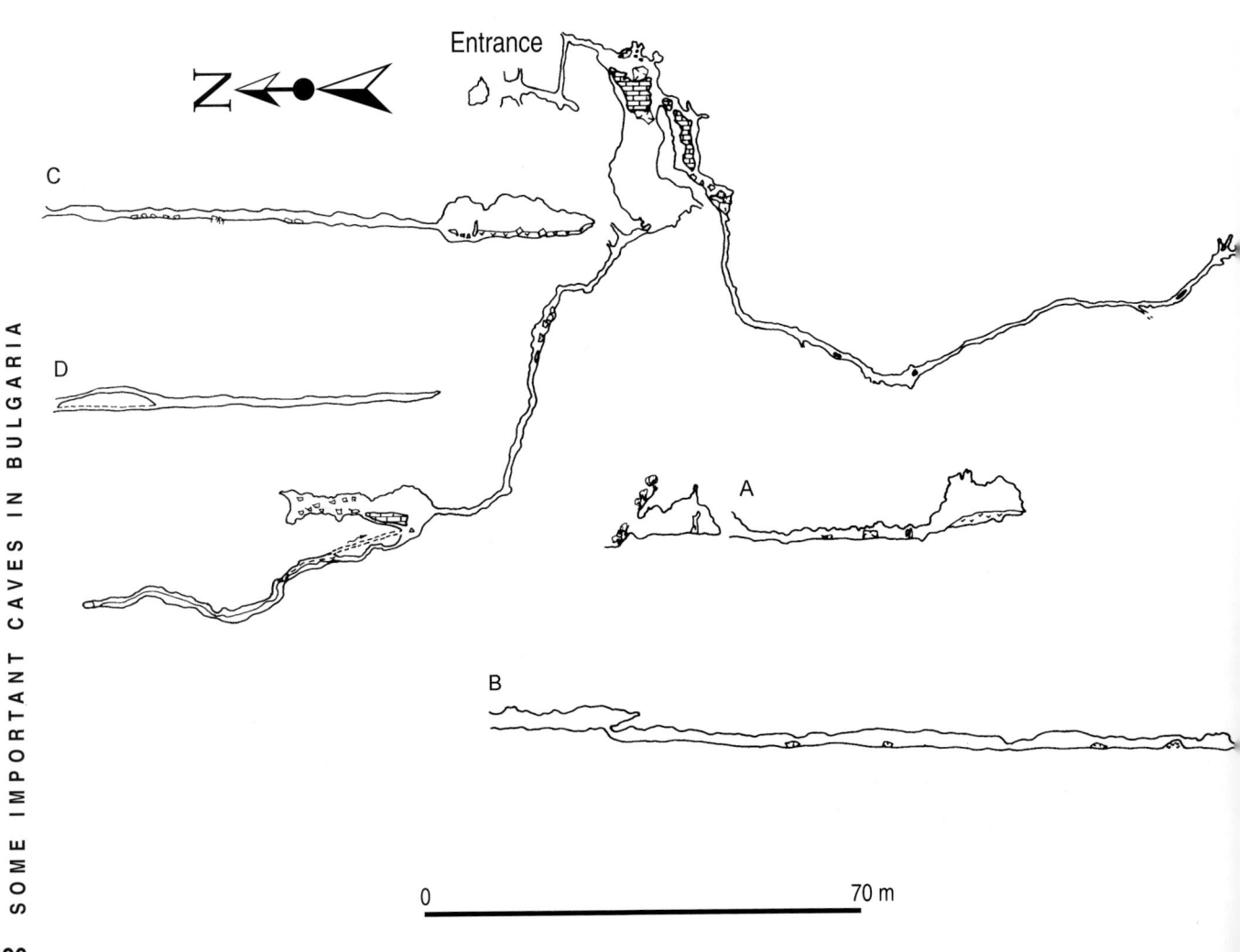

Name: **BRATANOVATA PESHTERA (257)**
Malko Tarnovo Town, Burgas Distr.
Length 348 m. Denivelation – 11 m
Coordinates: N 42° 00' 29.5" E 27° 26' 38.7"

A cave with two entrances in the locality Ai Dere. Wet, some parts with clay and small sinters. Air temperature 8.7–9.2 °C. Broken clay vessels, iron objects and a well preserved jar have been found in the cave.

The cave was studied and surveyed by cavers from Akademik Sofia Caving Club in 1976, the prolongation was surveyed in 1977 by Iv. Lichkov, P. Neykovski and A. Benderev.

Fauna (studied by Hr. Delchev, St. Andreev and P. Beron)): eight species known, including the troglobite woodlice *Trichoniscus valkanovi* and *T. beroni* (Isopoda).

Protected Natural Monument, together with 21.8 ha of adjacent land (Decree 206/23.03.1981).

Bratanovata peshtera

Malko Tarnovo Town

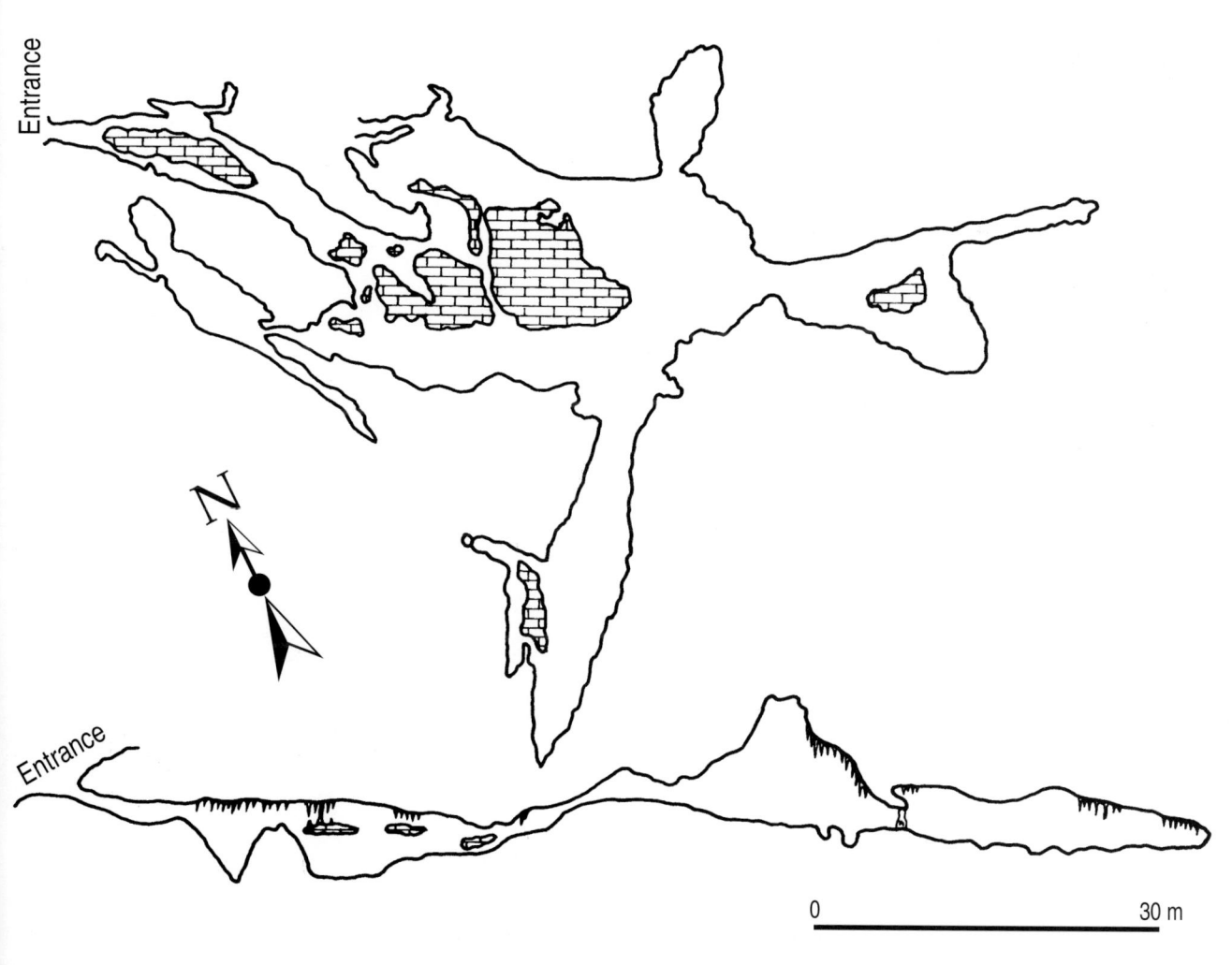

The entrance of Bratanovata Cave – photo Ivan Lichkov

The Candle – photo Ivan Lichkov

Coins from the time of Alexander the Great – photo Ivan Lichkov

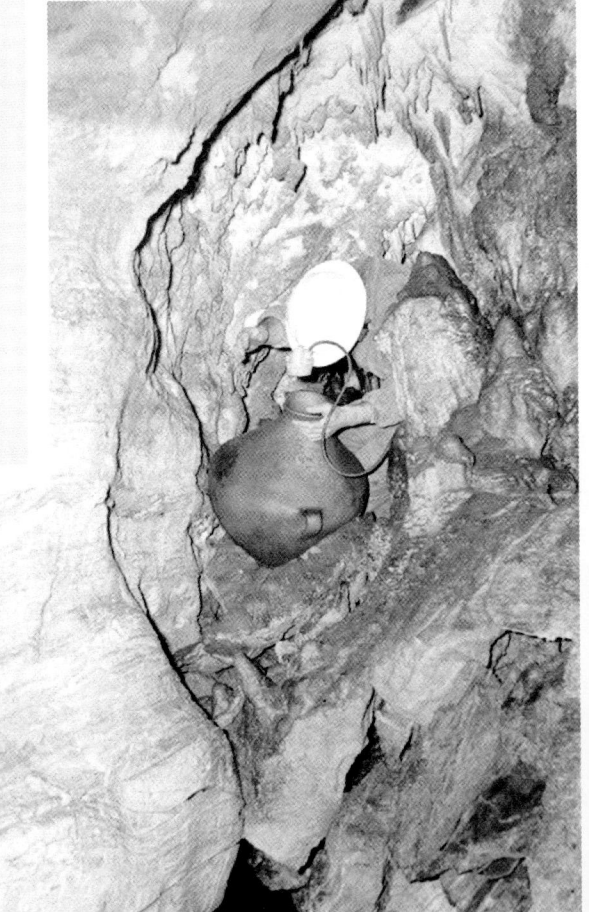

Neolithic pottery from the cave – photo Ivan Lichkov

Name: **DRÂNCHI DUPKA (253)**
Mramor Village, Yambol Distr.
Length 275 m. Denivelation 85 m
Coordinates: N 42° 02' 52.8" E 26° 32' 17.8"

A cave 3 km SE of Mramor, in Trias limestone of Sakar Anticline. A branched cave, going down in steps. Many boulders, dripping water.

Known to the local people. Partly explored on June 18, 1960 by cavers from Yambol, led by P. Karabadjakov. Surveyed in details by them in the 70s, and later by cavers from Plovdiv, led by Iv. Petrov and I. Savchev in 1986.

Fauna: nine species known, no troglobites.

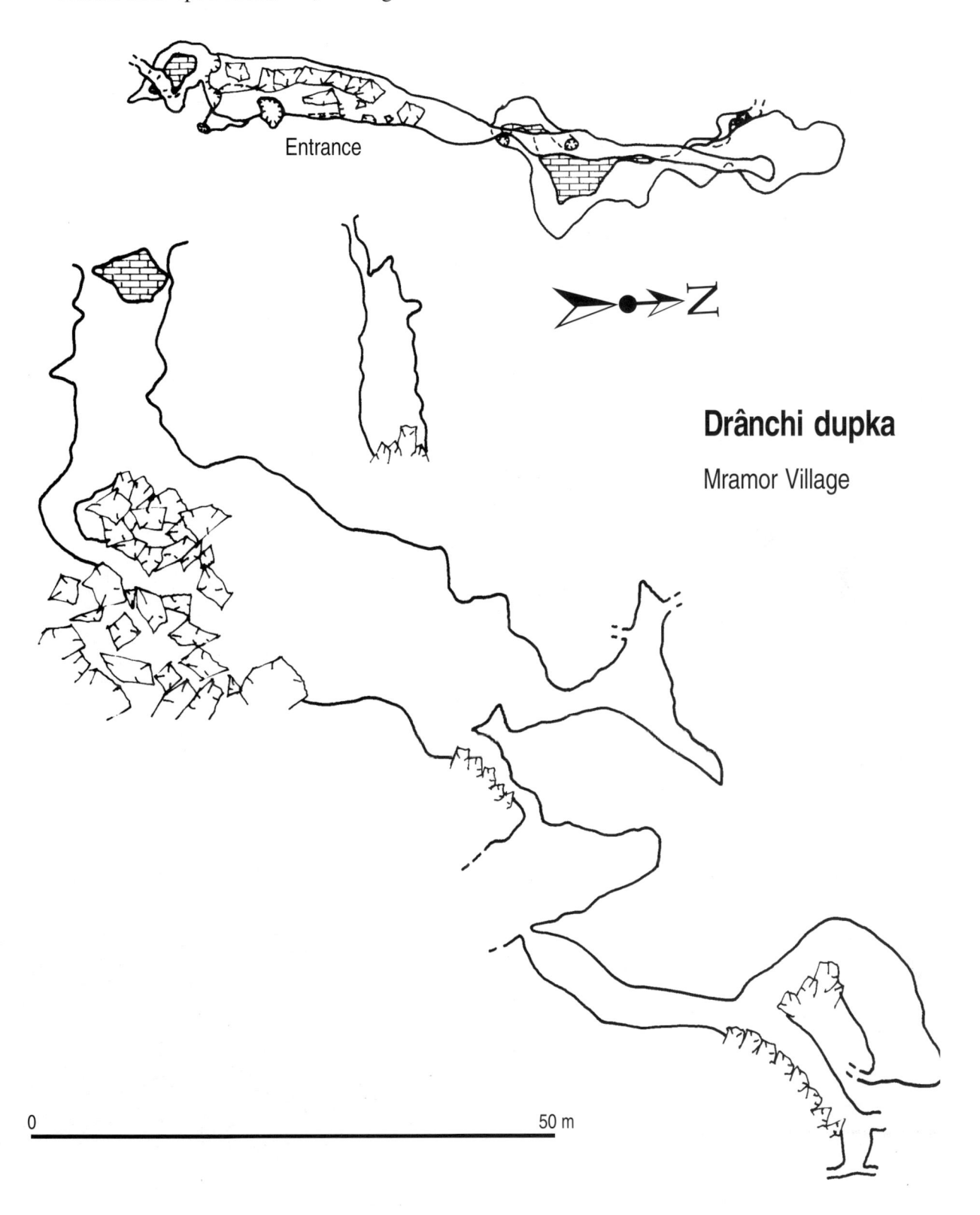

Drânchi dupka

Mramor Village

Name: **GEORGIEVATA (GYURGYOVSKATA) PESHTERA (259)**
Kosti Village, Burgas Distr.
Length 130 m. Denivelation -7 m

A cave in the area of Malkiya Budjak. Wet, horizontal, with a large entrance hall (20 x 10 m). From the hall start three galleries. Poor in formations, only at the end there are some stalactites. Floor covered with clay and blocks, fallen from the ceiling.

The cave has been explored and surveyed by St. Todorov, D. Dimitrova and A. Rachev from Strandja – Burgas and Akademik – Sofia caving clubs in 1975. Fauna collected by P. Beron (1975).

Fauna: six species, including the troglobite *Trichoniscus valkanovi* (Isopoda).

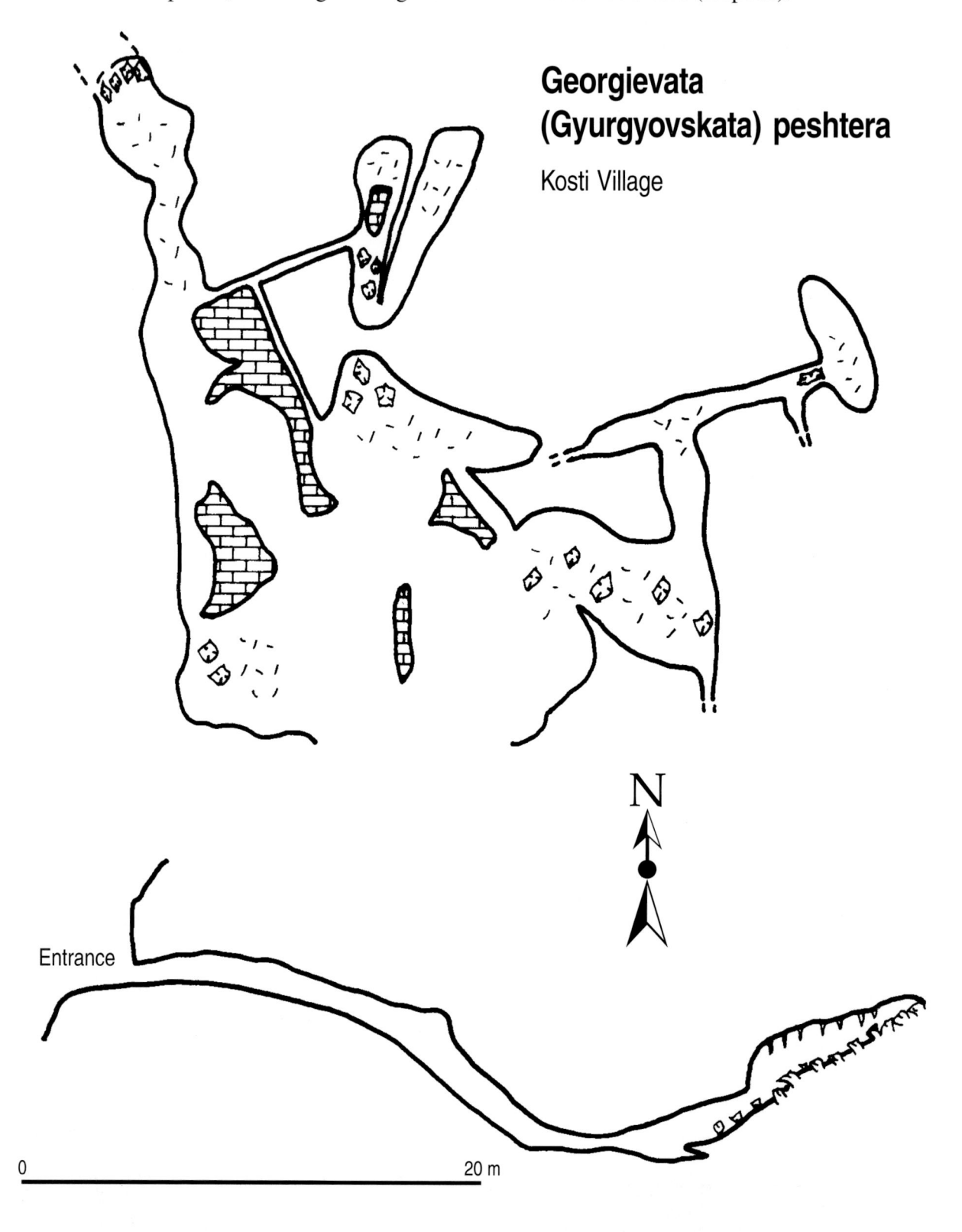

Georgievata (Gyurgyovskata) peshtera
Kosti Village

Name: **GOLYAMATA VÂPA (255)**
Stoilovo Village, Burgas Distr.
Altitude 320 m. Length 450 m. Denivelation -125 m

A cave pothole Golyamata Vapa in the surroundings of Petrova Niva (Bâzât), 1.5 km midway between Zvezdets and Stoilovo. Developed in Jurassic limestone. The largest chamber of the cave is 12 x 10 m and the height is more than 10 m.

The cave entrance is at the bottom of a funnel, which on its part is at the bottom of two valleys, dry in the summer. Danger of flooding. Golyamata Vapa is a descending cave pothole with several verticals, the highest being 20 m. Rich in many kinds of cave formations. Along most of the galleries, a small stream runs (1 l/sec), which in rainy times rises several times.

Golyamata Vapa was discovered and surveyed by the cavers from Akademik Sofia Caving Club P. Neykovski, A. Strezov, P. Delchev in 1975.

Fauna (collected by P. Beron in 1980, B. Petrov and T. Ivanova in 1991): 10 invertebrates known, no troglobites.

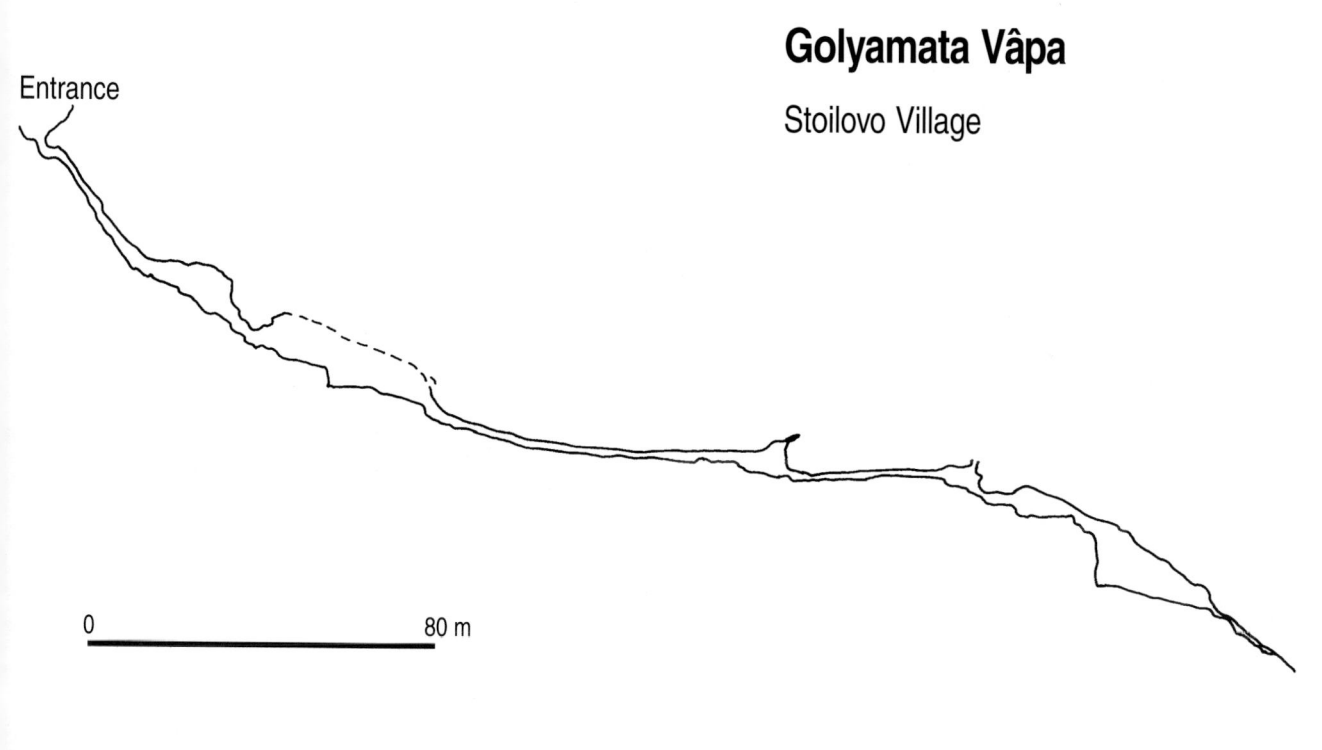

Golyamata Vâpa

Stoilovo Village

Name: **KALETO (254)**
Mladezhko Village, Burgas District
Length 302 m. Denivelation -15 m

A cave in the area of a Thracian fortress ("kale") in the locality Kaleto, SW of the village Mladezhko, Strandja Mt. Its entrance is a vertical well, 4.7 m deep. An inclined gallery follows, leading into a chamber with dimensions 10 x 7 m and an average height of 1.5 m. There are about 20 m of upper and lower, relatively short levels of the cave. The gallery reaches two walls of dry masonry of unclear purpose. After them, the main gallery enters a chamber, which is 10 m long and 8 m wide, then it goes down south. The length of the main gallery is 124 m, together with the two levels and the branches the total length of the cave becomes 302 m. In the cave, there are several flowstone formations, damaged by visitors. Formed in limestone from the Lower Trias (Anise – Nore).

Visited many times, explored especially in 1976, during an expedition of Akademik Sofia Caving Club. The actual map is due to Ivanka Hristova (Strandja Burgas Caving Club) and Tsvetan Lichkov (Akademik Sofia Caving Club).

Fauna (collected by P. Beron since 1963, later also by other Zoologists): 17 species known, but no troglobites.

Kaleto

Mladezhko Village

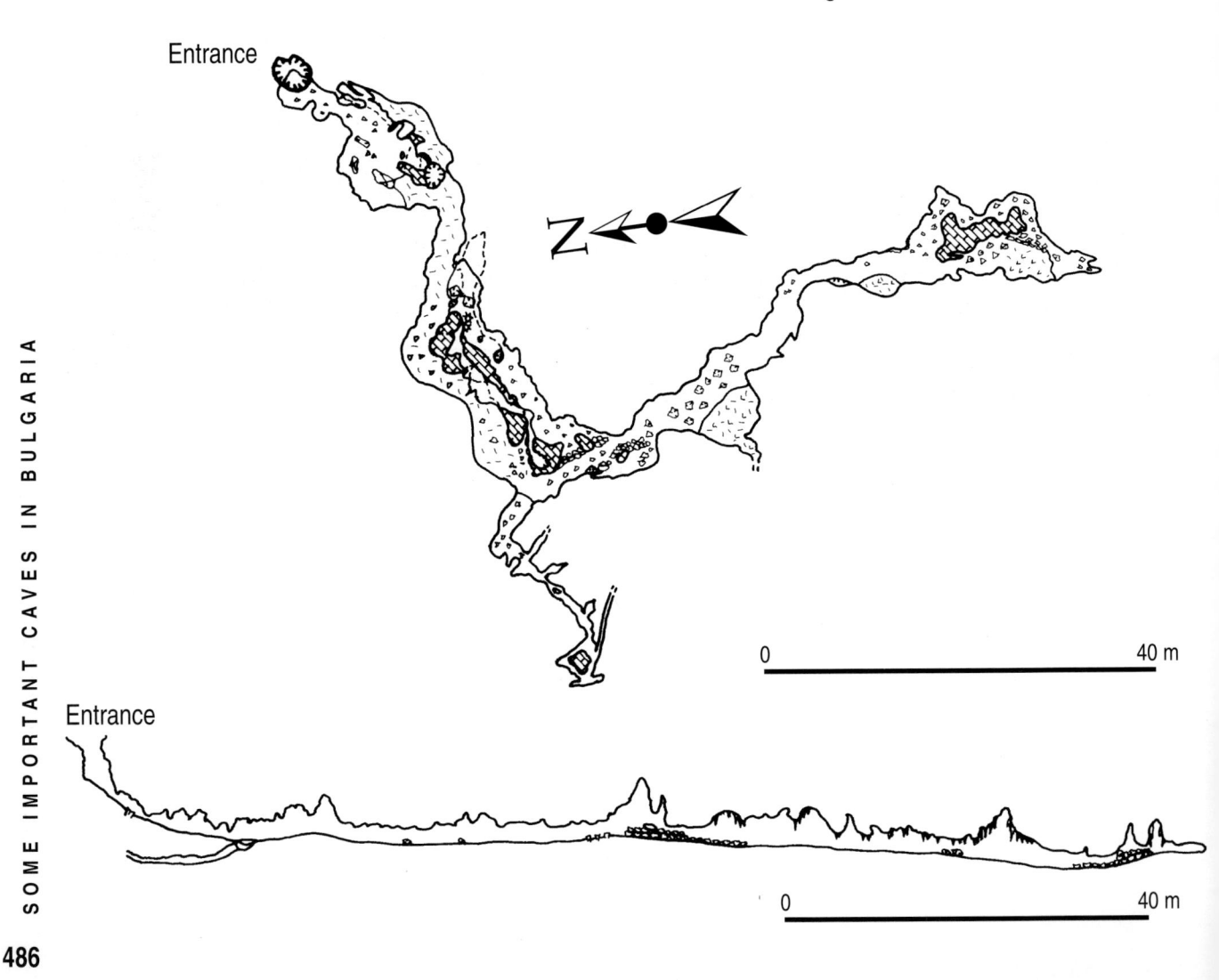

The entrance of Kaleto Cave – photo Ivan Lichkov

Name: **KIRECHNITSATA (KERECHNITSATA, GOLYAMATA MAHARA) (258)**
Kosti Village, Burgas Distr.
Total length 224 m. Denivelation -23 m
Coordinates: N 42° 00' 53.2" E 27° 49' 05.8"

A cave south of Kosti Village, in the locality of the same name. Wet, length on the main axis 143 m. Dimensions of entrance 7 x 6 m, nearby there is a pool with water. Inclined, branched. It opens in a ponor with a diam. of 14 m and depth 10 m. From the bottom of the ponor to the NE, a gallery (7 x 6 m) starts, near the entrance pool of water. At the 20th meter, two galleries start. The left one is 43 m long, the right has two branches, a total of ca. 80 m. The bottom of the gallery is covered with clay and fallen stones of different size. Surveyed in 1977 by P. Neykovski and P. Delchev (Akademik Sofia Caving Club).

Fauna: six species known, troglobite is *Lithobius bifidus* (Chilopoda).

Protected Natural Monument (together with 2 ha of adjacent land (Decree 4051/29.12.1973).

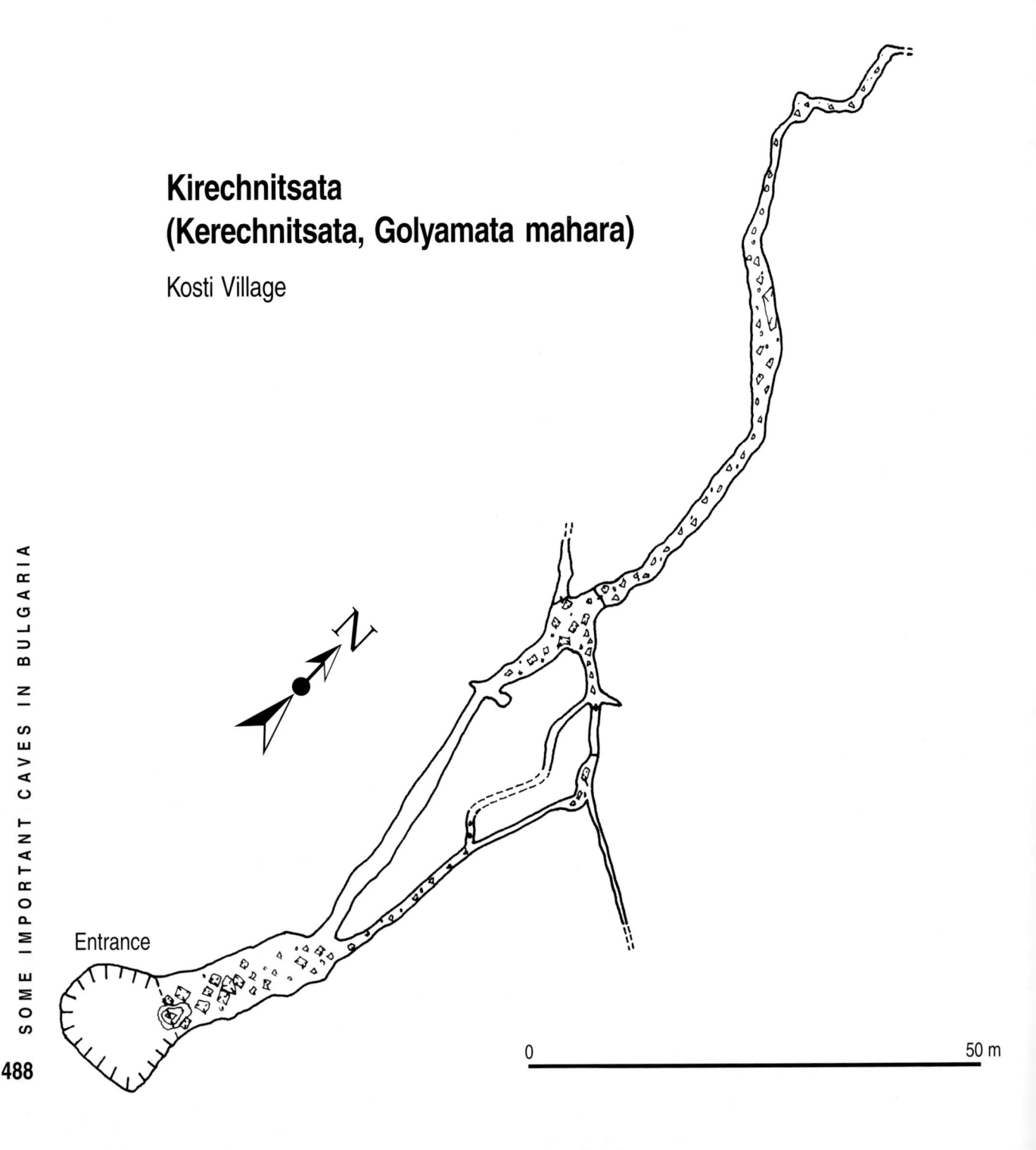

Kirechnitsata
(Kerechnitsata, Golyamata mahara)

Kosti Village

Entrance

Name: **PESHTERATA S DVATA VHODA (256)**
Stoilovo Village, Burgas Distr.
Length 208 m. Denivelation -29 m

A cave in the locality Bazat, not far from Petrova Niva. Formed in Jurassic limestone. The entrance is 5 m wide and 1 m high, facing west. A relatively spacious and humid cave. The entrance part is labyrinthic and dry. The floor is covered by clay and gravel material. Air temperature is 9 °C. In the cave there is a strong concentration of CO_2.

Explored and surveyed by Akademik – Sofia Caving Club in 1976, re-drawn and described in 1987 by I. Petrov, D. Angelov, M. Neychev, G. Shopov from the Speleology Circle at "S. Popov" Middle School (Akademik – Plovdiv Caving Club).

Fauna (collected by P. Beron in 1975): eight species known, troglobite is *Lithobius bifidus* (Chilopoda).

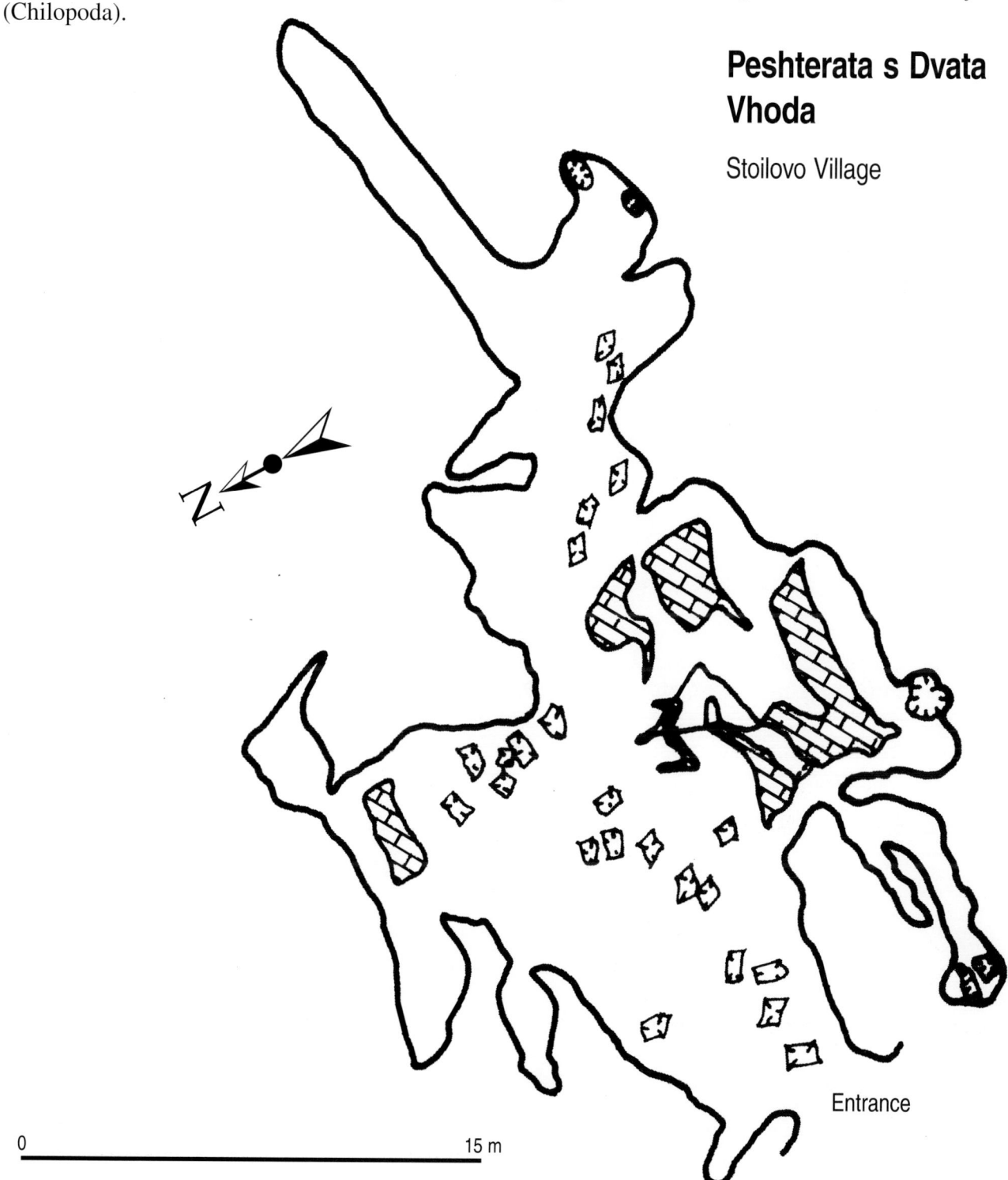

Peshterata s Dvata Vhoda

Stoilovo Village

Name: **STOYANOVATA PESHTERA (260)**
Kosti Village, Burgas Distr.
Length 150 m. Denivelation -27 m

A cave in the locality Malkiya Budjak, ca. 50 m from Rezovska reka. Elliptic shape of the entrance (1 x 2 m). As landmark serve 3 old beach trees near the entrance. Elliptic entrance (1 x 2 m). From it following the inclined (45°) floor we enter a small chamber with lot of fallen leaves. On the left there is a low brachyclasic gallery, after 15 m joining the main one. Descending diaclasic cave with length on the main axis 95 m. The main gallery is 1 to 7–8 m wide. There are two chambers – The White and The Big. Branching on the 30th m. The left branch leads to the White Chamber, full with flowstone formations, mostly moon milk. This chamber is connected with the Big Chamber, 50 m long and with inclination of 60°. The Big Chamber accounts for ca. 70% of the volume of the cave. The floor is covered with one meter deep layer of bat guano on moon milk. At the end of the chamber there is a 3 m vertical and a gallery 20 m long with onion-shaped stalactites at the end. Rights from the vertical, through a gallery 10 m long, intersected by verticals, is possible to reach the lowest point of the cave.

The cave is named after Stoyan – a local haydout (freedom fighter) and shepard. This cave has been explored, surveyed and described by P. Delchev and K. Burin from Akademik – Sofia Caving Club in 1975.

Fauna (studied by P. Beron in 1975 and 1980): spiders, opilionids, the troglobite millipede *Lithobius bifidus.*

Stoyanovata peshtera

Kosti Village

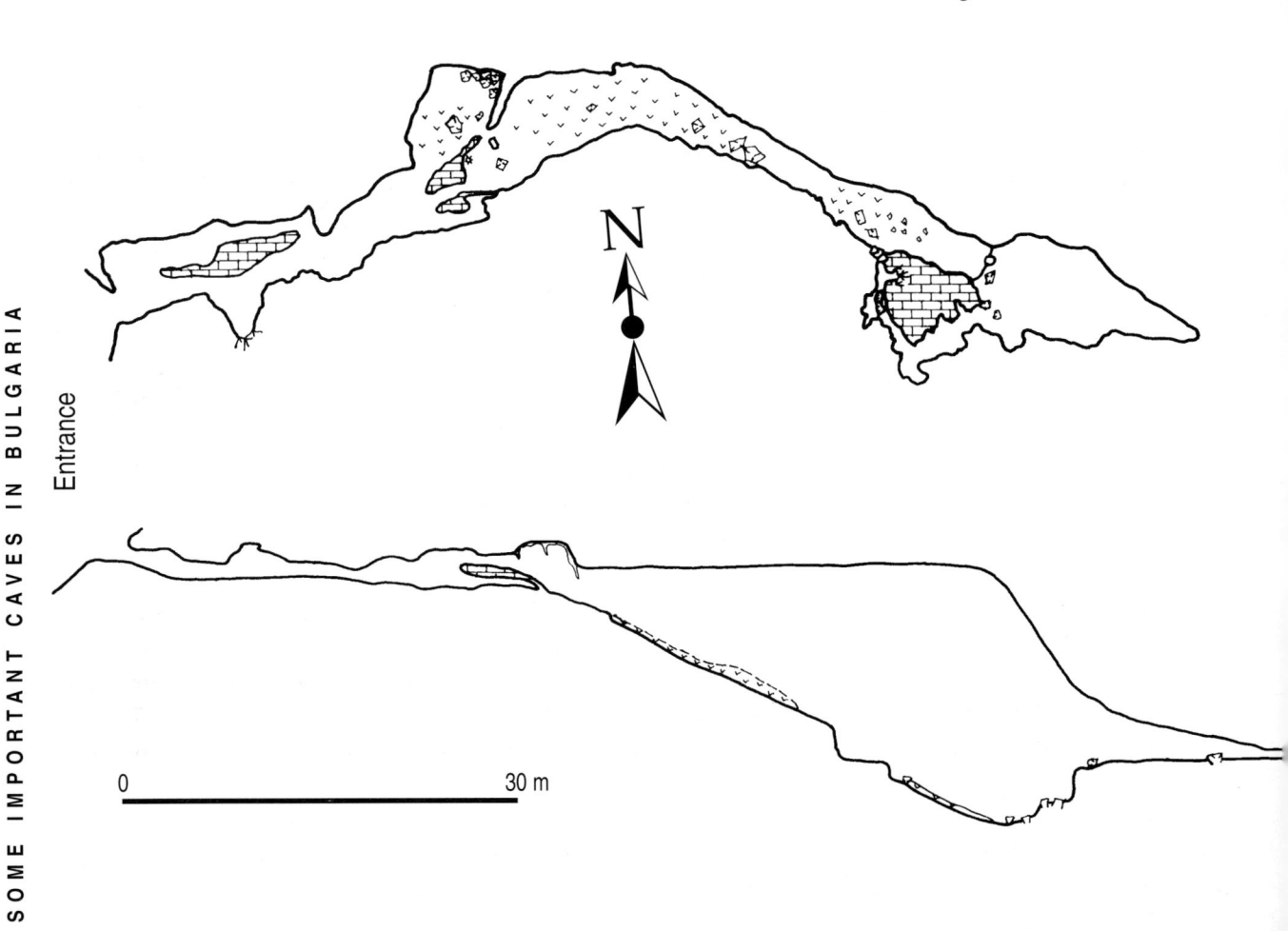

WHO IS WHO IN BULGARIAN SPELEOLOGY

Andreev, Dr Stoitse – born in 1937. Zoologist and Speleologist, explorer of caves in Bulgaria, Caucasus, Greece, Thailand and other countries. Assoc. Prof. in the National Museum of Natural History, Sofia (retired in 2002). Specialist in cave Isopoda and Amphipoda, author of many articles on the cave fauna of Bulgaria, Greece, Sarawak, Papua New Guinea, etc. Vice-President of the Republican Caving Commission (1968–1979).

Antov, Yasen – born in 1929. Journalist and writer, participant in the resurrection of the organized caving in Bulgaria. Leader of the Second International Expedition in the Rhodopes (1962). Author of many articles about caves and of the book "The Joy of the Searcher" (1967).

Atanassov, Dr Neno (1904–1996) – Zoologist, Director of the Natural History Museum in Sofia (1947–1962), Secretary of the Bulgarian Caving Society. Articles on the caves near Gintsi (Distr. Sofia), Byala (Distr. Sliven), Sâeva Dupka (Distr. Lovech) and others.

Balevski, Vassil – born in 1949. Caver, Photographer. Member of many caving expeditions in Bulgaria and abroad. Participant in the discovery, the exploration and the survey of many of the bigger caves in Troyan Balkan. First explorer of the deepest pothole in Bulgaria Raychova Dupka. Reached the bottom of Snezhna (Poland) through the highest-located entrance – Nad Kotlinami – 783 m. As a photographer he has spent many hours underground and shot many pictures of high quality, used by the Federation for its editions and for publicity on Speleology.

Benderev, Dr Alexey – born in 1954. Hydrogeologist and speleologist, Researcher in the Geological Institute – Bulgarian Academy of Sciences. Board member of BFS from 1992 to 1996. Participant in many caving expeditions in Bulgaria and abroad. In 1988, he took part in the National Caving Expedition in Cuba to study Guaso Plateau. Studies and publications in the field of Hydrogeology (Ponor Planina, Bosnek Area, Kameno Pole and other karstic areas). Co-author of the brochures about the deepest potholes in Bulgaria.

Beron, Dr Petar – born in 1940. Zoologist, Biospeleologist, Assoc. Professor, Director of the National Museum of Natural History in Sofia (since 1993), President of the Bulgarian Federation of Speleology (since April 12, 1985), one of the founders of Akademik Sofia Caving Club and of the Committee for Cave Tourism (1958). Leader of speleological expeditions to China, Vietnam, Cuba, participant in many other expeditions, explorer of caves in many countries in Europe, Asia, Africa, North and South America. Biologist of the British Speleological Expedition to Papua New Guinea (1975). Explorer of hundreds of caves in Bulgaria, author (partly with participation of V. Gueorguiev, P. Stoev and B. Petrov) of five parts of the Catalogue of Bulgarian Cave Fauna (1962, 1967, 1972, 1994, 2006). Author of books and other publications on the cave fauna of Corsica, Greece, South Asia, etc. Editor of the series of books "Tranteeva" and "Grottes bulgares". From 1993 to 1999, he was a Board Member (Adjoin Secretary) of U.I.S. Participant in the International Speleological Congresses in Barcelona, Budapest, Beijing, La Chaud-de-Font and Kalamas. Biospeleological specialisation in the Laboratoire Souterrain du C.N.R.S. in Moulis, France. President of the Balkan Speleological Union (since 2002).

Beshkov, Dr Vladimir – born in 1935. Zoologist, Herpetologist, Assoc. Professor in the Institute of Zoology (retired in 1995), explorer of caves and bats in Bulgaria and many other countries (Tanzania, Indonesia, Malaysia, Turkey, Greece, etc.). Author of papers on caves and bats.

Buresch, Dr Ivan (1885–1980). Zoologist and Biospeleologist, Member of the Bulgarian Academy of Sciences (since 1929, with an Academic Address devoted to the cave fauna of Bulgaria), initiator of the biological exploration of Bulgarian caves, Director of the Royal Institutions of Natural History in Sofia, co-founder of the Bulgarian Caving Society and its President (1940). Honorary President of the Bulgarian Federation of Speleology. He created the collection "Fauna cavernicola bulgarica" in the Royal Museum, in 1922 started the first systematic study of Bulgarian cave fauna. Published three important papers on this fauna (1924, 1926, 1936), as well as the first Catalogue of Bulgarian Cave Animals (reported at the International Congress in Budapest in 1927). Supervisor of the cave explorations in 1948–1949. He personally studied many of the caves in Bulgaria and organized his staff to study caves. Thanks to the materials, sent by him, many new taxa of cave animals have been described by prominent specialists.

Chapanov, Valentin (1962–1989). Caver and diver. Participant in expeditions for studying of many caves in Bulgaria, France (1984) and Spain (1986, 1987, 1988). In 1987, together with the divers I. Gunov and M. Dimitrov, he penetrated the bottom siphons of the pothole BU-56, overcame two new siphons and the 650-meter gallery after them, reaching a depth of 1408 m. Explorer of the siphons Popovata, Musinskata, Vodopada and others. Died during the exploration of the siphon in the cave Urushka Maara near Krushuna on May 26, 1989.

Daaliev, Trifon – born in 1947. Speleologist and alpinist, graduated in 1979 from the National Sport Academy. Secretary of BFS since 1979. Participant in many caving expeditions in Bulgaria, Cuba, Albania, Austria, France, Indonesia, Malaysia, China, Ukraine, Georgia, Poland, Turkey, Greece, etc. Author of publications about caves in many regions of Bulgaria and other countries. Participant in the exploration and the survey of many caves, including the deepest pothole explored so far by Bulgarians – S-1 in Austria. Head of the Unit for Cave Rescue at BFS. Co-author in the book "Bulgarians in the world's abysses" (1986). Compiled series of brochures with descriptions and maps of the most important caves and potholes in Bulgaria, also of the brochure "Preservation of Caves". Editor of the series "Grottes bulgares". During the XXII International Congress of Speleology in Switzerland (1997), he was elected First Vice-President of the Cave Rescue Commission of UIS.

Delchev, Dr Hristo – born in 1939. Zoologist and Biospeleologist, Assoc. Professor in the Institute of Zoology, specialist in cave and other spiders, former President of Akademik Sofia Tourist Society and Caving Club. Participant in the exploration of many caves in Bulgaria, Caucasus and other regions. Author of publications on cave spiders and of the manual "Speleology and Caving" (1979).

Delchev, Petar – born in 1949. Active caver, member of "Akademik" – Sofia Caving Club, leader and participant in many caving expeditions in Bulgaria, Austria, Vietnam, Cuba and other countries. Actively participated to the exploration of the pot holes in Pirin, the caves of Strandja and many others. Compiled the collection of papers "Pirin – caves and pot holes " (2002).

Denkov, Aleksander (1929–1972). Painter, founder of cave diving in Bulgaria. Head of the first group for cave diving in BFS, realized the first cave diving in a siphon in Bulgaria (Temnata Dupka, Lakatnik, 1959). Under his guidance and with his participation, the diving in the source Zhabokrek – Chiren was realized. Active member of the Committee for Caving in 1958–1972.

Dimitrov, Milen – born in 1963. Active caver and diver. He participated in expeditions to Albania, Spain, France. In 1986, he reached the bottom of BU – 56 and overcame three known siphons, helping his colleague I. Gunov to overcome a new one. In 1987, he participated (together with I. Gunov and V. Chapanov) in the penetration of the known one and of two new siphons and explored the 650-meter gallery, ending in another siphon at a 1408 m depth – thus BU-56 becoming the second deepest pot hole in the World at that time. Cave rescuer.

Dinev, Dr. Lyubomir (1911–1986) – Professor in Geography, President of the Bulgarian Federation of Speleology since its foundation in 1959 to 1985, then President of Honour. Author of articles on cave tourism. Board Member of U.I.S. from 1977 to 1981. Represented Bulgarian Speleology at UIS Congresses in Postoina, Stuttgart, Olomouc, Sheffield and Boulder Green.

Djambazov, Nikolay (1919–1982) – Archeologist, Associate Professor in Archeology, participant in excavations to the caves Devetashkata peshtera (1950–1952), Pesht (1951–1953), Lovech caves (1952–1956), Samuilitsa I and II near Kunino (1955–1960), Morovitsa (1955), Ochilata (1959–1960), Parnitsite (1959–1960), Orlova Chuka (1961), Bacho Kiro (1971–1976) and others. Author of many publications on cave archaeology. Retired in 1979. Books: "The caves in Bulgaria" (1958), "Devetashkata peshtera"(1960).

Dragandjikov, Ivan (1944–2000) – caver since 1957. Caving formation in 1961. Junior Instructor (1961). Founder of Zlosten – Kotel Caving Club (25 May 1963). Up to 1971 leader of the club. Participant to the National Expedition to Sniezhna (Poland, 1969). Instructor (1980). Contributed to declaring of Zlosten Protected Area.

Evtimov, Boris – caver, founder and President of Bessapara Caving Club at Kupena – Peshtera Tourist Society since it's creation in 1960 until the death of Evtimov. Instructor since 1976. Discoverer and first explorer of the cave Snezhanka (3 January 1961).

Garev, Borislav (1962–1994) – caver from Studenets – Pleven Caving Club. In 1980 became junior instructor. In 1981 created speleological section to the "Geo Milev" – Gymnasium in Pleven. Instructor (1982). Under his guidance were explored 56 caves. Contributed to the discovery of several new troglobites, two of which dedicated to him (the Isopod *Trichoniscus garevi* and the beatle *Duvalius garevi*). Founded the Youth Speleoclub "R. Popov" at the Pleven club. Participant to the expeditions to Jean Bernard (France) and to Bu-56 (Spain).

Gazdov, Senko – born in 1952. Speleologist, Secretary of Studenets Tourist Society in Pleven, leader and participant in many caving expeditions in Bulgaria, Spain, Albania, Austria, France, Italy and other countries. In 1984, he reached the bottom of Jean-Bernard (France, – 1358 m) and in Spain in 1986–1987 he reached the dry bottom of BU-56 at – 1325 m. Board member of BFS.

Genov, Nikolay – born in 1950. Caver and photographer. Author of many publications about caving, and of books about India, Egypt and other countries. Co-author of the book "Bulgarians in the abysses of the world" (1986). Initiator and technical leader of the expeditions Gouffre Berger – 1969 and Pierre – Saint – Martin – 1973. Participated actively in the exploration and surveying of the caves Morovitsa, Maliya Sovat, Lyastovitsa, Bacho Kiro and others.

Gladnishki, Nikolay – born in 1947. Speleologist, Junior Instructor in Speleology (1969), Senior Instructor (since 1976). Leader and participant in many expeditions in Cuba, China, Albania, Austria, Poland. Cave rescuer. He participated in the training of cave rescuers and in the surveying of many caves in Bulgaria and other countries, including the deepest pothole so far explored entirely by Bulgarians – S-1 in Austria. For many years Board Member of BFS (1974 -1990).

Grozdanov, Eng. Alexander – born in 1941. Explorer of many caves in Bulgaria and other countries. For many years actively working in "Akademik" – Sofia Caving Club. Co-author of the book "In the surroundings of Kotel " (1979).

Gueorguiev, Vassil (1935–1997) – Zoologist and Biospeleologist, Assoc. Prof. in the Institute of Zoology, author of many scientific publications, mainly on cave beetles, as well as of fundamental general papers on the cave fauna of Bulgaria and the Balkan Peninsula (La faune troglobie terrestre de la péninsule Balkanique. Origine, formation et zoogéographie, 1977), also of the popular books "Inhabitants of the eternal darkness" (1961) and "The secret of caves" (1968).

Gyaurov, Eng. Vesselin – born in 1945. Explorer of caves, junior instructor (1964), senior instructor (1976). One of the founders of "Edelweiss" – Sofia Caving Club, Board Member of the Counsel of BFS (1974-1989). Introducing the SRT in Bulgaria, for many years teaching in the caving courses of BFS. Constructor of the first jumars in Bulgaria ("Deltra") in Bulgaria.

Ikonomov, Dr Georgi – born in 1923. Medical Doctor, founder of the Caving Department of Russe District Museum and of the Caving Club in Russe (1958), author of the first map of the caves in Bulgaria (1958).

Ilandjiev, Dimitar – (1932–2000) – caver since 1956; caving formation in 1961, Instructor

(1964). Founding member and First President of Cherni Vrah – Sofia Caving Club (1960). Instructor of Ivan Vazov – Sofia Caving Club (1963). Discovered and explored many important caves.

Iliev, Zdravko – born in 1946. Caver since 1963, member of Edelweiss Sofia Caving Club. Participant in many caving expeditions, expert in Bulgarian caves, in charge of the Main Card Index of Bulgarian Caves. Publications: "The Karst of Karlukovo" (1997), "Kotlya" (2000), "40 year of organized caving in Shumen" (2001) and "Streshero" (2002). Honorary member of BFS.

Jalov, Alexey – born in 1953. Speleologist, graduated from the National Sport Academy in 1978, Permanent Vice-President of the Bulgarian Federation of Speleology (since 1979). Leader or participant in many speleological expeditions in Albania, China, Cuba, Vietnam and other countries. Author of many articles about caves in Bulgarian and foreign journals, as well as of the books "Days and nights underground" (1979), "Underground Notebook" (2000) and "Bulgarians in the abysses of the world" (1986, Editor).

Kolev, Boris (1933–2001) – Caver and Researcher, author of publications on the caves of the Eastern Rhodopes. One of the founders of the Caving Club in Haskovo (July 8, 1958). Honorary member of the Bulgarian Federation of Speleology.

Landjev, Nikola – born in 1953. Active caver from "Edelweiss" – Sofia Caving Club. Together with Rossen Vatev editor and publisher of the series "Speleopraktika" for more than 15 years. Leader and participant in many caving expeditions in Bulgaria, Austria and other countries. Together with Yordan Pavlov explored ice caves and pot holes in Pamir. Designed and prepared stretchers for cave rescue.

Lichkov, Eng. Ivan – born in 1946. Caver, active member of Akademik Sofia Caving Club, and explorer of many caves in Bulgaria, Austria, Georgia, Cuba, Vietnam, Antarctida and other countries and areas. Participant in many explorations of the caves of Strandja and of potholes in Pirin.

Lichkov, Eng. Tsvetan – born in 1940. Caver, participant in expeditions to the caves of Cuba in 1981 and 1988 and of many explorations of caves in Bulgaria, Germany, France, Ukraine, Vietnam and other countries.

Marinov, Bozhan (1909–1985) – caver from 1955, caver's formation in 1961. Junior Instructor (1961), Instructor (1964), Senior Instructor (1964). Founder and First President of "Madarski Konnik" – Shumen Caving Club (18.02.1961). Explored and surveyed some 35 caves in Shumen Area. Later moved to Varna, where he became President, than Honorary President of "Galata" – Varna Caving Club.

Markov, Vassil (1938–2000) – caver, founding member of Balkansko Eho Caving Club and it's Instructor since 1963. Junior Instructor (1963), Instructor (1964). Since 1965 member, than longtime President of Steneto Caving Club (Troyan). Senior Instructor (1976). Discovered and explored many caves in Troyan Balkan like Raychova Dupka, Trona, Golyamata Yama and others. Participants to many expeditions in Bulgaria and abroad.

Michev, Tanyo – born in 1939. Zoologist, caver and diver, Assoc. Professor in the Central Laboratory of General Ecology, Bulgarian Academy of Sciences, active in the ressurection of the organized caving in Bulgaria in the 50-ties. Participant to the study of many caves and pot holes in Bulgaria, as well as in some of the first siphon diving attemps.

Mikov, Vassil (1891–1970) – Archeologist, speleologist, explorer of the man-inhabited caves. Author of excavations of the caves near Kunino (1922), Dryanovo (1924), Karlukovo (1926), Devetashkata peshtera (1927, 1950), Loveshkite peshteri (1927), the caves near Karpachevo (1929), Rabishkata peshtera (1929), the caves near Dolni Lom (1929) etc. Author of more than 240 scientific and popular papers, among which: 1926. "Caves and pot holes between Iskar and Vit"; 1928. "Caves in Belogradchik Area" and others.

Mustakov, Vesselin – born in 1954. Caver, active member of Heliktit Caving Club – Sofia, participant in and leader of many caving expeditions in Bulgaria, Albania, Italy, Uzbekistan and other countries. Reached the bottom of KiIsi pot hole in 1990. For many years teaching in the courses for young cavers, instructors and rescuers in BFS.

Nedkov, Petko – born in 1934. Speleologist, Secretary of RCSCT (1967–1974), Vice President of BFS. Leader of the first Bulgarian expedition in a deep pothole (Snejna in Poland, 1969). Books: "Manual of Speleology" (1973), "ABC of the single rope" (1983). Head of the Cave Rescue Group at BFS (1974–1989). Honorary member of the Bulgarian Federation of Speleology.

Nedkov, Vassil (1949–1989). A prominent caver and alpinist. Participant to many caving expeditions in Bulgaria and other countries. In 1980, he reached the bottom of the pothole Mavro Skiadi on the island of Crete (-347 m). Many caving expeditions in Bulgaria, France, Italy, Greece, Cuba, China and other countries. Descended Pierre-Saint-Martin (-1171 m) in 1973, Spluga dela Preta (- 878 m) and Abisso Michele Gortani (-920 m) in 1979. In 1984, he reached the bottom of the Greek potholes Epos (- 451 m) and Provatina (- 407 m). In 1982, he explored caves in Austria – S-1 (-584 m), S-2 (-460 m), in Cuba – 1988 (Campanario, Sumidero del Guaso), China in 1989. Cave rescuer since 1977. Mountain climbing: 1985 – Eiger, North Face; 1988 – Peak Korzhenevska (7105 m). Board Member of BFS. Explorer of the siphons of the caves Katsite, Toplya, Mussinskata, Brashlyanskata, Temnata Dupka and others. Died during the penetration in the siphon in the cave Urushka Maara near the village Krushuna on May 26 th 1989.

Neykovski, Eng. Panayot (1936–1986). Caver, one of the founders of Akademik Sofia Caving Club. Participant and leader of expeditions in Ukraine, Cuba – 1981 and many expeditions in Bulgaria (Kotel, Strandja, Ponor Planina, Steneto, Pirin, Cherepish, etc.). Member of the management of Akademik Club and the Caving Federation. Author of many articles in journals and of the books "Troyanska Planina – Guidebook" – 1977, "Ponor Planina" – 1975, "Cherepish" – 1982.

Peltekov, Valery – born 1953. Caver and alpinist. Lecturer in Mountaineering. Junior Instructor since 1971, Instructor since 1975. Cave rescuer (1981). Participant in caving and climbing expeditions in Cuba, Albania, Greece, Austria, Poland, Caucasus, Himalaya, Pamir, Alps, Carpathians. Expert in Bulgarian mountaineering. Descended to the bottom of some of the biggest verticals: Epos – 451 m, Provatina – 407 m, Abisso Michele Gortani, Spluga della Preta. Co-author of the book "Bulgarians in the abysses of the world" (1986). For many years Board Member of BFS.

Penchev Simeon – born in 1938. Caver. Books: "Fairy Tale World" (1970). Participant in the exploration of several Bulgarian caves and potholes – Radolova Yama, Duhlata, Manailovata Peshtera, Vodopada and others.

Petkov, Krassimir – born in 1962. Caver and diver. Participant in many caving expeditions in Bulgaria and abroad. In France in 1984 he reached the bottom of Jean-Bernard (- 1358 m), in Spain in 1986 – 1987 he reached the dry bottom of BU-56 at – 1325 m. Explorer of the siphons in the caves Popovata, Vodopada, Glava Panega and others. In the siphon of Glava Panega he reached the deepest point of many diving attempts (- 52 m depth and 230 m length).

Petkov, Dr. Stephan (1866–1951) – Professor of Botany, first President of the Bulgarian Speleological Society (1929–1931 and 1933–1935). Studies cave algae.

Petkova, Angelina – born in 1941. Caver, President of the City Section for Caving and Speleology and of Cherni Vrâh Sofia Caving Club. Participant in the exploration of many important Bulgarian caves, drawing their maps, including the ones in this book.

Petrov, Eng. Pavel (1884–1984) – Hydroengeneer, Speleologist, Founder of the Bulgarian Speleological Society in 1929. Studied many Bulgarian caves, using a selfmade wooden boat to explore the lakes in Devetashkata Peshtera near Devetaki and Temnata Dupka near Lakatnik. Published articles on the caves explored.

Popov, Lyuben – born in 1935. Caver, President of the caving club at "Vitosha" Tourist Society, then of the caving club at "Aleko Konstantinov" – Sofia Tourist Society. Participant in the exploration of many important Bulgarian caves, also to caving expeditions in Poland, Russia, Italy. Articles in periodical editions.

Popov, Dr Rafail (1876–1940) – Archaeologist and Speleologist, Professor. Director of the National Archaeological Museum (1929–1938),

President of the Bulgarian Caving Society in 1937–1938 and in 1940 (until his death on August 15, the same year). Directed many excavations in Bulgarian caves, author of studies and books on cave archaeology and paleontology. As a result of the excavations in Temnata Dupka in Karlukovo, for the first time the presence of Paleolithic culture on the Balkan Peninsula was proved. Started the mapping of Bulgarian caves. Contributed to the study and the popularization of Bulgarian caves by publishing 47 scientific and popular papers.

Popov, Dr Vassil – born in 1954. Zoologist and Paleozoologist, Associate Professor in the Institute of Zoology, Bulgarian Academy of Sciences. Author of many articles on the paleontology of small mammals in caves. Participant in exploration of caves in Bulgaria and Vietnam.

Popov, Dr Vladimir (1912–1998) – Geographer, Geomorphologist, Karstologist, Assoc. Prof. in the Institute of Geography – Sofia (since 1965). Books: "Travel Underground" (1982), "Show caves in Bulgaria" (1987), "Ledenika", "Saeva Dupka", "Snezhanka", "Bacho Kiro" and many others. Author of precise measurements and survey by means of theodolith of many caves and of important morphological studies on the karst and caves. Also author of the map of Bulgarian cave regions (1:200 000) and of the first subdivision of Bulgarian caves into geographical regions.

Radev, Dr Jeko (1875–1934) – Geographer, Professor, author of the important study "Karstic forms in Western Stara Planina" (1915). The lake Jekovo Ezero in the cave Temnata Dupka near Lakatnik Railway Station bears his name.

Radev, Nenko (1899–1944) – explorer of caves, started publishing the first Catalogue of Bulgarian Caves (two parts published, in 1926 and 1928). Several cave animals are named after him (*Radevia, Paranemastoma radewi, Brachydesmus radewi, Pheggomisetes radevi*).

Radushev, Eng. Radush (1911–1991) – Speleologist, mining engineer, member of the First Bulgarian Caving Society and of BFS. In 1934, he carried out the first theodolith survey of a Bulgarian cave (Bacho Kiro). Participant in the cave brigade in 1948 and in the resurrection of the organized caving in Bulgaria in 1958. From 1964 to 1973, he was President of Edelweiss Sofia Caving Club. Since 1975 in charge of the Main Card Index of Bulgarian Caves. Participant in the International Speleological Congresses in Yugoslavia (1965) and Czechoslovakia (1973).

Rashkov, Ivan – born in 1946. Caver, participant in many expeditions to Bulgaria, France and other countries. In 1969, as technical leader, he reached 800 m in the pothole Gouffre Berger in France. President of Iskar Caving Club since 1979. Co-author of the book "Expeditions Gouffre Berger-69 and Pierre Saint Martin-73" (1977).

Raychev Georgi – born in 1953. Geographer, caver from the club in Chepelare (Rhodope). Explored many caves in the Rhodopes and other parts of Bulgaria, as well as in Albania and other countries. Together with his father D. Raychev, he compiled the jubilee edition "Fifty years of caving in Chepelare (1950–2000)"(2002).

Raychev, Dimitar (1922–2002). School teacher, veteran Speleologist, founder of a caving group (1950) and the caving club (1961) in Chepelare, as well as of the Museum of Speleology and Bulgarian Karst in Chepelare (Museum collection since 1968, inaugurated in 1980), together with his son Georgi Raychev and his daughter-in-law Yovka Raycheva. From 1964 to 1985 he was editor of the bulletin "Rodopski Peshternyak" (67 issues). Books: "Chepelare and its surroundings", "The Trigrad Gorge and Dyavolskoto Garlo", "The Karstic Wealth of Yagodina Village". Co-editor of the volume "50 Years of Caving in Chepelare (2002). Articles on Archaeology and Palaeontology in caves. Honorary member of the Bulgarian Federation of Speleology.

Sâbev, Dimitâr (1937–1980). Caver, Geologist, President of the City Section for Caving and Speleology – Sofia in 1979–1980. President of Edelweiss Sofia Caving Club in 1979–1980. One of the editors and promoter of the bulletin "Rodopski Peshternyak ". Author of the book "Rupchoskite Peshteri" and of many studies in the fields of Geology and the Speleology. Participant in expeditions in the Rhodopes and other parts of the country. Explored the karst and caves in the areas of Radyuva Planina, Yagodina, Gela, Zabardo, Rozhen and others.

Shanov, Dr Stefan – born in 1948. Geologist, Professor, Head of the Laboratory of Seismotectonics of the Institute of Geology – BAS. Karstologist, specialist in the tectonics of karstic areas. Author of more than 150 scientific publications, including more than 20 on the problems of karst. Caver (since 1969). From 1994 to 2002 President of Akademik Club, Sofia. Participant or Leader of many caving expeditions in Bulgaria, France, Cuba (1988), Albania (1994), etc.

Shopov, Yavor – born in 1962. Author of papers on the speleoluminescence and other studies on the cave minerals, etc. President of the Commission for Physicochemistry of Karst of UIS. Author of many publications on the luminescense, the cave minerals and others. Created the section "Speleology" and of Expeditionary yearbook. Studied caves and glaciers in Bulgaria, India and other countries.

Škorpil, Karel (1859–1944) and **Hermengild** (1858–1923) – brothers of Czeh origin, teachers – encyclopaedists, authors of the book "Krazski yavleniya v Bâlgariya" (Karst phenomena in Bulgaria) (1900) and other books on the karst and caves in Bulgaria.

Spassov, Eng. Konstantin – born in 1941. Hydrogeologist and Speleologist, Explorer of many caves in Bulgaria, ex- Vice President of the Bulgarian Federation of Speleology. Participant in expeditions in Poland (Snezhna, 1968), Ukraine, China, Cuba and other countries.

Stefanov, Dr Petar – born in 1954. Geographer, Geomorphologist, working in the Institute of Geography, BAS. Participant in many expeditions in Bulgaria, the expedition Cuba – 88, scientific conferences in Poland, etc.

Stoev, Dr Alexey – born in 1952. Astronomer, Director of the Astronomical Observatory in Stara Zagora. President of Sârnena Gora Tourist Society in Stara Zagora. Vice President of the Bulgarian Tourist Union, Board Member of the Bulgarian Federation of Speleology. Participant in many caving expeditions in Bulgaria, Cuba and other countries and of the International Congresses of Speleology. Author of articles of archaeoastronomy and speleo-climatology. Member of the governing body of the International Committee of Archaeoastronomy in Oxford and of the European Society for Astronomy and Culture.

Stoitsev, Vassil – born in 1941. Organizer in BFS from 1968 to 1978. Participant and leader of many caving expeditions in Bulgaria, Russia, Poland and other countries, also in explorations and mapping of Bulgarian caves. Deputy chairman of the Rescue Group at BFS in 1974 – 1979. Many articles in the journal Tourist, Echo Newspaper and other periodicals. Co-author of the books "Ponor Planina" (1975) and "In the surroundings of Kotel "(1979).

Strezov, Alexander – born in 1942. Caver from "Akademik" – Sofia Caving Club. Actively exploring for many years the longest cave in Bulgaria Duhlata and other caves in the region of Bosnek, as well as caves in Stranja, Devetaki and other karstic areas. Participant in expeditions in France and Great Britain. Publications in periodicals.

Taparkova-Pencheva, Ana – born in 1946. Caver and alpinist, President of Vitosha Sofia Caving Club. In 1969 she reached the bottom of Gouffre Berger in France (-1122 m, World record for women), also reached the bottom of Pierre Saint Martin (-1171 m) in 1973, of Snejnaja (- 1370 m) in Caucasus in 1986 (another women's world record) and of the Italian abysses. The first Bulgarian woman who crossed the -1000 m depth underground. Leader and member of many caving expeditions.

Tranteev, Petar (1924–1979) – Geographer and Speleologist, founder of the organized caving in Bulgaria after the 50s. Member of the First Bulgarian Caving Society, Secretary of the Committee for Cave Tourism and of the Republican Commission for Caving and Cave Tourism, Vice President of the Bulgarian Federation for Caving (1961–1966 and 1974–1979), Geographer in the Institute of Zoology of the Bulgarian Academy of Sciences (1966–1974). Scientific leader of the Expedition to Gouffre Berger in 1969, leader of many caving expeditions in Bulgaria, initiator of creating the Main Card Index of Bulgarian Caves. He educated hundreds of cavers and enjoyed a high international standing. Books: "Magura" (1962), "Caves – objects of tourism" (1965), "Karlukovo Gorge" (1966), "The Secret of the Caves" (1968, together with V. Gueorguiev), "The

Caves in Bulgaria" (1978, together with K. Kossev), "Surveying caves and potholes" (1981, together with Y. Velinov).

Vassilev, Viliyan (Vilitsata) (1957–1994). Caver since 1975. Board member of "Akademik" – Sofia Caving Club, instructor and coach of the club. Participant and leader of expeditions in Bulgaria, Vietnam'89; Georgia – "Pantyuhina'1989", Ukrayna – "Krim"1980. In France in 1987 did the traverse Felix Tromb – Hen Morte (Coquille – Pin Blanc) (length 10 km and denivelation -760 m, in 1989 reached the bottom of Gouffre Berger (-1122 m).

Velev, Georgi (1913–1989) – caver since 1958; caving formation in 1961; Junior Instructor (1961), Instructor – 1964, Senior Instructor - 1969. President of the caving club in Varna (1961–1971), President of the District Counsil of Speleology since 1979. Participant to the first caving visit to Romania in 1963.

SELECTED LITERATURE

BULGARIAN SPELEOLOGICAL SERIES:

Izvestiya na Bâlgarskoto peshterno druzhestvo (Bulletin de la Société Spéléologique de Bulgarie) – Vol. **1** (1936), **2** (1940)

Bâlgarski peshteri (Grottes bulgares) – Ed. Bulgarian Federation of Speleology – No **1-2** (1964, P. Beron, Y. Antov, T. Michev, Eds.), **3** (1985, P. Beron, T. Stoychev, A. Jalov, T. Daaliev, Eds.), **4** (1986, the same Eds.), **5** (1991, T. Daaliev, P. Nedkov, A. Jalov, Eds.), **6** (1999, P. Beron, T. Daaliev, A. Jalov, Eds.)

Rodopski peshternyak [Rhodope Caver] (published in Chepelare by D. Raychev from 1964 to 1985)

Godishnik na Akademik [Yearbook of "Akademik"] – **1** (1968, ed. 1969), **2** (1969-1970), **3** (1971–1972)

Kazanlâshki peshternyak [Kazanlâk Caver] (edition of the Caving Club in Kazanlâk) – **1** (1980), **2** (1983)

Vrachanski Peshternyak [Vratsa Caver](edition of the Caving Club "Veslets" in Vratsa) – 1964, 1986, 1987

Expeditionary Yearbook of Sofia University "Kliment Ohridski", **1**, 1985, Y. Shopov (Ed.), 146 p.; **3-4**, 1989, 128 p.; **5A**, 1990, 140 p.

Bâlgarska speleologiya [Bulgarian Speleology] – **1** (1989); **2** (1990) **3** (1991) Edition of BFSp, the Museum of Speleology and "Studenets" – Caving Club in Chepelare (Editors G. Raychev, A. Jalov, E. Stankova, B.Tranteev, D. Raychev, M. Barzakova, A. Baldjieva)

Speleopraktika – published since 1966 by the City Speleological Section – Sofia and later by the Association "Speleopraktika" (initiated by N. Landjev).

Infospeleo – Information Bulletin of BFSp., published since 2000 in Sofia (compiler A. Jalov)

PROCEEDINGS:

I National Conference of Speleology (February 1962)

II National Conference of Speleology (December 1976, Sofia). Proceedings (1977). National Conference of Speleology (L. Dinev, T. Panayotov, V. Popov, Eds.). Edition of BFSp., 122 p.

III National Conference of Speleology

IV National Conference of Speleology. Varna, 31.3.-3.4.1983. Topic: "Relations between the scientific and the practical Speleology". Proceedings of 63 p.

V National Conference of Speleology (Sofia,28-30.V.1987). Proceedings (T. Stoychev, Ed.) Ed. Medicina i Fizkultura, 1989, Sofia. 136 p.

VI National Conference of Speleology

VII National Conference of Speleology – National Scientific Conference on the problems of Karst and Speleology (Sofia, March 1999). Proceedings. Edition of "Heliktit" – Sofia. 122 p.

Scientific conference "Environment and cultural heritage in karst" (Sofia,10–11.10.2000) Proceedings (P. Delchev, S. Shanov, A. Benderev, Eds.). Edition of the Association "Environment and cultural heritage in karst", Sofia. 196 p.

Proceedings of Jubilee Scientific Conference "75 Years of organised speleology in Bulgaria" Sofia 4-5 April 2004. 2006, eds. A. Jalov & T. Daaliev, 161 p.

Almanah of "Edelweiss" – Sofia Caving Club. 40 years Yubilee. – Ed. "Edelweiss" – Sofia Caving Club, 2001, 87 p.

Yearbook of "Helictite" – Sofia Caving Club, 2000, Sofia, 40 p. (Comp. and Ed. A. Jalov)

Yearbook of "Helictite" – Sofia Caving Club, 2001, Sofia, 48 p. (Comp. and Ed. A. Jalov)

Yearbook of "Helictite" – Sofia Caving Club, 2002, Sofia, 77 p. (No 6)(Comp. and Ed. K. Stoichkov and J. Petrov)

40 Years Caving Club in Shumen 1961 – 2001. Authors M. Mircheva and Zdr. Iliev, 71 p., 45 maps

European Regional Conference on Speleology (Sofia 22-28.IX.1980) Proceedings. Vol.1 (342 p.); Vol.2 (588 p.) (Editorial Board: L. Dinev, P. Beron, A. Jalov, A. Manov, R. Radushev, K. Spassov): Edition of BFSp., 1983.

Scientific-practical conference on the tourism, alpinism, orientation, caving and protection of nature (Rousse-4-6.V.1979)(L. Dinev, Ed.), Sofia, 470 p.

A: CONCERNING KARST AND CAVES IN BULGARIA

Anon. 1977. [Terminology of karst and speleology in Bulgaria]. – Ed. BTS and BFPD, Sofia, 40 p. (in Bulgarian).

Arnaudov V. 1922. [The cave Ledenik in Vrachanska Planina]. – Estestvoznanie i Geografija, **1**: 142–156 (in Bulgarian).

Beron P. 1999. 70 years of organized caving in Bulgaria. – Grottes bulgares, **6**: 1–8.

Buresch Iv. 1930. [The caves in Bulgaria. Their beauty and the need of their preservation]. – Sofia, 44 p. (in Bulgarian).

Gospodinov D. 1991. [Haskovo and the caves]. – Ed. of the Caving Club at "Aida" – Haskovo Tourist Society, 39 p. (in Bulg.).

Daaliev T., A. Benderev. 1998. [The deepest pot holes in Trigrad Region]. – Ed. BFS, Sofia, (in Bulg.).

Daaliev T., A. Benderev. 2000. [The deepest pot holes in Troyan Subregion]. – Ed. BFS, Sofia, 32 p. (in Bulg.).

Daaliev T., A. Jalov, A. Benderev. 2000. [Brief information about the caves on the Plateau of Devetaki]. – Ed. BFS, Sofia, 22 p. (in Bulg.).

Dinev L., P. Beron. 1962. Entwicklung and Stand der Späläologie in Bulgarien. – Die Höhle, Wien, **13**(2): 45–48.

Djambazov N. 1958. [The caves in Bulgaria]. S., 132 p. (in Bulg.).

Iliev Z., P. Petkov. 2000. [Kotlya. Vrachanski Balkan. – Ed. Club "Streshero", 28 p.] (in Bulg.).

Jalov A. 1993. La Spéléologie en Bulgarie. – Spelunca, **52**: 23–26.

Leonidov A., P. Trifonov. 2005. Speleological Atlas of the Belogradchik region. – Publ. house "PAL", Sofia, 32 p. (Bulg. and Engl.).

Jalov A. 1999. The Caves in Bulgaria. Written sources from the period XII-XIX Century. – Nat. Sci. Conf. on the Problems of Karst and Speleology – Sofia, **99**: 5–10

Mikov V. 1926. [Caves and pot holes between Iskar and Vit]. – Estestvoznanie i Geografija, **10**(7–8): 236–249 (in Bulg.).

Mikov V. 1928. [Caves in Belogradchik Area]. – Balgarski Tourist, **20**(3): 43–44 (in Bulg.).

Mircheva M., Z. Iliev, K. Kostov. 2004. The caves of the Madara Plateau. – Sofia, 44 p. (Bulg., summ. Engl.).

Mishev K., Vl. Popov. 1958. [The karst in Vratchanska Planina. – Priroda, **7**: 7–13] (in Bulg.).

Pandev Z. K. 1993. [Dobrostan kartstic Massiv – caves and pot-holes, part 1: Dobrostan – Oreshets]. 128 p. (in Bulg.).

Pandev Z. K. 1994. [Dobrostan karstic Massiv – caves and pot-holes, part 2: Oreshets-Mostovo], 52 p.(Bulg., summ. Engl.).

Penchev S. 1970. [Fairy world]. – [Prirodonauchna biblioteka]. – Ed."Narodna prosveta", Sofia, 95 p. (in Bulg.).

Petrov Iv. 1977. [Sky of stone]. – Ed. "Medicina i Fizkultura", Sofia, 126 p. (in Bulg.).

Petroff P. 1928. La Grotte de Devettaqui. – Trav. Soc. Bulg. Sci. Nat., **13**: 193–208 (in Bulg., summ. Fr.).

Petrov P. 1929. [The Cave of Devetaki] – In: Lovech i Lovchansko, **1**: 23–32 (in Bulg.).

Popov V. 1967. [Snezhanka]. – Ed. Medicina i Fizkultura, S., (in Bulg.).

Popov V. 1968. [Ledenika]. – Ed. "Medicina i Fizkultura", Sofia, 42 p. (in Bulg.).

Popov V. 1972. [Bacho Kiro]. – Ed. Medicina i Fizkultura, S., (in Bulg.).

Popov V. 1979. [Saeva dupka]. – Ed. Medicina i Fizkultura, S., 28 p. (in Bulg.).

Popov V. 1982. [Travel underground]. – Ed. "Nauka i Izkustvo, Sofia, 152 p. (in Bulg.).

Popov V. 1987. [The show caves in Bulgaria]. – Ed. "Medicina i Fizkultura", Sofia, 75 p.] (in Bulg.).

Radev N. 1926. Materialien zur Erforschung der Höhlen Bulgariens – I. – Trav. Soc. Bulg. Sci. Nat., **12**: 151–182 (in Bulg., summ. in German).

Radev N. 1928. – Materialien zur Erforschung der Höhlen Bulgariens II. – Trav. Soc. Bulg. Sci. Nat., **13**: 115–130 (in Bulg., summ. in German).

Raychev D. 1979. [Trigradsko zhdrelo i Dyavolsko gârlo]. – Ed. Medicina i Fizkultura, 38 p. (in Bulg.).

Raychev D., G. Raychev. 1983. [Yagodina karstic region]. – Ed. Medicina i Fizkultura, 48 p. (in Bulg.).

Spassov A. 1986. [Biserna]. – Ed. Medicina i Fizkultura, Sofia, 30 p. (in Bulg.).
Stoitsev V., P. Neykovski. 1975. [Ponor planina]. – Malka turisticheska biblioteka, Sofia, 37 p. (in Bulg.).
Sabev D. 1967. [Rupchoskite peshteri]. – Malka turisticheska biblioteka, Sofia, 60 p. (in Bulg.).
Tranteev P. 1962. [Magura]. – Ed. Medicina i Fizkultura, Sofia, 52 p. (in Bulg.).
Trantcev P. 1964. Les dix gouffres les plus profonds en Bulgarie. – Grottes bulgares, **1-2**: 61–68 (in Bulg., summ. in French).
Tranteev P. 1965. [Caves objects of tourism]. – Ed. Medicina i Fizkultura, Sofia, 123 p. (in Bulg.).
Tranteev P., V. Gueorguiev. 1968. [The secret of the caves]. – Ed. Medicina i Fizkultura, Sofia, 230 p. (in Bulg.).
Tranteev P., K. Kosev. 1978. [The caves in Bulgaria] – Ed. Medicina i Fizkultura, Sofia, 94 p. (in Bulg.).
Prodeau M, S. Shanov and others. [1987]. Récit d'une amitié, suivi par: La Spéléologie en Bulgarie, La Division karstique de la Bulgarie. – Ed. Spéléo-Club de Saint-Hierblain, 40 p.
Yonge C.J. 1974. Bulgaria 1974. Sheffield University Speleological Society visit to Bulgaria (Summer 1974). – 18 p.

B: SOME PUBLICATIONS BY BULGARIANS ON THE KARST AND THE SPELEOLOGY IN THE WORLD

Aleksiev Iv. 2001. [Romania 1991]. – Collection of papers 40 years of "Akademik" – Caving Club – Sofia: 55–60 (in Bulg.).
Beron P. 1972. Aperçu sur la faune cavernicole de la Corse. – Publ. No. 3 du Lab. Souterrain du C.N.R.S. à Moulis, 55 pp.; summ. in Ann. de Spéléol., **27**(4): 807–810, Moulis.
Beron P. 1985. The contribution of Bulgarian Biospeleologists to the study of cave fauna of other countries. – Grottes bulgares, **3**: 25 – 31.
Beron P. 1986. [Five months in New Guinea]. – Ed. Zemizdat, Sofia, 172 p. (in Bulgarian).
Bonev I. 1989. Granite caves in the High Himalaya. – Proc. Int. Symp. Speleol. Tbilisi 1987: 181 – 183.
Bonev K. 1989. Geologic structural control of karst forming processes in the Plateau Guaso, Southeast Cuba. – Proceed. Intern. Congr. Speleol., Budapest, 1989, **2**: 669 (Abstract).

Daaliev T. 1990. Expedition Chine – 1989. – Spelunca, **38**: 8–9
Daaliev T. [1991]. Bulgarian – Cuban expedition "Guaso'88)". – Grottes bulgares, **5**: 25–35.
Daaliev T. [1991]. First Bulgarian – Chinese speleological expedition "Yunnan – 89". – Grottes bulgares, **5**: 38–45.
Daaliev T. 1992. Expedition Sino-Bulgare "Yunnan 89". – Regards, **10**: 29-30
Daaliev T. 1995. Explorations bulgares en Albanie. – Regards, **19**: 24
Daaliev T. 1999. Expédition "Indonésie – 95". – Grottes bulgares, **6**: 38–39.
Daaliev T. 2001a. [Cuba (Expedition "Guaso – 88")]. – Collection of papers 40 years of "Edelweiss" – Sofia Caving Club: 46–50 (in Bulg.).
Daaliev T. 2001b. [East of Bulgaria, south of the clouds]. – Collection of papers 40 years of "Edelweiss" – Sofia Caving Club: 51–54 (in Bulg.).
Daaliev T. 2001c. [Travel to the unforgetable]. – Collection of papers 40 years of "Edelweiss" – Sofia Caving Club: 64–67 (in Bulg.).
Garbev K. 1990. Syrie. – Spelunca, **37**: 20.
Gladnishky N., T. Daaliev [1991]. [Description of the cave El Campanario]. – Grottes bulgares, **5**: 29–35 (in Bulg.).
Gladnishky N., T. Daaliev. 1993. Cueva Del Campanario – International Caver, **8**: 27–32
Jalov A. 1985. [The important achievements of Bulgarian cavers in 1984]. – Grottes bulgares, **3**: 7–9 (in Bulg.).
Jalov A. (Editor). 1986. [Bulgarians in the abysses of the world]. – Ed. Medicina i Fizkultura, Sofia, 116 p. Authors: P. Nedkov, A. Taparkova-Pencheva, N. Genov, V. Peltekov, T. Daaliev, A. Jalov, S. Gazdov (in Bulg.).
Jalov A. 1988. Cuba. – Spelunca, **31**: 9–10.
Jalov A. 1989. Bulgarian-Cuban Expedition "Guaso'88". – Caves & Caving, **43**: 40.
Jalov A. 1990. Seconde expédition spéléologique sino-bulgare dans le Yunnan 1990.- Spelunca, **39**: 8-9.
Jalov A. 1990. [Bulgarian cavers around the world]. – Bâlgarska Speleologiya, **2**:1–5.
Jalov A. 1991. Vietnam. – Grottes bulgares, **5**: 46–50.
Jalov A. 1992. Albanie – une nouvelle région karstique s'offre aux expéditions. – Regards, **10**: 27–28.
Jalov A.1993. Albanian Alps'92 Expedition. – Caves & Caving, **61**: 35–36.

Jalov A. 1994a. Albania. – Caves & Caving, **63**: 33.

Jalov A. 1994b. Caving war in Albania. – Caves & Caving, **66**: 35.

Jalov A. 1996. Macedonia. Diving Record. – Caves & Caving, **71**: 7

Jalov A. 1997. Results from Bulgarian – Albanian Speleological researches in Albanian Alps from 1991 – 1996. – Proc. 12th Int. Congr. Speleol., La Chaux-de-Fonds, **4**: 25–28.

Jalov A. 1999a. Bulgarians in the caves of the world. – Grottes bulgares, **6**: 9–14.

Jalov A. 1999b. "Albania '96". – Grottes bulgares, **6**: 42–44.

Zhalov A. 2002. Results of Bulgarian-Albanian speleological researches in Albanian Alps from 1991-1996.- Svet, 1-2 (22-23) : 27-33.

Jalov A. 2003a. [The diary of an Albanian expedition "Golo bardo 2002"]. – Godishnik na peshteren club "Heliktit" za deynostta prez 2002 g., Sofia: 32–37 (in Bulg.).

Jalov A. 2003b. [In the gypsum labyrinths of Ozernaya]. – Godishnik na peshteren club "Heliktit" za deynostta prez 2002 g., Sofia: 38-42 (in Bulg.).

Jalov A. 2003c. Nuovo record di profondita in Albania: la BB 30 (-610 m). – Speleologia, **48**: 85.

Lichkov Z. 2006. Expenition "Cuba'81". Proc. Conf. "75 Years of organised speleology in Bulgaria": 43 – 46 (in Bulg.).

Kostov K. 2001. [International Caving Convention "Chatardag,99"]. – Collection of papers 40 years of "Akademik" – Sofia Caving Club: 68 – 69 (in Bulg.).

Mirchev V. 1987. [Jean Bernard'84].- Ed. Medicina i Fizkultura, Sofia, 108 p.

Rashkov Iv., A. Handjiyski. 1977. [Expedition Gouffre Berger and Pierre Saint-Martin– 69 – 73]. – Ed. Medicina i Fizkultura, Sofia, 60 p. (in Bulg.).

C: SELECTED SCIENTIFIC PUBLICATIONS AND GUIDEBOOKS

Biospeleology

In the field of Biospeleology general papers have been published by Buresch (1924, 1926, 1936), Guéorguiev (1966, 1977), Guéorguiev & Beron (1962), Beron & Guéorguiev (1967), Beron (1973, 1986, 1994, 2006a, 2006b), Beron, Petrov & Stoev (2006), Guéorguiev V., Hr. Deltshev, V. Golemansky (1994), Draganov & Dimitrova-Burin (1980). They contain full bibliography of the cave fauna of Bulgaria.

Beron P. 1964. Les grandes grottes à chauves – souris en Bulgarie. – Grottes bulgares, **1-2**: 37–43 (In Bulgarian, summ. in French).

Beron P. 1986. Dévelopement de la Biospéléologie en Bulgarie de 1977 à 1985 et vue d'ensemble sur les connaissances actuelles concernant la faune cavernicole bulgare. – Grottes bulgares, **4**: 53–63.

Beron P. 1972. Essai sur la faune cavernicole de Bulgarie. III. Résultats des recherches biospéologiques de 1966 à 1970. – Int. J. Speleol., **4**: 285–349.

Beron P. 1976. Subdivision zoogéographique de la Stara planina occidentale (Bulgarie) d'après sa faune cavernicole terrestre. – Acta zool. bulgarica, **4**: 30–37.

Beron P. 1994. Résultats des recherches biospéléologiques en Bulgarie de 1971 à 1994 et liste des animaux cavernicoles bulgares. – Série Tranteeva – 1, Sofia, 137 p. (contains bibliography of the cave fauna of Bulgaria until the end of 1994)

Beron P. 2005. Biodiversity of the cave fauna of Bulgaria. – In: Petrova A. (ed.), Actual state of the Biodiversity in Bulgaria – problems and perspectives. Ed. Drakon, Sofia: 397-420.

Beron P. 2006. Terrestrial cave animals in Bulgaria. – In: Fet V. and A. Popov (eds), Ecology and Zoogeography in Bulgaria (in print).

Beron P. & V. Guéorguiev. 1967. Essai sur la faune cavernicole de Bulgarie. II. Résultats des recherches biospéléologiques de 1961 à 1965. – Bull. Inst. Zool. Sofia, **24**: 151–212.

Beron P., B. Petrov, P. Stoev. 2004. Cave fauna in Eastern Rhodopes (Bulgaria and Greece). – In: P. Beron & A. Popov (Eds.), Biodiversity of Bulgaria. 2. Biodiversity of the Eastern Rhodopes (Bulgaria and Greece): 791–822.

Beron P., B. Petrov, P. Stoev. 2006. Cave fauna of Bulgaria. (in print).

Buresch Iv. 1924. Die Höhlenfauna Bulgariens. – Trav. Soc. Bulg. Sci. Nat., **11**: 143–166 (In Bulg., summ. in German).

Buresch Iv. 1926. Untersuchungen über die Höhlenfauna Bulgariens. II. – Trav. Soc. Bulg. Sci., **12**: 17 – 56 (In Bulg., summ. in German).

Buresch Iv. 1931. [Caves and cave fauna in Bulgaria]. – Letopis na BAN, **13**: 74–92. [Academic speech of Dr Buresch, in Bulgarian].

Buresch Iv. 1936. Übersicht der bisherigen Kentnisse und Erforschungen der Rezenten Höhlen-Fauna Bulgariens. – Bull. Soc. Spél. bulgare, **1**: 13–41 (In Bulgarian).

Guéorguiev V. 1961. [Inhabitants of the eternal darkness]. – Prirodonauchna biblioteka, Ed. "Narodna prosveta", Sofia, 63 p. (In Bulgarian).

Guéorguiev V. 1992. Subdivision zoogéographique de la Bulgarie d'après sa faune cavernicole terrestre. – Acta zool. bulgarica, **43**: 3–12 (in Bulg., summ. in French).

Guéorguiev V. & P. Beron. 1962. Essai sur la faune cavernicole de Bulgarie. – Ann. de Spéléologie; Toulouse, **17**(2-3): 285–441.

Guéorguiev V., Hr. Deltshev, V. Golemansky. 1994. Bulgarie. – In: Juberthie C. & V. Decu (Eds.), Encyclopaedia biospeologica, **1**, Moulis-Bucarest: 619–629.

Delchev Hr., V. Guéorguiev. 1977. [Review of the biospeleological studies in Bulgaria and their future problems]. – Sb. Materiali Speleologiya, Sofia: 82–89 (in Bulgarian).

Draganov S.J. 1977. [Findings related to the duckweed in our caves]. – Sbornik Speleologia, Sofia: 90–92 (in Bulg.).

Draganov S.J., E.D. Dimitrova – Burin. 1977. [Speleoalgological Research in Bulgaria] – Proc. 6th Intern. Congr. Speleology, Olomouc, **5**: 11–17 (in Russian).

Draganov S.J., E.D. Dimitrova – Burin. 1980. Speleo-Algological Studies in Bulgaria. – Ecologia, Sofia, **6**: 62–68.

Petkoff S. 1943. La flore des grottes souterraines, des grottes ouvertes et de leurs étangs environnants dans certains régions calcaires de Bulgarie. – Rev. Acad. Bulg. Sci., **68** (5): 109 – 188 (in Bulg., summ. Fr.).

Cvetkov L. 1963. [Life in underground water]. – Ed. Nauka i Iskustvo, Sofia, 100 p. (in Bulg.)

Karstology

Boyadjiev N. 1964. The karstic bassins in Bulgaria and their underground waters. – Bull. Inst. Hydrol. Meteorol., **2**.

Popov Vl. 1964. Morphology and genesis of the cave "Ledenika". – Bull. Inst. Géogr., **8**: 77–87.

Popov Vl. 1970a. Aire d'extension du calcaire (karst) en Bulgarie et quelques – unes de ses particularités. – Bull. Inst. Géogr., **13**: 5–19 (in Bulg., summ. Russ., Fr.).

Popov Vl. 1970b. Distribution of karst in the Stara Planina (Balkan) and some of its peculiarities. – Studia Geomorphologica Carpatho-Balcanica, Kraków, **4**: 237–247 (in Russian, summ. Engl., Pol.).

Popov Vl. 1976. [Cave regionalization in Bulgaria]. – Problemi na geografiyata, **2**(2): 14-24.

Popov Vl. 1977. Cave regionalization in Bulgaria. – Collection of papers Speleologiya, S., 15 – 31 (in Bulgarian).

Popov Vl., L. Zyapkov. 1969. Morphology and hydrology of karst in the northern part of the Predbalkan between the rivers Vit and Iskar. – Bull. Inst. Géogr., **12**.

Popov V. 1965. Rozšíreni krasu v Bulharsku. – ČSAV: **17**.

Radev Zh. 1915. Karstic forms in the Western Stara Planina. – Ann. Univ. Sofia, Fac. Hist. – Philol., **10-11**: 1–149.

Stefanov P. 1982. Morphometry and Morphography of the "Manailovska Dupka" Cave. – Problems of Geography, **3**: 51–59 (In Bulg., summ. Engl.).

Speleomineralogy

Bonev K. 1986. [On the origin of the Moonmilch]. – Grottes bulgares, **4**: 32 – 37 (in Bulg.).

Filipov A.F. 1983. [Modern problems and tasks of the speleomineralogy in Bulgaria]. – Proc. IVth Nat. Conf. of Speleology, Varna: 28–31 (in Bulg.).

Shopov Y. 1988. Bulgarian Cave Minerals. – The National Speleological Society Bull., **50**: 21–24.

Shopov Y. 1989. Genetic Classification of Cave Minerals. – Proc. 10th Int. Congress of Speleology, 13-20 August, Budapest, **1**: 101–105.

Shopov Y. 1990. [Development of the speleomineralogy in Bulgaria (1923-1986)]. – Bâlgarska speleologiya, **2**: 21-32.

Shopov Y. & A. Filipov. 2006. Cave minerals in Bulgaria. Proc. Conf. "75 Years of organised speleology in Bulgaria": 79 – 94 (in Bulg., abst. Engl.).

Caves and Man (Speleoarcheology, Speleofolklore, Onomastics, etc)

Antonov G. 1977. Sacred caves in Strandza mountain, S.E. Bulgaria. – Proc. 7th Intern. Speleological Congr., Sheffield, England, September, 1977: 2–5.

Boev P. 1963. Matériaux anthropologiques de la grotte Tabaškata peštera. – Bull. Inst. Archéol., **26**: 243–247.

Džambazov N. 1957. La grotte Pešt près de Staro selo, arrondissement de Vraca. – Bull. Inst. Archéol., **21**: 1-40.

Djambazov N. 1960a. Un nouveau site paléolithique de la vallée du Vit. – Archeologiya, **2**(1): 36–42.

Djambazov N. 1960b. Fouilles dans la Grande grotte près du village de Micré, arrondissement de Loveč. – Archeologiya, **2**(3): 54–61.

Djambazov N. 1960c. La grotte "Parnika" près du village Bezanovo, arrondissement de Lovetch. – Archeologiya, **2**(4): 54–58.

Djambazov N. 1963. Les grottes de Loveč. – Bull. Inst. Archéol., **26**: 195–241.

Džambazov N. 1970. Trouvailles paléolithiques dans la grotte Pešketo pres du village Liljače, dép. de Vratsa]. – Archeologiya, **12**(1): 58–66.

Djambazov N. 1977. [Archeological studies of caves in Bulgaria]. – Speleologia (Collection of papers at the speleological conference on 10.12.1976, Sofia): 102–111 (in Bulg.).

Djambazov N. 1979. Bulgarian Archeological science and the caves in 50 years. – II National conference on Speleology, Sofia.

Djambazov N. 1981. La grotte Samuilica II. – In: Cultures préhistoriques en Bulgarie, Bull. Inst. d'Archéologie, **36**: 5–62.

Filkoff L. 1929. The Drawings on the Walls of the Cave Magura. – Trav. Soc. Bulg. Sci. Nat., **14**: 143 – 150. (in Bulg., summ. Engl.).

Mikov V., N. Djambazov. 1960. [Devetashkata peshtera]. – Sofia, 199 p. (in Bulg.).

Mushmov N. 1924. [Sacred caves]. – Bâlgarski Tourist, **16**(2): 126 – 127 (in Bulg.).

Kolev B. 1983. Thracian sanctuaries in caves and rock niches in Eastern Rhodopes. – Proc. Europ. Reg. Conf. in Speleology, Sofia, **1**: 173 – 178.

Kozlowski J. K. 1982. Excavation in the Bacho Kiro Cave (Bulgaria).

Petrov Iv., D. Kostov. 1989. [Caves of historical and archeological importance in Stranja – Sakar]. – V Nat. Conf. Speleol.

Popov R. 1904. [Contribution to the prehistory of Bulgaria. The caves in Tarnovo gorge, the settlement near Madara and the caves above Shumen]. – Sb. Narodni Umotvoreniya i knizhnina, **20**: 1–27 (in Bulg.).

Popov R. 1913b. Die Ausgrabungen in der Höhle "Malkata Podlisza" beim Dorfe Beljakovez, umweit der Stadt Tirnovo (Nordbulgarien). – Praehistorischen Zeitschr., **5**(3–4): 450 – 458.

Popov R. 1921a. [Materials for the prehistory of Bulgaria. .....B. Speleological research in the area of Tarnovo, Belyakovets, Dryanovski Monastery and Karlukovo Village. Godishnik na Narodniya Muzey for 1920], **2**: 46–55.

Popov R. 1921b. [Tsarskata peshtera] – Estestvoznanie i Geografija, **6**(1): 28–33 (in Bulg.).

Popov R. 1921c. [The cave Morovitsa] – Estestvoznanie i Geografija, **6**: 434–460 (in Bulg.).

Popov R. 1921d. [The cave Golyama Podlistsa]. – Estestvoznanie i Geografija, **6**(1): 95–105 (in Bulg.).

Popov R. 1925. [The Plateau of Belyakovets – caves and prehistoric settlements (materials for the archeological map of Bulgaria)]. – 3. – Edition of the National Museum, Sofia, 1–58.

Popov R. 1931. [The cave Temnata dupka – new locality of the Paleolith in Bulgaria]. – Naroden Muzej, Sofia.

Popoff R. 1933. La grotte "Mirizlivka". Contribution à l'étude de la faune diluvienne et de la culture de l'homme quaternaire en Bulgarie. – Izdaniya na Narodniya Archeologicheski Muzej, Sofia, **26**: 74 p. (Bulg., summ. Fr.).

Sirakova S. 1986. Archeological studies of caves. – Grottes bulgares, **4**: 64–70 (in Bulg., summ. Engl.).

Spassov N. 1982. [Fossils from alpine capricorne and the giant deer in Bulgaria and the role of horns by the giant deer]. – Priroda, **31**(5): 21–28 (in Bulg.).

Stefanova M. 1986. Folklore themes on Bulgarian caves. – Grottes bulgares, **4**: 71–76 (in Bulg., summ. Engl.).

Stoev A., P. Muglova. 1992. "Moonarium" in the caves near the village of Bailovo – art gallery devoted to the Moon god or Moon callendar. – Actes de la Conf. Européenne de Spéléologie, Hélécine, Belgium, **1**: 149–152.

Stoïanov Il. 1904. Contribution à l'époque préhistorique en Bulgarie. – La grotte "Toplia" près du village Goliama Jeliazna. – Trav. Soc. Bulg. Sci. Nat., **2**: 103–171 (in Bulg., summ. Fr.).

Stoytchev T. 1986. A contribution of the Bulgarian Federation of Speleology to the study of caves from an Archeological and Historical aspect. – Grottes bulgares, **4**: 77–83 (in Bulg., summ. Engl.).

Teoclieva E. 1981. Archeological studies in the Strandja caves. – Scient.-Practical Conf. on Tourism, Alpinism, Tourist Orientation, Caving and Protection of Nature, Russe – 4-6 May 1979. – Proceedings, S.: 259-266 (in Bulgarian).

Todorov T. 2002. [The names of the caves and their classification]. – In: D. Raychev & G. Raychev, Eds., 50 Years of Caving in Chepelare: 53–57.

Zlatkova M. 1989. [Rock cave sanctuary at Baylovo Village, Sofia District. Methods of documentation]. – V Nat. Conf. of Speleology, Sofia, 28–30. 1987. Ed. Medicina i Fizkultura, 72–76 (in Bulg.).

Speleopaleonthology

Bocheński Z. 1982. Aves. – In: J. Kozlowski (Ed) Excavation in the Bacho Kiro Cave (Bulgaria), Final Report, Jagellonian Univ. Press, Krakow: 31–38.

Boev Z. 1994. The Upper Pleistocene birds. – In: Kozlowski J.K., H. Laville, B. Ginter (Eds.). 1994. Temnata cave. Excavations in Karlukovo Karst Area, Bulgaria. – Jagellonian Univ. Press, Krakow, **1**(2): 11–53.

Boev Z. 1997. The black grouse, *Tetrao tetrix* (L., 1758) (Tetraonidae, Aves), a disappeared species in Bulgaria (Paleolithic and Neolithic records). – Anthropozoologia, **25-26**: 643–646.

Boev Z. 1998a. A range fluctuation of Alpine swift (*Apus melba* [L., 1758]) (Apodidae – Aves) in Northern Balkan Peninsula in the Riss-Würm interglacial. – Biogeographia, **19**: 213–218.

Boev Z. 1998b. First fossil record of the Snowy Owl *Nyctea scandiaca* (Linnaeus, 1758) (Aves: Strigidae) from Bulgaria. – Hist. nat. bulg., **9**: 79–86.

Boev Z. 1999. The Middle Pleistocene avifauna of the Cave 16 (NW Bulgaria) and its palaeoecological implication. 8th Intern. Congress on the Zoogeogr. and Ecol. of Greece and Adjacent Regions, Kavala. 17-22 May 1999. Abstracts: 21–22. Hel. Z. Soc.

Boev Z. 2000a. Late Pleistocene avifauna of the Razhishkata Cave, Western Bulgaria. – Hist. nat. bulgarica, **12**: 71–87.

Boev Z. 2000b. Early Pleistocene and Early Holocene avifauna of the Cherdzhenitsa Cave, Northern Bulgaria. – Hist. nat. bulg., **11**: 107–116.

Boev Z. 2001a. Birds over the mammoth's head in Bulgaria. – In: G. Cavarretta, P. Gioia, M. Mussi, M.R. Palombo (eds.) The Worlds of Elephants. Proc. 1st Intern. Congress, Roma: 180–186.

Boev Z. 2001b. Late Pleistocene and Holocene avifauna from three caves in the vicinity of Tran (Pernik District – W Bulgaria). – Karst, Sofia, **1** (2000): 98–106.

Boev Z. 2001c. Late Pleistocene and Holocene avian finds from the vicinity of the Lakatnik r/w station (W Bulgaria). – Karst, Sofia, **1** (2000): 107–111.

Boev Z. 2001d. Late Pleistocene birds from the Kozarnika Cave (Montana District; NW Bulgaria). – Karst, Sofia, **1** (2000): 113–128.

Delpech F., J.-L. Guadelli. 1992. Les grands Mammifères gravettiens et aurignaciens de la grotte de Temnata. – In: J.K. Kozlowski, H. Laville, B. Ginter (eds.) Temnata cave. Excavations in Karlukovo Karst Area, Bulgaria. – Jagellonian Univ. Press, Krakow, **1**(1): 141–216.

Džambazov N. 1957. La grotte Pešt près de Staro selo, arrondissement de Vraca. – Bull. Inst. Archéol., **21**: 1–40 (Bulg., summ. Fr.).

Djambazov N. 1963. Les grottes de Loveč. – Bull. Inst. Archéol., **26**: 195–241 (Bulg., summ. Fr.).

Garrod D. 1939. Excavations in the cave of Batscho Kiro, North-East Bulgaria. – Bull. of the Amer. School of Prehistoric Research, **15**: 48–80.

Guadelli J.-L. & F. Delpech. 2000. Les grands Mammifères du debut du paléolithique supérieur à Temnata. – In: Ginter B., J.K. Kozlovski, J.-L. Guadelli, K. Laville (Eds) Temnata cave. Excavations in Karlukovo Karst Area, Bulgaria. – Jagellonian Univ. Press, Krakow, **2**(1): 53–158.

Kowalski K. 1982. Animal remains – general remarks. 66–71. – In: Excavation in Bacho Kiro Cave (Bulgaria) – final report.

Kowalski K., A. Nadachowski. 1982. Rodentia. In: Kozlowski J. (ed.) Excavation in Bacho Kiro Cave (Bulgaria). PWN, Warszawa: 45–51.

Kozlowski J.K. (Ed.) 1982. Excavation in the Bacho Kiro Cave (Bulgaria). Final Report. – PWN, Warszawa: 172 pp.

Markov G. 1963. Beitrag zur Untersuchung des Höhlenbären (*Ursus spelaeus* Blumenb.) in Bulgarien. – Bull. Inst. Zool. Mus. Sofia, **14**: 5–26 (Bulg., summ. Russ., Germ.).

Mlikovsky J. 1997. Late Pleistocene birds of Karlukovo, Bulgaria. – Hist. nat. bulg., **7**: 59–60.

Nikolov I. 1977. [Review of the fossil mammal fauna in Bulgarian caves and some problems of the future]. – Collection of papers Speleologiya, S., 98–101 (in Bulgarian).

Nikolov I. 1983. [Some notes on the cave fossil mammalian fauna in Bulgaria]. – European Regional Conference on Speleology, Sofia – Bulgaria, 22–28. IX. 1980, Proceedings, **1**: 215–218 (In Bulgarian).

Peshev Ts., Peshev D., Popov V. 2004. Mammalia. – Fauna Bulgarica 27. Ed. "Prof. Marin Drinov", 632 pp.

Petkow N. 1926. Beitrag zur Erforschung der neolitischen Säugetierfauna Bulgariens. – Trav. Soc. Bulg. Sci. Nat., **12**: 187–188 (in Bulg., summ. German).

Petkov [= Petkow] N. 1958. [Mechata Cave near the village Zhelen, Svoge District]. – Priroda, Sofia, **7**(1): 77–80 (in Bulgarian).

Popov [= Popoff=Popow] R. 1908. [Contribution to the neolithic mammalian fauna in Bulgaria]. – Sb. Narodni Umotvoreniya, **24**: 1–22 (in Bulgarian).

Popov R. 1913a. Beitrag zur Kenntnis der Diluvialfauna Bulgariens. – Spisanie na BAN, **7**: 115–142 (in Bulg., summ. Germ.).

Popov R. 1913b. Die Ausgrabungen in der Höhle "Malkata Podlisza" beim Dorfe Beljakovez, umweit der Stadt Tirnovo (Nordbulgarien). – Praehistorischen Zeitschr., **5**(3–4): 450–458.

Popow R. 1913c. Fouilles de la grotte "Morovitsa". – Bull. Soc. Arch. bulgare, **3**(1912–1913): 263–290.

Popov R. 1921. [Materials for the prehistory of Bulgaria. .....B. Speleological research in the area of Tarnovo, Belyakovets, Dryanovski Monasteri and Karlukovo Village]. Godishnik na Narodniya Muzey for 1920, **2**: 46–55.

Popov R. 1925. [The Plateau of Belyakovets – caves and prehistoric settlements. – Materials for the archeological map of Bulgaria III]. S.: 1–58.

Popov R. 1931. [The cave "Temnata dupka" – new locality from the Paleolith in Bulgaria]. – State Printing House, 143 p.

Popoff R. 1933. La grotte "Mirizlivka". Contribution à l'étude de la faune diluvienne et de la culture de l'homme quaternaire en Bulgarie. – Izdaniya na Narodniya Archeologicheski Muzej, Sofia, **26**: 74 p. (Bulg., summ. Fr.).

Popoff R. 1936. Fossile und Subfossile Reste aus den bis jetzt erforschenden Höhlen Bulgariens. – Bull. Soc. Spéléol. de Bulgarie, **1**: 1–11 (in Bulg.).

Popov R. 1939. The Animal remains from the cave of Bacho Kiro. – Bull. of the Amer. School of Prehistoric Research, **15**: 85–126.

Popov V. 1983. [Paleontological studies in the caves (importance, methods, results and directions of development)] – In: Proc. IVth Nat. Cof. of Speleology, Varna: 36–48 (in Bulg.).

Popov V. 1984. Small mammals (Mammalia: Insectivora, Lagomorpha, Rodentia) of the Upper Pleistocene deposits in the Mechata doupka Cave (Western Stara Planina Mts.) I. Taphonomy, paleoecological and zoogeographycal peculiarities. – Acta zool. bulgarica, **24**: 35–44.

Popov V. 1985. Small mammals (Mammalia: Insectivora, Lagomorpha, Rodentia) of the Upper Pleistocene deposits in the Mechata doupka Cave (Western Stara Planina Mts.) II. Description of species. – Acta zool. bulgarica, **26**: 23–49.

Popov V. 1986. Early Pleistocene Rodentia (Mammalia) from the "Temnata dupka" cave near Karlukovo (North Bulgaria). – Acta zool. bulgarica, **30**: 3–14.

Popov V. 1989. Middle Pleistocene small mammals (Insectivora, Lagomorpha, Rodentia) from Morovitsa cave (North Bulgaria). – Acta zool. cracov., **32** (13): 561–588.

Popov V. 1990. [Quaternary small mammals (Mammalia: Insectivora, Lagomorpha, Rodentia) from the Western Predbalkan. – Autoref., Thesis, 30 pp.]. In Bulgarian.

Popov V. 1994. Quaternary small mammals from deposits in Temnata – Prohodna Cave system. – In: Kozlowski J.K., H. Laville, B. Ginter (Eds.). 1994. Temnata cave. Excavations in Karlukovo Karst Area, Bulgaria. – Jagellonian Univ. Press, Krakow, **1**(2): 11–53.

Popov V. 2000. The small mammals (Mammalia: Insectivora, Chiroptera, Lagomorpha, Rodentia) from cave 16 (North Bulgaria) and the paleoenvironmental changes during the Late Pleistocene. – In: Ginter B., J.K. Kozlovski, J.-L. Guadelli, K. Laville (Eds) Temnata cave. Excavations in Karlukovo Karst Area, Bulgaria. – Jagellonian Univ. Press, Krakow, **2**(1): 159–240.

Raychev D. 2002. Discoveries of the cave club "Studenets", Chepelare. – In: D. Raychev &

G. Raychev, Eds., 50 Years of Caving in Chepelare: 50–53.
Raycheva Y. Tsv. 1983. [Localities of the bear of caverns (*Ursus spelaeus*) in Smolyan District]. – European Regional Conference on Speleology Sofia – Bulgaria, 22–28. IX. 1980, Proceedings, **1**: 219–227.
Spassov N. 1989. The position of jackals in the genus *Canis* and life history of the golden jackal (*Canis aureus* L.) in Bulgaria and on the Balkans. – Historia naturalis bulgarica, **1**: 44–54.
Stoïanov Il. 1904. Contribution à l'époque préhistorique en Bulgarie. – La grotte "Toplia" près du village Goliama Jeliazna. – Trav. Soc. Bulg. Sci. Nat., **2**: 103 – 171 (in Bulg., summ. Fr.).
Woloszyn B. W. 1982. Chiroptera: 40–45. – In: Kozlowski I. (ed.) Excavation in the Bacho Kiro Cave (Bulgaria). Warszawa, PWN.

Speleoclimatology

Mechkuev R., B. Todorov, I. Popivanova & S. Toshkov. 1983. [Basic microclimatic and chemical factors in the cave Magura]. – Collection of papers ERSK (in Bulgarian).
Muglova P., A. Stoev, A. Zhalov & A. Filipov. 2006. Investigations of the microclimat in Bulgaria: Scientific ideas, basic contributions and bibliography. Proc. Conf. "75 Years of organised speleology in Bulgaria": 111 – 119 (in Bulg., abst. Engl.).
Raychev G. 1991. [Microclimatic particularities of Yagodina cave system]. – Bulgarian Speleology, **3**.
Stoev A. 1981. [Microclimatic studies in some caves of Dobrostan Massif, based on particularities of the thermodynamics of their athmosphere].- In: Scientific-practical conference on tourism, alpinism, orientation, caving and protection of nature – Proceedings. Sofia: 267–292.
Stoev A. 1989. [Microclimatic studies in some caves of Karlukovo karstic region]. – In: Proc. V-th Nat. Conf. on Speleology. Ed. Medicina and Fizkultura, Sofia: 35–43.
Stoev A. 1999. Microclimatic subdivision of the caves in Bulgaria. – Nat. Conf. on the problems of Karst and Speleology, Sofia, 99, Proceedings: 82–86.

Cave Diving

Jalov A. 1999. Glava Panega – the longest siphon in Bulgaria. – Grottes bulgares, **6**: 29–30.
Jalov A. 1999. Achievements of cave divers. – Grottes bulgares, **6**: 31.
Zdravkov I. 1992. The exploration of Kotel's karst springs. – Grottes bulgares, **5**: 9–10.
Tranteev P., A. Denkov. 1960. [Days and nights underground].- Eho, **2** (7), 19.02.1960: 3 (in Bulgarian).
Tranteev P., T. Michev.1964. [Through lakes and siphons]. – Eho, **1**(235): 12 (in Bulgarian).

Protection of caves

Tranteev P. 1977. [Protection of caves]. – Speleologiya, Sofia: 112–118 (in Bulg.).

Equipment of the caver. Technique and tactics for exploring and surveying caves and pot holes

Delchev Hr. 1979. [Speleology and caving. A Manuel]. Ed. VIF, Sofia, 86 p. (in Bulg.).
Nedkov P. 1973. [Manuel of Speleology. Technique and tactics]. – Ed. "Medicina i Fizkultura", Sofia, 146 p. (in Bulgarian).
Nedkov P. 1983. [ABC of single rope technique]. – Ed. Medicina i Fizkultura, 96 p. (in Bulg.).
Tranteev P., Y. Velinov. 1981. [Surveying caves and pot holes]. – Ed. "Medicina i Fizkultura", Sofia, 95 p.] (in Bulgarian).

History of Bulgarian Speleology

Jalov A. 1990. [Data about the history of the caving in Bulgaria]. – Balgarska speleologiya, **2**: 6–7 (in Bulg.).
Jalov A.1994. Bibliographical data for caves and speleological activities in Bulgaria from XII Century to 1887. – In: Protostoria della Speleologia Edd.Nuova Prhomos. Citta di Castello, Italia: 331–334.
Jalov A. 2001. Historiography of speleological activity in Bulgaria from 1901 to the founding of Speleological society (1929) with bibliography of the published papers in that period. – In: Karst, vol.1. Proceedings of the First National Conference on Environment and Cultural Heritage in Karst, Sofia 10–11 November 2000, Earth and Man National Museum: 158–169.

See also the papers of Buresch (1926, 1928, 1936), Djambazov (1977, 1979). Guéorguiev & Beron (1962) in this Bibliography.